T0220811

Birkhäuser Advanced Texts Basler Lehrbücher

Series editors

Steven G. Krantz, Washington University, St. Louis, USA
Shrawan Kumar, University of North Carolina at Chapel Hill, Chapel Hill, USA
Jan Nekovář, Université Pierre et Marie Curie, Paris, France

More information about this series at http://www.springer.com/series/4842

Emmanuele DiBenedetto

Real Analysis

Second Edition

 Birkhäuser

Emmanuele DiBenedetto
Nashville, TN
USA

ISSN 1019-6242   ISSN 2296-4894   (electronic)
Birkhäuser Advanced Texts Basler Lehrbücher
ISBN 978-1-4939-8151-9   ISBN 978-1-4939-4005-9   (eBook)
DOI 10.1007/978-1-4939-4005-9

Mathematics Subject Classification (2010): 26A21, 26A45, 26A46, 26A48, 26A51, 26B05, 26B30, 28A33, 28C07, 31A15, 39B72, 46B25, 46B26, 46B50, 46C05, 46C15, 46E10, 46E25, 46E27, 46E30, 46E35, 46F05, 46F10

Printed on acid-free paper

This book is published under the trade name Birkhäuser
The registered company is Springer Science+Business Media LLC New York
(www.birkhauser-science.com)

Preface to the Second Edition

This is a revised and expanded version of my 2002 book on real analysis. Some topics and chapters have been rewritten (i.e., Chaps. 7–10) and others have been expanded in several directions by including new topics and, most importantly, considerably more practice problems. Noteworthy is the collection of problems in calculus with distributions at the end of Chap. 8. These exercises show how to solve algebraic equations and differential equations in the sense of distributions, and how to compute limits and series in $\mathcal{D}'$. Distributional calculations in most texts are limited to computing the fundamental solution of some linear partial differential equations. We have sought to give an array of problems to show the wide applicability of calculus in $\mathcal{D}'$. I must thank U. Gianazza and V. Vespri for providing me with most of these problems, taken from their own class notes. Chapter 9 has been expanded to include a proof of the Riesz convolution rearrangement inequality in N-dimensions. This is preceded by the topics on Steiner symmetrization as a supporting background. Chapter 11 is new, and it goes more deeply in the local fine properties of weakly differentiable functions by using the notion of p-capacity of sets in $\mathbb{R}^N$. It clarifies various aspects of Sobolev embedding by means of the isoperimetric inequality and the co-area formula (for smooth functions). It also links to measure theory in Chaps. 3 and 4, as the p-capacity separates the role of measures versus outer measure. In particular, while Borel sets are p-capacitable, Borel sets of positive and finite capacity are not measurable with respect to the measure generated by the outer measure of p-capacity. Thus, it also provides an example of nonmetric outer measures and non-Borel measure. As it stands, this book provides a background to more specialized fields of analysis, such as probability, harmonic analysis, functions of bounded variation in several dimensions, partial differential equations, and functional analysis. A brief connection to BV functions in several variables is offered in Sect. 7.2c of the Complements of Chap. 5.

The numbering of the sections of the Problems and Complements of each chapter follow the numbering of the section in that chapter. Exceptions are Chaps. 6 and 8. Most of the Problems and Complements of Chap. 8 are devoted to calculus with distributions, not directly related to the sections of that chapter.

Sections 20c–23c of the Complements of Chap. 6 are devoted to present the Vitali–Saks–Hahn theorem. The relevance of the theorem is in that it gives sufficient conditions on a set of integrable functions to be *uniformly integrable*. This in turn it permits one to connect the notions of weak and strong convergence to convergence in measure. In particular, as a consequence it gives necessary and sufficient conditions for a weakly convergent sequence in L^1 to be strongly convergent in L^1. As an application, in ℓ_1, weak and strong convergence coincide (Sects. 22c–23c of Chap. 6).

Over the years, I have benefited from comments and suggestions from several collaborators and colleagues including U. Gianazza, V. Vespri, U. Abdullah, Olivier Guibé, A. Devinatz[†], J. Serrin[†], J. Manfredi, and several current and former students, including Naian Liao, Colin Klaus Stockdale, Jordan Nikkel, and Zach Gaslowitz Special thanks go to Ugo Gianazza and Olivier Guibé for having read in detail large parts of the manuscript and for pointing out imprecise statements and providing valuable suggestions. To all of them goes my deep gratitude.

This work was partially supported by NSF grant DMS-1265548.

Preface to the First Edition

This book is a self-contained introduction to real analysis assuming only basic notions on limits of sequences in $\mathbb{R}^N$, manipulations of series, their convergence criteria, advanced differential calculus, and basic algebra of sets.

The passage from the setting in $\mathbb{R}^N$ to abstract spaces and their topologies is gradual. Continuous reference is made to the $\mathbb{R}^N$ setting where most of the basic concepts originated.

The first eight chapters contain material forming the backbone of a basic training in real analysis. The remaining three chapters are more topical, relating to maximal functions, functions of bounded mean oscillation, rearrangements, potential theory and the theory of Sobolev functions. Even though the layout of the book is theoretical, the entire book and the last chapters in particular have in mind applications of mathematical analysis to models of physical phenomena through partial differential equations.

The preliminaries contain a review of the notions of countable sets and related examples. We introduce some special sets, such as the Cantor set and its variants and examine their structure. These sets will be a reference point for a number of examples and counterexamples in measure theory (Chapter 3) and in the Lebesgue differentiability theory of absolute continuous functions (Chapter 5). This initial Chapter contains a brief collection of the various notions of *ordering*, the Hausdorff maximal principle, Zorn's Lemma, the well-ordering principle, and their fundamental connections.

These facts keep appearing in measure theory (Vitali's construction of a Lebesgue non-measurable set), topological facts (Tychonov's Theorem on the compactness of the product of compact spaces; existence of Hamel bases) and functional analysis (Hahn-Banach Theorem; existence of maximal orthonormal bases in Hilbert spaces).

Chapter 2 is an introduction to those basic topological issues that hinge upon analysis or that are, one way or another, intertwined with it. Examples include Uhryson's Lemma and the Tietze Extension Theorem, characterization of compactness and its relation to the Bolzano-Weierstrass property, structure of the compact sets in $\mathbb{R}^N$, and various properties of semi-continuous functions defined on compact sets. This analysis of compactness has in mind the structure of the compact subsets of the space of continuous functions (Chapter 5) and the characterizations of the compact subsets of the spaces $L^p(E)$ for all $1 \leq p < \infty$ (Chapter 6).

The Tychonov Theorem is proved with its application in mind in the proof of the Alaoglu Theorem on the weak* compactness of closed balls in a linear, normed space.

We introduce the notion of linear, topological vector spaces and that of linear maps and functionals and their relation to boundedness and continuity.

The discussion turns quickly to metric spaces, their topology, and their structure. Examples are drawn mostly from spaces of continuous or continuously differentiable functions or integrable functions. The notions and characterizations of compactness are rephrased in the context of metric spaces. This is preparatory to characterizing the structure of compact subsets of $L^p(E)$.

The structure of complete metric spaces is analyzed through Baire's Category Theorem. This plays a role in subsequent topics, such as an indirect proof of the existence of nowhere differentiable functions (Chapter 5), in the structure of Banach spaces (Chapter 6), and in questions of completeness and non-completeness of various topologies on $C_o^\infty(E)$ (Chapter 8).

Chapter 3 is a modern account of measure theory. The discussion starts from the structure of open sets in $\mathbb{R}^N$ as sequential coverings to construct measures and a brief introduction to the algebra of sets. Measures are constructed from outer measure by the Charathéodory process. The process is implemented in specific examples such as the Lebesgue-Stiltjes measures in $\mathbb{R}$ and the Hausdorff measure. The latter seldom appears in introductory textbooks in Real Analysis. We have chosen to present it in some detail because it has become, in the past two decades, an essential tool in studying the fine properties of solutions of partial differential equations and systems. The Lebesgue measure in $\mathbb{R}^N$ is introduced directly starting from the Euclidean measure of cubes rather than regarding it, more or less abstractly, as the N-product of the Lebesgue measure on $\mathbb{R}$. In $\mathbb{R}^N$, we distinguish between Borel sets and Lebesgue measurable sets, by cardinality arguments and by concrete counterexamples.

For general measures, emphasis is put on necessary and sufficient criteria of measurability in terms of $\mathcal{G}_\delta$ and $\mathcal{F}_\sigma$. In this, we have in mind the operation of measuring a set as an approximation process. From the applications point of view, one would like to approximate the measure of a set by the measure of measurable sets containing it and measurable sets contained into it. The notion is further expanded in the theory of Radon measures and their regularity properties.

It is also further expanded into the covering theorems, even though these represent an independent topic in their own right. The Vitali Covering Theorem is presented by the proof due to Banach. The Besicovitch covering is presented by emphasizing its value for general Radon measures in $\mathbb{R}^N$. For both, we stress the measure-theoretical nature of the covering as opposed to the notion of covering a set by inclusion.

Coverings have made possible an understanding of the local properties of solutions of partial differential equations, chiefly the Harnack inequality for non-negative solutions of elliptic and parabolic equations. For this reason, in the Complements of this chapter, we have included various versions of the Vitali and Besicovitch covering theorems.

Chapter 4 introduces the Lebesgue integral. The theory is preceded by the notions of measurable functions, convergence in measure, Egorov's Theorem on selecting almost everywhere convergent subsequences from sequences convergent in measure, and Lusin's Theorem characterizing measurability in terms of quasi-continuity. This theorem is given relevance as it relates measurability and local behavior of measurable functions. It is also a concrete application of the necessary and sufficient criteria of measurability of the previous chapter.

The integral is constructed starting from non-negative simple functions by the Lebesgue procedure. Emphasis is placed on convergence theorems and the Vitali's Theorem on the absolute continuity of the integral. The Peano-Jordan and Riemann integrals are compared to the Lebesgue integral by pointing out differences and analogies.

The theory of product measures and the related integral is developed in the framework of the Charathéodory construction by starting from measurable rect-angles. This construction provides a natural setting for the Fubini-Tonelli Theorem on multiple integrals.

Applications are provided ranging from the notion of convolution, the conver-gence of the Marcinkiewicz integral, to the interpretation of an integral in terms of the distribution function of its integrand.

The theory of measures is completed in this chapter by introducing the notion of signed measure and by proving Hahn's Decomposition Theorem. This leads to other natural notions of decompositions such as the Jordan and Lebesgue Decomposition Theorems. It also suggests naturally other notions of comparing two measures, such as the absolute continuity of a measure v with respect to another measure μ. It also suggests representing v, roughly speaking, as the integral of μ by the Radon-Nykodým Theorem.

Relating two measures finds application in the Besicovitch-Lebesgue Theorem, presented in the next chapter, and connecting integrability of a function to some of its local properties.

Chapter 5 is a collection of applications of measure theory to issues that are at the root of modern analysis. What does it mean for a function of one real variable to be differentiable? When can one compute an integral by the Fundamental Theorem of Calculus? What does it mean to take the derivative on an integral? These issues motivated a new way of measuring sets and the need for a new notion of integral.

The discussion starts from functions of bounded variation in an interval and their Jordan's characterization as the difference of two monotone functions. The notion of differentiability follows naturally from the definition of the four Dini's numbers. For a function of bounded variation, its Dini numbers, regarded as functions, are measurable. This is a remarkable fact due to Banach.

Functions of bounded variations are almost everywhere differentiable. This is a celebrated theorem of Lebesgue. It uses, in an essential way, Vitali's Covering Theorem of Chapter 3.

We introduce the notion of absolutely continuous functions and discuss similarities and differences with respect to functions of bounded variation. The Lebesgue theory of differentiating an integral is developed in this context. A natural related issue is that of the density of a Lebesgue measurable subset of an interval. Almost every point of a measurable set is a density point for that set. The proof uses a remarkable theorem of Fubini on differentiating term by term a series of monotone functions.

Similar issues for functions of N real variables are far more delicate. We present the theory of differentiating a measure v with respect to another μ by identifying precisely such a derivative in terms of the singular part and the absolutely continuous part of μ with respect to v. The various decompositions of measures of Chapter 4 find here their natural application, along with the Radon-Nykodým Theorem.

The pivotal point of the theory is the Besicovitch-Lebesgue Theorem asserting that the limit of the integral of a measurable function f when the domain of integration shrinks to a point x actually exists for almost all x and equals the value of f at x. The shrinking procedure is achieved by using balls centered at x, and the measure can be any Radon measure. This is the strength of the Besicovitch covering theorem. We discuss the possibility of replacing balls with domains that are, roughly speaking, comparable to a ball. As a consequence, almost every point of an N-dimensional Lebesgue-measurable set is a density point for that set.

The final part of the chapter contains an array of facts of common use in real analysis. These include basic facts on convex functions of one variable and their almost everywhere double differentiability. A similar fact for convex functions of several real variables (known as the Alexandrov Theorem) is beyond the scope of these notes. In the Complements, we introduce the Legendre transform and indicate the main properties and features.

We present the Ascoli-Arzelá Theorem, keeping in mind a description of compact subsets of spaces of continuous functions.

We also include a theorem of Kirzbraun and Pucci extending bounded, continuous functions in a domain into bounded, continuous functions in the whole $\mathbb{R}^N$ with the *same* upper bound and the *same* concave modulus of continuity. This theorem does not seem to be widely known.

The final part of the chapter contains a detailed discussion of the Stone-Weierstrass Theorem. We present first the Weierstrass Theorem (in N dimensions) as a pure fact of Approximation Theory. The polynomials approximating a continuous function f in the sup-norm over a compact set are constructed explicitly by means of the Bernstein polynomials. The Stone Theorem is then presented as a way of identifying the structure of a class of functions that can be approximated by polynomials.

Chapter 6 introduces the theory of L^p spaces for $1 \leq p \leq \infty$. The basic inequalities of Hölder and Minkowski are introduced and used to characterize the norm and the related topology of these spaces. A discussion is provided to identify elements of $L^p(E)$ as equivalence classes.

We introduce also the $L^p(E)$ spaces for $0 < p < 1$ and the related topology. We establish that there are not convex open sets except $L^p(E)$ itself and the empty set.

We then turn to questions of convergence in the sense of $L^p(E)$ and their completeness (Riesz-Fisher Theorem) as well as issues of separating such spaces by simple functions. The latter serves as a tool in the notion of weak convergence of sequences of functions in $L^p(E)$. Strong and weak convergence are compared and basic facts relating weak convergence and convergence of norms are stated and proved.

The Complements contain an extensive discussion comparing the various notions of convergence.

We introduce the notion of functional in $L^p(E)$ and its boundedness and continuity and prove the Riesz representation Theorem, characterizing the form of all the bounded linear functionals in $L^p(E)$ for $1 \leq p \leq \infty$. This proof is based on the Radon-Nykodým Theorem and as such is measure theoretical in nature.

We present a second proof of the same theorem based on the topology of L^p. The open balls that generate the topology of $L^p(E)$ are *strictly* convex for $1 < p < \infty$. This fact is proved by means of the Hanner and Clarkson's inequality, which while technical, are of interest in their own right.

The Riesz Representation Theorem permits one to prove that if E is a Lebesgue-measurable set in $\mathbb{R}^N$, then $L^p(E)$ for $1 \leq p < \infty$, are separable. It also permits one to select weakly convergent subsequences from bounded ones. This fact holds in general, reflexive, separable Banach spaces (Chapter 7). We have chosen to present it independently as part of the L^p theory. It is our point of view that a good part of functional analysis draws some of its key facts from concrete spaces, such as spaces of continuous functions, the L^p, space and the spaces ℓ_p.

The remainder of the chapter presents some technical tools regarding $L^p(E)$ for E, a Lebesgue-measurable set in $\mathbb{R}^N$, to be used in various parts of the later chapters. These include the continuity of the translation in the topology of $L^p(E)$, the Friedrichs mollifyiers, and the approximation of functions in $L^p(E)$ with $C^\infty(E)$ functions. It includes also a characterization of the compact subsets in $L^p(E)$.

Chapter 7 is an introduction to those aspects of functional analysis closely related to the Euclidean spaces $\mathbb{R}^N$, the spaces of continuous functions defined on some open set $E \subset \mathbb{R}^N$, and the spaces $L^p(E)$. These naturally suggest the notion of finite dimensional and infinite dimensional normed spaces. The difference between the two is best characterized in terms of the compactness of their closed unit ball. This is a consequence of a beautiful counterexample of Riesz.

The notions of maps and functionals is rephrased in terms of the norm topology. In $\mathbb{R}^N$, one thinks of a linear functional as an affine functions whose level sets are hyperplanes through the origin. Much of this analogy holds in general normed spaces with the proper rephrasing.

Families of pointwise equi-bounded maps are proven to be uniformly equi-bounded as an application of Baire's Category Theorem.

We also briefly consider special maps such as those generated by Riesz potential (estimates of these potentials are provided in Chapter 9), and related Fredholm integral equations.

A proof of the classical Open Mapping Theorem and Closed Graph Theorem are presented as a way of inverting continuous maps to identify isomorphisms out of continuous linear maps.

The Hahn-Banach Theorem is viewed in its geometrical aspects of separating closed convex sets in a normed space and of "drawing" tangent planes to a convex set.

These facts all play a role in the notion of weak topology and its properties. Mazur's Theorem on weak and strong closure of convex sets in a normed space is related to the weak topology of the $L^p(E)$ spaces. These provide the main examples, as convexity is explicit through Clarkson's inequalities.

The last part of the chapter gives an introduction to Hilbert spaces and their geometrical aspects through the parallelogram identity. We present the Riesz Representation Theorem of functionals through the inner product. The notion of basis is introduced and its cardinality is related to the separability of a Hilbert space. We introduce orthonormal systems and indicate the main properties (Bessel's inequality) and some construction procedures (Gram-Schmidt). The existence of a complete system is a consequence of the Hausdorff maximum principle. We also discuss various equivalent notions of completeness.

Chapter 8 is about spaces of real-valued continuous functions, differentiable functions, infinitely differentiable functions with compact support in some open set $E \subset \mathbb{R}^N$, and weakly differentiable functions.

Together with the $L^p(E)$ spaces, these are among the backbone spaces of real analysis.

We prove the Riesz Representation Theorem for continuous functions of compact support in $\mathbb{R}^N$. The discussion starts from positive functionals and their representation. Radon measures are related to positive functionals and bounded, signed Radon measures are related to bounded linear functionals. Analogous facts hold for the space of continuous functions with compact support in some open set $E \subset \mathbb{R}^N$.

We then turn to making precise the notion of a topology for $C_o^\infty(E)$. Completeness and non-completeness are related to metric topologies in a constructive way. We introduce the Schwartz topology and the notion of continuous maps and functionals with respect to such a topology. This leads to the theory of distributions and its related calculus (derivatives, convolutions etc. of distributions).

Their relation to partial differential equations is indicated through the notion of fundamental solution. We compute the fundamental solution for the Laplace operator also in view of its applications to potential theory (Chapter 9) and to Sobolev inequalities (Chapter 10).

The notion of weak derivative in some open set $E \subset \mathbb{R}^N$ is introduced as an aspect of the theory distributions. We outline their main properties and state and prove the by now classical Meyers-Serrin Theorem. Extension theorems and approximation by smooth functions defined in domains larger than E are provided. This leads naturally to a discussion of the smoothness properties of ∂E for these approximations and/or extensions to take place (cone property, segment property, etc.).

We present some calculus aspects of weak derivatives (chair rule, approximations by difference quotients, etc.) and turn to a discussion of $W^{1,\infty}(E)$ and its relation to Lipschitz functions. For the latter, we conclude the chapter by stating and proving the Rademaker Theorem.

Chapter 9 is a collection of topics of common use in real analysis and its applications. First is the Wiener version of the Vitali Covering Theorem (commonly referred to as the "simple version" of Vitali's Theorem). This is applied to the notion of maximal function, its properties, and its related strong type L^p estimates for $1 < p < \infty$. Weak estimates are also proved and used in the Marcinkiewicz Interpolation Theorem. We prove the by now classical Calderón-Zygmund Decomposition Theorem and its applications to the space functions of bounded mean oscillation (BMO) and the Stein-Fefferman L^p estimate for the sharp maximal function.

The space of BMO is given some emphasis. We give the proof of the John-Nirenberg estimate and provide its counterexample. We have in mind here the limiting case of some potential estimates (later in the chapter) and the limiting Sobolev embedding estimates (Chapter 10).

We introduce the notion of rearranging the values of functions and provide their properties and the related notion of equi-measurable function. The discussion is for functions of one real variable. Extensions to functions of N real variables are indicated in the Complements.

The goal is to prove the Riesz convolution inequality by rearrangements. The several proofs existing (Riesz, Zygmund, Hardy-Littlewood-Polya) all use, one way or another, the symmetric rearrangement of an integrable function.

We have reproduced here the proof of Hardy-Littlewood-Polya as appearing in their monograph [70]. In the process, we need to establish Hardy's inequality, of interest in its own right.

The Riesz convolution inequality is presented in several of its variants, leading to an N-dimensional version of it through an application of the continuous version of the Minkowski inequality.

Besides its intrinsic interest of these inequalities, what we have in mind here is to recover some limiting cases of potential estimates an their related Sobolev embedding inequalities.

The final part of the chapter introduces the Riesz potentials and their related L^p estimates, including some limiting cases. These are on one hand based on the previous Riesz convolution inequality, and on the other hand to Trudinger's version of the BMO estimates for particular functions arising as potentials.

Chapter 10 provides an array of embedding theorems for functions in Sobolev spaces. Their importance to analysis and partial differential equations cannot be underscored. Although good monographs exist ([1, 104]), I have found it laborious to extract the main facts, listed in a clean manner and ready for applications.

We start from the classical Gagliardo-Nirenberg inequalitites and proceed to Sobolev inequalities. We have made an effort to trace, in the various embedding inequalities, how the smoothness of the boundary enters in the estimates. For example, whenever the cone condition is required, we trace back in the various constant the dependence on the height and the angle of the cone. We present the Poincaré inequalities for bounded, convex domains E, and trace the dependence of the various constants on the "modulus of convexity" of the domain through the ratio of the radius of the smallest ball containing E and the largest ball contained in E. The limiting case $p = N$ of the Sobolev inequality builds of the limiting inequalities for the Riesz potentials, and is preceded by an introduction to Morrey spaces and their connection to BMO.

The characterization of the compact subsets of $L^p(E)$ (Chapter 6) is used to prove Reillich's Theorem on compact Sobolev inequalities.

We introduce the notion of trace of function in $W^{1,p}(\mathbb{R}^N \times \mathbb{R}^+)$ on the hyperplane $x_{N+1} = 0$. Through a partition of unity and a local covering, this provides the notion of trace of functions in $W^{1,p}(E)$ on the boundary ∂E, provided such a boundary is sufficiently smooth. Sharp inequalities relating functions in $W^{1,p}(E)$ with the integrability and regularity of their traces on ∂E are established in terms of fractional Sobolev spaces. Such inequalities are first established for E being a half-space and ∂E an hyperplane, and then extended to general domains E with sufficiently smooth boundary ∂E. In the Complements we characterize functions f defined and integrable on ∂E as traces on ∂E of functions in some Sobolev spaces $W^{1,p}(E)$. The relation between p and the order of integrability of f on ∂E is shown to be sharp. For special geometries, such as a ball, the inequality relating the integral of the traces and the Sobolev norm can be made explicit. This is indicated in the Complements.

The last part of the chapter contains a newly established *multiplicative* Sobolev embedding for functions in $W^{1,p}(E)$ that do not necessarily vanish on ∂E. The open set E is required to be convex. Its value is in its applicability to the asymptotic behavior of solutions to Neumann problems related to parabolic partial differential equations.

Acknowledgments

These notes have grown out of the courses and topics in real analysis I have taught over the years at Indiana University, Bloomington; Northwestern University; the University of Rome Tor Vergata, Italy; and Vanderbilt University.

My thanks go to the numerous students who have pointed out misprints and imprecise statements. Among them are John Renze, Ethan Pribble, Lan Yueheng, Kamlesh Parwani, Ronnie Sadka, Marco Battaglini, Donato Gerardi, Zsolt Macskasi, Tianhong Li, Todd Fisher, Liming Feng, Mikhail Perepelitsa, Lucas Bergman, Derek Bruff, David Peterson, and Yulya Babenko.

Special thanks go to Michael O'Leary, David Diller, Giuseppe Tommassetti, and Gianluca Bonuglia. The material of the last three chapters results from topical seminars organized, over the years, with these former students.

I am indebted to Allen Devinatz[†], Edward Nussbaum, Ethan Devinatz, Haskell Rhosental, and Juan Manfredi for providing me with some counterexamples and for helping me to make precise some facts related to unbounded linear functionals in normed spaces.

I would like also to thank Henghui Zou, James Serrin[†], Avner Friedman, Craig Evans, Robert Glassey, Herbert Amann, Enrico Magenes[†], Giorgio Talenti, Gieri Simonett, Vincenzo Vespri, and Mike Mihalik for reading, at various stages, portions of the manuscript and for providing valuable critical comments.

The input of Daniele Andreucci has been crucial and it needs to be singled out. He has read the entire manuscript and has made critical remarks and suggestions. He has also worked out in detail a large number of the problems and suggested some of his own. I am very much indebted to him.

These notes were conceived as a book in 1994, while teaching topics in real analysis at the School of Engineering of the University of Rome Tor Vergata, Italy. Special thanks go to Franco Maceri, former Dean of the School of Engineering of that university, for his vision of mathematics as the natural language of applied sciences and for fostering that vision.

I learned real analysis from Carlo Pucci[†] at the University of Florence, Italy in 1974–1975. My view of analysis and these notes are influenced by Pucci's teaching. According to his way of thinking, every theorem had to be motivated and had to go along with examples and counterexamples, i.e., had to withstand a scientific "critique."

Contents

Chapter 1
Preliminaries

1 Countable Sets

A set E is countable if it can be put in one-to-one correspondence with a subset of the natural numbers $\mathbb{N}$. Every subset of a countable set is countable.

Proposition 1.1 *The set S_E of the finite sequences of elements of a countable set E is countable.*

Proof Let $\{2, 3, 5, 7, 11, \ldots, m_j \ldots\}$ be the sequence of prime numbers. Every positive integer n has a unique factorization, of the type

$$n = 2^{\alpha_1} 3^{\alpha_2} \cdots m_j^{\alpha_j}$$

where the sequence $\{\alpha_1, \ldots, \alpha_j\}$ is finite and the α_i are nonnegative integers. Let now $\sigma = \{e_1, \ldots, e_j\}$ be an element of S_E. Since E is countable, to each element e_i there corresponds a unique positive integer α_i. Thus to σ there corresponds the unique positive integer given by the indicated factorization. ∎

Corollary 1.1 *The set of pairs $\{m, n\}$ of integers is countable.*

Corollary 1.2 *The set $\mathbb{Q}$ of the rational numbers is countable.*

Proof The rational numbers can be put in one-to-one correspondence with a subset of the ratios $\frac{m}{n}$ for two integers m, n with $n \neq 0$. ∎

Proposition 1.2 *The union of a countable collection of countable sets is countable.*

Proof Let $\{E_j\}$ be a countable collection of countable sets. Since each of the E_j is countable, their elements may be listed as

© Springer Science+Business Media New York 2016
E. DiBenedetto, *Real Analysis*, Birkhäuser Advanced
Texts Basler Lehrbücher, DOI 10.1007/978-1-4939-4005-9_1

$$E_1 = \{a_{11}\ a_{12}\ a_{13}\ \ldots\ a_{1n}\ \ldots\}$$
$$E_2 = \{a_{21}\ a_{22}\ a_{23}\ \ldots\ a_{2n}\ \ldots\}$$
$$\cdots = \{\cdots\ \ \ \cdots\ \ \ \cdots\ \ \ \cdots\ \ \ \cdots\ \ \cdots\}$$
$$E_m = \{a_{m1}\ a_{m2}\ a_{m3}\ \ldots\ a_{mn}\ \ldots\}$$
$$\cdots = \{\cdots\ \ \ \cdots\ \ \ \cdots\ \ \ \cdots\ \ \ \cdots\ \ \cdots\}$$

Thus the elements of $\cup E_j$ are in one-to-one correspondence with a subset of the ordered pairs $\{m, n\}$ of natural numbers. ∎

Proposition 1.3 (Cantor [22]) *The interval* $[0, 1]$ *as a subset of* $\mathbb{R}$, *is not countable.*

Proof The proof uses Cantor's diagonal process. If $[0, 1]$ were countable, its elements could be listed as the countable collection

$$x_j = 0.a_{1j}a_{2j}\cdots a_{mj}\cdots \quad \text{for } j \in \mathbb{N}$$

where a_{mj} are integers from 0 to 9. Now set $a_j = 1$ if a_{jj} is even and $a_j = 2$ if a_{jj} is odd. Then, the element $x = 0.a_1 a_2 \cdots$, is in $[0, 1]$ and is different from any one of the $\{x_j\}$. ∎

2 The Cantor Set

Divide the closed interval $[0, 1]$ into 3 equal subintervals and remove the central open interval $I_1 = (\frac{1}{3}, \frac{2}{3})$, so that $[0, 1] - I_1 = [0, \frac{1}{3}] \cup [\frac{2}{3}, 1]$. Subdivide each of these intervals in 3 equal parts and remove their central open interval. If I_2 is the set that has been removed

$$I_2 = \left(\tfrac{1}{3^2}, \tfrac{2}{3^2}\right) \cup \left(\tfrac{7}{3^2}, \tfrac{8}{3^2}\right)$$
$$[0, 1] - I_1 \cup I_2 = \left[0, \tfrac{1}{3^2}\right] \cup \left[\tfrac{2}{3^2}, \tfrac{3}{3^2}i\right] \cup \left[\tfrac{6}{3^2}, \tfrac{7}{3^2}\right] \cup \left[\tfrac{8}{3^2}, 1\right].$$

We subdivide each of the closed intervals making up $[0, 1] - I_1 \cup I_2$, into 3 equal subintervals and remove their central open interval. If I_3 is the set that has been removed

$$I_3 = \left(\tfrac{1}{3^3}, \tfrac{2}{3^3}\right) \cup \left(\tfrac{7}{3^3}, \tfrac{8}{3^3}\right) \cup \left(\tfrac{19}{3^3}, \tfrac{20}{3^3}\right) \cup \left(\tfrac{25}{3^3}, \tfrac{26}{3^3}\right)$$
$$[0, 1] - (I_1 \cup I_2 \cup I_3) = \left[0, \tfrac{1}{3^3}\right] \cup \left[\tfrac{2}{3^3}, \tfrac{3}{3^3}\right] \cup \left[\tfrac{6}{3^3}, \tfrac{7}{3^3}\right] \cup \left[\tfrac{8}{3^3}, \tfrac{9}{3^3}\right]$$
$$\cup \left[\tfrac{18}{3^3}, \tfrac{19}{3^3}\right] \cup \left[\tfrac{20}{3^3}, \tfrac{21}{3^2}\right] \cup \left[\tfrac{24}{3^3}, \tfrac{25}{3^3}\right] \cup \left[\tfrac{26}{3^3}, 1\right].$$

Proceeding in this fashion defines a sequence of disjoint open sets I_n, each being the finite, disjoint union of open intervals and satisfying

$$\text{meas}(I_n) = \frac{2^{n-1}}{3^n} \quad \text{and} \quad \sum \text{meas}(I_n) = 1.$$

The Cantor set $C = [0, 1] - \cup I_n$ is the set that remains after removing, the union of the I_n out of $[0, 1]$. The Cantor set C is closed and each of its point is an accumulation point of the extremes of the intervals I_n. Thus C coincides with the set of all its accumulation points. Set

$$
\mathcal{E} = \left\{ \begin{array}{l} \text{the collection of all sequences } \{\varepsilon_n\} \\ \text{where the numbers } \varepsilon_n \text{ are either } 0 \text{ or } 1 \end{array} \right\}.
$$

Every element $x \in C$ can be represented as (**2.1–2.2** of the Complements)

$$
x = \sum \frac{2}{3^j} \varepsilon_j \qquad \text{for some sequence } \{\varepsilon_n\} \in \mathcal{E}. \tag{2.1}
$$

Every element of C is associated to one and only one sequence $\{\varepsilon_n\} \in \mathcal{E}$ by the representation formula (2.1). For example

$$
\begin{array}{ll}
\frac{1}{3} \leftrightarrow \{0, 1, 1, 1, \ldots, 1, \ldots\} & \frac{2}{3} \leftrightarrow \{1, 0, 0, 0, \ldots, 0, \ldots\} \\
\frac{1}{9} \leftrightarrow \{0, 0, 1, 1, \ldots, 1, \ldots\} & \frac{2}{9} \leftrightarrow \{0, 1, 0, 0, \ldots, 0, \ldots\} \\
\frac{7}{9} \leftrightarrow \{1, 0, 1, 1, \ldots, 1, \ldots\} & \frac{8}{9} \leftrightarrow \{1, 1, 0, 0, \ldots, 0, \ldots\}.
\end{array}
$$

Vice versa any such a sequence identifies by (2.1) one and only one element of C. The set of all sequences in $\mathcal{E}$ has the cardinality of the real numbers in $[0, 1]$, being their binary representation. Thus C has the cardinality of $\mathbb{R}$ and therefore is uncountable. It also follows from (2.1) that the Cantor set could be defined alternatively as the set of those numbers in $[0, 1]$ whose ternary expansion has only the digits 0 and 2. The two definitions are equivalent.

3 Cardinality

Two sets X and Y have the same *cardinality* if there exists a one-to-one map f from X onto Y. In such a case one writes $\text{card}(X) = \text{card}(Y)$. If X is finite, then $\text{card}(X)$ is the number of elements of X. The formal inequality $\text{card}(X) \le \text{card}(Y)$ means that there exists a one-to-one map from X into Y. In particular if $X \subset Y$, then $\text{card}(X) \le \text{card}(Y)$. The formal inequality $\text{card}(X) \ge \text{card}(Y)$ means that there exists a one-to-one map from X onto Y. In particular if $X \supset Y$, then $\text{card}(X) \ge \text{card}(Y)$.

Proposition 3.1 (Schöder–Bernstein) *Let X and Y be any two sets. If* $\text{card}(X) \le \text{card}(Y)$ *and* $\text{card}(X) \ge \text{card}(Y)$, *then* $\text{card}(X) = \text{card}(Y)$.

Proof Let f and be a one-to-one function from X into Y and let g be a one-to-one function from Y into X. Partition X into the disjoint union of three sets X_o, X_1, X_2 by the following iterative procedure. If x is not in the range of g we say that $x \in X_1$. If

x is in the range of g form $g^{-1}(x) \in Y$. If $g^{-1}(x)$ is not in the range of f the process terminates and we say that $x \in X_2$. Otherwise form $f^{-1}(g^{-1}(x))$. If such an element is not in the range of g the process terminates and we say that $x \in X_1$. Proceeding in this fashion, either the process can be continued indefinitely or it terminates. If it terminates with an element not in the range of g we say that the starting element is in X_1. If it terminates with an element not in the range of f we say that the starting x is in X_2. If it can be continued indefinitely we say that $x \in X_o$. The three sets X_o, X_1, X_2 are disjoint and $X = X_o \cup X_1 \cup X_2$. Similarly $Y = Y_o \cup Y_1 \cup Y_2$ where the sets Y_j for $j = 0, 1, 2$ are constructed similarly. By construction f is a bijection from X_o onto Y_o and from X_1 onto Y_2. Similarly g is a bijection from Y_1 onto X_2. Thus the map $h : X \to Y$ defined by

$$h(x) = \begin{cases} f(x) & \text{if } x \in X_o \cup X_1 \\ g^{-1}(x) & \text{if } x \in X_2 \end{cases}$$

is a one-to-one map from X onto Y.   ∎

Corollary 3.1 $\operatorname{card}(X) = \operatorname{card}(Y)$ *if and only if* $\operatorname{card}(X) \le \operatorname{card}(Y)$ *and* $\operatorname{card}(X) \ge \operatorname{card}(Y)$.

The formal strict inequality $\operatorname{card}(X) < \operatorname{card}(Y)$ means that any one-to-one function $f : X \to Y$ is not a surjection, that is, roughly speaking, X contains strictly fewer elements than Y. For example $\operatorname{card}(\mathbb{N}) < \operatorname{card}(\mathbb{R})$.

3.1 Some Examples

A set X has the cardinality of $\mathbb{N}$ if it can be put in a one-to-one correspondence with $\mathbb{N}$. In such a case one writes $\operatorname{card}(X) = \operatorname{card}(\mathbb{N})$. For example $\operatorname{card}(\mathbb{Z}) = \operatorname{card}(\mathbb{Q}) = \operatorname{card}(\mathbb{N})$. A set X has the cardinality of $\mathbb{R}$ if it can be put in a one-to-one correspondence with $\mathbb{R}$. In such a case one writes $\operatorname{card}(X) = \operatorname{card}(\mathbb{R})$. For example if $\mathcal{C}$ is the Cantor set, $\operatorname{card}(\mathcal{C}) = \operatorname{card}(\mathbb{R})$. For a positive integer m denote by

$$X^m = \underbrace{X \times X \times \cdots \times X}_{m \text{ times}}$$

the collection of all m-tuples $(x_1, \ldots, x_m)$ of elements of X. Also, denote by 2^X the set of all subsets of a set X. Thus $2^{\mathbb{N}}$ is the collections of all subsets of $\mathbb{N}$ and $2^{\mathbb{R}}$ is the collection of all subsets of $\mathbb{R}$.

Proposition 3.2 $\operatorname{card}(\mathbb{N}^m) = \operatorname{card}(\mathbb{N})$ *and* $\operatorname{card}(2^{\mathbb{N}}) = \operatorname{card}(\mathbb{R})$. *Moreover for any* X *there holds* $\operatorname{card}(2^X) > \operatorname{card}(X)$.

Proof The first statement follows form Proposition 1.1. A non-empty subset $A \subset \mathbb{N}$, consists of an increasing sequence, finite or infinite, of positive integers, say for example $A = \{m_1, \ldots, m_n, \ldots\}$. Label by zero the elements of $\mathbb{N} - A$, and by 1 the elements of A, and keep their ordering within $\mathbb{N}$. This uniquely identifies the sequence

$$\varepsilon_A = \{0, \ldots, 1_{m_1}, \ldots, 0, \ldots, 1_{m_2}, \ldots\} \in \mathcal{E}.$$

If $A = \emptyset$ associate to $\emptyset$ the sequence $\varepsilon_\emptyset$ whose elements are all zero. Conversely, any sequence $\{\varepsilon_n\} \in \mathcal{E}$ identifies uniquely, by the inverse process, one and only one element of $2^\mathbb{N}$. Thus $2^\mathbb{N}$ is in one-to-one correspondence with the Cantor set $\mathcal{E}$, which, in turn, is in one-to-one correspondence with $\mathbb{R}$.

The last statement is established by proving that no function $f : X \to 2^X$ is a surjection. Let f be any such function and set $A_f = \{x \in X | x \notin f(x)\}$. The set A_f could be empty. If f were a surjection, there would exists $y \in X$ such that $f(y) = A_f$. By the definition of A_f, such a y cannot be neither in A_f nor in $X - A_f$. ∎

Corollary 3.2 *Given any non-empty set X, there exists a set Y containing X and of strictly larger cardinality.*

Corollary 3.3 $\text{card}(\mathbb{R}) < \text{card}(2^\mathbb{R})$.

4 Cardinality of Some Infinite Cartesian Products

For a positive integer m, the set X^m is the collection of all m-tuples of elements of X. Any such m-tuple $(x_1, \ldots, x_m)$ can be regarded as a function from the first m integers $(1, \ldots, m)$ into X. By analogy $X^\mathbb{N}$ is defined as the collection of all sequences of elements of X, and any such sequence $\{x_n\}$ can be regarded as a function from $\mathbb{N}$ into X. For example $\mathbb{R}^\mathbb{N}$ is the collection of all sequences of real numbers or equivalently is the collection of all functions $f : \mathbb{N} \to \mathbb{R}$. The product space $(0, 1)^\mathbb{N}$ is called the Hilbert cube and is the collection of all sequences $\{x_n\}$ of elements in $(0, 1)$. Let $\{0, 1\}$ denote any set consisting of only 2 elements, say for example 0 and 1. Then $\{0, 1\}^\mathbb{N}$ is in one-to-one correspondence with the Cantor set. Therefore $\text{card}(\{0, 1\}^\mathbb{N}) = \text{card}(\mathbb{R})$.

If A is any set, countable or not, the Cartesian product X^A is defined to be the collection of all functions $f : A \to X$. For example $(0, 1)^{(0,1)}$ is the collection of all functions $f : (0, 1) \to (0, 1)$. Likewise $\mathbb{R}^\mathbb{R}$ is the set of all real valued functions defined in $\mathbb{R}$.

Proposition 4.1 $\text{card}(2^\mathbb{N} \times 2^\mathbb{N}) = \text{card}(2^\mathbb{N}) = \text{card}(\mathbb{R})$.

Proof We exhibit a one-to-one correspondence between $(2^\mathbb{N} \times 2^\mathbb{N})$ and $2^\mathbb{N}$. For any two subsets A and B of $\mathbb{N}$, set

$$g(A) = \bigcup_{n \in A} \{2n\}, \qquad h(B) = \bigcup_{m \in B} \{2m - 1\}.$$

Then for any pair of sets $A, B \in 2^{\mathbb{N}}$ define

$$f(A, B) = g(A) \cup h(B) = C \in 2^{\mathbb{N}}.$$

By this procedure, every element $(A, B) \in 2^{\mathbb{N}} \times 2^{\mathbb{N}}$ is mapped into one and only one element of $2^{\mathbb{N}}$. Vice-versa, given any $C \in 2^{\mathbb{N}}$, by separating its even and odd numbers, identifies in a unique manner two sets A and B in $2^{\mathbb{N}}$, and hence a unique element $(A, B) \in 2^{\mathbb{N}} \times 2^{\mathbb{N}}$.   ∎

Corollary 4.1 $\mathrm{card}(\mathbb{R}^m) = \mathrm{card}(\mathbb{R})$, *for all $m \in \mathbb{N}$*.

Proof Let first $m = 2$. Then

$$\mathrm{card}(\mathbb{R}^2) = \mathrm{card}(\mathbb{R} \times \mathbb{R}) = \mathrm{card}(2^{\mathbb{N}} \times 2^{\mathbb{N}}) = \mathrm{card}(\mathbb{R}).$$

For general $m \in \mathbb{N}$ the statement follows by induction.   ∎

Proposition 4.2 *Let X, Y and Z be any triple of sets. Then*

$$\mathrm{card}\left(X^{Y \times Z}\right) = \mathrm{card}\left[\left(X^Y\right)^Z\right].$$

Proof Let $f \in X^{Y \times Z}$ so that $f(y, z) \in X$ for all pairs $(y, z) \in Y \times Z$. For a fixed $z \in Z$, set $h_z(y) = f(y, z)$. This gives an element of X^Y. Thus as z ranges over Z, the map $h_z(\cdot)$ uniquely identifies a function from Z into X^Y. Conversely any element $h_z \in (X^Y)^Z$ uniquely identifies an element $f \in X^{Y \times Z}$ by the same formula.   ∎

Corollary 4.2 $\mathrm{card}(\mathbb{R}^{\mathbb{N}}) = \mathrm{card}(\mathbb{R})$.

Proof Since $\mathbb{R}$ is in one-to-one correspondence with $\{0, 1\}^{\mathbb{N}}$

$$\mathrm{card}(\mathbb{R}^{\mathbb{N}}) = \mathrm{card}\left[(\{0, 1\}^{\mathbb{N}})^{\mathbb{N}}\right] = \mathrm{card}\left(\{0, 1\}^{\mathbb{N} \times \mathbb{N}}\right) = \mathrm{card}(\{0, 1\}^{\mathbb{N}}) = \mathrm{card}(\mathbb{R})$$

since $\mathbb{N} \times \mathbb{N}$ and $\mathbb{N}$ are in one-to-one correspondence.   ∎

Corollary 4.3 $\mathrm{card}\left(\mathbb{N}^{\mathbb{N}}\right) = \mathrm{card}(\mathbb{R})$.

Proof An element of $\mathbb{N}^{\mathbb{N}}$ is a sequence of elements of $\mathbb{N}$. In particular $\mathbb{N}^{\mathbb{N}}$ contains those sequences containing only two fixed elements of $\mathbb{N}$, say for example $\{1, 2\}$. Therefore $\{1, 2\}^{\mathbb{N}} \subset \mathbb{N}^{\mathbb{N}}$, and

$$\mathrm{card}\left(\mathbb{N}^{\mathbb{N}}\right) \geq \mathrm{card}\left(\{1, 2\}^{\mathbb{N}}\right) = \mathrm{card}(\mathbb{R}).$$

On the other hand $\mathbb{N}^{\mathbb{N}}$ is contained in $\mathbb{R}^{\mathbb{N}}$, and

$$\mathrm{card}\left(\mathbb{N}^{\mathbb{N}}\right) \leq \mathrm{card}\left(\mathbb{R}^{\mathbb{N}}\right) = \mathrm{card}(\mathbb{R}).$$ ∎

5   Orderings, the Maximal Principle, and the Axiom of Choice

A relation $\prec$ on a set X, is a *partial ordering* of X if it is transitive ($x \prec y$ and $y \prec z$ implies $x \prec z$) and antisymmetric ($x \prec y$ and $y \prec x$ implies $x = y$). The relation $\leq$ of *less than or equal to* is a partial ordering of $\mathbb{R}$. The *set inclusion* $\subseteq$ partially orders the set 2^X of all subsets of X.

A partial ordering $\prec$ on a set X is a *linear or total ordering* on X if for any two elements $x, y \in X$, either $x \prec y$ or $y \prec x$. The relation $\leq$ is a linear ordering on $\mathbb{R}$, whereas $\subseteq$ is not a linear ordering on 2^X.

Let X be a set partially ordered by a relation $\prec$ and let $E \subset X$. An *upper bound* of a subset $E \subset X$, is an element $x \in X$ such that $y \prec x$ for all $y \in E$. If $x \in E$, then x is a *maximal element* of E.

A linearly ordered subset $E \subset X$ is *maximal* if any linearly ordered subset of X containing E, coincides with E. Partial and linear ordering are meant with respect to the same ordering $\prec$. In particular if E is a proper subset of X, linearly ordered and maximal with respect to the same relation $\prec$ by which X is partially ordered, then for all $x \in X - E$, the set $E \cup \{x\} \subset X$ is not linearly ordered, by the same ordering $\prec$.

The Hausdorff Maximal Principle: *Every partially ordered set contains a maximal linearly ordered subset, by the same ordering.*

The Hausdorff Maximal Principle will be taken here as an *Axiom*. The maximality and ordering statements given below, that is, *Zorn's lemma*, the *Axiom of Choice*, and the *Well Ordering Principle*, will be proven to be a consequence of the Hausdorff Maximal Principle. Actually it can be proven that all these maximality and ordering statements are equivalent, that is, any one of them, taken as an *Axiom*, implies the remaining three ([80]).

Zorn's Lemma: *Let X be a partially ordered set X, such that every linearly ordered subset has an upper bound. Then X has a maximal element.*

Proof Let M be the maximal linearly ordered set claimed by the Maximal Principle. An upper bound for M is a maximal element of X.   ∎

The Axiom of Choice: *Let $\{X_\alpha\}$ be a collection of nonempty sets, as the index α ranges over some set A. There exists a function f defined in A, such that $f(\alpha) \in X_\alpha$. Equivalently one may choose an element x_α out of each set X_α. If the sets of the collection $\{X_\alpha\}$ are disjoint, one might select one and only one representative out of each X_α.*

Proof A function φ that maps the elements β of a subset $B \subset A$ into elements of X_β, may be regarded as a set of ordered pairs $\varphi = \cup\{(\beta, x_\beta)$ where $\beta \in B$ and $x_\beta \in X_\beta$, such that no two of such pairs have the same first coordinate. The set Φ of all such φ is partially ordered by inclusion. Let Ψ be a linearly ordered subset of Φ and set

$\psi = \cup\{\varphi | \varphi \in \Psi\}$. Such a ψ is union of pairs (α, x_α) for $\alpha \in A$ and $x_\alpha \in X_\alpha$. Since Ψ is linearly ordered, any two such pairs (α, x_α) and (β, x_β) must belong to some $\varphi \in \Psi$. Therefore $\alpha \neq \beta$ since φ is a function. This implies that ψ is a function in Φ and is an upper bound for Ψ. Thus every linearly ordered subset of Φ has an upper bound. By Zorn's lemma Φ has a maximal element f. To prove the axiom it remains to show that the domain of f is A. If not, we may select an element $\alpha \in A - \text{dom}\{f\}$ and associate to it an element $x_\alpha \in X_\alpha$, since X_α is not empty. Then the function $f' = f \cup (\alpha, x_\alpha)$ is in Φ and $f \subset f'$, against the maximality of f. ∎

Corollary 5.1 *Let X be a set. There exists a function $f : 2^X \to X$ such that $f(E) \in E$, for every $E \subset X$. Equivalently, one may choose an element out of every subset of X.*

6 Well Ordering

A linear ordering $\prec$ is a *well ordering* of X if every subset of X has a first element. The set $\mathbb{N}$ of positive integers is well ordered by $\leq$. The set $\mathbb{R}$ is not well ordered by the same ordering.

Well Ordering Principle ([177]): *Every set X can be well ordered, that is, there exists a relation $\prec$ that well orders X.*

Proof Let $f : 2^X \to X$ be a function as in Corollary 5.1, whose existence is guaranteed by the axiom of choice. Set

$$x_1 = f(X) \quad \text{and} \quad x_n = f\left(X - \bigcup_{j=1}^{n-1} x_j\right) \quad \text{for } n \geq 2.$$

The sequence $\{x_n\}$ can be given the ordering of $\mathbb{N}$ and, as such, is well ordered. Let D be a subset of X and let $\prec$ be a linear ordering defined on D. A subset $E \subset D$ is a *segment*, relative to $\prec$ if for any $x \in E$, all the elements $y \in D$ such that $y \prec x$, belong to E. The segments of $\{x_n\}$, relative to the ordering induced by $\mathbb{N}$, are the sets of the form $\{x_1, x_2, \ldots, x_m\}$ for some $m \in \mathbb{N}$. The union and intersection of two segments is a segment. The empty set is a segment relative to any linear ordering $\prec$. Denote by $\mathcal{F}$ the family of *linear* orderings $\prec$ defined on subsets $D \subset X$ and satisfying:

If $E \subset D$ is a segment, then the first element of $D - E$ is $f(X - E)$.   (∗)

Such a family is not empty since the ordering of $\mathbb{N}$ on the domain $D = \{x_n\}$ is in $\mathcal{F}$.

Lemma 6.1 *Every element of $\mathcal{F}$ is a well ordering on its domain.*

Proof Let $D \subset X$ be the domain of a linear ordering $\prec \in \mathcal{F}$. For a nonempty subset $A \subset D$ let

$$E = \{y \in D \mid y \prec x \text{ for all } x \in A \text{ and } y \neq x\}.$$

Then the first element of A is the first element of $D - E$ and the latter is $f(X - E)$ since E is a segment.   ∎

Lemma 6.2 *Let $\prec_1$ and $\prec_2$ be two elements in $\mathcal{F}$ with domains D_1 and D_2. Then, one of the two domains, say for example D_1 is a segment for the other, say for example D_2 with respect to the corresponding ordering $\prec_2$. Moreover $\prec_1$ and $\prec_2$ coincide on such a segment.*

Proof Let E denote the set of all x such that

$$\{y \in D_1 \mid y \prec_1 x\} = \{y \in D_2 \mid y \prec_2 x\}$$

and such that $\prec_1$ and $\prec_2$ agree on these two sets. By construction E is a segment for both D_1 and D_2 with respect to their orderings. The set E must coincide with at least one of D_1 and D_2. If not, the element $f(X - E)$ is the first element of both $D_1 - E$ and $D_2 - E$. By the definition of E, such an element is in E. This is however a contradiction since $f(X - E) \notin E$.   ∎

Let D_o be the union of the domains of the elements of $\mathcal{F}$. Let also $\prec_o$ be that ordering on D_o that coincides with the ordering $\prec$ in $\mathcal{F}$ on its domain D. By Lemma 6.2 this is a linear ordering on D_o and it satisfies the requirement (*) of the class $\mathcal{F}$. Therefore by Lemma 6.1, it is a well ordering on D_o. It remains to show that $D_o = X$. Consider the set $D_o' = D_o \cup \{f(X - D_o)\}$, and the ordering $\prec_o'$ that coincides with $\prec_o$ on D_o and by which $f(X - D_o)$ follows any element of D_o. One checks that $\prec_o'$ satisfies the requirements of the class $\mathcal{F}$ and its domain is D_o'. Therefore $D_o = D_o'$. This however is a contradiction unless $X - D_o = \emptyset$.   ∎

6.1   The First Uncountable

Let X be an uncountable set, well ordered by the ordering $\prec$. Without loss of generality we may assume that X has an upper bound, that is, there exists some $x^* \in X$ such that $x \prec x^*$ for all $x \in X - \{x^*\}$. Indeed if not we may add to X an element x^* and on $X \cup \{x^*\}$ define an ordering $\prec^*$ to coincide with $\prec$ on X and by which $x \prec x^*$ for all $x \in X$.

For $x \in X$, set $E_x = \{y \in X \mid y \prec x\}$. If E_x is a finite set, then x is called a *finite ordinal*. Consider now the set $E^o = \{x \in X \mid E_x \text{ is infinite}\}$. Since X is infinite and has an upper bound $E^o \neq \emptyset$. Since X is well ordered by $\prec$, the set E^o has a least element denoted by ω. The set E_ω is infinite and for each $x \prec \omega$ the set E_x is finite. For this reason ω is referred to as the *first infinite ordinal*. The set E_ω is the set of *finite ordinals* and is in one-to-one correspondence with the set $\mathbb{N}$ of the natural numbers ordered with the natural ordering of $\mathbb{N}$. If E_x is a countable set, then x is called a *countable ordinal*.

Set $E^1 = \{x \in X | E_x \text{ is uncountable}\}$. Since X is uncountable and has an upper bound, $E^1 \neq \emptyset$. Since X is well ordered by $\prec$, the set E^1 has a least element denoted by Ω. The set E_Ω is uncountable and for each $x \prec \Omega$ the set E_x is countable. For this reason Ω is referred to as the *first uncountable ordinal*.

Let $\mathbb{R}$ be well ordered by $\prec$. The cardinality of E_ω is denoted by $\aleph_o$. The cardinality of E_Ω is denoted by $\aleph_1$. The inclusion $E_\Omega \subset \mathbb{R}$ implies $\aleph_1 \leq \mathrm{card}(\mathbb{R})$. The *continuum hypothesis* states that the cardinality of $\mathbb{R}$ is $\aleph_1$, that is, roughly speaking, there are no sets, in the sense of cardinality, between E_Ω and $\mathbb{R}$.

Problems and Complements

1c   Countable Sets

1.1. A real number x is called *algebraic* if it is the root of a polynomial with rational coefficients. Prove that the set of algebraic numbers is countable.

1.2. Prove that the set of all the prime numbers is countable.

2c   The Cantor Set

2.1. Let $s \geq 2$ be a positive integer and let $\{\sigma_n\}$ be a sequence of positive integers that can take only the values $1, \ldots, s-1$. Any $x \in [0, 1]$ has the representation

$$x = \sum \frac{\sigma_n}{s^n} \quad \text{where } \{\sigma_n\} \text{ is one such sequence.} \qquad (2.1c)$$

The sequence $\{\sigma_n\}$ that identifies x is unique except if x is of the form $x = r/s^n$ for some positive integer r. In the latter case there are exactly 2 sequences that identify x. Conversely if $\{\sigma_n\}$ is any such sequence, then (2.1c) converges to a number $x \in [0, 1]$. For $s = 2, 3, 10$, this gives the binary, ternary or decimal representation of x.

2.2. Let L_n and R_n denote respectively the set of all the left and right hand points of the 2^{n-1} intervals removed out of $[0, 1]$, at the nth step of the construction of the Cantor set. For example $L_o = \{1\}$ and $R_o = \{0\}$, and

$$L_1 = \{\tfrac{1}{3}\} \; L_2 = \{\tfrac{1}{3^2}, \tfrac{7}{3^2}\} \; L_3 = \{\tfrac{1}{3^3}, \tfrac{7}{3^3}, \tfrac{19}{3^3}, \tfrac{25}{3^3}\} \quad \cdots$$

$$R_1 = \{\tfrac{2}{3}\} \; R_2 = \{\tfrac{2}{3^2}, \tfrac{8}{3^2}\} \; R_3 = \{\tfrac{2}{3^3}, \tfrac{8}{3^3}, \tfrac{20}{3^3}, \tfrac{26}{3^3}\} \quad \cdots.$$

By construction $\cup(L_n \cup R_n) \subset \mathcal{C}$ with strict inclusion. The elements of L_n and R_n can be constructed recursively as

$$L_n = \left\{ \text{numbers of the form } r + \frac{1}{3^n} \text{ for } r \in \bigcup_{j=0}^{n-1} R_j \right\}$$

$$R_n = \left\{ \text{numbers of the form } r + \frac{2}{3^n} \text{ for } r \in \bigcup_{j=0}^{n-1} R_j \right\}.$$

It follows that if $\beta_n \in R_n$, there exist a finite sequence of positive integers $j_1 < j_2 < \cdots < j_{n-1} \leq (n-1)$, such that

$$\beta_n = \frac{2}{3^{j_1}} + \frac{2}{3^{j_2}} + \cdots + \frac{2}{3^{j_{n-1}}} + \frac{2}{3^n}.$$

Therefore β_n has a ternary expansion with only digits 0 and 2, of the form

$$\beta_n = \{0, \ldots, 2_{j_1}, 0, \ldots, 2_{j_2}, 0, \ldots, 2_{j_{n-1}}, 0, \ldots, 2_n, 0_{n+1}, 0, 0, \ldots\}.$$

Likewise an element $\alpha_n \in L_n$ is of the form

$$\alpha_n = \frac{2}{3^{j_1}} + \frac{2}{3^{j_2}} + \cdots + \frac{2}{3^{j_{n-1}}} + \frac{1}{3^n}.$$

The ternary expansion of α_n has only digits 0 and 2 and is of the form

$$\alpha_n = \{0, \ldots, 2_{j_1}, 0, \ldots, 2_{j_2}, 0, \ldots, 2_{j_{n-1}}, 0, \ldots, 1_n, 0_{n+1}, 0, 0, \ldots\}$$
$$= \{0, \ldots, 2_{j_1}, 0, \ldots, 2_{j_2}, 0, \ldots, 2_{j_{n-1}}, 0, \ldots, 0_n, 2_{n+1}, 2, 2, \ldots\}.$$

If (α_n, β_n) is an interval removed out of $[0, 1]$ at the nth step of the process, then α_n and β_n have the same ternary expansion up to the terms of order $(n-1)$. The term of order n is 2 for the right end point and is 0 for the left hand point. The remaining terms in the expansion of α_n are all 2 and those in the expansion of β_n are all 0.

2.1c A Generalized Cantor Set of Positive Measure

Let $\alpha \in (0, 1)$ and out of the interval $[0, 1]$ remove the central open interval

$$I_1^\alpha = \left(\frac{1}{2} - \frac{\alpha}{6}, \frac{1}{2} + \frac{\alpha}{6} \right) \qquad \text{of length} \qquad \frac{\alpha}{3}.$$

Out of each of the remaining two intervals $(0, \frac{1}{2} - \frac{\alpha}{6})$ and $(\frac{1}{2} + \frac{\alpha}{6}, 1)$, remove the two central intervals each of length $\frac{1}{9}\alpha$ and denote by I_2^α their union. Proceeding in this fashion, define sequences of open sets I_n^α each being the finite union of disjoint open intervals and satisfying

$$\text{meas}(I_n^\alpha) = \alpha \frac{2^{n-1}}{3^n} \quad \text{and} \quad \sum \text{meas}(I_n^\alpha) = \alpha.$$

The generalized Cantor set is the set $\mathcal{C}_\alpha = [0, 1] - \cup I_n^\alpha$ that remains after removing the union of the I_n^α out of $[0, 1]$.

2.2c A Generalized Cantor Set of Measure Zero

Fix $\delta \in (0, \frac{1}{2})$. From $[0, 1]$ remove the open middle interval $I_o = (\delta, 1 - \delta)$, of length $(1 - 2\delta)$, so that

$$[0, 1] - I_o = J_{1,1} \cup J_{1,2}, \quad \text{where} \quad J_{1,1} = [0, \delta], \quad J_{1,2} = [1 - \delta, 1].$$

Next, out of $J_{1,1}$ and $J_{1,2}$ remove the open middle intervals $I_{1,1}$ and $I_{1,2}$, each of length $\delta - 2\delta^2$, and denote by I_1 their union. Thus

$$I_1 = I_{1,1} \cup I_{1,2}, \quad \text{meas}(I_1) = 2\delta - 4\delta^2$$

where

$$I_{1,1} = (\delta^2, \delta - \delta^2), \quad I_{1,2} = (1 - (\delta - \delta^2), 1 - \delta^2).$$

The set that remains is

$$[0, 1] - (I_o \cup I_1) = J_{2,1} \cup J_{2,2} \cup J_{2,3} \cup J_{2,4}$$

where
$$J_{2,1} = [0, \delta^2] \qquad\qquad J_{2,2} = [\delta - \delta^2, \delta]$$
$$J_{2,3} = [1 - \delta, 1 - (\delta - \delta^2)] \qquad J_{2,4} = [1 - \delta^2, 1].$$

From this

$$\text{meas}(I_o \cup I_1) = 1 - 4\delta^2 \quad \text{and} \quad \sum_{j=1}^{4} \text{meas}(J_{2,j}) = 4\delta^2.$$

Next out of each $J_{2,j}$ remove the open middle interval $I_{2,j}$, of length $\delta^2 - 2\delta^3$ and denote by I_3 their union. Thus

$$I_3 = \bigcup_{j=1}^{4} I_{2,j} \qquad \text{and} \qquad \text{meas}(I_3) = 4\delta^2 - 8\delta^3.$$

The set that remains is

$$[0, 1] - \bigcup_{i=0}^{3} I_i = \bigcup_{j=1}^{2^3} J_{3,j}$$

where $J_{3,j}$ are closed disjoint sub-intervals of $[0, 1]$, each of length δ^3. Proceeding in this fashion we generate sequences of open sets I_n, each being the finite union of disjoint, open intervals and of disjoint, closed intervals $J_{m,n} \subset [0, 1]$, such that

$$[0, 1] - \bigcup_{i=0}^{n} I_i = \bigcup_{j=1}^{2^n} J_{n,j}.$$

Moreover

$$\sum_{i=0}^{n} \text{meas}(I_i) = 1 - 2^n \delta^n \qquad \text{and} \qquad \sum_{j=1}^{2^n} \text{meas}(J_{n,j}) = 2^n \delta^n.$$

The generalized Cantor set $\mathcal{C}_\delta$ is defined as $\mathcal{C}_\delta = [0, 1] - \cup I_i$. For $\delta = \frac{1}{3}$ this coincides with the Cantor set $\mathcal{C}$ introduced in § 2. By construction, $\text{meas}(\mathcal{C}_\delta) = 0$, for all $\delta \in (0, \frac{1}{2})$.

Remark 2.1c For each $n \in \mathbb{N}$ the Cantor set $\mathcal{C}_\delta$ is covered by the union of the closed intervals $J_{n,j}$ for $j = 1, 2, \ldots, 2^n$.

2.3c   *Perfect Sets*

A non-empty set $E \subset \mathbb{R}^N$ is *perfect* if it coincides with the set of its accumulation points. The Cantor set is perfect. The set of the rational numbers is not perfect.

Proposition 2.1c *A perfect set is uncountable.*

Proof Assume that $E = \bigcup x_n$, for a sequence $\{x_n\}$ of elements in $\mathbb{R}^N$. Pick $y_1 \in (E - x_1)$. Since $\text{dist}\{x_1, y_1\} > 0$, there exists a closed cube Q_1, centered at y_1 that avoids x_1. Since E is perfect and Q_1 is closed and bounded $Q_1 \cap E$ is compact and avoids x_1. Next, remove x_2 out of E. The point y_1 as an element of E, is an accumulation point of elements in $E - x_2$. Therefore the intersection $Q_1^o \cap (E - x_2)$ is not empty and we may pick an element y_2 out of it. Since y_2 is different than x_1 and x_2, there exists a closed cube Q_2 centered at y_2, contained in Q_1 and not containing x_1 and x_2. For such a cube $Q_1 \cap E \supset Q_2 \cap E$ is compact and avoids x_1

and x_2. Proceeding in this fashion construct a nested sequence of closed bounded sets $Q_n \cap E$ satisfying

$$Q_{n-1} \cap E \supset Q_n \cap E \qquad \text{is compact and avoids } x_1, \ldots, x_n.$$

Thus $\bigcap (Q_n \cap E) = \emptyset$. This however is impossible by **2.4** below.   ∎

2.4. Let $\{E_n\}$ be a sequence of non-empty, closed sets in $\mathbb{R}^N$ such that $E_{n+1} \subset E_n$. If E_{n_o} is bounded for some index n_o, then $\cap E_n \neq \emptyset$. If all the E_n are unbounded, their intersection might be empty.

3c   Cardinality

Below is an equivalent but constructive proof of the Schröder-Bernstein Theorem. Define

$$X_o = X \quad Y_o = Y \qquad \text{and} \qquad X_1 = f(X_o) \quad Y_1 = g(Y_o)$$

and then inductively for all $n \in \mathbb{N}$

$$X_{2n} = g(X_{2n-1}) \qquad X_{2n+1} = f(X_{2n})$$
$$Y_{2n} = f(Y_{2n-1}) \qquad Y_{2n+1} = g(Y_{2n}).$$

By construction

$$X_o \supseteq Y_1 \supseteq X_2 \supseteq Y_3 \supseteq X_4 \supseteq Y_5 \cdots$$
$$Y_o \supseteq X_1 \supseteq Y_2 \supseteq X_3 \supseteq Y_4 \supseteq X_5 \cdots .$$

From this

$$\bigcap_{j=0}^{\infty} X_{2j} = \bigcap_{j=0}^{\infty} Y_{2j+1} \qquad \text{and} \qquad \bigcap_{j=0}^{\infty} Y_{2j} = \bigcap_{j=0}^{\infty} X_{2j+1}.$$

Set

$$X_\infty = \bigcap_{j=0}^{\infty} X_{2j} = \bigcap_{j=0}^{\infty} Y_{2j+1} \quad \text{and} \quad Y_\infty = \bigcap_{j=0}^{\infty} Y_{2j} = \bigcap_{j=0}^{\infty} X_{2j+1}.$$

The set X can be partitioned into the disjoint union of the sets

$$X_\infty, \quad X_o - Y_1, \quad Y_1 - X_2, \quad X_2 - Y_3, \quad \ldots.$$

Likewise Y can be partitioned into the disjoint union of the sets

$$Y_\infty, \quad Y_o - X_1, \quad X_1 - Y_2, \quad Y_2 - X_3, \quad \ldots.$$

A bijection from X onto Y is given by

$$h(x) = \begin{cases} f(x) & \text{if } x \in X_{2j} - Y_{2j+1} & \text{for some } j = 0, 1, 2, \ldots \\ g^{-1}(x) & \text{if } x \in Y_{2j+1} - X_{2j+2} & \text{for some } j = 0, 1, 2, \ldots \\ f(x) & \text{if } x \in X_{\infty}. \end{cases}$$

3.1. The first statement of Proposition 3.2 would be false if $\mathbb{N}$ is replaced by a finite set X. Give conditions on an infinite set X, for which $\text{card}(X^m) = \text{card}(X)$.

Chapter 2
Topologies and Metric Spaces

1 Topological Spaces

Let X be a set. A collection $\mathcal{U}$ of subsets of X defines a *topology* on X if:

i. the empty set $\emptyset$ and X belong to $\mathcal{U}$

ii. the union of any collection of sets in $\mathcal{U}$ is in $\mathcal{U}$

iii. the intersection of finitely many elements of $\mathcal{U}$ is in $\mathcal{U}$.

The pair $\{X; \mathcal{U}\}$, that is X endowed with the topology generated by $\mathcal{U}$, is a topological space. The elements $\mathcal{O}$ of $\mathcal{U}$ are the *open* sets of X. A set C in X is *closed* if $X - C$ is open. The empty set $\emptyset$ and X are both open and closed. It follows from the definitions that the finite union of closed sets is closed and the intersection of any collection of closed sets is closed.

An open neighborhood of a set $A \subset X$ is any open set that contains A. In particular a neighborhood of a singleton $x \in X$, is any open set $\mathcal{O}$ such that $x \in \mathcal{O}$. A subset $\mathcal{O} \subset X$ is open if and only if is an open neighborhood of any of its points.

A point $x \in A$ is an *interior* point of A if there exists an open set $\mathcal{O}$ such that $x \in \mathcal{O} \subset A$. The interior of A is the set of all its interior points. A set $A \subset X$ is open if and only if it coincides with its interior.

A point x is a point of *closure* of A if every open neighborhood of x intersects A. The closure $\bar{A}$ of A is the set of all the points of closure of A. A set A is closed if and only if $A = \bar{A}$. It follows that A is closed if and only if it is the intersection of all closed sets containing A.

Let $\{x_n\}$ be a sequence of elements of X. A point $x \in X$ is a *cluster* point for the sequence $\{x_n\}$ if every open set $\mathcal{O}$ containing x, contains infinitely many elements of $\{x_n\}$. The sequence $\{x_n\}$ converges to x if for every open set $\mathcal{O}$ containing x, there exists a positive integer $m(\mathcal{O})$ depending on $\mathcal{O}$, such that $x_n \in \mathcal{O}$ for all $n \geq m(\mathcal{O})$. Thus limit points for $\{x_n\}$ are cluster points. The converse is false. Indeed, there exist

© Springer Science+Business Media New York 2016
E. DiBenedetto, *Real Analysis*, Birkhäuser Advanced
Texts Basler Lehrbücher, DOI 10.1007/978-1-4939-4005-9_2

sequences $\{x_n\}$ with a cluster point x_o such that no subsequence of $\{x_n\}$ converges to x_o (**1.11** of the Complements).

Proposition 1.1 (Cauchy) *A sequence $\{x_n\}$ of elements of a topological space $\{X; \mathcal{U}\}$ converges to x if and only if every subsequence $\{x_{n'}\} \subset \{x_n\}$ contains in turn a subsequence $\{x_{n''}\} \subset \{x_{n'}\}$ converging to x.*

Let A and B be subsets of X. The set B is *dense* in A if $A \subset \bar{B}$. If also $B \subset A$, then $\bar{A} = \bar{B}$. The space $\{X; \mathcal{U}\}$ is *separable* if it contains a countable dense set.

Let X_o be a subset of X. The collection $\mathcal{U}$ induces a topology on X_o, by the family $\mathcal{U}_o = \{\mathcal{O} \cap X_o\}$. The pair $\{X_o; \mathcal{U}_o\}$ is a topological subspace of $\{X; \mathcal{U}\}$. A subspace of a separable topological space need not be separable (**4.9** of the Complements).

Let $\{X; \mathcal{U}\}$ and $\{Y; \mathcal{V}\}$ be any two topological spaces. A function $f : X \to Y$ is continuous at a point $x \in X$ if for every open set $O \in \mathcal{V}$ containing $f(x)$, there is an open set $\mathcal{O} \in \mathcal{U}$ containing x and such that $f(\mathcal{O}) \subset O$. A function $f : X \to Y$ is continuous if it is continuous at every $x \in X$. This implies that f is continuous if and only if the pre-image of every open set is open, that is if for every open set $O \in \mathcal{V}$, the set $f^{-1}(O)$ is an open set $\mathcal{O} \in \mathcal{U}$. Equivalently f is continuous if and only if the pre-image of a closed set is closed.

The restriction of a continuous function $f : X \to Y$ to a subset $X_o \subset X$ is continuous with respect to the induced topology of $\{X_o; \mathcal{U}_o\}$.

An *homeomorphism* between $\{X; \mathcal{U}\}$ and $\{Y; \mathcal{V}\}$ is a continuous one-to-one function f from X onto Y, with continuous inverse f^{-1}. If $f : X \to Y$ is a homeomorphism then $f(\mathcal{O}) \in \mathcal{V}$ for all $\mathcal{O} \in \mathcal{U}$.

Two homeomorphic topological spaces are equivalent in the sense that the elements of X are in one-to-one correspondence with the elements of Y and the open sets making up the topology of $\{X; \mathcal{U}\}$ are in one-to-one correspondence with the open sets making up the topology of $\{Y; \mathcal{V}\}$.

The collection 2^X of all subsets of X generates a topology on X called the *discrete topology*. Every function f from $\{X; 2^X\}$ into a topological space $\{Y; \mathcal{V}\}$ is continuous.

By a *real valued* function f defined on some $\{X; \mathcal{U}\}$, we mean a function from $\{X; \mathcal{U}\}$ into $\mathbb{R}$ endowed with the Euclidean topology.

The *trivial topology* on X is that for which the only open sets are X and $\emptyset$. The closure of any point $x \in X$ is X. All the continuous real-valued functions defined on X are constant.

As a short-hand notation, we denote by X a topological space, whenever a topology $\mathcal{U}$ is clear from the context, or whenever the specification of a topology $\mathcal{U}$ is immaterial.

1.1 Hausdorff and Normal Spaces

A topological space $\{X; \mathcal{U}\}$ is a *Hausdorff* space if it separates points, that is, if for any two distinct points $x, y \in X$, there exist disjoint open sets $\mathcal{O}_x$ and $\mathcal{O}_y$ such that $x \in \mathcal{O}_x$ and $y \in \mathcal{O}_x$.

Proposition 1.2 *Let $\{X; \mathcal{U}\}$ be a Hausdorff topological space. Then, the points $x \in X$ are closed.*

Proof Every point $y \in (X - x)$ is contained in some open set contained in $X - x$. Since $X - x$ is the union of all such open sets, it is open. Thus $\{x\}$ is closed. ∎

Remark 1.1 The converse is false. See § **4.2** of the Complements.

A topological space $\{X; \mathcal{U}\}$ is *normal* if it separates closed sets, that is, for any two disjoint closed sets C_1 and C_2, there exist disjoint open sets $\mathcal{O}_1$ and $\mathcal{O}_2$ such that $C_1 \subset \mathcal{O}_1$ and $C_2 \subset \mathcal{O}_2$.

A normal space need not be Hausdorff. For example the trivial topology is not Hausdorff but it is normal. However, if in addition the singletons $\{x\}$ are closed, then a normal space is Hausdorff. The converse is false as there exist Hausdorff spaces that do not separate closed sets (§ **1.19** of the Complements).

2 Urysohn's Lemma

Lemma 2.1 (Uryson [166]) *Let $\{X; \mathcal{U}\}$ be normal. Given any two closed, disjoint sets A and B in X, there exist a continuous function $f : X \to [0, 1]$ such that $f = 0$ on A and $f = 1$ on B.*

Proof We may assume that neither A nor B is empty. Indeed, for example, $A = \emptyset$, the function $f = 1$ satisfies the conclusion of the Lemma. Let t denote nonnegative, rational dyadic numbers in $[0, 1]$, that is of the form

$$t = \frac{m}{2^n} \qquad m = 0, 1, \ldots, 2^n; \qquad n = 0, 1, \ldots .$$

For each such t we construct an open set $\mathcal{O}_t$ in such a way that the family $\{\mathcal{O}_t\}$ satisfies

$$\mathcal{O}_o \supset A, \quad \mathcal{O}_1 = X - B, \quad \text{and} \quad \bar{\mathcal{O}}_\tau \subset \mathcal{O}_t \quad \text{whenever } \tau < t. \tag{2.1}$$

Since $\{X; \mathcal{U}\}$ is normal, there exists an open set $\mathcal{O}_o$ containing A and whose closure is contained in $X - B$. For $n = 0$ and $m = 0$, select such an open set $\mathcal{O}_o$. For $n = 0$ and $m = 1$ select $\mathcal{O}_1$ as in (2.1). To $n = 1$ and $m = 0, 1, 2$, there correspond sets $\mathcal{O}_o$, $\mathcal{O}_{\frac{1}{2}}$, and $\mathcal{O}_1$. The first and last have been selected and we select set $\mathcal{O}_{\frac{1}{2}}$ so that

$$\bar{\mathcal{O}}_o \subset \mathcal{O}_{\frac{1}{2}} \subset \bar{\mathcal{O}}_{\frac{1}{2}} \subset \mathcal{O}_1.$$

To $n = 2$ and $m = 0, 1, 2, 3, 4$ there correspond open sets $\mathcal{O}_{\frac{m}{2^2}}$ of which only $\mathcal{O}_{\frac{1}{4}}$ and $\mathcal{O}_{\frac{3}{4}}$ have to be selected. Since $\{X; \mathcal{U}\}$ is normal, there exist open sets $\mathcal{O}_{\frac{1}{4}}$ and $\mathcal{O}_{\frac{3}{4}}$ such that

$$\bar{\mathcal{O}}_o \subset \mathcal{O}_{\frac{1}{4}} \subset \bar{\mathcal{O}}_{\frac{1}{4}} \subset \mathcal{O}_{\frac{1}{2}} \quad \text{and} \quad \bar{\mathcal{O}}_{\frac{1}{2}} \subset \mathcal{O}_{\frac{3}{4}} \subset \bar{\mathcal{O}}_{\frac{3}{4}} \subset \mathcal{O}_1.$$

Proceeding by induction, if the open sets $\mathcal{O}_{m/2^{n-1}}$ have been selected we choose the sets $\mathcal{O}_{m/2^n}$ by first observing that the ones corresponding to m even, have already been selected. Therefore, we have only to choose those corresponding to m odd. For any such m fixed the open sets $\mathcal{O}_{(m-1)/2^n}$ and $\mathcal{O}_{(m+1)/2^n}$ have been selected. Since $\{X; \mathcal{U}\}$ is normal, there exists an open set $\mathcal{O}_{m/2^n}$ such that

$$\bar{\mathcal{O}}_{\frac{m-1}{2^n}} \subset \mathcal{O}_{\frac{m}{2^n}} \subset \bar{\mathcal{O}}_{\frac{m}{2^n}} \subset \mathcal{O}_{\frac{m+1}{2^n}}.$$

Define $f : X \to [0, 1]$ by setting $f(x) = 1$ for $x \in B$ and

$$f(x) = \inf\{t \mid x \in \mathcal{O}_t\} \quad \text{for } x \in X - B.$$

By construction $f(x) = 0$ on A and $f(x) = 1$ on B. It remains to prove that f is continuous. From the definition of f it follows that for all $s \in (0, 1]$

$$[f < s] = \bigcup_{t<s} \mathcal{O}_t \quad \text{and} \quad [f \leq s] = \bigcap_{t>s} \mathcal{O}_t.$$

Therefore $[f < s]$ is open. On the other hand by the last of (2.1) $\bar{\mathcal{O}}_\tau \subset \mathcal{O}_t$, whenever $\tau < t$. Therefore

$$[f \leq s] = \bigcap_{t>s} \mathcal{O}_t = \bigcap_{t>s} \bar{\mathcal{O}}_t$$

and $[f \leq s]$ is closed.   ∎

Corollary 2.1 *A Hausdorff space $\{X; \mathcal{U}\}$ is normal if and only if, for every pair of closed disjoint subsets C_1 and C_2 of X, there exists a continuous function $f : X \to \mathbb{R}$ such that $f = 1$ on C_1 and $f = 0$ on C_2.*

3 The Tietze Extension Theorem

Theorem 3.1 (Tietze [158]) *Let $\{X; \mathcal{U}\}$ be normal. A continuous function f from a closed subset C of X into $\mathbb{R}$ has a continuous extension on X, that is, there exists a continuous real-valued function f_* defined on the whole X, such that $f = f_*$ on C. Moreover if f is bounded, say*

$$|f(x)| \leq M \quad \text{for all } x \in C \text{ for some } M > 0$$

then f_* satisfies the same bound.

Proof Assume first that f is bounded and that $M \leq 1$. We will construct a sequence of real valued, continuous functions $\{g_n\}$, defined on the whole X, such that for all $n \in \mathbb{N}$

$$|g_n(x)| \leq \frac{1}{3}\left(\frac{2}{3}\right)^{n-1} \qquad \text{for all } x \in X$$

$$\left|f(x) - \sum_{j=1}^{n} g_j(x)\right| \leq \left(\frac{2}{3}\right)^{n} \qquad \text{for all } x \in C.$$

(3.1)

Assuming the sequence $\{g_n\}$ has been constructed, by virtue of the first of (3.1), the series $\sum g_n$ is uniformly convergent in X and $|\sum g_n| \leq 1$. Therefore, the functions $f_n = \sum_{j=1}^{n} g_j$ are continuous and form a sequence $\{f_n\}$, uniformly convergent on X to a continuous function f_*. From the second of (3.1) it follows that $f = f_*$ on C. It remains to construct the sequence $\{g_n\}$.

Since f is continuous, the two sets $[f \leq -\frac{1}{3}]$ and $[f \geq \frac{1}{3}]$ are closed and disjoint. By Urysohn's lemma, there exists a continuous function $g_1 : X \to [-\frac{1}{3}, \frac{1}{3}]$ such that

$$g_1 = \frac{1}{3} \quad \text{on} \quad [f \geq \frac{1}{3}] \quad \text{and} \quad g_1 = -\frac{1}{3} \quad \text{on} \quad [f \leq -\frac{1}{3}].$$

By construction $|f(x) - g_1(x)| \leq \frac{2}{3}$ for all $x \in C$. The function $h_1 = f - g_1$ is continuous and bounded on C. The two sets $[h_1 \leq -\frac{2}{9}]$ and $[h_1 \geq \frac{2}{9}]$ are closed and disjoint. Therefore by Urysohn's lemma there exists a continuous function $g_2 : X \to [-\frac{2}{9}, \frac{2}{9}]$ such that

$$g_2 = \frac{2}{9} \quad \text{on} \quad [h_1 \geq \frac{2}{9}] \quad \text{and} \quad g_2 = -\frac{2}{9} \text{ on } \quad [h_1 \leq -\frac{2}{9}].$$

By construction

$$|h_1 - g_2| = |f - (g_1 + g_2)| \leq \frac{4}{9} \quad \text{on } C.$$

The sequence $\{g_n\}$ is constructed inductively by this procedure. This proves Tietze's Theorem if f is bounded and $|f| \leq 1$. If f is bounded and $|f| \leq M$, the conclusion follows by replacing f with f/M. If f is unbounded, set

$$f_o = \frac{f}{1 + |f|}.$$

Since $f_o : C \to \mathbb{R}$ is continuous and bounded it has a continuous extension $g_o : X \to [-1, 1]$. In particular

$$g_o(x) = \frac{f(x)}{1 + |f(x)|} \in (-1, 1) \quad \text{for all } x \in C.$$

This implies that the set $[|g_o| = 1]$ is closed and disjoint from C. By the Uryshon Lemma there exists a continuous function $\eta : X \to \mathbb{R}$ such that $\eta = 1$ on C and $\eta = 0$ on $[|g_o| = 1]$. The function

$$g = \frac{\eta g_o}{1 - \eta |g_o|} : X \to \mathbb{R}$$

is continuous and coincides with f on C.   ∎

4 Bases, Axioms of Countability and Product Topologies

A family of open sets $\mathcal{B}$ is a *base* for the topology of $\{X; \mathcal{U}\}$, if for every open set $\mathcal{O}$ and every $x \in \mathcal{O}$, there exists a set $B \in \mathcal{B}$ such that $x \in B \subset \mathcal{O}$. A collection $\mathcal{B}_x$ of open sets, is a base at x if for each open set $\mathcal{O}$ containing x, there exists $B \in \mathcal{B}_x$ such that $x \in B \subset \mathcal{O}$. Thus if $\mathcal{B}$ is a base for the topology of $\{X; \mathcal{U}\}$, it is also a base for each of the points of X. More generally, $\mathcal{B}$ is a base if and only if it contains a base for each of the points of X. A set $\mathcal{O}$ is open if and only if for each $x \in \mathcal{O}$ there exists $B \in \mathcal{B}$ such that $x \in B \subset \mathcal{O}$.

Let $\mathcal{B}$ be a base for $\{X; \mathcal{U}\}$. Then:

i.   Every $x \in X$ belongs to some $B \in \mathcal{B}$
ii. For any two given sets B_1 and B_2 in $\mathcal{B}$ and every $x \in B_1 \cap B_2$, there exists some $B_3 \in \mathcal{B}$ such that $x \in B_3 \subset B_1 \cap B_2$.

The notion of base is induced by the presence of a topology generated by $\mathcal{U}$ on X. Conversely, if a collection of sets $\mathcal{B}$ satisfies (**i**) and (**ii**), then it permits one to construct a topology on X for which $\mathcal{B}$ is a base.

Proposition 4.1 *Let $\mathcal{B}$ be a collection of sets in X satisfying (**i**)–(**ii**). There exists a collection $\mathcal{U}$ of subsets of X, which generates a topology on X, for which $\mathcal{B}$ is a base.*

Proof Let $\mathcal{U}$ consist of the empty set $\emptyset$ and the collection of all subsets $\mathcal{O}$ of X, such that for every $x \in \mathcal{O}$ there exists an element $B \in \mathcal{B}$ such that $x \in B \subset \mathcal{O}$. Such a collection is not empty since $X \in \mathcal{U}$.

It follows from the definition that the union of any collection of elements in $\mathcal{U}$ remains in $\mathcal{U}$. Moreover $\mathcal{U}$ contains the empty set $\emptyset$ and X.

Let $\mathcal{O}_1$ and $\mathcal{O}_2$ be any two elements of $\mathcal{U}$ with nonempty intersection. For every $x \in \mathcal{O}_1 \cap \mathcal{O}_2$ there exist sets B_1, B_2 and B_3 in $\mathcal{B}$ such that $B_i \subset \mathcal{O}_i$, $i = 1, 2$ and

$$x \in B_3 \subset B_1 \cap B_2 \subset \mathcal{O}_1 \cap \mathcal{O}_2.$$

Therefore $\mathcal{O}_1 \cap \mathcal{O}_2 \in \mathcal{U}$. This implies that the collection $\mathcal{U}$ generates a topology on X for which $\mathcal{B}$ is a base.   ∎

A topological space $\{X; \mathcal{U}\}$ satisfies the *first axiom of countability* if each point $x \in X$ has a countable base. The space $\{X; \mathcal{U}\}$ satisfies the *second axiom of countability* if there exists a countable base for its topology.

Proposition 4.2 *Every topological space satisfying the second axiom of countability is separable.*

Proof Let $\{\mathcal{O}\}$ be a countable base for the topology of $\{X; \mathcal{U}\}$. For each $i \in \mathbb{N}$ select an element $x_i \in \mathcal{O}_i$. This generates a countable, dense subset of $\{X; \mathcal{U}\}$. ∎

4.1 Product Topologies

Let $\{X_1; \mathcal{U}_1\}$ and $\{X_2; \mathcal{U}_2\}$ be two topological spaces. The *product topology $\mathcal{U}_1 \times \mathcal{U}_2$* on the Cartesian product $X_1 \times X_2$ is constructed by considering the collection $\mathcal{B}$ of all products $\mathcal{O}_1 \times \mathcal{O}_2$ where $\mathcal{O}_i \in \mathcal{U}_i$ for $i = 1, 2$. These are called the *open rectangles* of the product topology. First, one verifies that they form a base in the sense of (**i**)-(**ii**). Then the product topological space $\{X_1 \times X_2; \mathcal{U}_1 \times \mathcal{U}_2\}$ is constructed by the procedure of Proposition 4.1. The symbol $\mathcal{U}_1 \times \mathcal{U}_2$ means the collection $\mathcal{U}$ of sets in $X_1 \times X_2$, constructed by the procedure of Proposition 4.1.

If $\{X_1, \mathcal{U}_1\}$ and $\{X_2; \mathcal{U}_2\}$ are Hausdorff spaces, then the topological product space is a Hausdorff space.

Let $X_1 \times X_2$ be endowed with the product topology $\mathcal{U}_1 \times \mathcal{U}_2$. Then the projections

$$\pi_j : X_1 \times X_2 \to X_j, \qquad j = 1, 2$$

are continuous. Moreover $\mathcal{U}_1 \times \mathcal{U}_2$ is the weakest topology on $X_1 \times X_2$ for which such projections are continuous. The procedure can be iterated to construct the product topology on the product of n topological spaces $\{X_i; \mathcal{U}_i\}_{i=1}^n$. Such a topology is the weakest topology for which the projections

$$\pi_j : \prod_{i=1}^n X_i \to X_j, \qquad j = 1, \dots, n$$

are continuous. More generally, given an infinite family of topological spaces $\{X_\alpha; \mathcal{U}_\alpha\}_{\alpha \in A}$, the product topology $\prod \mathcal{U}_\alpha$ on the Cartesian product $\prod X_\alpha$ is constructed as the weakest topology for which the projections

$$\pi_\beta : \prod X_\alpha \to X_\beta$$

are continuous for all $\beta \in A$. Such a topology must contain the collection

$$\mathcal{B} = \left\{ \begin{matrix} \text{finite intersections of the inverse images} \\ \pi_\alpha^{-1}(\mathcal{O}_\alpha) \text{ for } \alpha \in A \text{ and } \mathcal{O}_\alpha \in \mathcal{U}_\alpha \end{matrix} \right\}.$$

One verifies that $\mathcal{B}$ satisfies the conditions (**i**)–(**ii**) of a base. Then, the product topology is generated, starting from such a base, by the procedure of Proposition 4.1.

Let f be a function defined on A such that $f(\alpha) \in X_\alpha$ for all $\alpha \in A$.

The collection $\{f(\alpha)\}$ can be identified with a point in $\prod X_\alpha$. Conversely any point $x \in \prod X_\alpha$ can be identified with one such function. If $X_\alpha = X$ for some set X and all $\alpha \in A$, then $\prod X_\alpha$ is denoted by X^A and it is identified with the set of all functions defined on A and with values in X. If $A = \mathbb{N}$ then $X^{\mathbb{N}}$ is the set of all sequences $\{x_n\}$ of elements of X.

5   Compact Topological Spaces

A collection $\mathcal{F}$ of open sets $\mathcal{O}$ is an *open covering* of X if every $x \in X$ is contained in some $\mathcal{O} \in \mathcal{F}$. The covering is countable if it consists of countably many elements, and it is finite if it consists of a finite number of open sets.

It might occur that X is covered by a subfamily $\mathcal{F}'$ of elements of $\mathcal{F}$. Such a subfamily, if it exists, is an open sub-covering of X, relative to $\mathcal{F}$.

The topological space $\{X; \mathcal{U}\}$ is *compact* if every open covering $\mathcal{F}$ contains a finite sub-covering $\mathcal{F}'$. A set $X_o \subset X$ is compact if $\{X_o; \mathcal{U}_o\}$ is a compact topological space.

A collection $\mathcal{G}$ of closed subsets of X has the *finite intersection property* if the elements of any finite subcollection have nonempty intersection. Let $\mathcal{F}$ be an open covering for X and let $\mathcal{G}$ be the collection of the complements of the elements in $\mathcal{F}$. The elements of $\mathcal{G}$ are closed and if X is compact, $\mathcal{G}$ does not have the finite intersection property. More generally X is compact if and only if every collection of closed subsets of X with empty intersection, does not have the finite intersection property.

Proposition 5.1 (*i*)   *$\{X; \mathcal{U}\}$ is compact if and only if every collection $\mathcal{G}$ of closed sets with the finite intersection property has nonempty intersection.*

(*ii*)   *Let E be a closed subset of a compact space $\{X; \mathcal{U}\}$. Then E is compact.*

(*iii*)   *Let E be a compact subset of a Hausdorff topological space. Then E is closed.*

(*iv*)   *Let f from $\{X; \mathcal{U}\}$ into $\{Y; \mathcal{V}\}$ be continuous. If $\{X; \mathcal{U}\}$ is compact then $f(X) \subset Y$ is compact. If in addition f is one-to-one, and $\{Y; \mathcal{V}\}$ is Hausdorff, then f is a homeomorphism between $\{X; \mathcal{U}\}$ and $\{Y; \mathcal{V}\}$.*

Proof Part (i) follows from the previous remarks. To prove (ii), let $\mathcal{F}$ be any open covering for E. Then the collection $\{\mathcal{F}, (X - E)\}$ is an open covering for X. From this we may extract a finite sub-covering $\mathcal{F}'$ for X which, by possibly removing $X - E$, gives a finite sub-covering for E.

Turning to (iii), let $y \in (X - E)$ be fixed. Since X is a Hausdorff space, for every $x \in E$ there exist disjoint open sets $\mathcal{O}_x$ and $\mathcal{O}_{x,y}$ separating x and y. The collection $\{\mathcal{O}_x\}$ forms an open cover for E, from which we extract a finite one $\{\mathcal{O}_{x_1}, \ldots, \mathcal{O}_{x_n}\}$ for some positive integer n. The intersection $\bigcap_{j=1}^{n} \mathcal{O}_{x_j,y}$ is open and does not intersect

E. Therefore every element y of the complement of E contains an open neighborhood not intersecting E. Thus E is closed.

To establish (iv), let $\{X; \mathcal{U}\}$ be compact and let $\{Y; \mathcal{V}\}$ be the image of a continuous function f from X onto Y. Given an open covering $\{\Phi\}$ of Y, the collection $\mathcal{F} = \{f^{-1}(\Phi)\}$ is an open covering for X from which we may extract a finite one $\{f^{-1}(\Phi_1), \ldots, f^{-1}(\Phi_n)\}$. Then the finite collection $\{\Phi_1, \ldots, \Phi_n\}$ covers Y.

Let $f : X \to Y$ be continuous and one-to-one and let $\{Y; \mathcal{V}\}$ be Hausdorff. A closed subset $E \subset X$ is compact, and its image $f(E)$ is compact in Y and hence closed. Therefore f^{-1} is continuous. ∎

Remark 5.1 In (iii), the assumption that $\{X; \mathcal{U}\}$ be Hausdorff cannot be removed. Indeed if $\mathcal{U}$ is the trivial topology on X, every proper subset of X is compact and not closed.

A topological space $\{X; \mathcal{U}\}$ is *locally compact* if for each $x \in X$ there exists an open set $\mathcal{O}$ containing x and such that $\bar{\mathcal{O}}$ is compact. For example, $\mathbb{R}^N$ endowed with the Euclidean topology is locally compact but not compact.

5.1 Sequentially Compact Topological Spaces

A topological space $\{X; \mathcal{U}\}$ is *countably compact* if every countable open covering of X contains a finite sub-covering. If $\{X; \mathcal{U}\}$ is compact it is also countably compact. The converse is false (**5.7** of the Complements).

A topological space $\{X; \mathcal{U}\}$ has the Bolzano–Weierstrass property if every infinite sequence $\{x_n\}$ of elements of X has at least one cluster point.

A topological space $\{X; \mathcal{U}\}$ is *sequentially compact* if every infinite sequence $\{x_n\}$ of elements of X has a convergent subsequence.

Thus if $\{X; \mathcal{U}\}$ is sequentially compact it has the Bolzano–Weierstrass property. The converse is false.[1]

Proposition 5.2 (i) *The continuous image of a countably compact space is countably compact.*

(ii) *$\{X; \mathcal{U}\}$ is countably compact if and only if every countable family $\mathcal{G}$ of closed sets with the finite intersection property has nonempty intersection.*

(iii) *$\{X; \mathcal{U}\}$ has the Bolzano–Weierstrass property if and only if it is countably compact.*

(iv) *If $\{X; \mathcal{U}\}$ is sequentially compact then it is countably compact.*

(v) *If $\{X; \mathcal{U}\}$ is countably compact and if it satisfies the first axiom of countability, then it is sequentially compact.*

Proof Parts (i)–(ii) follows from the definitions and the proof of Proposition 5.1. The proof of (iii) uses the characterization of countable compactness stated in (ii).

---

[1] By part (v) of Proposition 5.2 a counterexample can be constructed starting from a space $\{X; \mathcal{U}\}$ that does not satisfy the first axiom of countability. See **5.7** of the Complements.

Let $\{X; \mathcal{U}\}$ be countably compact and let $\{x_n\}$ be a sequence of elements of X. The closed sets

$$B_n = \text{closure of } \{x_n, x_{n+1}, \dots, \}$$

satisfy the finite intersection property and therefore have nonempty intersection. Any element $x \in \cap B_n$ is a cluster point for $\{x_n\}$.

Conversely let $\{X; \mathcal{U}\}$ satisfy the Bolzano–Weierstrass property and let $\{B_n\}$ be a countable collection of closed subsets of X with the finite intersection property. Since for all $n \in \mathbb{N}$ the intersection $\bigcap_{j=1}^{n} B_j$ is nonempty we may select an element x_n out of it. The sequence $\{x_n\}$ has at least one cluster point x, which by construction, belongs to the intersection of all B_n.

Part (iv) follows from (iii), since sequential compactness implies the Bolzano–Weierstrass property.

To prove (v) let $\{x_n\}$ be an infinite sequence of elements of X and let x be in the closure of $\{x_n\}$. Since $\{X; \mathcal{U}\}$ satisfies the first axiom of countability, there exist a nested countable collection of open sets $\mathcal{O}_m \supset \mathcal{O}_{m+1}$, each containing x and each containing an element x_m out of the sequence $\{x_n\}$. The subsequence $\{x_m\}$ converges to x. ∎

Proposition 5.3 *Let $\{X; \mathcal{U}\}$ satisfy the second axiom of countability. Then every open covering of X contains a countable sub-covering.*

Proof Let $\{B_n\}$ a countable collection of open sets that forms a base for the topology of $\{X; \mathcal{U}\}$ and let $\mathcal{F}$ be an open covering of X. To each B_n we associate one and only one open set $\mathcal{O}_n \in \mathcal{F}$ that contains it. The countable collection $\mathcal{O}_n$ is a countable sub-covering of $\mathcal{F}$. ∎

Corollary 5.1 *For spaces satisfying the second axiom of countability, compactness, countable compactness and sequential compactness, are equivalent.*

6 Compact Subsets of $\mathbb{R}^N$

Let E be a subset of $\mathbb{R}^N$. We regard E as a topological space with the topology inherited from the Euclidean topology of $\mathbb{R}^N$.

Proposition 6.1 *The closed interval $[0, 1]$ is compact.*

Proof Let $\{I_\alpha\}$ be a collection of open intervals covering $[0, 1]$ and set

$$\mathcal{E} = \bigcup \left\{ \begin{array}{l} x \in [0, 1] \text{ such that the closed interval } [0, x] \text{ is} \\ \text{covered by finitely many elements out of } \{I_\alpha\} \end{array} \right\}.$$

Let $c = \sup\{x | x \in \mathcal{E}\}$. Since 0 belongs to some I_α we have $0 < c \leq 1$. Such an element c it is covered by some open set $I_{\alpha_c} \in \{I_\alpha\}$, and therefore, there exist

$\varepsilon > 0$ such that $(c - \varepsilon, c + \varepsilon) \subset I_{\alpha_c}$. By the definition of c, the interval $[0, c - \varepsilon]$ is covered by finitely many open sets $\{I_1, \ldots, I_n\}$ out of $\{I_\alpha\}$. Augmenting such a finite collection with I_{α_c} gives a finite covering of $[0, c + \varepsilon]$. Thus if $c < 1$, it is not the supremum of the set $\mathcal{E}$. ∎

Proposition 6.2 *The closed interval* $[0, 1]$ *has the Bolzano–Weierstrass property.*

Proof Let $\{x_n\}$ be an infinite sequence of elements of $[0, 1]$ without a cluster point in $[0, 1]$. Then, each of the open intervals $(x - \varepsilon, x + \varepsilon)$, for $x \in [0, 1]$ and $\varepsilon > 0$, contains at most finitely many elements of $\{x_n\}$. The collection of all such intervals forms an open covering of $[0, 1]$, from which we may select a finite one. This would imply $\{x_n\}$ is finite. ∎

Corollary 6.1 *Every sequence in* $[0, 1]$ *has a convergent subsequence.*

Corollary 6.2 *A bounded, closed subset* $E \subset \mathbb{R}^N$ *has the Bolzano–Weierstrass property.*

Proof Let $\{x_n\}$ be a sequence of elements in E and represent each of the x_n in terms of its coordinates, that is, $x_n = (x_{1,n}, \ldots, x_{N,n})$. Since $\{x_n\}$ is bounded, each of the sequences $\{x_{j,n}\}$ is contained in some closed interval $[a_j, b_j]$. Out of $\{x_{1,n}\}$ we extract a convergent subsequence $\{x_{1,n_1}\}$. Then out of $\{x_{2,n_1}\}$ we extract a convergent subsequence $\{x_{2,n_2}\}$. Proceeding in this fashion, the sequence $\{x_{n_N}\}$ has a limit. Since E is closed, such a limit is in E. ∎

Proposition 6.3 (Borel-Riesz ([18, 124])) *Let E be a bounded, closed subset of $\mathbb{R}^N$. Then, every open covering $\mathcal{U}$ of E contains a finite sub-covering $\mathcal{U}'$.*

Proof By Proposition 5.3, may assume the covering is countable, say $\mathcal{U} = \{\mathcal{O}_n\}$. We claim that $E \subset \bigcup_{i=1}^m \mathcal{O}_i$ for some $m \in \mathbb{N}$. Indeed if not, we may select for each positive integer n, an element $x_n \in E - \bigcup_{i=1}^n \mathcal{O}_n$ and select, out of the sequence $\{x_n\}$, a subsequence $\{x_{n'}\}$ convergent to some $x \in E$. Since the collection $\{\mathcal{O}_n\}$ covers E, there exists an index m such that $x \in \mathcal{O}_m$. Thus $x_{n'} \in \mathcal{O}_m$ for infinitely many n'. ∎

Proposition 6.4 (Heine-Borel) *Every compact subset of $\mathbb{R}^N$, endowed with the Euclidean topology, is closed and bounded.*

Proof Let $E \subset \mathbb{R}^N$ be compact. Since $\mathbb{R}^N$, endowed with the Euclidean topology, is a Hausdorff space, E is closed by (iii) of Proposition 5.1. The collection of balls $\{B_n\}$ centered at the origin and radius $n \in \mathbb{N}$ is an open covering for E. Since E is contained in the union of a finite sub-covering, it is bounded. ∎

Theorem 6.1 *A subset E of $\mathbb{R}^N$ is compact if and only if is bounded and closed.*

7   Continuous Functions on Countably Compact Spaces

Let f be a map from a topological space $\{X; \mathcal{U}\}$ into $\mathbb{R}$ and for $t \in \mathbb{R}$ set $[f < t] = \{x \in X | f(x) < t\}$. The sets $[f \leq t]$, $[f \geq t]$, and $[f > t]$ are defined analogously. A map f from a topological space $\{X; \mathcal{U}\}$ into $\mathbb{R}$, is *upper semi-continuous* if $[f < t]$ is open for all $t \in \mathbb{R}$, and it is *lower semi-continuous* if $[f > t]$ is open for all $t \in \mathbb{R}$. A map $f : X \to \mathbb{R}$ is continuous if and only if is both upper and lower semi-continuous.

Theorem 7.1 (Weierstrass-Baire) *Let $\{X; \mathcal{U}\}$ be countably compact and let $f : X \to \mathbb{R}$ be upper semi-continuous. Then f is bounded above in X and it achieves its maximum in X.*

Proof The collection of sets $\{[f < n]\}$ is a countable open covering of X, from which we extract a finite one, say, for example, $[f < n_1], \ldots, [f < n_N]$. Then $f \leq \max\{n_1, \ldots, n_N\}$. Thus f is bounded above. Next, let f_o denote the supremum of f on X. The sets $[f \geq f_o - \frac{1}{n}]$ are closed and form a family with the finite intersection property. Since $\{X; \mathcal{U}\}$ is countably compact, their intersection is nonempty. Therefore, there is an element $x_o \in [f \geq f_o - \frac{1}{n}]$ for all $n \in \mathbb{N}$. By construction $f(x_o) = f_o$.   ∎

Corollary 7.1 *(i)   A continuous real-valued function from a countably compact topological space $\{X; \mathcal{U}\}$ takes its maximum and minimum in X.*

(ii)   A continuous real-valued function from a countably compact topological space $\{X; \mathcal{U}\}$ is uniformly continuous.

Theorem 7.2 (Dini) *Let $\{X; \mathcal{U}\}$ be countably compact and let $\{f_n\}$ be a sequence of real-valued, upper semi-continuous functions such that $f_{n+1} \leq f_n$ for all $n \in \mathbb{N}$, and converging pointwise in X to a lower semi-continuous function f. Then $\{f_n\} \to f$ uniformly in X.*

Proof By possibly replacing f_n with $f_n - f$, we may assume that $\{f_n\}$ is a decreasing sequence of upper semi-continuous functions converging to zero pointwise in X. For every $\varepsilon > 0$, the collection of open sets $[f_n < \varepsilon]$ covers X and we extract a finite cover, say for example, up to a possible reordering

$$[f_{n_1} < \varepsilon], \; [f_{n_2} < \varepsilon], \cdots, [f_{n_\varepsilon} < \varepsilon], \quad n_1 < n_2 \cdots < n_\varepsilon.$$

Since $\{f_n\}$ is decreasing $[f_{n_\varepsilon} < \varepsilon] = X$. Thus $f_n(x) < \varepsilon$ for all $x \in X$ and all $n \geq n_\varepsilon$.   ∎

8   Products of Compact Spaces

Theorem 8.1 (Tychonov ([165])) *Let $\{X_\alpha; \mathcal{U}_\alpha\}$ be a family of compact spaces. Then $\prod X_\alpha$ endowed with the product topology, is compact.*

The proof is based on showing that every collection of closed sets with the finite intersection property, has nonempty intersection.

Lemma 8.1 *Let* $\{X; \mathcal{U}\}$ *be a topological space and let* $\mathcal{G}_o$ *be a collection of subsets of* X *with the finite intersection property. There exists a maximal collection* $\mathcal{G}$ *of subsets of* X *with the finite intersection property and containing* $\mathcal{G}_o$, *that is, if* $\mathcal{G}'$ *is another collection of subsets of* X *with the finite intersection property and containing* $\mathcal{G}$, *then* $\mathcal{G}' = \mathcal{G}$. *Moreover, the finite intersection of elements in* $\mathcal{G}$ *is in* $\mathcal{G}$ *and every subset of* X *that intersects each set of* $\mathcal{G}$ *is in* $\mathcal{G}$.[2]

Proof The family of all collections of sets with the finite intersection property and containing $\mathcal{G}_o$ is partially ordered by inclusion, so that by the Hausdorff principle, there is a maximal linearly ordered subfamily $\mathcal{F}$. We claim that $\mathcal{G}$ is the union of all the collections in $\mathcal{F}$.

Any n-tuple $\{E_1, \ldots, E_n\}$ of elements of $\mathcal{G}$ belongs to at most n collections $\mathcal{G}_j$. Since $\{\mathcal{G}_j\}$ is linearly ordered there is a collection $\mathcal{G}_n$ that contains the others. Therefore, $E_i \in \mathcal{G}_n$ for all $i = 1, \ldots, n$ and since $\mathcal{G}_n$ has the finite intersection property, $\cap E_i \neq \emptyset$. Thus $\mathcal{G}$ has the finite intersection property. The maximality of $\mathcal{G}$ follows by its construction.

The collection $\mathcal{G}'$ of all finite intersections of sets in $\mathcal{G}$ contains $\mathcal{G}$ and has the finite intersection property. Therefore $\mathcal{G}' = \mathcal{G}$ by maximality.

Let E be a subset of X that intersects all the sets in $\mathcal{G}$. Then, the collection $\mathcal{G} \bigcup \{E\}$ has the finite intersection property and contains $\mathcal{G}$. Therefore $E \in \mathcal{G}$, by maximality. ∎

Proof (of Tychonov's Theorem) Let $\mathcal{G}_o$ be a collection of closed sets in $\prod_\alpha X_\alpha$, with the finite intersection property and let $\mathcal{G}$ be the maximal collection constructed in Lemma 8.1. While the sets in $\mathcal{G}_o$ are closed, the elements of $\mathcal{G}$ need not be closed. We will establish that the intersection of the closure of all elements in $\mathcal{G}$ is not empty. For each $\alpha \in A$ let $\mathcal{G}_\alpha$ be the collection of the projection of $\mathcal{G}$ into X_α, that is

$$\mathcal{G}_\alpha = \{\text{collection of } \pi_\alpha(E)| \text{ for } E \in \mathcal{G}\}.$$

The sets in $\mathcal{G}_\alpha$ need not be closed nor open. However since $\mathcal{G}$ has the finite intersection property in $\prod X_\alpha$, the collection $\mathcal{G}_\alpha$ has the finite intersection property in X_α. Therefore the collection of their closures in $\{X_\alpha, \mathcal{U}_\alpha\}$

$$\overline{\mathcal{G}}_\alpha = \{\text{collection of } \overline{\pi_\alpha(E)}| \text{ for } E \in \mathcal{G}\}$$

has nonempty intersection, since each of $\{X_\alpha; \mathcal{U}_\alpha\}$ is compact. Select an element

$$x_\alpha \in \bigcap \{\overline{\pi_\alpha(E)}| \text{ for } E \in \mathcal{G}\} \subset X_\alpha.$$

We claim that the element $x \in \prod X_\alpha$, whose α-coordinate is x_α, belongs to the closure of all sets in $\mathcal{G}$. Let $\mathcal{O}$ be a set, open in the product topology that contains

---

[2]It is not claimed here that the elements of $\mathcal{G}$ are closed.

x. By the construction of the product topology, there exists finitely many indices $\alpha_1, \ldots, \alpha_n$ and finitely many sets $\mathcal{O}_{\alpha_j}$, open in X_{α_j}, such that

$$x \in \bigcap_{j=1}^{n} \pi_{\alpha_j}^{-1}(\mathcal{O}_{\alpha_j}) \subset \mathcal{O}.$$

For each j the projection x_{α_j} belongs to $\mathcal{O}_{\alpha_j}$. Since x_{α_j} belongs to the closure of all sets in $\mathcal{G}_{\alpha_j}$, the open set $\mathcal{O}_{\alpha_j}$ intersects all the sets in $\mathcal{G}_{\alpha_j}$. Therefore, $\pi_{\alpha_j}^{-1}(\mathcal{O}_{\alpha_j})$ intersects all the sets in $\mathcal{G}$ and by Lemma 8.1 it belongs to $\mathcal{G}$. Likewise the finite intersection $\bigcap_{j=1}^{n} \pi_{\alpha_j}^{-1}(\mathcal{O}_{\alpha_j})$ intersects every element in $\mathcal{G}$ and therefore it belongs to $\mathcal{G}$. Thus, an arbitrary open set $\mathcal{O}$ containing x, intersects all the sets in $\mathcal{G}$ and therefore x belongs to the closure of all such sets.   ∎

Remark 8.1 Tychonov's theorem provides a motivation for defining the topology on a product space $\prod X_\alpha$ as the weakest topology for which all the projection maps π_α are continuous. Indeed if all the topological spaces $\{X_\alpha; \mathcal{U}_\alpha\}$ are Hausdorff, the product topology is also a Hausdorff topology. But then any topology stronger than the product topology would violate Tychonov's theorem. This follows from Proposition 5.1c and **5.3** of the Complements.

9   Vector Spaces

A *linear space* consists of a set X, whose elements are called *vectors*, and a field $\mathcal{F}$, whose elements are called scalars, endowed with operations of sum $+ : X \times X \to X$, and multiplication by scalars $\bullet : \mathcal{F} \times X \to X$ satisfying the addition laws

$$
\begin{aligned}
x + y &= y + x \\
(x + y) + z &= x + (y + z),
\end{aligned}
\qquad \text{for all } x, y, z \in X
$$

$$\text{there exists } \Theta \in X \text{ such that } x + \Theta = x \text{ for all } x \in X$$

$$\text{for all } x \in X \text{ there exists } -x \in X \text{ such that } x + (-x) = \Theta$$

and the scalar multiplication laws

$$
\begin{aligned}
\lambda(x + y) &= \lambda x + \lambda y & \text{for all } x, y \in X \\
\lambda(\mu x) &= (\lambda \mu) x & \text{for all } \lambda, \mu \in \mathcal{F} \\
(\lambda + \mu) x &= \lambda x + \mu x & \text{for all } \lambda, \mu \in \mathcal{F} \\
1 x &= x & \text{where 1 is the unit element of } \mathcal{F}.
\end{aligned}
$$

It follows that $\lambda \Theta = \Theta$ for all $\lambda \in \mathcal{F}$ and if 0 is the zero-element of $\mathcal{F}$, then $0x = \Theta$ for all $x \in X$. Also, for all $x, y \in X$ and $\lambda \in \mathcal{F}$

$$(-1)x = -x, \qquad x - y = x + (-y), \qquad \lambda(x - y) = \lambda x - \lambda y.$$

A nonempty subset $X_o \subset X$ is a linear *subspace* of X if it is closed under the inherited operations of sum and multiplication by scalars. The largest linear subspace of X is X itself and the smallest is the null space $\{\Theta\}$. A *linear combination* of an n-tuple of vectors $\{x_1, \ldots, x_n\}$, is an expression of the form

$$y = \sum_{j=1}^{n} \lambda_j x_j \quad \text{where} \quad \{\lambda_1, \ldots, \lambda_n\} \text{ is an } n\text{-tuple of scalars.}$$

If $X_o \subset X$, the *linear span* of X_o is the set of all linear combinations of elements of X_o. It is a linear space, and it is the smallest linear subspace of X containing X_o, or *spanned* by X_o. An n-tuple $\{e_1, \ldots, e_n\}$ of vectors is *linearly independent* if

$$\sum_{j=1}^{n} \lambda_j e_j = 0 \quad \text{implies that} \quad \lambda_j = 0 \quad \text{for all } j = 1, \ldots, n.$$

A linear space X is of dimension n if it contains an n-tuple of linearly independent vectors whose span is the whole X. Any such n-tuple, say for example $\{e_1, \ldots, e_n\}$ is a *basis* in the sense that given $x \in X$ there exists an n-tuple of scalars $\{\lambda_1, \ldots, \lambda_n\}$ such that $x = \sum_{j=i}^{n} \lambda_j e_j$. For each $x \in X$, the n-tuple $\{\lambda_1, \ldots, \lambda_n\}$ is uniquely determined by the basis $\{e_1, \ldots, e_n\}$. While $\mathcal{F}$ could be any field we will consider $\mathcal{F} = \mathbb{R}$ and call X vector space over the reals.

Let A and B be subsets of a linear space X and let $\alpha, \beta \in \mathbb{R}$. Define the set operation

$$\alpha A + \beta B = \cup \{\alpha a + \beta b | a \in A, \ b \in B\}.$$

One verifies that the sum is commutative and associative, that is,

$$A + B = B + A \quad \text{and} \quad A + (B + C) = (A + B) + C.$$

Moreover
$$A + (B \cup C) = (A + B) \cup (A + C).$$

However $A + A \neq 2A$ and $A - A \neq \{\Theta\}$.

9.1 Convex Sets

A convex combination of two elements $x, y \in X$ is an element of the form $tx + (1 - t)y$ where $t \in [0, 1]$. As t ranges over $[0, 1]$ this describes the *line segment* of extremities x and y. The convex combination of n elements $\{x_1, \ldots, x_n\}$ of X is an element of the form

$$\sum_{j=1}^{n} \alpha_j x_j \quad \text{where} \quad \alpha_j \geq 0 \text{ and } \sum_{j=1}^{n} \alpha_j = 1.$$

A set $A \subset X$ is convex if for any pair $x, y \in A$ the elements $tx + (1 - t)y$ belong to A for all $t \in [0, 1]$. Alternatively, if the line segment of extremities x and y belongs to A.

The *convex hull* $c(A)$ of a set $A \subset X$ is the smallest convex set containing A. It can be characterized as either the intersection of all the convex sets containing A, or as the set of all convex combinations of n-tuples of elements in A, for any n.

The intersection of convex sets is convex; the union of convex sets need not be convex. Linear subspaces of X are convex.

9.2   Linear Maps and Isomorphisms

Let X and Y be linear spaces over $\mathbb{R}$. A map $T : X \to Y$ is linear if

$$T(\lambda x + \mu y) = \lambda T(x) + \mu T(y) \quad \text{for all } x, y \in X \text{ and } \lambda, \mu \in \mathbb{R}.$$

The image of T is $T(X) \subset Y$ and the kernel of T is $\ker\{T\} = T^{-1}\{0\}$. The image $T(X)$ is a linear subspace of Y and the kernel $\ker\{T\}$ is a linear subspace of X. A linear map $T : X \to Y$ is an *isomorphism* between X and Y if it is one-to-one and onto. The inverse of an isomorphism is an isomorphism and the composition of two isomorphisms is an isomorphism. If X and Y are finite-dimensional and are isomorphic, then they have the same dimension.

10   Topological Vector Spaces

A vector space X endowed with a topology $\mathcal{U}$ is a *topological vector space* over $\mathbb{R}$, if the operations of sum $+ : X \times X \to X$, and multiplication by scalars $\bullet : \mathbb{R} \times X \to X$ are continuous with respect to the product topologies of $X \times X$ and $\mathbb{R} \times X$.

Fix $x_o \in X$. The translation by x_o is defined by $T_{x_o}(x) = x_o + x$ for all $x \in X$. For a fixed $\lambda \in \mathbb{R} - \{0\}$, the dilation by λ is defined by $D_\lambda(x) = \lambda x$ for all $x \in X$. If $\{X; \mathcal{U}\}$ is a topological vector space, the maps T_{x_o} and D_λ are homeomorphisms from $\{X; \mathcal{U}\}$ onto itself. In particular if $\mathcal{O}$ is open then $x + \mathcal{O}$ is open for all fixed $x \in X$. Any topology with such a property is *translation invariant*.

Remark 10.1 This notion can be used to construct a vector topological space $\{X; \mathcal{U}\}$ for which the sum is not continuous. It suffices to construct a vector space endowed with a topology which is not translation invariant. For an example of a linear, topological vector space for which the product by scalars is not continuous, see **10.4** and **10.5** of the Complements.

Let $\{X; \mathcal{U}\}$ be a topological vector space. If $\mathcal{B}_\Theta$ is a base at the zero element Θ of X, then for any fixed $x \in X$, the collection $\mathcal{B}_x = x + \mathcal{B}_\Theta$ forms a base for the topology $\mathcal{U}$ at x. Thus a base $\mathcal{B}_\Theta$ at Θ determines the topology $\mathcal{U}$ on X. If the elements of the base $\mathcal{B}_\Theta$ are convex, the topology of $\{X;\mathcal{U}\}$ is called *locally convex*. An example of a topological vector space with a nonlocally convex topology, is in § 3.5c of the Complements of Chap. 6.

An open neighborhood of the origin $\mathcal{O}$ is symmetric if $\mathcal{O} = -\mathcal{O}$.

The next remarks imply that the topology of a topological vector space, while non-necessarily locally convex is, roughly speaking, ball-like and, while not necessarily Hausdorff is roughly speaking close to being Hausdorff.

Proposition 10.1 *Let $\{X;\mathcal{U}\}$ be a topological vector space. Then:*

(i) *The topology $\mathcal{U}$ is generated by a symmetric base $\mathcal{B}_\Theta$.*
(ii) *If $\mathcal{O}$ is an open neighborhood of the origin, then $X = \bigcup_{\lambda \in \mathbb{R}} \lambda \mathcal{O}$.*
(iii) *$\{X;\mathcal{U}\}$ is Hausdorff if and only if the points are closed.*
(iv) *$\{X;\mathcal{U}\}$ is Hausdorff if and only if $\bigcap\{\mathcal{O} \in \mathcal{B}_\Theta\} = \{\Theta\}$.*

Proof The continuity of the multiplication by scalars implies that if $\mathcal{O}$ is open, also $\lambda\mathcal{O}$ is open for all $\lambda \in \mathbb{R} - \{0\}$. If $\Theta \in \mathcal{O}$, then $\Theta \in \lambda\mathcal{O}$ for all $|\lambda| \leq 1$. In particular if $\mathcal{O}$ is a neighborhood of the origin also $-\mathcal{O}$ is a neighborhood of the origin. The set $A = -\mathcal{O} \cap \mathcal{O}$ is an open neighborhood the origin, and is symmetric since $A = -A$. One verifies that the collection of such symmetric sets is a base $\mathcal{B}_\Theta$ at the origin, for the topology of $\{X;\mathcal{U}\}$.

To prove (ii) fix $x \in X$ and let $\mathcal{O}$ be an open neighborhood of Θ. Since $0 \cdot x = \Theta$, by the continuity of the product by scalars, there exist $\varepsilon > 0$ and an open neighborhood $\mathcal{O}_x$ of x such that $\lambda \cdot y \in \mathcal{O}$ for all $|\lambda| < \varepsilon$ and all $y \in \mathcal{O}_x$. Thus $\delta \cdot x \in \mathcal{O}$ for some $0 < |\delta| < \varepsilon$ and $x \in \delta^{-1}\mathcal{O}$.

The direct part of (iii) follows from Proposition 1.1. For the converse, assume that Θ and $x \in (X - \Theta)$ are closed. Then there exists an open set $\mathcal{O}$ containing the origin Θ and not containing x. Since $\Theta + \Theta = \Theta$ and the sum $+ : (X \times X) \to X$ is continuous, there exists two open sets $\mathcal{O}_1$ and $\mathcal{O}_2$ such that $\mathcal{O}_1 + \mathcal{O}_2 \subset \mathcal{O}$. Set

$$\mathcal{O}_o = \mathcal{O}_1 \cap \mathcal{O}_2 \cap (-\mathcal{O}_1) \cap (-\mathcal{O}_2).$$

Then

$$\mathcal{O}_o + \mathcal{O}_o \subset \mathcal{O} \quad \text{and} \quad \mathcal{O}_o \cap (x + \mathcal{O}_o) = \emptyset.$$

The last statement is a consequence of (iii). ∎

Proposition 10.2 *Let $\{X;\mathcal{U}\}$ and $\{Y;\mathcal{V}\}$ be topological vector spaces. A linear map $T : X \to Y$ is continuous if and only if is continuous at the origin Θ of X.*

Proof Since T is linear, $T(\Theta) = \theta \in Y$, where θ is the origin of Y. Let $O \in \mathcal{V}$ be an open set containing θ. By assumption $T^{-1}(O)$ is an open set containing Θ. Let $x \in X$ be fixed. An open set in Y that contains $T(x)$ is of the form $T(x) + O$, where O is an open set containing θ. The pre-image $T^{-1}(T(x) + O)$ contains the open set $x + T^{-1}(O)$. ∎

10.1   Boundedness and Continuity

Let $\{X; \mathcal{U}\}$ be a topological vector space. A subset $E \subset X$ is bounded if for every open neighborhood $\mathcal{O}$ of the origin Θ, there exists $\mu > 0$ such that $E \subset \lambda\mathcal{O}$ for all $\lambda > \mu$. A map T from a topological vector space $\{X; \mathcal{U}\}$ into a topological vector space $\{Y; \mathcal{V}\}$ is *bounded* if it maps bounded subsets of X into bounded subsets of Y.

Proposition 10.3 *A linear, continuous map T from a topological vector space $\{X; \mathcal{U}\}$ into a topological vector space $\{Y; \mathcal{V}\}$, is bounded.*

Proof Let $E \subset X$ be bounded. For every neighborhood O of the origin θ of Y, open in the topology of $\{Y; \mathcal{V}\}$, the inverse image $T^{-1}(O)$ is a neighborhood of the origin Θ, open in the topology of $\{X; \mathcal{U}\}$. Since E is bounded, there exists some $\delta > 0$ such that $E \subset \delta T^{-1}(O)$. Therefore, $T(E) \subset \delta O$.   ∎

Remark 10.2 Linearity alone does not imply boundedness. An example of unbounded linear map between two topological vector spaces is in § 15. Further examples are in **3.4** and **3.5** of the Complements of Chap. 7.

Remark 10.3 For general topological vector spaces $\{X; \mathcal{U}\}$ and $\{Y; \mathcal{V}\}$, the converse of Proposition 10.3 is false; that is, linearity and boundedness do not imply continuity of T. See **10.3** of the Complements for a counterexample. However, the converse is true for linear, bounded maps T between *metric* vector spaces, as stated in Proposition 14.2.

11   Linear Functionals

If the target space Y is the field $\mathbb{R}$ endowed with the Euclidean topology, the linear map $T : X \to \mathbb{R}$ is called a *functional* on $\{X; \mathcal{U}\}$. A linear functional $T : \{X; \mathcal{U}\} \to \mathbb{R}$ is bounded in a neighborhood of Θ, if there exists an open set $\mathcal{O}$ containing Θ and a positive number k such that $|T(x)| < k$ for all $x \in \mathcal{O}$.

Proposition 11.1 *Let $T : \{X; \mathcal{U}\} \to \mathbb{R}$ be a not identically zero, linear functional on X. Then:*

(i)   If T is bounded in a neighborhood of the origin, then T is continuous.
(ii)   If $\ker\{T\}$ is closed then T is bounded in a neighborhood of the origin.
(iii) T is continuous if and only if $\ker\{T\}$ is closed.
(iv) T is continuous if and only if it is bounded in a neighborhood of the origin.

Proof Let $\mathcal{O}$ be an open neighborhood of the origin such that $|T(x)| \leq k$ for all $x \in \mathcal{O}$. For every $\varepsilon \in (0, k)$ the pre-image of the open interval $(-\varepsilon, \varepsilon)$, contains the open sets $\lambda\mathcal{O}$ for all $0 < \lambda < \varepsilon/k$. Thus T is continuous at the origin and therefore continuous by Proposition 10.2.

Turning to (ii), if $\ker\{T\}$ is closed there exist $x \in X$ and some open neighborhood $\mathcal{O}$ of Θ, such that $(x + \mathcal{O}) \cap \ker\{T\} = \emptyset$. By (i) of Proposition 10.1, we may assume, that $\mathcal{O}$ is symmetric and that $\lambda\mathcal{O} \subset \mathcal{O}$ for all $|\lambda| \leq 1$. This implies that $T(\mathcal{O})$ is a symmetric interval about the origin of $\mathbb{R}$. If such an interval is bounded, there is nothing to prove. If such an interval coincides with $\mathbb{R}$, then there exist $y \in \mathcal{O}$ such that $T(y) = T(x)$. Thus $(x - y) \in \ker\{T\}$ and $(x + \mathcal{O}) \cap \ker\{T\}$ is not empty since $y \in \mathcal{O}$. The contradiction proves (ii).

To prove (iii) observe that the origin $\{0\}$ of $\mathbb{R}$ is closed. Therefore if T is continuous $T^{-1}(0) = \ker\{T\}$ is closed.

The remaining statements follow from (i)–(ii). ∎

Proposition 11.2 *Let $\{T_1, \ldots, T_n\}$ be a finite collection of bounded linear functionals on a Hausdorff, linear, topological vector space $\{X; \mathcal{U}\}$, and set*

$$K = \bigcap_{j=1}^{n} \ker\{T_j\}.$$

If T is a bounded linear functional on $\{X; \mathcal{U}\}$ vanishing on K, then there exists a n-tuple $(\alpha_1, \ldots, \alpha_n) \in \mathbb{R}^n$ such that

$$T = \sum_{j=1}^{n} \alpha_j T_j.$$

Proof The map

$$X \ni x \to \big(T_1(x), \ldots, T_n(x)\big) \in \mathbb{R}^n$$

is bounded and linear, and its image $\mathbb{R}_o^n$ is a closed subspace $\mathbb{R}^n$. The map

$$\mathbb{R}_o^n \ni \big(T_1(x), \ldots, T_n(x)\big) \to \mathcal{T}_o\big(T_1(x), \ldots, T_n(x)\big) \overset{\text{def}}{\longrightarrow} T(x) \in \mathbb{R}$$

is well defined since T vanishes on K. The map $\mathcal{T}_o$ is bounded and linear, and it extends to a bounded linear map $\mathcal{T} : \mathbb{R}^n \to \mathbb{R}$, defined in the whole $\mathbb{R}^n$. The latter must be of the form

$$\mathbb{R}^n \ni (y_1, \ldots, y_n) \to \mathcal{T}(y_1, \ldots, y_n) = \sum_{j=1}^{n} \alpha_j y_j$$

for a fixed n-tuple $(\alpha_1, \ldots, \alpha_n) \in \mathbb{R}^n$. Since $\mathcal{T}$ agrees with $\mathcal{T}_o$ on $\mathbb{R}_o^n$

$$X \ni x \to T(x) = \sum_{j=1}^{n} \alpha_j T_j(x).$$ ∎

12   Finite Dimensional Topological Vector Spaces

The next proposition asserts that a n-dimensional Hausdorff topological vector space, can only be given, up to a homeomorphism, the Euclidean topology of $\mathbb{R}^n$.

Proposition 12.1 *Let $\{X; \mathcal{U}\}$ be a n-dimensional Hausdorff topological vector space over $\mathbb{R}$. Then $\{X; \mathcal{U}\}$ is homeomorphic to $\mathbb{R}^n$ equipped with the Euclidean topology.*

Proof Given a basis $\{\mathbf{e}_1, \ldots, \mathbf{e}_n\}$ for $\{X; \mathcal{U}\}$, the representation map

$$\mathbb{R}^n \ni (\lambda_1, \ldots, \lambda_n) \rightarrow T(\lambda_1, \ldots, \lambda_n) = \sum_{i=1}^{n} \lambda_i \mathbf{e}_i \in X$$

is linear, one-to-one, and onto. Let $\mathcal{O}$ be a neighborhood of the origin in X, which we may assume to be symmetric and such that $\alpha\mathcal{O} \subset \mathcal{O}$ for all $|\alpha| \leq 1$. By the continuity of the sum and multiplication by scalars the pre-image $T^{-1}(\mathcal{O})$ contains an open ball about the origin of $\mathbb{R}^n$. Thus T is continuous at the origin and therefore continuous.

To show that T^{-1} is continuous assume first that $n = 1$. In such a case $T(\lambda) = \lambda\mathbf{e}$ for some $\mathbf{e} \in (X - \Theta)$. The kernel of the inverse map $T^{-1} : X \rightarrow \mathbb{R}$ consist only of the zero element $\{\Theta\}$, which is closed since X is Hausdorff. Therefore T^{-1} is continuous by (iii) of Proposition 11.1. Proceeding by induction, assume that the representation map T is a homeomorphism between $\mathbb{R}^m$ and any m-dimensional Hausdorff space, for all $m = 1, \ldots, n-1$. Thus in particular any $(n-1)$-dimensional Hausdorff space is closed.

The inverse of the representation map has the form

$$X \ni x \rightarrow T^{-1}(x) = (\lambda_1(x), \ldots, \lambda_{n-1}(x), \lambda_n(x)).$$

Each of the n maps $\lambda_j(\cdot) : X \rightarrow \mathbb{R}$ is a linear functional on X, whose null-space is a $(n-1)$-dimensional subspace of X. Such a subspace is closed by the induction hypothesis. Thus each of the $\lambda_j(\cdot)$ is continuous. ∎

Corollary 12.1 *Every finite dimensional subspace of a Hausdorff topological vector space is closed.*

If $\{X; \mathcal{U}\}$ is n-dimensional and not Hausdorff, it is not homeomorphic to $\mathbb{R}^n$. An example is $\mathbb{R}^N$ with the trivial topology.

12.1   Locally Compact Spaces

A topological vector space $\{X; \mathcal{U}\}$ is locally compact if there exist an open neighborhood of the origin whose closure is compact.

Proposition 12.2 *Let $\{X;\mathcal{U}\}$ be a Hausdorff, locally compact topological vector space. Then X is of finite dimension.*

Proof Let $\mathcal{O}$ be a neighborhood of the origin, whose closure is compact. We may assume that $\mathcal{O}$ is symmetric and $\lambda\mathcal{O} \subset \mathcal{O}$ for all $|\lambda| \leq 1$. There exist at most finitely many points $x_1, \ldots, x_n \in \mathcal{O}$, such that

$$\overline{\mathcal{O}} \subset (x_1 + \tfrac{1}{2}\mathcal{O}) \cup (x_2 + \tfrac{1}{2}\mathcal{O}) \cup \cdots \cup (x_n + \tfrac{1}{2}\mathcal{O}).$$

The space $Y = \mathrm{span}\{x_1, \ldots, x_n\}$, is a closed, finite dimensional subspace of X. From the previous inclusion, $\tfrac{1}{2}\mathcal{O} \subset Y + \tfrac{1}{4}\mathcal{O}$. Therefore

$$\mathcal{O} \subset Y + \tfrac{1}{2}\mathcal{O} \subset 2Y + \tfrac{1}{4}\mathcal{O} = Y + \tfrac{1}{4}\mathcal{O}.$$

Thus, by iteration

$$\mathcal{O} \subset \bigcap \left(Y + \tfrac{1}{2^n}\mathcal{O}\right) = \bar{Y} = Y.$$

This implies that $\lambda\mathcal{O} \subset Y$ for all $\lambda \in \mathbb{R}$. Thus, by (ii) of Proposition 10.1

$$X = \bigcup \lambda\mathcal{O} \subset Y \subset X. \qquad \blacksquare$$

The assumption that $\{X;\mathcal{U}\}$ be Hausdorff cannot be removed. Indeed, any $\{X;\mathcal{U}\}$ with the trivial topology is compact, and hence locally compact. However, it is not Hausdorff and, in general, it is not of finite dimension.

13 Metric Spaces

A *metric* on a nonvoid set X is a function $d : X \times X \to \mathbb{R}$ satisfying the properties:

(i) $d(x, y) \geq 0$ for all pairs $(x, y) \in X \times X$
(ii) $d(x, y) = 0$ if and only if $x = y$
(iii) $d(x, y) = d(y, x)$ for all pairs $(x, y) \in X \times X$
(iv) $d(x, y) \leq d(x, z) + d(y, z)$ for all $x, y, z \in X$.

This last requirement is called the *triangle inequality*. The pair $\{X; d\}$ is a metric space. Denote by $B_\rho(x) = \{y \in X | d(y, x) < \rho\}$, the open ball centered at x and of radius $\rho > 0$. The collection $\mathcal{B}$ of all such balls, satisfies the conditions (i)–(ii) of § 4 and therefore, by Proposition 4.1, generates a topology $\mathcal{U}$ on $\{X; d\}$, called metric topology, for which $\mathcal{B}$ is a base. The notions of open or closed sets can be given in terms of the elements of $\mathcal{B}$. In particular, a set $\mathcal{O} \subset X$ is open if for every $x \in \mathcal{O}$ there exists some $\rho > 0$ such that $B_\rho(x) \subset \mathcal{O}$.

A point x is a point of closure for a set $E \subset X$ if $B_\varepsilon(x) \cap E \neq \emptyset$ for all $\varepsilon > 0$. A set E is closed if and only if it coincides with the set of all its points of closure. In particular points are closed.

Let $\{x_n\}$ be a sequence of elements of X. A point $x \in X$ is a cluster point for $\{x_n\}$ if for all $\varepsilon > 0$, the open ball $B_\varepsilon(x)$ contains infinitely many elements of $\{x_n\}$. The sequence $\{x_n\}$ converges to x if for every $\varepsilon > 0$ there exists n_ε such that $d(x, x_n) < \varepsilon$ for all $n \geq n_\varepsilon$. The sequence $\{x_n\}$ is a Cauchy sequence if for every $\varepsilon > 0$ there exists an index n_ε, such that $d(x_n, x_m) \leq \varepsilon$, for all $m, n \geq n_\varepsilon$.

A metric space $\{X; d\}$ is *complete* if every Cauchy sequence $\{x_n\}$ of elements of X converges to some element $x \in X$.

13.1 Separation and Axioms of Countability

The distance between two subsets A, B of X is defined by

$$d(A, B) = \inf_{x \in A; y \in B} d(x, y).$$

Proposition 13.1 *Let A be a subset of X. The function $x \to d(A, x)$ is continuous in $\{X; d\}$.*

Proof Let $x, y \in X$ and $z \in A$. By the requirement (iv) of a metric

$$d(z, x) \leq d(x, y) + d(z, y).$$

Taking the infimum of both sides for $z \in A$ gives

$$d(A, x) \leq d(x, y) + d(A, y).$$

Interchanging the role of x and y yields

$$|d(A, x) - d(A, y)| \leq d(x, y). \qquad \blacksquare$$

If E_1 and E_2 are two disjoint closed subsets of $\{X; d\}$, then the two sets

$$\mathcal{O}_1 = \{x \in X \mid d(x, E_1) < d(x, E_2)\}$$
$$\mathcal{O}_2 = \{x \in X \mid d(x, E_2) < d(x, E_1)\}$$

are open and disjoint. Moreover $E_1 \subset \mathcal{O}_1$ and $E_2 \subset \mathcal{O}_2$. Thus every metric space is normal. In particular every metric space is Hausdorff.

Every metric space satisfies the first axiom of countability. Indeed the collection of balls $B_\rho(x)$ as ρ ranges over the rational numbers of $(0, 1)$, is a countable base for the topology at x.

Proposition 13.2 *A metric space $\{X; d\}$ is separable if and only if it satisfies the second axiom of countability.*[3]

Proof Let $\{X; d\}$ be separable and let A be a countable, dense subset of $\{X; d\}$. The collection of balls centered at points of A and with rational radius forms a countable base for the topology of $\{X; d\}$. The converse follows from Proposition 4.2. ∎

Corollary 13.1 *Every subset of a separable metric space is separable.*

Proof Let $\{x_n\}$ be a countable dense subset. For a pair of positive integers (m, n), consider the balls $B_{1/m}(x_n)$ centered at x_n and radius $1/m$. If Y is a subset of X, the ball $B_{1/m}(x_n)$ must intersect Y for some pair (m, n). For any such pair, select an element $y_{n,m} \in B_{1/m}(x_n) \cap Y$. The collection of such $y_{m,n}$ is a countable, dense subset of Y. ∎

13.2 Equivalent Metrics

From a given metric d on X, one can generate other metrics. For example, for a given d, set

$$d_o(x, y) = \frac{d(x, y)}{1 + d(x, y)}. \tag{13.1}$$

One verifies that d_o satisfies the requirements (i)–(iii). To verify that d_o satisfies (iv) it suffice to observe that the function

$$t \to \frac{t}{1 + t} \qquad \text{for} \qquad t \geq 0$$

is nondecreasing. Thus d_o is a new metric on X and generates the metric spaces $\{X; d_o\}$. Starting from the Euclidean metric in $\mathbb{R}^N$, one may introduce a new metric by

$$d_*(x, y) = \left| \frac{x}{1 + |x|} - \frac{y}{1 + |y|} \right| \qquad x, y \in \mathbb{R}^N. \tag{13.2}$$

More generally, the same set X can be given different metrics, say for example d_1 and d_2, to generate metric spaces $\{X; d_1\}$ and $\{X; d_2\}$.

Two metrics d_1 and d_2 on the same set X are equivalent if they generate the same topology. Equivalently d_1 and d_2 are equivalent if they define the same open sets. In such a case, the identity map between $\{X; d_1\}$ and $\{X; d_2\}$ is a homeomorphism.

---

[3] An example of non separable metric space is in § 15.1 of Chap. 6. See also **15.2.** of the Complements of Chap. 6.

13.3 Pseudo Metrics

A function $d : (X \times X) \to \mathbb{R}$ is a pseudometric if it satisfies all but (ii) of the requirements of being a metric. For example $d(x, y) = ||x| - |y||$ is a pseudometric on $\mathbb{R}$. The open balls $B_\rho(x)$ are defined as for metrics and generate a topology on X, called the pseudometric topology. The space $\{X; d\}$ is a pseudo-metric space. The statements of Propositions 13.1, 13.2 and Corollary 13.1 continue to hold for pseudo-metric spaces.

14 Metric Vector Spaces

Let $\{X; d\}$ and $\{Y; \eta\}$ be metric spaces. The notion of continuity of a function from X into Y can be rephrased in terms of the metrics η and d. Precisely, a function $f : \{X; d\} \to \{Y; \eta\}$ is continuous at some $x \in X$, if and only if for every $\varepsilon > 0$ there exists $\delta = \delta(\varepsilon, x)$ such that $\eta\{f(x), f(y)\} < \varepsilon$ whenever $d(x, y) < \delta$. The function f is continuous if it is continuous at each $x \in X$ and it is uniformly continuous if the choice of δ depends on ε and is independent of x.

A homeomorphism f between two metric spaces $\{X; d\}$ and $\{Y; \eta\}$ is *uniform* if the map $f : X \to Y$ is one-to-one and onto, and if it is *uniformly* continuous and has uniformly continuous inverse.

An isometry between $\{X; d\}$ and $\{Y; \eta\}$ is a homeomorphism f between $\{X; d\}$ and $\{Y; \eta\}$ that preserves distances, that is, such that

$$\eta\{f(x), f(y)\} = d(x, y) \qquad \text{for all } x, y \in X.$$

Thus an isometry is a uniform homeomorphism between $\{X; d\}$ and $\{Y; \eta\}$.

Let $\{X_1; d_1\}$ and $\{X_2; d_2\}$ be metric spaces. The product metric $(d_1 \times d_2)$ on the Cartesian product $(X_1 \times X_2)$ is defined by

$$(d_1 \times d_2)\{(x_1, x_2), (y_1, y_2)\} = d_1(x_1, y_1) + d_2(x_2, y_2)$$

for all $x_1, y_1 \in X_1$ and $x_2, y_2 \in X_2$. One verifies that the topology generated by $(d_1 \times d_2)$ on $(X_1 \times X_2)$ coincides with the product topology of $\{X_1; d_1\}$ and $\{X_2; d_2\}$.

If X is a *vector* space, then $\{X; d\}$ is a topological vector space if the operations of sum $+ : X \times X \to X$ and product by scalars $\bullet : \mathbb{R} \times X \to X$, are continuous with respect to the topology generated by d on X and the topology generated by $(d \times d)$ on $X \times X$.

A metric d on a vector space X is translation invariant if

$$d(x + z, y + z) = d(x, y) \qquad \text{for all } x, y, z \in X.$$

If d is translation invariant, then the metric d_o in of (13.1) is translation invariant. The metric d_* in (13.2) is not translation invariant.

Proposition 14.1 *If d on a vector space X is translation invariant then the sum $+ : (X \times X) \to X$ is continuous.*

Proof It suffices to show that $X \times X \ni (x, y) \to x + y$ is continuous at an arbitrary point $(x_o, y_o) \in X \times X$. From the definition of product topology

$$\begin{aligned}
d(x + y, x_o + y_o) &= d(x - x_o, y_o - y) \\
&\leq d(x - x_o, \Theta) + d(y_o - y, \Theta) \\
&= d(x, x_o) + d(y, y_o) \\
&= (d \times d)\{(x, y), (x_o, y_o)\}.
\end{aligned}$$
∎

Translation invariant metrics generate translation invariant topologies. There exist nontranslation invariant metrics that generate translation invariant topologies.

Remark 14.1 The topology generated by a metric on a vector space X, need not be locally convex. A counterexample is in Corollary 3.1c of the Complements of Chap. 6.

Remark 14.2 In general the notion of a metric on a vector space X does not imply, alone, any continuity statement of the operations of sum or product by scalars. Indeed there exists metric spaces for which both operations are discontinuous.

To construct examples, let $\{X; d\}$ be a metric vector space and let h be a discontinuous bijection from X onto itself. Setting

$$d_h(x, y) \overset{\text{def}}{=} d(h(x), h(y)) \tag{14.1}$$

defines a metric in X. The bijection h can be chosen in such a way that for the metric vector space $\{X; d_h\}$ the sum and the multiplication by scalars are both discontinuous. One such a choice is in § 14c of the Complements.[4]

14.1 Maps Between Metric Spaces

The notion of maps between metric spaces and their properties, is inherited from the corresponding notions between topological vector spaces. In particular Propositions 10.1–10.3 and 11.1, continue to hold in the context of metric spaces. However for metric spaces Proposition 10.3 admits a converse.

Proposition 14.2 *Let $\{X; d\}$ and $\{Y; \eta\}$ be metric vector spaces. A bounded linear map $T : X \to Y$ is continuous.*

---

[4]This construction was suggested by Ethan Devinatz.

Proof For any ball $\mathcal{B}_r$ in $\{Y; \eta\}$, of radius r centered at the origin of Y, there exists a ball B_ρ in $\{X; d\}$, centered at origin of X such that $B_\rho \subset T^{-1}(\mathcal{B}_r)$. If not, for all $\delta > 0$ the ball $\delta^{-1}B_1$ is not contained in $T^{-1}(\mathcal{B}_r)$. Thus $T(B_1)$ is not contained in $\delta\mathcal{B}_r$ for any $\delta > 0$ against the boundedness of T. The contradiction implies T is continuous at the origin and, by linearity T is continuous everywhere. ∎

15 Spaces of Continuous Functions

Let E be a subset of $\mathbb{R}^N$, denote by $C(E)$ the collection of all continuous functions $f : E \to \mathbb{R}$ and set

$$d(f, g) = \sup_{x \in E} |f(x) - g(x)| \qquad f, g \in C(E). \tag{15.1}$$

If E is compact, this defines a metric in $C(E)$ by which $C(E)$ turns into a metric vector space. The metric in (15.1) generates a topology in $C(E)$ called the topology of *uniform convergence*. Cauchy sequences in $C(E)$ converge uniformly to a continuous function in E. In this sense $C(E)$ is *complete*.

If E is compact, $C(E)$ is separable (Corollary 16.1 of Chap. 5).

If E is open, a function $f \in C(E)$, while bounded on every compact subset of E, in general is not bounded in E.

If $E \subset \mathbb{R}^N$ is compact, then any $f \in C(E)$ is uniformly continuous in E. If E is open then $f \in C(E)$ does not imply uniform continuity even if f is bounded in E.

Let $\{E_n\}$ be a collection of bounded open sets invading E, that is, $\bar{E}_n \subset E_{n+1}$ for all n, and $E = \cup E_n$. For every $f, g \in C(E)$ set

$$d_n(f, g) = \sup_{x \in \bar{E}_n} |f(x) - g(x)|.$$

Each d_n, while a metric in $C(\bar{E}_n)$, is a pseudometric in $C(E)$. Setting

$$d(f, g) = \sum \frac{1}{2^n} \frac{d_n(f, g)}{1 + d_n(f, g)} \tag{15.2}$$

defines a metric in $C(E)$ by which $\{C(E); d\}$ is a metric vector space.

A sequence $\{f_n\}$ of functions in $C(E)$ converges to $f \in C(E)$, in the metric (15.2), if and only if $\{f_n\} \to f$ uniformly on every compact subset of E. Cauchy sequences in $C(E)$ converge uniformly over compact subsets of E, to a function in $C(E)$. In this sense, the space $C(E)$ with the topology generated by the metric (15.2) is complete.

Denote by $\mathcal{L}^1(E)$ the collection of functions in $C(E)$ whose Riemann integral over E is finite. Since $\mathcal{L}^1(E)$ is a linear subspace of $C(E)$, it can be given the metric (15.2) and the corresponding topology. This turns $\mathcal{L}^1(E)$ into a metric vector space. The linear functional

$$T(f) = \int_E f dx \; : \mathcal{L}^1(E) \to \mathbb{R}$$

is unbounded and hence discontinuous. As an example let $E = (0, 1)$. The functions $f_n(t) = t^{\frac{1}{n}-1}$ are all in $\mathcal{L}^1(0, 1)$ and the sequence $\{f_n\}$ is bounded in the topology of (15.2) since $d(f_n, 0) \leq 1$. However $T(f_n) = n$. If E is bounded, the linear functional

$$T(f) = \int_E f dx \; : C(\bar{E}) \to \mathbb{R}$$

is bounded and hence continuous.

15.1 Spaces of Continuously Differentiable Functions

Let E be an open subset of $\mathbb{R}^N$ and denote by $C^1(E)$ the collection of all continuously differentiable functions $f \; : E \to \mathbb{R}$. Denote by $C^1(\bar{E})$ the collection of functions in $C^1(E)$ whose derivatives f_{x_j} for $j = 1, \ldots, N$ admit a continuous extension to $\bar{E}$, which we continue to denote by f_{x_j}.

For $f, g \in C^1(\bar{E})$ set formally

$$d(f, g) = \sup_{x \in \bar{E}} |f(x) - g(x)| + \sum_{j=1}^{N} \sup_{x \in \bar{E}} |f_{x_j}(x) - g_{x_j}(x)|. \tag{15.3}$$

If E is bounded, so that $\bar{E}$ is compact, this defines a metric in $C^1(\bar{E})$ by which $C^1(\bar{E})$ turns into a metric vector space. Cauchy sequences in $C^1(\bar{E})$ converge to functions in $C^1(\bar{E})$. Therefore $C^1(\bar{E})$ is complete.

The space $C^1(\bar{E})$ can also be given the metric (15.1). This turns $C^1(\bar{E})$ into a metric space. The topology generated by such a metric in $C^1(\bar{E})$ is the same as the topology that $C^1(\bar{E})$ inherits as a subspace of $C(\bar{E})$. With respect to such a topology $C^1(\bar{E})$ is not complete. The linear map

$$T(f) = f_{x_j} : C^1(\bar{E}) \to C(\bar{E}) \quad \text{for some fixed } j \in \{1, \ldots, N\}$$

is bounded, and hence continuous, provided $C^1(\bar{E})$ is given the metric (15.3). It is unbounded, and hence discontinuous if $C^1(\bar{E})$ is given the metric (15.1).

As an example, let $\bar{E} = [0, 1]$. The functions $f_n(t) = t^n$ are in $C^1[0, 1]$ for all $n \in \mathbb{N}$ and the sequence $\{f_n\}$ is bounded in $C[0, 1]$ since $d(f_n, 0) = 1$. However $T(f_n) = nt^{n-1}$ is unbounded in $C[0, 1]$.

If E is open and $f \in C^1(E)$, the functions f and f_{x_j} for $j = 1, \ldots, N$, while bounded on every compact subset of E, in general are not bounded in E. A metric in $C^1(E)$ can be introduced along the lines of (15.2).

15.2 Spaces of Hölder and Lipschitz Continuous Functions

Let E be an open set in $\mathbb{R}^N$ and let $\alpha \in (0, 1]$ be fixed. A function bounded $f : E \to \mathbb{R}$ is said to be *Hölder continuous* with exponent α if there exists a constant $L_\alpha > 0$ such that

$$|f(x) - f(y)| \leq L_\alpha |x - y|^\alpha \quad \text{for all pairs } x, y \in E. \tag{15.4}$$

The best constant L_α is given by

$$[f]_{\alpha,E} = \sup_{x,y \in E} \frac{|f(x) - f(y)|}{|x - y|^\alpha}. \tag{15.5}$$

The collection of all Hölder continuous functions with exponent $\alpha \in (0, 1)$ is denoted by $C^\alpha(E)$. If $\alpha = 1$ these functions are called *Lipschitz continuous*, and their collection is denoted by Lip(E). Setting

$$d(f, g) = \sup_{x \in E} |f(x) - g(x)| + [f - g]_{\alpha,E}. \tag{15.6}$$

defines a translation invariant metric in $C^\alpha(E)$ or Lip(E) which turns these into metric topological vector spaces.

16 On the Structure of a Complete Metric Space

Let $\{X; \mathcal{U}\}$ be a topological space. A set $E \subset X$ is *nowhere dense* in X, if $\bar{E}^c$ is dense in X. If E is nowhere dense, then also $\bar{E}$ is nowhere dense. A closed set E is nowhere dense, if and only if it does not contain any open set. If E is nowhere dense, for any open set $\mathcal{O}$ the complement $\mathcal{O} - \bar{E}$ must contain an open set. Indeed if not $\bar{E}$ would contain the open set $\mathcal{O}$. If E is nowhere dense and open, $\bar{E} - E$ is nowhere dense. If E is nowhere dense and closed, $E - \overset{o}{E}$ is nowhere dense.

A finite subset of $[0, 1]$ is nowhere dense in $[0, 1]$. The Cantor set is nowhere dense in $[0, 1]$. Such a set is the *uncountable* union of if $\bar{E}^c$ is dense in X. If E is nowhere if $\bar{E}^c$ is dense in X. If E is nowhere nowhere dense sets. The rationals are not nowhere dense in $[0, 1]$. However, they are the *countable* union of nowhere dense sets in $[0, 1]$. Thus, the uncountable union of nowhere dense sets might be nowhere dense and the countable union of nowhere dense sets, might be dense.

A set $E \subset X$ is said to be *meager*, or of *first category* if is the countable union of nowhere dense sets. A set that is not of first category, is said to be of *second category*.

The complement of a set of first category is called a *residual* or *non-meager* set. The rationals in $[0, 1]$ are a set of first category. The Cantor set is of first category in $[0, 1]$.

A metric space $\{X; d\}$ is *complete* if every Cauchy sequence $\{x_n\}$ of elements of X converges to some element $x \in X$. An example of noncomplete metric space is the set of the rationals in $[0, 1]$ with the Euclidean metric. Every metric space can be completed as indicated in § 16.3c of the Complements. The completion of the rationals are the real numbers.

The *Baire Category Theorem* asserts that a complete metric space cannot be the countable union of nowhere dense sets, much the same way as $[0, 1]$ is not the union of the rationals.

Theorem 16.1 (Baire [7]) *A complete metric space is of second category.*

Proof If not, there exist a countable collection $\{E_n\}$ of nowhere dense subsets of X, such that $X = \cup E_n$. Pick $x_o \in X$ and consider the open ball $B_1(x_o)$ centered at x_o and radius one. Since E_1 is nowhere dense in X, the complement $B_1(x_o) - \bar{E}_1$ contains an open set. Select an open ball $B_{r_1}(x_1)$, such that

$$\bar{B}_{r_1}(x_1) \subset B_1(x_o) - \bar{E}_1 \subset \bar{B}_1(x_o).$$

The selection can be done so that $r_1 < \frac{1}{2}$. Since E_2 is nowhere dense the complement $B_{r_1}(x_1) - \bar{E}_2$ contains an open set so that we may select an open ball $B_{r_2}(x_2)$ such that

$$\bar{B}_{r_2}(x_2) \subset B_{r_1}(x_1) - \bar{E}_2 \subset \bar{B}_{r_1}(x_1).$$

The selection can be done so that $r_2 < \frac{1}{3}$. Proceeding in this fashion generates a sequence of points $\{x_n\}$ and a family of balls $\{B_{r_n}(x_n)\}$ such that

$$\bar{B}_{r_{n+1}}(x_{n+1}) \subset \bar{B}_{r_n}(x_n) \quad r_n \leq \frac{1}{n+1} \quad \text{and} \quad \bar{B}_{r_n}(x_n) \cap \bigcup_{j=1}^{n} \bar{E}_n = \emptyset$$

for all n. The sequence $\{x_n\}$ is Cauchy and we let x denote its limit. Now the element x must belong to all the closed ball $\bar{B}_{r_n}(x_n)$ and it does not belong to any of the $\bar{E}_n$. Thus $x \notin \cup \bar{E}_n$ and $X \neq \cup E_n$. ∎

Corollary 16.1 *A complete metric space $\{X; d\}$ does not contain open subsets of first category.*

16.1 The Uniform Boundedness Principle

Theorem 16.2 (Banach-Steinhaus [15]) *Let $\{X; d\}$ be a complete metric space and let $\mathcal{F}$ be a family of continuous, real-valued functions defined in X. Assume that the functions $f \in \mathcal{F}$ are pointwise equi-bounded, that is, for all $x \in X$, there exists a positive number $F(x)$ such that*

$$|f(x)| \leq F(x) \quad \text{for all} \quad f \in \mathcal{F}. \tag{16.1}$$

Then, there exists a nonempty open set $\mathcal{O} \in X$ and a positive number F, such that

$$|f(x)| \leq F \quad \text{for all} \quad f \in \mathcal{F} \quad \text{and all} \quad x \in \mathcal{O}. \tag{16.2}$$

Thus if the functions of the family $\mathcal{F}$ are pointwise equibounded in X, they are uniformly equibounded within some open subset of X. For this reason the theorem is also referred to as the *Uniform Boundedness Principle*.

Proof For $n \in \mathbb{N}$, let $E_{n,f}$ and E_n be subsets of X defined by

$$E_{n,f} = \{x \in X \mid |f(x)| \leq n\}, \qquad E_n = \bigcap_{f \in \mathcal{F}} E_{n,f}.$$

The sets $E_{n,f}$ are closed, since the functions f are continuous. Therefore also the sets E_n are closed. Since the functions f are pointwise equi-bounded, for each $x \in X$ there exists some integer n such that $|f(x)| \leq n$ for all $f \in \mathcal{F}$. Therefore each $x \in X$ belongs to some E_n, that is, $X = \cup E_n$.

Since $\{X; d\}$ is complete, by the Baire category theorem, at least one of the E_n must not be nowhere dense. Since E_n is closed, it must contain a nonempty open set $\mathcal{O}$. Such a set satisfies (16.2) with $F = n$.   ∎

The Baire category theorem, and related category arguments, are remarkable, as they afford function-theoretical conclusions from purely topological information.

17 Compact and Totally Bounded Metric Spaces

Since a metric space satisfies the first axiom of countability, sequential compactness, countable compactness and the Bolzano–Weierstrass property all coincide (Proposition 5.2).

A metric space $\{X; d\}$ is *totally bounded* if for each $\varepsilon > 0$ there exists a finite collection of elements of X, say $\{x_1, \ldots, x_m\}$ for some positive integer m depending upon ε, such that X is covered by the union of the balls $B_\varepsilon(x_i)$ of radius ε and centered at x_i. A finite sequence $\{x_1, \ldots, x_m\}$ with such a property is called a finite ε-net for X.

Proposition 17.1 *A countably compact metric space $\{X; d\}$ is totally bounded.*

Proof Proceeding by contradiction, assume that there exists some $\varepsilon > 0$ for which there is no finite ε-net. Then, for a fixed $x_1 \in X$ the ball $B_\varepsilon(x_1)$ does not cover X and we choose $x_2 \in X - B_\varepsilon(x_1)$. The union of the two balls $B_\varepsilon(x_1)$ and $B_\varepsilon(x_2)$ does not cover X and we select $x_3 \in X - B_\varepsilon(x_1) \cup B_\varepsilon(x_2)$. Proceeding in this fashion generates a sequence of points $\{x_n\}$ at mutual distance of at least ε. Such a sequence cannot have a cluster point, thus contradicting the Bolzano–Weierstrass property.   ∎

Corollary 17.1 *A countably compact metric space is separable.*

Proof For positive integers m and n, let $E_{n,m}$ be the finite ε-net of X corresponding to $\varepsilon = \frac{1}{n}$. The union $\cup E_{n,m}$ is a countable subset of X which is dense in X.   ∎

If $\{X; d\}$ is separable, every open covering of X has a countable sub-covering. Therefore, countable compactness implies compactness (Proposition 5.3 and Corollary 5.1). Thus for separable metric spaces, all the various notions of compactness are equivalent.

We next examine the relation between compactness and total boundedness.

If $\{X; d\}$ is compact it is also totally bounded. Indeed having fixed $\varepsilon > 0$, the balls $B_\varepsilon(x)$ centered at all points of X, form an open covering of X, from which one may extract a finite one. It turns out that total boundedness implies compactness, provided the metric space $\{X; d\}$ is complete.

Proposition 17.2 *A totally bounded and complete metric space $\{X; d\}$ is sequentially compact.*

Proof Let $\{x_n\}$ be a sequence of elements of X. The proof consists of selecting a Cauchy sequence $\{x_{n'}\} \subset \{x_n\}$. Since $\{X; d\}$ is complete, such a Cauchy subsequence would then converge to some $x \in X$ thereby establishing that $\{X; d\}$ is sequentially compact. Fix $\varepsilon = \frac{1}{2}$ and determine a corresponding $\frac{1}{2}$-net $\{y_{1,1}, \ldots, y_{1,m_1}\}$ for some positive integer m_1. The union of the balls $B_{\frac{1}{2}}(y_{1,j})$ for $j = 1, \ldots, m_1$, covers X. Therefore at least one of these balls, say for example $B_{\frac{1}{2}}(y_{1,j})$ contains infinitely many elements of $\{x_n\}$. Select these elements and relabel them, to form a sequence $\{x_{n_1}\}$. These elements satisfy $d(x_{n_1}, x_{m_1}) < 1$.

Next, let $\varepsilon = \frac{1}{2^2}$ and determine a corresponding $\frac{1}{4}$-net $\{y_{2,1}, \ldots, y_{2,m_2}\}$ for some positive integer m_2. There exist a ball $B_{\frac{1}{4}}(y_{2,j})$, for some $j \in \{1, \ldots, m_2\}$ that contains infinitely many elements of $\{x_{n_1}\}$. Select these elements and relabel them to form a sequence $\{x_{n_2}\}$. Thy satisfy $d(x_{n_2}, x_{m_2}) < \frac{1}{2}$.

Let $h \geq 2$ be a positive integer. If the subsequence $\{x_{n_{h-1}}\}$ has been selected, we let $\varepsilon = 2^{-h}$, and determine a corresponding 2^{-h}-net, say for example $\{y_{h,1}, \ldots, y_{h,m_h}\}$ for some positive integer m_h. There exist a ball $B_{\frac{1}{2^h}}(y_{h,j})$, for some $j \in \{1, \ldots, m_h\}$ that contains infinitely many elements of $\{x_{n_{h-1}}\}$. We select these elements and relabel them to form a sequence $\{x_{n_h}\}$, whose elements satisfy

$$d(x_{n_h}, x_{m_h}) < \frac{1}{2^{h+1}}.$$

The Cauchy subsequence $\{x_{n'}\}$ is selected by diagonalization, out of the sequences $\{x_{n_h}\}$.   ∎

Theorem 17.1 *A metric space $\{X; d\}$ is compact if and only if is totally bounded and complete.*

17.1   Pre-Compact Subsets of X

The various notions of compactness and their characterization in terms of total bound-edness, do not require that $\{X; d\}$ be a vector space. Thus, in particular they apply to any subset $K \subset X$, endowed with the metric d inherited from $\{X; d\}$, by regarding $\{K; d\}$ as a metric space in its own right.

Proposition 17.3 *A subset $K \subset X$ of a metric space $\{X; d\}$ is compact if and only if it is sequentially compact.*

A subset $K \subset X$ is *pre-compact* if its closure $\bar{K}$ is compact.

Proposition 17.4 *A subset K of a complete metric space $\{X; d\}$ is pre-compact if and only if is totally bounded.*

Problems and Complements

1c   Topological Spaces

1.1.   A countable union of open sets is open. A countable union of closed sets need not be closed.

1.2.   A countable intersection of closed sets is closed. A countable intersection of open sets need not be open.

1.3.   Let $\mathcal{U}_1$ and $\mathcal{U}_2$ be topologies on X. Then $\mathcal{U}_1 \cap \mathcal{U}_2$ is a topology on X; however $\mathcal{U}_1 \cup \mathcal{U}_2$ need not be a topology on X.

1.4.   The Euclidean topology on $\mathbb{R}$ induces a relative topology on $[0, 1)$. The sets $[0, \varepsilon)$ for $\varepsilon \in (0, 1)$ are open in the relative topology of $[0, 1)$ and not in the original topology of $\mathbb{R}$.

1.5.   Let $X = \mathbb{N} \cup \{\omega\}$, where ω is the first infinite ordinal. A set $\mathcal{O} \subset X$ is open if either is any subset of $\mathbb{N}$, or if it contains $\{\omega\}$ and all but finitely many elements of $\mathbb{N}$. The collection of all such sets, complemented with $\emptyset$ and X defines a topology on X. A function $f : X \to \mathbb{R}$ is continuous with respect to such a topology if and only if $\lim f(n) = f(\omega)$.

1.6.   Linear combinations of continuous functions are continuous. Let $g : \{X; \mathcal{U}\} \to \{Y; \mathcal{V}\}$ and $f : \{Y; \mathcal{V}\} \to \{Z; \mathcal{Z}\}$ be continuous. Then $f(g) : \{X; \mathcal{U}\} \to \{Z; \mathcal{Z}\}$ is continuous. The maximum or minimum of two real valued, continuous functions is continuous.

1.7.   Let $\mathcal{U}_1$ and $\mathcal{U}_2$ be topologies on X. The topology $\mathcal{U}_1$ is stronger or finer than $\mathcal{U}_2$ if $\mathcal{U}_2 \subset \mathcal{U}_1$, that is, roughly speaking, if $\mathcal{U}_1$ contains more open sets than $\mathcal{U}_2$. Equivalently if the identity map from $\{X; \mathcal{U}_1\}$ onto $\{X; \mathcal{U}_2\}$ is continuous.

1.8.   Let $\{f_n\}$ be a sequence of real valued, continuous functions from $\{X; \mathcal{U}\}$ into $\mathbb{R}$. If $\{f_n\} \to f$ uniformly, then f is continuous.

1.9. Let $C \subset X$ be closed and let $\{x_n\}$ be a sequence of points in C. Every cluster point of $\{x_n\}$ belongs to C.

1.10. Let $f : X \to Y$ be continuous and let $\{x_n\}$ be a sequence in X. If x is a cluster point of $\{x_n\}$, then $f(x)$ is a cluster point of $\{f(x_n)\}$.

1.11. Let X be the collection of pairs (m, n) of nonnegative integers. Any subset of X that does not contain $(0, 0)$ is declared to be open. A set $\mathcal{O}$ that contains $(0, 0)$ is open if and only if for all but a finite number of integers m, the set $\{n \in \mathbb{N} \cup \{0\} | (m, n) \notin \mathcal{O}\}$ is finite. For a fixed m the collection of (m, n) as n ranges over $\mathbb{N} \cup \{0\}$ can be regarded as a *column*. With this terminology, a set $\mathcal{O}$ containing $(0, 0)$ is open if and only if it contains all but a finite number of elements for all but a finite number of columns. This defines a Hausdorff topology on X. No sequence in X can converge to $(0, 0)$. The sequence (n, n) has $(0, 0)$ as a cluster point, but no subsequence of (n, n) converges to $(0, 0)$. This example is in [4].

1.12c Connected Spaces

A topological space $\{X; \mathcal{U}\}$ is connected if it is not the union of two disjoint open sets. A subset $X_o \subset X$ is connected if the space $\{X_o; \mathcal{U}_o\}$ is connected.

1.13. The continuous image of a connected space is connected.

1.14. Let $\{A_\alpha\}$ be a family of connected subsets of $\{X; \mathcal{U}\}$ with nonempty intersection. Then $\cup A_\alpha$ is connected.

1.15. **(Intermediate Value Theorem)**] Let f be a real valued continuous function on a connected space $\{X; \mathcal{U}\}$. Let $a, b \in X$ such that $f(a) < z < f(b)$ for some real number z. There exists $c \in X$ such that $f(c) = z$.

1.16. The discrete topology is a Hausdorff topology. If X is finite, then the discrete topology is the only one for which $\{X; \mathcal{U}\}$ is Hausdorff.

1.17. Let $\{X; \mathcal{U}\}$ be Hausdorff. Then $\{X; \mathcal{U}_1\}$ is Hausdorff for any stronger topology $\mathcal{U}_1$.

1.18. A Hausdorff space $\{X; \mathcal{U}\}$ is normal if and only if, for any closed set C and any open set $\mathcal{O}$ such that $C \subset \mathcal{O}$, there exists an open set O such that $C \subset O \subset \overline{O} \subset \mathcal{O}$.

1.19c Separation Properties of Topological Spaces

A topological space $\{X; \mathcal{U}\}$ is said to be *regular* if points are separated from closed sets, that is, for a given closed set $C \subset X$ and x not in C, there exist disjoint open sets $\mathcal{O}_C$ and $\mathcal{O}_x$ such that $C \subset \mathcal{O}_C$ and $x \in \mathcal{O}_x$.

A Hausdorff space is said to be of type (T_2). A regular space for which the singletons $\{x\}$ are closed is said to be of type (T_3). A normal space for which the singletons $\{x\}$ are closed is said to be of type (T_4). The separation properties of a topological space $\{X; \mathcal{U}\}$ are classified as follows:

T_o: Points are separated by open sets, that is, for any two given points $x, y \in X$ there exists an open set containing one of the two points, say for example y, but not the other.

T_1: The singletons $\{x\}$ are closed.

T_2: Hausdorff spaces.

T_3: Regular $+T_1$.

T_4: Normal $+T_1$.

From the definitions it follows that $(T_4) \Longrightarrow (T_3) \Longrightarrow (T_2) \Longrightarrow (T_1) \Longrightarrow ((T_o)$. The converse implications are false in general. In particular (T_o) does not imply (T_1). For example the space $X = \{x, y\}$ with the open sets $\{\emptyset, X, y\}$ is (T_o) and not (T_1). We have already observed that (T_1) does not imply Hausdorff. Hausdorff in turn does not imply normal. Counterexamples are rather specialized and can be found in [150, 41].

We will be concerned only with Hausdorff and normal spaces.

4c Bases, Axioms of Countability and Product Topologies

4.1. Let $\{X; \mathcal{U}\}$ satisfy the first axiom of countability and let $A \subset X$. For every $x \in \bar{A}$, there exists a sequence $\{x_n\}$ of elements of A converging to x. For every cluster point y of a sequence $\{x_n\}$ of points in X, there exists a subsequence $\{x_{n'}\} \to y$.

4.2. Let X be infinite and let $\mathcal{U}$ consist of the empty set and the collection of all subsets of X whose complement is finite. Then $\mathcal{U}$ is a topology on X. If X is uncountable $\{X; \mathcal{U}\}$ does not satisfy the first axiom of countability. The points are closed but $\{X; \mathcal{U}\}$ is not Hausdorff.

4.3. Let $\mathcal{B}$ be the collection of all intervals of the form $[\alpha, \beta)$. Then $\mathcal{B}$ is a base for a topology $\mathcal{U}$ on $\mathbb{R}$, constructed as in Proposition 4.1. The set $\mathbb{R}$ endowed with such a topology satisfies the first but not the second axiom of countability. The intervals $[\alpha, \beta)$ are both open and closed. This is called the *half-open interval* topology. The sequence $\{1 - \frac{1}{n}\}$ converges to 1 in the Euclidean topology and not in the half-open interval topology.

4.4. Let X be an uncountable set, well ordered by $\prec$ and let Ω be the first uncountable. Set $X_o = \{x \in X | x \prec \Omega\}$, $X_1 = X_o \cup \Omega$, and consider the collection $\mathcal{B}_o$ of sets

$$\{x \in X_o | x \prec \alpha\} \quad \text{for some } \alpha \in X_o$$
$$\{x \in X_o | \beta \prec x\} \quad \text{for some } \beta \in X_o$$
$$\{x \in X_o | \alpha \prec x \prec \beta\} \quad \text{for } \alpha, \beta \in X_o.$$

Define similarly a collection of sets $\mathcal{B}_1$ where the various elements are taken out of X_1.

i.   The collection $\mathcal{B}_o$ forms a base for a topology $\mathcal{U}_o$ on X_o. The resulting space $\{X_o; \mathcal{U}_o\}$ satisfies the first but not the second axiom of countability. Moreover $\{X_o; \mathcal{U}_o\}$ is separable.

ii. The collection $\mathcal{B}_1$ forms a base for a topology $\mathcal{U}_1$ on X_1. The resulting space $\{X_1; \mathcal{U}_1\}$ does not satisfy the first axiom of countability and is not separable.

4.5. The product of two connected topological spaces is connected.

4.6. The product of a family $\{X_\alpha; \mathcal{U}_\alpha\}$ of Hausdorff spaces is Hausdorff.

4.7. A sequence $\{x_n\}$ of elements of $\prod X_\alpha$ converges to some $x \in \prod X_\alpha$ if and only if the sequences of the projections $\{x_{\alpha,n}\}$ converge to the projections x_α of x.

4.8. The *countable* product of separable topological spaces is separable.

4.9. Let $\{X; \mathcal{U}\}$ satisfy the second axiom of countability. Every topological subspace of X is separable. If $\{X; \mathcal{U}\}$ is separable but it does not satisfy the second axiom of separability, a topological subspace of X might not be separable. The interval $[0, 1]$ with the half-open interval topology is separable. The Cantor set $C \subset [0, 1]$ with the inherited topology is not separable.

4.10c   The Box Topology

Let $\{X_\alpha; \mathcal{U}_\alpha\}$ be a family of topological spaces and set

$$\mathcal{B} = \bigcup \{\textstyle\prod \mathcal{O}_\alpha | \mathcal{O}_\alpha \in \mathcal{U}_\alpha\}.$$

Each set in $\mathcal{B}$ is an open rectangle, since it is the Cartesian product of open sets in $\mathcal{U}_\alpha$. The collection $\mathcal{B}$ forms a base for a topology in $\prod_\alpha X_\alpha$, called the box-topology. While the projections π_α are continuous with respect to such a topology, the box topology contains, roughly speaking, too many open sets.

As an example let $[0, 1]$ be endowed with the topology inherited from the Euclidean topology on $\mathbb{R}$. Then the Hilbert box $[0, 1]^\mathbb{N}$ can be endowed with either the product topology or the box-topology. The sequence $\{x_n\} = \{\frac{1}{n}, \dots, \frac{1}{n}, \dots\}$ converges to zero in the product topology and not in the box-topology. Indeed the neighborhood of the origin $\mathcal{O} = \prod [0, \frac{1}{j})$ does not contain any of the elements of $\{x_n\}$.

5c   Compact Topological Spaces

5.1. A Hausdorff and compact topological space is regular and normal.

5.2. If $\{X; \mathcal{U}\}$ is compact, then $\{X; \mathcal{U}_o\}$ is compact for any weaker topology $\mathcal{U}_o \subset \mathcal{U}$. The converse is false.

Proposition 5.1c *Let $f : \{X; \mathcal{U}\} \to \{Y; \mathcal{V}\}$ be continuous, one-to-one and onto. If $\{X; \mathcal{U}\}$ is compact and $\{Y; \mathcal{V}\}$ is Hausdorff then f is a homeomorphism.*

Proof The inverse f^{-1} is one-to-one and onto. It is continuous if for every closed set $C \subset X$, the image $f(C)$ is closed. If $C \subset X$ is closed, it is compact and its continuous image $f(C)$ is compact and hence closed since $\{Y; \mathcal{V}\}$ is Hausdorff.   ∎

5.3. Let $\{X; \mathcal{U}\}$ be Hausdorff and compact. Then the previous Proposition implies that:

    i. $\{X; \mathcal{U}_1\}$ is not compact for any stronger topology $\mathcal{U}_1$.

    ii. $\{X; \mathcal{U}_o\}$ is not Hausdorff for any weaker topology $\mathcal{U}_o$.

    iii. If $\{X; \mathcal{U}_1\}$ is compact for a stronger topology $\mathcal{U}_1 \supset \mathcal{U}$, then $\mathcal{U} = \mathcal{U}_1$.

Thus, the topological structure of a compact Hausdorff space $\{X; \mathcal{U}\}$ is rigid in the sense that one cannot strengthen its topology without loosing compactness and cannot weaken it without loosing the separation property.

5.4. Let $\|x\|$ be the Euclidean norm in $\mathbb{R}^N$ and consider the function

$$f(x) = \begin{cases} \dfrac{\max\{|x_1|, \ldots, |x_N|\}}{\|x\|} x & \text{for } x \neq 0 \\ 0 & \text{for } x = 0. \end{cases}$$

The function f maps cubes of wedge 2ρ in $\mathbb{R}^N$, onto balls of radius ρ in $\mathbb{R}^N$, it is continuous, one-to-one and onto. Thus f is a homeomorphism between $\mathbb{R}^N$ equipped with the topology generated by the cubes with faces parallel to the coordinate planes, and $\mathbb{R}^N$ equipped with the topology generated by the balls.

5.5. A space X consisting of more than one point and equipped with the trivial topology is compact and not Hausdorff.

5.6. Let $\{X; \mathcal{U}\}$ be locally compact. A subset $C \subset X$ is closed if and only if $C \cap K$ is closed, for every closed compact subset $K \subset X$.

5.7. Let $\{X_o; \mathcal{U}_o\}$ and $\{X_1; \mathcal{U}_1\}$ be the spaces introduced in **4.4**. The space $\{X_o; \mathcal{U}_o\}$ is sequentially compact but not compact. The space $\{X_1; \mathcal{U}_1\}$ is compact.

5.8c   The Alexandrov One-Point Compactification of $\{X; \mathcal{U}\}$ ([3])

Let $\{X; \mathcal{U}\}$ be a noncompact Hausdorff topological space. Having fixed $x_* \notin X$ consider the set $X_* = X \cup \{x_*\}$ and define a collection of sets $\mathcal{U}_*$ consisting of $\mathcal{U}, X_*$, and all subsets $\mathcal{O}_* \subset X_*$ containing x_* and such that $X_* - \mathcal{O}_*$ is compact in $\{X; \mathcal{U}\}$. Then $\mathcal{U}_*$ is a Hausdorff topology on X_*. Moreover $\{X_*; \mathcal{U}_*\}$ is compact, X is dense in X_* and the restriction of $\mathcal{U}_*$ to X, coincides with the original topology $\mathcal{U}$ on X.

5.9. The topological space of **1.5** is compact. It can be regarded as the Alexandrov one-point compactification of $\mathbb{N}$ equipped with the discrete topology.

5.10. The one-point compactification of $\mathbb{R}^N$ with its Euclidean topology, is home-omorphic to the unit sphere in $\mathbb{R}^{N+1}$ by stereographic projection.

5.11. The one-point compactification of $\{X_o; \mathcal{U}_o\}$ in **4.4** is $\{X_1; \mathcal{U}_1\}$.

7c   Continuous Functions on Countably Compact Spaces

7.1c   Upper-Lower Semi-continuous Functions

7.1. Characteristic functions of open(closed) sets in $\mathbb{R}^N$ are lower(upper) semi-continuous. A function f for an open set $E \subset \mathbb{R}^N$ into $\mathbb{R}^*$ is upper(lower) semi-continuous if and only is for each $x \in E$

$$\limsup_{y \to x} f(y) \leq f(x) \quad \left(\liminf_{y \to x} f(y) \geq f(x)\right).$$

7.2. Let $\{f_\alpha\}$ be a collection of upper(lower) semi-continuous functions on a topological space $\{X; \mathcal{U}\}$. Then $\inf(\sup) f_\alpha$ is upper(lower) semi-continuous.

7.3. The finite sum of nonnegative upper(lower) semi-continuous functions is upper(lower) semi-continuous.

7.4. Let $\{f_n\}$ be a sequence of nonnegative, lower semi-continuous function on $\{X; \mathcal{U}\}$. Then $\sum f_n$ is lower semi-continuous.

7.5. Let $\{f_n\}$ be a sequence of nonnegative, upper semi-continuous function on $\{X; \mathcal{U}\}$. Then $\sum f_n$ need not be upper semi-continuous. Give a counterexample.

7.6. **Modulus of Continuity**: For an arbitrary real valued function f defined on an open set $E \subset \mathbb{R}^N$, and for $\varepsilon > 0$, set

$$\eta(x, \varepsilon) = \sup\{|f(y) - f(z)| : y, z \in B_\varepsilon(x) \cap E\}$$
$$\eta(x) = \inf_\varepsilon \eta(x, \varepsilon).$$

Prove that $\eta(\cdot)$ is upper semi-continuous. Prove that f is continuous at x if and only if $\eta(x) = 0$ and therefore the points of continuity of any $f : E \to \mathbb{R}$ are countable intersection of open sets. The function

$$E \times \mathbb{R}^+ \ni (x, \varepsilon) \to \eta(x, \varepsilon)$$

is the *modulus of continuity* of f at x. The function f is Hölder continuous at x, with Hölder exponent $\alpha \in (0, 1]$, if there exists positive constant $\delta = \delta(x)$ and $C = C(x)$ depending upon such that $\eta(x, \varepsilon) \leq C(x)\varepsilon^\alpha$ for all $0 < \varepsilon \leq \delta(x)$. In such a case

$$|f(x) - f(y)| \leq C(x)|x - y|^\alpha \quad \text{for all } |x - y| \leq \delta(x).$$

The function f is Hölder continuous in E with exponent α, if the constants $\delta(x)$ and $C(x)$ are independent of $x \in E$. If $\alpha = 1$ then f is Lipschitz continuous at x and respectively in E.

7.7.   **Upper and Lower Envelope of a Function**: For an arbitrary real valued function f defined on an open set $E \subset \mathbb{R}^N$, and for $\varepsilon > 0$, set

$$\varphi(x) = \sup_{\varepsilon>0} \inf_{|x-y|<\varepsilon} f(y) \quad \text{lower envelope of } f \text{ at } x$$

$$\psi(x) = \inf_{\varepsilon>0} \sup_{|x-y|<\varepsilon} f(y) \quad \text{upper envelope of } f \text{ at } x.$$

Prove that φ is lower semi-continuous and ψ is upper semi-continuous; moreover $\varphi \le f \le \psi$.

7.2c   Characterizing Lower-Semi Continuous Functions in $\mathbb{R}^N$

Proposition 7.1c *Let E be an open subset of $\mathbb{R}^N$. A function $f : E \to \mathbb{R}^+$ is lower semi-continuous if and only if it is the pointwise limit of an increasing sequence of continuous functions defined in E.*

Proof ($\Longrightarrow$) For $n \in \mathbb{N}$ and $x \in E$ set

$$f_n(x) = \inf\{f(z) + n|x - z| : z \in E\}.$$

Prove that $|f_n(x) - f_n(y)| \le n|x - y|$.   ∎

7.3c   On the Weierstrass-Baire Theorem

7.8.   The set of discontinuities of a real valued function could be as diverse as possible. As an example consider the functions

$$f(x) = \begin{cases} 1 & \text{if } x \in \mathbb{Q} \\ -1 & \text{if } x \in [0, 1] - \mathbb{Q} \end{cases} \qquad g(x) = \begin{cases} x & \text{if } x \in \mathbb{Q} \\ -x & \text{if } x \in [0, 1] - \mathbb{Q}. \end{cases}$$

The first is everywhere discontinuous but its absolute value is continuous. The second is continuous only at $x = 0$.

7.9. There exists a function $f : (0, 1) \to \mathbb{R}$ continuous at the irrationals and discontinuous at the rationals of $(0, 1)$. To construct an example recall that a rational number $r \in (0, 1]$ can be written as the ratio m/n of two positive integers in *lowest terms*. That is, m and n are the smallest integers for which $r = \frac{m}{n}$. A rational number r is an equivalence class of ratios of the form $\frac{m}{n}$. Out of such an equivalence class we select the representative in *lowest terms*. Set

$$f(x) = \begin{cases} \frac{1}{n} & \text{if } x \in \mathbb{Q} \cap (0, 1] \\ 0 & \text{if } x \in (0, 1] - \mathbb{Q}. \end{cases} \tag{7.1c}$$

However there exists no function $f : (0, 1] \to \mathbb{R}$ continuous at the rationals and discontinuous at the irrationals of $(0, 1]$ (see Corollary 16.1c of the Complements).

The function in (7.1c) is everywhere upper semi-continuous in $(0, 1]$ since, for every $y \in (0,]$

$$\lim_{x \to y} \sup f(x) = 0 \le f(y).$$

7.10. There exists functions that are everywhere finite in their domain of definition and not bounded in every subset of their domain of definition. Continue to represent a rational number $r \in (0, 1)$ as the ratio m/n of two positive integers in *lowest terms*. Then set

$$f(x) = \begin{cases} n & \text{if } x \in \mathbb{Q} \cap (0, 1) \\ 0 & \text{if } x \in (0, 1) - \mathbb{Q}. \end{cases} \tag{7.2c}$$

Such a function is everywhere finite in $[0, 1]$ and unbounded in every subinterval of $[0, 1]$. Indeed let $I \subset [0, 1]$ be an interval. If f were bounded in I, then the denominator n of all rational numbers $\frac{m}{n} \in I$, would be bounded. This would imply that there are only finitely many rationals in I.

7.11. There exist real valued, bounded functions, defined on a compact set that do not take neither maxima or minima.

Continue to represent a rational number $r \in (0, 1)$ as the ratio m/n of two positive integers in *lowest terms*. Then set

$$f(x) = \begin{cases} (-1)^n \frac{n}{n+1} & \text{if } x \in \mathbb{Q} \cap (0, 1) \\ 0 & \text{if } x \in (0, 1) - \mathbb{Q}. \end{cases} \tag{7.3c}$$

About any point of $(0, 1)$ the values of f are arbitrarily close to ± 1. The function in (7.3c) is nowhere upper semi-continuous in $[0, 1]$. Indeed for every $y \in [0, 1]$

$$\lim_{x \to y} \sup f(x) = 1 > f(y).$$

Thus the assumption that f be upper semi-continuous cannot be relaxed in the Weierstrass-Baire Theorem. The function in (7.3c) is also nowhere monotone in $[0, 1]$.

7.4c On the Assumptions of Dini's Theorem

7.12. The assumption that the limit function f be lower semi-continuous cannot be removed from Dini's theorem. Indeed the sequence $\{x^n\}$ for $x \in [0, 1]$ is decreasing, each x^n is continuous in $[0, 1]$ but the limit f is not lower semi-continuous. Accordingly, the convergence $\{x^n\} \to f$ is not uniform in $[0, 1]$.

7.13. The assumption that each of the f_n be upper semi-continuous, cannot be removed from Dini's theorem. Set

$$f_n(x) = \begin{cases} 0 & \text{for } x = 0 \\[2mm] 1 & \text{for } 0 < x < \frac{1}{n} \\[2mm] 0 & \text{for } \frac{1}{n} \le x \le 1. \end{cases}$$

The sequence $\{f_n\}$ is decreasing, it converges to zero pointwise in $[0, 1]$, but the convergence is not uniform.

7.14. The requirement that the sequence $\{f_n\}$ be decreasing cannot be removed from Dini's Theorem. Set

$$f_n(x) = \begin{cases} 2n^2 x & \text{for } 0 \le x \le \frac{1}{2n} \\[2mm] n - 2n^2(x - \frac{1}{2n}) & \text{for } \frac{1}{2n} \le x \le \frac{1}{n} \\[2mm] 0 & \text{for } \frac{1}{n} \le x \le 1. \end{cases}$$

The functions f_n are continuous in $[0, 1]$ and converge to zero pointwise in $[0, 1]$. However the convergence is not uniform.

9c Vector Spaces

9.1. The element $\Theta \in X$ is unique.

9.2. Let A, B and C be subsets of a vector space X. Then:

(i) $A \cap B \ne \emptyset$ if and only if $\Theta \in A - B$.

(ii) $A \cap (B + C) \ne \emptyset$ if and only if $B \cap (A - C) \ne \emptyset$. Equivalently if and only if $C \cap (A - B) \ne \emptyset$.

(iii) X_o is a subspace of X if and only if $\alpha X_o = X_o$ for all $\alpha \in \mathbb{R} - \{0\}$ and $x + X_o = X_o$ for all $x \in X_o$.

(iv)   If X_o and X_1 are linear subspaces of X then $\alpha X_o + \beta X_1$ is a subspace of X.

(v)   If A and B are convex, then $A + B$ is convex and λA is convex for all $\lambda \in \mathbb{R}$.

9.3c   Hamel Bases

A collection $\{x_\alpha\}$ of elements of a vector space X is linearly independent if any *finite* subcollection of elements $\{x_\alpha\}$ is linearly independent.

A linearly independent collection $\{x_\alpha\}$ is a *Hamel basis* for a vector space X if $\mathrm{span}\{x_\alpha\} = X$. Equivalently, if every $x \in X$ has a unique representation as a *finite* linear combination of elements of $\{x_\alpha\}$, that is

$$x = \sum_{j=1}^m c_j x_{\alpha_j} \quad \text{for some finite } m, \quad c_j \in \mathbb{R}.$$

Proposition 9.1c *Every vector space X has a Hamel basis.*

Proof Let $\mathcal{L}$ be the collection of all subsets of X whose elements are linearly independent. This collection is partially ordered by inclusion. Every linearly ordered subcollection $\{B_\sigma\}$ of $\mathcal{L}$ has an upper bound given by $B = \cup B_\sigma$. Indeed the elements of B are linearly independent since any finitely many of them must belong to some B_σ, and the elements of B_σ are linearly independent. Therefore by Zorn's lemma $\mathcal{L}$ has a maximal element $\{x_\alpha\}$. The elements of $\{x_\alpha\}$ are linearly independent and every $x \in X$ can be written as a finite linear combination of them. Indeed if not, the collection $\{x_\alpha, x\}$ belongs to $\mathcal{L}$, contradicting the maximality of $\{x_\alpha\}$. ∎

9.4.   A Hamel basis for $\mathbb{R}^N$ is the usual Euclidean basis.

9.5.   Let ℓ denote the collection of all sequences $\{c_n\}$ of real numbers and consider the countable subcollection of ℓ

$$
\begin{aligned}
\mathbf{e}_1 &= \{1, 0, 0, \ldots, 0_m, 0, \ldots\} \\
\mathbf{e}_2 &= \{0, 1, 0, \ldots, 0_m, 0, \ldots\} \\
\cdots &= \cdots \quad \cdots \quad \cdots \quad \cdots \\
\mathbf{e}_m &= \{0, 0, 0, \ldots, 1_m, 0, \ldots\} \\
\cdots &= \cdots \quad \cdots \quad \cdots \quad \cdots
\end{aligned}
\tag{9.1c}
$$

Every $x \in \ell$ can be written as $x = \sum c_n \mathbf{e}_n$. However $\{\mathbf{e}_n\}$ is not a Hamel basis for ℓ.

9.6c   On the Dimension of a Vector Space

If the Hamel basis of a vector space X is of the form $\{x_n\}$ for $n \in \mathbb{N}$, the dimension of X is $\aleph_o$, that is, the cardinality of $\mathbb{N}$. More generally, if $\{x_\alpha\}$ for $\alpha \in A$ is a Hamel basis for a vector space X, then the dimension of X is the cardinality of A. This definition of dimension of X is independent of the choice of the Hamel basis.

9.7. Let ℓ_o denote the collection of all sequences of real numbers $\{c_n\}$ with only finitely many non zero elements. Then (9.1c) is a Hamel basis for ℓ_o and the dimension of ℓ_o is $\aleph_o$.

9.8. Let $\ell[0, 1]$ denote the collection of all sequences $\{c_n\}$ of real numbers in $[0, 1]$. The dimension of $\ell[0, 1]$ is no less than the cardinality of $\mathbb{R}$, since the collection $x_\alpha = \{\alpha, \alpha^2, \dots, \alpha^n, \dots\}$ for $\alpha \in (0, 1)$ is linearly independent.

9.9. A vector space with a countable Hamel basis is separable.

9.10. The pair $\{\mathbb{R}; \mathbb{Q}\}$, that is, the reals $\mathbb{R}$ over the field of the rationals $\mathbb{Q}$, is a vector space. If $x \in \mathbb{R}$ is not an algebraic number, then the elements $\{1, x, x^2, \dots\}$ are linearly independent. The dimension of $\{\mathbb{R}; \mathbb{Q}\}$ is not less than the cardinality of $\mathbb{R}$.

10c   Topological Vector Spaces

10.1. Let A and B be subsets of a topological vector space $\{X; \mathcal{U}\}$. Then:

(i)   If A and B are open, the $\alpha A + \beta B$ is open.

(ii) $\bar{A} + \bar{B} \subset \overline{A + B}$. The inclusion is in general strict unless either one of $\bar{A}$ or $\bar{B}$ is compact.

(iii) If $A \subset X$ is convex then $\bar{A}$ and $\overset{o}{A}$ are convex.

(iv)   The convex hull of an open set is open.

10.2. If $x \in \mathcal{O} \in \mathcal{B}_\Theta$, there exists an open set A such that $x + A \subset \mathcal{O}$.

10.3. The identity map from $\mathbb{R}$ equipped with the Euclidean topology, onto $\mathbb{R}$ equipped with the half-open interval topology of **4.3**, is bounded, linear but not continuous.

10.4. Let E be an open set in $\mathbb{R}^N$ and denote by $C(E)$ the linear vector space of all real valued continuous functions defined in E. In $C(E)$ introduce a topology as follows. For $g \in C(E)$ and $\rho > 0$, stipulate that the set,

$$\mathcal{O}_{g,\rho} = \Big\{ f \in C(E) \;\Big|\; \sup_E |f - g| < \rho \Big\}$$

is an open neighborhood of g. The collection of such $\mathcal{O}_{g,\rho}$ is a base for a topology in $C(E)$. The sum $+ : C(E) \times C(E) \to C(E)$ is continuous with respect to such a topology. However the multiplication by scalars $\bullet :$ $\mathbb{R} \times C(E) \to C(E)$, is not continuous.

10.5. For $x, y \in \mathbb{R}$ set

$$d(x, y) = \begin{cases} |x - y| + 1 & \text{if either } x = 0 \text{ or } y = 0 \text{ but not both} \\ |x - y| & \text{otherwise.} \end{cases}$$

Prove that $d(\cdot, \cdot)$ is a distance in $\mathbb{R}$ and that the resulting topological space, is not a linear topological vector space. The ball $B_{\frac{1}{2}}(0) = \{O\}$ is open, and its pre-image under translation need not be open.

13c Metric Spaces

13.1. Properties (i) and (iii) in the definition of a metric, follow from (ii) and (iv). Setting $x = y$ in the triangle inequality (iv), and using (ii) gives $2d(x, z) \geq 0$ for all $x, z \in X$. Setting $z = y$ in (iv) gives $d(x, y) \leq d(y, x)$, and by symmetry $d(y, x) \leq d(x, y)$. Thus a metric could be defined as a function $d : (X \times X) \to \mathbb{R}$ satisfying (ii) and (iv).

13.2. The identically zero pseudo-metric generates the trivial topology on X. The function $d(x, y) = 1$ if $x \neq y$ and $d(x, y) = 0$ if $x = y$ is the *discrete* metric on X and generates the discrete topology. With respect to such a metric the open balls $B_1(x)$ contain only the element x and their closure still coincides with x. Thus $\bar{B}_1(x) \neq \{y \in X | d(x, y) \leq 1\}$.

13.3. The function $(x, y) \to \min\{1; |x - y|\}$ is a metric on $\mathbb{R}$.

13.4. Let $A \subset X$. Then $\bar{A} = \cup\{x | d(A, x) = 0\}$.

13.5. A function $f : \{X; d\} \to \{Y; \eta\}$ is continuous at $x \in X$ if and only if $\{f(x_n)\} \to f(x)$, for every sequence $\{x_n\} \to x$.

13.6. Two metrics d_1 and d_2 on X are equivalent if and only if:

(i)   For every $x \in X$ and every ball $B_\rho^1(x)$ in the metric d_1, there exists a radius $r = r(\rho, x)$ such that the ball $B_r^2(x)$ in the metric d_2 is contained in $B_\rho^1(x)$.

(ii) For every $x \in X$ and every ball $B_r^2(x)$ in the metric d_2, there exists a radius $\rho = \rho(r, x)$ such that the ball $B_\rho^1(x)$ in the metric d_1 is contained in $B_r^2(x)$.

The two metrics are uniformly equivalent if the choices of r in (i) and the choice of ρ in (ii) are independent of $x \in X$. Equivalently, d_1 and d_2 are uniformly equivalent if and only if the identity map between $\{X; d_1\}$ and $\{X; d_2\}$ is a uniform homeomorphism.

13.7. In $\mathbb{R}^N$ the following metrics are uniformly equivalent

$$d_p(x, y) = \begin{cases} \left(\sum_{i=1}^{N} |x_i - y_i|^p \right)^{\frac{1}{p}} & \text{for } p \in [1, \infty) \\ \max_{1 \leq i \leq N} |x_i - y_i| & \text{for } p = \infty. \end{cases}$$

The discrete metric in $\mathbb{R}^N$ is not equivalent to any of the metrics d_p.

13.8.   The metric d_o in (13.1) is equivalent, but not uniformly equivalent, to the original metric d.

13.9.   A metric space $\{X; d\}$ is bounded if there exists and element $\Theta \in X$ and a number $M > 0$ such that $d(x, \Theta) < M$ for all $x \in X$. Boundedness depends only on the metric and it is neither an intrinsic property of X nor a topological property. In particular the same set X can be endowed with two equivalent metrics d and d_o in such a way that $\{X; d\}$ is not bounded and $\{X; d_o\}$ is bounded.

13.10c   The Hausdorff Distance of Sets

Let $\{X; d\}$ be a metric space. For $A \subset X$ and $\sigma > 0$ set

$$A_\sigma = \{x \in X \mid d(x, A) < \sigma\}.$$

The Hausdorff distance of two sets A and B in X is ([72], Chap. VIII)

$$d_{\mathcal{H}}(A, B) = \inf\{\sigma > 0 \text{ such that } A \subset B_\sigma \text{ and } B \subset A_\sigma\}.$$

If A and B have nonempty intersection their distance is zero but their Hausdorff distance might be positive. There exist distinct subsets A and B of X whose Hausdorff distance is zero. Thus $d_{\mathcal{H}}$ is a pseudo-metric on 2^X and generates the pseudo-metric space $\{2^X; d_{\mathcal{H}}\}$.

The identity map from $\{X; d\}$ to $\{2^X; d_{\mathcal{H}}\}$ is an isometry.

The topology on $\{2^X; d_{\mathcal{H}}\}$ is generated only by the original metric d, via the definition of $d_{\mathcal{H}}$, and not by the topology of $\{X; d\}$. Indeed there might exist metrics d_1 and d_2 that generate the same topology on X and such that the corresponding Hausdorff distances $d_{1,\mathcal{H}}$ and $d_{2,\mathcal{H}}$ generate different topologies on 2^X. As an example let $X = \mathbb{R}^+$ endowed with the two equivalent metrics

$$d_1(x, y) = \left| \frac{x}{1+x} - \frac{y}{1+y} \right|, \qquad d_2(x, y) = \min\{1; |x - y|\}.$$

The topologies of $\{2^{\mathbb{R}^+}; d_{1,\mathcal{H}}\}$ and $\{2^{\mathbb{R}^+}; d_{2,\mathcal{H}}\}$ are different. The set of natural numbers $\mathbb{N}$ is a point in $2^{\mathbb{R}^+}$. The ball $B_\varepsilon^1(\mathbb{N})$ centered at $\mathbb{N}$ and of radius $\varepsilon \in (0, 1)$, in the topology of $\{2^{\mathbb{R}^+}; d_{1,\mathcal{H}}\}$ contains infinitely many finite subsets of $\mathbb{R}^+$. The ball $B_\varepsilon^2(\mathbb{N})$ in the topology of $\{2^{\mathbb{R}^+}; d_{2,\mathcal{H}}\}$ does not contain any finite subset of $\mathbb{R}^+$.

13.11c Countable Products of Metric Spaces

Let $\{X_n; d_n\}$ be a countable collection of metric spaces. Then the product topology on $\prod X_n$ coincides with the topology generated by the metric

$$d(x, y) = \sum \frac{1}{2^n} \frac{d_n(x_n, y_n)}{1 + d_n(x_n, y_n)}. \tag{13.1c}$$

This will follow from the two inclusions:

(i)   Every neighborhood $\mathcal{O}_y$ of a point $y \in \prod X_n$ open in the product topology, contains a ball $B_\varepsilon(y)$ with respect to the metric in (13.1c).

(ii) Every ball $B_\varepsilon(y)$ with respect to the metric in (13.1c) contains a neighborhood of y, open in the product topology.

Elements $y \in \prod X_n$ are sequences $\{y_n\}$ such that $y_n \in X_n$. For a fixed $y \in \prod X_n$ a neighborhood $\mathcal{O}_y$ of y, open in the product topology contains an open set of the form

$$\mathcal{O}_{y,k} = \prod_{n=1}^{k} B_\varepsilon^{(n)}(y_n) \quad \text{for some finite } k \tag{13.2c}$$

where $B_\varepsilon^{(n)}(y_n)$ is the ball in $\{X_n; d_n\}$, centered at y_n and of radius ε.

There exists $\delta > 0$ sufficiently small depending on ε and k, such that the ball $B_\delta(y)$ in $\prod X_n$ is contained in $\mathcal{O}_y$. Indeed from

$$\sum \frac{1}{2^n} \frac{d_n(x_n, y_n)}{1 + d_n(x_n, y_n)} < \delta$$

it follows that, the number δ can be chosen so small that

$$d_n(x_n, y_n) < 2^{n+1}\delta \leq \varepsilon \quad \text{for all } n = 1, \ldots, k.$$

Thus $B_\delta(y) \subset \mathcal{O}_y$. Conversely, every ball $B_\delta(y)$ in $\prod X_n$ contains an open set of the form (13.2c). Indeed let k be a positive integer so large that

$$\sum_{n=k}^{\infty} \frac{1}{2^n} < \frac{\delta}{2}.$$

For such a k fixed the open set in (13.2c) with $\varepsilon = \frac{1}{2}\delta$ is contained in $B_\delta(y)$.

13.12.   The countable product of complete metric spaces is complete.

14c   Metric Vector Spaces

Referring back to (14.1), the discontinuity of the bijection h is meant with respect to the topology generated by the original metric, whereas the discontinuity of the sum $+ : X \times X \to X$, or the product by scalars $\bullet : \mathbb{R} \times X \to X$, should be proved with respect to the new metric d_h.

14.1.   Let $X = \mathbb{R}$ and let d be the usual Euclidean metric. Define

$$
h(x) = \begin{cases} 1 & \text{if } x = 0 \\ 0 & \text{if } x = 1 \\ x & \text{otherwise} \end{cases}
$$

and let $d_h = d(h)$ be defined as in (14.1). For $\varepsilon \in (0, 1)$ the ball $\mathcal{B}_\varepsilon(0)$, centered at 0 and radius ε, in the new metric d_h, consists of the singleton $\{0\}$ and the open ball $B_\varepsilon(1)$, in the original Euclidean metric, of radius ε and centered at 1 from which the singleton $\{1\}$ has been removed. Likewise the ball $\mathcal{B}_\varepsilon(1)$ of radius ε and centered at 1, in the new metric d_h, consists of the singleton $\{1\}$ and the and the open ball $B_\varepsilon(0)$, in the original Euclidean metric, of radius ε and centered at 0 from which the singleton $\{0\}$ has been removed. For such a metric d_h, both the sum and the multiplication by scalars are discontinuous.

14.2.   The half-open interval topology of **4.3** is not metrizable, that is, there exists no metric on $\mathbb{R}$ that generates the half-open interval topology. Combine **10.3** with Proposition 14.2.

14.3.   Let ℓ_∞ be the collection of all sequences $x = \{x_n\}$ or real numbers such that $\sup |x_n| < \infty$, endowed with with the metric

$$
d(x, y) = \sup_n |x_n - y_n|. \tag{14.1c}
$$

Show that ℓ_∞ is not complete nor separable.

14.4.   Let $t : \mathbb{R} \to [0, 1]$ be the *tent function* defined by

$$
t(s) = \begin{cases} 1 - 2|s| & \text{if } |s| \le \tfrac{1}{2}; \\ 0 & \text{otherwise.} \end{cases} \tag{14.2c}
$$

For $x \in \ell_\infty$ define
$$
T(x) = \sum x_i t(s - i).
$$

Prove that T is an isometry between ℓ_∞ and $T(\ell_\infty)$.

15c Spaces of Continuous Functions

15.1c *Spaces of Hölder and Lipschitz Continuous Functions*

15.1. Prove that the quantity $[f - g]_{\alpha,E}$ is a pseudo-metric in $C^\alpha(E)$.
15.2. Prove that $C^\alpha(E)$ is complete for the metric in (15.6).
15.3. Prove that $C^\beta(E) \subset C^\alpha(E)$ for all $\beta > \alpha$.
15.4. Prove that Hölder continuous functions in E are uniformly continuous.
15.5. Let E be a subset of $\mathbb{R}^N$ containing the origin, and let $\alpha \in (0, 1)$ be fixed. Prove that $E \ni x \to |x|^\alpha$ is in $C^\alpha(E)$. Prove also that for any $g \in C^1(\bar{E})$

$$d(|x|^\alpha, g) \geq 1 \quad \text{with respect to the metric } d(\cdot, \cdot) \text{ in (15.6).}$$

15.6. Prove that $C^\alpha(E)$ is not separable in its own metric topology.
15.7. Prove that $E \ni x \to |x| \in \mathrm{Lip}(E)$. Prove also that for any $g \in C^1(\bar{E})$, such that $g_{x_j}(x_o) = 0$ for $j = 1, \ldots, N$, for some $x_o \in E$,

$$d(|x|, g) \geq 1 \quad \text{with respect to the metric } d(\cdot, \cdot) \text{ in (15.6)}$$

15.8. In $\mathrm{Lip}(0, 1)$ consider the functions

$$(0, 1) \ni x \to |a - x|, \ |b - x| \quad \text{for fixed } a, b \in (0, 1).$$

Prove that if $a \neq b$ then

$$d(|a - x|, |b - x|) \geq 2 \quad \text{with respect to the metric } d(\cdot, \cdot) \text{ in (15.6).}$$

Deduce that $\mathrm{Lip}(E)$ is not separable.

16c On the Structure of a Complete Metric Space

16.1. Let d_1 and d_2 be two equivalent metrics on the same vector space X. The two metric spaces $\{X; d_1\}$ and $\{X; d_2\}$ have the same topology and the identity map is a homeomorphism. However, the identity map does not preserve completeness. As an example consider $\mathbb{R}$ with the Euclidean metric and the metric d_o given in (13.1) corresponding to the Euclidean metric.
16.2. Intersection Properties of a Complete Metric Space

Proposition 16.1c (Cantor) *Let $\{X; d\}$ be a complete metric space, and let $\{E_n\}$ be a countable collection of closed subsets of X such that $E_{n+1} \subset E_n$ and $\mathrm{diam}\{E_n\} \to 0$. Then $\cap E_n \neq \emptyset$.*

16.3c Completion of a Metric Space

Every metric space $\{X; d\}$ can be completed by the following procedure.

(a)   First, one defines X' as the set of all the Cauchy sequences $\{x_n\}$ of elements in X and verifies that such a set has the structure of a linear space. Then on X' one defines a distance function

$$(\{x_n\}; \{y_n\}) \to d'(\{x_n\}; \{y_n\}) = \lim d(x_n, y_n).$$

Since $\{x_n\}$ and $\{y_n\}$ are Cauchy sequences in X, the sequence $\{d(x_n, y_n)\}$ is a Cauchy sequence in $\mathbb{R}^+$. Thus the indicated limit exists. Since several pairs of Cauchy sequences might generate the same limit, this is not a metric on X'. One verifies however that it is a pseudo-metric.

(b)   In X' introduce an equivalence relation by which two sequences $\{x_n\}$ and $\{y_n\}$ are equivalent if $d'(\{x_n\}; \{y_n\}) = 0$. One verifies that such a relation is symmetric, reflexive and transitive and therefore generates equivalence classes. Define X^* as the set of equivalence classes of all Cauchy sequences of $\{X; d\}$. Any such class, contains only sequences at zero mutual pseudo-distance. For any two such equivalence classes x^* and y^* choose representatives $\{x_n\} \in x^*$ and $\{y_n\} \in y^*$ and set

$$d^*(x^*, y^*) = d'(\{x_n\}; \{y_n\}).$$

One verifies that the definition is independent of the choices of the representatives and that d^* defines a metric in X^*. The original metric space $\{X; d\}$ is embedded into $\{X^*; d^*\}$ by identifying elements of X with elements of X^* as constant Cauchy sequences. Such an embedding is an isometry.

(c)   The metric space $\{X^*; d^*\}$ is complete. Let $\{x_j^*\}$ be a Cauchy sequence in $\{X^*; d^*\}$ and select a representative $\{x_{j,n}\}$ out of each equivalence class x_j^*. By construction any such a representative is a Cauchy sequence in $\{X; d\}$. Therefore for each $j \in \mathbb{N}$, there exists an index n_j such that $d(x_{j,n}, x_{j,n_j}) \le \frac{1}{j}$ for all $n \ge n_j$. By diagonalization select now the sequence $\{x_{j,n_j}\}$ and verify that itself is a Cauchy sequence in $\{X; d\}$. Thus $\{x_{j,n_j}\}$ identifies an equivalence class $x^* \in X^*$. The Cauchy sequence $\{x_n^*\}$ converges to x^* in $\{X^*; d^*\}$. Finally the original metric space $\{X; d\}$, with the indicated embedding, is dense in $\{X^*; d^*\}$.

Remark 16.1c While every metric space can be completed, a deeper problem is that of characterizing the elements of the new space and its metric. A typical example is the completion of the rational numbers into the real numbers.

16.4c   Some Consequences of the Baire Category Theorem

The category theorem is equivalent to the following:

Proposition 16.2c *Let $\{X; d\}$ be a complete metric space. Then a countable collection $\{\mathcal{O}_n\}$ of open dense subsets of X, has nonempty intersection.*

16.5.   Let $\{X; d\}$ be a complete metric space. Then every closed, proper subset of X of first category, is nowhere dense.

16.6.   The countable union of sets of first category is of first category. However the countable union of nowhere dense sets need not be nowhere dense. Give an example.

16.7.   Let $\{E_n\}$ be a countable collection of closed subsets of a complete metric space $\{X; d\}$, such that $\cup E_n = X$. Then $\cup \overset{\mathrm{o}}{E}_n$ is dense in X.

16.8.   The rational numbers $\mathbb{Q}$ cannot be expressed as the countable intersection of open intervals.

16.9.   An infinite dimensional complete metric space $\{X; d\}$ cannot have a countable Hamel basis.

Proposition 16.3c *Let $f : \mathbb{R} \to \mathbb{R}$ be continuous on a dense subset E_o of $\mathbb{R}$. Then f is continuous on a set E of the second category.*

Proof For $x \in (0, 1)$

$$f'(x) = \sup_{\varepsilon}\ \inf_{|x-y|<\varepsilon}\ f(y), \qquad f''(x) = \inf_{\varepsilon}\ \sup_{|x-y|<\varepsilon}\ f(y).$$

For $n \in \mathbb{N}$ set also

$$\mathcal{O}_n = \left\{ x \in \mathbb{R} \,\big|\, f''(x) - f'(x) < \tfrac{1}{n} \right\}.$$

The set of continuity of f is the intersection of the $\mathcal{O}_n$. The sets $\mathcal{O}_n$ are open and dense in $\mathbb{R}$ and their complements $\mathcal{O}_n^c$ are nowhere dense in $\mathbb{R}$. Therefore $\cup \mathcal{O}_n^c$ is of the first category in $\mathbb{R}$ and $\cap \mathcal{O}_n$ is of the second category in $\mathbb{R}$. ∎

Corollary 16.1c *There exist no function $f : [0, 1] \to \mathbb{R}$ continuous only at the rationals of $[0, 1]$.*

See also the construction in **7.2** of the Complements.

17c   Compact and Totally Bounded Metric Spaces

Lemma 17.1c (The Lebesgue Number Lemma) *Let $\{X; d\}$ be a sequentially compact topological space. For every finite open covering $\{\mathcal{O}_m\}_{m=1}^k$ for some $k \in \mathbb{N}$,* *there exists a positive number σ such that every ball $B_\sigma(y) \subset X$ is contained in some $\mathcal{O}_m$. The number σ is called the Lebesgue number of the covering.*

Proof If such a $\sigma > 0$ does not exist, there exists a sequence of balls $\{B_{\frac{1}{n}}(x_n)\}_{n\in\mathbb{N}}$, of centers $\{x_n\}$ and radii $\frac{1}{n}$ each not contained in any of the open sets $\mathcal{O}_m$. There exists a subsequence $\{x_{n'}\} \subset \{x_n\}$ and $y \in X$ such that $\{x_{n'}\} \to y$. Since $\{\mathcal{O}_m\}$ is a covering $y \in \mathcal{O}_m$ for some m. Since $\mathcal{O}_m$ is open there exists $\sigma_m > 0$ such that $B_{\sigma_m}(y) \subset \mathcal{O}_m$. Then

$$B_{\frac{1}{n'}}(x_{n'}) \subset B_{\sigma_m}(y) \subset \mathcal{O}_m \quad \text{for } n' > \frac{2}{\sigma_m}. \qquad \blacksquare$$

17.1c   An Application of the Lebesgue Number Lemma

Let $\mathcal{C}_\delta$ be the Cantor set constructed in § 2.1c–§ 2.2c of Chap. 1, and let $J_{n,j}$ be the closed intervals left after the n-stage of removal of the middle open intervals $I_{n,j}$. Then for any given open covering $\{I_m\}$ of $\mathcal{C}_\delta$ there exists n sufficiently large such that each $J_{n,j}$ is contained in some $\mathcal{I}_m$.

Chapter 3
Measuring Sets

1 Partitioning Open Subsets of $\mathbb{R}^N$

Proposition 1.1 (Cantor [21]) *Every open subset E of $\mathbb{R}$ is the union of a countable collection of pairwise disjoint, open intervals.*

Proof For $x \in E$, let E_x be the union of all open intervals containing x and contained in E. By construction E_x is an interval. If x and y are two distinct elements of E, then either $E_x = E_y$ or $E_x \cap E_y = \emptyset$. Indeed if their intersection is not empty, their union is an interval containing both x and y and contained in E. Since E_x is an interval, it contains a rational number. Therefore, the collection of the intervals E_x that are distinct is countable and E is the union of such intervals. ∎

An open subset of $\mathbb{R}^N$ cannot be partitioned, in general, into countably many, mutually disjoint, open cubes. However, it can be partitioned into a countable collection of disjoint, $\frac{1}{2}$-closed dyadic cubes.

Let $\mathbf{q} = (q_1, \ldots, q_N) \in \mathbb{Z}^N$ denote a N-tuple of integers. For a positive integer p and some N-tuple $\mathbf{q}$, denote by $Q_{p,\mathbf{q}}$ the $\frac{1}{2}$-closed dyadic cube

$$Q_{p,\mathbf{q}} = \left\{ x \in \mathbb{R}^N \mid \frac{q_i - 1}{2^p} < x_i \le \frac{q_i}{2^p}; \quad i = 1, \ldots, N \right\}. \tag{1.1}$$

The whole $\mathbb{R}^N$ can be partitioned into disjoint, $\frac{1}{2}$-closed dyadic cubes, by slicing it with the hyperplanes $\{x_j = q_{\ell_j} 2^{-p}\}$ where for each $j = 1, \ldots, N$ the numbers q_{ℓ_j} range over the integers $\mathbb{Z}$. By this procedure, for all fixed $p \in \mathbb{N}$

$$\mathbb{R}^N = \bigcup_{\mathbf{q} \in \mathbb{Z}^N} Q_{p,\mathbf{q}} \quad \text{with} \quad Q_{p,\mathbf{q}} \cap Q_{p,\mathbf{q}'} = \emptyset \quad \text{for} \quad \mathbf{q} \ne \mathbf{q}'.$$

The collection of all $\frac{1}{2}$-closed dyadic cubes in $\mathbb{R}^N$ is denoted by $\mathcal{Q}_{\text{diad}}$.

Proposition 1.2 *An open set $E \subset \mathbb{R}^N$ is the union of a countable collection of $\frac{1}{2}$-closed disjoint, dyadic cubes.*

© Springer Science+Business Media New York 2016
E. DiBenedetto, *Real Analysis*, Birkhäuser Advanced
Texts Basler Lehrbücher, DOI 10.1007/978-1-4939-4005-9_3

Proof Consider the $\frac{1}{2}$-closed dyadic cubes $Q_{1,q}$. At most countably many of them are contained in E and we denote by $\mathcal{Q}_1$ their union, i.e.,

$$\mathcal{Q}_1 = \left\{ \bigcup Q_{1,q} \mid Q_{1,q} \subset E \right\}.$$

If the complement $E - \mathcal{Q}_1$ is nonempty, it contains at most countably many of the $\frac{1}{2}$-closed dyadic cubes $Q_{2,q}$, and we denote by $\mathcal{Q}_2$ their union, i.e.,

$$\mathcal{Q}_2 = \left\{ \bigcup Q_{2,q} \mid Q_{2,q} \subset E - \mathcal{Q}_1 \right\}.$$

Proceeding in this fashion define inductively

$$\mathcal{Q}_n = \left\{ \bigcup_q Q_{n,q} \mid Q_{n,q} \subset E - \bigcup_{j=1}^{n-1} \mathcal{Q}_j \right\} \qquad n = 2, 3, \ldots.$$

The union of the $\mathcal{Q}_n$ consist of countably many $\frac{1}{2}$-closed, disjoint, dyadic cubes of the type (1.1). Moreover, $E = \cup \mathcal{Q}_n$. Indeed, since E is open, for every $x \in E$ there exists a $\frac{1}{2}$-closed dyadic cube $Q_{p,q}$ contained in E and containing x. Such a cube must be contained in some of the $\mathcal{Q}_n$ for some n.   ∎

Remark 1.1 The decomposition of an open set E into cubes is not unique. For example, by analogous arguments, an open set E could be decomposed into a countable union of closed dyadic cubes with pairwise disjoint interior. Another decomposition is in § 1c of the Complements. Having determined one, reorder and relabel the cubes as $\{Q_n\}$, and write $E = \cup Q_n$.

Remark 1.2 The collection $\mathcal{Q}_{\text{diad}}$ of all $\frac{1}{2}$-closed dyadic cubes in $\mathbb{R}^N$, has the following set algebraic properties:

- The intersection of any two elements in $\mathcal{Q}_{\text{diad}}$, if nonempty, is an element of $\mathcal{Q}_{\text{diad}}$
- The mutual, relative complement of any two elements in $\mathcal{Q}_{\text{diad}}$ is the finite disjoint union of elements in $\mathcal{Q}_{\text{diad}}$.

2 Limits of Sets, Characteristic Functions, and σ-Algebras

Let $\{E_n\}$ be a countable collection of sets. The upper and lower limits of the sequence $\{E_n\}$, are defined as

$$\limsup E_n = \bigcap_{n=1}^{\infty} \bigcup_{j=n}^{\infty} E_j \qquad \liminf E_n = \bigcup_{n=1}^{\infty} \bigcap_{j=n}^{\infty} E_j. \qquad (2.1)$$

The sequence $\{E_n\}$ is convergent if

$$\limsup E_n = \liminf E_n.$$

Such a limit exists if the collection $\{E_n\}$ is *monotone increasing*, i.e., if $E_n \subset E_{n+1}$ for all $n \in \mathbb{N}$, or if $\{E_n\}$ is *monotone decreasing*, i.e., if $E_{n+1} \subset E_n$ for all $n \in \mathbb{N}$. However, for the limit to exist the sequence $\{E_n\}$ need not be monotone. It might also occur that a sequence $\{E_n\}$ of nonvoid sets, and whose cardinality tends to infinity, has a limit and the limit is the empty set. For example, this occurs if E_n is the set of positive integers between n and $2n$.

The characteristic function χ_E of a set E is defined as

$$\chi_E(x) = \begin{cases} 1 & \text{if } x \in E \\ 0 & \text{if } x \notin E. \end{cases}$$

It follows from the definition that for any two sets E and F, and their symmetric difference $E \Delta F = (E - F) \cup (F - E)$

$$\chi_{E-F} = \chi_E (1 - \chi_F) \quad \text{and} \quad \chi_{E \Delta F} = |\chi_E - \chi_F|.$$

It also follows from the definition that for a sequence of sets $\{E_n\}$

$$\chi_{\bigcup_{j=n}^{\infty} E_j} = \sup_{j \geq n} \chi_{E_j} \qquad \chi_{\bigcap_{j=n}^{\infty} E_j} = \inf_{j \geq n} \chi_{E_j}.$$

Thus the notion of upper and lower limits of a sequence of sets can be equivalently rephrased in terms of upper and lower limits of the corresponding characteristic functions.

A set $E \subset \mathbb{R}^N$ is of the type of $\mathcal{F}_\sigma$ if it is the union of a countable collection of closed subsets of $\mathbb{R}^N$. It is of the type of $\mathcal{G}_\delta$ if it is the intersection of a countable collection of open subsets of $\mathbb{R}^N$. The set of the rational numbers is of the type of $\mathcal{F}_\sigma$ and the set of the irrational numbers is of the type of $\mathcal{G}_\delta$.

Similarly, one may define sets of type $\mathcal{F}_{\sigma\delta}$ and $\mathcal{G}_{\delta\sigma} \ldots$ as

$$\mathcal{F}_{\sigma\delta} = \{ \text{countable intersection of sets of the type } \mathcal{F}_\sigma \},$$
$$\mathcal{G}_{\delta\sigma} = \{ \text{countable union of sets of the type } \mathcal{G}_\delta \}.$$

Let X be a set. A collection $\mathcal{A}$ of subset of X is an *algebra* if it contains X and is closed under *finite* union and complements. This implies that $\emptyset \in \mathcal{A}$.

The collection $\mathcal{A}$ is a σ-*algebra* of subsets of X, if it is an algebra and if in addition is closed under *countable* union [72].

There are algebras that are not σ-algebras (**3.1** of the Complements).

The collection 2^X of all subsets of X is a σ-algebra called the *discrete* σ-algebra of subsets of X. The collection $\{X; \emptyset\}$ is a σ-algebra, called the *trivial* σ-algebra of subsets of X.

Proposition 2.1 *Given any collection $\mathcal{O}$ of subsets of X, there exists a smallest σ-algebra $\mathcal{A}_o$ that contains $\mathcal{O}$.*

Proof Let $\mathcal{F}$ be the collection of all the σ-algebras containing $\mathcal{O}$. Such a collection is nonvoid since it contains discrete σ-algebra. Set

$$\mathcal{A}_o = \bigcap \{\mathcal{A} \mid \mathcal{A} \in \mathcal{F}\}.$$

Any two sets in $\mathcal{A}_o$ belong to all the σ-algebras in $\mathcal{F}$. Therefore, their union is in $\mathcal{A}_o$ since it must be in all the $\mathcal{A} \in \mathcal{F}$. Analogously, one proves that the complement of a set in $\mathcal{A}_o$ remains in $\mathcal{A}_o$ and the countable union of sets in $\mathcal{A}_o$ remains in $\mathcal{A}_o$. Therefore, $\mathcal{A}_o$ is a σ-algebra. If $\mathcal{A}'$ is any σ-algebra containing $\mathcal{O}$, then it must be in the family $\mathcal{F}$. Thus $\mathcal{A}_o \subset \mathcal{A}'$.   ∎

Let $\mathcal{B}$ denote the smallest σ-algebra containing the open subsets of $\mathbb{R}^N$. The elements of $\mathcal{B}$ are the Borel sets and $\mathcal{B}$ is the Borel σ-algebra [72].

Open and closed subsets of $\mathbb{R}^N$ are Borel sets. Sets of the type of $\mathcal{F}_\sigma$, $\mathcal{G}_\delta$, $\mathcal{F}_{\sigma\delta}$, $\mathcal{G}_{\delta\sigma}, \dots$ are Borel sets.

3 Measures

Denote by $\mathbb{R}^* = \{-\infty\} \cup \mathbb{R} \cup \{+\infty\}$ the set of the *extended* real numbers, with the formal ordering $-\infty < c < +\infty$, and formal operations $\pm\infty \pm c = \pm\infty$ for all $c \in \mathbb{R}$, and $(\pm\infty)c = (\pm\infty)$ sign c for all $c \in \mathbb{R} - \{0\}$. If $c = 0$ somewhat arbitrarily one sets $0 \cdot \infty = 0$. The operation $\infty - \infty$ is not defined,

Let X be a set and let $\mathcal{A}$ be a σ-algebra of subsets of X. A set function μ defined on $\mathcal{A}$ with values on the extended reals $\mathbb{R}^*$, is *countably subadditive* if for a countable collections $\{E_n\}$ of elements of $\mathcal{A}$

$$\mu\left(\bigcup E_n\right) \leq \sum \mu(E_n).$$

The set function μ is *countably additive* if for a countable collections $\{E_n\}$ of *disjoint* elements of $\mathcal{A}$

$$\mu\left(\bigcup E_n\right) = \sum \mu(E_n) \qquad \left(E_i \cap E_j = \emptyset \text{ for } i \neq j\right).$$

The triple $\{X, \mathcal{A}, \mu\}$ is a measure space and μ is a measure on X if

(i)   the domain of μ is a σ − algebra $\mathcal{A}$ (ii) μ is nonnegative on $\mathcal{A}$
(iii) μ is countably additive   (iv) $\mu(E) < \infty$ for some $E \in \mathcal{A}$.

Proposition 3.1 *Let* $\mu : \mathcal{A} \to \mathbb{R}^*$ *be a measure and let* $A, B \in \mathcal{A}$. *Then:*

(a). μ is monotone, that is $\mu(A) \leq \mu(B)$, whenever $A \subset B$.

(b). *If $A \subset B$ and if $\mu(B) < \infty$*

$$\mu(B - A) = \mu(B) - \mu(A). \tag{3.1}$$

(c). *If $\mu(A \cup B) < \infty$*

$$\mu(A \cup B) = \mu(A) + \mu(B) - \mu(A \cap B). \tag{3.2}$$

(d). *The set function μ is countably subadditive. Moreover for a countable collection $\{E_n\}$ of elements of $\mathcal{A}$[1]*

$$\liminf \mu(E_n) \geq \mu\,(\liminf E_n). \tag{3.3}$$

(e). *Finally, if $\mu(\cup E_n) < \infty$, then*

$$\limsup \mu(E_n) \leq \mu\,(\limsup E_n). \tag{3.4}$$

Proof Write $B = A \cup (B - A)$. Since A and $B - A$ are disjoint, by (iii)

$$\mu(B) = \mu(A) + \mu(B - A).$$

This proves the monotonicity of μ, since $\mu(B - A) \geq 0$. It also proves (3.1) if $\mu(B) < \infty$. To prove (3.2) write $A \cup B$ as the disjoint union

$$A \cup B = A \cup (B - A \cap B)$$

and apply (iii) and (3.1). Let $\{E_n\}$ be a countable collection of sets in $\mathcal{A}$ and set

$$B_1 = E_1 \quad \text{and} \quad B_n = E_n - \bigcup_{j=1}^{n-1} E_j \quad \text{for} \quad n = 2, 3, \ldots.$$

The sets B_n are mutually disjoint and their union coincides with the union of the E_n. Therefore, by (iii) and the monotonicity of μ

$$\mu\left(\bigcup E_n\right) = \mu\left(\bigcup B_n\right) = \sum \mu(B_n) = \sum \mu\left(E_n - \bigcup_{j=1}^{n-1} E_j\right) \leq \sum \mu(E_n).$$

Thus μ is countably subadditive. To prove (3.3) write

$$\liminf E_n = \cup D_n, \quad \text{where} \quad D_n = \cap_{j \geq n} E_j.$$

Since $D_n \subset D_{n+1}$

$$\bigcup D_n = D_1 \cup \bigcup (D_{n+1} - D_n).$$

---

[1] This is a version of Fatou's lemma (§ 8.1 of Chap. 4).

May assume that $\mu(D_n) < \infty$ for all n, otherwise the conclusion is trivial. Since the sets $D_{n+1} - D_n$ are disjoint, by countable additivity and (3.1)

$$\begin{aligned}
\mu\,(\liminf E_n) = \mu\left(\bigcup D_n\right) &= \mu\left[D_1 \cup \bigcup (D_{n+1} - D_n)\right] \\
&= \mu(D_1) + \sum \mu(D_{n+1} - D_n) \\
&= \mu(D_1) + \sum \left[\mu(D_{n+1}) - \mu(D_n)\right] \\
&= \lim \mu(D_n) \leq \liminf \mu(E_n).
\end{aligned}$$

To prove (3.4), consider the increasing family of sets

$$\bigcup E_n - \bigcup_{j \geq n} E_j,$$

and compute

$$\begin{aligned}
\mu\left[\lim \left(\bigcup E_n - \bigcup_{j \geq n} E_j\right)\right] &= \mu\left(\bigcup E_n - \limsup_{j \geq n} \bigcup E_j\right) \\
&= \mu\left(\bigcup E_n\right) - \mu\left(\limsup_{j \geq n} \bigcup E_j\right).
\end{aligned}$$

On the other hand, by (3.3)

$$\begin{aligned}
\mu\left[\lim \left(\bigcup E_n - \bigcup_{j \geq n} E_j\right)\right] &\leq \liminf \mu\left(\bigcup E_n - \bigcup_{j \geq n} E_j\right) \\
&= \mu\left(\bigcup E_n\right) - \limsup \mu\left(\bigcup_{j \geq n} E_j\right).
\end{aligned}$$

Combining these two inequalities, proves (3.4).   ∎

Remark 3.1 Let $E \in \mathcal{A}$ be a set for which (iv) holds. Then (3.1) implies that $\mu(\emptyset) = \mu(E - E) = 0$.

3.1 Finite, σ-Finite, and Complete Measures

The measure μ is finite if $\mu(X) < \infty$. It is σ-finite if there exists a countable collection $\{E_n\}$ of subsets of X such that $X = \cup E_n$ and $\mu(E_n) < \infty$ for all n. An example of a non σ-finite measure is in **3.2** of the Complements.

A measure space $\{X, \mathcal{A}, \mu\}$ is complete if every subset of a set of measure zero is in the σ-algebra $\mathcal{A}$. From the monotonicity of μ it follows that if $\{X, \mathcal{A}, \mu\}$ is complete, the measure of every subset of a set of measure zero is zero. An example of a not complete measure is in § 14.3. Every measure can be completed (§ 3.1c of the Complements).

3.2 Some Examples

Let X be a set. The set function $\mu(\emptyset) = 0$ and $\mu(X) = \infty$ is a measure on the trivial σ-algebra $\{X; \emptyset\}$. Let X be a set and let E be a subset of X. Define $\mu(E)$ to be the number of elements in E if E is finite, and infinity otherwise. The set function μ is a measure on 2^X called the *counting* measure on X.

Let $X = \{x_n\}$ be a sequence and let $\{\alpha_n\}$ be a sequence of nonnegative numbers. The set function

$$X \supset E \longrightarrow \mu(E) = \sum \{\alpha_n \mid x_n \in E\}$$

is a σ-finite measure on X.

Let X be infinite and for every subset $E \subset X$ define $\mu(E) = 0$ if E is countable and $\mu(E) = \infty$ otherwise. This defines a measure on 2^X.

The sum of two measures defined on the same σ-algebra is a measure.

Let $\{\mu_n\}$ be a sequence of measures on X defined on the same σ-algebra $\mathcal{A}$. Then $\mu = \sum \mu_n$, is a measure on X defined on $\mathcal{A}$.

Let $\{X, \mathcal{A}, \mu\}$ be a measure space and for a fixed set $B \in \mathcal{A}$ define,

$$\mathcal{A}_B = \{ \text{ the collection of sets } A \cap B \text{ for } A \in \mathcal{A}\}.$$

Then $\mathcal{A}_B$ is a σ-algebra and the restriction of μ to $\mathcal{A}_B$ is a measure.

Let $\mathcal{A}$ be the discrete σ-algebra of all subset of $\mathbb{R}^N$. Fix $x \in \mathbb{R}^N$ and define

$$\mu(E) = \begin{cases} 1 & \text{if } x \in E, \\ 0 & \text{if } x \notin E. \end{cases} \tag{3.5}$$

One verifies that μ is a measure defined on $2^{\mathbb{R}^N}$. It is called the *Dirac delta measure* in $\mathbb{R}^N$ with mass concentrated at x, and it is denoted by δ_x.

4 Outer Measures and Sequential Coverings

A set function μ_e from the subsets of X into $\mathbb{R}^*$ is an *outer measure* if,

(i) μ_e is defined on all subsets of X (ii) μ_e is monotone
(iii) μ_e is nonnegative and $\mu_e(\emptyset) = 0$ (iv) μ_e is countably subadditive.

A collection $\mathcal{Q}$ of subsets of X is a *sequential covering* for X if it contains the empty set and if for every $E \subset X$ there exists a countable collection $\{Q_n\}$ of elements of $\mathcal{Q}$ such that $E \subset \cup Q_n$. For example, the collection of the $\frac{1}{2}$-closed dyadic cubes $\mathcal{Q}_{\text{diad}}$ is a sequential covering for $\mathbb{R}^N$.

Outer measures can be constructed from a sequential covering $\mathcal{Q}$ by the following procedure. Let λ be a given nonnegative set function defined on $\mathcal{Q}$, taking values on $\mathbb{R}^*$ and such that $\lambda(\emptyset) = 0$. For every $E \subset X$, set

$$\mu_e(E) = \inf\left\{ \sum \lambda(Q_n) \mid Q_n \in \mathcal{Q} \text{ and } E \subset \bigcup Q_n \right\}. \tag{4.1}$$

It follows from the definition that, if $\mu_e(E) < \infty$, for every $\varepsilon > 0$ there exists a countable collection $\{Q_{\varepsilon,n}\}$ of elements of $\mathcal{Q}$ such that

$$E \subset \bigcup Q_{\varepsilon,n} \quad \text{and} \quad \sum \lambda(Q_{\varepsilon,n}) \le \mu_e(E) + \varepsilon. \tag{4.2}$$

Proposition 4.1 *The set function μ_e defined by (4.1) is an outer measure.*

Proof The requirements (i)–(iii) are a direct consequence of the definitions. Let $\{E_n\}$ be a countable collection of subsets of X. In proving (iv) may assume that $\mu_e(E_n) < \infty$ for all n. Fix $\varepsilon > 0$. For each $n \in \mathbb{N}$ there exists a countable collection $\{Q_{j_n}\}$ of elements of $\mathcal{Q}$ such that

$$E_n \subset \bigcup Q_{j_n} \quad \text{and} \quad \sum \lambda(Q_{j_n}) \le \mu_e(E_n) + \varepsilon 2^{-n}.$$

The union of the sets Q_{j_n} as both n and j_n range over $\mathbb{N}$, covers the union of the E_n. Therefore,

$$\mu_e\left(\bigcup E_n\right) \le \sum_n \sum_{j_n} \lambda(Q_{j_n}) \le \sum \mu_e(E_n) + \varepsilon. \qquad \blacksquare$$

The outer measure μ_e generated by the sequential covering $\mathcal{Q}$ and the nonnegative set function λ need not coincide with λ on elements of $\mathcal{Q}$. By construction

$$\mu_e(Q) \le \lambda(Q) \quad \text{for all } Q \in \mathcal{Q} \tag{4.3}$$

and strict inequality might occur (§ 4.2 and § 5 below).

4.1 The Lebesgue Outer Measure in $\mathbb{R}^N$

Let $\mathcal{Q}_{\text{diad}}$ denote the collection of the $\frac{1}{2}$-closed dyadic cubes in $\mathbb{R}^N$ and let λ be the Euclidean measure of cubes, i.e.,

$$\text{for all } Q \in \mathcal{Q}_{\text{diad}} \quad \lambda(Q) = \left(\frac{\text{diam}\,Q}{\sqrt{N}}\right)^N.$$

The Lebesgue outer measure of a set $E \subset \mathbb{R}^N$ is defined by

$$\mu_e(E) = \inf \left\{ \sum \left(\frac{\operatorname{diam} Q_n}{\sqrt{N}} \right)^N \mid E \subset \bigcup Q_n, \quad Q_n \in \mathcal{Q}_{\text{diad}} \right\}. \tag{4.4}$$

4.2 The Lebesgue–Stieltjes Outer Measure [89, 154]

Let $f : \mathbb{R} \to \mathbb{R}$ be monotone nondecreasing and right continuous. For an open interval $(a, b) \subset \mathbb{R}$ define $\lambda(a, b) = f(b) - f(a)$. The collection of open intervals (a, b) forms a sequential covering of $\mathbb{R}$. The corresponding outer measure $\mu_{f,e}$ is the Lebesgue–Stieltjes outer measure on $\mathbb{R}$ generated by f. By construction

$$\mu_{f,e}(a, b] = f(b) - f(a).$$

However, it might occur that $\mu_{f,e}(a, b) < f(b) - f(b)$. Thus for such an outer measure, (4.3) might hold with strict inequality.

5 The Hausdorff Outer Measure in $\mathbb{R}^N$ [71]

For $\varepsilon > 0$ let $\mathcal{E}_\varepsilon$ be the sequential covering of $\mathbb{R}^N$, consisting all subsets E of $\mathbb{R}^N$ whose diameter is less than ε. Fix $\alpha > 0$ and set $\lambda(\emptyset) = 0$ and

$$\mathcal{E}_\varepsilon \ni E \longrightarrow \lambda(E) = (\operatorname{diam} E)^\alpha.$$

This defines a nonnegative set function on $\mathcal{E}_\varepsilon$ which in turn generates the outer measure

$$\mathcal{H}_{\alpha,\varepsilon}(E) = \inf \left\{ \sum (\operatorname{diam} E_n)^\alpha \mid E \subset \bigcup E_n, \quad E_n \in \mathcal{E}_\varepsilon \right\}. \tag{5.1}$$

For such an outer measure the inequality in (4.3) might be strict. Indeed if Q is a cube of unit edge in $\mathbb{R}^N$

$$\lambda(Q) = (\sqrt{N})^\alpha \quad \text{and} \quad \mathcal{H}_{\alpha,1}(Q) = 0 \quad \text{for all } \alpha > N.$$

If $\varepsilon' < \varepsilon$, then $\mathcal{H}_{\alpha,\varepsilon} \le \mathcal{H}_{\alpha,\varepsilon'}$. Therefore, the limit

$$\mathcal{H}_\alpha(E) = \lim_{\varepsilon \to 0} \mathcal{H}_{\alpha,\varepsilon}(E)$$

exists and defines a nonnegative set function $\mathcal{H}_\alpha$ on the subsets of $\mathbb{R}^N$.

Proposition 5.1 $\mathcal{H}_\alpha$ *is an outer measure on* $\mathbb{R}^N$. *Moreover,*

$$\mathcal{H}_\alpha(E) < \infty \quad \text{implies} \quad \mathcal{H}_\beta(E) = 0 \quad \text{for all } \beta > \alpha;$$
$$\mathcal{H}_\alpha(E) > 0 \quad \text{implies} \quad \mathcal{H}_\beta(E) = \infty \quad \text{for all } \beta < \alpha.$$

Proof The first statement is proved as in Proposition 4.1. Let $\{E_n\}$ be countable collection of elements of $\mathcal{E}_\varepsilon$ such that $E \subset \bigcup E_n$. For $\beta > \alpha$

$$\mathcal{H}_{\beta,\varepsilon}(E) \leq \sum (\operatorname{diam} E_n)^\beta \leq \varepsilon^{\beta-\alpha} \sum (\operatorname{diam} E_n)^\alpha.$$

Thus $\mathcal{H}_{\beta,\varepsilon}(E) \leq \varepsilon^{\beta-\alpha} \mathcal{H}_{\alpha,\varepsilon}(E)$ for all $\varepsilon > 0$.   ∎

Proposition 5.2 *Let* $\mu_e(\cdot)$ *denote the Lebesgue outer measure in* $\mathbb{R}^N$ *defined in* (4.4). *There exists two positive constants* $c_N \leq C_N$ *depending only upon* N *such that, for every set* $E \subset \mathbb{R}^N$

$$c_N \mathcal{H}_N(E) \leq \mu_e(E) \leq C_N \mathcal{H}_N(E).$$

Moreover, $c_1 = C_1 = 1$. *In particular, for* $N = 1$ *the Lebesgue outer measure on* $\mathbb{R}$ *coincides with the Hausdorff outer measure* $\mathcal{H}_1$.

Proof May assume that both $\mu_e(E)$ and $\mathcal{H}_N(E)$ are finite.

Having fixed $\varepsilon > 0$ there exists a countable collection of $\frac{1}{2}$-closed dyadic cubes $\{Q_n\}$ whose union covers E and

$$\mu_e(E) \geq \sum \left(\frac{\operatorname{diam} Q_n}{\sqrt{N}}\right)^N - \varepsilon.$$

By possibly subdividing the cubes Q_n and using the finite additivity of the Euclidean measure of cubes, may assume that $\operatorname{diam} Q_n < \varepsilon$. Then

$$\mu_e(E) \geq \frac{1}{N^{N/2}} \sum (\operatorname{diam} Q_n)^N - \varepsilon \geq \frac{1}{N^{N/2}} \mathcal{H}_{N,\varepsilon}(E) - \varepsilon$$

for all $\varepsilon > 0$. This proves the left inequality with $c_N = \sqrt{N^{-N}}$.

For the upper bound on $\mu_e(E)$, having fixed $\varepsilon > 0$, there exists a countable collection $\{E_n\}$ of elements of $\mathcal{E}_\varepsilon$ such that

$$\mathcal{H}_{N,\varepsilon}(E) \geq \sum (\operatorname{diam} E_n)^N - \varepsilon. \tag{5.2}$$

Each of the E_n can be included in a $\frac{1}{2}$-closed dyadic cube Q_n of edge not exceeding $2\operatorname{diam} E_n$. Therefore,

$$(\operatorname{diam} E_n)^N \geq \frac{1}{2^N}\{\text{volume of } Q_n\} \quad \text{and} \quad E_n \subset Q_n.$$

From this and (4.4)

$$\mathcal{H}_{N,\varepsilon}(E) \geq \frac{1}{2^N} \sum \left(\frac{\mathrm{diam}\, Q_n}{\sqrt{N}}\right)^N - \varepsilon \geq \frac{1}{2^N} \mu_e(E) - \varepsilon$$

for all $\varepsilon > 0$. This establishes the upper bound with $C_N = 2^N$.

If $N = 1$ each of the sets E_n in (5.2) can be included in the finite union $(\alpha_n, \beta_n]$ of $\frac{1}{2}$-closed, dyadic intervals, in such a way that

$$\mathrm{diam}\, E_n \geq \text{length of } (\alpha_n, \beta_n] - 2^{-n}\varepsilon.$$

From this and (4.4)

$$\mathcal{H}_{1,\varepsilon}(E) \geq \sum \text{length of } (\alpha_n, \beta_n] - 2\varepsilon \geq \mu_e(E) - 2\varepsilon. \qquad \blacksquare$$

The Hausdorff outer measure is additive on sets that are at mutual positive distance.

Proposition 5.3 *Let E and F be subsets of $\mathbb{R}^N$ such that $\mathrm{dist}\{E; F\} = \delta$ for some $\delta > 0$. Then $\mathcal{H}_\alpha(E \cup F) = \mathcal{H}_\alpha(E) + \mathcal{H}_\alpha(F)$, for all $\alpha > 0$.*

Proof Since $\mathcal{H}_\alpha$ is subadditive, it suffices to prove the statement with equality replaced by $\geq$. Also may assume that $\mathcal{H}_\alpha(E \cup F)$ is finite.

Having fixed $\varepsilon < \frac{1}{2}\delta$, there exists a collection of sets $\{G_n\}$ whose union contains $E \cup F$ and each of diameter less than ε, such that

$$\mathcal{H}_{\alpha,\varepsilon}(E \cup F) \geq \sum (\mathrm{diam}\, G_n)^\alpha - \varepsilon.$$

If a point $x \in E$ is covered by some G_n, such a set does not intersect F. Likewise, if $y \in F$ is covered by G_m, then $G_m \cap E = \emptyset$. Therefore, the collection $\{G_n\}$ can be separated into two subcollections $\{E_n\}$ and $\{F_n\}$. The union of the E_n contains E and the union of the F_n contains F. From this

$$\mathcal{H}_{\alpha,\varepsilon}(E \cup F) \geq \sum (\mathrm{diam}\, E_n)^\alpha + \sum (\mathrm{diam}\, F_n)^\alpha - \varepsilon$$
$$\geq \mathcal{H}_{\alpha,\varepsilon}(E) + \mathcal{H}_{\alpha,\varepsilon}(F) - \varepsilon. \qquad \blacksquare$$

5.1 Metric Outer Measures

An outer measure μ_e in $\mathbb{R}^N$ is a *metric* outer measure if for any two sets E and F at positive mutual distance, $\mu(E \cup F) = \mu_e(E) + \mu_e(F)$. The Lebesgue outer measure in $\mathbb{R}^N$ is metric. The Lebesgue–Stieltjes outer measure on $\mathbb{R}$ is a metric outer measure. The Hausdorff outer measure $\mathcal{H}_\alpha$ is metric. An example of a nonmetric outer measure is in § 14.2 of Chap. 11.

6 Constructing Measures from Outer Measures [26]

Let μ_e be an outer measure on X and let A and E be any two subsets of X. Then the set identity $A = (A \cap E) \cup (A - E)$, implies

$$\mu_e(A) \leq \mu_e(A \cap E) + \mu_e(A - E). \tag{6.1}$$

Consider the collection $\mathcal{A}$ of those sets $E \subset X$ satisfying (6.1) with equality, for all sets $A \subset X$, i.e., the collection of the sets $E \subset X$ such that [25 26]

$$\mu_e(A) \geq \mu_e(A \cap E) + \mu_e(A - E) \tag{6.2}$$

for all sets $A \subset X$. Sets $E \in \mathcal{A}$ are said μ_e-measurable.

Proposition 6.1 (i) *The empty set is in $\mathcal{A}$*
(ii) If E is a set of outer measure zero, then $E \in \mathcal{A}$
(iii) If $E \in \mathcal{A}$, its complement E^c is in $\mathcal{A}$
(iv) If E_1 and E_2 are in $\mathcal{A}$, then $E_1 \cup E_2$, $E_1 - E_2$, $E_1 \cap E_2$ are in $\mathcal{A}$
(v) Let $\{E_n\}$ be a collection of disjoint sets in $\mathcal{A}$. Then

$$\mu_e \left(A \cap \bigcup E_n \right) = \sum \mu_e \left(A \cap E_n \right) \quad \textit{for every } A \subset X.$$

(vi) The countable union of sets in $\mathcal{A}$ is in $\mathcal{A}$.

Proof To establish that a set E is in the collection $\mathcal{A}$, it suffices to verify (6.2) for all sets A of finite outer measure. Statements (i) and (ii) follow from (6.2) and the monotonicity of μ_e. Statement (iii) follows from the set identities

$$A \cap E^c = A - E \qquad A - E^c = A \cap E.$$

To prove the first statement in (iv) let E_1 and E_2 be elements of $\mathcal{A}$ and write (6.2) for the pair A, E_1 and for the pair $(A - E_1)$, E_2,

$$\mu_e(A) \geq \mu_e(A \cap E_1) + \mu_e(A - E_1);$$
$$\mu_e(A - E_1) \geq \mu_e\big((A - E_1) \cap E_2\big) + \mu_e\big((A - E_1) - E_2\big).$$

Add these inequalities and use the subadditivity of μ_e and the set identities

$$(A - E_1) - E_2 = A - (E_1 \cup E_2)$$
$$\big((A - E_1) \cap E_2\big) \cup \big(A \cap E_1\big) = A \cap (E_1 \cup E_2).$$

This gives

$$\mu_e(A) \geq \mu_e\big(A \cap (E_1 \cup E_2)\big) + \mu_e\big(A - (E_1 \cup E_2)\big).$$

The remaining statements in (vi) follow from the set identities

$$E_1 - E_2 = \left(E_1^c \cup E_2\right)^c \qquad E_1 \cap E_2 = \left(E_1^c \cup E_2^c\right)^c.$$

We first establish (v) for a finite collection of disjoint sets. That is, if

$$B_n = \bigcup_{j=1}^{n} E_j \qquad \text{and} \qquad E_i \cap E_j = \emptyset \text{ for } i \neq j$$

then for every set $A \subset X$

$$\mu_e(A \cap B_n) = \sum_{j=1}^{n} \mu_e(A \cap E_j).$$

The statement is obvious for $n = 1$. Assuming it holds for n, we show it continues to hold for $n + 1$. By (iv), B_n is in $\mathcal{A}$, and may write (6.2) for the pair $(A \cap B_{n+1})$ and B_n. This gives

$$\mu_e(A \cap B_{n+1}) = \mu_e(A \cap B_{n+1} \cap B_n) + \mu_e(A \cap B_{n+1} - B_n)$$
$$= \mu_e(A \cap B_n) + \mu_e(A \cap E_{n+1}).$$

Let now $\{E_n\}$ be a countable collection of disjoint sets in $\mathcal{A}$. By the subadditivity and monotonicity of the outer measure μ_e

$$\sum \mu_e(A \cap E_n) \geq \mu_e\left(A \cap \bigcup E_n\right) \geq \mu_e\left(A \cap \bigcup_{j=1}^{m} E_j\right) = \sum_{j=1}^{m} \mu_e(A \cap E_j)$$

To prove (vi) assume first that the sets of the collection $\{E_n\}$ are mutually disjoint. For every $A \subset X$ and every $m \in \mathbb{N}$

$$\mu_e(A) = \mu_e\left(A \cap \bigcup_{j=1}^{m} E_j\right) + \mu_e\left(A - \bigcup_{j=1}^{m} E_j\right)$$
$$\geq \sum_{j=1}^{m} \mu_e(A \cap E_j) + \mu_e\left(A - \bigcup E_n\right).$$

Letting $m \to \infty$ and using the countable subadditivity of μ_e, shows that the union of the E_n satisfies (6.2) and therefore is in $\mathcal{A}$.

For a general countable collection $\{E_n\}$ of elements of $\mathcal{A}$ write $D_1 = E_1$ and $D_n = E_n - \bigcup_{j=1}^{n-1} E_j$. The sets D_n are in $\mathcal{A}$, they are mutually disjoint and their union coincides with the union of the E_n.   ∎

Proposition 6.2 *The restriction of* μ_e *to* $\mathcal{A}$ *is a complete measure.*

Proof By Proposition 6.1, $\mathcal{A}$ is a σ-algebra and the restriction $\mu_e|_{\mathcal{A}}$ satisfies the requirements of a complete measure. In particular, the countable additivity follows from (v) with A replaced with the union of the E_n and the completeness follows from (ii). ∎

7 The Lebesgue–Stieltjes Measure on $\mathbb{R}$

The Lebesgue–Stieltjes outer measure $\mu_{f,e}$ induced by an increasing, right continuous function $f : \mathbb{R} \to \mathbb{R}$, generates a σ-algebra $\mathcal{A}_f$ of subset of $\mathbb{R}$ and a measure μ_f defined on $\mathcal{A}_f$.

Proposition 7.1 $\mathcal{A}_f$ *contains the Borel sets on* $\mathbb{R}$.

Proof It suffices to verify that intervals of the type $(\alpha, \beta]$ belong to $\mathcal{A}_f$, i.e., that for every subset $E \subset \mathbb{R}$

$$\mu_{f,e}(E) \geq \mu_{f,e}(E \cap (\alpha, \beta]) + \mu_{f,e}(E - (\alpha, \beta]). \tag{7.1}$$

Lemma 7.1 *Let* λ_f *be the Lebesgue–Stieltjes set function defined on open intervals of* $\mathbb{R}$, *from which* $\mu_{f,e}$ *is constructed. Then for any open interval* I *and any interval of the type* $(\alpha, \beta]$,

$$\lambda_f(I) \geq \mu_{f,e}(I \cap (\alpha, \beta]) + \mu_{f,e}(I - (\alpha, \beta]).$$

Proof Let $I = (a, b)$ and assume that $(\alpha, \beta] \subset (a, b)$. Denote by ε and η positive numbers. By the right continuity of f and the definition of $\mu_{f,e}$,

$$\begin{aligned}
\lambda_f(I) &= f(b) - f(a) \\
&= \lim_{\varepsilon \to 0} \big(f(b) - f(\beta + \varepsilon)\big) \\
&\quad + \lim_{\varepsilon \to 0} \lim_{\eta \to 0} \big(f(\beta + \varepsilon) - f(\alpha + \eta)\big) + \lim_{\eta \to 0} \big(f(\alpha + \eta) - f(a)\big) \\
&\geq \mu_{f,e}(\beta, b) + \mu_{f,e}(\alpha, \beta] + \mu_{f,e}(a, \alpha] \\
&\geq \mu_{f,e}(\alpha, \beta] + \mu_{f,e}\big(I - (\alpha, \beta]\big).
\end{aligned}$$

The cases $\alpha \leq a < \beta < b$ and $a \leq \alpha < b < \beta$ are handled similarly. ∎

Returning to the proof of (7.1), may assume that $\mu_{f,e}(E)$ is finite. Having fixed $\varepsilon > 0$, there exists a collection of open intervals $\{I_n\}$ whose union covers E, such that

$$\mu_{f,e}(E) + \varepsilon \geq \sum \lambda_f(I_n)$$
$$\geq \sum \mu_{f,e}(I_n \cap (\alpha, \beta]) + \sum \mu_{f,e}(I_n - (\alpha, \beta])$$
$$\geq \mu_{f,e}\left(\bigcup I_n \cap (\alpha, \beta]\right) + \mu_{f,e}\left(\bigcup I_n - (\alpha, \beta]\right)$$
$$\geq \mu_{f,e}\left(E \cap (\alpha, \beta]\right) + \mu_{f,e}\left(E - (\alpha, \beta]\right) \qquad \blacksquare$$

7.1 Borel Measures

A measure μ in $\mathbb{R}^N$ is a Borel measure if the σ-algebra of its domain of definition, contains the Borel sets. By Proposition 7.1, the Lebesgue–Stieltjes measure on $\mathbb{R}$ is a Borel measure.

8 The Hausdorff Measure on $\mathbb{R}^N$

For a fixed $\alpha > 0$, the Hausdorff outer measure $\mathcal{H}_\alpha$ generates a σ-algebra $\mathcal{A}_\alpha$ and a measure $\mathcal{H}^\alpha$ called the Hausdorff measure on $\mathbb{R}^N$.

Proposition 8.1 $\mathcal{A}_\alpha$ contains the Borel sets in $\mathbb{R}^N$.

Proof It suffices to verify that closed sets $E \subset \mathbb{R}^N$ belong to $\mathcal{A}_\alpha$, i.e., that for every subset $A \subset \mathbb{R}^N$,

$$\mathcal{H}_\alpha(A) \geq \mathcal{H}_\alpha(A \cap E) + \mathcal{H}_\alpha(A - E) \quad \text{for } E \subset \mathbb{R}^N \text{ closed.}$$

May assume that $\mathcal{H}_\alpha(A)$ is finite. For $n \in \mathbb{N}$, set

$$E_n = \left\{ x \in \mathbb{R}^N \mid \text{dist}\{x; E\} \leq \frac{1}{n} \right\}$$

and estimate below

$$\mathcal{H}_\alpha(A) = \mathcal{H}_\alpha\big[(A \cap E_n) \cup (A - E_n)\big]$$
$$\geq \mathcal{H}_\alpha\big[(A \cap E) \cup (A - E_n)\big]$$
$$= \mathcal{H}_\alpha(A \cap E) + \mathcal{H}_\alpha(A - E_n).$$

The last inequality follows from Proposition 5.3, since

$$\text{dist}\left\{ (A \cap E); (A - E_n) \right\} \geq \frac{1}{n}.$$

From this

$$\mathcal{H}_\alpha(A) \geq \mathcal{H}_\alpha(A \cap E) + \mathcal{H}_\alpha(A - E) - \mathcal{H}_\alpha\big[A \cap (E_n - E)\big].$$

Lemma 8.1 $\lim \mathcal{H}_\alpha\big[A \cap (E_n - E)\big] = 0.$

Proof For $j = n, n+1, \ldots,$ set

$$F_j = \left\{ x \in A \mid \frac{1}{j+1} < \text{dist}\{x; E\} \leq \frac{1}{j} \right\}.$$

Then $A \cap (E_n - E) = \bigcup_{j=n}^{\infty} F_j$, and

$$\mathcal{H}_\alpha\big(A \cap (E_n - E)\big) \leq \sum_{j=n}^{\infty} \mathcal{H}_\alpha(F_j).$$

The conclusion would follow from this, if the series $\sum \mathcal{H}_\alpha(F_j)$ were convergent. To this end regroup the F_j into those whose index j is even and those whose index is odd. By construction $\text{dist}\{F_j; F_i\} > 0$ if $i \neq j$ and if either both i and j are even, or if both are odd. For any two such sets

$$\mathcal{H}_\alpha(F_j) + \mathcal{H}_\alpha(F_i) = \mathcal{H}_\alpha(F_j \cup F_i)$$

by virtue of Proposition 5.3. Therefore, for all finite m

$$\sum_{j=1}^{m} \mathcal{H}_\alpha(F_j) \leq \sum_{h=1}^{m} \mathcal{H}_\alpha(F_{2h}) + \sum_{h=0}^{m} \mathcal{H}_\alpha(F_{2h+1})$$
$$= \mathcal{H}_\alpha\Big(\bigcup_{h=1}^{m} F_{2h} \Big) + \mathcal{H}_\alpha\Big(\bigcup_{h=0}^{m} F_{2h+1} \Big)$$
$$\leq 2\mathcal{H}_\alpha(F_1) \leq 2\mathcal{H}_\alpha(A). \qquad \blacksquare$$

Corollary 8.1 *For all $\alpha > 0$, the Hausdorff measure $\mathcal{H}^\alpha$ is a Borel measure.*

Remark 8.1 The definition (5.1) valid for $\alpha > 0$ and the previous remarks suggest we define $\mathcal{H}_o$ to be the counting measure.

Remark 8.2 The proof of Proposition 8.1 only used that $\mathcal{H}_\alpha$ is a metric outer measure. Therefore, the measure generated by any metric outer measure in $\mathbb{R}^N$ is a Borel measure. This is known as the Carathéodory sufficient condition for a measure in $\mathbb{R}^N$ to be a Borel measure [26]. An example of a nonmetric outer measure is in § 14.2 of Chap. 11.

9 Extending Measures from Semi-algebras to σ-Algebras

A measure space $\{X, \mathcal{A}, \mu\}$ can be constructed starting from a sequential covering $\mathcal{Q}$ of X and a nonnegative set function $\lambda : \mathcal{Q} \to \mathbb{R}^*$. First, one constructs an outer measure μ_e by the procedure of § 4 and then, by the procedure of § 6 such an outer measure μ_e, generates a σ-algebra $\mathcal{A}$ of subsets of X and a measure μ defined on $\mathcal{A}$. The elements of the originating sequential covering $\mathcal{Q}$ need not be measurable with respect to the resulting measure μ. Even if the elements of $\mathcal{Q}$ are μ_e-measurable, the set function λ and the outer measure μ_e, might disagree on $\mathcal{Q}$. Examples can be constructed using the Lebesgue–Stieltjes outer measure in $\mathbb{R}$ or the Hausdorff outer measure in $\mathbb{R}^N$.

The next proposition provides sufficient conditions both on $\mathcal{Q}$ and λ for the elements of $\mathcal{Q}$ to be measurable and for λ to coincide with μ_e on $\mathcal{Q}$.

The collection $\mathcal{Q}$ is said to be a *semi-algebra* if

(i). The intersection of any two elements in $\mathcal{Q}$ is in $\mathcal{Q}$
(ii). For any two elements Q_1 and Q_2 in $\mathcal{Q}$, the difference $Q_1 - Q_2$ is the *finite disjoint* union of elements in $\mathcal{Q}$.

The collection of all the half-closed intervals of the type $(a, b]$ for $a, b \in \mathbb{R}$, is a semi-algebra. The collection of the open intervals on $\mathbb{R}$ is not a semi-algebra. The collection $\mathcal{Q}_{\text{diad}}$ of the $\frac{1}{2}$-closed dyadic cubes in $\mathbb{R}^N$ is a semi-algebra. The sequential covering $\mathcal{E}_\varepsilon$ from which the Hausdorff outer measures are constructed, is a semi-algebra.

The set function $\lambda : \mathcal{Q} \to \mathbb{R}^*$ is finitely additive on $\mathcal{Q}$, if for any finite collection $\{Q_1, \ldots, Q_n\}$ of disjoint elements of $\mathcal{Q}$, whose union is in $\mathcal{Q}$,[2]

$$\lambda\Big(\bigcup_{j=1}^n Q_j \Big) = \sum_{j=1}^n \lambda(Q_j), \qquad \Big(\bigcup_{j=1}^n Q_j \in \mathcal{Q} \Big).$$

The set function λ is countably subadditive on $\mathcal{Q}$, if for any countable collection $\{Q_n\}$ of elements of $\mathcal{Q}$, whose union is in $\mathcal{Q}$

$$\lambda\Big(\bigcup Q_j \Big) \le \sum \lambda(Q_n), \qquad \Big(\bigcup Q_n \in \mathcal{Q} \Big).$$

The Euclidean measure of parallepipeds is finitely additive. The Lebesgue–Stieltjes set function λ_f is finitely additive. The Hausdorff set function is not finitely additive.

Proposition 9.1 *Assume that $\mathcal{Q}$ is a semi-algebra, and the set function λ is finitely additive on $\mathcal{Q}$. Then $\mathcal{Q} \subset \mathcal{A}$. If in addition λ is countably subadditive, then λ agrees with μ_e on $\mathcal{Q}$.*

---

[2]The notion of sequential covering $\mathcal{Q}$ does not require that $\mathcal{Q}$ be closed under finite union, even if it is a semi-algebra.

Proof Let $Q \in \mathcal{Q}$ be fixed and select $A \subset X$. If $\mu_e(A) = \infty$ then (6.2) holds with $E = Q$. If $\mu_e(A) < \infty$, having fixed $\varepsilon > 0$, there exists a countable collection $\{Q_{\varepsilon,n}\}$ of elements of $\mathcal{Q}$ such that

$$\varepsilon + \mu_e(A) \geq \sum \lambda(Q_{\varepsilon,n}), \qquad \text{and} \qquad A \subset \bigcup Q_{\varepsilon,n}. \tag{9.1}$$

For each fixed n write

$$Q_{\varepsilon,n} = (Q_{\varepsilon,n} \cap Q) \cup (Q_{\varepsilon,n} - Q).$$

The intersection $Q_{\varepsilon,n} \cap Q$ is in $\mathcal{Q}$, by virtue of (i), whereas by (ii)

$$Q_{\varepsilon,n} - Q = \bigcup_{j_n=1}^{m_n} Q_{j_n}$$

where the Q_{j_n} are disjoint elements of $\mathcal{Q}$. Therefore, each $Q_{\varepsilon,n}$ can be written as the finite union of disjoint elements of $\mathcal{Q}$. Since $\lambda(\cdot)$ is finitely additive

$$\lambda(Q_{\varepsilon,n}) = \lambda(Q_{\varepsilon,n} \cap Q) + \sum_{j_n=1}^{m_n} \lambda(Q_{j_n}). \tag{9.2}$$

The elements of the collection $\{Q_{\varepsilon,n} \cap Q\}$ are in $\mathcal{Q}$ and their union covers $A \cap Q$. Likewise, the elements of the collection $\{Q_{j_n}\}$ as j_n ranges over $\{1, \ldots, m_n\}$ and n ranges over $\mathbb{N}$, are in $\mathcal{Q}$ and their union covers $A - Q$. Therefore, by the definition of outer measure

$$\sum_n \lambda(Q_{\varepsilon,n}) = \sum_n \lambda(Q_{\varepsilon,n} \cap Q) + \sum_n \sum_{j_n=1}^{m_n} \lambda(Q_{j_n})$$
$$\geq \mu_e(A \cap Q) + \mu_e(A - Q).$$

This proves that Q is μ_e-measurable. To prove that $\mu_e(Q) = \lambda(Q)$, may assume that $\mu_e(Q) < \infty$. Fix an arbitrary $\varepsilon > 0$ and let $\{Q_{\varepsilon,n}\}$ be a countable collection of elements in $\mathcal{Q}$ such that

$$\varepsilon + \mu_e(Q) \geq \sum \lambda(Q_{\varepsilon,n}) \qquad \text{and} \qquad Q \subset \bigcup Q_{\varepsilon,n}. \tag{9.3}$$

From (9.2), for each fixed n, estimate below

$$\lambda(Q_{\varepsilon,n}) \geq \lambda(Q_{\varepsilon,n} \cap Q).$$

The elements of the collection $\{Q_{\varepsilon,n} \cap Q\}$ are in $\mathcal{Q}$ by (i). Moreover, their union is in $\mathcal{Q}$ since, by the second of (9.3), $Q = \bigcup(Q \cap Q_{\varepsilon,n})$. Therefore, since $\lambda(\cdot)$ is countably subadditive

$$\varepsilon + \mu_e(Q) \geq \sum \lambda(Q_{\varepsilon,n} \cap Q) \geq \lambda(Q). \qquad \blacksquare$$

Remark 9.1 For the inclusion $Q \subset A$ it suffices to require that λ be finitely additive on Q. If λ is both finitely additive and countably subadditive, then it is also countably additive.

A set function λ on a semi-algebra Q is said to be a *measure on* Q if it satisfies the requirements (i)–(iv) of § 3. The countable additivity (iii) is required to hold for any countable collection $\{Q_n\}$ of sets in Q whose union is in Q. The assumptions of Proposition 9.1 are verified if λ is a measure on a semi-algebra Q. In such a case $Q \subset A$ and μ_e agrees with λ on Q. This way the measure μ, restriction of μ_e to A, can be regarded as an extension of λ from Q to A (see **4.2**. of the Complements).

9.1 On the Lebesgue–Stieltjes and Hausdorff Measures

The assumptions of Proposition 9.1 are not satisfied, on different accounts, for neither the Lebesgue–Stieltjes measure in $\mathbb{R}$ nor the Hausdorff measure in $\mathbb{R}^N$. For the Lebesgue–Stieltjes measure, Q is the collection of all open intervals $(a, b) \subset \mathbb{R}$. Such a collection is not a semi-algebra. The set function λ_f is finitely additive and countably additive. The Lebesgue–Stieltjes measurability of the open intervals must be established by a different argument (Proposition 7.1). For the Hausdorff measure in $\mathbb{R}^N$, the sequential covering Q is the collection of all subsets of $\mathbb{R}^N$ whose diameter is less than some $\varepsilon > 0$. Such a collection is a semi-algebra. However, the set function $\lambda(E) = (\text{diam} E)^\alpha$ is not finitely additive for all $\alpha > 0$. The $\mathcal{H}_\alpha$-measurability of the open sets is established by an independent argument (Proposition 8.1).

10 Necessary and Sufficient Conditions for Measurability

Let $\{X, A, \mu\}$ be the measure space generated by the pair $\{Q; \lambda\}$ where Q is a semi-algebra of subsets of X and λ is a measure on Q. Denote by Q_σ the collection of all sets E_σ, that are the countable union of elements of Q. Also denote by $Q_{\sigma\delta}$ the collection of sets $E_{\sigma\delta}$ that are the countable intersection of elements of Q_σ.

Proposition 10.1 *Let $E \subset X$ be of finite outer measure. For every $\varepsilon > 0$ there exists a set $E_{\sigma,\varepsilon} \in Q_\sigma$ such that*

$$E \subset E_{\sigma,\varepsilon} \quad \text{and} \quad \mu_e(E) \geq \mu(E_{\sigma,\varepsilon}) - \varepsilon. \tag{10.1}$$

Moreover, there exists $E_{\sigma\delta} \in Q_{\sigma\delta}$, such that $E \subset E_{\sigma\delta}$ and $\mu_e(E) = \mu(E_{\sigma\delta})$.

Proof By (4.2), for every $\varepsilon > 0$ there exists $E_{\sigma,\varepsilon} = \bigcup Q_n \in Q_\sigma$, such that

$$\mu_e(E) + \varepsilon \geq \sum \lambda(Q_n) = \sum \mu(Q_n) \geq \mu\left(\bigcup Q_n\right) = \mu(E_{\sigma,\varepsilon}).$$

This proves (10.1). As a consequence, for each $n \in \mathbb{N}$ there exists $E_{\sigma,\frac{1}{n}} \in \mathcal{Q}_\sigma$ such that

$$\mu(E_{\sigma,\frac{1}{n}}) - \frac{1}{n} \leq \mu_e(E) \leq \mu(E_{\sigma,\frac{1}{n}}).$$

Thus $E_{\sigma\delta} = \bigcap E_{\sigma,\frac{1}{n}}$ is one possible choice for such a claimed set. ∎

Remark 10.1 Proposition 10.1 continues to hold if μ is any measure that agrees with λ on $\mathcal{Q}$.

Proposition 10.2 *Let* $\{X, \mathcal{A}, \mu\}$ *be the measure space generated by a measure* λ *on a semi-algebra* $\mathcal{Q}$. *A set* $E \subset X$ *of finite outer measure, is* μ-*measurable if and only if for every* $\varepsilon > 0$ *there exists a set* $E_{\sigma,\varepsilon} \in \mathcal{Q}_\sigma$, *such that*

$$E \subset E_{\sigma,\varepsilon} \quad and \quad \mu_e(E_{\sigma,\varepsilon} - E) \leq \varepsilon. \tag{10.2}$$

Proof The necessary condition follows from Proposition 10.1. Indeed if E is μ-measurable

$$\mu(E_{\sigma,\varepsilon}) \leq \mu_e(E) + \varepsilon \quad \Longleftrightarrow \quad \mu(E_{\sigma,\varepsilon} - E) \leq \varepsilon.$$

For the sufficient condition, assuming (10.2) holds, we verify that E satisfies (6.2) for all $A \subset X$. Since $E_{\sigma,\varepsilon}$ is μ-measurable

$$\begin{aligned}
\mu_e(A) &= \mu_e(A \cap E_{\sigma,\varepsilon}) + \mu_e(A - E_{\sigma,\varepsilon}) \\
&\geq \mu_e(A \cap E) + \mu_e(A - E) - \mu_e\big(A \cap (E_{\sigma,\varepsilon} - E)\big) \\
&\geq \mu_e(A \cap E) + \mu_e(A - E) - \varepsilon.
\end{aligned}$$ ∎

Remark 10.2 The sufficient condition of Proposition 10.2 does not require that E be of finite outer measure.

Proposition 10.3 *Let* $\{X, \mathcal{A}, \mu\}$ *be a measure space generated by a measure* λ *on a semi-algebra* $\mathcal{Q}$. *A set* $E \subset X$ *of finite outer measure, is* μ-*measurable if and only if there exists* $E_{\sigma\delta} \in \mathcal{Q}_{\sigma\delta}$, *such that* $E \subset E_{\sigma\delta}$ *and* $\mu_e(E_{\sigma\delta} - E) = 0$.

11 More on Extensions from Semi-algebras to σ-Algebras

Theorem 11.1 [48, 61] *Every measure* λ *on a semi-algebra* $\mathcal{Q}$ *generates a measure space* $\{X, \mathcal{A}, \mu\}$, *where* $\mathcal{A}$ *is a* σ-*algebra containing* $\mathcal{Q}$ *and* μ *is a measure on* $\mathcal{A}$ *which agrees with* λ *on* $\mathcal{Q}$. *Moreover, if* $\mathcal{Q}_o$ *is the smallest* σ-*algebra containing* $\mathcal{Q}$, *the restriction of* μ *to* $\mathcal{Q}_o$ *is an extension of* λ *to* $\mathcal{Q}_o$. *If* λ *is* σ-*finite on* $\mathcal{Q}$, *such an extension is unique.*

Proof There is only to prove the uniqueness whenever λ is σ-finite. Assume that μ_1 and μ_2 are both extensions of λ and let μ_e be the outer measure generated by $\{Q; \lambda\}$. Since Q is a semi-algebra, every element of Q_σ can be written as the disjoint union of elements of Q. Since μ_1 and μ_2 agree on Q, they also agree on Q_σ. Next we show that both μ_1 and μ_2 agree with the outer measure μ_e on sets of Q_o of finite outer measure.

Let $E \in Q_o$ be of finite outer measure. By (10.1) for every $\varepsilon > 0$ there exists $E_{\sigma,\varepsilon} \in Q_\sigma$, such that

$$\mu_1(E_{\sigma,\varepsilon}) \leq \mu_e(E) + \varepsilon.$$

Since $E \subset E_{\sigma,\varepsilon}$ and $\varepsilon > 0$ is arbitrary, this implies $\mu_1(E) \leq \mu_e(E)$, for all $E \in Q_o$ of finite outer measure. Also, since μ_e is a measure on Q_o

$$\mu_e(E_{\sigma,\varepsilon} - E) = \mu_e(E_{\sigma,\varepsilon}) - \mu_e(E) = \mu_1(E_{\sigma,\varepsilon}) - \mu_e(E) \leq \varepsilon.$$

From this, since both μ_e and μ_1 are measures on Q_o and $E \subset E_{\sigma,\varepsilon}$

$$\mu_e(E) \leq \mu_e(E_{\sigma,\varepsilon}) = \mu_1(E_{\sigma,\varepsilon}) = \mu_1(E) + \mu_1(E_{\sigma,\varepsilon} - E)$$
$$\leq \mu_1(E) + \mu_e(E_{\sigma,\varepsilon} - E) \leq \mu_1(E) + \varepsilon.$$

Therefore, $\mu_1(E) = \mu_e(E)$ on sets $E \in Q_o$ of finite outer measure. Interchanging the role of μ_1 and μ_2, one concludes that $\mu_1(E) = \mu_2(E) = \mu_e(E)$, for every $E \in Q_o$ of finite outer measure.

Fix now $E \in Q_o$ not necessarily of finite outer measure. Since λ is σ-finite on Q, there exists a sequence of sets $Q_n \in Q$ such that $E = \bigcup Q_n \cap E$ and each of the intersections $Q_n \cap E$ is in Q_o and is of finite outer measure. Since Q is a semi-algebra, may assume that the Q_n are mutually disjoint. Then

$$\mu_1(E) = \sum \mu_1(Q_n \cap E) = \sum \mu_2(Q_n \cap E) = \mu_2(E). \qquad \blacksquare$$

The requirement that λ be σ-finite is essential to insure a unique extension (**11.7** of § 10.2c of the Complements).

12   The Lebesgue Measure of Sets in $\mathbb{R}^N$

The collection Q of the $\frac{1}{2}$-closed dyadic cubes, including the empty set, is a semi-algebra of subsets of $\mathbb{R}^N$. The Euclidean measure λ of cubes in $\mathbb{R}^N$, provides a nonnegative, finitely additive set function defined on Q from which one may construct the outer measure μ_e as indicated in (4.1). The restriction of μ_e to the σ-algebra $\mathcal{M}$ generated by μ_e, is the Lebesgue measure in $\mathbb{R}^N$, and sets in $\mathcal{M}$ are said to be Lebesgue measurable.

The collection of $\frac{1}{2}$-closed dyadic cubes and the Euclidean measure on them satisfy the assumption of Proposition 9.1. Therefore, the elements in Q are Lebesgue measurable. As a consequence, open and closed sets, and sets of the type $\mathcal{F}_\sigma$, $\mathcal{G}_\delta$, $\mathcal{F}_{\sigma\delta}$, $\mathcal{G}_{\delta\sigma}$..., are Lebesgue measurable.

The outer measure of a singleton $\{y\} \in \mathbb{R}^N$ is zero, since $\{y\}$ may be included into cubes or arbitrarily small measure. From this and the countable subadditivity of μ_e it follows that any countable set in $\mathbb{R}^N$ has outer measure zero. Therefore, by (iii) of Proposition 6.1, any countable set in $\mathbb{R}^N$ is Lebesgue measurable and has Lebesgue measure zero. In particular, the set $\mathbb{Q}$ of the rational numbers is measurable and has measure zero. Analogously, the set of points in $\mathbb{R}^N$ of rational coordinates is measurable and has measure zero.

Every Borel set is Lebesgue measurable. The converse is false as the inclusion $\mathcal{B} \subset \mathcal{M}$ is strict. In § 14 we exhibit a measurable subset of $[0, 1]$ which is not a Borel set.

Remark 12.1 By virtue of Theorem 11.1, the restriction of the Lebesgue measure to the Borel σ-algebra is the unique extension of the Euclidean measure of cubes, from Q into $\mathcal{B}$.

12.1 A Necessary and Sufficient Condition of Measurability

Let μ be the Lebesgue measure in $\mathbb{R}^N$. For a subset E of $\mathbb{R}^N$ define

$$\mu'_e(E) = \inf \left\{ \mu(\mathcal{O}) \mid \text{where } \mathcal{O} \text{ is open and } E \subset \mathcal{O} \right\}.$$

The definition is analogous to that of the outer measure (4.1) except for the class of sets where the infimum is taken.

Since every open set is the countable, disjoint union of $\frac{1}{2}$-closed cubes, the class of the open sets containing E is contained in the class of the countable unions of $\frac{1}{2}$-closed cubes, containing E. Thus $\mu_e(E) \le \mu'_e(E)$.

Proposition 12.1 $\mu_e(E) = \mu'_e(E)$.

Proof One may assume that $\mu_e(E) < \infty$. Having fixed $\varepsilon > 0$, let $\{Q_{\varepsilon,n}\}$ be a countable collection of $\frac{1}{2}$-closed, dyadic cubes, whose union contains E and satisfying (4.2). For each n there exists an open cube $Q'_{\varepsilon,n}$ congruent to the interior of $Q_{\varepsilon,n}$ such that

$$Q_{\varepsilon,n} \subset Q'_{\varepsilon,n} \quad \text{and} \quad \mu(Q'_{\varepsilon,n} - Q_{\varepsilon,n}) \le \frac{1}{2^n}\varepsilon.$$

The union of the $Q'_{\varepsilon,n}$ is open and contains E. Therefore,

$$\mu'_e(E) \le \sum \mu(Q'_{\varepsilon,n}) \le \sum \mu(Q_{\varepsilon,n}) + \sum 2^{-n}\varepsilon \le \mu_e(E) + 2\varepsilon. \qquad \blacksquare$$

Proposition 12.2 *A set $E \subset \mathbb{R}^N$ such that $\mu_e(E) < \infty$, is Lebesgue measurable if and only if for every $\varepsilon > 0$ there exists an open set $E_{o,\varepsilon}$ such that*

$$E \subset E_{o,\varepsilon} \quad and \quad \mu_e(E_{o,\varepsilon} - E) \leq \varepsilon. \tag{12.1}$$

Equivalently, a set $E \subset \mathbb{R}^N$ of finite Lebesgue outer measure, is Lebesgue measurable if and only if there exists a set E_δ of the type of a $\mathcal{G}_\delta$, such that

$$E \subset E_\delta \quad and \quad \mu_e(E_\delta - E) = 0. \tag{12.2}$$

Proof The sufficient condition follows from Propositions 10.2 and Remark 10.1. Since the Lebesgue measure in $\mathbb{R}^N$ is σ-finite, the necessary condition follows from Propositions 10.1–10.3, using that $\mu_e(E) = \mu'_e(E)$. ∎

Proposition 12.3 *A set $E \subset \mathbb{R}^N$ such that $\mu_e(E) < \infty$, is Lebesgue measurable if and only if for every $\varepsilon > 0$ there exists a closed set $E_{c,\varepsilon}$ such that*

$$E_{c,\varepsilon} \subset E \quad and \quad \mu_e(E - E_{c,\varepsilon}) \leq \varepsilon. \tag{12.3}$$

Equivalently, a set $E \subset \mathbb{R}^N$ of finite Lebesgue outer measure, is Lebesgue measurable if and only if there exists a set E_σ of the type of a $\mathcal{F}_\sigma$, such that

$$E_\sigma \subset E \quad and \quad \mu_e(E - E_\sigma) = 0. \tag{12.4}$$

Proof (Sufficient Condition) Assume that for all $\varepsilon > 0$ there exists a closed set $E_{c,\varepsilon}$ satisfying (12.3). Then $E_{c,\varepsilon}^c$ is open, it contains E^c and

$$\mu_e(E_{c,\varepsilon}^c - E^c) = \mu_e(E - E_{c,\varepsilon}) \leq \varepsilon.$$

Therefore by Proposition 12.2, the set $E_{c,\varepsilon}^c$ is Lebesgue measurable. Hence $E_{c,\varepsilon}$ is Lebesgue measurable. ∎

Proof (Necessary Condition) Assume first that E is bounded and Lebesgue measurable, and let Q be a closed cube containing E in its interior. The set $Q - E$ is Lebesgue measurable and of finite measure. Therefore, by Proposition 12.2, for every $\varepsilon > 0$ there exists an open set $E_{o,\varepsilon}$ such that

$$Q - E \subset E_{o,\varepsilon} \quad and \quad \mu\big[E_{o,\varepsilon} - (Q - E)\big] \leq \varepsilon.$$

The set $E_{c,\varepsilon} = E_{o,\varepsilon}^c \cap Q$ is closed, is contained in E and it satisfies (12.3). For $n \in \mathbb{N}$ set $E_n = E \cap [|x| \leq n]$. If E is Lebesgue measurable and of finite measure, having fixed $\varepsilon > 0$ there exists n so that $\mu(E \cap [|x| > n]) < \frac{1}{2}\varepsilon$. The set E_n is bounded and measurable. Hence there exists a closed set $E_{c,\varepsilon} \subset E_n$, such that $\mu(E_n - E_{c,\varepsilon}) < \frac{1}{2}\varepsilon$. The set $E_{c,\varepsilon}$ is closed, is contained in E and

$$\mu(E - E_{c,\varepsilon}) \leq \mu(E_n - E_{c,\varepsilon}) + \mu(E \cap [|x| > n]) \leq \varepsilon. \quad ∎$$

Remark 12.2 The sufficient part of Propositions 12.2–12.3, does not require that $\mu_e(E) < \infty$. This follows from the proof of Proposition 10.2 and Remark 10.2.

13   Vitali's Nonmeasurable Set [168]

The following construction, due to Vitali, exhibits a subset of $[0, 1]$ which is not Lebesgue measurable. Let $\bullet : [0, 1) \times [0, 1) \to [0, 1)$ be the addition mod-1 acting on pairs $x, y \in [0, 1)$, that is

$$x \bullet y = \begin{cases} x + y & \text{if } x + y < 1; \\ x + y - 1 & \text{if } x + y \geq 1. \end{cases}$$

If E is a Lebesgue measurable subset of $[0, 1)$, then for every fixed $y \in [0, 1)$, the set

$$E \bullet y \overset{\text{def}}{=} \{x \bullet y \mid x \in E\}$$

is Lebesgue measurable and $\mu(E \bullet y) = \mu(E)$. Next introduce an equivalence relation $\sim$ in $[0, 1)$ by $x \sim y$ if $x - y$ is rational. Such a relation identifies equivalence classes in $[0, 1)$. If $\mathcal{E}$ is one such class, then any two elements of $\mathcal{E}$ differ by a rational number. In particular, the rational numbers in $[0, 1)$ all belong to one such equivalence class. Select one and only one element out of each class, to form a set E which by this procedure, contains one and only one element from each of these equivalence classes. In particular, any two distinct elements $x, y \in E$ are not equivalent. Such a selection is possible by the axiom of choice. Let now $r_o = 0$ and let $\{r_n\}$ denote the sequence of rational numbers in $(0, 1)$, and set $E_n = E \bullet r_n$. The sets E_n are pairwise disjoint. Indeed if $x \in E_n \cap E_m$, there exist two elements $x_n, x_m \in E$, and two rational numbers r_n and r_m, such that

$$x_n \bullet r_n = x_m \bullet r_m.$$

Therefore $x_n - x_m$ is rational, and $x_n \sim x_m$. This however contradicts the definition of E, unless $m = n$. Next observe that each element of $[0, 1)$ belongs to some E_n. Indeed every $x \in [0, 1)$ must belong to some equivalence class and therefore there must exist some $y \in E$ such that $x - y$ is rational. If $x - y \geq 0$, then $x = y + r_n$ for some r_n and hence $x \in E_n$. If $x - y < 0$, then $x = y - r_m$ for some r_m. This can be rewritten as

$$x = y + (1 - r_m) - 1 \qquad \text{or equivalently as} \qquad x = y \bullet (1 - r_m).$$

Thus in either case, $x \in E_n$ for some n. Therefore, $[0, 1) = \bigcup E_n$. If E were Lebesgue measurable, also E_n would be Lebesgue measurable and it would have the same measure. Since the sets E_n are mutually disjoint,

$$\mu\big([0, 1)\big) = \sum \mu(E_n).$$

This however is a contradiction since the right-hand side is either zero or infinity.

14   Borel Sets, Measurable Sets, and Incomplete Measures

Proposition 14.1 *There exists a subset E of $[0, 1]$ which is Lebesgue measurable but it is not a Borel set.*

Proposition 14.2 *The restriction of the Lebesgue measure on $\mathbb{R}$ to the σ-algebra of the Borel sets in $\mathbb{R}$, is not a complete measure.*

The next sections prepare for the proof of these propositions which is given in § 14.3.

14.1   A Continuous Increasing Function $f : [0, 1] \to [0, 1]$

Construct inductively a nonincreasing sequence of functions $\{f_n\}$ defined in $[0, 1]$ by setting $f_o(x) = x$

$$f_1(x) = \begin{cases} \frac{1}{2}x & \text{if } 0 \le x \le \frac{1}{3} \\[2mm] 2x - \frac{1}{2} & \text{if } \frac{1}{3} \le x \le \frac{1}{2} \\[2mm] \frac{1}{2}x + \frac{1}{4} & \text{if } \frac{1}{2} \le x \le \frac{5}{6} \\[2mm] 2x - 1 & \text{if } \frac{5}{6} \le x \le 1. \end{cases}$$

The function f_1 has been constructed by dividing first $[0, 1]$ into the two subintervals $[0, \frac{1}{2}]$ and $[\frac{1}{2}, 1]$. The first subinterval is subdivided in turn into the two subintervals $[0, \frac{1}{3}]$ and $[\frac{1}{3}, \frac{1}{2}]$. In the first of these f_1 is affine and has derivative $\frac{1}{2}$, whereas in the second f_1 is affine and has derivative 2. The second subinterval $[\frac{1}{2}, 1]$ is divided into two intervals $[\frac{1}{2}, \frac{5}{6}]$ and $[\frac{5}{6}, 1]$. In the first of these f_1 is affine and has derivative $\frac{1}{2}$, whereas in the second f_1 is affine and has derivative 2. The resulting function f_1 is continuous and increasing in $[0, 1]$. Moreover, $f_1 \le f_o$ and $f_1 = f_o$ at each of the end points of the initial subdivision. The functions f_n for $n \ge 2$ are constructed inductively to satisfy:

i.   Each f_n is continuous and increasing in $[0, 1]$. Moreover, $[0, 1]$ is subdivided into 4^n subintervals in such a way that f_n is affine on each of them and has derivative either 2^{-n} or 2^n.

ii. $f_{n+1}(x) \leq f_n(x)$ for all $n \in \mathbb{N}$ and all $x \in [0, 1]$.

iii.   If α is anyone of the end points of the 4^n intervals into which $[0, 1]$ has been subdivided, then $f_m(\alpha) = f_n(\alpha)$ for all $m \geq n$.

iv.   If $[\alpha, \beta]$ is an interval where f_n is affine, then $(\beta - \alpha) \leq 4^{-n}$ and $f_n(\beta) - f_n(\alpha) \leq 2^{-n}$.

Let f_n be constructed and let $[\alpha, \beta]$ be one of the intervals where f_n is affine. Then f_{n+1} restricted to $[\alpha, \beta]$ can be constructed by the following a graphical procedure. Set

$$A = \big(\alpha, f_n(\alpha)\big), \qquad B = \big(\beta, f_n(\beta)\big)$$

and let C be the midpoint of the segment $\overline{AB}$. Next let D be the unique point below the segment $\overline{AC}$ such that the slope of $\overline{AD}$ is 2^{-n-1} and the slope of $\overline{DC}$ is 2^{n+1}. Likewise, let E be the unique point below the segment $\overline{CB}$ such that the slope of $\overline{CE}$ is 2^{-n-1} and the slope of $\overline{CB}$ is 2^{n+1}. Then the polygonal $ADCEB$ is the graph of f_{n+1} within $[\alpha, \beta]$.

Since the f_n are continuous and strictly increasing in $[0, 1]$, their inverses f_n^{-1} are also continuous and strictly increasing in $[0, 1]$. Moreover by construction, such inverses satisfy properties (i)–(iv) except (ii) where the inequality is reversed. For a fixed $n \in \mathbb{N}$, any fixed $x \in [0, 1]$ belongs to at least one of the 4^n closed subintervals where f_n is affine. If for example $x \in [\alpha, \beta]$, then for all $m \geq n$

$$0 \leq f_n(x) - f_m(x) \leq f_n(\beta) - f_n(\alpha) \leq 2^{-n}.$$

Since $x \in [0, 1]$ is arbitrary the sequence $\{f_n\}$ converges uniformly in $[0, 1]$ to a nondecreasing, uniformly continuous function f in $[0, 1]$.

Having fixed $x < y$ in $[0, 1]$, there exists n so large that one of the intervals $[\alpha, \beta]$ where f_n is affine, is contained in $[x, y]$. Therefore,

$$f(x) \leq f(\alpha) = f_n(\alpha) < f_n(\beta) = f(\beta) \leq f(y).$$

Thus f is strictly increasing in $[0, 1]$ and has a continuous strictly increasing inverse f^{-1} in $[0, 1]$. Set

$$A_n = \bigcup \big\{ \text{intervals } [\alpha, \beta] \text{ where } f_n \text{ is affine and } f_n' = 2^n \big\};$$
$$B_n = \bigcup \big\{ \text{intervals } [\alpha, \beta] \text{ where } f_n \text{ is affine and } f_n' = 2^{-n} \big\}.$$

From the definition

$$[\alpha, \beta] \in A_n \implies f(\beta) - f(\alpha) = 2^n(\beta - \alpha)$$
$$[\alpha, \beta] \in B_n \implies f(\beta) - f(\alpha) = 2^{-n}(\beta - \alpha).$$

Therefore, adding over all such intervals

$$\mu\big(f(A_n)\big) = 2^n \mu(A_n) \qquad\qquad \mu(A_n) + \mu(B_n) = 1$$
$$\mu\big(f(B_n)\big) = 2^{-n} \mu(B_n) \qquad\qquad \mu\big(f(A_n)\big) + \mu\big(f(B_n)\big) = 1.$$

From this compute

$$\mu(A_n) = \frac{2^n - 1}{2^{2n} - 1} \qquad\qquad \mu(B_n) = 2^n \frac{2^n - 1}{2^{2n} - 1}$$
$$\mu\big(f(A_n)\big) = 2^n \frac{2^n - 1}{2^{2n} - 1} \qquad\qquad \mu\big(f(B_n)\big) = \frac{2^n - 1}{2^{2n} - 1}.$$

Set

$$S_n = \bigcup_{j=n}^{\infty} A_j \qquad\qquad S = \bigcap_{n=1}^{\infty} S_n.$$

The set S is measurable and compute

$$0 \le \mu(S) = \lim \mu(S_n) \le \lim \sum_{j=n}^{\infty} \mu(A_j) = 0.$$

The sets $f(A_n)$ are measurable being the finite union of intervals. Therefore, the sets

$$f(S_n) = \bigcup_{j=n}^{\infty} f(A_j) \qquad\qquad f(S) = \bigcap_{n=1}^{\infty} f(S_n)$$

are also measurable. Then compute

$$1 \ge \mu\big(f(S)\big) = \lim \mu\big(f(S_n)\big) \ge \lim \mu\big(f(A_n)\big) = 1.$$

Therefore, f maps the set $S \subset [0, 1]$ of measure zero onto $f(S) \subset [0, 1]$ of measure 1. Likewise, f maps the set $[0, 1] - S$ of measure 1, onto $[0, 1] - f(S)$ of measure zero.

14.2 On the Preimage of a Measurable Set

Since f is continuous, the preimage of an open or closed subset of $[0, 1]$ is open or closed and hence Lebesgue measurable. More generally one might consider the family

$$\mathcal{F} = \left\{ \begin{array}{l} \text{the collection of the subsets } E \text{ of } [0, 1] \\ \text{such that } f^{-1}(E) \text{ is Lebesgue measurable} \end{array} \right\}. \qquad (14.1)$$

Since f is strictly increasing, the complement of any set in $\mathcal{F}$ is in $\mathcal{F}$.

If $\{E_n\}$ is a countable collection of elements of $\mathcal{F}$, then

$$f^{-1}\left(\bigcup E_n\right) = \bigcup f^{-1}(E_n) \quad \text{and} \quad f^{-1}\left(\bigcap E_n\right) = \bigcap f^{-1}(E_n).$$

Therefore, the countable union or intersection of elements in $\mathcal{F}$ remains in $\mathcal{F}$. Thus $\mathcal{F}$ is a σ-algebra of subsets of $[0, 1]$. It follows that $\mathcal{F}$ must contain the Borel sets $\mathcal{B}$ of $[0, 1]$, since they form the smallest σ-algebra containing the open sets. In particular, the preimage of a Borel set is measurable.

Since f is continuous and increasing, the same argument shows that the preimage of a Borel set is a Borel set.

14.3   Proof of Propositions 14.1 and 14.2

Since the Lebesgue measure is complete, every subset of $\mathcal{S}$ is measurable and has measure zero. Likewise, every subset of $[0, 1] - f(\mathcal{S})$ is measurable and has measure zero. Let E be the Vitali nonmeasurable subset of $[0, 1]$. Then $E - \mathcal{S}$ is also nonmeasurable. Indeed if it were measurable E would be the disjoint union of the measurable sets $E - \mathcal{S}$ and $E \cap \mathcal{S}$. The set $\mathcal{D} = f(E - \mathcal{S})$ is contained in $[0, 1] - f(\mathcal{S})$. Therefore, $\mathcal{D}$ is measurable and it has measure zero. The preimage of $\mathcal{D}$ is not measurable. The measurable set $\mathcal{D}$ is not a Borel set, for otherwise $f^{-1}(\mathcal{D})$ would be measurable. Since $\mathcal{D}$ is measurable and has measure zero, by (12.2) of Proposition 12.2, there exists a set $\mathcal{D}_\delta$ of the type of $\mathcal{G}_\delta$ such that $\mathcal{D} \subset \mathcal{D}_\delta$ and $\mu(\mathcal{D}_\delta) = 0$. Since $\mathcal{D}$ is not a Borel set, the restriction of the Lebesgue measure to the σ-algebra of the Borel sets, is not complete.   ■

Remark 14.1 Returning to the family $\mathcal{F}$ defined in (14.1), the same example shows that $\mathcal{F}$ does not contain the σ-algebra $\mathcal{M}$ of the Lebesgue measurable subsets of $[0, 1]$, as there exists measurable sets whose preimage is not measurable. By interchanging the role of f and f^{-1} shows that in general $\mathcal{F}$ is not contained in $\mathcal{M}$.

15   Borel Measures

A feature of the Lebesgue measure in $\mathbb{R}^N$ is that the measure of a measurable set $E \subset \mathbb{R}^N$ of finite measure, can be approximated by the measure of open sets containing E or closed sets contained in E. This is the content of Propositions 12.2–12.3.

A Borel measure μ in $\mathbb{R}^N$ is a measure defined on a σ-algebra containing the Borel sets $\mathcal{B}$.[3] Such a requirement alone does not guarantee that the μ-measure of a Borel set $E \subset \mathbb{R}^N$ of finite μ-measure can be approximated by the μ-measure of open sets containing E and closed sets contained in E.

---

[3] Some authors define it as a measure μ whose domain of definition is *exactly* $\mathcal{B}$.

As an example consider the counting measure on $\mathbb{R}$. A single point is a Borel set with counting measure one and every open set that contains it has counting measure infinity. However, if μ is a finite Borel measure in $\mathbb{R}^N$ and E is a Borel set, then $\mu(E)$ can be approximated by the μ-measure of closed sets included in E or by the μ-measure of open sets containing E.

Proposition 15.1 *Let μ be a finite Borel measure in $\mathbb{R}^N$ and let E be a Borel set. For every $\varepsilon > 0$ there exists a closed set $E_{c,\varepsilon} \subset E$, such that*

$$E_{c,\varepsilon} \subset E \quad\text{and}\quad \mu(E - E_{c,\varepsilon}) \le \varepsilon. \tag{15.1}$$

Moreover, for every $\varepsilon > 0$ there exists an open set $E_{o,\varepsilon}$, such that

$$E \subset E_{o,\varepsilon} \quad\text{and}\quad \mu(E_{o,\varepsilon} - E) \le \varepsilon. \tag{15.2}$$

Proof (of (15.1)) Let $\mathcal{A}$ be the σ-algebra where μ is defined and set

$$\mathcal{C}_o = \left\{ \begin{array}{l} \text{the collection of sets } E \in \mathcal{A} \text{ such that for every } \varepsilon > 0 \\ \text{there exists a closed set } C \subset E, \text{ such that } \mu(E - C) \le \varepsilon \end{array} \right\}.$$

Such a collection is not empty since the closed sets are in $\mathcal{C}_o$. Let $\{E_n\}$ be a countable collection of sets in $\mathcal{C}_o$ and, having fixed $\varepsilon > 0$, select closed sets $C_n \subset E_n$ such that $\mu(E_n - C_n) \le 2^{-n}\varepsilon$ for all $n \in \mathbb{N}$. Then

$$\mu\left(\bigcap E_n - \bigcap C_n\right) \le \mu\left[\bigcup(E_n - C_n)\right] \le \sum \mu(E_n - C_n) \le \varepsilon.$$

Since $\bigcap C_n$ is closed, the intersection $\bigcap E_n$ belongs to $\mathcal{C}_o$. Next, by (3.3)–(3.4) of Proposition 3.1, since μ is finite

$$\lim_{m\to\infty} \mu\left(\bigcup E_n - \bigcup_{n=1}^{m} C_n\right) = \mu\left(\bigcup E_n - \bigcup C_n\right) \le \mu\left[\bigcup(E_n - C_n)\right] \le \varepsilon.$$

Therefore, there exists a positive integer m_ε such that

$$\mu\left(\bigcup E_n - \bigcup_{n=1}^{m_\varepsilon} C_n\right) \le 2\varepsilon.$$

Since the union $\bigcup_{n=1}^{m_\varepsilon} C_n$ is closed, the union $\bigcup E_n$ belong to $\mathcal{C}_o$.

The collection $\mathcal{C}_o$ contains trivially the closed sets, and in particular the closed dyadic cubes in $\mathbb{R}^N$. Since every open set is the countable union of such cubes, $\mathcal{C}_o$ contains also the open sets (Remark 1.1). Set

$$\mathcal{C} = \left\{\text{the collection of sets } E \in \mathcal{C}_o \text{ such that } (\mathbb{R}^N - E) \in \mathcal{C}_o\right\}.$$

If $E \in \mathcal{C}$ then the definition implies that $(\mathbb{R}^N - E)$ belongs to $\mathcal{C}$. Thus $\mathcal{C}$ is closed under taking complements. In particular $\mathcal{C}$ contains all the open sets and all the closed sets. Let $\{E_n\}$ be a countable collections of sets in $\mathcal{C}$, i.e., both E_n and $(\mathbb{R}^N - E_n)$ belong to $\mathcal{C}_o$ for all n. Then $\cup E_n \in \mathcal{C}_o$ and

$$\mathbb{R}^N - \bigcup E_n = \bigcap (\mathbb{R}^N - E_n) \in \mathcal{C}_o.$$

Analogously $\cap E_n \in \mathcal{C}_o$ and

$$\mathbb{R}^N - \bigcap E_n = \bigcup (\mathbb{R}^N - E_n) \in \mathcal{C}_o.$$

Thus $\mathcal{C}$ is a σ-algebra. Since $\mathcal{C}$ contains the open sets it contains the σ-algebra of the Borel sets.   ∎

Proof (of (15.2)) Let $E \in \mathcal{B}$. Then $(\mathbb{R}^N - E)$ is a Borel set and by (15.1), having fixed $\varepsilon > 0$ there exists a closed set $C \subset (\mathbb{R}^N - E)$ such that

$$\mu\big[(\mathbb{R}^N - E) - C\big] = \mu\big((\mathbb{R}^N - C) - E\big) \le \varepsilon.$$

Since $\mathbb{R}^N - C$ is open (15.2) follows.   ∎

Corollary 15.1 *Let μ be a finite Borel measure in $\mathbb{R}^N$ and let $E \in \mathcal{B}$.*
There exists $E_\sigma \in \mathcal{F}_\sigma$, such that $E_\sigma \subset E$ and $\mu(E - E_\sigma) = 0$.
There exists $E_\delta \in \mathcal{G}_\delta$, such that $E \subset E_\delta$ and $\mu(E_\delta - E) = 0$.

16   Borel, Regular, and Radon Measures

The approximation with closed sets contained in E continues to hold for Borel measures that are not necessarily finite, provided E is of finite measure.

Proposition 16.1 *Let μ be a Borel measure in $\mathbb{R}^N$ and let E be a Borel set of finite measure. For every $\varepsilon > 0$ there exists a closed set $E_{c,\varepsilon} \subset E$, such that (15.1) holds.*

Proof Let $\mathcal{A}$ be the σ-algebra where μ is defined. The set E being fixed, set $\mu_E(A) = \mu(E \cap A)$ for all $A \in \mathcal{A}$. Then μ_E is a finite Borel measure in $\mathbb{R}^N$.   ∎

Statement (15.2) is false for general Borel measures even if $\mu(E) < \infty$, as indicated by the counting measure on $\mathbb{R}$. However, it continues to hold for Borel measures that are finite on bounded sets.

Proposition 16.2 *Let μ be a Borel measure in $\mathbb{R}^N$ which is finite on bounded sets, and let E be a Borel set of finite measure. For every $\varepsilon > 0$ there exists an open set $E_{o,\varepsilon} \supset E$, such that (15.2) holds.*

Proof Let Q_n be the open cube centered at the origin, of edge n and with faces parallel to the coordinate planes. Since E is a Borel set, $Q_n - E$ is also a Borel set and $\mu(Q_n - E) < \infty$. By Proposition 16.1, having fixed $\varepsilon > 0$, there exists a closed set $C_n \subset (Q_n - E)$ such that,

$$\mu\big[(Q_n - E) - C_n\big] = \mu\big[(Q_n - C_n) - E\big] \le 2^{-n}\varepsilon.$$

The set $Q_n - C_n$ is open and contains $Q_n \cap E$. Therefore,

$$E = \bigcup Q_n \cap E \subset \bigcup (Q_n - C_n) = E_{o,\varepsilon}.$$

The set $E_{o,\varepsilon}$ is open, contains E and

$$\mu(E_{o,\varepsilon} - E) = \mu\big[\bigcup (Q_n - C_n) - E\big] \le \sum 2^{-n}\varepsilon. \qquad \blacksquare$$

16.1 Regular Borel Measures

Let μ be a Borel measure in $\mathbb{R}^N$. The statements of Proposition 15.1 and Propositions 16.1–16.2, give conditions for the measure of a Borel set E to be approximated by the measure of closed sets contained in E or open sets containing E. Those Borel measures for which the indicated approximation holds for all measurable sets E, define a subclass of measures called *regular*.

A Borel measure μ in $\mathbb{R}^N$ is *outer regular* if for every measurable set $E \subset \mathbb{R}^N$ of finite measure,

$$\mu(E) = \inf\{\mu(\mathcal{O}) \text{ where } \mathcal{O} \text{ is open and } E \subset \mathcal{O}\}. \qquad (16.1)$$

A Borel measure μ in $\mathbb{R}^N$ is *inner regular* if for every measurable set $E \subset \mathbb{R}^N$ of finite measure,

$$\mu(E) = \sup\{\mu(C) \text{ where } C \text{ is closed and } C \subset E\}. \qquad (16.2)$$

A Borel measure μ in $\mathbb{R}^N$ is *regular* if it is both outer and inner regular.

The the counting measure on $\mathbb{R}$ is not outer regular, but it is inner regular.

The Hausdorff measure $\mathcal{H}^\alpha$ in $\mathbb{R}^N$, for $0 \le \alpha < N$ is inner regular and not outer regular.

By Proposition 15.1, a finite Borel measure in $\mathbb{R}^N$ defined exactly on $\mathcal{B}$ is regular. By Propositions 16.1–16.2 a Borel measure defined exactly on $\mathcal{B}$, and finite on bounded, measurable subsets of $\mathbb{R}^N$ is regular.

By Propositions 12.2–12.3, the Lebesgue measure in $\mathbb{R}^N$ is regular.

If μ is outer regular, then for all measurable sets E of finite measure, there exists a Borel set $E_\delta \in \mathcal{G}_\delta$ such that $E \subset E_\delta$ and $\mu(E) = \mu(E_\delta)$.

If μ is inner regular, then for all measurable sets E of finite measure, there exists a Borel set $E_\sigma \in \mathcal{F}_\sigma$ such that $E_\sigma \subset E$ and $\mu(E) = \mu(E_\sigma)$.

16.2 Radon Measures

A Radon measure in $\mathbb{R}^N$ is a Borel measure which is finite on bounded subsets of $\mathbb{R}^N$. The Lebesgue measure in $\mathbb{R}^N$ is a Radon measure. The Lebesgue–Stieltjes measure on $\mathbb{R}$ is a Radon measure. The Dirac measure δ_x with mass concentrated at x is a Radon measure. The counting measure on $\mathbb{R}$ is a Borel measure but not a Radon measure. The Hausdorff measure $\mathcal{H}^\alpha$ is a Borel measure in $\mathbb{R}^N$ but not a Radon measure for all $\alpha \in [0, N)$.

Let μ be a Radon measure in $\mathbb{R}^N$ and for every set $E \subset \mathbb{R}^N$, set

$$\mu_e(E) = \inf \left\{\mu(\mathcal{O}) \text{ where } \mathcal{O} \text{ is open and } E \subset \mathcal{O}\right\}. \tag{16.3}$$

By the previous remarks, this is an outer measure which coincides with μ on the Borel sets.

Proposition 16.3 *A Radon measure μ in $\mathbb{R}^N$ generates, by the formula (16.3), an outer measure μ_e which coincides with μ on the Borel sets. Moreover for every set $E \subset \mathbb{R}^N$ such that $\mu_e(E) < \infty$, there exists a set E_δ of the type of a $\mathcal{G}_\delta$ such that $E \subset E_\delta$ and $\mu_e(E) = \mu(E_\delta)$.*

17 Vitali Coverings

Let $\{X, \mathcal{A}, \mu\}$ be $\mathbb{R}^N$ endowed with the Lebesgue measure, and let $\mathcal{F}$ denote a family of closed, nontrivial cubes in $\mathbb{R}^N$. A family $\mathcal{F}$ is a *fine* Vitali covering for a set $E \subset \mathbb{R}^N$, if for every $x \in E$ and every $\varepsilon > 0$, there exist a cube $Q \in \mathcal{F}$ such that $x \in Q$ and $\text{diam}\, Q < \varepsilon$.

The collection of N-dimensional closed dyadic cubes of diameter not exceeding some given positive number, is an example of a fine Vitali covering for any set $E \subset \mathbb{R}^N$. Other examples are in § 2, 3, 5 of Chap. 5.

Theorem 17.1 (Vitali [168, 10]) *Let E be a bounded, Lebesgue measurable set in $\mathbb{R}^N$ and let $\mathcal{F}$ be a fine Vitali covering for E. There exists a countable collection $\{Q_n\}$ of cubes $Q_n \in \mathcal{F}$, with pairwise disjoint interior, such that*

$$\mu(E - \cup Q_n) = 0. \tag{17.1}$$

Remark 17.1 The theorem does not claim that $\cup Q_n$ covers all points of E. Rather, $\cup Q_n$ covers E in a measure-theoretical sense. Construction of pointwise coverings are in § 17.1c of the Complements.

Remark 17.2 The theorem is more general in that E need not be Lebesgue measurable. See **17.1**–**17.3** of § 17c the Complements. If $\mathcal{F}$ is a covering of E, but not necessarily a fine Vitali covering, a similar statement holds in a weaker form (§ 1 of Chap. 9).

Proof Without loss of generality may assume that E and the cubes making up the family $\mathcal{F}$ are all included in some larger cube Q. Label by $\mathcal{F}_o$ the family $\mathcal{F}$, and out of $\mathcal{F}_o$ select a cube Q_o. If Q_o covers E then the theorem is proven. Otherwise introduce the family of cubes

$$\mathcal{F}_1 = \left\{ \begin{array}{l} \text{the collection of cubes } Q \in \mathcal{F}_o \text{ whose interior is} \\ \text{disjoint from the interior of } Q_o, \text{ i.e., } \overset{\circ}{Q} \cap \overset{\circ}{Q}_o = \emptyset \end{array} \right\}.$$

If Q_o does not cover E, such a family is not empty. Introduce also the number

$$d_1 = \{\text{the supremum of the diameters of the cubes } Q \in \mathcal{F}_1\}.$$

Then out of $\mathcal{F}_1$ select a cube Q_1 whose diameter is larger than $\frac{1}{2}d_1$. If $Q_o \cup Q_1$ covers E then the theorem is proven. Otherwise introduce the family of cubes

$$\mathcal{F}_2 = \left\{ \begin{array}{l} \text{the collection of cubes } Q \in \mathcal{F}_1 \text{ whose interior is} \\ \text{disjoint from the interior of } Q_1, \text{ i.e., } \overset{\circ}{Q} \cap \overset{\circ}{Q}_1 = \emptyset \end{array} \right\}$$

and the number

$$d_2 = \{\text{the supremum of the diameters of the cubes } Q \in \mathcal{F}_2\}.$$

Then out of $\mathcal{F}_2$ select a cube Q_2 whose diameter is larger than $\frac{1}{2}d_2$. Proceeding in this fashion, define inductively families $\{\mathcal{F}_n\}$, positive numbers $\{d_n\}$ and cubes $\{Q_n\}$, by the recursive procedure

$$\mathcal{F}_n = \left\{ \begin{array}{l} \text{the collection of cubes } Q \in \mathcal{F}_{n-1} \text{ whose interior is} \\ \text{disjoint from the interior of } Q_{n-1}, \text{ i.e., } \overset{\circ}{Q} \cap \overset{\circ}{Q}_{n-1} = \emptyset \end{array} \right\};$$

$$d_n = \{\text{the supremum of the diameters of the cubes } Q \in \mathcal{F}_n\};$$

$$Q_n = \{\text{a cube selected out of } \mathcal{F}_n, \text{ such that } \operatorname{diam} Q_n > \tfrac{1}{2}d_n\}.$$

The cubes $\{Q_n\}$ have pairwise disjoint interior and they are all included in some larger cube Q. Therefore,

$$\sum \left(\frac{\operatorname{diam} Q_n}{\sqrt{N}}\right)^N = \sum \mu(Q_n) < \infty. \tag{17.2}$$

The convergence of the series implies that $\lim \operatorname{diam} Q_n = 0$. To prove (17.1) proceed by contradiction, by assuming that

$$\mu\left(E - \bigcup Q_n\right) \geq 2\varepsilon \qquad \text{for some } \varepsilon > 0. \tag{17.3}$$

First, for each Q_n construct a larger cube Q'_n of diameter

$$\operatorname{diam} Q'_n = (4\sqrt{N} + 1)\operatorname{diam} Q_n \tag{17.4}$$

with same center as Q_n and faces parallel to the faces of Q_n. By the convergence of the series in (17.2) there exists some $n_\varepsilon \in \mathbb{N}$ such that

$$\mu\left(\bigcup_{n=n_\varepsilon+1}^{\infty} Q'_n\right) \leq \sum_{n=n_\varepsilon+1}^{\infty} \mu(Q'_n) \leq \varepsilon.$$

Using this inequality and (17.3) estimate

$$\mu\left[\left(E - \bigcup_{n=1}^{n_\varepsilon} Q_n\right) - \bigcup_{n=n_\varepsilon+1}^{\infty} Q'_n\right] \geq \mu\left(E - \bigcup Q_n\right) - \mu\left(\bigcup_{n=n_\varepsilon+1}^{\infty} Q'_n\right) \geq \varepsilon.$$

This implies that there exists an element

$$x \in \left(E - \bigcup_{n=1}^{n_\varepsilon} Q_n\right) - \bigcup_{n=n_\varepsilon+1}^{\infty} Q'_n. \tag{17.5}$$

Such an element, must have positive distance 2σ from the union of the first n_ε cubes. Indeed such a finite union is closed and x does not belong to any of the cubes Q_n for $n = 1, \ldots, n_\varepsilon$. By the definition of fine Vitali covering, given such a σ, there exists a cube $Q_\delta \in \mathcal{F}$ of positive diameter $0 < \delta \leq \sigma$ that covers x. By construction Q_δ does not intersect the interior of any of the first n_ε cubes Q_n. It follows that Q_δ belongs to the family $\mathcal{F}_{n_\varepsilon+1}$. Next we claim that

$$Q_\delta \cap \overset{\circ}{Q_n} \neq \emptyset \qquad \text{for some } n = n_\varepsilon + 1, \ n_\varepsilon + 2, \ldots.$$

Indeed, if Q_δ did not intersect the interior of any such cubes, it would belong to all the families $\mathcal{F}_n$. This however would imply that

$$0 < \delta = \operatorname{diam} Q_\delta \leq d_n \to 0 \qquad \text{as } n \to \infty.$$

Let then $m \geq (n_\varepsilon + 1)$ be the smallest positive integer for which $Q_\delta \cap \overset{\circ}{Q_m} \neq \emptyset$. In particular,

$$Q_\delta \notin \mathcal{F}_{m+1} \qquad Q_\delta \in \mathcal{F}_m \qquad \text{and} \qquad \delta \leq d_m.$$

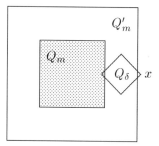

Fig. 1 About the proof of Vitali's covering theorem

By the selection (17.5), the element x does not belong to Q'_m. Therefore, the intersection $Q_\delta \cap \overset{\circ}{Q}_m$ can be not empty only if (Fig. 1)

$$\delta = \operatorname{diam} Q_\delta > \frac{1}{2\sqrt{N}} \left(\operatorname{diam} Q'_m - \operatorname{diam} Q_m \right).$$

This and (17.4) yield the contradiction $d_m \geq \delta > d_m$.   ∎

Corollary 17.1 *Let E be a bounded, Lebesgue measurable set in $\mathbb{R}^N$ and let $\mathcal{F}$ be a fine Vitali covering for E. For every $\varepsilon > 0$, there exists a finite collection $\{Q_1, \ldots, Q_{n_\varepsilon}\}$ of cubes in $\mathcal{F}$, with pairwise disjoint interior, such that*

$$\sum \mu(Q_n) - \varepsilon \leq \mu(E) \leq \mu\Big(\bigcup_{n=1}^{n_\varepsilon} E \cap Q_n \Big) + \varepsilon.$$

Proof Having fixed $\varepsilon > 0$, let $E_{o,\varepsilon}$ be an open set containing E and satisfying (12.1). Introduce the subfamily $\mathcal{F}_\varepsilon$ of the cubes in $\mathcal{F}$ that are contained in $E_{o,\varepsilon}$, and out of $\mathcal{F}_\varepsilon$ select a countable collection of closed cubes $\{Q_n\}$, with pairwise disjoint interior, satisfying (17.1). By construction

$$\sum \mu(Q_n) \leq \mu(E_{o,\varepsilon}) \leq \mu(E) + \varepsilon.$$

This implies that there exists a positive integer n_ε such that $\sum_{n_\varepsilon+1}^{\infty} \mu(Q_n) \leq \varepsilon$. From this and (17.1)

$$\mu(E) = \mu\Big(\bigcup E \cap Q_n \Big) \leq \mu\Big(\bigcup_{n=1}^{n_\varepsilon} E \cap Q_n \Big) + \varepsilon.$$ ∎

18 The Besicovitch Covering Theorem

Let E be a subset of $\mathbb{R}^N$. A collection $\mathcal{F}$ of closed balls in $\mathbb{R}^N$ is a Besicovitch covering for E if each $x \in E$ is the center of a nontrivial ball $B(x)$ belonging to $\mathcal{F}$.

Theorem 18.1 (Besicovitch [16]) *Let E be a bounded subset of $\mathbb{R}^N$ and let $\mathcal{F}$ be a Besicovitch covering for E such that*

$$\{\text{the supremum of the radii of the balls in } \mathcal{F}\} \overset{\text{def}}{=} R < \infty. \qquad (18.1)$$

There exists a countable collection $\{x_n\}$ of points in E and a corresponding collection of balls $\{B_n\}$ in $\mathcal{F}$

$$B_n = B_{\rho_n}(x_n) \quad \text{balls centered at } x_n \text{ and radius } \rho_n$$

such that $E \subset \bigcup B_n$. Moreover, there exists a positive integer c_N depending only upon the dimension N, and independent of E, the covering $\mathcal{F}$, and R, such that the balls $\{B_n\}$ can be organized into c_N subcollections

$$\mathcal{B}_1 = \{B_{n_1}\}, \quad \mathcal{B}_2 = \{B_{n_2}\}, \dots, \mathcal{B}_{c_N} = \{B_{n_{c_N}}\}$$

in such a way that the balls $\{B_{n_j}\}$ of each subcollection $\mathcal{B}_j$ are disjoint.

Remark 18.1 The theorem continues to hold, by essentially the same proof, if the balls making up the Besicovitch covering $\mathcal{F}$ are replaced by cubes with faces parallel to the coordinate planes [112].

Proof Since E is bounded and $R < \infty$, may assume that E and the balls making up the family $\mathcal{F}$ are all included in some large ball B_o centered at the origin. Set $E_1 = E$ and

$$\mathcal{F}_1 = \{ \text{ the collection of balls } B(x) \in \mathcal{F} \text{ whose center is in } E_1\}$$
$$r_1 = \{\text{the supremum of the radii of the balls in } \mathcal{F}_1\}.$$

There exists $x_1 \in E_1$ and a ball

$$B_1 = B_{\rho_1}(x_1) \in \mathcal{F}_1 \qquad \text{of radius } \rho_1 > \tfrac{3}{4}r_1.$$

If $E_1 \subset B_1$, the process terminates. Otherwise set $E_2 = E_1 - B_1$ and

$$\mathcal{F}_2 = \{ \text{ the collection of balls } B(x) \in \mathcal{F} \text{ whose center is in } E_2\}$$
$$r_2 = \{\text{the supremum of the radii of the balls in } \mathcal{F}_2\}.$$

There exists $x_2 \in E_2$ and a ball

$$B_2 = B_{\rho_2}(x_2) \in \mathcal{F}_2 \qquad \text{of radius } \rho_2 > \tfrac{3}{4}r_2.$$

Proceeding recursively, define countable collections of sets E_n, balls B_n, families $\mathcal{F}_n$ and positive numbers r_n by

$$E_n = E - \bigcup_{j=1}^{n-1} B_j, \qquad x_n \in E_n$$

$\mathcal{F}_n = \{$ the collection of balls $B(x) \in \mathcal{F}$ whose center is in $E_n\}$

$r_n = \{$the supremum of the radii of the balls in $\mathcal{F}_n\}$

$B_n = B_{\rho_n}(x_n) \in \mathcal{F}_n \qquad$ of radius $\rho_n > \frac{3}{4}r_n$.

By construction, if $m > n$

$$\rho_n > \tfrac{3}{4}r_n \geq \tfrac{3}{4}r_m \geq \tfrac{3}{4}\rho_m. \tag{18.2}$$

This implies the balls $B_{\frac{1}{3}\rho_n}(x_n)$ are disjoint. Indeed since $x_m \notin B_n$

$$|x_n - x_m| > \rho_n = \tfrac{1}{3}\rho_n + \tfrac{2}{3}\rho_n > \tfrac{1}{3}\rho_n + \tfrac{1}{3}\rho_m. \tag{18.3}$$

The balls $B_{\frac{1}{3}\rho_n}(x_n)$ are all contained in B_o and are disjoint. Therefore, $\{\rho_n\} \to 0$ as $n \to \infty$. The union of the balls $\{B_n\}$ covers E. If not, select $x \in E - \cup B_n$ and a nontrivial ball $B_\rho(x)$ centered at x and radius $\rho > 0$. Such a ball exists since $\mathcal{F}$ is a Besicovitch covering. By construction $B_\rho(x)$ must belong to all the families $\mathcal{F}_n$. Therefore, $0 < \rho \leq r_n \to 0$.

The proof of the last statement regarding the subcollections $\mathcal{B}_j$, is based on the following geometrical fact, whose proof is postponed to the next section.

Proposition 18.1 *There exists $c_N \in \mathbb{N}$ depending only on N and independent of E, such that, for every $k \in \mathbb{N}$, at most c_N balls out of $\{B_1, \ldots, B_k\}$ intersect B_k.*

The collections $\mathcal{B}_j$ are constructed by regarding them initially as empty boxes, to be filled with disjoint balls, taken out of $\{B_n\}$. Each element of $\{B_n\}$ is allocated to some of the boxes $\mathcal{B}_j$ as follows.

First, for $j = 1, \ldots, c_N$, put B_j into $\mathcal{B}_j$. Consider next the ball B_{c_N+1}. By Proposition 18.1, at least one of the first c_N balls does not intersect B_{c_N+1}, say for example, B_1. Then allocate B_{c_N+1} to $\mathcal{B}_1$.

Consider the subsequent ball B_{c_N+2}. At least 2 of the first $(c_N + 1)$ balls do not intersect B_{c_N+2}. If one of the B_j for $j = 2, \ldots, c_N$, say for example, B_2, does not intersect B_{c_N+2}, allocate B_{c_N+2} to $\mathcal{B}_2$. If all the balls B_j for $j = 2, \ldots, c_N$ intersect B_{c_N+2} then, B_1 and B_{c_N+1} do not intersect B_{c_N+2}, since at least 2 of the first $(c_N + 1)$ balls do not intersect B_{c_N+2}. Then allocate B_{c_N+2} to $\mathcal{B}_1$ which now would contain 3 disjoint balls.

Proceeding recursively, assume all the balls

$$B_1, \ldots, B_{c_N}, \ldots, B_{c_N+n-1} \quad \text{for some } n \in \mathbb{N}$$

have been allocated, so that at the $(n - 1)^{\text{th}}$ step of the process, each of the $\mathcal{B}_j$ contains at most n disjoint balls. To allocate B_{c_N+n} observe that by Proposition 18.1, at least n of the first $(c_N + n - 1)$ balls must be disjoint from B_{c_N+n}. This implies that the elements of at least one of the boxes $\mathcal{B}_j$ for $j = 1, \ldots, c_N$, are all disjoint from $B_{c_N + n}$. Allocate $B_{c_N + n}$ to one such a box and proceed inductively. ∎

19 Proof of Proposition 18.1

Fix some positive integer k, consider those balls B_j for $j = 1, \ldots, k$ that intersect $B_k = B_{\rho_k}(x_k)$, and divide them into two sets

$$\mathcal{G}_1 = \left\{ \begin{array}{c} \text{the collection of balls } B_j = B_{\rho_j}(x_j) \text{ for } j = 1, \ldots, k \\ \text{that intersect } B_k \text{ and such that } \rho_j \leq \tfrac{3}{4} M \rho_k \end{array} \right\}$$

$$\mathcal{G}_2 = \left\{ \begin{array}{c} \text{the collection of balls } B_j = B_{\rho_j}(x_j) \text{ for } j = 1, \ldots, k \\ \text{that intersect } B_k \text{ and such that} \rho_j > \tfrac{3}{4} M \rho_k \end{array} \right\}$$

where $M > 3$ is a positive integer to be chosen.

Lemma 19.1 *The number of balls in $\mathcal{G}_1$ does not exceed $4^N (M + 1)^N$.*

Proof Let $\#\{\mathcal{G}_1\}$ be the number of elements in $\mathcal{G}_1$. The balls $\{B_{\frac{1}{3}\rho_j}(x_j)\}$ are disjoint and are contained in $B_{(M+1)\rho_k}(x_k)$. Indeed, since $B_j \cap B_k \neq \emptyset$

$$|x_j - x_k| \leq \rho_j + \rho_k \leq \left(\tfrac{3}{4}M + 1\right)\rho_k.$$

Moreover for any $x \in B_{\frac{1}{3}\rho_j}(x_j)$,

$$|x - x_k| \leq |x - x_j| + |x_j - x_k| \leq \tfrac{1}{3}\rho_j + \left(\tfrac{3}{4}M + 1\right)\rho_k \leq (M + 1)\rho_k.$$

From this, denoting by κ_N the volume of the unit ball in $\mathbb{R}^N$

$$\sum_{j: B_j \in \mathcal{G}_1} \kappa_N \left(\tfrac{1}{3}\rho_j\right)^N \leq \kappa_N (M + 1)^N \rho_k^N.$$

Since $j < k$, it follows from (18.2) that $\tfrac{1}{3}\rho_j > \tfrac{1}{4}\rho_k$. Therefore,

$$\#\{\mathcal{G}_1\}\kappa_N \left(\tfrac{1}{4}\rho_k\right)^N \leq \kappa_N (M + 1)^N \rho_k^N. \quad \blacksquare$$

An upper estimate of the number of balls in $\mathcal{G}_2$ is derived by counting the number of rays originating from the center x_k of B_k to each of the centers x_j of $B_j \in \mathcal{G}_2$. We first establish that the angle between any two such rays is not less than an absolute angle θ_o. Then we estimate the number of rays originating from x_k and mutually forming an angle of at least θ_o.

Let $B_{\rho_n}(x_n)$ and $B_{\rho_m}(x_m)$ be any two balls in $\mathcal{G}_2$ and set

$$\theta = \{\text{angle between the rays from } x_k \text{ to } x_n \text{ and } x_m\}.$$

Lemma 19.2 *The number M can chosen so that $\theta > \theta_o = \arccos \frac{5}{6}$.*

Proof Assume $n < m < k$. By construction $x_m \notin B_{\rho_n}(x_n)$, i.e.,

$$|x_n - x_m| > \rho_n. \tag{19.1}$$

Also $x_k \notin B_{\rho_n}(x_n) \cup B_{\rho_m}(x_m)$, i.e.,

$$\rho_n < |x_n - x_k| \qquad \text{and} \qquad \rho_m < |x_m - x_k|.$$

Since both $B_{\rho_n}(x_n)$ and $B_{\rho_m}(x_m)$ intersect B_k and are in $\mathcal{G}_2$

$$\begin{aligned}
\tfrac{3}{4}M\rho_k < \rho_n \le |x_n - x_k| \le \rho_n + \rho_k \\
\tfrac{3}{4}M\rho_k < \rho_m \le |x_m - x_k| \le \rho_m + \rho_k.
\end{aligned} \tag{19.2}$$

By elementary trigonometry

$$\cos \theta = \frac{|x_n - x_k|^2 + |x_m - x_k|^2 - |x_n - x_m|^2}{2|x_n - x_k||x_m - x_k|}.$$

Assuming $\cos \theta > 0$ and using (19.1) and (19.2) estimate

$$\begin{aligned}
\cos \theta &\le \frac{(\rho_n + \rho_k)^2 + (\rho_m + \rho_k)^2 - \rho_n^2}{2\rho_n \rho_m} \\
&\le \frac{\rho_m^2 + 2\rho_k^2 + 2\rho_k(\rho_n + \rho_m)}{2\rho_n \rho_m} \\
&\le \frac{1}{2}\frac{\rho_m}{\rho_n} + \frac{\rho_k}{\rho_n}\frac{\rho_k}{\rho_m} + \frac{\rho_k}{\rho_m} + \frac{\rho_k}{\rho_n} \\
&\le \frac{1}{2}\frac{\rho_m}{\rho_n} + \left(\frac{4}{3}\right)^2 \frac{1}{M^2} + 2\frac{4}{3}\frac{1}{M}.
\end{aligned}$$

Since $m > n$, from (18.2) it follows that $\rho_n > \frac{3}{4}\rho_m$. Therefore,

$$\cos \theta \le \frac{2}{3} + \frac{4}{3}\frac{1}{M}\left(\frac{4}{3}\frac{1}{M} + 2\right).$$

Now choose M so large that the $\cos \theta \le \frac{5}{6}$.   ∎

If $N = 2$, the number of rays originating from the origin and mutually forming an angle $\theta > \theta_o$ does not exceed $2\pi/\theta_o$. If $N \ge 3$ let $\mathcal{C}(\theta_o)$ be a circular cone in $\mathbb{R}^N$ with vertex at the origin, whose axial cross section with a 2-dimensional hyperplane, forms an angle $\frac{1}{2}\theta_o$. Denote by $\sigma_N(\theta_o)$ the solid angle corresponding to $\mathcal{C}(\theta_o)$. Then the number of rays originating from the origin and mutually forming an angle $\theta > \theta_o$ does not exceed $\omega_N/\sigma_N(\theta_o)$.

The number c_N claimed by Proposition 18.1 is estimated by,

$$c_N = \#\{\mathcal{G}_1\} + \#\{\mathcal{G}_2\} \le 4^N (M+1)^N + \frac{\omega_N}{\sigma_N(\theta_o)}.$$

20 The Besicovitch Measure-Theoretical Covering Theorem

A family $\mathcal{F}$ of closed balls in $\mathbb{R}^N$ is a *fine* Besicovitch covering for a set $E \subset \mathbb{R}^N$, if for every $x \in E$ and every $\varepsilon > 0$, there exists a ball $B_\rho(x) \in \mathcal{F}$ centered at x and of radius $\rho < \varepsilon$.

The next measure theoretical covering, called the Besicovitch covering theorem, holds for any Radon measure μ and its associated outer measure μ_e (§ 16.1). The set E to be covered in a measure theoretical sense, is not required to be μ-measurable.

Theorem 20.1 (Besicovitch [16]) *Let E be a bounded set in $\mathbb{R}^N$ and let $\mathcal{F}$ be a fine Besicovitch covering for E. Let μ be a Radon measure in $\mathbb{R}^N$ and let μ_e be the outer measure associated with it.*

There exists a countable collection $\{B_n\}$ of disjoint balls $B_n \in \mathcal{F}$ such that

$$\mu_e\,(E - \cup B_n) = 0. \tag{20.1}$$

Remark 20.1 It is not claimed that $E \subset \bigcup B_n$. The collection $\{B_n\}$ forms a measure-theoretical covering of E in the sense of (20.1).

Remark 20.2 A fine Besicovitch covering of a set $E \subset \mathbb{R}^N$ differs from a fine Vitali covering, in that each $x \in E$ is required to be the center of a ball of arbitrarily small radius. In this respect it is less flexible than the fine Vitali covering. However, Besicovitch covering theorem applies to any Radon measure and in this respect is more flexible than the Vitali covering theorem.

Proof May assume $\mu_e(E) > 0$ otherwise the statement is trivial. Since E is bounded, may assume that both E and all the balls making up the covering $\mathcal{F}$ are contained in some larger ball B_o.

Let $\mathcal{B}_j$ for $j = 1, \ldots, c_N$ be the subcollections of disjoint balls claimed by Theorem 18.1. Since $E \subset \bigcup_{j=1}^{c_N} \bigcup_{n_j=1}^{\infty} B_{n_j}$

$$\mu_e\Big(E \cap \bigcup_{j=1}^{c_N} \bigcup_{n_j=1}^{\infty} B_{n_j}\Big) = \mu_e(E) > 0.$$

Therefore, there exists some index $j \in \{1, \ldots, c_N\}$ for which

$$\mu_e\Big(E \cap \bigcup_{n_j=1}^{\infty} B_{n_j}\Big) \ge \frac{1}{c_N}\mu_e(E).$$

Since all the balls B_{n_j} are disjoint and are all included in B_o

$$\mu_e\left(E \cap \bigcup_{n_j=1}^{\infty} B_{n_j}\right) \leq \sum_{n_j=1}^{\infty} \mu(B_{n_j}) \leq \mu(B_o) < \infty.$$

Therefore, there exists some index m_1 such that

$$\mu_e\left(E \cap \bigcup_{n_j=1}^{m_1} B_{n_j}\right) \geq \frac{1}{2c_N}\mu_e(E). \tag{20.2}$$

The finite union of balls is μ-measurable. Therefore, by the Carathéodory criterion of measurability (6.2) and the lower estimate in (20.2)

$$\mu_e(E) = \mu_e\left(E \cap \bigcup_{n_j=1}^{m_1} B_{n_j}\right) + \mu_e\left(E - \bigcup_{n_j=1}^{m_1} B_{n_j}\right)$$

$$\geq \frac{1}{2c_N}\mu_e(E) + \mu_e\left(E - \bigcup_{n_j=1}^{m_1} B_{n_j}\right).$$

Therefore,

$$\mu_e\left(E - \bigcup_{n_j=1}^{m_1} B_{n_j}\right) \leq \eta\mu_e(E) \qquad \eta = 1 - \frac{1}{2c_N} \in (0,1).$$

Set now

$$E_1 = E - \bigcup_{n_j=1}^{m_1} B_{n_j}.$$

If $\mu_e(E_1) = 0$ the process terminates and the theorem is proven. Otherwise let $\mathcal{F}_1$ denote the collection of balls in $\mathcal{F}$ that do not intersect any of the balls B_{n_j} for $n_j = 1, \ldots, m_1$. Since $\mathcal{F}$ is a fine Besicovitch covering for E, the family $\mathcal{F}_1$ is not empty and it is a fine Besicovitch covering for E_1. Repeating the previous selection process for the set E_1 and the Besicovitch covering $\mathcal{F}_1$, yields a finite number m_2 of closed disjoint balls B_{n_ℓ} in $\mathcal{F}_1$ such that

$$\mu_e\left(E_1 - \bigcup_{n_\ell=1}^{m_2} B_{n_\ell}\right) \leq \eta\mu_e(E_1) = \eta\mu_e\left(E - \bigcup_{n_j=1}^{m_1} B_{n_j}\right) \leq \eta^2\mu_e(E).$$

Relabeling the balls B_{n_j} and B_{n_ℓ} yields a finite number s_2 of closed, disjoint balls B_n in $\mathcal{F}$ such that

$$\mu_e\left(E - \bigcup_{n=1}^{s_2} B_n\right) \leq \eta^2 \mu_e(E).$$

Repeating the process k times gives a collection of s_k closed disjoint balls in $\mathcal{F}$ such that

$$\mu_e\left(E - \bigcup_{n=1}^{s_k} B_n\right) \leq \eta^k \mu_e(E). \tag{20.3}$$

If for some $k \in \mathbb{N}$

$$\mu_e\left(E - \bigcup_{n=1}^{s_k} B_n\right) = 0$$

the process terminated and the theorem is proven. Otherwise, (20.3) holds for all $k \in \mathbb{N}$. Letting $k \to \infty$ proves (20.1).   ∎

Problems and Complements

1c   Partitioning Open Subsets of $\mathbb{R}^N$

The dyadic cubes covering an open set E can be chosen so that their diameter is proportional to their distance from the boundary of E.

Proposition 1.1c   (Whitney [174]) *Every open set $E \subset \mathbb{R}^N$ can be partitioned into the countable union of closed dyadic cubes $\{Q_n\}$, with pairwise disjoint interior and satisfying,*

$$diam Q_j \leq dist\{Q_j; \partial E\} \leq 4 diam Q_j \qquad \text{for all } j.$$

2c   Limits of Sets, Characteristic Functions and σ-Algebras

2.1.   From the definition (2.1) it follows that $E' \subset E''$. There are sequences of sets $\{E_n\}$ for which the inclusion is strict.

2.2.   Prove that for a sequence of sets $\{E_n\}$

$$\limsup \chi_{E_n} = \chi_{E''} \qquad\qquad \liminf \chi_{E_n} = \chi_{E'}$$
$$\left(\limsup E_n\right)^c = \liminf E_n^c \quad \left(\liminf E_n\right)^c = \limsup E_n^c.$$

Set $D_1 = E_1$ and $D_{n+1} = D_n \Delta E_{n+1}$ for $n \in \mathbb{N}$. Prove that $\lim D_n$ exists if and only if $\lim E_n = \emptyset$.

2.3.   Let $\mathcal{A}$ be the collection of subsets $E \subset X$ such that either E or E^c is finite. Then if X is not finite $\mathcal{A}$ is an algebra but not a σ-algebra.

2.4.   Construct the smallest σ-algebra generated by two elements of X.

2.5.   Construct the smallest σ-algebra generated by the collection of all finite subsets of X.

2.6.   Prove that an algebra $\mathcal{A}$ is a σ-algebra if and only if for every nondecreasing (nonincreasing) sequence of sets $\{E_n\} \subset \mathcal{A}$, $\lim E_n \in \mathcal{A}$.

2.7.   The smallest σ-algebra $\mathcal{A}_o$ containing a collection of sets $\mathcal{O} \subset 2^X$, is the union of all the smallest σ-algebras containing a countable collection of elements of $\mathcal{O}$.

2.8.   An infinite σ-algebra contains a countably infinite collection of disjoint non-empty sets.

2.9.   If a σ-algebra $\mathcal{A}$ contains infinitely many sets, then card $\mathcal{A} \geq$ card $\mathbb{R}$.

2.10.   Let $\{X; \mathcal{U}\}$ be a topological space such that the collection of its open sets is an algebra. Characterize $\{X; \mathcal{U}\}$.

2.11.   The intersection of any collection of algebras is an algebra. The union of algebras need not be an algebra.

2.12.   For a countable collection of sets $\{E_n\}$ characterize $\limsup E_n$ as the set of all x lying in infinitely many E_n.

3c Measures

3.1.   Let X be an infinite set and let $\mathcal{A}$ be the collection of subsets of X that are either finite or have a finite complement. Let also $\mu : \mathcal{A} \to \mathbb{R}^*$ be a set function defined by, $\mu(E) = 0$ if E is finite and $\mu(E) = \infty$ if E has finite complement. The collection $\mathcal{A}$ is not a σ-algebra and μ is not countably additive.

3.2.   Let X be an uncountable set and let $\mathcal{A}$ be the collection of subsets of X that are either countable or have a countable complement. Let also $\mu : \mathcal{A} \to \mathbb{R}^*$ be a set function defined by, $\mu(E) = 0$ if E is countable and $\mu(E) = \infty$ if E has countable complement. The collection $\mathcal{A}$ is a σ-algebra and μ is a measure.

3.3.   **Semifinite Measures:** A measure μ on a σ-algebra $\mathcal{A}$ is semifinite if every measurable set of infinite measure contains a measurable set of positive and finite measure. Measures that are σ-finite are semifinite. The converse is false. Give an example of a non-semifinite measure.

Prove that if μ is semifinite, every measurable set of infinite measure contains a measurable set of arbitrarily large measure.

3.4.   For a measure μ on a σ-algebra $\mathcal{A}$ set

$$\mathcal{A} \ni E \to \mu_o(E) = \sup\{\mu(A) \mid \mathcal{A} \ni A \subset E, \text{ and } \mu(A) < \infty\}.$$

Prove that μ_o is semifinite and that if μ is semifinite, then $\mu = \mu_o$.

3.5.   **Locally Measurable Sets:** For a measure space $\{X, \mathcal{A}, \mu\}$, define the collection $\mathcal{A}_{\text{loc}}$

$$\mathcal{A}_{\text{loc}} = \left\{ \begin{array}{c} \text{collection of all } E \subset X, \text{ such that } E \cap A \in \mathcal{A} \\ \text{for all } A \in \mathcal{A} \text{ such that } \mu(A) < \infty \end{array} \right\}.$$

Prove that if μ is σ-finite, then $\mathcal{A} = \mathcal{A}_{\text{loc}}$. Give an example showing that the inclusion $\mathcal{A} \subset \mathcal{A}_{\text{loc}}$ might be strict. Prove that $\mathcal{A}_{\text{loc}}$ is a σ-algebra and construct a measure μ_{loc} on $\mathcal{A}_{\text{loc}}$ that coincides with μ on $\mathcal{A}$.

3.6.   **Extension of $\{X, \mathcal{A}, \mu\}$ to $\{X, \mathcal{A}_{\text{loc}}, \mu_{\text{loc}}\}$:** Prove that the set function

$$\mathcal{A}_{\text{loc}} \ni E \rightarrow \mu_{\text{loc}}(E) = \begin{cases} \mu(E) & \text{if } E \in \mathcal{A} \\ \infty & \text{otherwise} \end{cases}$$

is a measure on $\mathcal{A}_{\text{loc}}$.

3.7.   **Measure Theoretical Distance of Sets.** Let μ be a measure on a σ-algebra $\mathcal{A}$. Prove that $\mu(A \Delta B) = 0$, for $A, B \in \mathcal{A}$, implies $\mu(A) = \mu(B)$. Prove that the relation

$$A \sim B \quad \text{if and only if} \quad \mu(A \Delta B) = 0$$

is an equivalence relation on $\mathcal{A}$. A distance $d(\cdot; \cdot)$ between any two sets $E, F \in \mathcal{A}$ can be defined by setting $d(E; F) = \mu(E \Delta F)$. Prove that

$$d(E; F) \geq 0, \quad d(E; F) = d(F; E), \quad d(E; F) \leq d(E; G) + d(G; F)$$

for any $G \in \mathcal{A}$. Moreover, if $E \sim E'$ and $F \sim F'$, then $d(E; F) = d(E'; F')$. The set operations

$$\mathcal{A} \ni E, F \longrightarrow E \cup F, \quad E \cap F, \quad E^c$$

are continuous with respect to $d(\cdot; \cdot)$. Specifically, for all $\varepsilon > 0$, there exists δ, depending on ε and $\mu(X)$, such for every pair of sets

$$\mathcal{A} \ni E_i, F_i \text{ for } i = 1, 2, \text{ such that } d(E_1; E_2) < \delta \text{ and } d(F_1; F_2) < \delta,$$

there holds

$$d\big((E_1 \cup F_1); (E_2 \cup F_2)\big) < \varepsilon;$$
$$d\big((E_1 \cap F_1); (E_2 \cap F_2)\big) < \varepsilon;$$
$$d\big(E_1^c; E_2^c\big) < \varepsilon.$$

This follows from the finite additivity of μ, and the set identities

$$(E_1 \cup F_1)\Delta(E_2 \cup F_2) = (E_1 \Delta E_2)\Delta(F_1 \Delta F_2)\Delta E_1$$
$$\cap (F_1 \Delta F_2)\Delta F_2 \cap (E_1 \Delta E_2);$$

$$(E_1 \cap F_1)\Delta(E_2 \cap F_2) = E_1 \cap (F_1 \Delta F_2)\Delta F_2 \cap (E_1 \Delta E_2);$$
$$E_1^c \Delta E_2^c = E_1 \Delta E_2;$$
$$(E_1 \Delta F_1)\Delta(E_2 \Delta F_2) = (E_1 \Delta E_2)\Delta(F_1 \Delta F_2).$$

3.8. **Finitely and Countably Additive Measures:** A set function μ satisfying the requirements (i)–(iv) of § 3 of a measure, is also called a countably additive measure. A finite, and finitely additive measure on a σ-algebra $\mathcal{A}$ is a nonnegative set function $\mu : \mathcal{A} \to \mathbb{R}^*$ such $\mu(X) < \infty$, and for every finite collection $\{E_1, \ldots, E_n\}$ of disjoint elements of $\mathcal{A}$

$$\mu(E_1 \cup \cdots \cup E_n) = \mu(E_1) + \cdots + \mu(E_n).$$

Prove that a finite, and finitely additive measure μ is countably additive if and only if for every nondecreasing (nonincreasing) sequence of sets $\{E_n\} \subset \mathcal{A}$,

$$\mu(\lim E_n) = \lim \mu(E_n).$$

3.9. Let μ be a measure on a σ-algebra $\mathcal{A}$ and let $\{E_\alpha\}$ be a collection of disjoint sets in $\mathcal{A}$. Prove that for every measurable set E of positive σ-finite measure, $\mu(E \cap E_\alpha) > 0$ for at most countably many E_α.

3.10. Let $\{X, \mathcal{A}, \mu\}$ be a measure space and let $\{E_n\} \subset \mathcal{A}$ satisfy $\sum \mu(E_n) < \infty$. Prove that the set of points belonging to infinitely many E_n has measure zero (see **2.12**).

3.1c Completion of a Measure Space

If $\{X, \mathcal{A}, \mu\}$ is not complete it can be completed as follows. First define

$$\mathcal{A}_{\text{compl}} = \left\{ \begin{array}{c} \text{the collection of sets of the type } E \cup N \text{ where } E \in \mathcal{A} \\ \text{and } N \text{ is a subset of a set in } \mathcal{A} \text{ of measure zero} \end{array} \right\}.$$

Then set

$$\mathcal{A}_{\text{compl}} \ni E \cup N \to \mu_{\text{compl}}(E \cup N) = \mu(E).$$

The definition does not depend on the choices of E and N identifying the same element $E \cup N \in \mathcal{A}_{\text{compl}}$. Indeed if $E_1 \cup N_1 = E_2 \cup N_2 \in \mathcal{A}_{\text{compl}}$ then $E_1 \subset E_2 \cup N_2'$ and $E_2 \subset E_1 \cup N_1'$, where N_1' and N_2' are subsets of sets in $\mathcal{A}$ of measure zero. Therefore,

$$\mu_{\text{compl}}(E_1 \cup N_1) = \mu_{\text{compl}}(E_2 \cup N_2).$$

Prove that $\mathcal{A}_{\text{compl}}$ is a σ-algebra and μ_{compl} is a complete measure.

4c Outer Measures

4.1. If $\mu_e(N) = 0$ then $\mu_e(E) = \mu_e(E \cup N)$ for every $E \subset X$.

4.2. A finitely additive outer measure is a measure.

4.3. The countable sum of outer measures is an outer measure.

4.4. Construct the Lebesgue–Stieltjes outer measure corresponding to the functions

$$f(x) = e^x, \quad f(x) = \begin{cases} [\![x]\!] & \text{for } x \geq 0, \\ 0 & \text{for } x < 0, \end{cases} \quad f(x) = \begin{cases} 1 & \text{for } x \geq 0, \\ -1 & \text{for } x < 0. \end{cases}$$

4.5. Let Q consist of X, $\emptyset$ and all the singletons in X. Define set functions λ_1 and λ_2, from Q into $\mathbb{R}^*$, by

$$\lambda_1(X) = \infty, \ \lambda_1(\emptyset) = 0, \ \lambda_1(E) = 1, \ \text{for all } E \in Q \text{ other than } X \text{ and } \emptyset;$$
$$\lambda_2(X) = 1, \quad \lambda_2(\emptyset) = 0, \ \lambda_2(E) = 0, \ \text{for all } E \in Q \text{ such that } E \neq X.$$

Each of these is a set function on a sequential covering of X. Describe the outer measures they generate.

4.6. The Lebesgue outer measure on $\mathbb{R}$ coincides with the Lebesgue–Stieltjes outer measure generated by $f(x) = x$.

5c The Hausdorff Outer Measure in $\mathbb{R}^N$

5.1c The Hausdorff Dimension of a Set $E \subset \mathbb{R}^N$

By Propositions 5.1 and 5.2, if $\mu_e(E)$ is finite, then $\mathcal{H}_{N+\eta}(E) = 0$ for all $\eta > 0$. If $E \subset \mathbb{R}^2$ is a segment, then $\mu_e(E) = \mathcal{H}_2(E) = 0$. Moreover,

$$\mathcal{H}_{1+\eta}(E) = 0 \quad \text{for all } \eta > 0 \quad \text{and} \quad \mathcal{H}_1(E) = \{\text{length of } E\}.$$

The Hausdorff dimension of a set $E \subset \mathbb{R}^N$ is defined by

$$\dim_{\mathcal{H}}(E) = \inf \{\alpha \mid \mathcal{H}_\alpha(E) = 0\}. \tag{5.1c}$$

5.1. Let $\{E_n\}$ be a countable collection of sets in $\mathbb{R}^N$ with the same Hausdorff dimension d. Then their union has the same Hausdorff dimension d.

5.2. The Hausdorff dimension of a point in $\mathbb{R}^N$ is zero. The Hausdorff dimension of a countable set in $\mathbb{R}^N$ is zero.

5.2c   The Hausdorff Dimension of the Cantor Set is ln 2/ln 3

For $\delta \in (0, 1)$, let $\mathcal{C}_\delta$ be the generalized Cantor set introduced in § 2.2c of the Complements of Chap. 1. For $n \in \mathbb{N}$, consider the intervals $\{J_{n,j}\}$, $j = 1, 2 \ldots, 2^n$ introduced in the same section. For each $n \in \mathbb{N}$ they form a finite sequence of disjoint closed intervals covering $\mathcal{C}_\delta$ and each of length δ^n. Therefore, by the definition of the outer measure $\mathcal{H}_\alpha$, for $\alpha > 0$,

$$\mathcal{H}_\alpha(\mathcal{C}_\delta) \le \lim_{n\to\infty} \sum_{j=1}^{2^n} (\mathrm{diam} J_{n,j})^\alpha = \lim_{n\to\infty} 2^n \delta^{\alpha n}. \tag{5.2c}$$

Therefore if $\alpha > \log_{1/\delta} 2$, then $\mathcal{H}_\alpha(\mathcal{C}_\delta) = 0$. It follows from (5.1c) that the Hausdorff dimension of $\mathcal{C}_\delta$ does not exceed $\log_{1/\delta} 2$. If $\delta = \frac{1}{3}$ then $\mathcal{C}_\delta$ coincides with the standard Cantor set $\mathcal{C}$. Thus

$$\dim_{\mathcal{H}}(\mathcal{C}) \le \frac{\ln 2}{\ln 3}.$$

To prove the converse inequality we may assume that $\mathcal{H}_\alpha(\mathcal{C}_\delta) < \infty$. Then given $\varepsilon > 0$ there exists a countable collection of sets $\mathcal{I}_m$ of diameter not exceeding ε, whose union covers $\mathcal{C}_\delta$, such that

$$\sum_{m=1}^{\infty} (\mathrm{diam} \mathcal{I}_m)^\alpha \le \mathcal{H}_\alpha(\mathcal{C}_\delta) + \varepsilon.$$

By possibly changing ε into 2ε on the right-hand side, may assume that $\mathcal{I}_m$ are open intervals. Since $\mathcal{C}_\delta$ is compact, may also assume that the collection $\mathcal{I}_m$ is finite. Since $\{I_m\}$ is a finite covering of $\mathcal{C}_\delta$ consisting of open intervals, there exists $n \in \mathbb{N}$ sufficiently large, such that each of the intervals $J_{n,j}$ must be contained in some $\mathcal{I}_m$ (§ 17.1c of Chap. 2).

Lemma 5.1c *For each* $m \in \mathbb{N}$ *let* n_m *be the smallest positive integer such that* $J_{n_m, j} \subset \mathcal{I}_m$ *for some* $j = 1, \ldots, 2^{n_m}$. *Then at most 4 of the intervals* $J_{n_m, j}$ *intersect* $\mathcal{I}_m$, *say for example* J_{n_m, j_ℓ}, *for* $\ell = 1, \ldots, \ell_{m,n}$ *with* $\ell_{n,m} \le 4$.

Proof It follows from the construction of the $J_{n,j}$ in § 2.2c of the Complements of Chap. 1. Indeed if not $J_{n_m-1, j} \subset \mathcal{I}_m$ for some $j = 1, \ldots 2^{n_m-1}$. ∎

Corollary 5.1c *For each* $m \in \mathbb{N}$

$$\mathrm{diam} \mathcal{I}_m \ge \frac{1}{4} \sum_{\ell=1}^{\ell_{m,n}} \mathrm{diam} J_{n_m, j_\ell}.$$

Let $n = \max\{n_m\}$. Then

$$\varepsilon + \mathcal{H}_\alpha(\mathcal{C}_\delta) \geq \sum (\operatorname{diam}\mathcal{I}_m)^\alpha \geq \frac{1}{4^\alpha} \sum_m \sum_{\ell=1}^{\ell_{m,n}} (\operatorname{diam}J_{n_m,\ell})^\alpha$$

$$\geq \frac{1}{4^\alpha} \sum_{j=1}^{2^n} (\operatorname{diam}J_{n,j})^\alpha = \frac{1}{4^\alpha} 2^n \delta^{\alpha n}. \qquad \blacksquare$$

Thus the Hausdorff dimension of $\mathcal{C}_\delta$ is $\ln 2 / \ln 1/\delta$.

8c   The Hausdorff Measure in $\mathbb{R}^N$

8.1c   Hausdorff Outer Measure of the Lipschitz Image of a Set

A function $f : \mathbb{R}^N \to \mathbb{R}^m$ for $N, m \in \mathbb{N}$ is Lipschitz continuous if there exists a constant L such that

$$|f(x_1) - f(x_2)|_m \leq L|x_1 - x_2|_N \quad \text{for all } x_1, x_2 \in \mathbb{R}^N. \tag{8.1c}$$

Here $|\cdot|_m$ and $|\cdot|_N$ denote the Euclidean distance in $\mathbb{R}^m$ and $\mathbb{R}^N$, respectively. The constant L is the Lipschitz constant of f.

Proposition 8.1c *Let $f : \mathbb{R}^N \to \mathbb{R}^m$ be Lipschitz continuous with Lipschitz constant L and let $E \subset \mathbb{R}^N$. Then, for all $0 \leq \alpha < \infty$,*

$$\mathcal{H}_\alpha\big[f(E)\big] \leq L^\alpha \mathcal{H}_\alpha(E). \tag{8.2c}$$

Proof The statement is obvious for $\alpha = 0$. Assuming $\alpha > 0$, fix $\varepsilon > 0$ and let $\{E_n\}$ be a countable collection of subsets of $\mathbb{R}^N$, each of diameter not exceeding ε, and whose union covers E. Then $f(E) \subset \bigcup f(E_n)$ and

$$\operatorname{diam} f(E_n) \leq L \operatorname{diam} E_n \leq L\varepsilon.$$

From this
$$\mathcal{H}_{\alpha, L\varepsilon}\big[f(E)\big] \leq L^\alpha \sum (\operatorname{diam} E_n)^\alpha. \qquad \blacksquare$$

Corollary 8.1c *A Lipschitz function $f : \mathbb{R}^N \to \mathbb{R}^m$ maps sets of α-Hausdorff measure zero, into sets of α-Hausdorff measure zero.*

Corollary 8.2c *For $m \leq N$, let $P_{N,m}$ be the projection of sets in $\mathbb{R}^N$ into $\mathbb{R}^m$. Then for all $E \subset \mathbb{R}^N$ and all $0 \leq \alpha < \infty$*

$$\mathcal{H}_\alpha\big[P_{N,m}(E)\big] \leq \mathcal{H}_\alpha(E). \tag{8.3c}$$

As a consequence

$$\{Hausdorff\ dimension\ of\ P_{N,m}(E)\} \le \{Hausdorff\ dimension\ of\ E\}. \qquad (8.4c)$$

8.2c Hausdorff Dimension of Graphs of Lipschitz Functions

The graph of a function f from a set $E \subset \mathbb{R}^N$ into $\mathbb{R}$ is the set

$$\mathcal{G}_{f;E} = \{(x, f(x))\ \text{for}\ x \in E\} \subset \mathbb{R}^{N+1}. \qquad (8.5c)$$

Proposition 8.2c *Let $f : \mathbb{R}^N \to \mathbb{R}$ be Lipschitz and let $E \subset \mathbb{R}^N$ be Lebesgue measurable and of positive Lebesgue measure. Then the*

$$Hausdorff\ dimension\ of\ \mathcal{G}_{f;E} = N. \qquad (8.6c)$$

Proof By Corollary 8.2c the Hausdorff dimension of $\mathcal{G}_{f;E}$ is no less than N. For the converse inequality, use an argument similar to the proof of Proposition 5.1 to prove that

$$\mathcal{H}_\alpha(\mathcal{G}_{f;E}) = 0 \quad \text{for all}\ \alpha > N. \qquad \blacksquare$$

9c Extending Measures from Semi-algebras to σ-Algebras

Given a nonnegative set function $\lambda : \mathcal{Q} \to \mathbb{R}^*$ on a semi-algebra $\mathcal{Q}$ one might ask under what conditions λ is actually a measure on $\mathcal{Q}$. It turns out that more stringent conditions need to be imposed both on λ and $\mathcal{Q}$. The collection $\mathcal{Q}$ is required to be an *algebra* and λ is required to be regular on $\mathcal{Q}$ in the following sense.

9.1c Inner and Outer Continuity of λ on Some Algebra $\mathcal{Q}$

A nonnegative set function λ on some algebra $\mathcal{Q}$ is *inner continuous*, if for every $Q \in \mathcal{Q}$ and all increasing sequences $\{Q_n\} \subset \mathcal{Q}$ such that $\bigcup Q_n = Q$, there holds $\lim \lambda(Q_n) = \lambda(Q)$.

The set function λ is *outer continuous* if for every $Q \in \mathcal{Q}$ and all decreasing sequences $\{Q_n\} \subset \mathcal{Q}$ such that $\bigcap Q_n = Q$, there holds $\lim \lambda(Q_n) = \lambda(Q)$.

A measure λ on some algebra $\mathcal{Q}$ is both inner and outer continuous. The next proposition asserts that the converse is true, provided λ is finitely additive and finite.

Proposition 9.1c *A nonnegative, finite, finitely additive set function λ on some algebra Q is countably additive on Q, if and only if is either inner or outer continuous.*

Proof Let $\{Q_n\}$ be a disjoint sequence of sets in Q, whose union is $Q \in Q$. If λ is inner continuous

$$\lambda(Q) = \lim \lambda\Big(\bigcup_{j=1}^{n} Q_j \Big) = \lim \sum_{j=1}^{n} \lambda(Q_j) = \sum \lambda(Q_n).$$

If λ is outer continuous

$$\lambda(Q) = \lambda\Big(\bigcup_{j=1}^{n} Q_j \Big) + \lambda\Big(Q - \bigcup_{j=1}^{n} Q_j \Big)$$

$$= \lim \sum_{j=1}^{n} \lambda(Q_j) + \lim \lambda\Big(Q - \bigcup_{j=1}^{n} Q_j \Big) \qquad \blacksquare$$

Corollary 9.1c *A nonnegative, finite, finitely additive set function λ on some algebra Q is countably additive on Q, if and only if for every decreasing sequence $\{Q_n\} \subset Q$ with empty intersection, $\lim \lambda(Q_n) = 0$.*

10c   More on Extensions from Semi-algebras to σ-Algebras

10.1c   Self-extensions of Measures

Let $\{X, \mathcal{A}, \mu\}$ be a measure space, not necessarily generated by an outer measure. Using $\mathcal{A}$ as a sequential covering of X and $\mu : \mathcal{A} \to \mathbb{R}^*$ as a set function, one may construct an outer measure μ_{ee}, by the procedure indicated in (4.1), and a measure $\mu\mu$ defined on a σ-algebra $\mathcal{A}_{\mu\mu}$.

The measure $\mu\mu$ is a self-extension of μ.

11.1.   Prove that if μ is σ-finite, then $\mu\mu$ is the completion of μ. In particular, if $\{X, \mathcal{A}, \mu\}$ is σ-finite and complete then $\mu = \mu\mu$.

11.2.   Let $\{X, \mathcal{A}_{\text{compl}}, \mu_{\text{compl}}\}$ denote the completion of $\{X, \mathcal{A}, \mu\}$ (§ 3.1c). Prove that $\mathcal{A}_{\mu\mu} = \mathcal{A}_{\text{compl;loc}}$ and $\mu\mu = \mu_{\text{compl;loc}}$ (see **3.5–3.6** of § 3c of the Complements).

11.3.   Assume $\mu(X) < \infty$ and let $E \subset X$ be such that $\mu_{ee}(E) = \mu(X)$. Prove that if $A, B \in \mathcal{A}$ satisfy $A \cap E = B \cap E$, then $\mu(A) = \mu(B)$. Define

$$\{A \cap E : A \in \mathcal{A}\} = \mathcal{A}_E \ni A \cap E \to \mu_E(A \cap E) = \mu(A).$$

Prove that $\mathcal{A}_E$ is a σ-algebra on E and that μ_E is a well defined measure on $\mathcal{A}_E$.

In the problems below assume that $\{X, \mathcal{A}, \mu\}$ itself is generated by an outer measure μ_e. By construction $\mu_e \le \mu_{ee}$.

11.4. Prove that if $\{X, \mathcal{A}, \mu\}$ is generated by a measure λ on a semi-algebra Q then $\mu_{ee} = \mu_e$. Moreover, $\mathcal{A}_{\mu\mu} = \mathcal{A}_{\text{loc}}$ and $\mu\mu = \mu_{\text{loc}}$.

11.5. Prove that $\mu_{ee}(E) = \mu_e(E)$ if and only if there exists a μ-measurable set $A \supset E$ such that $\mu(A) = \mu_e(E)$.

11.6. Give an example of $\{X, \mathcal{A}, \mu\}$ and $E \subset X$ for which $\mu_e(E) < \mu_{ee}(E)$.

An example can be constructed by the following steps. First divide all subsets of $\mathbb{R}$ into three classes

$$(2^{\mathbb{R}})_o = \{\text{the collection of all countable subsets of } \mathbb{R}\}$$

$$(2^{\mathbb{R}})_1 = \{\text{the collection of all } uncountable \text{ subsets of } \mathbb{R}$$
$$\text{with } uncountable \text{ complement}\}$$

$$(2^{\mathbb{R}})_2 = \{\text{the collection of all } uncountable \text{ subsets of } \mathbb{R}$$
$$\text{with } countable \text{ complement}\}.$$

Then define a set function $\mu_e : 2^{\mathbb{R}} \to \mathbb{R}^+$, by

$$2^{\mathbb{R}} \ni E \to \mu_e(E) = \begin{cases} 0 \text{ if } E \in (2^{\mathbb{R}})_o \\ 1 \text{ if } E \in (2^{\mathbb{R}})_1 \\ 2 \text{ if } E \in (2^{\mathbb{R}})_2 \end{cases}$$

and verify that μ_e is an outer measure on $2^{\mathbb{R}}$. Let $\{\mathbb{R}, \mathcal{A}, \mu\}$ be the measure space generated by μ_e. Since μ_e is not finitely additive $\mathcal{A}$ is not expected to coincide with $2^{\mathbb{R}}$ (Proposition 9.1).

Lemma 10.1c *Sets in $(2^{\mathbb{R}})_o$ and in $(2^{\mathbb{R}})_2$ are measurable, whereas sets in $(2^{\mathbb{R}})_1$ are not measurable.*

Proof The first two statements are established by the Carathéodory measurability condition (6.2). If $E \in (2^{\mathbb{R}})_1$ then E^c is uncountable and it can be separated into the disjoint union of two uncountable sets $B, C \in (2^{\mathbb{R}})_1$. In the Carathéodory condition (6.2) take $A = E \cup B \in (2^{\mathbb{R}})_1$. Then

$$\mu_e(A) = 1, \quad \mu_e(A \cap E) = \mu_e(E) = 1, \quad \mu_e(A - E) = \mu(B) = 1. \qquad \blacksquare$$

If $E \in (2^{\mathbb{R}})_2$, for any $\varepsilon > 0$, there is no measurable set $A \supset E$ such that

$$1 = \mu_e(E) \geq \mu_e(A) - \varepsilon, \qquad \text{since } \mu_e(A) \text{ is either 0 or 2.}$$

10.2c *Nonunique Extensions of Measures λ on Semi-algebras*

11.7. Let $X = \mathbb{Q} \cap [0, 1]$ and let Q be the semi-algebra of the finite unions of sets of the type $(a, b] \cap X$ where a, b are real numbers and $0 \leq a < b \leq 1$. The set function

$$\lambda\{(a,b]\cap X\} = \infty \quad \text{and} \quad \lambda(\emptyset) = 0$$

is a non σ-finite measure on $\mathcal{Q}$. Prove that $\mathcal{Q}_o = 2^{\mathcal{Q}}$ and verify that

$$\mu_1 = \{\text{the counting measure on } \mathcal{Q}_o\}$$
$$\mu_2 = \{\mu_2(E) = \infty \text{ for every } E \in \mathcal{Q}_o \text{ and } \mu(\emptyset) = 0\}$$

are extensions of λ to $2^{\mathcal{Q}}$.

11.8. Let $\{X, \mathcal{A}, \mu\}$ be generated by a finite measure λ on a semi-algebra $\mathcal{Q}$. Define the *inner measure* of a set $E \subset X$ by

$$\mu_i(E) = \mu(X) - \mu_e(X - E).$$

Prove that E is measurable if and only if $\mu_e(E) = \mu_i(E)$.

12c The Lebesgue Measure of Sets in $\mathbb{R}^N$

In this section "measure" means the Lebesgue measure in $\mathbb{R}^N$ and μ_e is the outer measure form which the Lebesgue measure is constructed.

12.1. The Lebesgue measure of a polyhedron coincides with its Euclidean measure.

12.2. The Lebesgue outer measure and measure are translation invariant.

12.3. The interval $[0, 1]$ is not countable for otherwise it would have measure zero. The Cantor set provides an example of a measurable uncountable set of measure zero.

12.4. There exist unbounded sets with finite measure.

12.5. Let $0 < k < N$ be an integer. A $\mathbb{R}^{N-k}$ hyperplane in $\mathbb{R}^N$ has measure zero.

12.6. The boundary of a ball in $\mathbb{R}^N$ has measure zero. However, there exist open sets in $\mathbb{R}^N$ whose boundary has positive measure. For example, the complement of the generalized Cantor set. Such a set also provides an example of an open set $E \subset [0, 1]$ which is dense in $[0, 1]$, its measure is less than 1 and for every interval $I \subset [0, 1]$, the measure of $I \cap E$ is positive.

12.7. The set of the rational numbers has Lebesgue measure zero and its boundary has infinite measure.

12.8. Let E be a bounded measurable set in $\mathbb{R}^N$. There exists a set E_δ of the type of $\mathcal{G}_\delta$, containing E and such that

$$\mu_e(A \cap E) = \mu_e(A \cap E_\delta) \quad \text{for all} \quad A \subset \mathbb{R}^N.$$

This is false if E is not measurable.

12.9. Let $E \in \mathcal{M}$ be of positive measure. For every $\varepsilon \in (0, 1)$ there exists a cube Q_ε such that $\mu(E \cap Q_\varepsilon) > \varepsilon\mu(Q_\varepsilon)$.

Proof May assume that $\mu(E) < \infty$. For every $\varepsilon \in (0, 1)$ there exists an open set $E_o \supset E$ such that $\varepsilon\mu(E_o) < \mu(E)$. Since E_o is open it is the countable union of dyadic, disjoint cubes $\{Q_n\}$. Therefore,

$$\varepsilon\mu(E_o) = \varepsilon\mu\left(\bigcup Q_n\right) = \varepsilon\sum\mu(Q_n)$$
$$< \mu(E) = \mu\left(\bigcup E \cap Q_n\right) = \sum\mu(E \cap Q_n).$$

Hence $\mu(E \cap Q_n) > \varepsilon\mu(Q_n)$ for some n. ∎

12.10. Let $E \subset \mathbb{R}$ be a bounded, Lebesgue measurable set of positive measure. Prove that the set
$$E - E = \{x - y; x, y, \in E\}$$

contains an interval about the origin.

Proof Having fixed $\varepsilon \in (0, 1)0$, let $I = (a, b)$ be an open interval for which $\mu(E \cap I) > \varepsilon\mu(I)$. Such an interval exists by **12.9**. Let $\alpha \in (0, 1)$ and consider the interval

$$\left(-\alpha\mu(I), \alpha\mu(I)\right) = J.$$

The numbers $\alpha, \varepsilon \in (0, 1)$ can be chosen so that

$$(I \cap E) \cap [(I \cap E) + x] \neq \emptyset \quad \text{for all fixed } x \in J. \tag{*}$$

This would establish the claim. Indeed, for each $x \in J$ there exist $y, z \in (I \cap E)$ such that $y = z + x$, i.e., $x = y - z$ for some $y, z \in E$. To prove (*) proceed by contradiction. By construction

$$(I \cap E) \cup [(I \cap E) + x] \subset \left(a, b + \alpha\mu(I)\right).$$

If (*) is violated

$$2\varepsilon\mu(I) < 2\mu(I \cap E) = \mu(I \cap E) + \mu[(I \cap E) + x] \leq (1 + \alpha)\mu(I).$$

Choose $\alpha = \frac{1}{2}$ nd $\varepsilon = \frac{3}{4}$. ∎

12.11. For $\varepsilon > 0$ and $x_n \in \mathbb{Q} \cap [0, 1]$ construct the open interval $I_{\varepsilon,n}$, about x_n of length $2^{-n}\varepsilon$, and set $E_\varepsilon = \bigcup I_{\varepsilon,n}$ and $E = \bigcap E_{1/m}$. Prove that $E_\varepsilon \subset [0, 1]$ with strict inclusion, for all $\varepsilon \in (0, 1)$.

12.12 Let $E \subset [0, 1]$ be Lebesgue measurable and of Lebesgue measure zero. Prove that the set $E^2 = \{x^2 : x \in E\}$ is measurable and has measure zero.

12.13. The Lebesgue measure on $\mathbb{R}$ coincides with the Lebesgue–Stieltjes measure generated by $f(x) = x$.

12.1c Inner Measure and Measurability

Let μ_e be the Lebesgue outer measure in $\mathbb{R}^N$. Define the Lebesgue inner measure $\mu_i(E)$ of a bounded set $E \subset \mathbb{R}^N$ as

$$\mu_i(E) = \sup\{\mu_e(C) \mid \text{where } C \text{ is closed and } C \subset E\}. \tag{12.1c}$$

If E is of finite inner measure, for every $\varepsilon > 0$ there exists a closed set $E_{c,\varepsilon}$ such that $E_{c,\varepsilon} \subset E$ and $\mu_i(E) \le \mu(E_{c,\varepsilon}) + \varepsilon$.

Proposition 12.1c *A bounded set $E \subset \mathbb{R}^N$ is Lebesgue measurable if and only if $\mu_i(E) = \mu_e(E)$.*

12.2c The Peano–Jordan Measure of Bounded Sets in $\mathbb{R}^N$

Let E be a bounded set in $\mathbb{R}^N$ and construct the two classes of sets

$$\mathcal{O}_E = \left\{ \begin{array}{l} \text{sets that are the finite union of open} \\ \text{cubes in } \mathbb{R}^N \text{ and that contain } E \end{array} \right\}$$

$$\mathcal{I}_E = \left\{ \begin{array}{l} \text{sets that are the finite union of closed} \\ \text{cubes in } \mathbb{R}^N \text{ and that are contained in } E \end{array} \right\}.$$

The Peano–Jordan outer and inner measure of E are defined as

$$\mu^e_{\mathcal{P}-\mathcal{J}}(E) = \inf_{O \in \mathcal{O}_E} \mu(O), \qquad \mu^i_{\mathcal{P}-\mathcal{J}}(E) = \sup_{I \in \mathcal{I}_E} \mu(I). \tag{12.2c}$$

A bounded set $E \subset \mathbb{R}^N$ is Peano–Jordan measurable if its Peano–Jordan outer and inner measures coincide. From the definition of Lebesgue outer and inner measure it follows that

$$\mu^i_{\mathcal{P}-\mathcal{J}}(E) \le \mu_i(E) \le \mu_e(E) \le \mu^e_{\mathcal{P}-\mathcal{J}}(E).$$

Thus a Peano–Jordan measurable set is Lebesgue measurable. The converse is false. The set $\mathbb{Q} \cap [0, 1]$ is Lebesgue measurable and its measure is zero. However,

$$\mu^e_{\mathcal{P}-\mathcal{J}}(\mathbb{Q} \cap [0, 1]) = 1, \qquad \mu^i_{\mathcal{P}-\mathcal{J}}(\mathbb{Q} \cap [0, 1]) = 0.$$

Thus $\mathbb{Q} \cap [0, 1]$ is not Peano–Jordan measurable. This last example shows that the Peano–Jordan measure is not a measure in the sense of § 3, since its domain is not a σ-algebra.

12.3c   Lipschitz Functions and Measurability

A continuous function from $\mathbb{R}^N$ into itself need not preserve measurability as shown by the continuous function $f : [0, 1] \to [0, 1]$ constructed in § 14. It is then natural to ask what properties one may require on $f : \mathbb{R}^N \to \mathbb{R}^N$, to map Lebesgue measurable set in $\mathbb{R}^N$ into Lebesgue measurable sets in $\mathbb{R}^N$. The next proposition asserts that this occurs if f is Lipschitz continuous in the sense of (8.1c).

Proposition 12.2c *A Lipschitz continuous function $f : \mathbb{R}^N \to \mathbb{R}^N$, maps Lebesgue measurable sets in $\mathbb{R}^N$ into Lebesgue measurable sets in $\mathbb{R}^N$.*

Prove the proposition by the following steps:

i. f maps compact sets of $\mathbb{R}^N$ into compact sets in $\mathbb{R}^N$. Hence it maps bounded, closed sets into bounded closed sets.

ii. f maps countable unions of closed dyadic cubes into measurable sets. Hence f maps open sets into measurable sets.

iii. f maps sets of measure zero in $\mathbb{R}^N$ into sets of measure zero of $\mathbb{R}^N$. See Corollary 8.1c.

iv.   If $E \subset \mathbb{R}^n$ is bounded and measurable, there exists a set $E_\sigma \subset E$ of the type of $\mathcal{F}_\sigma$, and a measurable set $\mathcal{E} \subset E$ of measure zero, such that $E = E_\sigma \cup \mathcal{E}$. Then $f(E) = f(E_\sigma) \cup f(\mathcal{E})$.

v.   If E is unbounded write

$$E = \bigcup E \cap \{ j \le |x| < j + 1 \}.$$

Remark 12.1c Let f be a Lipschitz map from $\mathbb{R}^N$ into $\mathbb{R}^m$ for some $N, m \in \mathbb{N}$. The conclusion of the proposition is false, in general, if $N \ne m$. Give counterexamples and identify which of the previous points fails.

12.3.1c   Linear Maps, Measurability and Volumes

Let T be a linear bijection in $\mathbb{R}^N$ (i.e., a linear nonsingular transformation of $\mathbb{R}^N$ onto itself). Such a map is Lipschitz continuous and therefore it maps Lebesgue measurable sets into Lebesgue measurable sets.

Proposition 12.3c *Let $T : \mathbb{R}^N \to \mathbb{R}^N$ linear and nonsingular. Then for every Lebesgue measurable set $E \subset \mathbb{R}^N$,*

$$\mu(TE) = |\det T| \mu(E). \tag{12.3c}$$

Prove the proposition by the following steps:

i. T maps parallelepipeds in $\mathbb{R}^N$ into parallelepipeds in $\mathbb{R}^N$.

ii.   Let P be a parallelepiped in $\mathbb{R}^N$ with edges identified by the vectors $\mathbf{x}_i$ for $i = 1, \ldots, N$. Then

$$\mu(P) = |\det(x_{ij})|, \quad \text{and} \quad \mu(TP) = |\det T|\mu(P)$$

where (x_{ij}) is the matrix whose rows are the vectors $\mathbf{x}_i$.

iii. If $E \subset \mathbb{R}^N$ is open, then (12.3c) holds. An open set is the countable disjoint union of dyadic cubes.

iv. If $E \subset \mathbb{R}^N$ is bounded and measurable, for every $\varepsilon > 0$ there exists a countable collection $\{Q_n\}$ of disjoint dyadic cubes such that

$$E \subset \bigcup Q_n \quad \text{and} \quad \sum \mu(Q_n) \le \mu(E) + \varepsilon.$$

Now $TE \subset \bigcup TQ_n$ and hence

$$\mu(TE) \le |\det T|\mu(E) + |\det T|\varepsilon.$$

v. If $E \subset \mathbb{R}^N$ is bounded and measurable, for every $\varepsilon > 0$ there exists an open set E_o such that
$$E \subset E_o \quad \text{and} \quad \mu(E_o - E) \le \varepsilon.$$

Now $TE \subset TE_o$ and

$$|\det T|\mu(E) \le |\det T|\mu(E_o) = \mu(TE_o)$$
$$= \mu(TE) + \mu[T(E_o - E)] \le \mu(TE) + |\det T|\varepsilon.$$

vi. If E is measurable and unbounded, write it as the disjoint union of bounded measurable sets.

Thus in particular the Lebesgue measure is invariant by rotations and translations.

13c Vitali's Nonmeasurable Set

13.1. Every measurable subset A of the nonmeasurable set E has measure zero. Indeed the sets $A_n = A \bullet r_n \subset E_n$ for $r_n \in \mathbb{Q} \cap [0, 1]$ are disjoint, have each measure equal to the measure of A, and their union is contained in $[0, 1]$. Thus

$$\sum \mu(A_n) = \mu\left(\bigcup A_n\right) \le 1.$$

13.2. Every set $A \subset [0, 1]$ of positive outer measure, contains a nonmeasurable set. Indeed at least one of the intersections $A \cap E_n$ is nonmeasurable. If all such intersections were measurable then by the subadditive property of the outer measure,

$$0 < \mu_e(A) \le \sum \mu(A \cap E_n) = 0$$

since all the $A \cap E_n$ are measurable subsets of E.

13.3. Let $E \subset [0, 1]$ be the Vitali nonmeasurable set, and let $\mu_e(\cdot)$ be the Lebesgue outer measure. Prove that $\mu_e(E) > 0$. Moreover for all $\varepsilon > 0$, there exists a nonneasurable set $E_\varepsilon \subset [0, 1]$, such that $0\mu_e(E_\varepsilon) < \varepsilon$.

13.4. The Lebesgue measure on $[0, 1]$ is a set function defined of the Lebesgue measurable subsets of $[0, 1]$ satisfying:

(a). μ is nonnegative
(b). μ is countably additive
(c). μ is translation invariant
(d). If $I \subset [0, 1]$ is an interval then $\mu(I)$ coincides with the Euclidean measure if I.

It is impossible to define a set function μ satisfying (a)–(d) and defined on all subsets of $[0, 1]$.

The construction of the nonmeasurable set $E \subset [0, 1]$ uses only the properties (a)–(c) of the function μ and it is independent of the particular construction of the Lebesgue measure. The final contradiction argument uses property (d). If a function μ satisfying (a)–(d) and defined in all the subsets of $[0, 1]$ were to exist, the same construction would imply that $\mu([0, 1])$ is either zero or infinity. The requirement (d) cannot be removed from these remarks. Indeed the counting measure or the identically zero measure would satisfy (a)–(c) but not (d).

14c Borel Sets, Measurable Sets and Incomplete Measures

The strict inclusion $\mathcal{B} \subset \mathcal{M}$ can be established by an indirect cardinality argument.

14.1. Let $\mathcal{B}_o$ denote the collection of all open intervals of $[0, 1]$. Prove that card $(\mathcal{B}_o) \leq$ card $(\mathbb{R})$.

14.2. Define inductively $\mathcal{B}_n$ for all $n \in \mathbb{N}$, as the collection of all countable unions, countable intersections and complements of elements of $\mathcal{B}_{n-1}$. Prove that card $(\mathcal{B}_n) \leq$ card $(\mathbb{R}^\mathbb{N}) =$ card $(\mathbb{R})$.

14.3. Let Ω be the first uncountable. For each $\alpha < \Omega$ define $\mathcal{B}_\alpha$ as the collection of all countable unions, countable intersection and complements of elements of $\mathcal{B}_\beta$ for all ordinal numbers $\beta < \alpha$. By definition of first uncountable, the cardinality of $\mathcal{B}_\alpha$ does not exceed card $(\mathbb{R}^\mathbb{N}) =$ card $(\mathbb{R})$.

14.4. The smallest σ-algebra containing the open sets of $[0, 1]$ can be constructed by this procedure by setting $\mathcal{B} = \bigcup_{\alpha < \Omega} \mathcal{B}_\alpha$. Hence the cardinality of $\mathcal{B}$ does not exceed card $(\mathbb{R})$.

14.5. Since the Lebesgue measure is complete, every subset of the Cantor set is measurable and has measure zero. Since the cardinality of the Cantor set is the cardinality of $\mathbb{R}$, the cardinality of all the subset of the Cantor sets is card $(2^\mathbb{R})$.

14.6. The cardinality of the Lebesgue measurable subsets of $[0, 1]$ is not less than card $(2^\mathbb{R})$, and the cardinality of the Borel subsets of $[0, 1]$ does not exceed card $(\mathbb{R})$.

16c Borel, Regular and Radon Measures

16.1c Regular Borel Measures

16.1. Prove that in the Proposition 15.1, the closed sets $E_{c,\varepsilon}$ can be taken to be bounded.

16.2. prove that the counting measure on $\mathbb{R}$ is not outer regular, but it is inner regular.

16.3. prove that the Hausdorff measure in $\mathbb{R}^N$ generated by the outer measure $\mathcal{H}_{N-\varepsilon}$, for $0 < \varepsilon < N$ is inner regular and not outer regular.

16.2c Regular Outer Measures

Let μ be a Borel measure in $\mathbb{R}^N$, and let μ_e be an outer measure in $\mathbb{R}^N$ that coincides with μ on the Borel sets. The outer measure μ_e is *outer regular* with respect to μ if for every set $E \subset \mathbb{R}^N$ of finite outer measure, there exists a set E_δ of the type of a $\mathcal{G}_\delta$ such that $E \subset E_\delta$ and $\mu_e(E) = \mu(E_\delta)$. An example of such μ_e is the one generated from a Radon measure μ by formula (16.3).

16.4. Prove that if μ_e is generated by a nonnegative set function λ on the collection of open sets, then μ_e is outer regular with respect to μ. In particular, the Lebesgue–Stieltjes outer measure $\mu_{f,e}$ is outer regular with respect to the Lebesgue–Stieltjes measure μ_f.

16.5. Prove that the Hausdorff outer measure $\mathcal{H}_\alpha$ for $\alpha > 0$, is outer regular with respect to the Hausdorff measure it generates.

Let $E \subset \mathbb{R}^N$ be of finite $\mathcal{H}_\alpha$ outer measure. For $m \in \mathbb{N}$ fixed, There exists a countable collection of sets $\{E_{m,n}\}$, each of diameter less than $\frac{1}{m}$, whose union contains E and satisfying

$$\mathcal{H}_{\alpha,\frac{1}{m}}(E) \geq \sum_n \operatorname{diam}(E_{m,n})^\alpha - \tfrac{1}{m}.$$

The sets

$$\mathcal{O}_{m,n} = \{x \in \mathbb{R}^N \mid \operatorname{dist}\{x; E_{m,n}\} < \tfrac{1}{m}\operatorname{diam}(E_{m,n})\}$$

are open and satisfy

$$\left(1 + \tfrac{2}{m}\right)^\alpha \mathcal{H}_\alpha(E) \geq \sum \operatorname{diam}(\mathcal{O}_{m,n})^\alpha - \tfrac{1}{m} \geq \mathcal{H}_{\alpha,\varepsilon_m}(E_\delta) - \tfrac{1}{m}$$

where

$$E_\delta = \bigcap_m \bigcup_n \mathcal{O}_{m,n}, \quad \text{and} \quad \varepsilon_m = \left(1 + \tfrac{1}{m}\right)\tfrac{1}{m}.$$

17c Vitali Coverings

17.1.   The notion of Vitali's covering of a set $E \subset \mathbb{R}^N$ is independent of the measurability of E. Let μ_e be the outer measure associated to the Lebesgue measure in $\mathbb{R}^N$ and generated by (16.1). Extend the Vitali covering theorem to the case of a set E of finite outer measure $\mu_e(E)$.

17.2.   Let E be a bounded subset of $\mathbb{R}^N$ of finite outer measure which admits a fine Vitali covering with cubes contained in E. Then E is Lebesgue measurable. The requirement that the cubes making up the Vitali covering, be contained in E is essential. Give a counterexample.

17.3.   Let $\{Q_\alpha\}$ be an uncountable family of closed, nontrivial cubes in $\mathbb{R}^N$. Then $\bigcup Q_\alpha$ is Lebesgue measurable.

17.1c Pointwise and Measure-Theoretical Vitali Coverings

It has been remarked that the collection $\{Q_n\}$ claimed by Theorem 17.1 covers E only in a measure-theoretical sense. Here we indicate how the covering can be realized to be pointwise.

Proposition 17.1c *Let $\mathcal{F}$ be a family of closed, nontrivial cubes $Q(d) \subset \mathbb{R}^N$ of uniformly bounded diameter d, i.e.,*

$$\sup\{d \mid Q(d) \in \mathcal{F}\} = D < \infty$$

There exists a countable collection of disjoint cubes $\{Q_n(d_n)\} \subset \mathcal{F}$ such that

$$\bigcup\{Q(d) \in \mathcal{F}\} \subset \bigcup Q_n(3d_n). \tag{17.1c}$$

Proof For $j = 1, 2, \ldots$ define

$$\mathcal{F}_j = \left\{ Q(d) \in \mathcal{F} \mid \frac{D}{2^j} < d \le \frac{D}{2^{j-1}} \right\}.$$

Select $\{Q_{n_1}(d_{n_1})\}$ as any maximal collection of disjoint cubes in $\mathcal{F}_1$. Then introduce the collection of cubes

$$\left\{ Q(d) \in \mathcal{F}_2 \mid Q(d) \cap \bigcup Q_{n_1}(d_{n_1}) = \emptyset \right\}$$

and out of this collection select any maximal subcollection of disjoint cubes $\{Q_{n_2}(d_{n_2})\}$. Proceeding in this fashion, select recursively $\{Q_{n_j}(d_{n_j})\}$ as any maximal subcollection of disjoint cubes out of

$$\{Q(d) \in \mathcal{F}_j \mid Q(d) \cap \bigcup_{k=1}^{j-1}\{Q_{n_k}(d_{n_k})\} = \emptyset\}$$

The countable, disjoint collection claimed by the proposition is

$$\{Q_n(d_n)\} = \bigcup_{j\in\mathbb{N}}\{Q_{n_j}(d_{n_j})\}$$

A cube $Q(d) \in \mathcal{F}$ belongs to some $\mathcal{F}_j$. By maximality of $\{Q_{n_j}(d_{n_j})\}$ there exists $Q_j(d_j) \in \{Q_{n_j}(d_{n_j})\}$, such that $Q(d) \cap Q_j(d_j) \neq \emptyset$. By construction

$$\frac{D}{2^j} < \mathrm{diam}\, Q(d) \leq \frac{D}{2^{j-1}} \quad \text{and} \quad \frac{D}{2^j} < \mathrm{diam}\, Q_j(d_j).$$

Therefore, $d < 2d_j$.   ∎

Proposition 17.2c *Let $\mathcal{F}$ be a fine Vitali covering of a set $E \subset \mathbb{R}^N$. There exists a countable collection of cubes $\{Q_n(d_n)\} \subset \mathcal{F}$ such that for any finite collection $\{Q_1, \ldots, Q_m\} \subset \mathcal{F}$*

$$E - \bigcup_{\ell=1}^m Q_\ell \subset \bigcup_{Q_n(d_n)\in\{Q_n(d_n)\}-\{Q_1,\ldots,Q_m\}} Q_n(3d_n).$$

Proof Since $\mathcal{F}$ is a fine Vitali covering for E, may assume, without loss of generality, that the diameters of the cubes in $\mathcal{F}$ are uniformly bounded. Then select $\{Q_n(d_n)\}$ as in the previous proposition and fix the finite collection $\{Q_1, \ldots, Q_m\} \subset \mathcal{F}$. If E is covered by the union of the Q_ℓ there is nothing to prove. Otherwise take

$$x \in E - \bigcup_{\ell=1}^m Q_\ell$$

and select a cube $Q_x \in \mathcal{F}$ containing x and not intersecting any of the Q_ℓ. Such a cube exists since $\mathcal{F}$ is a fine Vitali covering and the cubes are closed. By the previous proof there exists $Q_j(d_j) \in \{Q_n(d_n)\}$ such that $Q_x \subset Q_j(3d_j)$.   ∎

18c The Besicovitch Covering Theorem

18.1c *The Besicovitch Theorem for Unbounded E*

The boundedness of E insures that, for the balls $B_{\rho_n}(x_n)$, the radii $\{\rho_n\} \to 0$ as $n \to \infty$. This information only enters in the proof in a qualitative way. The number c_N, claimed by the theorem, is independent of $\mathcal{F}$, the number R in (18.1), the set E and its boundedness. These remarks permit one to extend the Besicovitch covering theorem to the case of E unbounded. For $n \in \mathbb{N}$ set

$$E_n = \{2R(n-1) \le |x| < 2Rn\}$$

and apply the Besicovitch theorem to each of the E_n to obtain finite collections of countable collections of disjoint balls $\{\mathcal{B}_1^n, \ldots, \mathcal{B}_{c_N}^n\}$. By construction, the balls of each of the collections $\mathcal{B}_j^3$, do not intersect any of the balls of $\mathcal{B}_j^1$. Hence they can be incorporated into $\mathcal{B}_j^1$ to form a larger subcollection of disjoint balls. Likewise, the balls of each of the collections $\mathcal{B}_j^5$ do not intersect any of the balls of $\mathcal{B}_j^3$ and thy can be incorporated into $\mathcal{B}_j^1 \cup \mathcal{B}_j^3$ to form a larger subcollection of disjoint balls. Proceeding by induction set

$$\mathcal{B}_j^{\text{odd}} = \bigcup_{n \text{ odd}} \mathcal{B}_j^n \qquad \text{for } j = 1, \ldots, c_N.$$

Each of $\mathcal{B}_j^{\text{odd}}$ is a countable collection of disjoint balls. A similar argument holds for the collections $\mathcal{B}_j^n$ of even index n. Set

$$\mathcal{B}_j^{\text{even}} = \bigcup_{n \text{ even}} \mathcal{B}_j^n \qquad \text{for } j = 1, \ldots, c_N.$$

The finite collection of countable collections of disjoint balls claimed by the theorem can be taken to be

$$\mathcal{B}_1^{\text{even}}, \ldots, \mathcal{B}_{c_N}^{\text{even}}, \mathcal{B}_1^{\text{odd}}, \ldots, \mathcal{B}_{c_N}^{\text{odd}}.$$

Thus the theorem continues to hold with c_N replaced by $2c_N$.

18.2c The Besicovitch Measure-Theoretical Inner Covering of Open Sets $E \subset \mathbb{R}^N$

Proposition 18.1c *Let $E \subset \mathbb{R}^N$ be open. There exists a countable collection of disjoint, closed balls $\{B_n\}$, contained in E, such that $\mu(E - \cup E_n) = 0$.*

18.3c A Simpler Form of the Besicovitch Theorem

The next is a covering statement, based only on the geometry of cubes in $\mathbb{R}^N$ and independent of measures.

Theorem 18.1c (Besicovitch) *Let E be a bounded subset of $\mathbb{R}^N$ and let $\mathcal{F}$ be a collection of cubes in $\mathbb{R}^N$ with faces parallel to the coordinate planes and such that each $x \in E$ is the center of a nontrivial cube $Q(x)$ belonging to $\mathcal{F}$.*

There exists a countable collection $\{x_n\}$ of points $x_n \in E$, and a corresponding collection of cubes $\{Q(x_n)\}$ in $\mathcal{F}$ such that

$$E \subset \bigcup Q(x_n) \quad \text{and} \quad \sum \chi_{Q(x_n)} \le 4^N. \tag{18.1c}$$

Remark 18.1c The second of (18.1c) asserts that each point $x \in \mathbb{R}^N$ is covered by at most 4^N cubes out of $\{Q(x_n)\}$. Equivalently at most 4^N cubes overlap at each given point.

It is remarkable that the largest number of possible overlaps of the cubes $Q(x_n)$, at each given point, is independent of the set E and the covering $\mathcal{F}$, and depends only on the geometry of the cubes in $\mathbb{R}^N$.

Lemma 18.1c *Let $\{Q(x_n)\}$ be a countable collection of cubes in $\mathbb{R}^N$ with centers at x_n and satisfying*

$$\text{If } n < m \text{ then } x_m \notin Q(x_n) \text{ and } \mu(Q(x_m)) \le 2\mu(Q(x_n)). \tag{18.2c}$$

Then each point $x \in \mathbb{R}^N$ is covered by at most 4^N cubes out of $\{Q(x_n)\}$.

Proof Assume first $N = 2$. Having fixed $x \in \mathbb{R}^2$ may assume up to a translation that x is the origin. Denote by $2\rho_n$ the edge of the cube $Q(x_n)$ and let $\{Q_j\}$ be the collection of squares containing the origin, whose center x_j is in the first quadrant, and of edge $2\rho_j$. Starting from Q_1 and the corresponding edge $2\rho_1$ consider the 4 closed squares,

$$S_1 = [0, \rho_1] \times [0, \rho_1] \qquad S_2 = [0, \rho_1] \times [\rho_1, 2\rho_1]$$
$$S_3 = [\rho_1, 2\rho_1] \times [\rho_1, 2\rho_1] \qquad S_4 = [\rho_1, 2\rho_1] \times [0, \rho_1].$$

Let also
$$S_o = [0, 2\rho_1] \times [0, 2\rho_1] = S_1 \cup S_2 \cup S_3 \cup S_4.$$

By construction Q_1 covers S_1 and by the second of (18.2c) the center x_j of each of the Q_j, for $j = 2, 3, \ldots$, cannot lie outside S_o. Indeed if it did, since Q_j contains the origin, $\rho_j > 2\rho_1$ and $\mu(Q_j) > 4\mu(Q_1)$.

Thus the centers x_j of the Q_j, for $j = 2, 3, \ldots$, must belong to some of the squares S_1, S_2, S_3, S_4. Now x_j cannot belong to S_1 because otherwise the first of (18.2c) would be violated, since $S_1 \subset Q_1$.

If some $x_j \in S_2$, then since Q_j contains the origin, $S_2 \subset Q_j$. Therefore, by the first of (18.2c), no other center x_ℓ, for $\ell > j$, belongs to S_2. This implies that S_2 contains at most one center x_j.

By a similar argument S_3 and S_4 contain each at most one center x_j of a cube Q_j. Thus the collection $\{Q_j\}$ contains at most 4 cubes.

Defining analogously the collections of cubes containing the origin and whose center is, respectively, in the second, third and fourth quadrant, we conclude that each of these collections contains at most 4 cubes. Thus the origin is covered by at most 16 cubes out of the collection $\{Q(x_n)\}$.

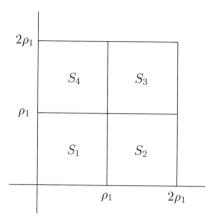

Fig. 1c Constructing the sequence of cubes $Q(x_n)$

A similar argument for $N = 3$ gives that each point is covered by at most 64 cubes. The general case follows by induction (Fig. 1c). ∎

Proof (of Theorem 18.1c) Set

$$E_1 = E \quad \text{and} \quad \lambda_1 = \sup \{\mu(Q(x)) \mid x \in E\}.$$

If $\lambda_1 = \infty$ there exists cubes $Q(x) \in \mathcal{F}$ of arbitrarily large edge and centered at some $x \in E$. Since E is bounded we select one such a cube. If $\lambda_1 < \infty$ select $x_1 \in E_1$ and a cube $Q(x_1)$ such that $\mu(Q(x_1)) > \frac{1}{2}\lambda_1$. Then set

$$E_2 = E_1 - Q(x_1) \quad \text{and} \quad \lambda_2 = \sup \{\mu(Q(x)) \mid x \in E_2\}.$$

If $E_1 \subset Q(x_1)$ the process terminates. Otherwise $\lambda_2 > 0$ and we select $x_2 \in E_2$ and a cube $Q(x_2)$ such that $\mu(Q(x_2)) > \frac{1}{2}\lambda_2$.

Proceeding recursively, define countable collections of sets E_n, points $x_n \in E_n$, corresponding cubes $Q(x_n)$ and positive numbers λ_n by

$$E_n = E - \bigcup_{j=1}^{n-1} Q(x_j) \quad \lambda_n = \sup \{\mu(Q(x)) \mid x \in E_n\}, \quad \mu(Q(x_n)) \geq \frac{1}{2}\lambda_n.$$

By construction $\{\lambda_n\}$ is a decreasing sequence and

$$\mu(Q(x_m)) \leq \lambda_m \leq \lambda_n \leq 2\mu(Q(x_n)) \quad \text{for } n < m$$

as long as $\lambda_m > 0$. Therefore by the previous lemma, at most 4^N of the cubes $\{Q(x_n)\}$ overlap at each $x \in \mathbb{R}^N$. It remains to prove that $E \subset \bigcup Q(x_n)$.

If $E_n = \emptyset$ for some n, then, $E \subset \bigcup_{j=1}^{n-1} Q(x_j)$ and the process terminates. If $E_n \neq \emptyset$ for all n, we claim that $\lim \lambda_n = 0$. To this end compute

$$\limsup \mu(Q(x_n)) = \delta.$$

Let x_o be a cluster point of $\{x_n\}$. If $\delta > 0$, then x_o would be covered by infinitely many cubes $Q(x_n)$. Therefore, $\delta = 0$. The relations

$$\tfrac{1}{2}\lambda_n \leq \mu(Q(x_n)) \leq \lambda_n, \qquad \lambda_{n+1} \leq \lambda_n$$

now imply that $\lim \lambda_n = 0$. If $x \in E - \bigcup Q(x_n)$, there exists a nontrivial cube $Q(x) \in \mathcal{F}$ such that $\mu(Q(x)) < \lambda_n$ for all n. Therefore, $\mu(Q(x)) = 0$ and $Q(x)$ is a trivial cube.   ∎

18.4c   Another Besicovitch-Type Covering

Proposition 18.2c *Let E be a bounded subset of $\mathbb{R}^N$ and let $x \to \rho(x)$ be a function from E into $(0,1)$.*[4] *There exists a countable collection $\{x_n\}$ of points in E, such that the closed balls $B(x_n, \rho(x_n))$, with center at x_n and radius $\rho(x_n)$, are pairwise disjoint, and*

$$E \subset \bigcup B(x_n, 3\rho(x_n)). \tag{18.3c}$$

Proof Let $\mathcal{F}_1$ be the collection of pairwise disjoint balls $B(x, \rho(x))$ such that $\tfrac{1}{2} \leq \rho(x) < 1$. Since E is bounded, $\mathcal{F}_1$ contains at most a finite number of such balls, say for example,

$$B(x_1, \rho(x_1)), \ B(x_2, \rho(x_2)), \ldots, B(x_{n_1}, \rho(x_{n_1})) \tag{18.4c}_1$$

for some $n_1 \in \mathbb{N}$. If their union covers E, the construction terminates. If not, let $\mathcal{F}_2$ be the collection of pairwise disjoint balls $B(x, \rho(x))$, not intersecting any of the balls selected in $(18.4c)_1$, and such that $2^{-2} \leq \rho(x) < 2^{-1}$. Since E is bounded, $\mathcal{F}_2$, if not empty, contains at most a finite number of such balls, say for example

$$B(x_{n_1+1}, \rho(x_{n_1+1})), \ B(x_{n_1+2}, \rho(x_{n_1+2})), \ldots, B(x_{n_2}, \rho(x_{n_2}))$$

for some $n_2 \in \mathbb{N}$. If the union of the balls

$$B(x_i, \rho(x_i)) \quad \text{for} \quad i = 1, \ldots, n_1, n_1 + 1, \ldots, n_2 \tag{18.4c}_2$$

covers E, the construction terminates. If $\mathcal{F}_2$ is empty, or if the union of these balls does not cover E, construct the collection $\mathcal{F}_3$ of pairwise disjoint balls $B(x, \rho(x))$,

---

[4]Neither E nor $x \to \rho(x)$ are required to be measurable.

not intersecting any of balls selected in $(18.4c)_2$, and such that $2^{-3} \le \rho(x) < 2^{-2}$. Proceeding recursively, assume we have selected a finite collection of pairwise disjoint balls

$$B\big(x_1, \rho(x_1)\big), \; B\big(x_2, \rho(x_2)\big), \; \ldots, \; B\big(x_{n_j}, \rho(x_{n_j})\big) \qquad (18.4c)_j$$

for some $n_j \in \mathbb{N}$. If their union does not cover E, construct the family $\mathcal{F}_{j+1}$ of pairwise disjoint balls $B\big(x, \rho(x)\big)$, not intersecting any of balls selected in $(18.4c)_j$, and such that $2^{-(j+1)} \le \rho(x) < 2^{-j}$. Since E is bounded, $\mathcal{F}_{j+1}$, if not empty, contains at most a finite number of such balls, which we add to the collection in $(18.4c)_j$. If for some $j \in \mathbb{N}$, the union of the balls in $(18.4c)_j$ covers E the process terminates. Otherwise this recursive procedure generates a countable collection of pairwise disjoint balls $\{B\big(x_n, \rho(x_n)\big)\}$. It remains to prove that such a collection satisfies (18.3c). Fix $x \in E$. By construction

$$B\big(x, \rho(x)\big) \cap B\big(x_j, \rho(x_j)\big) \neq \emptyset \quad \text{for some } j.$$

Therefore, $\rho(x) \le 2\rho(x_j)$. For one such x_j fixed

$$|x - x_j| \le \rho(x) + \rho(x_j) \le 3\rho(x_j).$$

Thus $x \in B\big(x_j, 3\rho(x_j)\big)$.   ∎

Chapter 4
The Lebesgue Integral

1 Measurable Functions

Let $\{X, \mathcal{A}, \mu\}$ be a measure space and $E \in \mathcal{A}$. For a function $f : E \to \mathbb{R}^*$ and $c \in \mathbb{R}$ set

$$[f > c] = \{x \in E \mid f(x) > c\}.$$

The sets $[f \geq c], [f < c], [f \leq c]$, are defined similarly, and are linked by

$$[f \geq c] = \bigcap \left[f > c - \frac{1}{n}\right] \quad [f > c] = \bigcup \left[f \geq c + \frac{1}{n}\right]$$

$$[f \leq c] = E - [f > c] \qquad [f < c] = E - [f \geq c].$$

Therefore, if any one of the four sets

$$[f > c], \qquad [f \geq c], \qquad [f < c], \qquad [f \leq c] \tag{1.1}$$

is measurable for all $c \in \mathbb{R}$, the remaining three are measurable for all $c \in \mathbb{R}$. A function $f : E \to \mathbb{R}^*$ is *measurable* if at least one of the sets in (1.1) is measurable for all $c \in \mathbb{R}$.

Remark 1.1 The notion of measurable function depends only on the σ-algebra $\mathcal{A}$ and is independent of the measure μ defined on $\mathcal{A}$.

Proposition 1.1 *A function $f : E \to \mathbb{R}^*$ is measurable if and only if at least one of the sets in (1.1) is measurable for all $c \in \mathbb{Q}$.*

Proof Assume for example that the third of the sets in (1.1) is measurable for all $c \in \mathbb{Q}$. Having fixed some $c \in \mathbb{R} - \mathbb{Q}$ let $\{q_n\}$ be a sequence of rational numbers decreasing to c. Then $[f > c] = \bigcup [f \geq q_n]$ is measurable. ∎

© Springer Science+Business Media New York 2016
E. DiBenedetto, *Real Analysis*, Birkhäuser Advanced
Texts Basler Lehrbücher, DOI 10.1007/978-1-4939-4005-9_4

Proposition 1.2 *Let* $f : E \to \mathbb{R}^*$ *be measurable and let* $\alpha \in \mathbb{R} - \{0\}$. *Then*

(i)   The functions $|f|$, αf, $\alpha + f$, f^2, *are measurable*
(ii) If $f \neq 0$ *then also* $1/f$ *is measurable*
(iii) For any measurable subset $E' \subset E$, *the restriction* $f \big|_{E'}$ *is measurable.*

Proof The statements (i) and (ii) follow from the set identities:

$$[|f| > c] = \begin{cases} [f > c] \cup [f < -c] & \text{if } c \geq 0 \\ \\ E & \text{if } c < 0 \end{cases}$$

$$[\alpha + f > c] = [f > c - \alpha]$$

$$[\alpha f > c] = \begin{cases} \left[f > \dfrac{c}{\alpha}\right] & \text{if } \alpha > 0 \\ \\ \left[f < \dfrac{c}{\alpha}\right] & \text{if } \alpha < 0 \end{cases}$$

$$[f^2 > c] = \begin{cases} \left[f > \sqrt{c}\right] \cup \left[f < -\sqrt{c}\right] & \text{if } c \geq 0 \\ \\ E & \text{if } c < 0 \end{cases}$$

$$\left[\frac{1}{f} > c\right] = \begin{cases} [f > 0] \cap \left[f < \dfrac{1}{c}\right] & \text{if } c > 0 \\ \\ [f > 0] & \text{if } c = 0 \\ \\ [f > 0] \cup \left[f < \dfrac{1}{c}\right] & \text{if } c < 0. \end{cases}$$

To prove (iii) observe that $\left[f \big|_{E'} > c\right] = [f > c] \cap E'$.   ∎

Proposition 1.3 *Let* $f : E \to \mathbb{R}^*$ *and* $g : E \to \mathbb{R}$ *be measurable. Then*

(i)   the set $[f > g]$ *is measurable*
(ii) the functions $f \pm g$ *are measurable*
(iii) the function fg *is measurable*
(iv) if $g \neq 0$, *the function* f/g *is measurable.*

Proof Let $\{q_n\}$ be the sequence of the rational numbers. Then

$$[f > g] = \bigcup [f \geq q_n] \cap [g < q_n].$$

This proves (i). To prove (ii) observe that for all $c \in \mathbb{R}$

$$[f \pm g > c] = [f > \mp g + c]$$

and the latter is measurable in view of (i). By the previous proposition $(f \pm g)^2$ are measurable and this implies (iii) in view of the identity

$$4fg = (f + g)^2 - (f - g)^2.$$

Finally (iv) follows from (iii) and (ii) of the previous proposition. ∎

Proposition 1.4 *Let* $\{f_n\}$ *be a sequence of measurable functions in* E. *Then the functions*

$$\varphi = \sup f_n, \quad \psi = \inf f_n, \quad f'' = \limsup f_n, \quad f' = \liminf f_n$$

are measurable.

Proof For every $c \in \mathbb{R}$,

$$[\varphi > c] = \bigcup [f_n > c] \quad \text{and} \quad [\psi \ge c] = \bigcap [f_n \ge c]. \qquad ∎$$

Let f and g be two functions defined in E. We say that $f = g$ almost everywhere (a.e.) in E if there exists a set $\mathcal{E} \subset E$ of measure zero such that

$$f(x) = g(x) \quad \text{for all } x \in E - \mathcal{E} \quad \text{and} \quad \mu(\mathcal{E}) = 0.$$

More generally, a property of real-valued functions defined on a measure space $\{X, \mathcal{A}, \mu\}$ is said to hold *almost everywhere* (a.e.), if it does hold for all $x \in X$ except possibly for a measurable set $\mathcal{E}$ of measure zero.

Lemma 1.1 *Let* $\{X, \mathcal{A}, \mu\}$ *be complete. If* f *is measurable and* $f = g$ *a.e. in* E, *then also* g *is measurable.*

Corollary 1.1 *Let* $\{X, \mathcal{A}, \mu\}$ *be complete and let* $\{f_n\}$ *be a sequence of measurable functions defined in* $E \in \mathcal{A}$ *and taking values in* $\mathbb{R}^*$. *A function* $f : E \to \mathbb{R}^*$ *such that* $f = \lim f_n$ *a.e. in* E *is measurable.*

2 The Egorov–Severini Theorem [39, 145]

Let $\{f_n\}$ be a sequence of functions defined in a measurable set E with values in $\mathbb{R}^*$ and set

$$f''(x) = \limsup f_n(x) \qquad f'(x) = \liminf f_n(x) \qquad x \in E. \qquad (2.1)$$

The functions f'' and f' are defined in E and take values in $\mathbb{R}^*$. We will assume throughout that they are a.e. finite in E, i.e., there exist measurable sets $\mathcal{E}''$ and $\mathcal{E}'$ contained in E such that

$$
\begin{aligned}
f''(x) \in \mathbb{R} \quad &\text{for all } x \in E - \mathcal{E}'' \quad \text{and} \quad \mu(\mathcal{E}'') = 0 \\
f'(x) \in \mathbb{R} \quad &\text{for all } x \in E - \mathcal{E}' \quad \text{and} \quad \mu(\mathcal{E}') = 0.
\end{aligned}
\tag{2.2}
$$

The upper limit in (2.1) is uniform, if for every $\varepsilon > 0$ there exists an index n_ε such that

$$
f_n(x) < f''(x) + \varepsilon \quad \text{for all } n \geq n_\varepsilon \quad \text{and for all } x \in E - \mathcal{E}''.
$$

Similarly the lower limit in (2.1) is uniform, if for every $\varepsilon > 0$, there exists an index n_ε such that

$$
f_n(x) \geq f'(x) - \varepsilon \quad \text{for all } n \geq n_\varepsilon \quad \text{and for all } x \in E - \mathcal{E}'.
$$

Proposition 2.1 *Let* $\{f_n\}$ *be a sequence of measurable functions defined on a measurable set* $E \in \mathcal{A}$ *of finite measure, and with values in* $\mathbb{R}^*$.

Assume that, for example, the second(first) of (2.2) holds. Then for every $\eta > 0$ *there exists a measurable set* $E_\eta \subset E$, *such that* $\mu(E - E_\eta) \leq \eta$ *and the lower(upper) limit in (2.1) is uniform in* E_η.

Proof The statement is only proved for the lower limit, the arguments for the upper limit being analogous. Fix $m, n \in \mathbb{N}$, and introduce the sets

$$
E_{m,n} = \bigcap_{j=n}^{\infty} \left\{ x \in (E - \mathcal{E}') \mid f_j(x) \geq f'(x) - \frac{1}{m} \right\}.
$$

For $m \in \mathbb{N}$ fixed, the sets $E_{m,n}$ are measurable and expanding. By the definitions of f' and $\mathcal{E}'$

$$
E - \mathcal{E}' = \bigcup_{n=1}^{\infty} E_{m,n} \quad \text{and} \quad \mu(E) = \lim_n \mu(E_{m,n}).
$$

Therefore, having fixed $\eta > 0$, there exists an index $n(m, \eta)$ such that

$$
\mu(E - E_{m,n(m,\eta)}) \leq \frac{1}{2^m} \eta.
$$

The set claimed by the proposition is

$$
E_\eta = \bigcap_{m=1}^{\infty} E_{m,n(m,\eta)}.
\tag{2.3}
$$

Indeed E_η is measurable and by construction

$$\mu(E - E_\eta) = \mu\Big(\bigcup_{m=1}^{\infty} (E - E_{m,n(m,\eta)})\Big) \leq \sum_{m=1}^{\infty} \mu(E - E_{m,n(m,\eta)}) \leq \eta.$$

Fix an arbitrary $\varepsilon > 0$ and let m_ε be the smallest positive integer such that $\varepsilon m_\varepsilon \geq 1$. From the inclusion $E_\eta \subset E_{m_\varepsilon, n(m_\varepsilon, \eta)}$ and the definition of the sets $E_{m,n}$, it follows that

$$f_n(x) \geq f'(x) - \varepsilon \quad \text{for all } n \geq n_\varepsilon \equiv n(m_\varepsilon, \eta) \quad \text{and for all } x \in E_\eta. \qquad \blacksquare$$

Theorem 2.1 (Egorov-Severini [39, 145]) *Let $\{f_n\}$ be a sequence of measurable functions defined in a measurable set E of finite measure, and with values in $\mathbb{R}^*$. Assume that the sequence converges a.e., in E, to a function $f : E \rightarrow \mathbb{R}^*$, which is finite a.e. in E. Then, for every $\eta > 0$ there exists a measurable set $E_\eta \subset E$, such that $\mu(E - E_\eta) \leq \eta$ and the limit is uniform in E_η.*

Remark 2.1 The Egorov–Severini theorem is in general false if E is not of finite measure (**2.2** of the Complements).

2.1 The Egorov–Severini Theorem in $\mathbb{R}^N$

Proposition 2.2 *Let $\{X, \mathcal{A}, \mu\}$ be $\mathbb{R}^N$ endowed with a inner regular Borel measure μ (in the sense of (16.2) of Chap. 2). Let $\{f_n\}$ be a sequence of μ measurable functions defined on a μ-measurable set $E \subset \mathbb{R}^N$, of finite measure, and with values in $\mathbb{R}^*$. Assume that, for example, the second of (2.2) holds. Then for every $\eta > 0$ there exists a closed set $E_{c,\eta} \subset E$, such that $\mu(E - E_{c,\eta}) \leq \eta$ and the lower limit in (2.1) is uniform in $E_{c,\eta}$.*

Proof The set E_η in (2.3) is a μ-measurable subset of $\mathbb{R}^N$ of finite measure. Since μ is inner regular, there exists a closed set $E_{c,\eta} \subset E_\eta$ such that $\mu(E_\eta - E_{c,\eta}) \leq \eta$. $\blacksquare$

The proposition holds in particular if μ is the Lebesgue measure in $\mathbb{R}^N$, since the latter is inner regular (Proposition 12.3 of Chap. 3).

3 Approximating Measurable Functions by Simple Functions

A function f from a measurable set E with values in $\mathbb{R}$ is *simple* if it is measurable and if it takes a finite number of values. The characteristic function of a measurable set is simple. Let f be simple in E, let $\{f_1, \dots, f_n\}$ be the distinct values taken by f in E and set

$$E_i = \{x \in E \mid f(x) = f_i\}. \tag{3.1}$$

The sets E_i are measurable and disjoint, and f can be written in its *canonical form*

$$f = \sum_{i=1}^{n} f_i \chi_{E_i}. \tag{3.2}$$

Given measurable sets $E_1, \ldots, E_n$ and real numbers $f_1, \ldots, f_n$, the expression in (3.2) defines a simple function, but not in general its canonical form. This occurs only if the E_i are disjoint and the f_i are distinct.

The sum and the product of simple functions are simple functions.

If f and g are simple functions written in their canonical form (3.2), then $f + g$ and fg are simple but not necessarily in their canonical form.

Proposition 3.1 *Let* $f : E \to \mathbb{R}^*$ *be a nonnegative measurable function. There exists a sequence of simple functions* $\{f_n\}$, *such that* $f_n \le f_{n+1}$ *and*

$$f(x) = \lim f_n(x) \quad \text{for all } x \in E.$$

Proof For a fixed $n \in \mathbb{N}$ set

$$f_n(x) = \begin{cases} n & \text{if } f(x) \ge n \\ \dfrac{j}{2^n} & \text{if } \dfrac{j}{2^n} \le f(x) < \dfrac{j+1}{2^n} \\ & \text{for } j = 0, 1, \ldots, n2^n - 1. \end{cases} \tag{3.3}$$

By construction $f_n \le f_{n+1}$. Since f is measurable, the sets

$$\left[f \ge \frac{j}{2^n} \right] - \left[f \ge \frac{j+1}{2^n} \right], \quad j = 0, 1, \ldots, n2^n - 1$$

and the set $[f \ge n]$, are measurable and disjoint. Thus the f_n are simple.

Fix some $x \in E$. If $f(x)$ is finite, by choosing $n_o \ge f(x)$,

$$0 \le f(x) - f_n(x) \le \frac{1}{2^n} \quad \text{for all } n \ge n_o.$$

If $f(x) = \infty$, then $f_n(x) = n$ for all positive integers n and the conclusion holds in either case. ∎

A function f from a set E into $\mathbb{R}^*$ can be decomposed as

$$f = f^+ - f^- \quad \text{where} \quad f^{\pm} = \tfrac{1}{2}(|f| \pm f). \tag{3.4}$$

Corollary 3.1 *Let $f : E \to \mathbb{R}^*$ be a measurable function. There exists a sequence of simple functions $\{f_n\}$, such that $f(x) = \lim f_n(x)$ for all $x \in E$.*

4 Convergence in Measure (Riesz [125], Fisher [46])

Let $\{f_n\}$ be a sequence of measurable functions from a measurable set E of finite measure, into $\mathbb{R}^*$ and let $f : E \to \mathbb{R}^*$ be measurable and a.e. finite in E. The sequence $\{f_n\}$ converges *in measure* to f, if for any $\eta > 0$

$$\lim \mu \{x \in E \mid |f_n(x) - f(x)| > \eta\} = 0.$$

Proposition 4.1 *The convergence in measure identifies the limit uniquely up to a set of measure zero, i.e., if $\{f_n\}$ converges in measure to f and g, then $f = g$ a.e. in E.*

Proof For $n \in \mathbb{N}$ and a.e. $x \in E$,

$$|f(x) - g(x)| \leq |f(x) - f_n(x)| + |f_n(x) - g(x)|.$$

Therefore, for all $\eta > 0$

$$[|f - g| > \eta] \subset \left[|f - f_n| > \tfrac{1}{2}\eta\right] \cup \left[|f_n - g| > \tfrac{1}{2}\eta\right].$$

Take the measure of both sides and let $n \to \infty$. ∎

Proposition 4.2 *Let $\{X, \mathcal{A}, \mu\}$ be a measure space and let $E \in \mathcal{A}$ be of finite measure. If $\{f_n\}$ converges a.e. in E to a function $f : E \to \mathbb{R}^*$ which is finite a.e. in E, then $\{f_n\}$ converges to f in measure.*

Proof Having fixed an arbitrary $\varepsilon > 0$, by the Egorov–Severini theorem, there exists a measurable set $E_\varepsilon \subset E$, such that $\mu(E - E_\varepsilon) \leq \varepsilon$ and $\{f_n\}$ converges to f uniformly in E_ε. Therefore, for any $\eta > 0$,

$$\lim \sup \mu\{x \in E \mid |f_n(x) - f(x)| > \eta\} \leq \varepsilon. \qquad \blacksquare$$

Remark 4.1 The proposition is in general false if E is not of finite measure.

Remark 4.2 Convergence in measure does not imply a.e. convergence as shown by the following example. For $m, n \in \mathbb{N}$ and $m \leq n$, let

$$\varphi_{nm}(x) = \begin{cases} 1 & \text{for } x \in [\dfrac{m-1}{n}, \dfrac{m}{n}] \\ 0 & \text{for } x \in [0, 1] - [\dfrac{m-1}{n}, \dfrac{m}{n}]. \end{cases} \qquad (4.1)$$

Then construct a sequence of functions $f_n : [0, 1] \to \mathbb{R}$ by setting

$$f_1 = \varphi_{11}, \; f_2 = \varphi_{21}, \; f_3 = \varphi_{22}, \; f_4 = \varphi_{31}, \; f_5 = \varphi_{32},$$
$$f_6 = \varphi_{33}, \; f_7 = \varphi_{41}, \; f_8 = \varphi_{42}, \; f_9 = \varphi_{43}, \quad \cdots \cdots .$$

The sequence $\{f_n\}$ converges in measure to zero in $[0, 1]$. However $\{f_n\}$ does not converge to zero anywhere in $[0, 1]$. Indeed for any fixed $x \in [0, 1]$ there exist infinitely many indices $m, n \in \mathbb{N}$ such that

$$\frac{m-1}{n} \leq x \leq \frac{m}{n} \qquad \text{and hence} \qquad \varphi_{nm}(x) = 1.$$

Even though the sequence $\{f_n\}$ does not converge to zero anywhere in $[0, 1]$, it contains a subsequence $\{f_{n'}\} \subset \{f_n\}$ converging to zero a.e. in $[0, 1]$. For example one might select $\{f_{n'}\} = \{\varphi_{n1}\}$.

Proposition 4.3 (Riesz [126]) *Let $\{X, \mathcal{A}, \mu\}$ be a measure space and let $E \in \mathcal{A}$. Let $\{f_n\}$ and f be measurable functions from E into $\mathbb{R}^*$, a.e. finite in E. If $\{f_n\}$ converges in measure to f, there exists a subsequence $\{f_{n'}\} \subset \{f_n\}$ converging to f a.e. in E.*

Proof For $m, n \in \mathbb{N}$, arguing as in Proposition 4.1

$$\mu\left([|f_n - f_m| > \eta]\right) \leq \mu\left(\left[|f_n - f| > \tfrac{1}{2}\eta\right]\right) + \mu\left(\left[|f_m - f| > \tfrac{1}{2}\eta\right]\right).$$

From this and the definition of convergence in measure

$$\lim_{n,m \to \infty} \mu\left([|f_n - f_m| > \eta]\right) = 0. \tag{4.2}$$

We will establish the proposition under the assumption (4.2). It implies that for every $j \in \mathbb{N}$, there exists a positive integer n_j such that

$$\mu\left(\left[|f_n - f_m| > \frac{1}{2^j}\right]\right) \leq \frac{1}{2^{j+1}} \qquad \text{for all } m, n \geq n_j.$$

The numbers n_j may be chosen so that $n_j < n_{j+1}$. Setting

$$E_{n_j} = \left[|f_{n_j} - f_{n_{j+1}}| \leq \frac{1}{2^j}\right]$$

one estimates

$$\mu\left(E - E_{n_j}\right) \leq \frac{1}{2^{j+1}} \qquad \text{and} \qquad \mu\left(E - \bigcap_{j \geq k} E_{n_j}\right) \leq \frac{1}{2^k}$$

for every fixed, positive integer k. The subsequence $\{f_{n_j}\}$ selected out of $\{f_n\}$, is convergent for all $x \in \bigcap_{j=k}^{\infty} E_{n_j}$. Indeed for any such x, and any pair of indices

$k \leq n_j < n_{j+\ell}$

$$|f_{n_j}(x) - f_{n_{j+\ell}}(x)| \leq \sum_{i=j}^{j+\ell-1} |f_{n_i}(x) - f_{n_{i+1}}(x)| \leq \frac{1}{2^{j-1}}.$$

Since $k \in \mathbb{N}$ is arbitrary, $\{f_{n_j}\}$ converges a.e. in E. ∎

The next proposition can be regarded as a Cauchy-type criterion for a sequence $\{f_n\}$ to converge in measure.

Proposition 4.4 *Let $\{X, \mathcal{A}, \mu\}$ be a measure space and let $E \in \mathcal{A}$ be of finite measure. Let $\{f_n\}$ be a sequence of measurable functions from a measurable set E of finite measure, into $\mathbb{R}^*$. The sequence $\{f_n\}$ converges in measure if and only if (4.2) holds for all $\eta > 0$.*

Proof The necessary condition has been established in the first part of the proof of Proposition 4.3, leading to (4.2). To prove its sufficiency, let (4.2) hold for all $\eta > 0$ and let $\{f_{n'}\}$ be a subsequence, selected out of $\{f_n\}$ and convergent a.e. in E. The limit $f = \lim f_{n'}$, a.e. in E, defines a measurable function $f : E \to \mathbb{R}^*$, which is finite a.e. in E. Having fixed positive numbers η and ε, by virtue of (4.2), there exists an index n_ε such that

$$\mu\left(\left[|f_n - f_m| > \tfrac{1}{2}\eta\right]\right) \leq \tfrac{1}{2}\varepsilon \qquad \text{for all } n, m \geq n_\varepsilon.$$

Since $\{f_{n'}\} \to f$ a.e. in E, by the Egorov–Severini theorem, there exists a measurable set $E_\varepsilon \subset E$, and an index n'_ε, such that $\mu(E - E_\varepsilon) \leq \tfrac{1}{2}\varepsilon$, and

$$|f_{n'}(x) - f(x)| \leq \tfrac{1}{2}\eta \quad \text{for all } x \in E_\varepsilon \quad \text{and for all } n' \geq n'_\varepsilon.$$

From this and the inclusion

$$[|f_n - f| > \eta] \subset \left[|f_n - f_{n'}| > \tfrac{1}{2}\eta\right] \bigcup \left[|f_{n'} - f| > \tfrac{1}{2}\eta\right]$$

it follows

$$\mu\left([|f_n - f| > \eta]\right) \leq \varepsilon \quad \text{for all} \quad n \geq \max\{n_\varepsilon; n'_\varepsilon\}. \qquad ∎$$

5 Quasicontinuous Functions and Lusin's Theorem

Let μ be a Borel, regular measure in $\mathbb{R}^N$, in the sense of (16.1)–(16.2) of Chap. 3. Let $E \subset \mathbb{R}^N$ be measurable and of finite measure. A function $f : E \to \mathbb{R}^*$ is *quasi-continuous* if for every $\varepsilon > 0$ there exists a closed set $E_{c,\varepsilon} \subset E$, such that

$$\mu(E - E_{c,\varepsilon}) \leq \varepsilon \quad \text{and the restriction of } f \text{ to } E_{c,\varepsilon} \text{ is continuous.} \tag{5.1}$$

Proposition 5.1 *A simple function defined in a measurable set $E \subset \mathbb{R}^N$ of finite measure is quasi-continuous.*

Proof Let $f : E \to \mathbb{R}$ be simple and let $\{f_1, \ldots, f_n\}$ be its range. Since the sets E_i, defined in (3.1) are measurable, having fixed $\varepsilon > 0$, there exists closed sets $E_{c,i} \subset E_i$ such that

$$\mu(E_i - E_{c,i}) \leq \frac{1}{n}\varepsilon \quad \text{for } i = 1, \ldots, n.$$

The set $E_{c,\varepsilon} = \bigcup_{i=1}^{n} E_{c,i}$ is closed and $\mu(E - E_{c,\varepsilon}) \leq \varepsilon$. The sets $E_{c,i}$ being closed and disjoint, are at positive mutual distance. Since f is constant on each of them, it is continuous in $E_{c,\varepsilon}$. ∎

Theorem 5.1 (Lusin [99]) *Let E be a μ-measurable set in $\mathbb{R}^N$ of finite measure. A function $f : E \to \mathbb{R}$ is measurable if and only if it is quasi-continuous.*

Proof (Necessity) Assume first that E is bounded. By Proposition 3.1 there exists a sequence of simple functions $\{f_n\}$ that converges to f pointwise in E. Since each of the f_n is quasi-continuous, having fixed $\varepsilon > 0$ there exist closed sets $E_{c,n} \subset E$ such that

$$\mu(E - E_{c,n}) \leq \frac{1}{2^{n+1}}\varepsilon$$

and the restriction of f_n to $E_{c,n}$ is continuous. By the Egorov–Severini theorem in $\mathbb{R}^N$, there exists a closed set $E_{c,o} \subset E$ such that

$$\mu(E - E_{c,o}) \leq \tfrac{1}{2}\varepsilon \quad \text{and } f_n \text{ converges uniformly to } f \text{ in } E_{c,o}.$$

The set $E_{c,\varepsilon} = \cap E_{c,n}$ is closed and

$$\mu(E - E_{c,\varepsilon}) = \mu\left(E - \cap E_{c,n}\right)$$
$$= \mu\left[\cup(E - E_{c,n})\right] \leq \sum_{n=0}^{\infty} \mu(E - E_{c,n}) \leq \varepsilon.$$

Since the functions f_n are continuous in $E_{c,\varepsilon}$ and converge to f uniformly in $E_{c,\varepsilon}$, also f is continuous in E_ε.

If E is unbounded, having fixed $\varepsilon > 0$ there exists $n \in \mathbb{N}$ such that

$$\mu(E \cap [|x| > n]) \leq \tfrac{1}{2}\varepsilon.$$

The set $E \cap [|x| \leq n]$ is bounded and there exists a closed set

$$E_{c,\varepsilon} \subset E \cap [|x| \leq n] \quad \text{such that} \quad \mu(E \cap [|x| \leq n] - E_{c,\varepsilon}) \leq \tfrac{1}{2}\varepsilon$$

and the restriction of f to $E_{c,\varepsilon}$ is continuous. The set $E_{c,\varepsilon}$ is closed, is contained in E and it satisfies (5.1). ∎

Proof (Sufficiency) Let f be quasi-continuous in E. Having fixed $\varepsilon > 0$, let $E_{c,\varepsilon}$ be a closed set satisfying (5.1). To show that $[f \geq c]$ is measurable write

$$[f \geq c] = ([f \geq c] \cap E_{c,\varepsilon}) \cup ([f \geq c] \cap (E - E_{c,\varepsilon})).$$

Since the restriction of f to $E_{c,\varepsilon}$ is continuous, $[f \geq c] \cap E_{c,\varepsilon}$ is closed. Moreover

$$\mu_e([f \geq c] \cap (E - E_{c,\varepsilon})) \leq \mu(E - E_{c,\varepsilon}) \leq \varepsilon.$$

Therefore, $[f \geq c]$ is measurable by Proposition 12.3 of Chap. 3. ∎

6 Integral of Simple Functions ([87])

Let $\{X, \mathcal{A}, \mu\}$ be a measure space and let $E \in \mathcal{A}$. For a measurable set A and $\alpha \in \mathbb{R}$ define,

$$\int_E \alpha\chi_A d\mu = \begin{cases} \alpha\mu(E \cap A) & \text{if } \alpha \neq 0 \\ 0 & \text{if } \alpha = 0. \end{cases}$$

Since $\mu(E \cap A) \in \mathbb{R}^*$, the first of these is well defined, as an element of $\mathbb{R}^*$, for all $\alpha \in \mathbb{R} - \{0\}$. Let $f : E \to \mathbb{R}$ be a nonnegative simple function, with canonical representation

$$f = \sum_{i=1}^{n} f_i \chi_{E_i}, \tag{6.1}$$

where $\{E_1, \ldots, E_n\}$ is a finite collection of mutually disjoint measurable sets exhausting E and $\{f_1, \ldots, f_n\}$ is a finite collection of mutually distinct, nonnegative numbers. The Lebesgue integral of a nonnegative simple function f is defined by

$$\int_E f d\mu = \sum_{i=1}^{n} \int_E f_i \chi_{E_i} d\mu = \sum_{i=1}^{n} f_i \mu(E_i). \tag{6.2}$$

This could be finite or infinite. If it is finite, f is said to be *integrable* in E.

Remark 6.1 If $f : E \to \mathbb{R}^*$ is nonnegative, simple and integrable, the set $[f > 0]$ has finite measure.

Remark 6.2 The integral of a nonnegative simple function is independent of the representation of f. In particular if the representation in (6.1) is not canonical, the integral of f is still given by (6.2).

Let $f, g : E \to \mathbb{R}$ be nonnegative simple functions. Then

$$f \geq g \quad \text{a.e. in } E \quad \Longrightarrow \quad \int_E f d\mu \geq \int_E g d\mu.$$

If f and g are both nonnegative, simple and integrable

$$\int_E (\alpha f + \beta g) d\mu = \alpha \int_E f d\mu + \beta \int_E g d\mu \qquad \text{for all } \alpha, \beta \in \mathbb{R}^+.$$

7 The Lebesgue Integral of Nonnegative Functions

Let $f : E \to \mathbb{R}^*$ be measurable and nonnegative and let $\mathcal{S}_f$ denote the collection of all nonnegative simple functions $\zeta : E \to \mathbb{R}$ such that $\zeta \leq f$. Since $\zeta \equiv 0$ is one such function, the class $\mathcal{S}_f$ is not empty.

The Lebesgue integral of f over E is defined by

$$\int_E f d\mu = \sup_{\zeta \in \mathcal{S}_f} \int_E \zeta d\mu. \qquad (7.1)$$

This could be finite or infinite. The elements $\zeta \in \mathcal{S}_f$ are not required to vanish outside a set of finite measure. For example if f is a positive constant on a measurable set of infinite measure, its integral is well defined by (7.1) and is infinity. The key new idea of this notion of integral is that the *range* of a nonnegative function f is partioned, as opposed to its *domain*, as in the notion of Riemann integral (see also § 7.2c of the Complements).

If $f : E \to \mathbb{R}^*$ is measurable and nonpositive, define

$$\int_E f d\mu = - \int_E (-f) d\mu \qquad (f \leq 0). \qquad (7.2)$$

A nonnegative measurable function $f : E \to \mathbb{R}^*$ is said to be *integrable* if (7.1) defines a finite number. As an example, if μ is the counting measure on $\mathbb{N}$, a nonnegative function $f : \mathbb{N} \to \mathbb{R}$ is integrable if and only if $\sum f(n) < \infty$.

If $f, g : E \to \mathbb{R}^*$ are measurable, nonnegative and $f \leq g$ a.e. in E, then $\mathcal{S}_f \subset \mathcal{S}_g$. Thus

$$\int_E f d\mu \leq \int_E g d\mu.$$

A measurable function $f : E \to \mathbb{R}^*$ is said to be integrable if $|f|$ is integrable. From the decomposition (3.4) it follows that $f^\pm \leq |f|$. Therefore, if f is integrable, also $f^\pm$ are integrable. If f is integrable, its integral is defined by

$$\int_E f d\mu = \int_E f^+ d\mu - \int_E f^- d\mu. \tag{7.3}$$

If $E' \subset E$ is measurable and $f : E \to \mathbb{R}^*$ is integrable, then also $f\chi_{E'}$ is integrable and

$$\int_{E'} f d\mu = \int_E f\chi_{E'} d\mu. \tag{7.4}$$

The integral of a nonnegative, measurable function $f : E \to \mathbb{R}^*$ is always defined, finite or infinite, by (7.1). More generally, for a measurable function $f : E \to \mathbb{R}^*$, we set

$$\int_E f d\mu = \begin{cases} +\infty & \text{if} \quad \int_E f^+ d\mu = \infty \quad \text{and} \quad \int_E f^- dx < \infty \\ -\infty & \text{if} \quad \int_E f^+ d\mu < \infty \quad \text{and} \quad \int_E f^- dx = \infty. \end{cases} \tag{7.5}$$

If f^+ and f^- are both not integrable, then the integral of f is not defined.

8 Fatou's Lemma and the Monotone Convergence Theorem

Given a measure space $\{X, \mathcal{A}, \mu\}$ and a measurable set E, let $\{f_n\}$ denote a sequence of measurable functions from E with values in $\mathbb{R}^*$.

Lemma 8.1 (Fatou [43]) *Let $\{f_n\}$ be a sequence of measurable and a.e. nonnegative functions in E. Then*

$$\int_E \liminf f_n d\mu \le \liminf \int_E f_n d\mu. \tag{8.1}$$

Proof Set $f = \liminf f_n$ and select a nonnegative simple function $\zeta \in \mathcal{S}_f$. Assume first that ζ be integrable, so that it vanishes outside a measurable set $F \subset E$, of finite measure. For fixed $x \in F$ and $\varepsilon > 0$ there exists an index $n_\varepsilon(x)$ such that

$$f_n(x) \ge \zeta(x) - \varepsilon \qquad \text{for all} \quad n \ge n_\varepsilon(x).$$

By the Egorov–Severini theorem as in Proposition 2.1, having fixed $\eta > 0$, there exists a set $F_\eta \subset F$ such that $\mu(F - F_\eta) \le \eta$ and this inequality holds uniformly in F_η, i.e., for every fixed $\varepsilon > 0$, there exists n_ε such that

$$f_n(x) \ge \zeta(x) - \varepsilon \quad \text{for all} \quad n \ge n_\varepsilon \quad \text{and for all} \quad x \in F_\eta.$$

From this, for $n \ge n_\varepsilon$

$$\int_E f_n d\mu \ge \int_{F_\eta} f_n d\mu \ge \int_{F_\eta} (\zeta - \varepsilon) d\mu$$

$$\geq \int_F \zeta d\mu - \int_{F-F_\eta} \zeta d\mu - \varepsilon\mu(F)$$

$$\geq \int_E \zeta d\mu - \eta \sup \zeta - \varepsilon\mu(F).$$

Since $\mu(F)$ is finite, this implies

$$\liminf \int_E f_n d\mu \geq \int_E \zeta d\mu \qquad \text{for all integrable } \zeta \in \mathcal{S}_f. \tag{8.2}$$

If ζ is not integrable, it equals some positive number δ, on a measurable set $F \subset E$ of infinite measure. Fix $\varepsilon \in (0, \delta)$ and set

$$F_n = \{x \in E \mid f_j(x) \geq \delta - \varepsilon \text{ for all } j \geq n\}.$$

From the definition of lower limit $F_n \subset F_{n+1}$ and $F \subset \cup F_n$. Therefore,

$$\liminf \mu(F_n) \geq \mu(\liminf F_n) = \infty.$$

by (3.3) of Proposition 3.1 of Chap. 3. From this

$$\liminf \int_E f_n d\mu \geq \liminf \int_{F_n} f_n d\mu \geq (\delta - \varepsilon) \liminf \mu(F_n).$$

Thus in either case (8.2) holds for all $\zeta \in \mathcal{S}_f$.   ∎

In the conclusion (8.1) of Fatou's lemma, equality does not hold in general. For example in $\mathbb{R}$ with the Lebesgue measure, the sequence

$$f_n(x) = \begin{cases} 1 & \text{for } x \in [n, n+1] \\ 0 & \text{otherwise} \end{cases}$$

satisfies (8.1) with strict inequality. This raises the issue of when (8.1) holds with equality, or equivalently when one can pass to the limit under integral.

Theorem 8.1 (Monotone Convergence) *Let $\{f_n\}$ be a monotone increasing sequence of measurable, nonnegative functions in E, i.e.,*

$$0 \leq f_n(x) \leq f_{n+1}(x) \qquad \text{for all } x \in E \text{ and for all } n \in \mathbb{N}.$$

Then

$$\lim \int_E f_n d\mu = \int_E \lim f_n d\mu.$$

Remark 8.1 The integrals are meant in the sense of (7.1). In particular both sides could be infinite.

Proof The sequence $\{f_n\}$ converges for all $x \in E$ to a measurable function $f : E \to \mathbb{R}^*$. Therefore, by Fatou's lemma

$$\int_E f d\mu = \int_E \lim f_n d\mu \leq \lim \int_E f_n d\mu \leq \int_E f d\mu.$$ ∎

9 More on the Lebesgue Integral

Proposition 9.1 *Let* $f, g : E \to \mathbb{R}$ *be integrable. Then for all* $\alpha, \beta \in \mathbb{R}$

$$\int_E (\alpha f + \beta g) d\mu = \alpha \int_E f d\mu + \beta \int_E g d\mu. \tag{9.1}$$

If $f \geq g$ *a.e. in* E, *then*

$$\int_E f d\mu \geq \int_E g d\mu. \tag{9.2}$$

Also for every integrable function f

$$\left| \int_E f d\mu \right| \leq \int_E |f| d\mu. \tag{9.3}$$

If E' *is a measurable subset of* E, *then*

$$\int_E f d\mu = \int_{E-E'} f d\mu + \int_{E'} f d\mu. \tag{9.4}$$

Proof For $\alpha \geq 0$ denote by $\alpha \mathcal{S}_f$ the collection of functions of the form $\alpha \zeta$ where $\zeta \in \mathcal{S}_f$. If $\alpha \geq 0$ and $f \geq 0$ then $\alpha \mathcal{S}_f = \mathcal{S}_{\alpha f}$. Therefore,

$$\int_E \alpha f d\mu = \sup_{\eta \in \mathcal{S}_{\alpha f}} \int_E \eta d\mu = \alpha \sup_{\zeta \in \mathcal{S}_f} \int_E \zeta d\mu = \alpha \int_E f d\mu.$$

Similarly, if $\alpha < 0$ use (7.2) and conclude that for every nonnegative measurable function $f : E \to \mathbb{R}^*$

$$\int_E \alpha f d\mu = \alpha \int_E f d\mu. \tag{9.5}$$

If $\alpha > 0$ and f is integrable and of variable sign, then (9.5) continues to hold in view of (7.3) and the decomposition $\alpha f = (\alpha f)^+ - (\alpha f)^-$. A similar argument applies if $\alpha < 0$ and we conclude that (9.5) holds true for every integrable function and every $\alpha \in \mathbb{R}$. Therefore, it suffices to prove (9.1) for $\alpha = \beta = 1$.

Assume first that both f and g are nonnegative. There exist monotone increasing sequences of simple functions $\{\zeta_n\}$ and $\{\xi_n\}$ converging pointwise in E, to f and g, respectively. By the monotone convergence theorem

$$\int_E (f+g)d\mu = \lim \int_E (\zeta_n + \xi_n)d\mu$$

$$= \lim \int_E \zeta_n d\mu + \lim \int_E \xi_n d\mu = \int_E f d\mu + \int_E g d\mu.$$

Next we assume that $f \geq 0$ and $g \leq 0$. First observe that $f+g$ is integrable since $|f+g| \leq |f| + |g|$. From the decomposition

$$(f+g)^+ - g = (f+g)^- + f$$

and (9.1) proven for the sum of two nonnegative functions

$$\int_E (f+g)^+ d\mu + \int_E -g d\mu = \int_E (f+g)^- d\mu + \int_E f d\mu.$$

This and the definitions (7.2)–(7.3) prove (9.1) for $f \geq 0$ and $g \leq 0$. If f and g are integrable with no further sign restriction

$$\int_E (f+g)d\mu = \int_E [(f^+ + g^+) - (f^- + g^-)]d\mu = \int_E f d\mu + \int_E g d\mu.$$

To prove (9.2) observe that from $f - g \geq 0$ and (9.1)

$$0 \leq \int_E (f-g)d\mu = \int_E f d\mu - \int_E g d\mu.$$

Inequality (9.3) follows from (9.2) and

$$-|f| \leq f \leq |f|.$$

Finally, (9.4) follows from (9.1) upon writing

$$f = f\chi_{E'} + f\chi_{E-E'}. \qquad \blacksquare$$

Corollary 9.1 *Let* $f : E \to \mathbb{R}^*$ *be integrable and let E be of finite measure. Then*

$$\mu(E) \inf_E f \leq \int_E f d\mu \leq \mu(E) \sup_E f.$$

10 Convergence Theorems

The properties of the Lebesgue integral, permit one to formulate various versions of Fatou's lemma and of the monotone convergence theorem. For example the conclusion of Fatou's lemma continues to hold if the functions f_n are of variable sign, provided they are uniformly bounded below by some integrable function g.

Proposition 10.1 *Let $g : E \to \mathbb{R}^*$ be integrable and assume that $f_n \geq g$ a.e. in E for all $n \in \mathbb{N}$. Then*

$$\liminf \int_E f_n d\mu \geq \int_E \liminf f_n d\mu.$$

Proof Since $f_n - g \geq 0$, the sequence $\{f_n - g\}$ satisfies the assumptions of Fatou's lemma. Thus

$$\liminf \int_E f_n d\mu \geq \int_E g d\mu + \int_E (\liminf f_n - g) d\mu. \qquad \blacksquare$$

Proposition 10.2 *Let $\{f_n\}$ be a sequence of nonnegative, measurable, functions on E. Then*

$$\sum \int_E f_n d\mu = \int_E \sum f_n d\mu.$$

Proof The sequence $\{\sum_{i=1}^n f_i\}$ is a monotone sequence of nonnegative, measurable functions. $\qquad \blacksquare$

Remark 10.1 It is not required that the f_n be integrable, nor that $\sum f_n$ be integrable. The integral of measurable, nonnegative functions, finite or infinite, is well defined by (7.1).

Theorem 10.1 (Dominated Convergence) *Let $\{f_n\}$ be a dominated and convergent sequence of integrable functions in E, i.e.,*

$$\lim f_n(x) = f(x) \quad \text{for all } x \in E$$

and there exists an integrable function $g : E \to \mathbb{R}^*$ such that

$$|f_n| \leq g \quad \text{a.e. in } E \quad \text{for all } n \in \mathbb{N}.$$

Then the limit function $f : E \to \mathbb{R}^*$ is integrable and

$$\lim \int_E f_n d\mu = \int_E \lim f_n d\mu.$$

Proof The limit function f is measurable and by Fatou's lemma

$$\int_E |f|d\mu \le \liminf \int_E |f_n|d\mu \le \int_E gd\mu < \infty.$$

Thus f is integrable. Next

$$g - f_n \ge 0 \quad \text{and} \quad f_n + g \ge 0 \quad \text{for all } n \in \mathbb{N}.$$

Therefore, by Fatou's lemma

$$\int_E fd\mu \le \liminf \int_E f_n \le \limsup \int_E f_nd\mu \le \int_E fd\mu. \qquad \blacksquare$$

11   Absolute Continuity of the Integral

Theorem 11.1 (Vitali [169]) *Let E be measurable and let $f : E \to \mathbb{R}^*$ be integrable. For every $\varepsilon > 0$ there exists $\delta > 0$ such that for every measurable subset $\mathcal{E} \subset E$ of measure less than δ*

$$\int_{\mathcal{E}} |f|d\mu < \varepsilon.$$

Proof May assume that $f \ge 0$. For $n \in \mathbb{N}$ consider the functions

$$E \ni x \to f_n(x) = \begin{cases} f(x) & \text{if } f(x) < n \\ n & \text{if } f(x) \ge n. \end{cases}$$

Since $\{f_n\}$ is increasing

$$\int_E f_nd\mu \le \int_E fd\mu \quad \text{and} \quad \lim \int_E f_nd\mu = \int_E fd\mu.$$

Having fixed $\varepsilon > 0$ there exists some index n_ε such that

$$\int_E f_{n_\varepsilon}d\mu > \int_E fd\mu - \frac{1}{2}\varepsilon.$$

Choose $\delta = \varepsilon/2n_\varepsilon$. Then for every measurable set $\mathcal{E} \subset E$ of measure $\mu(\mathcal{E}) < \delta$

$$\int_{\mathcal{E}} fd\mu \le \int_{\mathcal{E}} f_{n_\varepsilon}d\mu + \int_{E-\mathcal{E}}(f_{n_\varepsilon} - f)d\mu + \frac{1}{2}\varepsilon \le n_\varepsilon\mu(\mathcal{E}) + \frac{1}{2}\varepsilon < \varepsilon. \qquad \blacksquare$$

12 Product of Measures

Let $\{X, \mathcal{A}, \mu\}$ and $\{Y, \mathcal{B}, \nu\}$ be two measure spaces. Any pair of sets $A \subset X$ and $B \subset Y$, generates a subset $A \times B$ of the Cartesian product $X \times Y$ called a generalized rectangle. There are subsets of $X \times Y$ that are not rectangles.

The intersection of any two rectangles is a rectangle, by the formula

$$(A_1 \times B_1) \cap (A_2 \times B_2) = (A_1 \cap A_2) \times (B_1 \cap B_2).$$

The mutual complement of any two rectangles, while not a rectangle, can be written as the disjoint union of two rectangles, by the decomposition

$$(A_2 \times B_2) - (A_1 \times B_1) = (A_2 - A_1) \times B_2 \cup (A_1 \cap A_2) \times (B_2 - B_1).$$

Thus, the collection $\mathcal{R}$ of all rectangles is a semi-algebra. If $A \in \mathcal{A}$ and $B \in \mathcal{B}$, the rectangle $A \times B$ is called a *measurable rectangle*. The collection of all measurable rectangles is denoted by $\mathcal{R}_o$. By the previous remarks $\mathcal{R}_o$ is a semi-algebra. Since $X \times Y \in \mathcal{R}_o$, such a collection forms a sequential covering of $X \times Y$. The semi-algebra $\mathcal{R}_o$ can be endowed with the nonnegative set function

$$\lambda(A \times B) = \mu(A)\nu(B) \tag{12.1}$$

for all measurable rectangles $A \times B$.

Proposition 12.1 *Let $\{A_n \times B_n\}$ be a countable collection of disjoint, measurable rectangles whose union is a measurable rectangle $A \times B$. Then*

$$\lambda(A \times B) = \sum \lambda(A_n \times B_n).$$

Proof For each $x \in A$

$$B = \bigcup \{B_j \mid (x, y) \in A_j \times B_j; \; y \in B\}.$$

Since for each $x \in A$ fixed this is a disjoint union

$$\nu(B)\chi_A(x) = \sum \nu(B_j)\chi_{A_j}(x).$$

Integrating in $d\mu$ over A and using Proposition 10.2 now gives

$$\mu(A)\nu(B) = \sum \mu(A_n)\nu(B_n). \qquad \blacksquare$$

Thus λ is unambiguously defined, since the measure of a measurable rectangle does not depend on its partitions into countably many, pairwise disjoint measurable rectangles. The proposition also implies that λ is a measure on the semi-algebra

$\mathcal{R}_o$. Therefore, λ can be extended to a complete measure $(\mu \times \nu)$ on $X \times Y$, which coincides with λ on $\mathcal{R}_o$ (Theorem 11.1 of Chap. 3).

Theorem 12.1 *Every pair $\{X, \mathcal{A}, \mu\}$ and $\{Y, \mathcal{B}, \nu\}$ of measure spaces generates a complete, product measure space*

$$\{X \times Y, (\mathcal{A} \times \mathcal{B}), (\mu \times \nu)\}$$

where $(\mathcal{A} \times \mathcal{B})$ is a σ-algebra containing $\mathcal{R}_o$ and $(\mu \times \nu)$ is a measure on $(\mathcal{A} \times \mathcal{B})$ that coincides with (12.1) on measurable rectangles.

13 On the Structure of $(\mathcal{A} \times \mathcal{B})$

Denote by $(\mathcal{A} \times \mathcal{B})_o$ the smallest σ-algebra generated by the collection of all measurable rectangles. Set also

$$\mathcal{R}_\sigma = \{\text{countable unions of elements of } \mathcal{R}_o\}$$
$$\mathcal{R}_{\sigma\delta} = \{\text{countable intersections of elements of } \mathcal{R}_\sigma\}.$$

By construction

$$\mathcal{R}_o \subset \mathcal{R}_\sigma \subset \mathcal{R}_{\sigma\delta} \subset (\mathcal{A} \times \mathcal{B})_o \subset (\mathcal{A} \times \mathcal{B}).$$

For $E \subset X \times Y$ the two sets

$$Y \supset E_x = \{y \mid (x, y) \in E\} \quad \text{for a fixed } x \in X$$
$$X \supset E_y = \{x \mid (x, y) \in E\} \quad \text{for a fixed } y \in Y$$

are, respectively, the X-section and the Y-section of E.

Proposition 13.1 *Let $E \in (\mathcal{A} \times \mathcal{B})_o$. Then for every $y \in Y$ the Y-section E_y is in $\mathcal{A}$ and for every $x \in X$ the X-section E_x is in $\mathcal{B}$.*

Proof The collection $\mathcal{F}$ of all sets $E \in (\mathcal{A} \times \mathcal{B})$ such that $E_x \in \mathcal{B}$ for all $x \in X$ is a σ-algebra. Since $\mathcal{F}$ contains all the measurable rectangles, it must contain the smallest σ-algebra generated by the measurable rectangles. ∎

Remark 13.1 There exist nonmeasurable sets $E \subset X \times Y$ such that all the x and y sections are measurable (**13.4** of the Complements).

Remark 13.2 There exist $(\mu \times \nu)$-measurable rectangles $A \times B$ that are not measurable rectangles. To construct an example, let $A_o \subset X$ be not μ-measurable but included into a measurable set of finite μ-measure. Let also $B_o \in \mathcal{B}$ be of zero ν-measure. The rectangle $A_o \times B_o$ is $(\mu \times \nu)$-measurable and has measure zero.

For each $\varepsilon > 0$, there exists a measurable rectangle R_ε containing $A_o \times B_o$ and of measure less than ε. Therefore, $A_o \times B_o$ is $(\mu \times \nu)$-measurable, by the criterion of measurability of Proposition 10.2 of Chap. 3. This last example implies that Proposition 13.1 does not hold if $(\mathcal{A} \times \mathcal{B})_o$ is replaced by $(\mathcal{A} \times \mathcal{B})$. In particular the inclusion $(\mathcal{A} \times \mathcal{B})_o \subset (\mathcal{A} \times \mathcal{B})$ is strict.

Remark 13.3 While $\{X \times Y, (\mathcal{A} \times \mathcal{B}), (\mu \times \nu)\}$ is complete, the restriction of $(\mu \times \nu)$ to $(\mathcal{A} \times \mathcal{B})_o$ in general is not complete. For example let $\mu = \nu$ be the Lebesgue measure on $X = Y = \mathbb{R}$. The segment $(0, 1) \times \{0\}$ is measurable in the product measure and has measure zero. However, if $E \in (0, 1)$ is the Vitali nonmeasurable set, the set $E \times \{0\}$ is contained in $(0, 1) \times \{0\}$, is measurable in $(\mu \times \nu)$ but is not in $(\mathcal{A} \times \mathcal{B})_o$.

Proposition 13.2 *Assume that $\{X, \mathcal{A}, \mu\}$ and $\{Y, \mathcal{B}, \nu\}$ are complete measure spaces, and let $E \in \mathcal{R}_{\sigma\delta}$ be of finite $(\mu \times \nu)$-measure. Then the function $x \to \nu(E_x)$ is μ-measurable and the function $y \to \mu(E_y)$ is ν-measurable. Moreover*

$$\int_{X \times Y} \chi_E d(\mu \times \nu) = \int_X \nu(E_x) d\mu = \int_Y \mu(E_y) d\nu.$$

Proof The conclusion holds true if E is a measurable rectangle. If $E \in \mathcal{R}_\sigma$, it can be decomposed into the countable union of disjoint measurable rectangles E_n. The functions

$$x \to \nu(E_x) = \sum \nu(E_{n,x}), \qquad y \to \mu(E_y) = \sum \mu(E_{n,y})$$

are, respectively, μ and ν measurable. By monotone convergence

$$\int_{X \times Y} \chi_E d(\mu \times \nu) = \sum \int_{X \times Y} \chi_{E_n} d(\mu \times \nu)$$
$$= \sum \int_X \nu(E_{n,x}) d\mu = \sum \int_Y \mu(E_{n,y}) d\nu$$
$$= \int_X \sum \nu(E_{n,x}) d\mu = \int_Y \sum \mu(E_{n,y}) d\nu$$
$$= \int_X \nu(E_x) d\mu = \int_Y \mu(E_y) d\nu.$$

If $E \in \mathcal{R}_{\sigma\delta}$ there exists a countable collection $\{E_n\}$ of elements of $\mathcal{R}_\sigma$, each of finite measure, such that $E_{n+1} \subset E_n$, and $E = \cap E_n$. Since E is of finite measure, may assume that $(\mu \times \nu)(E_1) < \infty$. Then, since $E_1 \in \mathcal{R}_\sigma$,

$$(\mu \times \nu)(E_1) = \int_{E_1} d(\mu \times \nu) = \int_X \nu(E_{1,x}) d\mu = \int_Y \mu(E_{1,y}) d\nu < \infty.$$

The sequence of sets $\{E_{n,x}\}$ and $\{E_{n,y}\}$ are decreasing, for all $x \in X$ and all $y \in Y$, respectively, and have limits

$$E_x = \lim E_{n,x}, \qquad E_y = \lim E_{n,y}.$$

The sets E_x and E_y are ν-, and μ-measurable, respectively. Moreover $\cup E_{n,x}$ has finite ν-measure for μ-a.e. $x \in X$, and $\cup E_{n,y}$ has finite μ-measure for ν-a.e. $y \in Y$. Therefore, by (d)–(e) of Proposition 3.1 of Chap. 3,

$$\nu(E_x) = \lim \nu(E_{n,x}) \quad \text{for } \mu\text{-a.e. } x \in X;$$
$$\mu(E_y) = \lim \mu(E_{n,y}) \quad \text{for } \nu\text{-a.e. } y \in Y.$$

The functions $x \to \nu(E_{n,x})$ are μ-measurable. Since $\{X, \mathcal{A}, \mu\}$ is complete, their μ-a.e. limit $x \to \nu(E_x)$ is also μ-measurable. Likewise, since $\{Y, \mathcal{B}, \nu\}$ is complete, $y \to \mu(E_y)$ is ν-measurable. Moreover

$$\nu(E_x) \le \nu(E_{n,x}) \le \nu(E_{1,x}) \quad \text{for } \mu\text{-a.e. } x \in X,$$
$$\mu(E_y) \le \mu(E_{n,y}) \le \mu(E_{1,y}) \quad \text{for } \nu\text{-a.e. } y \in Y, \qquad \text{for all } n \in \mathbb{N}.$$

Then, by dominated convergence

$$\int_{X \times Y} \chi_E d(\mu \times \nu) = \lim \int_{X \times Y} \chi_{E_n} d(\mu \times \nu)$$
$$= \lim \int_X \nu(E_{n,x}) d\mu = \lim \int_Y \mu(E_{n,y}) d\nu$$
$$= \int_X \lim \nu(E_{n,x}) d\mu = \int_Y \lim \mu(E_{n,y}) d\nu$$
$$= \int_X \nu(E_x) d\mu = \int_Y \mu(E_y) d\nu. \qquad \blacksquare$$

Remark 13.4 If $E \in \mathcal{R}_\sigma$, the proof is based on the Monotone Convergence Theorem 8.1, and in view of Remark 8.1, it does not require that E be of finite measure. If $E \in \mathcal{R}_{\sigma\delta}$ the proof is based on the Dominated Convergence Theorem 10.1. The assumptions that E is of finite measure provides integrable upper bounds that permit one to pass to the limit under integral.

Proposition 13.3 *Assume that* $\{X, \mathcal{A}, \mu\}$ *and* $\{Y, \mathcal{B}, \nu\}$ *are complete measure spaces, and Let* $E \in (\mathcal{A} \times \mathcal{B})$ *be of* $(\mu \times \nu)$*-measure zero. Then*

$$E_x \text{ are } \nu\text{-measurable and } \nu(E_x) = 0 \quad \mu\text{-a.e. in } X$$
$$E_y \text{ are } \mu\text{-measurable and } \mu(E_y) = 0 \quad \nu\text{-a.e. in } Y.$$

Proof If $E \in \mathcal{R}_{\sigma\delta}$ the conclusion follows from the previous proposition. If E is $(\mu \times \nu)$-measurable, there exist a set $E_{\sigma\delta} \in \mathcal{R}_{\sigma\delta}$ such that $E \subset E_{\sigma\delta}$ and $(\mu \times \nu)(E_{\sigma\delta} - E) = 0$ (Proposition 10.3 of Chap. 3). Then $E_x \subset E_{\sigma\delta,x}$ and $E_y \subset E_{\sigma\delta,y}$, and the conclusion follows since μ and ν are complete. $\blacksquare$

Proposition 13.4 *Assume that* $\{X, \mathcal{A}, \mu\}$ *and* $\{Y, \mathcal{B}, \nu\}$ *are complete measure spaces and let* $E \in (\mathcal{A} \times \mathcal{B})$ *be of finite* $(\mu \times \nu)$-*measure. Then*

> E_x *are* ν-*measurable for* μ-*a.e.* $x \in X$ *and* $x \to \nu(E_x)$ *is integrable*
>
> E_y *are* μ-*measurable for* ν-*a.e.* $y \in Y$ *and* $y \to \mu(E_y)$ *is integrable.*

Moreover

$$\int_{X \times Y} \chi_E d(\mu \times \nu) = \int_X \nu(E_x) d\mu = \int_Y \mu(E_y) d\nu.$$

Proof There exists $E_{\sigma\delta} \in \mathcal{R}_{\sigma\delta}$, such that $E \subset E_{\sigma\delta}$ and $E_{\sigma\delta} - E = \mathcal{E}$ has $(\mu \times \nu)$-measure zero. Therefore, by Proposition 13.3 the sets $E_x = E_{\sigma\delta,x} - \mathcal{E}_x$ are ν-measurable for μ-almost all $x \in X$, and $\nu(E_x) = \nu(E_{\sigma\delta,x})$ for μ-almost all $x \in X$. A similar statement holds for ν-almost all E_y. Since E and $\mathcal{E}$ are disjoint

$$\begin{aligned}
\int_{X \times Y} \chi_E d(\mu \times \nu) &= \int_{X \times Y} \chi_{E_{\sigma\delta}} d(\mu \times \nu) \\
&= \int_X \nu(E_{\sigma\delta,x}) d\mu = \int_Y \mu(E_{\sigma\delta,y}) d\nu \\
&= \int_X \nu(E_x) d\mu = \int_Y \mu(E_y) d\nu.
\end{aligned}$$

∎

14 The Theorem of Fubini–Tonelli

Theorem 14.1 (Fubini [52]) *Let* $\{X, \mathcal{A}, \mu\}$ *and* $\{Y, \mathcal{B}, \nu\}$ *be two complete measure spaces and let*

$$X \times Y \ni (x, y) \longrightarrow f(x, y) \quad \text{be integrable in } X \times Y.$$

Then

> $X \ni x \longrightarrow f(x, y)$ *is* μ-*integrable in* X *for* ν-*almost all* $y \in Y$
>
> $Y \ni y \longrightarrow f(x, y)$ *is* ν-*integrable in* Y *for* μ-*almost all* $x \in X$.

Moreover

$$X \ni x \longrightarrow \int_Y f(x, y) d\nu \quad \text{is } \mu\text{-integrable in } X$$

$$Y \ni y \longrightarrow \int_X f(x, y) d\mu \quad \text{is } \nu\text{-integrable in } Y$$

(14.1)

and

$$\int_{X\times Y} f(x, y)d(\mu \times \nu) = \int_X \left(\int_Y f(x, y)d\nu\right)d\mu$$

$$= \int_Y \left(\int_X f(x, y)d\mu\right)d\nu.$$

(14.2)

Proof By the decomposition (3.4) we may assume that $f \geq 0$. By Proposition 13.4, the statement holds true if f is the characteristic function of a measurable set E of finite measure. If f is nonnegative and integrable there exists a sequence $\{f_n\}$ of nonnegative integrable simple functions such that $\{f_n\} \nearrow f$ a.e. in $(X \times Y)$. Since each of the f_n is integrable it vanishes outside a set of finite measure. Therefore, Proposition 13.4 holds for each of such f_n. Then by monotone convergence

$$\int_{X\times Y} f(x, y)d(\mu \times \nu) = \lim \int_{X\times Y} f_n(x, y)d(\mu \times \nu)$$

$$= \int_X \lim \left(\int_Y f_n(x, y)d\nu\right)d\mu = \int_Y \lim \left(\int_X f_n(x, y)d\mu\right)d\nu$$

$$= \int_X \left(\int_Y f(x, y)d\nu\right)d\mu = \int_Y \left(\int_X f(x, y)d\mu\right)d\nu.$$

■

14.1   The Tonelli Version of the Fubini Theorem

The double integral formula (14.2) requires that f be integrable in the product measure $(\mu \times \nu)$. Tonelli observed that if f is nonnegative the integrability requirement can be relaxed provided $(\mu \times \nu)$ is σ-finite.

Theorem 14.2 (Tonelli [160]) *Let $\{X, \mathcal{A}, \mu\}$ and $\{Y, \mathcal{B}, \nu\}$ be complete and σ-finite, and let $f : (X \times Y) \to \mathbb{R}^*$ be measurable and nonnegative. Then the measurability statements in (14.1) and the double integral formula (14.2) hold. The integrals in (14.2) could be either finite or infinite.*

Proof The integrability requirement in the Fubini theorem was used to insure the existence of a sequence $\{f_n\}$ of integrable functions each vanishing outside a set of finite measure and converging to f. The positivity of f and the σ-finiteness in the Tonelli theorem provide a similar information.   ■

If f is integrable in $d(\mu \times \nu)$ then Fubini's theorem holds and equality occurs in (14.2). If f is not integrable and nonnegative then the left-hand side of (14.2) is infinite. Tonelli's theorem asserts that in such a case also the right-hand side is well defined and is infinity, provided $(\mu \times \nu)$ is σ-finite.

In particular, Tonelli's theorem could be used to establish whether a nonnegative, measurable function $f : (X \times Y) \to \mathbb{R}^*$ is integrable, through the equality of the two right-hand sides of (14.2).

The requirement that $(\mu \times \nu)$ be σ-finite cannot be removed as shown by the example in **14.4** of the Complements.

15 Some Applications of the Fubini–Tonelli Theorem

15.1 Integrals in Terms of Distribution Functions

Let $f : E \to \mathbb{R}^*$ be measurable and nonnegative. The *distribution function* of f relative to E is defined as

$$\mathbb{R}^+ \ni t \longrightarrow \mu([f > t]).$$

This is a nonincreasing function of t and if f is finite a.e. in E, then

$$\lim_{t \to \infty} \mu([f > t]) = 0 \qquad \text{unless} \qquad \mu([f > t]) \equiv \infty.$$

If f is integrable, such a limit can be given a quantitative form. Indeed

$$t\mu([f > t]) = \int_E t \chi_{[f > t]} d\mu \leq \int_E f d\mu < \infty.$$

Proposition 15.1 *Let $\{X, \mathcal{A}, \mu\}$ be complete and σ-finite and let $f : E \to \mathbb{R}^*$ be measurable and nonnegative. Let also ν be a complete and σ-finite measure on $\mathbb{R}^+$ such that $\nu([0, t)) = \nu([0, t])$ for all $t > 0$. Then*

$$\int_E \nu([0, f]) d\mu = \int_0^\infty \mu([f > t]) d\nu. \tag{15.1}$$

In particular if $\nu([0, t]) = t^p$ for some $p > 0$, then

$$\int_E f^p d\mu = p \int_0^\infty t^{p-1} \mu([f > t]) dt \tag{15.2}$$

where dt is the Lebesgue measure on $\mathbb{R}^+$.

Proof The function $f : E \to \mathbb{R}^*$, when regarded as a function from $E \times \mathbb{R}^+$ into $\mathbb{R}^*$, is measurable in the product measure $(\mu \times \nu)$. Likewise, the function $g(t) = t$ from $\mathbb{R}^+$ into $\mathbb{R}^*$, when regarded as a function from $E \times \mathbb{R}$ into $\mathbb{R}^*$, is measurable in the product measure $(\mu \times \nu)$. Therefore, the difference $f - t$ is measurable in the product measure $(\mu \times \nu)$. This implies that the set $[f - t > 0] = [f > t]$ is measurable in the product measure $(\mu \times \nu)$. Therefore, by the Tonelli theorem

$$\int_0^\infty \mu([f > t])d\nu = \int_0^\infty \left(\int_E \chi_{[f>t]}d\mu \right) d\nu$$

$$= \int_E \left(\int_0^\infty \chi_{[f>t]}d\nu \right) d\mu = \int_E \left(\int_0^f d\nu \right) d\mu$$

$$= \int_E \nu([0, f])d\mu. \qquad \blacksquare$$

Both sides of (15.1) could be infinity and the formula could be used to verify whether $\nu([0, f])$ is μ-integrable over E.

Corollary 15.1 *Let E be an open set in $\mathbb{R}^N$ and let f be continuous in E. Then for all $x \in E$,*

$$f(x) = \int_0^\infty \chi_{[f(x)>0]}dt. \qquad (15.3)$$

Proof Apply (15.2) with $\mu = \delta_x$. $\qquad\qquad\qquad\qquad\qquad\qquad\qquad\qquad\blacksquare$

Corollary 15.2 *Let E be an open set in $\mathbb{R}^N$ and let f and h be nonnegative Lebesgue measurable functions defined in E. Then for all $p > 0$*

$$\int_E f^p h dx = \int_0^\infty p t^{p-1} \left(\int_{[f>t]} h dx \right) dt \qquad (15.4)$$

where dx is the Lebesgue measure in $\mathbb{R}^N$.

Proof Apply (15.2) with $d\mu = hdx$. $\qquad\qquad\qquad\qquad\qquad\qquad\qquad\qquad\blacksquare$

If f and h are of variable sign, with f integrable and h bounded, then

$$\int_E f h d\mu = \int_0^\infty \left(\int_{[f^+>t]} h\, dx \right) dt - \int_0^\infty \left(\int_{[f^->t]} h\, dx \right) dt. \qquad (15.5)$$

In the next two applications in § 15.2 and § 15.3, the measure space $\{X, \mathcal{A}, \mu\}$ is $\mathbb{R}^N$ with the Lebesgue measure.

15.2   Convolution Integrals

Lemma 15.1 *Let $f : \mathbb{R}^N \to \mathbb{R}^*$ be measurable. Then the function*

$$\mathbb{R}^{2N} \ni (x, y) \to f(x - y)$$

is measurable with respect to the product measure of $\mathbb{R}^{2N}$.

Proof Consider the change of variables

$$\mathbb{R}^{2N} \ni \begin{pmatrix} x \\ y \end{pmatrix} \longrightarrow T\begin{pmatrix} x \\ y \end{pmatrix} = \begin{pmatrix} x - y \\ x + y \end{pmatrix} = \begin{pmatrix} \xi \\ \eta \end{pmatrix} \in \mathbb{R}^{2N}.$$

This an invertible, Lipschitz map from $\mathbb{R}^{2N}$ into itself. By Proposition 12.2c of the Complements of Chap. 3, Lebesgue measurable sets in $\mathbb{R}^{2N}$ in the (ξ, η) coordinates are mapped, by T^{-1}, into Lebesgue measurable sets in $\mathbb{R}^{2N}$ in the (x, y) coordinates. Now for all $c \in \mathbb{R}$

$$\{(x, y) \in \mathbb{R}^{2N} \mid f(x - y) > c\} = \{\xi \in \mathbb{R}^N \mid f(\xi) > c\} \times \mathbb{R}^N.$$

Since $f : \mathbb{R}^N \to \mathbb{R}^*$ is measurable, the latter is a measurable rectangle. ∎

Given any two nonnegative measurable functions $f, g : \mathbb{R}^N \to \mathbb{R}^*$ their *convolution* is defined as

$$x \longrightarrow (f * g)(x) = \int_{\mathbb{R}^N} g(y) f(x - y) dy. \tag{15.6}$$

Since f and g are both nonnegative, the right-hand side, finite or infinite, is well defined for all $x \in \mathbb{R}^N$.

Proposition 15.2 *Let f and g be nonnegative and integrable in $\mathbb{R}^N$. Then $(f * g)(x)$ is finite for a.e. $x \in \mathbb{R}^N$, the function $(f * g)$ is integrable in $\mathbb{R}^N$ and*

$$\int_{\mathbb{R}^N} (f * g) dx = \int_{\mathbb{R}^N} f dx \int_{\mathbb{R}^N} g dx.$$

Proof The function $(x, y) \to g(y) f(x - y)$ is nonnegative and measurable with respect to the product measure. Therefore, by the Tonelli theorem

$$\iint_{\mathbb{R}^{2N}} g(y) f(x - y) dx dy = \int_{\mathbb{R}^N} \left(\int_{\mathbb{R}^N} g(y) f(x - y) dx \right) dy$$
$$= \int_{\mathbb{R}^N} f dx \int_{\mathbb{R}^N} g dx. \qquad ∎$$

The convolution of any two integrable functions f and g is defined as in (15.6). Since

$$|g(y) f(x - y)| \leq |g(y)| |f(x - y)|$$

the convolution $(f * g)$ is well defined as an integrable function over $\mathbb{R}^{2N}$.

15.3   The Marcinkiewicz Integral ([101, 102])

Let $E \subset \mathbb{R}^N$ be non void, and for $x \in \mathbb{R}^N$ let $\delta(x)$ denote the distance from x to E. By the definition of distance, $\delta_E(x) = 0$ for all $x \in E$.

Lemma 15.2 *Let E be a nonempty set in $\mathbb{R}^N$. Then the distance function $x \to \delta(x)$ is Lipschitz continuous with Lipschitz constant one.*

Proof Fix x and y in $\mathbb{R}^N$ and assume that $\delta_E(x) \geq \delta_E(y)$. By definition of $\delta_E(y)$, having fixed $\varepsilon > 0$ there exists $z' \in E$ such that $\delta_E(y) \geq |y - z'| - \varepsilon$. Then estimate

$$0 \leq \delta_E(x) - \delta_E(y) \leq \inf_{z \in E} |x - z| - |y - z'| + \varepsilon$$

$$\leq |x - z'| - |y - z'| + \varepsilon \leq |x - y| + \varepsilon. \qquad \blacksquare$$

Let E be a bounded, closed set in $\mathbb{R}^N$. Fix a positive number λ and a cube Q containing E. The Marcinkiewicz integral relative to E and λ is the function

$$\mathbb{R}^N \ni x \longrightarrow M_{E,\lambda}(x) = \int_Q \frac{\delta_E^\lambda(y)}{|x - y|^{N+\lambda}} dy.$$

The right-hand side is well defined as the integral of a measurable, nonnegative function.

Proposition 15.3 *The Marcinkiewicz integral $M_{E,\lambda}(x)$ is finite for a.e. $x \in E$. Moreover the function $x \to M_{E,\lambda}(x)$ is integrable in E and*

$$\int_E M_{E,\lambda} dx \leq \frac{\omega_N}{\lambda} \mu(Q - E).$$

where ω_N is the measure of the unit sphere in $\mathbb{R}^N$.

Proof Since $\delta_E(y) = 0$ for all $y \in E$, by the Tonelli theorem

$$\int_E M_{E,\lambda}(x) dx = \int_Q \delta_E^\lambda(y) \left(\int_E \frac{dx}{|x - y|^{N+\lambda}} \right) dy$$

$$= \int_{Q-E} \delta_E^\lambda(y) \left(\int_E \frac{dx}{|x - y|^{N+\lambda}} \right) dy. \qquad (15.7)$$

For $y \in (Q - E)$ and $x \in E$, since E is closed, $|x - y| \geq \delta_E(y) > 0$. Therefore,

$$\int_E \frac{dx}{|x - y|^{N+\lambda}} \leq \int_{|x-y| \geq \delta_E(y)} \frac{d(x - y)}{|x - y|^{N+\lambda}}$$

$$\leq \omega_N \int_{\delta_E(y)}^\infty \frac{ds}{s^{1+\lambda}} = \frac{\omega_N}{\lambda \delta_E^\lambda(y)}.$$

Using this estimate in (15.7) establishes the proposition. ∎

16 Signed Measures and the Hahn Decomposition

Let μ_1 and μ_2 be two measures both defined on the same σ-algebra $\mathcal{A}$. If one of them is finite, the set function

$$\mathcal{A} \ni E \longrightarrow \mu(E) = \mu_1(E) - \mu_2(E)$$

is well defined and countably additive on $\mathcal{A}$. However since it is not necessarily nonnegative it is called a *signed measure*. Signed measures are also generated by an integrable function f on a measure space $\{X, \mathcal{A}, \mu\}$ by the formula

$$\mathcal{A} \ni E \longrightarrow \int_E f d\mu = \int_E f^+ d\mu - \int_E f^- d\mu. \tag{16.1}$$

More generally a *signed measure* on X is a set function μ satisfying:

(i) the domain of μ is a σ-algebra $\mathcal{A}$ (ii) $\mu(\emptyset) = 0$
(iii) μ takes at most one of $\pm\infty$ (iv) μ is countably additive.

The last property is intended in the sense of convergent or divergent series.

Any linear, real combination of measures defined on the same σ-algebra is a signed measure provided all but one are finite.

Let $\{X, \mathcal{A}, \mu\}$ be a measure space for a signed measure μ. A measurable set $E \subset \mathcal{A}$ is said to be *positive (negative)* if $\mu(A) \geq (\leq)0$, for all measurable subsets $A \subset E$. The difference and the union of two positive (negative) sets is positive (negative). Since μ is countably additive, the countable, disjoint union of positive (negative) sets is positive (negative). From this it follows that any countable union of positive (negative) sets is positive (negative).

Lemma 16.1 *Let $\{X, \mathcal{A}, \mu\}$ be a measure space for a signed measure μ. Let $E \subset X$ be measurable and such that $|\mu(E)| < \infty$. Then every measurable set $A \subset E$ satisfies $|\mu(A)| < \infty$.*

Proof Assume for example that μ does not take the value $+\infty$. Let $A \subset E$. If $\mu(A) > 0$ then $\mu(A) < \infty$. If $\mu(A) < 0$, taking the measure of the disjoint union $E = (E - A) \cup A$ gives $\mu(E) = \mu(E - A) + \mu(A)$, which in turn implies

$$0 < -\mu(A) = \mu(E - A) - \mu(E) < \infty. \quad ∎$$

Proposition 16.1 *Let $\{X, \mathcal{A}, \mu\}$ be a measure space for a signed measure μ. Every measurable set E of positive, finite measure contains a positive subset A of positive measure.*

Proof If E is positive we take $A = E$. Otherwise E contains a measurable set of negative measure. Let n_1 be the smallest positive integer for which there exists a measurable set $B_1 \subset E$ such that

$$\mu(B_1) \leq -\frac{1}{n_1}.$$

If $A_1 = E - B_1$ is positive then we take $A = A_1$. Otherwise A_1 contains a measurable set of negative measure. We then let n_2 be the smallest positive integer for which there exist a measurable set $B_2 \subset E - B_1$ of negative such that

$$\mu(B_2) \leq -\frac{1}{n_2}.$$

Proceeding in this fashion, if for some finite m the set

$$A_m = E - \bigcup_{j=1}^{m} B_j$$

is positive, the process terminates. Otherwise the indicated procedure generates the sequences of sets $\{B_j\}$ and $\{A_m\}$. We establish that the set

$$A = \cap A_m = E - \cup B_j$$

is positive by showing that every measurable subset $C \subset A$ has nonnegative measure. Since $A \subset E$ and E is of finite measure $|\mu(A)| < \infty$, by Lemma 16.1. By construction the sets B_j and A are measurable and disjoint. Since μ is countably additive

$$0 < \mu(E) = \mu(A) + \sum \mu(B_j) \leq \mu(A) - \sum \frac{1}{n_j}.$$

This implies that the series $\sum n_j^{-1}$ is convergent and therefore $n_j \to \infty$ as $j \to \infty$. It also implies that $\mu(A) > 0$. Let C be a measurable subset of A. Since A belongs to all A_j, by construction

$$\mu(C) \geq -\frac{1}{n_j - 1} \longrightarrow 0 \quad \text{as } j \to \infty. \qquad \blacksquare$$

Theorem 16.1 (Hahn Decomposition [58], Vol. 1) *Let $\{X, \mathcal{A}, \mu\}$ be a measure space for a signed measure μ. Then X can be decomposed into a positive set X^+ and a negative set X^-.*

Proof Assume for example that μ does not take the value $+\infty$ and set

$$M = \{\sup \mu(A) \text{ where } A \in \mathcal{A} \text{ is positive}\}.$$

Let $\{A_n\}$ be a sequence of positive sets such that $\mu(A_n)$ increases to M and set $A = \cup A_n$. The set A is positive and, by construction $\mu(A) \leq M$. On the other hand $A = (A - A_n) \cup A_n$ for all n. Since this is a disjoint union and A is positive

$$\mu(A) = \mu(A - A_n) + \mu(A_n) \geq \mu(A_n) \quad \text{for all } n.$$

Thus $\mu(A) = M$ and $M < \infty$. The complement $X - A$ is a negative set. For otherwise it would contain a set E of positive measure, which in turn would contain a positive set A_o of positive measure. Then A and A_o are disjoint and $A \cup A_o$ is a positive set. Therefore,

$$\mu(A \cup A_o) = \mu(A) + \mu(A_o) > M$$

contradicting the definition of M. The Hahn decomposition is realized by taking $X^+ = A$ and $X^- = X - X^+$. ∎

A set $E \in \mathcal{A}$ is a *null set* if every measurable subset of E has measure zero. There exist measurable sets of zero measure that are not null sets. By removing out of X^+ a null set $\mathcal{E}$ and adding it to X^-, the set $(X^+ - \mathcal{E})$ remains a positive set and $(X^- \cup \mathcal{E})$ remains negative. Moreover

$$X = (X^+ - \mathcal{E}) \cup (X^- \cup \mathcal{E}).$$

Thus the Hahn decomposition is not unique. However it can be determined up to null sets.

17   The Radon-Nikodým Theorem

Let μ and ν be two measures defined on the same σ-algebra $\mathcal{A}$. The measure ν is *absolutely continuous* with respect to μ if $\mu(E) = 0$ implies $\nu(E) = 0$, and in such a case write $\nu \ll \mu$. Let $\{X, \mathcal{A}, \mu\}$ be a measure space and let $f : X \to \mathbb{R}^*$ be measurable and nonnegative. The set function

$$\mathcal{A} \ni E \longrightarrow \nu(E) = \int_E f \, d\mu \tag{17.1}$$

is a measure defined on $\mathcal{A}$ and absolutely continuous with respect to μ.

Theorem 17.1 (Radon-Nikodým)[1] *Let $\{X, \mathcal{A}, \mu\}$ be σ-finite and let ν be a measure on the same σ-algebra, absolutely continuous with respect to μ ($\nu \ll \mu$). There exists*

---

[1]Although referred to as the Radon-Nikodým theorem, the first version of this theorem, in the context of a measure in $\mathbb{R}^N$ absolutely continuous with respect to the Lebesgue measure, is in Lebesgue [90]. Radon extended it to Radon measures in [121], and Nikodým to general measures in [115, 116].

a nonnegative μ-measurable function $f : X \to \mathbb{R}^*$ such that ν has the representation (17.1). Such a f is unique up to a set of μ-measure zero.

The function f in the representation (17.1) is called the Radon-Nikodým derivative of ν with respect to μ, since formally $d\nu = f\,d\mu$. It is not asserted that f is μ-integrable. This would occur if and only if ν is finite. The assumption that μ be σ-finite cannot be removed as shown by counterexamples in **17.1** and **17.2** of the Complements.

17.1 Sublevel Sets of a Measurable Function

Let $\{X, \mathcal{A}, \mu\}$ be a measure space and let $f : X \to \mathbb{R}^*$ be measurable. For $t \in \mathbb{R}$, any measurable set E_t such that

$$[f < t] \subset E_t \subset [f \le t]$$

is a sublevel set for f at the value t. The next proposition asserts that for any increasing collection of measurable sets $\{E_t\}$, as t ranges over a countable index, there exists a measurable function $f : X \to \mathbb{R}^*$, which admits E_t as sublevel sets.

Proposition 17.1 (von Neumann [114]) *For a countable index t, let $\{E_t\}$ be a collection of measurable sets such that $E_s \subset E_t$ for $s < t$. There exists a measurable function $f : X \to \mathbb{R}^*$ such that*

$$f \le t \text{ in } E_t, \quad \text{and} \quad f \ge t \text{ in } X - E_t. \tag{17.2}$$

Proof Define

$$X \ni x \to f(x) = \begin{cases} \inf\{t \mid x \in E_t\} & \text{if } x \in \bigcup E_t \\ +\infty & \text{otherwise.} \end{cases}$$

If $x \in E_t$, then $f(x) \le t$. If $x \notin E_t$, then x does not belong to E_s for any $s < t$, and hence $f(x) \ge t$. Such a function is measurable, since for all $c \in \mathbb{R}$

$$[f < c] = \bigcup_{t < c} E_t. \qquad \blacksquare$$

The assumptions imply that if $t > s$, then $\mu(E_s - E_t) = 0$. The proposition continues to hold if the monotonicity of $\{E_t\}$ is replaced by such a weaker, measure theoretical notion of monotonicty. In such a case however, the conclusion (17.2) holds up to a set of measure zero.

Corollary 17.1 *For a countable index t, let $\{E_t\}$ be a collection of measurable sets such that*

$$\mu(E_s - E_t) = 0 \quad whenever \ \ s < t.$$

There exists a measurable function $f : X \to \mathbb{R}^*$

$$f \leq t \ \ a.e. \ in \ E_t, \quad and \quad f \geq t \ \ a.e. \ in \ X - E_t. \tag{17.3}$$

Proof Set

$$\mathcal{E} = \bigcup_{s<t} (E_s - E_t) \quad and \quad E_t' = E_t \cup \mathcal{E}.$$

The collection $\{E_t'\}$ is monotone and, by Proposition 17.1, there exists a measurable function $f : X \to \mathbb{R}^*$ satisfying (17.2) with E_t replaced by E_t'. Since the index t is countable $\mu(\mathcal{E}) = 0$ and hence (17.3) holds except possibly for a set of measure zero. ∎

17.2 Proof of the Radon-Nikodým Theorem

Assume first that μ is finite. For nonnegative, rational t consider the signed measure $\nu - t\mu$ on $\{X, \mathcal{A}\}$, and the corresponding Hahn's decomposition $\{X_t^+, X_t^-\}$ up to a null set $\mathcal{E}_t$. If $s < t$,

$$(\nu - s\mu)(X_s^- - X_t^-) \leq 0 \quad and \quad (\nu - t\mu)(X_s^- - X_t^-) \geq 0.$$

Therefore,

$$\mu(X_s^- - X_t^-) \leq 0 \quad for \quad s < t.$$

By Corollary 17.1 there exists a measurable function $f : X \to \mathbb{R}^*$

$$f \leq t \ \ \mu\text{-a.e. in } X_t^-, \quad and \quad f \geq t \ \ \mu\text{-a.e. in } X - X_t^-.$$

Moreover $f \geq 0$ a.e. in X, since $X_t^- = \emptyset$, for all $t \leq 0$. For a measurable set $E \subset X$, and positive integers j, n, set

$$E_{n,j} = E \cap \left(X_{\frac{j+1}{n}}^- - X_{\frac{j}{n}}^- \right), \qquad E_n = E - \bigcup_{j\in\mathbb{N}} E_{n,j}$$

and, for fixed $n \in \mathbb{N}$, write E as the disjoint union

$$E = E_n \cup \bigcup_{j\in\mathbb{N}} E_{n,j}.$$

From this for any fixed $n \in \mathbb{N}$, and all $E \in \mathcal{A}$

$$\nu(E) = \nu(E_n) + \sum_{j \in \mathbb{N}} \nu(E_{n,j}).$$

By the properties of the Hahn decomposition

$$E_{n,j} \subset X^-_{\frac{j+1}{n}} \cap X^+_{\frac{j}{n}}.$$

Therefore,

$$\left(\nu - \tfrac{j+1}{n}\mu\right)(E_{n,j}) \leq 0 \quad \text{and} \quad \left(\nu - \tfrac{j}{n}\mu\right)(E_{n,j}) \geq 0.$$

From this

$$\nu(E_{n,j}) - \frac{1}{n}\mu(E_{n,j}) \leq \frac{j}{n}\mu(E_{n,j}) \leq \frac{j+1}{n}\mu(E_{n,j}) \leq \nu(E_{n,j}) + \frac{1}{n}\mu(E_{n,j}).$$

By construction

$$\frac{j}{n} \leq f \leq \frac{j+1}{n} \quad \text{on } E_{n,j}$$

which implies

$$\frac{j}{n}\mu(E_{n,j}) \leq \int_{E_{n,j}} f \, d\mu \leq \frac{j+1}{n}\mu(E_{n,j})$$

and hence

$$\nu(E_{n,j}) - \frac{1}{n}\mu(E_{n,j}) \leq \int_{E_{n,j}} f \, d\mu \leq \nu(E_{n,j}) + \frac{1}{n}\mu(E_{n,j}).$$

By construction, $E_n \subset X^+_{j/n}$ for all $j \in \mathbb{N}$, and hence

$$\left(\nu - \tfrac{j}{n}\mu\right)(E_n) \geq 0 \quad \Longrightarrow \quad \nu(E_n) \geq \tfrac{j}{n}\mu(E_n) \quad \text{for all } j \in \mathbb{N}.$$

If $\mu(E_n) > 0$ then $\nu(E_n) = \infty$ and if $\mu(E_n) = 0$ then $\nu(E_n) = 0$ since $\nu \ll \mu$. By the construction of f one has $f = \infty$ on E_n. Therefore, in either case

$$\nu(E_n) = \int_{E_n} f \, d\mu.$$

Adding this and the previous inequalities yields

$$\nu(E) - \frac{1}{n}\mu(E) \leq \int_E f \, d\mu \leq \nu(E) + \frac{1}{n}\mu(E).$$

Since n is arbitrary and μ is finite this implies the conclusion (17.1).

If $g : X \to \mathbb{R}^*$ is another nonnegative measurable function by which the measure ν can be represented as in (17.1), let

$$A_n = \left[f - g \geq \tfrac{1}{n} \right].$$

Then for all $n \in \mathbb{N}$

$$\frac{1}{n}\mu(A_n) \leq \int_{A_n} (f - g)d\mu = \nu(A_n) - \nu(A_n) = 0.$$

Thus $f = g$ μ-a.e. in X. Assume next that μ is σ-finite and let X_n be a sequence of expanding sets such that

$$\mu(X_n) \leq \mu(X_{n+1}) < \infty \quad \text{and} \quad X = \bigcup X_n.$$

Denote by μ_n the restriction of μ to $\mathcal{A} \cap X_n$, and let f_n be the unique function claimed by the Radon-Nikodým theorem for the pair of measures $\{\mu_n; \nu\}$ on the σ-algebra $\mathcal{A} \cap X_n$. While defined in X_n we regard f_n as defined in the whole X by setting it to be zero in $X - X_n$. By construction

$$f_{n+1}\big|_{X_n} = f_n \quad \text{for all } n.$$

The function f claimed by the theorem is

$$f = \sup f_n = \lim f_n.$$

Indeed if $E \in \mathcal{A}$, by monotone convergence

$$\nu(E) = \lim \nu(E \cap X_n) = \lim \int_E f_n d\mu = \int_E f d\mu.$$

The uniqueness of f is proved as in the case of μ finite.   ∎

18 Decomposing Measures

Two measures μ and ν on the same space $\{X, \mathcal{A}\}$ are *mutually singular* if X can be decomposed into two measurable, disjoint sets X_μ and X_ν such

$$\mu(E \cap X_\nu) = 0 \quad \text{and} \quad \nu(E \cap X_\mu) = 0 \quad \text{for all } E \in \mathcal{A}.$$

An example of mutually singular measures is given by (16.1). If μ and ν are mutually singular we write $\nu \perp \mu$. If $\nu \ll \mu$ and $\nu \perp \mu$, then $\nu = 0$.

18.1   The Jordan Decomposition

Given a measure space $\{X, \mathcal{A}, \mu\}$ for a signed measure μ, let $X = X^+ \cup X^-$ be the corresponding Hahn decomposition of X. For every $E \in \mathcal{A}$ set

$$\mu^+(E) = \mu(E \cap X^+) \quad \text{and} \quad \mu^-(E) = -\mu(E \cap X^-).$$

The set functions $\mu^\pm$ are measures on $\mathcal{A}$ and (i) at least one of them is finite, (ii) they are independent of the particular Hahn decomposition, (iii) they are mutually singular. Moreover for every $E \in \mathcal{A}$

$$\mu(E) = \mu^+(E) - \mu^-(E). \tag{18.1}$$

Theorem 18.1   (Jordan [79], Vol 1) *Let $\{X, \mathcal{A}, \mu\}$ be a measure space for a signed measure μ. There exists a unique pair (μ^+, μ^-) of mutually singular measures, one of which is finite such that $\mu = \mu^+ - \mu^-$.*

Proof Let $X = X^+ \cup X^-$ be the Hahn decomposition of X relative to μ and determined up to a null set. The existence of $\mu^\pm$ follows from (18.1). If $\mu = \nu^+ - \nu^-$ is another such decomposition, there exists disjoint sets Y^+ and Y^- such that $X = Y^+ \cup Y^-$, and

$$\nu^+(E \cap Y^-) = \nu^-(E \cap Y^+) = 0 \quad \text{for all } E \in \mathcal{A}.$$

Since Y^+ is positive and Y^- is negative with respect to the same measure μ, $X^+ = Y^+$ and $X^- = Y^-$ up to null sets. Therefore, for all $E \in \mathcal{A}$

$$\mu^\pm(E) = \mu(E \cap X^\pm \cap Y^\pm) = \nu^\pm(E). \qquad \blacksquare$$

The two measures $\mu^\pm$ are the *upper* and *lower* variation of μ. The measure $|\mu| = \mu^+ + \mu^-$ is the *total variation* of μ. Both measures $\mu^\pm$ are absolutely continuous with respect to $|\mu|$. Moreover for every $E \in \mathcal{A}$

$$-\mu^-(E) \le \mu(E) \le \mu^+(E) \quad \text{and} \quad |\mu(E)| \le |\mu|(E).$$

18.2   The Lebesgue Decomposition

Two signed measures μ and ν on the same space $\{X, \mathcal{A}\}$ are mutually singular, denoted by $\nu \perp \mu$, if the measures $|\mu|$ and $|\nu|$ are mutually singular. The signed measure ν is absolutely continuous with respect to μ, denoted by $\nu \ll \mu$, if $|\mu(E)| = 0$ implies $|\nu|(E) = 0$.

Theorem 18.2 (Lebesgue) *Let* $\{X, \mathcal{A}, \mu\}$ *be a σ-finite measure space for a signed measure* μ *and let* ν *be a σ-finite signed measure defined on* $\mathcal{A}$. *There exists a unique pair* (ν_o, ν_1) *of σ-finite, signed measures defined on the same σ-algebra* $\mathcal{A}$ *such that*

$$\nu = \nu_o + \nu_1 \quad \text{and} \quad \nu_o \perp \mu \quad \nu_1 \ll \mu.$$

Proof If $\nu = \nu_o' + \nu_1'$ is another such decomposition, then

$$\nu_o - \nu_o' = \nu_1' - \nu_1.$$

This implies that $\nu_o - \nu_o'$ is both singular and absolutely continuous with respect to μ and therefore identically zero. Analogously $\nu_1 = \nu_1'$.

The notions of mutually singular signed measures and absolute continuity of a signed measure ν with respect to a signed measure μ are set in terms of the same notions for their total variations $|\nu|$ and $|\mu|$. Therefore, we may assume that μ is a measure. If $\nu^{\pm}$ is the Jordan decomposition of ν, by treating ν^+ and ν^- separately, we may assume that also ν is a measure.

Set $\lambda = \mu + \nu$. Both μ and ν are absolutely continuous with respect to λ. Therefore, by the Radon-Nikodým Theorem, there exist measurable, nonnegative functions f and g such that

$$\mu(E) = \int_E f \, d\lambda \quad \text{and} \quad \nu(E) = \int_E g \, d\lambda \quad \text{for all } E \in \mathcal{A}.$$

Define ν_o and ν_1 by

$$\nu_o(E) = \nu(E \cap [f = 0]) \quad \nu_1(E) = \nu(E \cap [f > 0]).$$

By construction $\nu = \nu_o + \nu_1$. The measure ν_o is singular with respect to μ since $X = [f > 0] \cup [f = 0]$ and for every measurable set E

$$\nu_o(E \cap [f > 0]) = \mu(E \cap [f = 0]) = 0.$$

The measure ν_1 is absolutely continuous with respect to μ. Indeed $\mu(E) = 0$ implies that $f = 0$, λ-a.e. on E. Therefore,

$$0 = \lambda(E \cap [f > 0]) \geq \nu(E \cap [f > 0]) = \nu_1(E). \qquad \blacksquare$$

18.3 A General Version of the Radon-Nikodým Theorem

Theorem 18.3 *Let* $\{X, \mathcal{A}, \mu\}$ *be a σ-finite measure space and let* ν *be a σ-finite, signed measure on the same space* $\{X, \mathcal{A}\}$. *If* $\nu \ll \mu$, *there exists a measurable* *function $f : X \to \mathbb{R}^*$ such that*

$$\nu(E) = \int_E f d\mu \quad \text{for all } E \in \mathcal{A}. \tag{18.2}$$

The function f need not be integrable, however at least one of f^+ or f^- must be integrable. Precisely, if the signed measure ν does not take the value $+\infty$ $(-\infty)$, then the upper(lower) variation ν^+ (ν^-) of ν is finite and f^+ (f^-) is integrable. If f^+ is integrable and f^- is not, the integral in (18.2) is well defined in the sense of (7.5).

Such a function f is unique up to a set of μ-measure zero.

Proof Determine the Hahn decomposition $X = X^+ \cup X^-$ up to a null set, and the corresponding Jordan decomposition $\nu = \nu^+ - \nu^-$.

The upper and lower variations $\nu^\pm$ are absolutely continuous with respect to μ. Applying the Radon-Nikodým Theorem to the pairs $(\nu_\pm, \mu)$, determines nonnegative μ-measurable functions $f_\pm$ such that

$$\nu_\pm(E) = \nu(E \cap X^\pm) = \int_E f_\pm d\mu \quad \text{for all } E \in \mathcal{A}.$$

One verifies that $f_\pm$ vanish μ-a.e. in $X^\mp$ and that the function claimed by the theorem is $f = f_+ - f_-$. ∎

Problems and Complements

1c Measurable Functions

In the problems **1.1–1.12** $\{X, \mathcal{A}, \mu\}$ is a measure space and $E \in \mathcal{A}$.

1.1. The characteristic function of a set E is measurable if and only if E is measurable.

1.2. A function f is measurable if and only if its restriction to any measurable subset of its domain is measurable.

1.3. A function f is measurable if and only if f^+ and f^- are both measurable.

1.4. Let $\{X, \mathcal{A}, \mu\}$ be complete. A function defined on a set of measure zero is measurable.

1.5. Let $\{X, \mathcal{A}, \mu\}$ not be complete. Then there exists a measurable set A of measure zero that contains a nonmeasurable set B. The two functions $f = \chi_A$ and $g = \chi_{A-B}$ differ on a set of outer measure zero. However f is measurable and g is not.

1.6. If f is measurable then $[f = c]$ is measurable for all c in the range of f. The converse is false.

1.7. Let f be measurable. Then also $|f|^{p-1}f$ is measurable for all $p > 0$.

1.8. $|f|$ measurable does not imply that f is measurable. Likewise f^2 measurable does not imply that f is measurable.

1.9. A function f is measurable if and only if its preimage of a Borel set is measurable.

1.10. Let $\{f_n\}$ be a sequence of measurable functions from E into $\mathbb{R}^*$. The subset of E where $\lim f_n$ exists is measurable.

1.11. The supremum(infimum) of an uncountable family $\{f_\alpha\}$ of measurable functions, need not be measurable.

1.12. Upper(lower) semi-continuous functions $f : E \to \mathbb{R}^*$ are measurable.

In the problems **1.13–1.19**, $\{\mathbb{R}^N, \mathcal{M}, \mu\}$ is $\mathbb{R}^N$ with the Lebesgue measure and $E \in \mathcal{M}$.

1.13. There exists a nonmeasurable function $f : \mathbb{R} \to \mathbb{R}$ such that $f^{-1}(y)$ is measurable for all $y \in \mathbb{R}$. **Hint:** given a nonmeasurable set $E \subset \mathbb{R}$, define $f(x) = x$ for $x \in \mathbb{R} - E$ and $f(x) = -x$ for $x \in E$.

1.14. A monotone function f in some interval $(a, b) \subset \mathbb{R}$ is measurable.

1.15. Let $f : \mathbb{R} \to \mathbb{R}$ be measurable and let $g : \mathbb{R} \to \mathbb{R}$ be continuous. The composition $g(f) : E \to \mathbb{R}$ is measurable. However the composition $f(g) : E \to \mathbb{R}$ in general is not measurable. To construct a counterexample consider the function of § 14 of Chap. 3.

1.16. Let $f : [0, 1] \to \mathbb{R}^*$ be measurable. Then $\chi_{\mathbb{Q} \cap [0,1]}(f)$, is measurable.

1.17. Let $f : \mathbb{R}^2 \to \mathbb{R}^*$ be such that $f(\cdot, y)$ is continuous for all $y \in \mathbb{R}$ and $f(x, \cdot)$ is measurable for all $x \in \mathbb{R}$. Prove that f is measurable.

We indicate two approaches to this statement. The first is based on the following lemma.

Lemma 1.1c *Let $c \in \mathbb{R}$ and let $y \in \mathbb{R}$ be fixed. Then $f(x, y) \geq c$ if and only if, for all $n \in \mathbb{N}$ there exists a rational number r_m such that*

$$|x - r_m| < \tfrac{1}{n} \quad \text{and} \quad f(r_m, y) > c - \tfrac{1}{n}.$$

Proof Since $g = f(\cdot, y)$ is continuous $g^{-1}(c - \tfrac{1}{n}, \infty)$ is open and we may select $r_m \in \mathbb{Q}$ with the indicated properties. Conversely, let x satisfy the property of the lemma, and assume by contradiction that $g(x) < c$. Since g is continuous, there exists $k \in \mathbb{N}$ such that $g(x) < c - \tfrac{1}{k}$. Since $x \in g^{-1}(-\infty, c - \tfrac{1}{k})$, and since this set is open, there exists $h \in \mathbb{N}$ such that

$$(x - \tfrac{1}{h}, x + \tfrac{1}{h}) \subset g^{-1}(-\infty, c - \tfrac{1}{k}).$$

Take $n = \max\{k; h\}$ and select any $r \in \mathbb{Q}$ such that

$$|x - r| < \tfrac{1}{m} \quad \text{and} \quad r \in g^{-1}(-\infty, c - \tfrac{1}{k}).$$

For such choices

$$g(r) = f(r, y) < c - \tfrac{1}{n} \le c - \tfrac{1}{k}. \qquad\qquad \blacksquare$$

Using the lemma

$$[f \ge c] = \bigcap_n \bigcup_m \left(\left[f(r_m, \cdot) > c - \tfrac{1}{m} \right] \times B_{\frac{1}{n}}(r_m) \right).$$

The second approach is based on the following construction. For $n \in \mathbb{N}$ and $j \in \mathbb{Z}$ set $a_i = i/n$ and

$$f_n(x, \cdot) = \frac{f(a_{j+1}, \cdot)(x - a_j) - f(a_j, \cdot)(x - a_{j+1})}{a_{j+1} - a_j}, \qquad \text{for } a_{j+1} \le x \le a_j.$$

1.18. Find an example of a function $f : \mathbb{R}^2 \to \mathbb{R}^*$ measurable in each of its variables separately, and not measurable. **Hint:** see the Sierpinsky example **13.4.** of § 13c.

1.19. Let T be a linear nonsingular transformation of $\mathbb{R}^N$ onto itself and let $f : \mathbb{R}^N \to \mathbb{R}^*$ be measurable. Prove that $f(T) : \mathbb{R}^N \to \mathbb{R}^*$ is measurable. **Hint:** By § 12.3c of Chap. 3 it suffices to show that

$$[f(T) > c] = T^{-1}[f > c] \quad \text{for all } c \in \mathbb{R}.$$

1.1c   Sublevel Sets

Let $\{X, \mathcal{A}, \mu\}$ be a measure space and let $f : X \to \mathbb{R}$ be measurable. For $t \in \mathbb{R}$, the sublevel set of f at t, is defined as $E_t = [f \le t]$. By the definition

$$E_s \subset E_t \text{ for } s < t; \quad \bigcup E_t = X; \quad \bigcap E_t = \emptyset; \quad \bigcap_{s<t} E_t = E_s. \qquad (1.1c)$$

Conversely, given a collection of measurable sets $\{E_t\}_{t\in\mathbb{R}}$ in Ω, satisfying (1.1c) there exists a measurable function $f : X \to \mathbb{R}$ for which $E_t = [f \le t]$. Verify that (von Neumann [114])

$$X \ni x \to f(x) = \inf\{t : x \in E_t\}$$

is such a function and prove that it is unique.

2c The Egorov–Severini Theorem

2.1. Let $\{X, \mathcal{A}, \mu\}$ be a measure space and let $E \in \mathcal{A}$ be of finite measure. Let $\{f_n\}$ be a sequence of measurable functions from E into $\mathbb{R}^*$. Assume that for a.e. $x \in E$ the set $\{f_n(x)\}$ is bounded. For every $\varepsilon > 0$ there exists a measurable set $E_\varepsilon \subset E$ and a positive number k_ε such that

$$\mu(E - E_\varepsilon) \le \varepsilon \quad \text{and} \quad |f_n| \le k_\varepsilon \quad \text{on } E_\varepsilon \quad \text{for all } n \in \mathbb{N}.$$

2.2. Let μ be the counting measure on $2^\mathbb{N}$. Define f_n as the characteristic function of $\{1, \ldots, n\}$. The sequence $\{f_n\}$ converges to 1 everywhere in $\mathbb{N}$ but not in measure.

Proposition 2.1c *Let* $\{X, \mathcal{A}, \mu\}$ *be a measure space and let* $E \in \mathcal{A}$. *Let* $\{f_n\}$ *and* f *be measurable functions from* E *into* $\mathbb{R}^*$. *Assume that* f *is finite a.e. in* E. *Then* $\{f_n\} \to f$ *a.e. in* E *if and only if for all* $\eta > 0$

$$\lim \mu\Big(\bigcup_{j=n}^{\infty} [|f_j - f| \ge \eta] \Big) = 0. \tag{2.1c}$$

Hint: Denoting by A the set where $\{f_n\}$ is not convergent

$$A = \bigcup_m \limsup \big[|f_n - f| \ge \tfrac{1}{m}\big].$$

Then $\mu(A) = 0$ if and only if (2.1c) holds.

3c Approximating Measurable Functions by Simple Functions

Proposition 3.1c *Let* $\{X, \mathcal{A}, \mu\}$ *be a measure space and let* $f : X \to \mathbb{R}^*$ *be non-negative and measurable. There exists a countable collection* $\{E_j\} \subset \mathcal{A}$ *such that* $X = \bigcup E_j$ *and*

$$f = \sum_{j \in \mathbb{N}} \frac{1}{j} \chi_{E_j}. \tag{3.1c}$$

Proof Let $E_1 = [f \ge 1]$ and $f_1 = \chi_{E_1}$. Then, for $j \ge 2$ define recursively

$$E_j = \Big[f \ge \frac{1}{j} + f_{j-1} \Big] \quad \text{where} \quad f_j = \sum_{i=1}^{j} \frac{1}{i} \chi_{E_i}.$$

By construction $f \ge f_j$ for all j. If $f(x) = \infty$ then $x \in E_j$ for all j. Hence $f(x)$ has the representation (3.1c). If $f(x) < \infty$ then $x \notin E_j$ for infinitely many j. For such an x

$$0 \le f(x) - f_j(x) \le \frac{1}{j+1} \quad \text{for infinitely many } j.$$ ∎

The functions f_j are simple and hence the proposition gives an alternative proof of Proposition 3.1. However simple functions, even in canonical form, cannot be written, in general, as a finite sum of the type of (3.1c). As an example:

3.1. Find the representation (3.1c) of a positive, real multiple of the characteristic function of the interval $(0, 1)$.

3.2. Find the representation (3.1c) of a simple function taking only two positive values on distinct, measurable sets.

4c Convergence in Measure

In the problems **4.1–4.7**, $\{X, \mathcal{A}, \mu\}$ is complete measure space, and $E \in \mathcal{A}$ is of finite measure.

4.1. Let $f : E \to \mathbb{R}^*$ be measurable and assume that $|f| > 0$ a.e. on E. For every $\varepsilon > 0$ there exists a measurable set $E_\varepsilon \subset E$ and a positive number δ_ε such that $\mu(E - E_\varepsilon) \le \varepsilon$ and $|f| > \delta_\varepsilon$ on E_ε.

4.2. Let $\{f_n\} : E \to \mathbb{R}^*$ be a sequence of measurable functions converging to f a.e. in E. Assume that $|f| > 0$ and $|f_n| > 0$ a.e. on E for all $n \in \mathbb{N}$. For every $\varepsilon > 0$ there exists a measurable set $E_\varepsilon \subset E$ and a positive number δ_ε such that

$$\mu(E - E_\varepsilon) \le \varepsilon \quad \text{and} \quad |f_n| > \delta_\varepsilon \quad \text{on } E_\varepsilon \quad \text{for all } n \in \mathbb{N}.$$

4.3. Let μ be the counting measure on the rationals of $[0, 1]$. Then convergence in measure is equivalent to uniform convergence.

4.4. Let $\{f_n\} : E \to \mathbb{R}^*$ be a sequence of measurable and a.e. finite functions. There exists a sequence of positive numbers $\{k_n\}$ such that $f_n k_n^{-1} \to 0$ a.e. in E.

4.5. Let $\{f_n\}, \{g_n\} : E \to \mathbb{R}^*$ be sequences of measurable functions converging in measure to f and g, respectively, and let $\alpha, \beta \in \mathbb{R}$. Then

$$\{\alpha f_n + \beta g_n\}, \quad \{|f_n|\}, \quad \{f_n g_n\} \quad \longrightarrow \quad \alpha f + \beta g, \quad |f|, \quad fg$$

in measure. Moreover if $f \ne 0$ a.e. on E and $f_n \ne 0$ a.e. on E for all n, then $1/f_n$ converges to $1/f$ in measure. (*Hint*: Use **4.1–4.2**).

4.6. Let $\{X, \mathcal{A}, \mu\}$ be a measure space and let $\{E_n\}$ be a sequence of measurable sets. The sequence $\{\chi_{E_n}\}$ converges in measure if and only if $d(E_n; E_m) \to 0$ as $n, m \to \infty$ (see **2.2.** and **3.7.** of the Complements of Chap. 3).

4.7. Let $\{f_n\} : E \to \mathbb{R}^*$ be a sequence of measurable and a.e. finite functions. Prove that $\{f_n\} \to f$ in measure if and only if every subsequence of $\{f_n\}$ contains in turn a subsequence converging to f in measure.

7c The Lebesgue Integral of Nonnegative Measurable Functions

7.1c Comparing the Lebesgue Integral with the Peano-Jordan Integral

Let $E \subset \mathbb{R}^N$ be bounded and PeanoJordan measurable, and denote by $\mathcal{P} = \{E_n\}$ a finite partition of E into pairwise disjoint PeanoJordan measurable sets. For a bounded function $f : E \to \mathbb{R}$ set

$$h_n = \inf_{x \in E_n} f(x) \quad \mathcal{F}_{\mathcal{P}}^- = \sum h_n \mu_{\mathcal{P}-\mathcal{J}}(E_n)$$
$$k_n = \sup_{x \in E_n} f(x) \quad \mathcal{F}_{\mathcal{P}}^+ = \sum k_n \mu_{\mathcal{P}-\mathcal{J}}(E_n).$$

A bounded function $f : E \to \mathbb{R}$ is PeanoJordan integrable if for every $\varepsilon > 0$, there exists a partition $\mathcal{P}_\varepsilon$ of E into PeanoJordan measurable sets E_n, such that

$$\mathcal{F}_{\mathcal{P}}^+ - \mathcal{F}_{\mathcal{P}}^- \leq \varepsilon.$$

The sets E_n are also Lebesgue measurable. Therefore,

$$\sum h_n \mu_{\mathcal{P}-\mathcal{J}}(E_n) \leq \int_E f \, d\mu \leq \sum k_n \mu_{\mathcal{P}-\mathcal{J}}(E_n).$$

Thus if a bounded function f is PeanoJordan integrable it is also Lebesgue integrable. The converse is false. Indeed the characteristic function of the rationals $\mathbb{Q}$ of $[0, 1]$ is Lebesgue integrable and not PeanoJordan integrable.

This is not longer the case however if f is not bounded. Following Riemann's notion of improper integral, the function

$$f(x) = \frac{1}{x} \sin \frac{1}{x} \qquad \text{for} \quad x \in (0, 1]$$

is PeanoJordan integrable in $(0, 1)$ but not Lebesgue integrable.

In the problems **7.1–7.7**, $\{X, \mathcal{A}, \mu\}$ is a measure space and $E \in \mathcal{A}$.

7.1. Let $f : E \to \mathbb{R}^*$ be measurable and nonnegative. If $f = \infty$ in a set $\mathcal{E} \subset E$ of measure zero

$$\int_E f \, d\mu = \int_{E - \mathcal{E}} f \, d\mu.$$

7.2. Let $f : X \to \mathbb{R}^*$ be measurable and nonnegative. Then

$$\int_E f \, d\mu = 0 \quad \text{for all } E \in \mathcal{A} \quad \text{implies} \quad f = 0 \text{ a.e. in } X.$$

7.3. Let $f : E \to \mathbb{R}^*$ be integrable. If the integral of f over every measurable subset $A \subset E$ is nonnegative, then $f \geq 0$ a.e. on E.

7.4. Let $f : \mathbb{R} \to \mathbb{R}$ be Lebesgue integrable. Then for every $h \in \mathbb{R}$ and every interval $[a, b] \subset \mathbb{R}$

$$\int_{[a,b]} f d\mu = \int_{[a+h,b+h]} f(x - h) d\mu.$$

7.5. Construct the Lebesgue integral of a nonnegative μ-measurable function $f : \mathbb{R}^N \to \mathbb{R}$, when μ is the Dirac delta measure δ_x concentrated at some $x \in \mathbb{R}^N$.

7.6. Let E be of finite measure. A measurable function $f : E \to \mathbb{R}^*$ is integrable if and only if $\sum \mu([|f| \geq n]) < \infty$. **Hint:**

$$\sum \mu([|f| \geq n]) = \sum n\mu([n \leq |f| < n + 1]).$$

7.7. Let $\{X, \mathcal{A}, \mu\}$ be $\mathbb{R}^N$ with the Lebesgue measure. Let T be a linear nonsingular transformation of $\mathbb{R}^N$ onto itself and let $f : \mathbb{R}^N \to \mathbb{R}^*$ be Lebesgue integrable. Prove that

$$\int_E f(x) d\mu = \frac{1}{|\det T|} \int_{TE} f(y) d\mu.$$

Hint: § 12.3.1c of the Complements of Chap. 3 and Problem **1.19**.

7.2c On the Definition of the Lebesgue Integral

The original definition of Lebesgue was based only on the Lebesgue measure in $\mathbb{R}^N$. The integral of a measurable, nonnegative function f was defined as in (7.1), where the supremum was taken over the class of simple functions $\zeta \leq f$, and vanishing outside a set of finite measure. Denote by Φ_f such a class of simple functions and observe that

$$\int_E f d\mu = \sup_{\varphi \in \Phi_f} \int_E \varphi d\mu.$$

Such a definition, while adequate for the Lebesgue measure in $\mathbb{R}^N$, is not adequate for a general measure space $\{X, \mathcal{A}, \mu\}$. For example let $\{X, \mathcal{A}, \mu\}$ be the measure space of **3.2** of Chap. 3. Then

$$\infty = \int_X 1 d\mu = \sup_{\varphi \in \Phi_1} \int_X \varphi d\mu = 0.$$

9c More on the Lebesgue Integral

Let $\{X, \mathcal{A}, \mu\}$ be a measure space with $\mu(X) = 1$ and let $f : X \to \mathbb{R}$ be μ-measurable. In particular, for all Borel sets $B \in \mathbb{R}$, the set $f^{-1}(B)$ is μ-measurable. Set

$$\mu_f(B) = \mu([f^{-1}(B)]) \quad \text{for all Borel sets } B \subset \mathbb{R}. \tag{9.1c}$$

9.1. **Distribution Measure of a Measurable Function:** Verify that μ_f is a measure on $\mathcal{B}$. Verify that if $f = \chi_E$ for some $E \in \mathcal{A}$, then for a Borel set $B \subset \mathbb{R}$

$$\mu_f(B) = \chi_B(1)\mu(E) + \chi_B(0)\mu(X - E).$$

9.2. A function $h : \mathbb{R} \to \mathbb{R}^*$ is Borel measurable if the preimage of a Borel set in $\mathbb{R}$ is a Borel set in $\mathbb{R}$. Let h be Borel measurable and integrable with respect to μ_f. Prove that $h(f)$ is integrable in $\{X, \mathcal{A}, \mu\}$ and

$$\int_X h(f)d\mu = \int_{\mathbb{R}} h(t)d\mu_f. \tag{9.2c}$$

Hint: Given μ_f, compute the integral of both sides by assuming first that $h = \chi_B$ where $B \subset \mathbb{R}$ is a Borel set. Then in view of (9.1c), the integrals equal $\mu_f(B) = \mu\{f^{-1}(B)\}$. Next assume that h is nonnegative. Compute both sides of (9.2c) by using the definition (7.1) and show that both sides are equal. The general case follows by linear combinations and limiting processes.

9.3. For a μ-measurable function f on $\{X, \mathcal{A}, \mu\}$ prove that

$$\int_X |f|^p d\mu = \int_0^\infty t^p d\mu_f$$

9.4. **Equidistributed Measurable Functions:** Two measurable function $f_1, f_2 : X \to \mathbb{R}$ are equidistributed if

$$\mu[f_1^{-1}(B)] = \mu[f_2^{-1}(B)] \quad \text{for every Borel set } B \subset \mathbb{R}.$$

Give an example of equidistributed measurable functions $f_1 \neq f_2$. Prove that if f_1 and f_2 are equi-distributed, then

$$\int_X h(f_1)d\mu = \int_X h(f_2)d\mu \quad \text{for all } h(\cdot) \text{ as in } \mathbf{9.2}.$$

9.5. **Expectation and Variance of a Measurable Function:** The expectation $E(f)$ and the variance $\sigma^2(f)$ of a measurable function f are defined as

$$E(f) = \int_X f d\mu = \int_{\mathbb{R}} t d\mu_f$$

$$\sigma^2(f) = \int_X (f - E(f))^2 d\mu = \int_{\mathbb{R}} (t - E(f))^2 d\mu_f. \tag{9.3c}$$

The quantity $\sigma(f)$, is the *standard deviation* of f. Prove that $\sigma^2(f) = 0$ if and only if there exists $t \in \mathbb{R}$ such that $f = 1$ a.e. in X and $t = E(f)$.

9.6.   Let f_1 and f_2 be equidistributed measurable functions on $\{X, \mathcal{A}, \mu\}$. Then $\sigma^2(f_1) = \sigma^2(f_2)$.

10c   Convergence Theorems

In the problems **10.1–10.10**, $\{X, \mathcal{A}, \mu\}$ is a measure space and $E \in \mathcal{A}$.

10.1.   Let $f : E \to \mathbb{R}^*$ be integrable and let $\{E_n\}$ be a countable collection of measurable, disjoint subsets of E such that $E = \cup E_n$. Then

$$\int_E f d\mu = \sum \int_{E_n} f d\mu.$$

10.2.   Let μ be the counting measure on the positive rationals $\{r_1, r_2, \dots\}$, and let

$$f_n(r_j) = \begin{cases} \dfrac{1}{j} & \text{if } j < n \\[2mm] 0 & \text{if } j \geq n. \end{cases}$$

The sequence $\{f_n\}$ converges uniformly to a nonintegrable function.

10.3.   Let μ be a finite measure and let $\{f_n\}$ be a sequence of integrable functions converging uniformly in X. The limiting function f is integrable and one can pass to the limit under integral.

10.1c   Another Version of Dominated Convergence

Theorem 10.1c *Let $\{f_n\}$ be a sequence of integrable functions in E converging a.e. in E to some f. Assume that there exists a sequence of integrable functions $\{g_n\}$ converging a.e. in E to an integrable function g and such that*

$$\lim \int_E g_n d\mu = \int_E g d\mu.$$

Assume moreover that $|f_n| \le g_n$ a.e. in E for all $n \in \mathbb{N}$. Then f is integrable and

$$\lim \int_E f_n d\mu = \int_E f d\mu.$$

10.4. Let $\{f_n\} : E \to \mathbb{R}^*$ be a sequence of integrable functions converging to an integrable function f a.e. in E. Then

$$\lim \int_E |f_n - f| d\mu = 0 \quad \text{if and only if} \quad \lim \int_E |f_n| d\mu = \int_E |f| d\mu.$$

10.5. Let $\{f_n\} : E \to \mathbb{R}^*$ be a sequence of measurable functions satisfying

$$\sum \int_E |f_n| d\mu < \infty.$$

Then $\sum f_n$ defines, a.e. in E and integrable function and

$$\int_E \sum f_n d\mu = \sum \int_E f_n d\mu.$$

10.6. Let $\{f_n\} : E \to \mathbb{R}^*$ be a sequence of measurable functions satisfying $|f_n| \le |f|$ for an integrable function $f : E \to \mathbb{R}$. Then

$$\int_E \liminf f_n d\mu \le \liminf \int_E f_n d\mu \le \limsup \int_E f_n d\mu \le \int_E \limsup f_n d\mu.$$

10.7. Let E be of finite measure and let $\{f_n\} : E \to \mathbb{R}^*$ be a sequence of measurable functions satisfying $|f_n| \le |g|$ for an integrable function $g : E \to \mathbb{R}$. If $\{f_n\} \to f$ in measure, then

$$\int_E f d\mu = \lim \int_E f_n d\mu \quad \text{and} \quad \lim \int_E |f_n - f| d\mu = 0.$$

10.8. Let E be of finite measure and let $\{f_n\} : E \to \mathbb{R}^*$ be a sequence of nonnegative measurable functions converging in measure to f. Then

$$\int_E f d\mu \le \liminf \int_E f_n d\mu.$$

10.9. Prove that in the Egorov-Severini Theorem 2.1 the assumption $\mu(E) < \infty$ can be replaced by $|f_n| \le g$ for an integrable function $g : E \to \mathbb{R}^*$.

10.10. Let $\{f_n\} : E \to \mathbb{R}^*$ be a nonincreasing sequence of nonnegative, measurable functions. Give an example to show that in general

$$\lim \int_E f_n d\mu \neq \int_E \lim f_n d\mu.$$

Thus in the Monotone Convergence Theorem 8.1, the monotonicity assumption $f_n \leq f_{n+1}$, cannot be replaced with the monotonicity assumption $f_n \geq f_{n+1}$.

In the problems **10.11–10.16**, $\{X, \mathcal{A}, \mu\}$ is $\mathbb{R}^N$ with the Lebesgue measure.

10.11. Let sequences $\{f_n\}$ be defined by

$$f_n(x) = \begin{cases} n & \text{if } x \in [0, \frac{1}{n}] \\ 0 & \text{otherwise,} \end{cases} \qquad f_n = \begin{cases} \frac{1}{n} & \text{for } 0 \leq x \leq n \\ 0 & \text{for } x > n. \end{cases}$$

In either case

$$1 = \lim \int_{\mathbb{R}^+} f_n dx \neq \int_{\mathbb{R}^+} \lim f_n dx = 0.$$

10.12. Let $f : \mathbb{R} \to \mathbb{R}$ be Lebesgue measurable, nonnegative, and locally bounded. Assume that f is Riemann-integrable on $\mathbb{R}$. Then f is Lebesgue integrable on $\mathbb{R}$ and

$$\int_{\mathbb{R}} f d\mu = \lim \int_{-n}^{n} f d\mu.$$

10.13. Let $\{f_n\}$ be the sequence of nonnegative integrable functions defined on $\mathbb{R}^N$ by

$$f_n(x) = \begin{cases} n^N \exp\left\{ \dfrac{-1}{1 - n^2|x|^2} \right\} & \text{if } |x| < \frac{1}{n} \\ 0 & \text{if } |x| \geq \frac{1}{n}. \end{cases}$$

The sequence $\{f_n\}$ converges to zero a.e. in $\mathbb{R}^N$ and each f_n is integrable with uniformly bounded integral. However

$$\lim \int_{\mathbb{R}^N} f_n dx \neq \int_{\mathbb{R}^N} \lim f_n dx.$$

10.14. Let $\{f_n\}$ be the sequence defined in **7.7** of the Complements of Chap. 2. The assumptions of the dominated convergence theorem fail and

$$\lim \int_0^1 f_n(x) dx = \frac{1}{2} \quad \text{and} \quad \int_0^1 \lim f_n(x) dx = 0.$$

10.15. The two sequences $\{x^n\}$ and $\{nx^n\}$ converge to zero in $(0, 1)$. For the first one can pass to the limit under integral and for the second one cannot.

10.16. **Integrability and Boundedness:** There exist positive, Lebesgue integrable functions on $\mathbb{R}$, that are infinity at all the rationals. Let $\{r_n\}$ be the sequence of the rationals in $\mathbb{R}^+$ and set

$$f(x) = \sum \frac{1}{2^n} \frac{e^{-x}}{\sqrt{|x - r_n|}}.$$

Prove that f is integrable over $\mathbb{R}^+$ (Proposition 10.2).

11c   Absolute Continuity of the Integral

The proof of Theorem 11.1 shows that for an integrable function f, given $\varepsilon > 0$, the corresponding δ claimed by the theorem, depends upon ε and f.

A collection Φ of integrable functions $f : E \to \mathbb{R}^*$ is *uniformly integrable*, if for all $\varepsilon > 0$ there exists $\delta = \delta(\varepsilon) > 0$ such that the conclusion of Theorem 11.1 holds for all $f \in \Phi$, for all measurable sets $\mathcal{E} \subset E$ such that $\mu(\mathcal{E}) < \delta$ (Vitali [169]).

11.1. If Φ is finite then it is uniformly integrable.
11.2. Let $\{f_n\}$ be a sequence of integrable functions such that $|f_n| \le g$ for an integrable function g. Then $\{f_n\}$ is uniformly integrable.
11.3. Let E be of finite measure and let $\{f_n\}$ be a sequence of uniformly integrable functions converging a.e. in E to an integrable function f. Prove that (**Hint:** Egorov–Severini theorem)

$$\lim \int_E |f_n - f| d\mu = 0. \qquad (11.1c)$$

This gives an alternate proof of the Lebesgue dominated convergence theorem when $\mu(E) < \infty$. Give an example showing that the conclusion is false if E is not of finite measure.
11.4. Let $\{f_n\}$ be a sequence of integrable functions in E, satisfying (11.1c) for an integrable f. Then $\{f_n\}$ is uniformly integrable.
11.5. Show that the following sequences of functions defined in $(-1, 1)$, are not uniformly Lebesgue integrable.

$$f_n(x) = \sqrt{n}e^{-x^2}n; \qquad f_n(x) = \frac{x}{\sqrt{n}}e^{-x^2}n;$$

$$\qquad\qquad\qquad\qquad\qquad\qquad (11.2c)$$

$$f_n(x) = \frac{n}{n^2x^2 + 1}; \qquad f_n(x) = \frac{n^2x}{(n^2x^2 + 1)^2}.$$

11.6. Let E be of finite measure and let $\{f_t\} : E \to \mathbb{R}^*$ be a family of functions uniformly integrable in E and such that $f_t(x)$ is continuous in $t \in (0, 1)$ for every fixed $x \in E$. Prove that

$$(0, 1) \ni t \to \int_E f_t d\mu \quad \text{is continuous.}$$

12c Product of Measures

12.1c Product of a Finite Sequence of Measure Spaces

Given a finite sequence of measure spaces $\{X_j, \mathcal{A}_j, \mu_j\}_{j=1}^n$ for some $n \in \mathbb{N}$ their product space is constructed by the following steps:

12.1.   **Measurable n-Rectangles.** A measurable n-rectangle is a set of the form

$$E = \prod_{j=1}^n E_j \quad \text{for} \quad E_j \in \mathcal{A}_j \quad \text{for all} \quad j = 1, \dots, n.$$

Denote by $\mathcal{R}_o^n$ be the collection of all measurable n-rectangles and prove that it forms a semi-algebra.

12.2.   **Measuring Measurable Rectangles.** On $\mathcal{R}_o^n$ introduce the set function

$$\mathcal{R}_o^n \ni E \to \lambda_n(E) = \prod_{j=1}^n \mu_j(E_j). \tag{12.1c}$$

Prove that λ_n is countably additive on $\mathcal{R}_o^n$. For this use a n-dimensional version of Proposition 12.1. As a consequence λ_n is well defined on $\mathcal{R}_o^n$ in the sense that, for $Q \in \mathcal{R}_o^n$, the value $\lambda_n(Q)$ is independent of a particular partition of elements in $\mathcal{R}_o^n$ making up Q.

12.3.   **Constructing the Product Measure Space.** The set function λ_n on $\mathcal{R}_o^n$ generates an outer measure μ_e^n on $\prod_{j=1}^n X_j$. The latter in turn generates the measure space $\{Y_n, \mathcal{B}_n, \nu_n\}$, where

$$Y_n = \prod_{j=1}^n X_j, \quad \mathcal{B}_n = \Big(\prod_{j=1}^n \mathcal{A}_j\Big), \quad \nu_n = \Big(\prod_{j=1}^n \mu_j\Big).$$

Moreover $\mathcal{R}_o^n \subset \mathcal{B}_n$ and $\nu_n(E) = \lambda_n(E)$ for all $E \in \mathcal{R}_o^n$.

12.1.1c   Alternate Constructions

Fix an integer $1 \le m < n$ and construct first, by the indicated procedure, the two measure spaces $\{Y_m, \mathcal{B}_m, \nu_m\}$ and $\{Y^m, \mathcal{B}^m, \nu^m\}$, where

$$Y_m = \prod_{j=1}^{m} X_j, \qquad \mathcal{B}_m = \left(\prod_{j=1}^{m} \mathcal{A}_j \right), \qquad \nu_m = \left(\prod_{j=1}^{m} \mu_j \right)$$

$$Y^m = \prod_{j=m+1}^{n} X_j, \quad \mathcal{B}^m = \left(\prod_{j=m+1}^{n} \mathcal{A}_j \right), \quad \nu^m = \left(\prod_{j=m+1}^{n} \mu_j \right).$$

Then construct their product

$$\{Y_m \times Y^m, (\mathcal{B}_m \times \mathcal{B}^m), (\nu_m \times \nu^m)\}. \tag{12.2c}$$

Denote by $\mathcal{A}_o^n$ the smallest σ-algebra containing $\mathcal{R}_o^n$.

12.4. Prove that $\mathcal{A}_o^n$ is contained in $(\mathcal{B}_m \times \mathcal{B}^m)$ for all $1 \le m < n$.

12.5. Prove that for all $1 \le m < n$, the restriction of $(\nu_m \times \nu^m)$ to $\mathcal{R}_o^n$ coincides with the function λ_n introduced in (12.1c).

12.6. Prove that for all $1 \le m < n$, the product space in (12.2c) is generated by the same outer measure μ_e^n, generated by λ_n on $\mathcal{R}_o^n$. Conclude that

$$\{Y_n, \mathcal{B}_n, \nu_n\} = \{Y_m \times Y^m, (\mathcal{B}_m \times \mathcal{B}^m), (\nu_m \times \nu^m)\} \tag{12.3c}$$

and thus the finite product of measure spaces is associative.

12.7. Prove that the Lebesgue measure in $\mathbb{R}^N$ coincides with the product of N copies of the Lebesgue measure in $\mathbb{R}$.

13c On the Structure of $(\mathcal{A} \times \mathcal{B})$

13.1. Let $\{X, \mathcal{A}, \mu\}$ and $\{Y, \mathcal{B}, \nu\}$ be complete measure spaces. Let $A \subset X$ be μ-measurable and let $B \subset Y$ be not ν-measurable. Denote by $(\mathcal{A} \times \mathcal{B})_e$ the outer measure on $\mathcal{A} \times \mathcal{B}$ that generates the product measure space $\{X \times Y, (\mathcal{A} \times \mathcal{B}), (\mu \times \nu)\}$. Prove that:

i.   If $(\mathcal{A} \times \mathcal{B})_e (A \times B) = 0$ then $A \times B$ is $(\mathcal{A} \times \mathcal{B})$-measurable.

ii. If $0 < (\mathcal{A} \times \mathcal{B})_e (A \times B) < \infty$ then $A \times B$ is not $(\mathcal{A} \times \mathcal{B})$-measurable. If $(\mathcal{A} \times \mathcal{B})_e (A \times B) = \infty$ then $A \times B$ might be $(\mu \times \nu)$-measurable. Give an example or an argument.

iii. If $\{X, \mathcal{A}, \mu\}$ and $\{Y, \mathcal{B}, \nu\}$ are σ-finite and $(\mathcal{A} \times \mathcal{B})_e (A \times B) > 0$, then then $A \times B$ is not $(\mathcal{A} \times \mathcal{B})$-measurable.

13.2. Let $\{X, \mathcal{A}, \mu\}$ be non complete and let $A \subset X$ be non μ-measurable, but included in a μ-measurable set A' of measure zero. For every ν-measurable set $B \subset Y$, the rectangle $A \times B$ is $(\mu \times \nu)$-measurable. As a consequence, the assumption that both $\{X, \mathcal{A}, \mu\}$ and $\{Y, \mathcal{B}, \nu\}$ be complete, cannot be removed from Proposition 13.4.

13.3.   Let $E \subset [0, 1]$ be the Vitali non measurable set. The diagonal set $\mathcal{E} = \{(x, x) | x \in E\}$ is Lebesgue measurable in $\mathbb{R}^2$ and has measure zero. The rectangle $E \times E$ is not Lebesgue measurable in $\mathbb{R}^2$.

13.4.   **(Sierpinski [146]).** Let $\mathbb{R}$ be well ordered by $\prec$, denote by Ω the first uncountable and let

$$X = Y = E_\Omega = \{x \in \mathbb{R} | x \prec \Omega\}.$$

Let $\mathcal{A}$ be the σ-algebra of the subsets of X that are either countable or their complement is countable. For $E \in \mathcal{A}$, let $\mu(E) = 0$ if E is countable and $\mu(E) = 1$ otherwise. Consider the set

$$E = \{(x, y) \in X \times X \mid x \prec y\}.$$

All the x and y sections of E are measurable. However E is not $(\mu \times \mu)$-measurable. Indeed if E were measurable, it would have finite measure, since

$$E \subset X \times X \quad \text{and} \quad (\mu \times \mu)(X \times X)) = \mu(X)\mu(X) = 1.$$

Therefore, it would have to satisfy Proposition 13.4. However

$$\int_Y \mu(E_y) d\nu = 0 \quad \text{and} \quad \int_X \mu(E_x) d\mu = 1.$$

13.1c   Sections and Their Measure

Given a finite sequence $\{X_j, \mathcal{A}, \mu_j\}_{j=1}^n$ of measure spaces, let $\{Y_n, \mathcal{B}_n, \nu_n\}$ be their product measure space as constructed either in (12.1c)–(12.3c).

For a set $E \subset Y_n$ and $y_m \in Y_m$, and $y^m \in Y^m$ for some $1 \leq m < n$, the y_m-section E_{y_m} of E and the y^m-section E_{y^m} of E are defined as

$$Y^m \ni E_{y_m} = \{y^m \mid (y_m, y^m) \in E \text{ for a fixed } y_m \in Y_m\}$$
$$Y_m \ni E_{y^m} = \{y_m \mid (y_m, y^m) \in E \text{ for a fixed } y^m \in Y^m\}. \tag{13.4c}$$

13.5.   Prove that for all $E \in \mathcal{A}_o^n$ and all $1 \leq m < n$ the sections $E_{y_m} \in \mathcal{B}^m$ and $E_{y^m} \in \mathcal{B}_m$.

13.6.   Prove that for all $E \in \mathcal{A}_o^n$ of finite ν_n measure and for all $1 \leq m < n$

$$\nu_n(E) = \int_{Y_n} \chi_E d\nu_n = \int_{Y_m} \nu^m(E_{y_m}) d\nu_m = \int_{Y^m} \nu_m(E_{y^m}) d\nu^m. \tag{13.5c}$$

13.7.   State and prove a version of Fubini's Theorem when all the measure spaces $\{X_j, \mathcal{A}_j, \mu_j\}$, for $j = 1, \ldots, n$, are complete.

13.8. State and prove a version of Tonelli's Theorem when all the measure spaces $\{X_j, \mathcal{A}_j, \mu_j\}$, for $j = 1, \ldots, n$, are complete and σ-finite.

14c The Theorem of Fubini–Tonelli

14.1. Let ω_N be the measure of the unit sphere in $\mathbb{R}^N$. Then

$$\omega_{N+1} = 2\omega_N \int_0^{\pi/2} (\sin t)^{N-1} dt.$$

14.2. By the Fubini–Tonelli theorem

$$\int_{\mathbb{R}^N} e^{-|x|^2} dx = \prod_{i=1}^{N} \int_{\mathbb{R}} e^{-x_i^2} dx_i = \pi^{N/2}.$$

14.3. Let $\{X, \mathcal{A}, \mu\}$ be $[0, 1]$ with the Lebesgue measure and let $\{Y, \mathcal{B}, \nu\}$ be the rationals in $[0, 1]$ with the counting measure. The function $f(x, y) = x$ is integrable on $\{X, \mathcal{A}, \mu\}$ and not integrable on the product space.

14.4. Let $[0, 1]$ be equipped with the Lebesgue measure μ and the counting measure ν both acting on the same σ-algebra of the Lebesgue measurable subsets of $[0, 1]$. The corresponding product measure space is not σ-finite since ν is not σ-finite. The diagonal set $E = \{x = y\}$ is of the type of $\mathcal{R}_{\sigma\delta}$, and hence $(\mu \times \nu)$-measurable and

$$\nu(E_x) = 1 \quad \forall x \in [0, 1] \quad \text{and} \quad \mu(E_y) = 0 \quad \forall y \in [0, 1].$$

Therefore,

$$\int_{[0,1]} \nu(E_x) d\mu = 1 \quad \text{and} \quad \int_{[0,1]} \mu(E_y) d\nu = 0.$$

Moreover

$$\iint_{[0,1] \times [0,1]} \chi_E d(\mu \times \nu) = \infty.$$

14.5. Let f and g be integrable functions on complete measure spaces $\{X, \mathcal{A}, \mu\}$ and $\{Y, \mathcal{B}, \nu\}$, respectively. Then $F(x, y) = f(x)g(y)$ is integrable on the product space $(X \times Y)$.

14.6. Let $\{X, \mathcal{A}, \mu\}$ be any measure space and let $\{Y, \mathcal{B}, \nu\}$ be $\mathbb{N}$ with the counting measure. Prove that the conclusion of Fubini's theorem continues to hold even i $\{X, \mathcal{A}, \mu\}$ is not complete, and that the conclusion of Tonelli's theorem continues to hold even if $\{X, \mathcal{A}, \mu\}$ is neither complete nor σ-finite.

14.7. Let $\{X, \mathcal{A}, \mu\}$ and $\{Y, \mathcal{B}, \nu\}$ be both $\mathbb{N}$ with the counting measure and consider the function

$$f(m, n) = \begin{cases} 1 & \text{if } m = n \\ -1 & \text{if } m = n + 1 \\ 0 & \text{otherwise.} \end{cases}$$

Prove that f is not integrable in the product space $\{X \times Y, (\mathcal{A} \times \mathcal{B}), (\mu \times \nu)\}$ and the conclusion of Fubini's theorem fails.

14.8.   Let $\{X, \mathcal{A}, \mu\}$ and $\{Y, \mathcal{B}, \nu\}$ be both $(0, 1)$ with the Lebesgue measure. Let $f : (0, 1) \to \mathbb{R}$ be measurable and assume that

$$(0, 1) \times (0, 1) \ni (x, y) \to f(x) - f(y)$$

is integrable in the product space $\{X \times Y, (\mathcal{A} \times \mathcal{B}), (\mu \times \nu)\}$. Prove that f is integrable in $(0, 1)$.

15c   Some Applications of the Fubini–Tonelli Theorem

15.1c   Integral of a Function as the "Area Under the Graph"

Let $\{X, \mathcal{A}, \mu\}$ be complete, and σ-finite, and let $f : X \to \mathbb{R}^*$ be a nonnegative and integrable. The graph of f on E is the set of points $\{x : (x, f(x))\}$ and the set of points "under the graph" is

$$U_f = \{(x, y) \mid x \in X : 0 \le y < f(x)\}.$$

Prove that U_f is measurable in the product measure space of $\{X, \mathcal{A}, \mu\}$ and $\mathbb{R}$ with the Lebesgue measure dx, and

$$\{\text{measure of } U_f\} = \int_X f(x) d\mu.$$

Prove that the graph of f as a subset of $X \times \mathbb{R}$ is measurable in the product measure and has measure zero.

15.2c   Distribution Functions

Let $\{X, \mathcal{A}, \mu\}$ be a measure space and let $f : X \to \mathbb{R}^*$ and $\{f_n\} : X \to \mathbb{R}^*$ be nonnegative and integrable. Prove:

Proposition 15.1c *Assume that $\mu([f > t]) \not\equiv \infty$. Then*

$$\lim_{\varepsilon \to 0} \mu([f > t + \varepsilon]) = \mu([f > t])$$

$$\lim_{\varepsilon \to 0} \mu([f > t - \varepsilon]) = \mu([f \geq t]).$$

Therefore, the distribution function $t \to \mu([f > t])$ is right-continuous, and it is continuous at a point t if only if $\mu([f = t]) = 0$.

Proposition 15.2c *Let $\{f_n\} \to f$ in measure. Then for every $\varepsilon > 0$*

$$\limsup \mu([f_n > t]) \leq \mu([f > t - \varepsilon])$$

$$\liminf \mu([f_n > t]) \geq \mu([f > t + \varepsilon]).$$

Therefore, $\mu([f_n > t]) \to \mu([f > t])$ at those t where the distribution function of f is continuous.

15.1. For a measurable function f on $\{X, \mathcal{A}, \mu\}$ set

$$\mathbb{R}^* \ni t \to f_*(t) = \mu([f \leq t]) \tag{15.1c}$$

prove that f_* is nondecreasing, right-continuous, and $f_*(\infty) = \mu(X)$.

15.2. Give an example of f and $\{X, \mathcal{A}, \mu\}$ for which f_X is right-continuous and not continuous.

15.3. The function f_* generates the Lebesgue–Stiltjies measure μ_{f_*} on $\mathbb{R}$. Let μ_f be the measure defined by (9.1c). Prove that $\mu_{f_*} = \mu_f$ on the Borel sets of $\mathbb{R}$ and that μ_{f_*} is the completion of μ_f.

17c The Radon-Nikodým Theorem

17.1. Let ν be the Lebesgue measure in $[0, 1]$ and let μ be the counting measure on the same σ-algebra of the Lebesgue measurable subsets of $[0, 1]$. Then ν is absolutely continuous with respect to μ, but it does not exist a nonnegative, μ-measurable function $f : [0, 1] \to \mathbb{R}^*$ for which ν can be represented as in (17.1).

17.2. In $[0, 1]$ let $\mathcal{A}$ be the σ-algebra of the sets that are, either countable or have countable complement. Let μ be the counting measure on $\mathcal{A}$ and let $\nu : \mathcal{A} \to \mathbb{R}^*$ be defined by $\nu(E) = 0$ if E is countable and $\nu(E) = 1$ otherwise. Then ν is absolutely continuous with respect to μ but it does not have a Radon-Nikodým derivative.

17.3. **Changing the Variables of Integration:** Let the assumptions of the Radon-Nikodým theorem hold. If g is ν-integrable, $g(d\nu/d\mu)$ is μ-integrable and for every measurable set E

$$\int_E g \, d\nu = \int_E g \frac{d\nu}{d\mu} d\mu. \tag{17.1c}$$

17.4. **Linearity of the Radon-Nikodým Derivative:** Let μ and ν_1 and ν_2 be σ-finite measures defined on the same σ-algebra $\mathcal{A}$. If ν_1 and ν_2 are absolutely continuous with respect to μ

$$\frac{d(\nu_1 + \nu_2)}{d\mu} = \frac{d\nu_1}{d\mu} + \frac{d\nu_2}{d\mu} \qquad \mu\text{-a.e..} \tag{17.2c}$$

17.5. **The Chain Rule:** Let μ, ν, and η be σ-finite measures on X defined on the same σ-algebra $\mathcal{A}$. Assume that μ is absolutely continuous with respect to η and that ν is absolutely continuous with respect to μ. Then

$$\frac{d\nu}{d\eta} = \frac{d\nu}{d\mu}\frac{d\mu}{d\eta} \qquad \text{a.e. with respect to } \eta. \tag{17.3c}$$

17.6. **Derivative of the Inverse:** Let μ and ν be two not identically zero, σ-finite measures defined on the same σ-algebra $\mathcal{A}$ and mutually absolutely continuous. Then

$$\frac{d\nu}{d\mu} \neq 0 \quad \text{and} \quad \frac{d\mu}{d\nu} = \left(\frac{d\nu}{d\mu}\right)^{-1}. \tag{17.4c}$$

Hint: For all nonnegative ν-measurable functions g

$$\int_E g d\nu = \int_E g \frac{d\nu}{d\mu} d\mu.$$

This is true for simple functions and hence for nonnegative ν-measurable functions g. Now for all $E \in \mathcal{A}$ of finite μ-measure

$$\mu(E) = \int_E 1 d\mu = \int_E \frac{d\mu}{d\nu} d\nu = \int_E \frac{d\mu}{d\nu}\frac{d\nu}{d\mu} d\mu.$$

The inverse formula (17.4c) follows from the uniqueness of the Radon-Nikodým derivative.

17.7. Let $\{X, \mathcal{A}, \mu\}$ be $[0, 1]$ with the Lebesgue measure. Let $\{E_n\}$ be a measurable partition of $[0, 1]$ and let $\{\alpha_n\}$ be a sequence of positive numbers such that $\sum \alpha_n < \infty$. Find the Radon-Nikodým derivative of the measure

$$\mathcal{A} \ni E \longrightarrow \nu(E) = \sum \alpha_n \mu(E \cap E_n)$$

with respect to the Lebesgue measure.

Proposition 17.1c *Let μ and ν be two measures defined on the same σ-algebra $\mathcal{A}$. Assume that ν is finite. Then ν is absolutely continuous with respect to μ if and only if for every $\varepsilon > 0$ there exists $\delta > 0$ such that $\nu(E) < \varepsilon$ for every set $E \in \mathcal{A}$ such that $\mu(E) < \delta$.*

Proof (Necessity) If not, there exist $\varepsilon > 0$ and a sequence of measurable sets $\{E_n\}$, such that $\nu(E_n) \geq \varepsilon$ and $\mu(E_n) \leq 2^{-n}$. Let $E = \limsup E_n$ and compute

$$\nu(E) = \nu(\limsup E_n) \geq \limsup \nu(E_n) \geq \varepsilon.$$

On the other hand for all $n \in \mathbb{N}$

$$\mu(E) \leq \mu\left(\bigcup_{j=n}^{\infty} E_j \right) \leq \sum_{j=n}^{\infty} \mu(E_j) \leq \frac{1}{2^{n-1}}. \qquad \blacksquare$$

17.8.   The proposition might fail if ν is not finite. Let $X = \mathbb{N}$ and for every subset $E \subset \mathbb{N}$, set

$$\mu(E) = \sum_{n \in E} \frac{1}{2^n} \qquad \nu(E) = \sum_{n \in E} 2^n.$$

17.9.   **More on $\nu \ll \mu$:** There exist σ-finite measures ν on the Lebesgue measurable sets of $\mathbb{R}$, absolutely continuous with respect to the Lebesgue measure μ on $\mathbb{R}$ and such that $\nu(E) = \infty$ for every measurable set E with nonempty interior. Let $\{r_n\}$ be the sequence of the rationals in $\mathbb{R}$ and set

$$g(x) = \sum \frac{1}{2^n} \frac{e^{-|x|}}{|x - r_n|} \quad \text{and} \quad \nu(E) = \int_E g \, d\mu$$

for all Lebesgue measurable set $E \subset \mathbb{R}$.

17.10.   Let $\{X, \mathcal{A}, \mu\}$ be a measure space and let ν be a signed measure on $\mathcal{A}$ of finite total variation $|\nu|$. Then $\nu \ll \mu$ if and only if for every $\varepsilon > 0$ there exists $\delta > 0$ such that $|\nu|(E) < \varepsilon$ for all $E \in \mathcal{A}$ such that $\mu(E) < \delta$. Give an counterexample if $|\nu|$ is not finite.

18c   A Proof of the Radon-Nikodým Theorem When Both μ and ν Are σ-Finite

A more constructive proof of Theorem 17.1 can be given if both μ and ν are σ-finite. Assume first that both μ and ν are finite. Let Φ be the family of all measurable nonnegative functions $\varphi : X \to \mathbb{R}^*$ such that

$$\int_E \varphi \, d\mu \leq \nu(E) \quad \text{for all } E \in \mathcal{A}.$$

Since $0 \in \Phi$ such a class is not empty. For two given functions φ_1 and φ_2 in Φ the function $\max\{\varphi_1; \varphi_2\}$ is in Φ. Indeed for any $E \in \mathcal{A}$

$$\int_E \max\{\varphi_1; \varphi_2\}d\mu = \int_{E\cap[\varphi_1\geq\varphi_2]} \varphi_1 d\mu + \int_{E\cap[\varphi_2>\varphi_1]} \varphi_2 d\mu$$

$$\leq \nu(E\cap[\varphi_1\geq\varphi_2]) + \nu(E\cap[\varphi_2<\varphi_1]) = \nu(E).$$

Since ν is finite

$$M = \sup_{\varphi\in\Phi}\int_X \varphi d\mu \leq \nu(X) < \infty.$$

Let $\{\varphi_n\}$ be a sequence of functions in Φ such that

$$\lim \int_X \varphi_n d\mu = M$$

and set $f_n = \max\{\varphi_1, \cdots, \varphi_n\}$. The sequence $\{f_n\}$ is nondecreasing, and μ-a.e. convergent to a function f that belongs to Φ. Indeed by monotone convergence, for every measurable set E

$$\int_E f d\mu = \lim \int_E f_n d\mu \leq \nu(E).$$

Such a limiting function is the f claimed by the theorem. For this it suffices to establish that the measure

$$\mathcal{A} \ni E \longrightarrow \eta(E) = \nu(E) - \int_E f d\mu$$

is identically zero. If not, there exists a set $A \in \mathcal{A}$ such that $\eta(A) > 0$. Since both ν and η are absolutely continuous with respect to μ, for such a set, $\mu(A) > 0$. Also, since μ is finite, there exists $\varepsilon > 0$ such that

$$\xi(A) = \eta(A) - \varepsilon\mu(A) > 0.$$

The set function

$$\mathcal{A} \ni E \longrightarrow \xi(E) = \eta(E) - \varepsilon\mu(E)$$

is a signed measure on $\mathcal{A}$. Therefore, by Proposition 16.1 the set A contains a positive subset A_o of positive measure. In particular

$$\eta(E\cap A_o) - \varepsilon\mu(E\cap A_o) \geq 0 \quad \text{for all } E \in \mathcal{A}.$$

From this and the definition of η

$$\varepsilon\mu(E\cap A_o) \leq \nu(E\cap A_o) - \int_{E\cap A_o} f d\mu \quad \text{for all } E \in \mathcal{A}.$$

The function $(f + \varepsilon \chi_{A_o})$ belongs to Φ. Indeed, for every measurable set E

$$\int_E (f + \varepsilon \chi_{A_o}) d\mu = \int_{E - A_o} f d\mu + \int_{E \cap A_o} (f + \varepsilon) d\mu$$
$$\leq \nu(E - A_o) + \nu(E \cap A_o) = \nu(E).$$

This however contradicts the definition of M since

$$\int_X (f + \varepsilon \chi_{A_o}) d\mu = M + \varepsilon \mu(A_o) > M.$$

If $g : X \to \mathbb{R}^*$ is another nonnegative measurable function by which the measure ν can be represented, let $A_n = [f - g \geq \frac{1}{n}]$. Then for all $n \in \mathbb{N}$

$$\frac{1}{n} \mu(A_n) \leq \int_{A_n} (f - g) d\mu = \nu(A_n) - \nu(A_n) = 0.$$

Thus $f = g$ μ-a.e. in X.

Assume next that μ is σ-finite and ν is finite. Let E_n be a sequence of measurable, expanding sets such that

$$\mu(E_n) \leq \mu(E_{n+1}) < \infty \quad \text{and} \quad X = \bigcup E_n$$

and denote by μ_n the restriction of μ to E_n. Let f_n be the unique function claimed by the Radon-Nikodým theorem for the pair of finite measures $\{\mu_n; \nu\}$. By construction $f_{n+1} \big|_{E_n} = f_n$ for all n. The function f claimed by the theorem is $f = \sup f_n$. Indeed if $E \in \mathcal{A}$, by monotone convergence

$$\nu(E) = \lim \nu(E \cap E_n) = \lim \int_E f_n d\mu = \int_E f d\mu.$$

The uniqueness of such a f is proved as in the case of μ finite. Finally a similar argument, establishes the theorem when also ν is σ-finite. ∎

Chapter 5
Topics on Measurable Functions of Real Variables

1 Functions of Bounded Variation ([78])

Let f be a real-valued function defined and bounded in some interval $[a, b] \subset \mathbb{R}$. Denote by $P = \{a = x_o < x_1 < \cdots < x_n = b\}$ a partition of $[a, b]$ and set

$$V_f[a, b] = \sup_P \sum_{i=1}^{n} |f(x_i) - f(x_{i-1})|.$$

This number, finite or infinite, is called the *total variation* of f in $[a, b]$. If $V_f[a, b]$ is finite the function f is said to be of *bounded variation* in $[a, b]$ and one writes $f \in BV[a, b]$. If f is monotone in $[a, b]$, then $f \in BV[a, b]$, and

$$V_f[a, b] = |f(b) - f(a)|.$$

More generally, if f is the difference of two monotone functions in $[a, b]$, then $f \in BV[a, b]$. If f is Lipschitz continuous in $[a, b]$ with Lipschitz constant L, then $f \in BV[a, b]$, and $V_f[a, b] \le L(b - a)$.

Continuity does not imply bounded variation. The function

$$f(x) = \begin{cases} x \cos \dfrac{\pi}{x} & \text{for } x \in (0, 1] \\ 0 & \text{for } x = 0 \end{cases} \tag{1.1}$$

is continuous in $[0, 1]$ and not of bounded variation on $[0, 1]$. Consider the partition of $[0, 1]$

$$P_n = \left\{0 < \frac{1}{n} < \frac{1}{n-1} < \cdots < \frac{1}{n-(n-1)} = 1\right\}$$

© Springer Science+Business Media New York 2016
E. DiBenedetto, *Real Analysis*, Birkhäuser Advanced
Texts Basler Lehrbücher, DOI 10.1007/978-1-4939-4005-9_5

and estimate

$$V_f[0, 1] \geq \lim_{n \to \infty} \left\{ \left| \frac{\cos n\pi}{n} \right| + \sum_{j=1}^{n-1} \left| \frac{\cos \pi(n-j)}{n-j} - \frac{\cos \pi(n-j+1)}{n-j+1} \right| \right\}$$

$$= \lim_{n \to \infty} \left\{ \frac{1}{n} + \sum_{j=1}^{n-1} \left(\frac{1}{n-j} + \frac{1}{n-j+1} \right) \right\} \geq \lim_{n \to \infty} \sum_{i=1}^{n} \frac{1}{i}.$$

Bounded variation does not imply continuity. The function

$$f(x) = \begin{cases} 0 & \text{for } x \in [-1, 1] - \{0\} \\ 1 & \text{for } x = 0 \end{cases}$$

is discontinuous and of bounded variation and $V_f[-1, 1] = 2$.

Proposition 1.1 *Let f and g of bounded variation in $[a, b]$ and let α and β be real numbers. Then $(\alpha f + \beta g)$ and (fg) are of bounded variation in $[a, b]$. If $|g| \geq \varepsilon$ for some $\varepsilon > 0$, then (f/g) is of bounded variation in $[a, b]$.*

Proposition 1.2 *Let f be of bounded variation in $[a, b]$. Then f is of bounded variation in every closed subinterval of $[a, b]$. Moreover for every $c \in [a, b]$*

$$V_f[a, b] = V_f[a, c] + V_f[c, b]. \tag{1.2}$$

The *positive* and *negative* variations of f in $[a, b]$ are defined by

$$V_f^+[a, b] = \sup_P \sum_{j=1}^{n} [f(x_j) - f(x_{j-1})]^+$$

$$V_f^-[a, b] = \sup_P \sum_{j=1}^{n} [f(x_j) - f(x_{j-1})]^-.$$

Proposition 1.3 *Let f be of bounded variation in $[a, b]$. Then*

$$V_f[a, b] = V_f^+[a, b] + V_f^-[a, b]$$
$$f(b) - f(a) = V_f^+[a, b] - V_f^-[a, b].$$

In particular for every $x \in [a, b]$, there holds the Jordan decomposition

$$f(x) = f(a) + V_f^+[a, x] - V_f^-[a, x]. \tag{1.3}$$

Since $x \to V_f^\pm[a, x]$ are both non-decreasing, a function f of bounded variation can be written as the difference of two non-decreasing functions. We have already observed that the difference of two monotone functions in $[a, b]$ is of bounded variation in $[a, b]$. Thus a function f is of bounded variation in $[a, b]$ if and only if it is the difference of two monotone functions in $[a, b]$. The Jordan decomposition also implies that a function of bounded variation in $[a, b]$ is measurable in $[a, b]$.

Proposition 1.4 *A function f of bounded variation in $[a, b]$ has at most countably many jump discontinuities in $[a, b]$.*

Proof May assume that f is monotone increasing. Then, for every $c \in (a, b)$, the limits

$$\lim_{x \to c^+} f(x) = f(c^+), \qquad \lim_{x \to c^-} f(x) = f(c^-)$$

exist and are finite. If $f(c^+) > f(c^-)$, we select one and only one rational number out of the interval $(f(c^-), f(c^+))$. This way the set of jump discontinuities of f is put in one-to-one correspondence with a subset of $\mathbb{N}$. ∎

2 Dini Derivatives ([37])

Let f be a real-valued function defined in $[a, b]$. For a fixed $x \in [a, b]$ set

$$D_\pm f(x) = \liminf_{h \to 0^\pm} \frac{f(x+h) - f(x)}{h}$$
$$D^\pm f(x) = \limsup_{h \to 0^\pm} \frac{f(x+h) - f(x)}{h}. \tag{2.1}$$

These are the four *Dini numbers* or the four *Dini derivatives* of f at x. If f is differentiable at x these four numbers all coincide with $f'(x)$.

Proposition 2.1 (Banach-Sierpiński [9, 147]) *Let f be real-valued and non-decreasing in $[a, b]$. Then the functions $D_\pm f$ and $D^\pm f$ are measurable.*

Proof We prove that $D^+ f$ is measurable, the arguments for the remaining ones being similar. For $n \in \mathbb{N}$ set

$$u_n(x) = \sup_{0 < h < \frac{1}{n}} \frac{f(x+h) - f(x)}{h}.$$

Since $D^+ f = \lim u_n$, it suffices to prove that the u_n are measurable. By monotonicity f is measurable and, for h fixed, the difference quotients in the definition of u_n are measurable. We prove that the supremum of such difference quotients for $h \in (0, \frac{1}{n}]$ is be realized for h ranging over a countable subset of $(0, \frac{1}{n}]$. This way u_n would be the supremum of a countable collection of measurable functions. Set

$$v_n(x) = \sup_{h \in \mathbb{Q} \cap (0, \frac{1}{n}]} \frac{f(x+h) - f(x)}{h}.$$

By construction $v_n(x) \le u_n(x)$. To establish the reverse inequality, having fixed $\varepsilon > 0$, there exists $\tau \in (0, \frac{1}{n}]$ such that

$$\frac{f(x+\tau) - f(x)}{\tau} > u_n(x) - \varepsilon.$$

Having fixed such a τ, there exist $h \in \mathbb{Q} \cap (0, \frac{1}{n}]$, and $h \ge \tau$, such that

$$\frac{1}{\tau} < \frac{1}{h} + \frac{\varepsilon}{|f(x+\tau)| + |f(x)| + 1}.$$

Therefore, since $f(x+\tau) \le f(x+h)$

$$\frac{f(x+h) - f(x)}{h} + \varepsilon \ge \frac{f(x+\tau) - f(x)}{\tau} > u_n(x) - \varepsilon.$$

Thus $v_n(x) \ge u_n(x) - 2\varepsilon$ for all $\varepsilon > 0$.   ∎

For a function f defined in $[a, b]$ set

$$D'' f = \max\{D^- f; D^+ f\}, \qquad D' f = \min\{D_- f; D_+ f\}.$$

If f is non-decreasing the two functions $D'' f$ and $D' f$ are both measurable.

Proposition 2.2 *Let f be a real-valued, non-decreasing function in $[a, b]$. Then*

$$f(b) - f(a) \ge t\mu([D'' f > t]) \qquad \text{for all } t \in \mathbb{R}.$$

Proof The assertion is trivial if $\mu([D'' f > t]) = 0$ or if $t \le 0$. Assuming that

$$\mu([D'' f > t]) > 0 \quad \text{and} \quad t > 0$$

let $\mathcal{F}$ denote the family of all closed intervals $[\alpha, \beta] \subset [a, b]$ such that at least one of the extremes α or β is in $[D'' f > t]$ and such that

$$\frac{f(\beta) - f(\alpha)}{\beta - \alpha} > t.$$

By the definition of $D'' f$, having fixed $x \in [D'' f > t]$ and $\delta > 0$, there exists some interval $[\alpha, \beta] \in \mathcal{F}$ of length less than δ and such that $x \in [\alpha, \beta]$. Therefore $\mathcal{F}$ is a fine Vitali covering for $[D'' f > t]$. By Corollary 17.1 of Chap. 3, for any fixed $\varepsilon > 0$ there exist a finite collection of intervals $[\alpha_i, \beta_i] \in \mathcal{F}$ for $i = 1, \ldots, n$, with pairwise disjoint interior, such that

$$\sum_{i=1}^{n} (\beta_i - \alpha_i) > \mu([D'' f > t]) - \varepsilon.$$

From this and the definition of $\mathcal{F}$

$$f(b) - f(a) \geq \sum_{i=1}^{n} [f(\beta_i) - f(\alpha_i)]$$

$$> \sum_{i=1}^{n} t(\beta_i - \alpha_i) > t\mu([D''f > t]) - t\varepsilon \qquad \blacksquare$$

Corollary 2.1 *Let f be a real-valued, non-decreasing function defined in $[a, b]$. Then $D''f$ and $D'f$ are a.e. finite in $[a, b]$.*

Proof For all $t > 0$

$$\mu([D'f = \infty]) \leq \mu([D''f = \infty]) \leq \mu([D''f > t]) \leq \frac{f(b) - f(a)}{t}. \qquad \blacksquare$$

3   Differentiating Functions of Bounded Variation

A real valued function f defined in $[a, b]$ is differentiable at some $x \in (a, b)$ if and only if $D''f(x)$ is finite at x and $D''f(x) = D'f(x)$. The function f is a.e. differentiable in $[a, b]$ if and only if $D''f$ is a.e. finite in $[a, b]$ and $\mu([D''f > D'f]) = 0$.

Theorem 3.1 (Lebesgue [91])[1] *A real-valued, non-decreasing function f in $[a, b]$ is a.e. differentiable in $[a, b]$.*

Proof By Corollary 2.1, $D''f$ and $D'f$ are a.e. finite in $[a, b]$. Assume that $\mu([D''f > D'f]) > 0$ and, for $p, q \in \mathbb{N}$, set

$$E_{p,q} = \left[D'f < \frac{p}{q} < \frac{p+1}{q} < D''f(x) \right].$$

Since

$$[D''f > D'f] = \bigcup E_{p,q}$$

there exists a pair p, q of positive integers such that $\mu(E_{p,q}) > 0$. Let $\mathcal{F}$ be the family of closed intervals $[\alpha, \beta] \subset [a, b]$ such that at least one of the extremes α and β belongs to $E_{p,q}$ and such that

$$\frac{f(\beta) - f(\alpha)}{\beta - \alpha} < \frac{p}{q}.$$

Having fixed $x \in E_{p,q}$ and some $\delta > 0$, there exists an interval $[\alpha, \beta] \in \mathcal{F}$ of length less than δ, such that $x \in [\alpha, \beta]$. Therefore $\mathcal{F}$ is a fine Vitali covering of $E_{p,q}$.

---

[1] Also in [122]. A proof independent of measure theory is in [134], pp. 5–9.

By Corollary 17.1 of Chap. 3, having fixed an arbitrary $\varepsilon > 0$, one may extract out of $\mathcal{F}$, a finite collection of intervals $[\alpha_i, \beta_i]$ for $i = 1, \ldots, n$, with pairwise disjoint interior, such that

$$\sum_{i=1}^{n} (\beta_i - \alpha_i) - \varepsilon < \mu(E_{p,q}) < \mu\left(E_{p,q} \cap \bigcup_{i=1}^{n} [\alpha_i, \beta_i]\right) + \varepsilon.$$

Therefore by the construction of the family $\mathcal{F}$

$$\sum_{i=1}^{n} [f(\beta_i) - f(\alpha_i)] < \frac{p}{q} \sum_{i=1}^{n} (\beta_i - \alpha_i) < \frac{p}{q} \mu(E_{p,q}) + \frac{p}{q} \varepsilon.$$

By Proposition 2.2 applied to f restricted to the interval $[\alpha_i, \beta_i]$ we derive

$$f(\beta_i) - f(\alpha_i) > \frac{p+1}{q} \mu(E_{p,q} \cap [\alpha_i, \beta_i]).$$

Adding these inequalities for $i = 1, \ldots, n$, gives

$$\sum_{i=1}^{n} [f(\beta_i) - f(\alpha_i)] > \frac{p+1}{q} \sum_{i=1}^{n} \mu(E_{p,q} \cap [\alpha_i, \beta_i])$$

$$\geq \frac{p+1}{q} \mu\left(E_{p,q} \cap \bigcup_{i=1}^{n} [\alpha_i, \beta_i]\right) > \frac{p+1}{q} \mu(E_{p,q}) - \frac{p+1}{q} \varepsilon.$$

Combining the inequalities involving $\mu(E_{p,q})$

$$\frac{p}{q} \mu(E_{p,q}) + \frac{p}{q} \varepsilon > \frac{p+1}{q} \mu(E_{p,q}) - \frac{p+1}{q} \varepsilon.$$

From this $\mu(E_{p,q}) < (2p+1)\varepsilon$, for all $\varepsilon > 0$.   ∎

Corollary 3.1 *A real-valued function f of bounded variation in $[a, b]$ is a.e. differentiable in $[a, b]$.*

4 Differentiating Series of Monotone Functions

Theorem 4.1 (Fubini [53])[2] *Let $\{f_n\}$ be a sequence of real-valued, non-decreasing functions in $[a, b]$ and assume that the series $\sum f_n$ is convergent in $[a, b]$ to a real-valued function f defined in $[a, b]$. Then f is a.e. differentiable in $[a, b]$ and*

$$f'(x) = \sum f_n'(x) \qquad \text{for a.e. } x \in [a, b].$$

---

[2] Also in [161].

Proof By possibly replacing f_n with $f_n - f_n(a)$, we may assume that $f_n(a) = 0$ and $f_n \geq 0$. For $n \in \mathbb{N}$ write

$$f = \sum_{i=1}^{n} f_i + R_n \qquad \text{where} \qquad R_n = \sum_{j=n+1}^{\infty} f_j.$$

The functions R_n are non-decreasing and hence a.e. differentiable in $[a, b]$. The difference $R_n - R_{n+1} = f_{n+1}$, is also non-decreasing and a.e. differentiable in $[a, b]$. Therefore $R'_n - R'_{n+1} = f'_{n+1} \geq 0$, a.e. in $[a, b]$. This implies that the sequence $\{R'_n\}$ is decreasing, and has a limit $g = \lim R'_n$ a.e. in $[a, b]$. The sum f is non-decreasing and hence a.e. differentiable in $[a, b]$. Therefore

$$f' = \sum_{i=1}^{n} f'_i + R'_n \qquad \text{a.e. in } [a, b].$$

To prove the theorem it suffices to show that $g = 0$ a.e. in $[a, b]$. Apply Proposition 2.2 to the function R_n, and take into account that $R_n \geq g$ a.e. in $[a, b]$. This gives

$$R_n(b) \geq t\mu([R'_n > t]) \geq t\mu([g > t]) \qquad \text{for all } n \in \mathbb{N}.$$

Since the series is convergent everywhere in $[a, b]$ the left-hand side goes to zero as $n \to \infty$. Thus $\mu([g > t]) = 0$ for all $t > 0$. ∎

5 Absolutely Continuous Functions ([91, 169])

A real valued function f defined in $[a, b]$ is *absolutely continuous* in $[a, b]$ if for every $\varepsilon > 0$ there exists a positive number δ such that, for every finite collection of disjoint intervals $(a_j, b_j) \subset [a, b]$, $j = 1, \ldots, n$ of total length not exceeding δ

$$\sum_{j=1}^{n} |f(b_j) - f(a_j)| < \varepsilon \qquad \left(\sum_{j=1}^{n} (b_j - a_j) < \delta\right). \tag{5.1}$$

If g is integrable in $[a, b]$ then the function

$$x \to \int_{a}^{x} g(t)dt$$

is absolutely continuous in $[a, b]$. This follows from the absolute continuity of the integral. If f is Lipschitz continuous in $[a, b]$ it is absolutely continuous in $[a, b]$. The converse is false. A counterexample can be constructed using **5.1** of the Problems and Complements. Absolute continuity implies continuity but the converse is false. The function in (1.1) is continuous and not absolutely continuous.

The linear combination of two absolutely continuous functions f and g in $[a, b]$ as well as their product fg are absolutely continuous. Their quotient f/g is absolutely continuous if $|g| \geq c_o > 0$ in $[a, b]$.

Proposition 5.1 *Let f be absolutely continuous in $[a, b]$. Then f is of bounded variation in $[a, b]$.*

Proof Having fixed $\varepsilon > 0$ and the corresponding δ, partition $[a, b]$ by points $a = x_o < x_1 < \cdots < x_n = b$ such that

$$\tfrac{1}{2}\delta < x_i - x_{i-1} < \delta \qquad i = 1, \ldots, n.$$

The number n of intervals of this partition does not exceed $2(b - a)/\delta$. In each of them the variation of f is less than ε. Then by (1.2) of Proposition 1.2

$$\mathcal{V}_f[a, b] = \sum_{i=1}^{n} \mathcal{V}_f[x_{i-1}, x_i] \leq 2(b - a)\frac{\varepsilon}{\delta}. \qquad \blacksquare$$

Corollary 5.1 *Let f be absolutely continuous in $[a, b]$. Then f is a.e. differentiable in $[a, b]$.*

Proposition 5.2 *Let f be absolutely continuous in $[a, b]$. If $f' \geq 0$ a.e. in $[a, b]$ then f is non-decreasing in $[a, b]$.*

Proof Fix $[\alpha, \beta] \subset [a, b]$ and set $E = [f' \geq 0] \cap [\alpha, \beta]$. By the assumption $\mu(E) = \beta - \alpha$. Having fixed $\varepsilon > 0$, let δ be a corresponding positive number claimed by the absolute continuity of f. For every $x \in E$, there exist some $\sigma_x > 0$ such that

$$f(x + h) - f(x) > -\varepsilon h \qquad \text{for all } h \in (0, \sigma_x).$$

The collection of intervals $[x, x + h]$ for x ranging over E and $h \in (0, \sigma_x)$, is a fine Vitali covering for E. Therefore, in correspondence of the previously fixed $\delta > 0$, we may extract a finite collection of intervals $[\alpha_i, \beta_i]$ for $i = 1, \ldots, n$, with pairwise disjoint interior, such that

$$\mu(E) - \delta \leq \mu\Big[E \cap \bigcup_{i=1}^{n} (\alpha_i - \beta_i)\Big] \quad \text{and} \quad f(\beta_i) - f(\alpha_i) > -\varepsilon(\beta_i - \alpha_i).$$

The complement

$$[\alpha, \beta] - \bigcup_{i=1}^{n} (\alpha_i, \beta_i)$$

consists of finitely many disjoint intervals $[a_j, b_j]$ for $j = 1, \ldots, m$ of total length not exceeding δ. Therefore (5.1) holds for such a finite collection, and

$$f(\beta) - f(\alpha) = \sum_{i=1}^{n} [f(\beta_i) - f(\alpha_i)] + \sum_{j=1}^{m} [f(b_j) - f(a_j)]$$

$$\geq -\varepsilon \sum_{i=1}^{n} (\beta_i - \alpha_i) - \varepsilon \qquad \text{for all } \varepsilon > 0. \qquad \blacksquare$$

Corollary 5.2 *Let f be absolutely continuous in $[a, b]$. If $f' = 0$ a.e. in $[a, b]$ then f is constant in $[a, b]$.*

Remark 5.1 The conclusions of Proposition 5.2 and Corollary 5.2 are false if f is of bounded variation and not absolutely continuous. A counterexample is the function of the jumps introduced in § 1.1c (see **2.4** of the Complements).

Remark 5.2 The assumption of absolute continuity cannot be relaxed to the mere continuity as shown by the Cantor ternary function and its variants (§ 5.1c–§ 5.2c of the Problems and Complements). The same examples also show that bounded variation and continuity do not imply absolute continuity.

6 Density of a Measurable Set

Let $E \subset [a, b]$ be Lebesgue measurable. The set density functions

$$x \to d_E(x) = \int_a^x \chi_E(t)dt, \qquad d_{[a,b]-E}(x) = \int_a^x \chi_{[a,b]-E}(t)dt$$

are absolutely continuous and non-decreasing in $[a, b]$. Moreover

$$d_E(x) + d_{[a,b]-E}(x) = x - a$$
$$d_E'(x) + d_{[a,b]-E}'(x) = 1 \qquad \text{for a.e. } x \in [a, b].$$

Proposition 6.1 (Lebesgue [91]) *Let $E \subset [a, b]$ be Lebesgue measurable. Then*

$$d_E' = 1 \text{ a.e. in } E \quad \text{and} \quad d_E' = 0 \text{ a.e. in } [a, b] - E.$$

Proof It suffices to prove the first of these. If E is open then $d_E' = 1$ in E. Assume now that E is of the type of a $\mathcal{G}_\delta$, i.e., there exists a countable collection of open sets $\{E_n\}$ such that $E_{n+1} \subset E_n$ and $E = \cap E_n$. By dominated convergence

$$d_E(x) = \int_a^x \chi_E(t)dt = \lim \int_a^x \chi_{E_n}(t)dt = \lim d_{E_n}(x).$$

Therefore

$$d_E = d_{E_1} + \sum (d_{E_{n+1}} - d_{E_n}).$$

Each of the term of the series is non-increasing since $d'_{E_{n+1}} \leq d'_{E_n}$. Therefore by the Fubini theorem

$$d'_E = d'_{E_1} + \sum (d_{E_{n+1}} - d_{E_n})', \qquad \text{a.e. in } [a, b].$$

If $x \in E$ then $d'_{E_n} = 1$ for all $n \in \mathbb{N}$, and the assertion follows.

If E is a measurable subset of $[a, b]$, there exists a set E_δ of the type of a $\mathcal{G}_\delta$, such that $E \subset E_\delta$ and $\mu(E_\delta - E) = 0$. This implies that $d_E = d_{E_\delta}$ and thus $d'_E = d'_{E_\delta} = 1$ a.e. in E. ∎

7 Derivatives of Integrals

We have observed that if f is Lebesgue integrable in $[a, b]$ then the function

$$x \longrightarrow F(x) = \int_a^x f(t)dt$$

is absolutely continuous and hence a.e. differentiable in $[a, b]$.

Proposition 7.1 *Let f be Lebesgue integrable in $[a, b]$. Then $F' = f$, a.e. in $[a, b]$.*

Proof Assume first that f is simple. Then if $\{\lambda_1, \ldots, \lambda_n\}$ are the distinct values taken by f, there exist disjoint, measurable sets $E_i \subset [a, b], i = 1, \ldots, n$ such that

$$f = \sum_{i=1}^n \lambda_i \chi_{E_i} \qquad \text{and} \qquad F = \sum_{i=1}^n \lambda_i d_{E_i}.$$

Therefore the assertion follows from Proposition 6.1.

Assume next that f is integrable and nonnegative in $[a, b]$. There exists a sequence of simple functions $\{f_n\}$ such that

$$f_n \leq f_{n+1} \qquad \text{and} \qquad \lim f_n(x) = f(x) \quad \text{for all } x \in [a, b].$$

By dominated convergence

$$F(x) = \lim F_n(x) \qquad \text{where} \qquad F_n(x) = \int_a^x f_n(t)dt.$$

Therefore

$$F = F_1 + \sum (F_{n+1} - F_n).$$

The terms of the series are non-decreasing since

$$(F_{n+1} - F_n)' = f_{n+1} - f_n \geq 0 \quad \text{a.e. in } [a, b].$$

Therefore by the Fubini theorem

$$F' = \lim F_n' = \lim f_n = f \qquad \text{a.e. in } [a, b].$$

A general integrable f is the difference of two nonnegative integrable functions. ∎

Proposition 7.2 (Lebesgue [91]) *Let f be absolutely continuous in $[a, b]$. Then f' is integrable and*

$$f(x) = f(a) + \int_a^x f'(t)dt. \qquad (7.1)$$

Proof Assume first that f is non-decreasing so that $f' \geq 0$ a.e. in $[a, b]$. Defining $f(x) = f(b)$ for $x \geq b$, the limit

$$f'(x) = \lim \frac{f\left(x + \dfrac{1}{n}\right) - f(x)}{\dfrac{1}{n}}$$

exists a.e. in $[a, b]$. Then by Fatou's lemma

$$\int_a^b f'dx \leq \liminf n \int_a^b \left[f\left(x + \frac{1}{n}\right) - f(x)\right]dx$$

$$= \liminf n \left(\int_b^{b+\frac{1}{n}} f(x)dx - \int_a^{a+\frac{1}{n}} f(x)dx\right)$$

$$\leq \liminf n \frac{f(b) - f(a)}{n} = f(b) - f(a).$$

Since f is absolutely continuous, it is of bounded variation and by the Jordan decomposition is the difference of two non-decreasing functions. Thus f' is integrable in $[a, b]$. The function

$$g(x) = f(a) + \int_a^x f'(t)dt$$

is absolutely continuous and $g' = f'$ a.e. in $[a, b]$. Thus $(g - f)' = 0$ a.e. in $[a, b]$ and by Corollary 5.2, $g = f + \text{const}$ in $[a, b]$. Since $g(a) = f(a)$ the conclusion follows. ∎

Remark 7.1 The proof of Proposition 7.2 contains the following

Corollary 7.1 *Let f be of bounded variation in $[a, b]$. Then f' is integrable in $[a, b]$.*

Remark 7.2 Proposition 7.2 is false if f is only of bounded variation in $[a, b]$. A counterexample is given by a non-constant, non-decreasing simple function. For such a function the representation (7.1) does not hold.

The proposition continues to be false even by requiring that f be continuous. The Cantor ternary function (§ 5.1c of the Problems and Complements) is of bounded variation and continuous in $[0, 1]$ and its derivative is integrable. However it is not absolutely continuous and (7.1) does not hold. A similar conclusion holds for the function in § 5.2c of the Problems and Complements.

8 Differentiating Radon Measures

Let f be a nonnegative, Lebesgue measurable, real-valued function defined in $\mathbb{R}^N$, integrable on compact subsets of $\mathbb{R}^N$ and let $B_\rho(x)$ denote the closed ball in $\mathbb{R}^N$ centered at x and radius ρ. If μ is the Lebesgue measure in $\mathbb{R}^N$, the notion of differentiating the integral of f at some point $x \in \mathbb{R}^N$ is replaced by

$$\lim_{\rho \to 0} \frac{1}{\mu(B_\rho(x))} \int_{B_\rho(x)} f \, d\mu = \lim_{\rho \to 0} \frac{\nu(B_\rho(x))}{\mu(B_\rho(x))} \quad \text{where} \quad d\nu = f \, d\mu$$

provided the limit exists. More generally, given any two Radon measures μ and ν in $\mathbb{R}^N$, set

$$D_\mu^+ \nu(x) = \limsup_{\rho \to 0} \frac{\nu(B_\rho(x))}{\mu(B_\rho(x))}, \qquad D_\mu^- \nu(x) = \liminf_{\rho \to 0} \frac{\nu(B_\rho(x))}{\mu(B_\rho(x))} \qquad (8.1)$$

provided $\mu(B_\rho(x)) > 0$ for all $\rho > 0$. Set also $D_\mu^\pm \nu(x) = \infty$ if $\mu(B_\rho(x)) = 0$ for some $\rho > 0$. If for some $x \in \mathbb{R}^N$ the upper and lower limits in (8.1) are equal and finite, we set

$$D_\mu^+ \nu(x) = D_\mu^- \nu(x) = D_\mu \nu(x)$$

and say that the Radon measure ν is differentiable at x, with respect to the Radon measure μ.

Proposition 8.1 *Let μ and ν be two Radon measures in $\mathbb{R}^N$ and let μ_e and ν_e be their associated outer measures. For every $t > 0$ and every set*

$$E \subset [D_\mu^+ \nu \geq t] \quad \text{there holds} \quad \mu_e(E) \leq \frac{1}{t} \nu_e(E). \qquad (8.2)$$

Analogously, for every $t > 0$ and every set

$$E \subset [D_\mu^- \nu \leq t] \quad \text{there holds} \quad \mu_e(E) \geq \frac{1}{t} \nu_e(E). \qquad (8.3)$$

Proof In proving (8.2) assume first that the set E is bounded. Fix $t > 0$ and $\varepsilon \in (0, t)$. By the definition of $D_\mu^+ \nu(x)$, for every $x \in E$, there exists a ball $B_\rho(x)$ centered at x and of arbitrarily small radius ρ such that

$$(t - \varepsilon)\mu(B_\rho(x)) < \nu(B_\rho(x)). \tag{8.4}$$

Let $\mathcal{O}$ be an open set containing E and set

$$\mathcal{F} = \left\{ \begin{array}{c} \text{collection of closed balls } B_\rho(x) \text{ for } x \in E \\ \text{satisfying (8.4) and contained in } \mathcal{O} \end{array} \right\}.$$

Since $\mathcal{O}$ is open and ρ is arbitrarily small, such a collection is not empty and forms a fine Besicovitch covering for E. By the Besicovitch measure-theoretical covering theorem, there exists a countable collection $\{B(x_n)\}$ of disjoint, closed balls in $\mathcal{F}$, such that $\mu_e(E - \cup B_n) = 0$. From this and (8.4)

$$\mu_e(E) \le \sum \mu(B_n) < \frac{1}{t - \varepsilon} \sum \nu(B_n) \le \frac{1}{t - \varepsilon} \nu(\mathcal{O}).$$

Since ν is regular, there exists a set E_δ of the type of a $\mathcal{G}_\delta$ and containing E, such that $\nu_e(E) = \nu(E_\delta)$. Therefore

$$\nu_e(E) = \nu(E_\delta) = \inf\{\nu(\mathcal{O}) \text{ where } \mathcal{O} \text{ is open and contains } E_\delta\}.$$

Thus

$$\mu_e(E) \le \frac{1}{t - \varepsilon} \nu_e(E) \qquad \text{for all } \varepsilon \in (0, t).$$

This proves (8.2) if E is bounded. If not, construct a countable collection $\{E_n\}$ of bounded sets such that $E_n \subset E_{n+1}$ whose union if E. Then apply (8.2) to each of the E_n to obtain

$$\mu_e(E_n) \le \frac{1}{t} \nu_e(E) \qquad \text{for all } n \in \mathbb{N}.$$

For each n let $E_{n,\delta}$ be a set of the type of $\mathcal{G}_\delta$ such that $E_n \subset E_{n,\delta}$ and $\mu_e(E_n) = \mu(E_{n,\delta})$. By construction $E \subset \liminf E_{n,\delta}$. Therefore

$$\mu_e(E) \le \mu_e(\liminf E_{n,\delta}) = \mu_e(\liminf E_{n,\delta})$$
$$\le \liminf \mu(E_{n,\delta}) = \liminf \mu_e(E_n).$$

The proof of (8.3) is analogous.   ∎

9   Existence and Measurability of $D_\mu \nu$

The next proposition asserts that ν is differentiable with respect to μ for μ-almost all $x \in \mathbb{R}^N$. Equivalently $D_\mu \nu(x)$ exists μ-a.e. in $\mathbb{R}^N$.

Proposition 9.1   *There exists a Borel set* $\mathcal{E} \subset \mathbb{R}^N$ *of* μ-*measure zero, such that* $D_\mu^\pm \nu$ *is finite, and* $D_\mu \nu = D_\mu^\pm \nu$ *in* $\mathbb{R}^N - \mathcal{E}$.

Proof   Assume first that both μ and ν are finite and set $\mathcal{E}_\infty = [D_\mu^+ \nu = \infty]$. By (8.2)

$$\mu_e([D_\mu^+ \nu > t]) \leq \frac{1}{t} \nu(\mathbb{R}^N) \qquad \text{for all } t > 0.$$

Since $\mathcal{E}_\infty \subset [D_\mu^+ \nu > t]$, for all $t > 0$, one has $\mu_e(\mathcal{E}_\infty) = 0$. There exists a Borel set $\mathcal{E}_{\infty,\delta}$ of the type of a $\mathcal{G}_\delta$, such that $\mathcal{E}_\infty \subset \mathcal{E}_{\infty,\delta}$, and $\mu_e(\mathcal{E}_\infty) = \mu(\mathcal{E}_{\infty,\delta}) = 0$. Next, for positive integers p, q, set

$$E_{p,q} = \left[D_\mu^- \nu < \frac{p}{q} < \frac{p+1}{q} < D_\mu^+ \nu \right] - \mathcal{E}_{\infty,\delta}.$$

By (8.2)–(8.3)

$$\frac{p+1}{q} \mu_e(E_{p,q}) \leq \nu_e(E_{p,q}) \leq \frac{p}{q} \mu_e(E_{p,q}).$$

Therefore $\mu_e(E_{p,q}) = 0$. From this

$$\mu_e([D_\mu^- \nu < D_\mu^+ \nu]) \leq \mu_e(\bigcup E_{p,q}) \leq \sum \mu_e(E_{p,q}) = 0.$$

There exists a Borel set $[D_\mu^- \nu < D_\mu^+ \nu]_\delta$ containing $[D_\mu^- \nu < D_\mu^+ \nu]$ and such that

$$\mu_e([D_\mu^- \nu < D_\mu^+ \nu]) = \mu([D_\mu^- \nu < D_\mu^+ \nu]_\delta) = 0.$$

Setting

$$\mathcal{E} = \mathcal{E}_{\infty,\delta} \cup [D_\mu^- \nu < D_\mu^+ \nu]_\delta$$

proves the proposition if μ and ν are finite. For the general case, one first considers the restrictions μ_n and ν_n of μ and ν to the ball B_n centered at the origin and radius n, and then lets $n \to \infty$.   ∎

Henceforth we regard $D_\mu \nu$ as defined in the whole $\mathbb{R}^N$ by setting it to be zero on the Borel set $\mathcal{E}$ claimed by Proposition 9.1. For $\rho > 0$ set

$$f_\rho(x) = \begin{cases} \dfrac{\nu(B_\rho(x))}{\mu(B_\rho(x))} & \text{for } x \in \mathbb{R}^N - \mathcal{E} \\ 0 & \text{for } x \in \mathcal{E}. \end{cases} \qquad (9.1)$$

The limit as $\rho \to 0$ exists everywhere in $\mathbb{R}^N$ and equals $D_\mu \nu$. Taking such a limit along a countable collection $\{\rho_n\} \to 0$

$$D_\mu \nu = \lim f_{\rho_n} \qquad \text{for all } x \in \mathbb{R}^N. \tag{9.2}$$

Proposition 9.2 *For all* $t \in \mathbb{R}$ *the sets* $[D_\mu \nu \geq t]$ *are Borel sets. In particular* $D_\mu \nu$ *is Borel measurable.*

The proof hinges upon the following lemma.

Lemma 9.1 *For all fixed* $\rho > 0$, *the two functions*

$$\mathbb{R}^N \ni x \to \mu(B_\rho(x)), \quad \mathbb{R}^N \ni x \to \nu(B_\rho(x))$$

are upper semi-continuous.

Proof The statement for $x \to \mu(B_\rho(x))$ reduces to

$$\limsup_{y \to x} \mu(B_\rho(y)) \leq \mu(B_\rho(x)) \quad \text{for all } x \in \mathbb{R}^N.$$

Let $\{x_n\}$ be a sequence of points in $\mathbb{R}^N$ converging to x. Then

$$\limsup \chi_{B_\rho(x_n)} \leq \chi_{B_\rho(x)} \qquad \text{pointwise in } \mathbb{R}^N.$$

Equivalently

$$\liminf(1 - \chi_{B_\rho(x_n)}) \geq (1 - \chi_{B_\rho(x)}) \qquad \text{pointwise in } \mathbb{R}^N.$$

By Fatou's lemma

$$\begin{aligned}
\mu(B_{2\rho}(x)) - \mu(B_\rho(x)) &= \int_{B_{2\rho}} (1 - \chi_{B_\rho(x)}) d\mu \\
&\leq \int_{B_{2\rho}} \liminf(1 - \chi_{B_\rho(x_n)}) d\mu \\
&\leq \liminf \int_{B_{2\rho}} (1 - \chi_{B_\rho(x_n)}) d\mu \\
&= \liminf \{\mu(B_{2\rho}(x)) - \mu(B_\rho(x_n))\} \\
&= \mu(B_{2\rho}(x)) - \limsup \mu(B_\rho(x_n)).
\end{aligned}$$

∎

9.1 Proof of Proposition 9.2

If $t \leq 0$, then $[D_\mu \nu \geq t] = \mathbb{R}^N$. Therefore it suffices to consider $t > 0$. From (9.1)–(9.2), $D_\mu \nu = \inf \varphi_n$ where $\varphi_n = \sup_{j \geq n} f_{\rho_j}$. Since

$$[D_\mu \nu \geq t] = \bigcap_n [\varphi_n \geq t] = \bigcap_n \bigcap_k \left[\varphi_n > t - \frac{1}{k} \right] = \bigcap_n \bigcap_k \bigcup_{j \geq n} \left[f_{\rho_j} > t - \frac{1}{k} \right]$$

it suffices to show that $[f_\rho > \tau]$ are Borel sets for all $\rho, \tau > 0$. From (9.1)

$$[f_\rho > \tau] = \left\{ x \in \mathbb{R}^N \mid [\nu(B_\rho(x)) > \tau \mu(B_\rho(x))] \right\} - \mathcal{E}.$$

Since $\mathcal{E}$ is a Borel set, it suffices to show that $[\nu(B_\rho) > \tau \mu(B_\rho)]$ is a Borel set. Let $\{q_n\}$ denote the sequence of rational numbers. For a fixed $\tau > 0$

$$[\nu(B_\rho) > \tau \mu(B_\rho)] = \bigcup [\nu(B_\rho) \geq q_n] \cap [\tau \mu(B_\rho) < q_n].$$

Since the two functions $x \to \mu(B_\rho(x))$, $\nu(B_\rho(x))$ are upper semi-continuous, the sets $[\tau \mu(B_\rho) < q_n]$ are open and the sets $[\nu(B_\rho) \geq q_n]$ are closed. ∎

10 Representing $D_\mu \nu$

In representing $D_\mu \nu$ assume that the two Radon measures μ and ν are defined on the same σ-algebra $\mathcal{A}$. The measurable function $D_\mu \nu$ can be identified by considering separately the cases when ν is absolutely continuous or singular with respect to μ. For general Radon measures, $D_\mu \nu$ is identified by combining these two cases and applying the Lebesgue decomposition theorem of ν into two measures ν_o and ν_1 where the first is absolutely continuous and the second is singular with respect to μ (Theorem 18.2 of Chap. 4).

10.1 Representing $D_\mu \nu$ for $\nu \ll \mu$

Lemma 10.1 Let $\nu \ll \mu$. Then $\nu([D_\mu \nu = 0]) = 0$.

Proof Let $\mathcal{E}$ be the Borel set claimed by Proposition 9.1 and appearing in (9.1). Then for all $t > 0$, $[D_\mu \nu = 0] \subset \mathcal{E} \cup [D_\mu^- \nu < t]$. From this, (8.3), and the absolute continuity of ν with respect to μ

$$\nu([D_\mu \nu = 0]) \leq \nu([D_\mu^- \nu < t]) \leq t \mu([D_\mu^- \nu < t]).$$

If μ is finite, the conclusion follows by letting $t \to 0$. If not, restrict first μ and ν to balls B_n centered at the origin and radius n, and then let $n \to \infty$. ∎

It follows from Lemma 10.1 that

$$\nu([D_\mu\nu = 0]) = \int_{[D_\mu\nu=0]} D_\mu\nu d\mu = 0.$$

The next proposition asserts that such a formula actually holds with the set $[D_\mu\nu = 0]$ replaced by any μ-measurable set E.

Proposition 10.1 *Assume ν is absolutely continuous with respect to μ. Then for every μ-measurable set E*

$$\nu(E) = \int_E D_\mu\nu d\mu. \tag{10.1}$$

Proof Let $E \subset \mathbb{R}^N$ be μ-measurable and for $t > 1$ and $n \in \mathbb{Z}$ set

$$E_n = E \cap [t^n < D_\mu\nu \le t^{n+1}].$$

By construction $\bigcup_{n\in\mathbb{Z}} E_n \subset E$, and

$$E - \bigcup_{n\in\mathbb{Z}} E_n \subset [D_\mu\nu = 0] \quad \Longrightarrow \quad \nu\left(E - \bigcup_{n\in\mathbb{Z}} E_n\right) = 0.$$

From this and (8.3)

$$\nu(E) = \sum_{n\in\mathbb{Z}} \nu(E_n) \le \sum_{n\in\mathbb{Z}} t^{n+1}\mu(E_n)$$

$$= t \sum_{n\in\mathbb{Z}} t^n \mu(E_n) \le t \sum_{n\in\mathbb{Z}} \int_{E_n} D_\mu\nu d\mu = t \int_E D_\mu\nu d\mu.$$

Similarly using (8.2)

$$\nu(E) = \sum_{n\in\mathbb{Z}} \nu(E_n) \ge \sum_{n\in\mathbb{Z}} t^n \mu(E_n)$$

$$= \frac{1}{t} \sum_{n\in\mathbb{Z}} t^{n+1}\mu(E_n) \ge \frac{1}{t} \sum_{n\in\mathbb{Z}} \int_{E_n} D_\mu\nu d\mu = \frac{1}{t} \int_E D_\mu\nu d\mu.$$

Therefore

$$\frac{1}{t} \int_E D_\mu\nu d\mu \le \nu(E) \le t \int_E D_\mu\nu d\mu$$

for all $t > 1$. Letting $t \to 1$ proves (10.1). ∎

10.2   Representing $D_\mu \nu$ for $\nu \perp \mu$

Continue to assume that μ and ν are two Radon measures in $\mathbb{R}^N$ defined on the same σ-algebra $\mathcal{A}$.

Proposition 10.2 *Assume ν is singular with respect to μ. There exists a Borel set $\mathcal{E}_\perp$ of μ-measure zero, such that $D_\mu \nu = 0$ in $\mathbb{R}^N - \mathcal{E}_\perp$.*

Proof Since μ and ν are singular, $\mathbb{R}^N$ can be partitioned into two disjoint sets $\mathbb{R}^N_\mu$ and $\mathbb{R}^N_\nu$, such that for every $E \in \mathcal{A}$ (§ 18 of Chap. 4).

$$\mu(E \cap \mathbb{R}^N_\nu) = 0 \quad \text{and} \quad \nu(E \cap \mathbb{R}^N_\mu) = 0.$$

By (8.2), for all $t > 0$

$$\mu([D_\mu \nu > t]) = \mu([D_\mu \nu > t] \cap \mathbb{R}^N_\mu) \leq \frac{1}{t} \nu([D_\mu \nu > t] \cap \mathbb{R}^N_\mu) = 0.$$

Denoting by $\{t_n\}$ the positive rational numbers, $[D_\mu \nu > 0] = \bigcup [D_\mu \nu > t_n]$ and

$$\mu([D_\mu \nu > 0]) \leq \sum \mu([D_\mu \nu > t_n]) = 0. \qquad \blacksquare$$

11   The Lebesgue-Besicovitch Differentiation Theorem

Let μ be a Radon measure in $\mathbb{R}^N$ defined on a σ-algebra $\mathcal{A}$. A function $f : \mathbb{R}^N \to \mathbb{R}^*$ measurable with respect to $\mathcal{A}$, is *locally μ-integrable* in $\mathbb{R}^N$ if

$$\int_E |f| d\mu < \infty \quad \text{for every bounded set } E \in \mathcal{A}.$$

If f is nonnegative, the formula

$$\mathcal{A} \ni E \longrightarrow \nu(E) = \int_E f d\mu$$

defines a Radon measure ν in $\mathbb{R}^N$, absolutely continuous with respect to μ, whose Radon-Nykodým derivative with respect to μ is f. Moreover, such an f is unique, up to a set of μ-measure zero. Therefore by Proposition 10.1

$$D_\mu \nu = \frac{d\nu}{d\mu} = f \quad \mu - \text{a.e. in } \mathbb{R}^N.$$

Let now f be locally μ-integrable in $\mathbb{R}^N$ and of variable sign. By writing $f = f^+ - f^-$ and applying the same reasoning separately to $f^\pm$, proves a N-dimensional version of the Lebesgue differentiation theorem for general Radon measures.

Theorem 11.1 (Lebesgue-Besicovitch [16]) *Let μ be a Radon measure in $\mathbb{R}^N$ and let $f : \mathbb{R}^N \to \mathbb{R}^*$ be locally μ-integrable. Then*

$$\lim_{\rho \to 0} \frac{1}{\mu(B_\rho(x))} \int_{B_\rho(x)} f d\mu = f(x) \quad \text{for } \mu - a.e.x \in \mathbb{R}^N. \tag{11.1}$$

If μ is the Lebesgue measure in $\mathbb{R}$ and f is locally Lebesgue integrable, the limit in (11.1) takes the form

$$\lim_{h \to 0} \frac{1}{2h} \int_{x-h}^{x+h} f(y) dy = f(x) \quad \text{for a.e. } x \in \mathbb{R}. \tag{11.1}_{N=1}$$

In this sense, (11.1) can be regarded as a N-dimensional notion of taking the derivative of an integral at a fixed point $x \in \mathbb{R}^N$.

11.1 Points of Density

Let $E \subset \mathbb{R}^N$ be μ measurable. Applying (11.1) with $f = \chi_E$, gives

$$\lim_{\rho \to 0} \frac{\mu(E \cap B_\rho(x))}{\mu(B_\rho(x))} = \chi_E(x) \quad \mu - \text{a.e. in } E.$$

A point $x \in E$ for which such a limit is one, is a *point of density* of E.

Corollary 11.1 *Almost every point of a μ-measurable set $E \subset \mathbb{R}^N$ is a point of density for E.*

11.2 Lebesgue Points of an Integrable Function

Let μ be a Radon measure in $\mathbb{R}^N$ and let f be locally μ-integrable. The points $x \in \mathbb{R}^N$ where (11.1) holds form a set called the *set of differentiability* of f. A point x is a *Lebesgue point* for f if

$$\lim_{\rho \to 0} \frac{1}{\mu(B_\rho(x))} \int_{B_\rho(x)} |f(y) - f(x)| d\mu = 0. \tag{11.2}$$

A Lebesgue point is a differentiability point for f. The converse is false.

Theorem 11.2 *Let μ be a Radon measure in $\mathbb{R}^N$ and let f be locally μ-integrable. There exists a μ-measurable set $\mathcal{E} \subset \mathbb{R}^N$ of μ-measure zero, such that (11.2) holds for all $x \in \mathbb{R}^N - \mathcal{E}$. Equivalently, μ-a.e. $x \in \mathbb{R}^N$ is a Lebesgue point for f.*

Proof Let r_n be a rational number. The function $|f - r_n|$ is locally μ-integrable. Therefore there exists a Borel set $\mathcal{E}_n \subset \mathbb{R}^N$ of μ-measure zero, such that

$$\lim_{\rho \to 0} \frac{1}{\mu(B_\rho(x))} \int_{B_\rho(x)} |f - r_n| d\mu = |f(x) - r_n| \quad \text{for all } x \in \mathbb{R}^N - \mathcal{E}_n.$$

Since f is locally μ-integrable in $\mathbb{R}^N$, there exists $\mathcal{E}_o \subset \mathbb{R}^N$, of μ-measure zero, such that $f(x)$ is finite for all $x \in \mathbb{R}^N - \mathcal{E}_o$. The set $\mathcal{E} = \bigcup_{n=0}^\infty \mathcal{E}_n$, has μ-measure zero. Then for all $x \in \mathbb{R}^N - \mathcal{E}$, by the triangle inequality

$$\lim_{\rho \to 0} \frac{1}{\mu(B_\rho(x))} \int_{B_\rho(x)} |f - f(x)| d\mu \le 2|f(x) - r_n|$$

for all rational numbers $\{r_n\}$. Since $f(x)$ is finite, there exists a sequence $\{r_{n'}\} \subset \{r_n\}$ converging to $f(x)$. Thus (11.2) holds for all $x \in \mathbb{R}^N - \mathcal{E}$.   ∎

12   Regular Families

Let μ be a Radon measure in $\mathbb{R}^N$. For a fixed $x \in \mathbb{R}^N$ a family $\mathcal{F}_x$ of μ-measurable subsets of $\mathbb{R}^N$ is said to be *regular* at x if:

 (i) For every $\varepsilon > 0$ there exists $S \in \mathcal{F}_x$ such that $\operatorname{diam} S \le \varepsilon$
 (ii) There exists a constant $c \ge 1$, such that for each $S \in \mathcal{F}_x$

$$\mu(B(x)) \le c\mu(S). \tag{12.1}$$

where $B(x)$ is the smallest ball in $\mathbb{R}^N$ centered at x and containing S.

The first of these asserts, roughly speaking, that the sets $S \in \mathcal{F}_x$ shrink to x, even though x is not required to be in any of the sets $S \in \mathcal{F}_x$. The second says that each S is, roughly speaking, comparable to a ball centered at x.

If μ is the Lebesgue measure in $\mathbb{R}^N$, examples of regular families $\mathcal{F}_o$ at the origin, include the collection of cubes, ellipsoids or regular polygons centered at the origin. An example of a regular family $\mathcal{F}_o$ whose sets S don't contain the origin is the collection of spherical annuli $\frac{1}{2}\rho < |x| < \rho$.

The sets in $\mathcal{F}_x$ have no symmetry restrictions.

The sets shrinking to a point x, in Theorem 11.2 need not be balls, provided they shrink to x along a regular family $\mathcal{F}_x$.

Proposition 12.1 *Let f locally μ-integrable in $\mathbb{R}^N$. Then if x is a Lebesgue point for f and $\mathcal{F}_x$ is a regular family at x*

$$\lim_{\substack{diam\, S \to 0 \\ S \in \mathcal{F}_x}} \frac{1}{\mu(S)} \int_S |f(y) - f(x)| d\mu = 0. \tag{12.2}$$

In particular

$$f(x) = \lim_{\substack{diam\, S \to 0 \\ S \in \mathcal{F}_x}} \frac{1}{\mu(S)} \int_S f d\mu. \tag{12.3}$$

Proof Having fixed $S \in \mathcal{F}_x$ let $B(x)$ be the ball satisfying (12.1). Then

$$\frac{1}{\mu(S)} \int_S |f - f(x)| d\mu \le \frac{c}{\mu(B(x))} \int_{B(x)} |f - f(x)| d\mu. \qquad \blacksquare$$

Referring back to $(11.1)_{N=1}$, these remarks imply that for locally Lebesgue integrable functions of one variable

$$\lim_{h \to 0} \frac{1}{h} \int_x^{x+h} f d\mu = f(x) \qquad \text{for a.e. } x \in \mathbb{R}. \tag{$11.1)'_{N=1}$}$$

13   Convex Functions

A function f from an open interval (a, b) into $\mathbb{R}^*$ is convex if for every pair $x, y \in (a, b)$ and every $t \in [0, 1]$

$$f(tx + (1 - t)y) \le tf(x) + (1 - t)f(y).$$

A function f is concave if $-f$ is convex. The set

$$\mathcal{G}_f = \left\{ (x, y) \in \mathbb{R}^2 \mid x \in (a, b), \quad y \ge f(x) \right\}$$

is the *epigraph* of f. The function f is convex if and only if its epigraph is convex. The positive linear combination of convex functions is convex and the pointwise limit of a sequence of convex functions is convex.

Proposition 13.1 *Let $\{f_\alpha\}$ be a family of convex functions defined in (a, b). Then the function $f = \sup f_\alpha$ is convex in (a, b).*

Proof Fix $x, y \in (a, b)$ and $t \in [0, 1]$ and assume first that $f(tx + (1 - t)y)$ is finite. Having fixed an arbitrary $\varepsilon > 0$ there exists α such that

$$f(tx + (1 - t)y) \le f_\alpha(tx + (1 - t)y) + \varepsilon$$
$$\le tf_\alpha(x) + (1 - t)f_\alpha(y) + \varepsilon \le tf(x) + (1 - t)f(y) + \varepsilon.$$

If $f(tx + (1 - t)y) = \infty$, having fixed an arbitrarily large number k, there exists α, such that $k \le t f_\alpha(x) + (1 - t) f_\alpha(y)$.   ∎

Proposition 13.2 *Let f be a real-valued, convex function in some interval $(a, b) \subset \mathbb{R}$. Then the function*

$$y \longrightarrow \mathcal{F}(x; y) = \frac{f(x) - f(y)}{x - y} \qquad x, y \in (a, b); \quad x \ne y$$

is non-decreasing.

Proof Assume $y > x$. It suffices to show that

$$\mathcal{F}(x; z) \le \mathcal{F}(x; y) \quad \text{for} \quad z = tx + (1 - t)y \quad \text{for all} \quad t \in (0, 1).$$

By the convexity of f

$$\mathcal{F}(x; z) = \frac{f(tx + (1 - t)y) - f(x)}{(1 - t)(y - x)} \le \frac{(1 - t)[f(y) - f(x)]}{(1 - t)(y - x)} = \mathcal{F}(x; y).$$

∎

By symmetry, the function $x \to \mathcal{F}(x; y)$ is also non-decreasing.

Proposition 13.3 *Let f be a real-valued, convex function in some interval $[a, b] \subset \mathbb{R}$. Then f is locally Lipschitz continuous in (a, b).*

Proof Fix a subinterval $[c, d] \subset (a, b)$. Then for all $x, y \in [c, d]$

$$\mathcal{F}(c; a) = \frac{f(c) - f(a)}{c - a} \le \frac{f(x) - f(y)}{x - y} \le \frac{f(b) - f(d)}{b - d} = \mathcal{F}(b; d)$$

If $\mathcal{F}(x; y)$ is nonnegative, also $\mathcal{F}(b; d)$ is nonnegative. Therefore

$$|f(x) - f(y)| \le \mathcal{F}(b; d)|x - y|.$$

If the difference quotient $\mathcal{F}(x; y)$ is negative, then

$$|f(x) - f(y)| \le -\mathcal{F}(a; c)|x - y|.$$

∎

Proposition 13.4 *Let f be a real-valued convex function in (a, b). Then f is a.e. differentiable in (a, b). Moreover the right and left derivatives $D_\pm f(x)$ exist and are finite at each $x \in (a, b)$ and are both monotone non-decreasing functions. Also $D_- f(x) \le D_+ f(x)$ for all $x \in (a, b)$.*

Proof For each $x \in (a, b)$ fixed, the function $(h, k) \to \mathcal{F}(x + h; x + k)$ is non-decreasing in both variables. Therefore there exist and are finite the limits

$$D_- f(x) = \lim_{h \nearrow 0} \mathcal{F}(x + h; x) \leq \lim_{k \searrow 0} \mathcal{F}(x; x + k) = D_+ f(x).$$

Since f is absolutely continuous in every closed subinterval of (a, b), it is a.e. differentiable in (a, b) and $D_- f(x) = D_+ f(x)$ for a.e. $x \in (a, b)$. If $x < y$

$$D_+ f(x) = \lim_{h \searrow 0} \mathcal{F}(x + h; x) \leq \lim_{h \searrow 0} \mathcal{F}(y + h; y) = D_+ f(y).$$

Thus $D_\pm f$ are both non-decreasing. ∎

14 The Jensen's Inequality

Let φ be a real-valued, convex function in some interval (a, b). For a fixed $x \in (a, b)$ consider the set

$$\partial_x \varphi = [D_- \varphi(x), D_+ \varphi(x)].$$

If φ is differentiable at x, then $\partial_x \varphi = \varphi'(x)$. Otherwise $\partial_x \varphi$ is an interval. Fix $\alpha \in (a, b)$ and $m \in \partial_\alpha \varphi$. Since φ is convex the line through $(\alpha, \varphi(\alpha))$ and slope m lies below the epigraph of φ. In particular

$$\varphi(\alpha) + m(\eta - \alpha) \leq \varphi(\eta) \quad \text{for all } \eta \in (a, b). \tag{14.1}$$

Proposition 14.1 (Jensen [76]) *Let E be a measurable set of finite measure, and let $f : E \to \mathbb{R}$ be integrable in E. Then, for every real-valued, convex function φ defined in $\mathbb{R}$*

$$\varphi\left(\frac{1}{\mu(E)} \int_E f d\mu\right) \leq \frac{1}{\mu(E)} \int_E \varphi(f) d\mu. \tag{14.2}$$

Proof Applying (14.1) for the choices

$$\alpha = \frac{1}{\mu(E)} \int_E f d\mu \qquad \eta = f(x) \qquad \text{for a.e. } x \in E$$

yields

$$\varphi\left(\frac{1}{\mu(E)} \int_E f d\mu\right) + m\left(f(x) - \frac{1}{\mu(E)} \int_E f d\mu\right) \leq \varphi(f(x)).$$

Integrate over E and divide by the measure of E. ∎

15   Extending Continuous Functions

Let f be a continuous function defined on a set $E \subset \mathbb{R}^N$ with values in $\mathbb{R}$ and with modulus of continuity

$$\omega_f(s) = \sup_{\substack{|x-y| \le s \\ x,y \in E}} |f(x) - f(y)| \qquad s > 0.$$

The function $s \to \omega_f(s)$ is nonnegative and non-decreasing in $[0, \infty)$. The function f is uniformly continuous in E if and only if $\omega_f(s) \to 0$ as $s \to 0$.

15.1   The Concave Modulus of Continuity of f

Assume that $\omega_f(\cdot)$ is dominated in $[0, \infty)$ by some increasing, affine function $\ell(\cdot)$, i.e.,

$$\omega_f(s) \le as + b \quad \text{for all } s \in [0, \infty) \quad \text{for some } a, b \in \mathbb{R}^+. \tag{15.1}$$

Denote by $s \to c_f(s)$ the concave modulus of continuity of f, i.e., the smallest concave function in $[0, \infty)$ whose graph lies above the graph of $s \to \omega_f(s)$. If $\omega_f(\cdot)$ satisfies (15.1), then $c_f(\cdot)$ can be constructed as

$$c_f(s) = \inf\{\ell(s) \mid \ell \text{ is affine and } \ell \ge \omega_f \text{ in } [0, \infty)\}.$$

It follows from the definitions that

$$c_f(|x - y|) - |f(x) - f(y)| \ge 0 \quad \text{for all } x, y \in E. \tag{15.2}$$

Theorem 15.1 (Kirzbraun-McShane-Pucci)[3] *Let f be a real-valued, uniformly continuous function on a set $E \subset \mathbb{R}^N$ with modulus of continuity ω_f satisfying (15.1). There exists a continuous function $\tilde{f}$ defined on $\mathbb{R}^N$, which coincides with f on E. Moreover f and $\tilde{f}$ have the same concave modulus of continuity c_f and*

$$\sup_{\mathbb{R}^N} \tilde{f} = \sup_E f; \qquad \inf_{\mathbb{R}^N} \tilde{f} = \inf_E f.$$

---

[3] If the modulus of continuity is of Lipschitz type, the theorem is in M.D. Kirzbraun [83]. The proof of Kirzbraun is rather general as it does include vector-valued functions. A simpler proof for scalar functions, is in McShane [106]. Pucci observed that the concavity of the modulus of continuity is sufficient to construct the extension. The proof has been taken from the 1974 lectures on Real Analysis by C. Pucci, at the Univ. of Florence, Italy.

Proof For each $x \in \mathbb{R}^N$, set

$$g(x) = \inf_{y \in E} \{f(y) + c_f(|x - y|)\}.$$

The required extension is

$$\tilde{f}(x) = \min\{g(x); \sup_E f\}.$$

If $x \in E$, by (15.2)

$$f(y) + c_f(|x - y|) \geq f(x) + c_f(|x - y|) - |f(x) - f(y)| \geq f(x)$$

for all $y \in E$. Therefore $g = f$ within E. Next for all $x \in \mathbb{R}^N$ and all $y \in E$

$$\inf_E f + \inf_{y \in E} c_f(|x - y|) \leq g(x) \leq f(y) + c_f(|x - y|).$$

Therefore

$$\inf_{\mathbb{R}^N} g = \inf_E f \quad \text{and} \quad \sup_{\mathbb{R}^N} \tilde{f} = \sup_E f.$$

To prove that f and $\tilde{f}$ have the same concave modulus of continuity, it suffices to prove that g has the same concave modulus of continuity as f. Fix $x_1, x_2 \in \mathbb{R}^N$ and $\varepsilon > 0$. There exists $y \in E$ such that

$$g(x_1) \geq f(y) + c_f(|x_1 - y|) - \varepsilon.$$

Therefore for such $y \in E$

$$g(x_1) - g(x_2) \geq c_f(|x_1 - y|) - c_f(|x_2 - y|) - \varepsilon.$$

If $|x_2 - y| \leq |x_1 - x_2|$

$$g(x_1) - g(x_2) \geq -c_f(|x_1 - x_2|) - \varepsilon.$$

Otherwise

$$|x_1 - y| > |x_2 - y| - |x_1 - x_2| > 0.$$

Since $s \to c_f(s)$ is concave, $-c_f(\cdot)$ is convex, and by Proposition 13.2

$$\frac{c_f(|x_1 - y|) - c_f(0)}{|x_1 - y|} \geq \frac{c_f(|x_1 - y| + |x_2 - x_1|) - c_f(|x_1 - x_2|)}{|x_1 - y| + |x_2 - x_1| - |x_1 - x_2|}$$

$$\geq \frac{c_f(|x_2 - y|) - c_f(|x_1 - x_2|)}{|x_1 - y|}.$$

From this, taking into account that $c_f(0) = 0$

$$c_f(|x_1 - y|) - c_f(|x_2 - y|) \geq -c_f(|x_1 - x_2|).$$

Thus in either case

$$g(x_1) - g(x_2) \geq -c_f(|x_1 - x_2|) - \varepsilon.$$

Interchanging the role of x_1 and x_2 and taking into account that $\varepsilon > 0$ is arbitrary, gives

$$|g(x_1) - g(x_2)| \leq c_f(|x_1 - x_2|). \qquad \blacksquare$$

16   The Weierstrass Approximation Theorem

Theorem 16.1 (Weierstrass [172]) *Let f be a real-valued, uniformly continuous function defined on a bounded set $E \subset \mathbb{R}^N$. There exists a sequence of polynomials $\{P_j\}$ such that*

$$\sup_E |f - P_j| \to 0 \quad as \ j \to \infty.$$

Proof By Theorem 15.1 we may regard f as defined in the whole $\mathbb{R}^N$ with modulus of continuity ω_f. After a translation and dilation, we may assume that $\bar{E}$ is contained in the interior of the unit cube Q centered at the origin of $\mathbb{R}^N$ and with faces parallel to the coordinate planes.

For $x \in \mathbb{R}^N$ and $\delta > 0$, we let $Q_\delta(x)$ denote the cube of edge 2δ centered at x and congruent to Q. For $j \in \mathbb{N}$, set

$$p_j(x) = \frac{1}{\alpha_j^N} \prod_{i=1}^{N} (1 - x_i^2)^j \qquad \alpha_j = \int_{-1}^{1} (1 - t^2)^j \, dt,$$

These are polynomials of degree $2jN$ satisfying

$$\int_{Q_1(x)} p_j(x - y) dy = 1 \qquad \text{for all } j \in \mathbb{N} \text{ and all } x \in \mathbb{R}^N.$$

The approximating polynomials claimed by the theorem are

$$P_j(x) = \int_Q f(y) p_j(x - y) dy. \qquad (16.1)$$

These are called the Stieltjes polynomials relative to f. For $x \in E$ compute

$$P_j(x) - f(x) = \int_Q f(y) p_j(x - y) dy - \int_{Q_1(x)} f(x) p_j(x - y) dy.$$

Let $\delta > 0$ be so small that $Q_\delta(x) \subset Q$. Then

$$
\begin{aligned}
|P_j(x) - f(x)| &\leq \int_{Q_\delta(x) \cap Q} |f(x) - f(y)| p_j(x - y) dy \\
&+ \left| \int_{Q - Q_\delta(x)} f(y) p_j(x - y) dy \right| \\
&+ \left| \int_{Q_1(x) - Q_\delta(x)} f(x) p_j(x - y) dy \right| \\
&\leq \omega_f(2\sqrt{N}\delta) + \sup_E |f| \int_{Q - Q_\delta(x)} p_j(x - y) dy \\
&+ \sup_E |f| \int_{Q_1(x) - Q_\delta(x)} p_j(x - y) dy.
\end{aligned}
$$

To estimate the last two integrals we observe that for $y \notin Q_\delta(x)$, for at least one index $i \in \{1, \ldots, N\}$, there holds $|x_i - y_i| > \delta$. Therefore

$$
p_j(x - y) \leq \alpha_j^{-N}(1 - \delta^2)^j.
$$

Moreover from the definition of α_j

$$
\alpha_j \geq 2 \int_0^1 (1 - t)^j dt = \frac{2}{j + 1}.
$$

Combining these calculations we estimate

$$
\sup_E |f - P_j| \leq \omega(\delta) + 2^{1-N} \sup_E |f| (j + 1)^N (1 - \delta^2)^j. \qquad \blacksquare
$$

Corollary 16.1 *Let E be a compact subset of $\mathbb{R}^N$. Then $C(E)$ endowed with the topology of the uniform convergence, is separable.*

Proof The collection of polynomials in the real variables $x_1, \ldots, x_N$, with rational coefficients is a countable, dense subset of $C(E)$. $\qquad \blacksquare$

17 The Stone-Weierstrass Theorem

Let $\{X; \mathcal{U}\}$ be a compact Hausdorff space, and denote by $C(X)$ the collection of all real-valued, continuous functions defined in X. Setting

$$
d(f, g) = \sup_{x \in X} |f(x) - g(x)| \quad f, g \in C(X) \tag{17.1}
$$

defines a complete metric in $C(X)$. We continue to denote by $C(X)$ the resulting metric space. The sum of two functions in $C(X)$ is in $C(X)$ and the product of a function in $C(X)$ by a real number is an element of $C(X)$. Thus $C(X)$ is a vector space. One verifies that the operations of sum and product by real numbers

$$+ : C(X) \times C(X) \to C(X) \qquad \bullet : \mathbb{R} \times C(X) \to C(X)$$

are continuous with respect to the corresponding product topologies. Thus $C(X)$ is a topological vector space. The space $C(X)$ is also an *algebra* in the sense that the product of any two functions in $C(X)$ remains in $C(X)$. More generally a subset $\mathcal{F} \subset C(X)$ is an algebra if it is closed under the operations of sum, product, and product by real numbers. For example, the collection of functions $f \in C(X)$ that vanish at some fixed point $x_o \in X$, is an algebra. The intersection of all algebras containing a given subset of $C(X)$ is an algebra. Since $\{X; \mathcal{U}\}$ is Hausdorff its points are closed. Therefore, having fixed $x \neq y$ in X, there exists a continuous function $f : X \to [0, 1]$ such that $f(x) = 0$ and $f(y) = 1$ (Urysohn's lemma, § 2 of Chap. 2). Thus there exists an element of $C(X)$ that distinguishes any two fixed, distinct points in X. More generally an algebra $\mathcal{F} \subset C(X)$ *separates* points of X if for any pair of distinct points $x, y \in X$, there exists a function $f \in \mathcal{F}$, such that $f(x) \neq f(y)$. For example if E is a bounded, open subset of $\mathbb{R}^N$ the collection of all polynomials in the coordinate variables forms an algebra $\mathcal{P}$ of functions in $C(\bar{E})$. Such an algebra trivially separates points.

The classical Weierstrass theorem asserts that every $f \in C(\bar{E})$ can be approximated by elements of $\mathcal{P}$, in the metric of (17.1). Equivalently, $C(\bar{E})$ is the closure of $\mathcal{P}$ in the metric (17.1). The proof was based on constructing explicitly the approximating polynomials to a given $f \in C(\bar{E})$.

Stone's theorem identifies the structure that a subset of $C(X)$ must possess to be dense in $C(X)$.

Theorem 17.1 (Stone [155]) *Let $\{X; \mathcal{U}\}$ be a compact Hausdorff space and let $\mathcal{F} \subset C(X)$ be an algebra that separates points and that contains the constant functions. Then $\bar{\mathcal{F}} = C(X)$.*

18 Proof of the Stone-Weierstrass Theorem

Proposition 18.1 *Let $\{X; \mathcal{U}\}$ be a compact Hausdorff space and let $\mathcal{F} \subset C(X)$ be an algebra. Then*

  (i) *The closure $\bar{\mathcal{F}}$ in $C(X)$, is an algebra*
  (ii) *If $f \in \mathcal{F}$ then $|f| \in \bar{\mathcal{F}}$*
  (iii) *If f and g are in $\mathcal{F}$, then $\max\{f; g\}$ and $\min\{f; g\}$ are in $\bar{\mathcal{F}}$.*

Proof The first statement follows from the structure of an algebra and the notion of closure in the metric (17.1). To prove (ii) we may assume, without loss of generality, that $|f| \leq 1$. Regard f as a variable ranging over $[-1, 1]$. By the classical Weierstrass approximation theorem, applied to the function $[-1, 1] \ni f \to |f|$, having fixed $\varepsilon > 0$, there exists a polynomial $\mathcal{P}_\varepsilon(f)$ in the variable f, such that

$$\sup_{f \in [-1,1]} \left| |f| - \mathcal{P}_\varepsilon(f) \right| \leq \varepsilon.$$

This in turn implies that

$$\sup_{x \in X} \left| |f(x)| - \mathcal{P}_\varepsilon\big(f(x)\big) \right| \leq \varepsilon.$$

Since $\mathcal{F}$ is an algebra, $\mathcal{P}_\varepsilon(f) \in \mathcal{F}$. Thus $|f|$ is in the closure of $\mathcal{F}$. The last statements follow from (ii) and the identities

$$\max\{f; g\} = \tfrac{1}{2}(f + g) + \tfrac{1}{2}|f - g|$$
$$\min\{f; g\} = \tfrac{1}{2}(f + g) - \tfrac{1}{2}|f - g|.$$

■

18.1 Proof of Stone's Theorem

Having fixed $f \in C(X)$ and $\varepsilon > 0$, we exhibit a function $\varphi \in \bar{\mathcal{F}}$ such that $d(f, \varphi) \leq \varepsilon$. Since $\mathcal{F}$ separates points of X, for any two distinct points $\xi, \eta \in X$, there exists $h \in \mathcal{F}$ such that $h(\xi) \neq h(\eta)$. Since $\mathcal{F}$ contains the constants, there exist numbers λ and μ, such that the function $\varphi_{\xi\eta} = \lambda h + \mu$ is in $\mathcal{F}$ and $\varphi_{\xi\eta}(\xi) = f(\xi)$ and $\varphi_{\xi\eta}(\eta) = f(\eta)$. By keeping ξ fixed, regard $\varphi_{\xi\eta}$ as a family of continuous functions, parameterized with $\eta \in X$. Since $\varphi_{\xi\eta}$ and f coincide at η and are both continuous, for each $\eta \in X$, there exists an open set $\mathcal{O}_\eta$ containing η and such that $\varphi_{\xi\eta} < f + \varepsilon$ in $\mathcal{O}_\eta$. The collection of open sets $\mathcal{O}_\eta$, as η ranges over X is an open covering for X from which we extract a finite one $\{\mathcal{O}_{\eta_1}, \ldots, \mathcal{O}_{\eta_n}\}$, for some finite n. Set

$$\varphi_\xi = \min \left\{ \varphi_{\xi\eta_1}, \ldots, \varphi_{\xi\eta_n} \right\}.$$

By Proposition 18.1–(iii), $\varphi_\xi \in \bar{\mathcal{F}}$. Moreover by construction

$$\varphi_\xi \leq f + \varepsilon \text{ in } X \quad \text{and} \quad \varphi_\xi(\xi) = f(\xi) \text{ for all } \xi \in X.$$

Since φ_ξ and f coincide at ξ and they are both continuous, for each $\xi \in X$ there exists an open set $\mathcal{O}_\xi$ containing ξ and such that $\varphi_\xi > f - \varepsilon$ in $\mathcal{O}_\xi$. The collection of open sets $\mathcal{O}_\xi$, as ξ ranges over X is an open covering for X, from which we extract a finite one, for example $\{\mathcal{O}_{\xi_1}, \ldots, \mathcal{O}_{\xi_m}\}$ for some finite m. Set

$$\varphi = \max \left\{ \varphi_{\xi_1}, \ldots, \varphi_{\xi_m} \right\}.$$

By Proposition 18.1–(iii), $\varphi \in \bar{\mathcal{F}}$, and by construction $|f - \varphi| \leq \varepsilon$ in X.   ∎

19   The Ascoli-Arzelà Theorem

Let $E \subset \mathbb{R}^N$ be open and let $x \in E$. A sequence of functions $\{f_n\}$ from E into $\mathbb{R}$ is equibounded at x if there exists $M(x) > 0$ such that

$$|f_n(x)| \leq M(x) \qquad \text{for all } n \in \mathbb{N}. \tag{19.1}$$

The sequence $\{f_n\}$ is equi-continuous at x if there exists a continuous increasing function $\omega_x : \mathbb{R}^+ \to \mathbb{R}^+$ with $\omega_x(0) = 0$, such that for all $y \in E$

$$|f_n(x) - f_n(y)| \leq \omega_x(|x - y|) \qquad \text{for all } n \in \mathbb{N}. \tag{19.2}$$

Theorem 19.1 (Ascoli-Arzelà)[4] *Let $\{f_n\}$ be a sequence of pointwise equibounded and equi-continuous functions in E. There exists a subsequence $\{f_{n'}\} \subset \{f_n\}$ converging to a continuous function f in E. Moreover for all $x \in E$*

$$|f(x) - f(y)| \leq \omega_x(|x - y|) \quad \text{for all } y \in E$$

and the convergence is uniformly on compact subsets $K \subset E$.

Proof Let $\mathbb{Q}$ denote the set of points of $\mathbb{R}^N$ whose coordinates are rational. Such a set is countable and dense in E. Let $x_1 \in \mathbb{Q} \cap E$. Since the sequence of numbers $\{f_n(x_1)\}$ is bounded, we may select a subsequence $\{f_{n_1}(x_1)\}$ convergent to some real number that we denote with $f(x_1)$. If $x_2 \in \mathbb{Q} \cap E$, the sequence of numbers $\{f_{n_1}(x_2)\}$ is bounded, and we may select a convergent subsequence $\{f_{n_{1,2}}(x_2)\} \to f(x_2)$. Proceeding in this fashion we may select out of $\{f_n\}$, by a diagonalization process, a subsequence $\{f_{n'}\}$ such that

$$f_{n'}(x) \longrightarrow f(x) \quad \text{for all } x \in \mathbb{Q} \cap E.$$

Next, fix $x \in E - \mathbb{Q}$. Since $\mathbb{Q}$ is dense in E, for each $\varepsilon > 0$ there exist $x_\ell \in \mathbb{Q} \cap E$ such that $|x - x_\ell| < \varepsilon$. Therefore by the assumption of equi-continuity at x,

---

[4]First proved by Ascoli in [6] for equi-Lipschitz functions, and extended by Arzelà in [5] to a general family of equi-continuous functions.

$$|f_{n'}(x) - f_{m'}(x)| \le |f_{n'}(x) - f_{n'}(x_\ell)|$$
$$+ |f_{m'}(x) - f_{m'}(x_\ell)|$$
$$+ |f_{n'}(x_\ell) - f_{m'}(x_\ell)|$$
$$\le 2\omega_x(\varepsilon) + |f_{n'}(x_\ell) - f_{m'}(x_\ell)|.$$

Since $\{f_{n'}(x_\ell)\}$ is convergent, there exists a positive integer $m(x_\ell)$ large enough that

$$|f_{n'}(x_\ell) - f_{m'}(x_\ell)| \le \varepsilon \quad \text{for all} \quad n', m' > m(x_\ell).$$

Therefore for all such n' and m'

$$|f_{n'}(x) - f_{m'}(x)| \le \varepsilon + 2\omega_x(\varepsilon).$$

Hence $\{f_{n'}(x)\}$ is a Cauchy sequence with limit $f(x)$. For $x, y \in E$

$$|f(x) - f(y)| = \lim |f_{n'}(x) - f_{n'}(y)| \le \omega_x(|x - y|).$$

Let $K \subset E$ be compact and let $\varepsilon > 0$ be fixed. For each $x \in K$ there exists $\delta_x > 0$, depending on ε, such that

$$\omega_x(|x - y|) < \tfrac{1}{3}\varepsilon \quad \text{for all} \quad y \in B_{\delta_x}(x).$$

The collection of open balls $\{B_{\frac{1}{2}\delta_x}(x)\}_{x \in K}$ covers K and we select a finite one, for the radii $\{\frac{1}{2}\delta_{x_1}, \ldots \frac{1}{2}\delta_{x_m}\}$ for some finite m. Since $\mathbb{Q}$ is dense in E, the points x_j can be chosen in $\mathbb{Q}$, and in such a way that the collection of balls $\{B_{\delta_{x_j}}(x_j)\}_{j=1}^m$ covers K. From the convergence of $\{f_{n'}\} \to f$ there exists an integer $n_{m,\varepsilon}$ such that

$$|f(x_j) - f_{n'}(x_j)| < \tfrac{1}{3}\varepsilon \quad \text{for} \quad n' \ge n_{m,\varepsilon} \text{ and } j = 1, \ldots, m.$$

Each $x \in K$ is contained in some ball $B_{\delta_{x_j}}(x_j)$. Therefore for $n' \ge n_{m,\varepsilon}$

$$|f_{n'}(x) - f(x)| \le |f_{n'}(x) - f_{n'}(x_j)|$$
$$+ |f_{n'}(x_j) - f(x_j)|$$
$$+ |f(x) - f(x_j)| < \varepsilon. \qquad \blacksquare$$

19.1 Pre-compact Subsets of $C(\bar{E})$

Let E be a bounded, open subset of $\mathbb{R}^N$ and denote by $C(\bar{E})$ the collection of all real-valued, continuous functions defined in E, with the metric (17.1).

Proposition 19.1 *A subset $\mathcal{K} \subset C(\bar{E})$ is pre-compact in $C(\bar{E})$, if and only if the elements of K are pointwise equibounded and equi-continuous in $\bar{E}$.*

Proof Since $C(E)$ is metric and separable, compactness and sequential compactness are equivalent (Proposition 17.3 of Chap. 2). Then, the sufficient condition follows from Theorem 19.1. For the necessary part recall that if $\mathcal{K}$ is pre-compact in $C(\bar{E})$, then $\bar{\mathcal{K}}$ is totally bounded, and hence it admits an ε-net, for all $\varepsilon > 0$ (Theorem 17.1 of Chap. 2). ∎

Problems and Complements

1c Functions of Bounded Variations

1.1. The continuous function on $[0, 1]$

$$f(x) = \begin{cases} x^2 \cos \dfrac{\pi}{x} & \text{for } x \in (0, 1] \\ 0 & \text{for } x = 0 \end{cases}$$

is of bounded variation in $[0, 1]$.

1.2. Let f be the continuous function defined in $[0, 1]$ by

$$f(0) = 0, \quad f\left(\tfrac{1}{2n+1}\right) = 0, \quad f\left(\tfrac{1}{2n}\right) = \tfrac{1}{2n} \text{ for all } n \in \mathbb{N}$$
$$f \text{ is affine on the intervals } \left[\tfrac{1}{m+1}, \tfrac{1}{m}\right] \text{ for all } m \in \mathbb{N}.$$

Such an f is not of bounded variation in $[0, 1]$.

1.3. Prove Propositions 1.1–1.3.

1.4. Let $\{f_n\}$ be a sequence of functions in $[a, b]$ converging pointwise in $[a, b]$ to f. Then $V_f[a, b] \leq \liminf V_{f_n}[a, b]$, and strict inequality may occur as shown by the sequence

$$f_n(x) = \begin{cases} 0 & \text{for } x = 0 \\ \tfrac{1}{2} & \text{for } x \in (0, \tfrac{1}{n}] \\ 0 & \text{for } x \in (\tfrac{1}{n}, 1]. \end{cases}$$

1.5. Let f be continuous and of bounded variation in $[a, b]$. Then the functions $x \to V_f[a, x]$, $V_f^+[a, x]$, $V_f^-[a, x]$ are continuous in $[a, b]$.

1.6. Prove that the distribution function $f_*(\cdot)$ of a measurable function f on a measure space $\{X, \mathcal{A}, \mu\}$, as defined by (15.1c) of Chap. 4, is of bounded variation in every interval $[a, b] \subset \mathbb{R}$.

1.1c   The Function of The Jumps

Let f be of bounded variation in $[a, b]$ and regard it as defined in $\mathbb{R}$, by extending f to be equal to $f(a)$ on $(-\infty, a]$ and equal to $f(b)$ in $[b, \infty)$. Prove that the limits

$$f(x^{\pm}) \overset{\text{def}}{=} \lim_{h \to 0} f(x \pm h) \quad \text{for } h > 0$$

exist for all $x \in [a, b]$. The function of the jumps of f is defined by

$$J_f(x) = \sum_{a \le c_j \le x} \left[f(c_j^+) - f(c_j^-) \right].$$

The difference $f - J_f$ is continuous in $[a, b]$. Also J_f is of bounded variation in $[a, b]$ and

$$\mathcal{V}_f[a, b] = \mathcal{V}_{f - J_f}[a, b] + \mathcal{V}_{J_f}[a, b].$$

Therefore a function f of bounded variation in $[a, b]$ can be decomposed into the continuous function $f - J_f$ and J_f. The latter bears the possible discontinuities of f in $[a, b]$.

1.7. Construct a non-decreasing function in $[0, 1]$ which is discontinuous at all the rational points of $[0, 1]$.

1.2c   The Space $BV[a, b]$

Let $[a, b] \subset \mathbb{R}$ be a finite interval and denote by $BV[a, b]$ the collection of all functions $f : [a, b] \to \mathbb{R}$ of bounded variation in $[a, b]$. One verifies that $BV[a, b]$ is a linear vector space. Also setting

$$d(f, g) = |f(a) - g(a)| + \mathcal{V}_{f-g}[a, b] \quad \text{for } f, g \in BV[a, b] \tag{1.1c}$$

defines a distance in $BV[a, b]$ by which $\{BV[a, b]; d\}$ is a metric space.

For any two functions $f, g \in BV[a, b]$

$$\sup_{[a,b]} |f - g| \le |f(a) - g(a)| + \mathcal{V}_{f-g}[a, b]. \tag{1.2c}$$

Therefore a Cauchy sequence in $BV[a, b]$ is also Cauchy in the sup-norm. The converse is false as illustrated in Fig. 1c. The sequence $\{f_n\}$ generated as in Fig. 1c is not a Cauchy sequence in the topology of $BV[0, 1]$, while it is a Cauchy sequence in the topology of $C[0, 1]$.

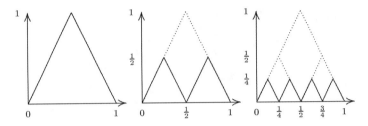

Fig. 1c Cauchy sequence in the sup-norm and not in the BV-norm

1.2.1c   Completeness of $BV[a, b]$

Let $\{f_n\}$ be a Cauchy sequence in $BV[a, b]$. There exists $f \in BV[a, b]$ such that $\{f_n\} \to f$ in the topology of $BV[a, b]$.

The proof is in two steps. First one uses (1.2c) to identify the limit f. Then one proves that such an f is actually in $BV[a, b]$ by using that $\{f_n\}$ is Cauchy in $BV[a, b]$. As a consequence $BV[a, b]$ is a complete metric space.

2c   Dini Derivatives

2.1. Compute $D^+f(0)$ and $D_+f(0)$ for the function in (1.1).

2.2. Let f have a maximum at some $c \in (a, b)$. Then $D^-f(c) \geq 0$.

2.3. Let f be continuous in $[a, b]$. If $D^+f \geq 0$ in $[a, b]$, then f is non-decreasing in $[a, b]$. The assumption that f be continuous cannot be removed.

Hint: Fix $[\alpha, \beta] \subset (a, b)$, and by continuity extend the function f to be equal to $f(\beta)$ for $x \geq \beta$ and to be equal to $f(\alpha)$ for $x \leq \alpha$. Having fixed $\varepsilon > 0$, the assumption implies that for each $x \in [\alpha, \beta]$ there exists

$$0 < h_x = h(x, \varepsilon) < \tfrac{1}{2}(b - \beta)$$

such that

$$f(x) < f(x + h_x) + h\varepsilon.$$

By continuity, such inequality continues to hold for all $y \in (x - \delta_x, x + \delta_x)$ for some $\delta_x > 0$, which without loss of generality we may assume not to exceed $\frac{1}{4}h_x$. The collection of these open intervals covers $[\alpha, \beta]$. From this select a finite sub-collection, say

$$(x_j - \delta_j, x_j + \delta_j) \quad \text{for } j = 1, \ldots, n.$$

To these are associated positive numbers $h_j < \frac{1}{2}(\beta - b)$ such that

$$f(y) < f(y + h_j) + h_j \varepsilon \quad \text{for all} \ \ y \in (x_j - \delta_j, x_j + \delta_j).$$

Starting with $y = \alpha$ and iterating this inequality $m \le n$ times gives

$$f(\alpha) < f\left(\alpha + \sum_{j=1}^{m} h_j\right) + \sum_{j=1}^{m} h_j \varepsilon.$$

By construction, there exists $m \le n$ such that

$$\alpha + \sum_{j=1}^{m} h_j \ge \beta \quad \text{and} \quad \sum_{j=1}^{m} h_j \le b - a.$$

Therefore

$$f(\alpha) < f(\beta) + \varepsilon(b - a) \quad \text{for all} \ \ \varepsilon > 0.$$

2.4. Let $f \in BV[a, b]$ and let J_f the function of the jumps of f introduced in § 1.1c. Prove that $J_f' = 0$ a.e. in $[a, b]$.

Hint: Assume that f is non-decreasing, so that J_f is non decreasing. Denoting by $\{c_n\}$ the sequence of the jumps of f

$$J_f(b) - J_f(a) = \sum_{a < c_j \le b} \left[f(c_j^+) - f(c_j^-) \right]. \tag{2.1c}$$

Fix $m \in \mathbb{N}$ and let $\varepsilon > 0$ be so small that the intervals

$$(\alpha_j, \beta_j] = \left(c_j - \frac{\varepsilon}{2m}, \ c_j + \frac{\varepsilon}{2m}\right] \cap (a, b] \quad \text{for} \ \ j = 1, \ldots, m$$

are disjoint. The complement of their union is the finite union of disjoint intervals $[\alpha_j', \beta_j')$. For $t > 0$ assume that $\mu([DJ_f > t]) > 0$, and choose

$$\varepsilon < \frac{1}{2}\mu([DJ_f > t]).$$

For such a choice estimate

$$\mu\left([DJ_f > t] \cap \bigcup_{j=1}^{m'} [\alpha_j', \beta_j']\right) > \frac{1}{2}\mu([DJ_f > t]) > 0$$

for a positive integer m'. By Proposition 2.2

$$J_f(\beta_j') - J_f(\alpha_j') \ge t\mu\left([DJ_f > t] \cap [\alpha_j', \beta_j']\right).$$

Therefore

$$J_f(b) - J_f(a) = \sum_{j=1}^{m} \left[J_f(\beta_j) - J_f(\alpha_j) \right] + \sum_{j=1}^{m+1} \left[J_f(\beta'_j) - J_f(\alpha'_j) \right]$$

$$\geq \sum_{j=1}^{m} \left[f(c_j^+) - f(c_j^-) \right] + \tfrac{1}{2}\mu([DJ_f > t]).$$

Letting $m \to \infty$ and taking into account (2.1c) implies $\mu([DJ_f > t]) = 0$ for all $t > 0$. For a different approach use Theorem 4.1.

2.1c A Continuous, Nowhere Differentiable Function ([167])

For a real number x, denote by $\{x\}$ the distance from x to its nearest integer and set

$$f(x) = \sum_{n=0}^{\infty} \frac{\{10^n x\}}{10^n}. \qquad (2.2c)$$

Each term of the series is continuous. Moreover the series is uniformly convergent being majorized by the geometric series $\sum 10^{-n}$. Therefore f is continuous. Since $f(x) = f(x + j)$ for every integer j and all $x \in \mathbb{R}$, it suffices to consider $x \in [0, 1)$. Any such x has a decimal expansion of the form $x = 0.a_1 a_2 \ldots a_n \ldots$, where a_i are integers from 0 to 9. By excluding the case when $a_i = 9$ for all i larger than some m, such a representation is unique.

For $n \in \mathbb{N}$ fixed compute

$$\{10^n x\} = 0.a_{n+1} a_{n+2} \ldots \qquad \text{if } 0.a_{n+1} a_{n+2} \cdots \leq \tfrac{1}{2}$$
$$\{10^n x\} = 1 - 0.a_{n+1} a_{n+2} \ldots \qquad \text{if } 0.a_{n+1} a_{n+2} \cdots > \tfrac{1}{2}.$$

Having fixed $x \in [0, 1)$ choose increments

$$h_m = \begin{cases} -10^{-m} & \text{if either } a_m = 4 \text{ or } a_m = 9 \\ +10^{-m} & \text{otherwise.} \end{cases}$$

Then form the difference quotients of f at x

$$\frac{f(x + h_m) - f(x)}{h_m} = 10^m \sum_{n=0}^{\infty} \pm \frac{\{10^n (x \pm 10^{-m})\} - \{10^n x\}}{10^n}.$$

The numerators of the terms of this last series, all vanish for $n \geq m$, whereas for $n = 0, 1, \ldots, (m - 1)$ they are equal to $\pm 10^{n-m}$. Therefore the difference quotient reduces to the sum of m terms each of the form ± 1. Such a sum is an integer, positive or negative, which has the same parity of m. Thus the limit as $h_m \to 0$ of the difference ratios does not exists.

Remark 2.1c The function in (2.2c) is not of bounded variation in any interval $[a, b] \subset \mathbb{R}$.

2.2c   An Application of the Baire Category Theorem

The existence of a continuous and nowhere differentiable function, can be established indirectly, by a category-type argument. The Dini derivatives $D_+ f$ and $D^+ f$ introduced in (2.1), are called the right Dini numbers.

Proposition 2.1c (Banach [12]) *There exists a real-valued, continuous function in* $[0, 1]$ *such that its Dini's numbers* $|D_+ f|$ *and* $|D^+ f|$ *are infinity at every point of* $(0, 1)$.

Proof For $n \in \mathbb{N}$ let E_n denote the collection of all functions $f \in C[0, 1]$, for which there exists at least one point $t \in [0, 1 - \frac{1}{n}]$ for which

$$\left| \frac{f(t + h) - f(t)}{h} \right| \le n \quad \text{for all } h \in (0, 1 - t).$$

Each E_n is closed and nowhere dense in $C[0, 1]$. Both statements are meant with respect to the topology of the uniform convergence in $C[0, 1]$. To prove that E_n is nowhere dense in $C[0, 1]$ observe that any continuous function in $[0, 1]$ can be approximated in the sup-norm by continuous functions with polygonal graph of arbitrarily large Lipschitz constant. Then the complement $C[0, 1] - \cup E_n$ is non-empty. ∎

4c   Differentiating Series of Monotone Functions

4.1. Let $\{f_n\}$ be a sequence of functions of bounded variation in $[a, b]$ such that the series $\sum f_n(x)$ and $\sum V_{f_n}[a, x]$ are both convergent in $[a, b]$. Then the sum f of the first series is of bounded variation in $[a, b]$ and the derivative can be computed term by term, a.e. in $[a, b]$.

5c   Absolutely Continuous Functions

5.1. Let f be absolutely continuous in $[a, b]$. Then f is Lipschitz continuous in $[a, b]$ if and only if f' is a.e. bounded in $[a, b]$.

5.2. The function

$$f(x) = \begin{cases} x^{1+\varepsilon} \sin \dfrac{1}{x} & \text{for } x \in (0, 1] \\ 0 & \text{for } x = 0 \end{cases}$$

is absolutely continuous in $[0, 1]$ for all $\varepsilon > 0$.

5.1c   The Cantor Ternary Function ([23])

Set $f(0) = 0$ and $f(1) = 1$. Divide the interval $[0, 1]$ into 3 equal subintervals and on the central interval $[\frac{1}{3}, \frac{2}{3}]$ set $f = \frac{1}{2}$, i.e., f is defined to be the average of its values at the extremes of the parent interval $[0, 1]$. Next divide the interval $[0, \frac{1}{3}]$ into 3 equal subintervals, and on the central interval $[\frac{1}{3^2}, \frac{2}{3^2}]$ set $f = \frac{1}{4}$, i.e., f is defined to be the average of its values at the extremes of the parent interval $[0, \frac{1}{3}]$. Likewise divide the interval $[\frac{2}{3}, 1]$ into 3 equal subintervals, and on the central interval $[\frac{7}{3^2}, \frac{8}{3^2}]$ set $f = \frac{3}{4}$, i.e., f is defined to be the average of its values at the extremes of the parent interval $[\frac{2}{3}, 1]$.

Proceeding in this fashion we define f in the whole $[0, 1]$, by successive averages. By construction f is non-constant, non-decreasing, and continuous in $[0, 1]$. Since it is constant on each of the intervals making up the complement of the Cantor set $\mathcal{C}$, its derivative vanishes in $[0, 1]$ except on $\mathcal{C}$. Thus $f' = 0$ a.e. on $[0, 1]$.

5.1.1c   Another Construction of the Cantor Ternary Function

The same function can be defined by an alternate procedure that uses the ternary expansion of the elements of the Cantor set. For $x \in \mathcal{C}$ let $\{\epsilon_j\}$ be sequence, with entries only 0 or 1, corresponding to the ternary expansion of x, as in (2.1) of Chap. 1. Then define

$$f(x) = f\left(\sum_{j=1}^{\infty} \frac{2}{3^j} \epsilon_i x, j \right) \overset{\text{def}}{=} \sum_{j=1}^{\infty} \frac{1}{2^j} \epsilon_{x,j}. \tag{5.1c}$$

Let (α_n, β_n) be an interval removed in the nth step of the construction of the Cantor set. The extremes α_n and β_n belong to $\mathcal{C}$ and their ternary expansion is described in **2.2** of the Complements of Chap. 1. From the form of such expansion compute

$$f(\alpha_n) - f(\beta_n) = \frac{1}{2^n} - \sum_{j=n+1}^{\infty} \frac{1}{2^j} = 0.$$

If (α_n, β_n) is an interval in $[0, 1] - \mathcal{C}$ we set

$$f(x) = f(\alpha_n) \qquad \text{for all } x \in [\alpha_n, \beta_n].$$

In such a way f is defined in the whole $[0, 1]$ and $f' = 0$ a.e. in $[0, 1]$. The right-hand side of (5.1c) is the decimal representation of the number in $[0, 1]$ whose binary representation is the sequence ϵ_x. As x ranges over C, the sequences ϵ_x, with only entries 0 or 1, range over all such sequences. Therefore f maps C onto $[0, 1]$. To show that f is continuous, observe that f is monotone and finite. Therefore its possible discontinuity points are discrete jumps. If f were not continuous then it would not be a surjection over $[0, 1]$.

The Cantor ternary function is continuous, of bounded variation, but not absolutely continuous. This can be established indirectly by means of Corollary 5.2. Give a direct proof.

5.2c   A Continuous Strictly Monotone Function with a.e. Zero Derivative

The Cantor ternary function is piecewise constant on the complement of the Cantor set. This accounts for $f' = 0$ a.e. in $[0, 1]$. We next exhibit a continuous strictly increasing function in $[0, 1]$, whose derivative vanishes a.e. in $[0, 1]$.

Let $t \in (0, 1)$ be fixed and define $f_o(x) = x$ and

$$f_1(x) = \begin{cases} (1+t)x & \text{for } 0 \le x \le \tfrac{1}{2} \\ (1-t)x + t & \text{for } \tfrac{1}{2} \le x \le 1. \end{cases}$$

The function f_1 is constructed by dividing $[0, 1]$ into two equal subintervals, by setting $f_1 = f_o$ at the end points of $[0, 1]$, by setting

$$f_1\left(\frac{1}{2}\right) = \frac{1-t}{2} f_o(0) + \frac{1+t}{2} f_o(1) = \frac{1+t}{2}$$

and by defining f_1 to be affine in the intervals $[0, \tfrac{1}{2}]$ and $[\tfrac{1}{2}, 1]$.

This procedure permits one to construct an increasing sequence $\{f_n\}$ of strictly increasing functions in $[0, 1]$. Precisely if f_n has been defined, it must be affine in each of the subintervals

$$\left[\frac{j}{2^n}, \frac{j+1}{2^n}\right] \qquad j = 0, 1, \ldots, 2^n - 1.$$

Subdivide each of these into two equal subintervals, and define f_{n+1} to be affine on each of these with values at the end points, given by

$$f_{n+1}\left(\frac{j}{2^n}\right) = f_n\left(\frac{j}{2^n}\right)$$

$$f_{n+1}\left(\frac{j+1}{2^n}\right) = f_n\left(\frac{j+1}{2^n}\right)$$

$$f_{n+1}\left(\frac{2j+1}{2^{n+1}}\right) = \frac{1-t}{2} f_n\left(\frac{j}{2^n}\right) + \frac{1+t}{2} f_n\left(\frac{j+1}{2^n}\right).$$

By construction $\{f_n\}$ is increasing and

$$f_m\left(\frac{j}{2^n}\right) = f_n\left(\frac{j}{2^n}\right) \quad \text{for all } m \geq n, \quad j = 0, 1, \ldots, 2^n - 1. \tag{5.2c}$$

The limit function f is non-decreasing. We show next that it is continuous and strictly increasing in $[0, 1]$. Every fixed $x \in [0, 1]$ is included into a sequence of nested and shrinking intervals $[\alpha_n, \beta_n]$ of the type

$$\alpha_n = \frac{m_{n,x}}{2^n} \quad \beta_n = \frac{m_{n,x} + 1}{2^n} \quad \text{for some } m_{n,x} \in \mathbb{N} \cup \{0\}.$$

By the construction of f_{n+1}, if the parent interval of $[\alpha_n, \beta_n]$ is $[\alpha_n, \beta_{n-1}]$

$$f_{n+1}(\beta_n) - f_{n+1}(\alpha_n) = \frac{1+t}{2}[f_n(\beta_{n-1}) - f_n(\alpha_n)].$$

Likewise if the parent interval of $[\alpha_n, \beta_n]$ is $[\alpha_{n-1}, \beta_n]$

$$f_{n+1}(\beta_n) - f_{n+1}(\alpha_n) = \frac{1-t}{2}[f_n(\beta_n) - f_n(\alpha_{n-1})].$$

Therefore by (5.2c) either

$$f_{n+1}(\beta_n) - f_{n+1}(\alpha_n) = \frac{1+t}{2}[f_n(\beta_{n-1}) - f_n(\alpha_{n-1})] \tag{5.2c$_+$}$$

or

$$f_{n+1}(\beta_n) - f_{n+1}(\alpha_n) = \frac{1-t}{2}[f_n(\beta_{n-1}) - f_n(\alpha_{n-1})]. \tag{5.2c$_-$}$$

From this by iteration

$$f_{n+1}(\beta_n) - f_{n+1}(\alpha_n) = \prod_{i=1}^{n} \frac{1 + \varepsilon_i t}{2} \quad \text{where} \quad \varepsilon_i = \pm 1.$$

Since for all $m \geq n + 1$

$$f_m(\beta_n) = f_{n+1}(\beta_n) \quad \text{and} \quad f_m(\alpha_n) = f_{n+1}(\alpha_n)$$

the previous equality implies

$$f(\beta_n) - f(\alpha_n) = \prod_{i=1}^{n} \frac{1 + \varepsilon_i t}{2} \quad \text{where} \quad \varepsilon_i = \pm 1.$$

For each fixed n the right-hand side is strictly positive. Thus $f(\beta_n) > f(\alpha_n)$, i.e., f is strictly monotone. On the other hand $(5.2c)_\pm$ imply also

$$f(\beta_n) - f(\alpha_n) \le \left(\frac{1+t}{2}\right)^n \longrightarrow 0 \quad \text{as} \quad n \to \infty.$$

Thus f is continuous in $[0, 1]$. Still from $(5.2c)_\pm$ we compute

$$\frac{f(\beta_n) - f(\alpha_n)}{\beta_n - \alpha_n} = \prod_{i=1}^{n} (1 + \varepsilon_i t).$$

As $n \to \infty$ the right-hand side either converges to zero, or diverges to infinity or the limit does not exists. However, since f is monotone it is a.e. differentiable. Therefore the limit exists for a.e. $x \in [0, 1]$ and is zero. By Corollary 5.2 such a function is not absolutely continuous. Give a direct proof.

5.3. The function f constructed in § 14 of Chap. 3 is not absolutely continuous.

5.4. Let μ be a Radon measure on $\mathbb{R}$ defined on the same σ-algebra of the Lebesgue measurable sets in $\mathbb{R}$, and absolutely continuous with respect to the Lebesgue measure on $\mathbb{R}$. Then set

$$f(x) = \begin{cases} +\mu([\alpha, x]) & \text{for } x \in [\alpha, \infty) \\ -\mu([x, \alpha]) & \text{for } x \in (-\infty, \alpha]. \end{cases}$$

The function f is locally absolutely continuous, i.e., its restriction to any bounded interval is absolutely continuous. The function f can be used to generate the Lebesgue-Stieltjes measure μ_f. The measure μ_f coincides with μ on the Lebesgue measurable sets.

5.3c *Absolute Continuity of the Distribution Function of a Measurable Function*

For a measurable function f on a measure space $\{X, \mathcal{A}, \mu\}$, let μ_f and f_* be respectively, the distribution measure and the distribution function of f, as defined (9.1) and (15.1c of the Complements of Chap. 4).

Proposition 5.1c *The distribution function f_* is absolutely continuous, in every closed subinterval of $\mathbb{R}$, if and only if the distribution measure μ_f is absolutely continuous with respect to the Lebesgue measure on $\mathbb{R}$. In such a case*

$$f_*(t) = \int_{-\infty}^{t} \varphi(s)\,ds$$

where φ is the Radon-Nikodým derivative of μ with respect to the Lebesgue measure ds.

5.5. Give an example of $\{X, \mathcal{A}, \mu\}$ and f for which f_* is continuous and not absolutely continuous.

7c Derivatives of Integrals

7.1. Construct a measurable set $E \subset (-1, 1)$ such that $d'_E(0) = \frac{1}{2}$.
For $n \in \mathbb{N}$ set[5]

$$I_n = \left[\frac{1}{2^n}, \frac{1}{2^{n-1}}\right) \quad \text{so that} \quad (0, 1) = \bigcup_{n \in \mathbb{N}} I_n.$$

Divide each I_n into 2^n equal subintervals, each of length 2^{-2n} and retain only those $\frac{1}{2}$-closed intervals of even parity, to obtain sets

$$E_n = \bigcup_{j=0}^{2^{n-1}-1} \left[\frac{2^n + 2j}{2^{2n}}, \frac{2^n + 2j + 1}{2^{2n}}\right), \quad \text{and} \quad E^+ = \bigcup E_n.$$

Verify that $\mu(E_n) = \frac{1}{2}\mu(I_n)$ and $\mu(E^+) = \frac{1}{2}$. having fixed $h \in (0, 1)$ there exists $n_h, i_h \in \mathbb{N}$, such that

$$h \in \left[\frac{1}{2^{n_h}}, \frac{1}{2^{n_h-1}}\right] \quad \text{and} \quad h \in \left[\frac{2^{n_h} + i_h}{2^{2n_h}}, \frac{2^{n_h} + (i_h + 1)}{2^{2n_h}}\right].$$

The first of these implies that $h = O(2^{-n_h})$. Next compute

$$(0, h] \cap E^+ = \left(\bigcup_{\ell=n_h+1}^{\infty} E_\ell\right) \cup \left(\bigcup_{j=0}^{2^{n_h-1}-1} (0, h] \cap \left(\frac{2^{n_h} + 2j}{2^{2n_h}}, \frac{2^{n_h} + 2j + 1}{2^{2n_h}}\right)\right).$$

From this

$$\mu\big[(0, h] \cap E\big] = \frac{1}{2}h + O(2^{-2n_h}) \quad \Longrightarrow \quad \mu\big[(0, h] \cap E^+\big] = \frac{1}{2}h + O(h^2).$$

Thus

$$\lim_{h \to 0} \frac{1}{h} \int_0^h \chi_{E^+} dx = \frac{1}{2}.$$

Let E^- be the symmetric of E^+ in $(-1, 0]$ and set $E = E^- \cup E^+$.

7.2. Let f be absolutely continuous in $[a, b]$. Then the function $x \to V_f[a, x]$ is also absolutely continuous in $[a, b]$. Moreover

$$V_f[a, x] = \int_a^x |f'(t)| dt \quad \text{for all } x \in [a, b].$$

---

[5]This construction was suggested by V. Vespri and U. Gianazza.

7.3. Let f be of bounded variation in $[a, b]$. The singular part of f is the function ([90])

$$\sigma_f(x) = f(x) - f(a) - \int_a^x f'(t)dt. \tag{7.1c}$$

The singular part of f is of bounded variation and $\sigma' = 0$ a.e. in $[a, b]$. It has the same singularities as f, and $(f - \sigma)$ is absolutely continuous.

Thus every function f of bounded variation in $[a, b]$, can be decomposed into the sum of an absolutely continuous function in $[a, b]$ and a singular function. Compare the σ_f with the functions of the jumps J_f given in § 1.1c of the Complements.

7.4. Let f be Lebesgue integrable in the interval $[a, b]$ and let F denote a primitive of f. Then for every absolutely continuous function g defined in $[a, b]$

$$\int_a^b fg\,dx = F(b)g(b) - F(a)g(a) - \int_a^b Fg'\,dx.$$

7.5. Let $f, g : [a, b] \to \mathbb{R}$ be absolutely continuous. Then

$$\int_a^b fg'\,dx + \int_a^b f'g\,dx = f(a)g(a) - f(b)g(b).$$

7.6. Let $h : [a, b] \to [c, d]$ be absolutely continuous, increasing and such that $h(a) = c$ and $h(b) = d$. Then for every nonnegative, Lebesgue measurable function $f : [c, d] \to \mathbb{R}$, the composition $f(h)$ is measurable and

$$\int_c^d f(s)ds = \int_a^b f\big(h(t)\big)h'(t)dt.$$

This is established sequentially for f the characteristic function of an interval, the characteristic function of an open set, the characteristic function of a measurable set, for a simple function.

Proposition 7.1c *Let f be absolutely continuous in $[a, b]$. Then for every measurable $E \subset [a, b]$ of measure zero, $f(E) \subset \mathbb{R}$ is a set of measure zero.*

Proof Combine Proposition 7.2 with and Vitali's absolute continuity of the integral (Theorem 11.1 of Chap. 4). ∎

The converse of Proposition 7.1c is in false. The characteristic function of the rationals maps any set into a set of measure zero. Such a function however is not continuous.

7.7. Prove by a counterexample that the converse of Proposition 7.1c is false, even if f is assumed to be continuous. **Hint:** The function f in (1.1) is continous in $[0, 1]$ and not $BV[0, 1]$. To show that it maps measurable sets of measure zero, into sets of measure zero, observe that $f \in AC[\varepsilon, 1]$ for all $\varepsilon > 0$.

Proposition 7.2c *Let* $f : [a, b] \rightarrow \mathbb{R}$ *be continuous and monotone and such that for every measurable* $E \subset [a, b]$ *of measure zero,* $f(E) \subset \mathbb{R}$ *is a set of measure zero. Then* f *is absolutely continuous in* $[a, b]$.

Proof Assuming f non-decreasing the function $h(x) = x + f(x)$ is strictly increasing. The set function $\nu(E) = \mu(h(E))$ for all Lebesgue measurable sets in $[a, b]$ is a measure on the same σ-algebra, satisfying $\nu \ll \mu$. Apply the Radon-Nykodým theorem. ∎

7.1c   Characterizing $BV[a, b]$ Functions

Denote by $C_o^1[a, b]$ the collection of continuously differentiable functions φ of compact support in $[a, b]$.

Proposition 7.3c *Let* $f \in BV[a, b]$. *Then*

$$\sup_{\substack{\varphi \in C_o^1(\mathbb{R}) \\ |\varphi| \leq 1}} \int_a^b f\varphi' dx \leq V_f[a, b]. \tag{7.2c}$$

Proof One may assume that f is monotone increasing and nonnegative. There exists an increasing sequence of simple functions such that $\{f_n\} \rightarrow f$ everywhere in $[a, b]$ (Proposition 3.1 of Chap. 4). By the monotonicity of f, the construction of $\{f_n\}$ identifies a partition

$$P = \{a = x_o < x_1 < \cdots < x_n = b\} \tag{7.3c}$$

of $[a, b]$ such that

$$f_n(x) = f(x_{j-1}) \quad \text{in the interval} \quad [x_{j-1}, x_j] \quad \text{for} \quad j = 1, \ldots, n.$$

For $0 < \delta \ll 1$ and n fixed construct the Lipschitz continuous functions

$$f_{n,\delta}(x) = \sum_{j=1}^n \left\{ f(x_{j-1})\chi_{[x_{j-1}, x_j]} + [f(x_j) - f(x_{j-1})]\chi_{[x_j-\delta, x_j]} \frac{x - x_j + \delta}{\delta} \right\}.$$

One verifies that

$$\int_a^b f\varphi' dx = \lim_{n \to \infty} \lim_{\delta \to 0} \int_a^b f_{n,\delta}\varphi' dx.$$

Since $f_{n,\delta}$ are absolutely continuous in $[a, b]$, integration by parts is justified. Hence for every $\varphi \in C_o^1[a, b]$,

$$\left| \int_a^b f_{n,\delta}\varphi' dx \right| = \left| \sum_{j=1}^n \int_{x_{j-1}}^{x_j} f_{n,\delta}\varphi' dx \right| \leq V_f[a, b]. \qquad ∎$$

The converse of (7.2c) is in general false. For example $\chi_{\{\frac{1}{2}\}} \in BV[0, 1]$ with variation 2, whereas the left-hand side of (7.2c) is zero. The latter only requires that f be integrable, whereas the notion of bounded variation requires that f be defined at every point of $[a, b]$. However if f is integrable in $[a, b]$ then it is unambiguously defined by (11.1), or (11.1)$'_{N=1}$, everywhere in $[a, b]$ except for a set of measure zero. Given f integrable in $[a, b]$ introduce partitions P_f as in (7.3c), where however x_j are differentiability points of f. Then define the *essential* variation of f in (a, b) as

$$\text{ess} - \mathcal{V}_f(a, b) = \sup_{P_f} \sum_{j=1}^{n} |f(x_j) - f(x_{j-1})|.$$

Proposition 7.4c *Let f be integrable in $[a, b]$. Then*

$$\text{ess} - \mathcal{V}_f(a, b) \leq \sup_{\substack{\varphi \in C_0^1[a,b] \\ |\varphi| \leq 1}} \int_a^b f\varphi' dx. \tag{7.4c}$$

Proof Assume the right-hand side of (7.4c) is finite, and fix a partition P_f. Without loss of generality may assume that a and b are points of P_f, and construct the polygonal of vertices $(x_j, f(x_j))$. Assume momentarily that such a polygonal changes its monotonicity at each of its veritice, i.e.,

$$\begin{aligned} &\text{if } f(x_{j-1}) < f(x_j) \text{ then } f(x_j) > f(x_{j+1}); \\ &\text{if } f(x_{j-1}) > f(x_j) \text{ then } f(x_j) < f(x_{j+1}). \end{aligned} \tag{7.5c}$$

Assume $f(a) < f(x_1)$ and set

$$\varphi_\delta(x) = \begin{cases} \dfrac{a-x}{\delta} & \text{for } a \leq x < a+\delta, \\ -1 & \text{for } a+\delta \leq x \leq x_1 - \delta, \\ \dfrac{x-x_1+\delta}{\delta} - 1 & \text{for } x_1 - \delta \leq x < x_1. \end{cases}$$

Then, for $j = 1, \ldots, n$, define φ recursively as

$$\varphi_\delta(x) = \begin{cases} \dfrac{x_{j-1} - x}{\delta} & \text{for } x_{j-1} \leq x < x_{j-1} + \delta, \\ -1 & \text{for } x_{j-1} + \delta \leq x \leq x_j - \delta, \quad \text{if } f(x_{j-1}) < f(x_j); \\ \dfrac{x - x_j + \delta}{\delta} - 1 & \text{for } x_j - \delta \leq x < x_j, \end{cases}$$

$$\varphi(x)_\delta = \begin{cases} \dfrac{x - x_{j-1}}{\delta} & \text{for } x_{j-1} \leq x < x_{j-1} + \delta, \\ 1 & \text{for } x_{j-1} + \delta \leq x \leq x_j - \delta, \quad \text{if } f(x_{j-1}) > f(x_j). \\ \dfrac{x_j - \delta - x}{\delta} + 1 & \text{for } x_j - \delta \leq x < x_j, \end{cases}$$

Assuming momentarily that such a choice is admissible, compute

$$\sup_{\substack{\varphi \in C_o^1[a,b] \\ |\varphi| \le 1}} \int_a^b f\varphi' dx \ge \limsup_{\delta \to 0} \left| \int_a^b f\varphi'_\delta dx \right|$$

$$= \limsup_{\delta \to 0} \left| \frac{\text{sign } \varphi'_\delta}{\delta} \int_a^{a+\delta} f dx \right.$$

$$+ \sum_{j=1}^{n-1} \frac{\text{sign } \varphi'_\delta}{\delta} \int_{x_j-\delta}^{x_j+\delta} f dx + \frac{\text{sign } \varphi'_\delta}{\delta} \int_{b-\delta}^b f dx \right|$$

$$\ge \sum_{j=1}^n |f(x_j) - f(x_{j-1})|,$$

since x_j are differentiability points of f.   ∎

Complete the proof by the following steps:

i. Prove that in (7.4c) the supremum can be taken over all Lipschitz continuous functions $\varphi \in C_o[a, b]$.

ii. Remove the assumptions and that P_f satisfies (7.5c), to establish the inequality for all partitions P_f of $[a, b]$.   ∎

Corollary 7.1c *An integrable function f in $[a, b]$ is of essentially bounded variation in $[a, b]$ if and only if the right-hand side of (7.4c) is finite.*

7.2c   Functions of Bounded Variation in N Dimensions [55]

The characterization of Corollary 7.1c suggests a notion bounded variations in several dimensions. A function f locally integrable in $\mathbb{R}^N$ is of bounded variation in a Lebesgue measurable set $E \subset \mathbb{R}^N$ if there exists a constant C such that

$$\left| \int_E f \text{ div } \varphi dx \right| \le C \tag{7.6c}$$

for all vector valued functions

$$\varphi = (\varphi_1, \ldots, \varphi_N) \in [C_o^1(\mathbb{R}^N)]^N \quad \text{such that } |\varphi| \le 1.$$

The smallest constant C for which (7.6c) holds is the variation of f in E and is denoted by $\|Df\|(E)$.

7.2.1c Perimeter of A Set

Let E be a bounded measurable set in $\mathbb{R}^N$ with smooth boundary ∂E. Prove that $\chi_E \in BV(\mathbb{R}^N)$ and that $V_{\chi_E}(\mathbb{R}^N) = \mathcal{H}^{N-1}(\partial E)$, where $\mathcal{H}^{N-1}$ is the $(N-1)$-dimensional Hausdorff measure of sets in $\mathbb{R}^N$.

If ∂E is not smooth, (7.6c) suggests defining the "measure of ∂E" or, roughly speaking, the perimeter of E, by

$$\text{Per}(E) = \sup_{\varphi \in [C_o^1]^N, |\varphi| \le 1} \int_{\mathbb{R}^N} \chi_E \operatorname{div} \varphi \, dx, \qquad (7.7c)$$

provided the right-hand side is finite. Sets $E \subset \mathbb{R}^N$ for which $\text{Per}(E) < \infty$ are called of *finite perimeter*. They correspond, roughly speaking, to sets for which the Gauss-Green theorem holds.

13c Convex Functions

13.1. Give an example of a bounded, discontinuous, convex function in $[a, b]$. Give an example of a convex function unbounded in (a, b).

13.2. A continuous function f in (a, b) is convex if and only if

$$f\left(\frac{x+y}{2}\right) \le \frac{f(x) + f(y)}{2} \qquad \text{for all } x, y \in (a, b).$$

13.3. Let f be convex, non-decreasing and non-constant in $(0, \infty)$. Then $f(x) \to \infty$ as $x \to \infty$.

13.4. Let f be convex in $[0, \infty)$. Then the limit of $x^{-1} f(x)$ as $x \to \infty$, exists finite or infinite.

13.5. Let $\{f_n\}$ be a sequence of convex functions in (a, b) converging to some real-valued function f. Then the convergence is uniform within any closed subinterval of (a, b). The conclusion is false if f is permitted to take values in $\mathbb{R}^*$, as shown by the sequence $\{x^n\}$ for $x \in (0, 2)$.

13.6. Let $f \in C^2(a, b)$. Then f is convex in (a, b) if and only if $f''(x) \ge 0$ for each $x \in (a, b)$. *Proof:* Having fixed $x < y$ it suffices to prove that

$$[0, 1] \ni t \to \varphi(t) = f(tx + (1 - t)y) - tf(x) - (1 - t)f(y) \qquad (13.1c)$$

is nonpositive in $[0, 1]$. Such a function vanishes at the end points of $[0, 1]$ and its extrema are minima since

$$\varphi''(t) = (x - y)f''(tx + (1 - t)y) \le 0.$$

Proposition 13.1c *A continuous function f in (a, b) is convex if and only if either one of the two one-sided derivatives $D_{\pm}f$ is non-decreasing.*

Proof Assume for example that D_+f is non-decreasing. If the function φ in (13.1c) has a positive maximum $\varphi(t_o) > 0$, at some $t_o \in (0, 1)$, then

$$D_+\varphi(t_o) = (x - y)D_+f(t_ox + (1 - t_o)y) + f(y) - f(x) \le 0.$$

Therefore, since D_+f is non-decreasing, $D_+\varphi(t)$ is non-positive in $[0, t_o]$. Thus φ is non-increasing in $[0, t_o]$ and $\varphi(t_o) \le 0$. ∎

13.7. The function $f(x) = |x|^p$ is convex for $p \ge 1$ and concave for $p \in (0, 1)$.

13.8c   Convex Functions in $\mathbb{R}^N$

Let E be a convex subset of $\mathbb{R}^N$. A function $f : E \to \mathbb{R}^*$ is convex if for every pair of points x and y in E and every $t \in [0, 1]$

$$f(tx + (1 - t)y) \le tf(x) + (1 - t)f(y).$$

The $(N + 1)$-dimensional set

$$\mathcal{G}_f = \left\{(x, x_{N+1}) \in \mathbb{R}^{N+1} \mid x \in E, \quad x_{N+1} \ge f(x)\right\}$$

is the epigraph of f. The function f is convex if and only if its epigraph is convex.

13.9. Let $E \subset \mathbb{R}^N$ be open and convex. A function $f \in C^2(E)$ is convex if and only if $\sum_{i,j=1}^{N} f_{x_ix_j}\xi_i\xi_j \ge 0$ for all $\xi \in \mathbb{R}^N$.

Hint: Fix $B_\rho(x) \subset E$ and ξ in the unit sphere of $\mathbb{R}^N$. The function $(-\rho, \rho) \ni t \to \varphi(t) = f(x + t\xi)$ is convex.

13.10. Construct a non convex function $f \in C^2(\mathbb{R}^2)$ such that f_{xx} and f_{yy} are both nonnegative.

13.11. Let $E \subset \mathbb{R}^N$ be open and convex, and let f be convex and real-valued in E. Then f is continuous on E. Moreover for every $x \in E$ there exist the left and right directional derivatives

$$D_{\mathbf{u}}^{\pm}f(x) = D_t^{\pm}f(x + t\mathbf{u})\Big|_{t=0} \qquad \text{for all } |\mathbf{u}| = 1.$$

Moreover $D_{\mathbf{u}}^-f \le D_{\mathbf{u}}^+f$. In particular for each $x \in E$, there exist the left and right derivatives $D_{x_j}^{\pm}f$, along the coordinate axes and $D_{x_j}^-f \le D_{x_j}^+f$.

13.12. Let $E \subset \mathbb{R}^N$ be convex. A function f defined in E is convex if and only if $f(x) = \sup \pi(x)$, where $\pi \le f$ is affine.

13.13. Let $f : \mathbb{R}^N \to \mathbb{R}$ be convex. There exist a positive number k such that $\lim \inf_{|x| \to \infty} |x|^{-1} f(x) \geq -k$.

13.14c The Legendre Transform ([92])

The Legendre transform f^* of a convex function $f : \mathbb{R}^N \to \mathbb{R}^*$ is defined by

$$f^*(x) = \sup_{y \in \mathbb{R}^N} \{x \cdot y - f(y)\}. \tag{13.2c}$$

Proposition 13.2c f^* *is convex in* $\mathbb{R}^N$ *and* $f^{**} = f$.

Proof The convexity of f follows from **13.12**. From the definition (13.2c), $f(y) + f^*(x) \geq y \cdot x$, for all $x, y \in \mathbb{R}^N$. Therefore

$$f(y) \geq \sup_{x \in \mathbb{R}^N} \{y \cdot x - f^*(x)\} = f^{**}(y).$$

Also, still from (13.2c)

$$f^{**}(x) = \sup_{y \in \mathbb{R}^N} \left\{x \cdot y - \sup_{z \in \mathbb{R}^N} \{y \cdot z - f(z)\}\right\}$$

$$= \sup_{y \in \mathbb{R}^N} \inf_{z \in \mathbb{R}^N} \{y \cdot (x - z) + f(z)\}.$$

Since f is convex, for a fixed $x \in \mathbb{R}^N$, there exists a vector $\mathbf{m}$ such that

$$f(z) - f(x) \geq \mathbf{m} \cdot (z - x) \qquad \text{for all } z \in \mathbb{R}^N.$$

Combining these inequalities yields

$$f^{**}(x) \geq f(x) + \sup_{y \in \mathbb{R}^N} \inf_{z \in \mathbb{R}^N} (z - x) \cdot (\mathbf{m} - y) = f(x). \qquad \blacksquare$$

13.15c Finiteness and Coercivity

The Legendre transform f^*, as defined by (13.2c), could be infinite even if f is finite in $\mathbb{R}^N$. For example in $\mathbb{R}$

$$|x|^* = \begin{cases} 0 & \text{if } |x| \leq 1 \\ \infty & \text{if } |x| > 1. \end{cases}$$

A convex function $f : \mathbb{R}^N \to \mathbb{R}$ is coercive at infinity if

$$\lim_{|x| \to \infty} \frac{f(x)}{|x|} = \infty.$$

Proposition 13.3c *If f is coercive at infinity, then f^* is finite in $\mathbb{R}^N$. If f is finite, then f^* is coercive at infinity.*

Proof Assume f is coercive at infinity. If the sup in (13.2c) is achieved for $y = 0$ the assertion is obvious. Otherwise

$$f^*(x) = \sup_{y \in \mathbb{R}^N - \{0\}} |y| \left\{ x \cdot \frac{y}{|y|} - \frac{f(y)}{|y|} \right\}.$$

Therefore the supremum is achieved for some finite y and $f^*(x)$ is finite. To prove the converse statement, fix $\lambda > 0$ and write

$$f^*(x) = \sup_{y \in \mathbb{R}^N} \{x \cdot y - f(y)\} \geq \{x \cdot y - f(y)\}\big|_{y = \lambda x/|x|}$$

$$= \lambda|x| - f\left(\lambda \frac{x}{|x|}\right) \geq \lambda|x| - \sup_{|u|=\lambda} |f(u)|.$$

Therefore, since $x \in \mathbb{R}^N - \{0\}$ is arbitrary

$$\lim_{|x| \to \infty} \frac{f^*(x)}{|x|} \geq \lambda \qquad \text{for all } \lambda > 0. \qquad \blacksquare$$

13.16c   The Young's Inequality

Prove that the Legendre transform of the convex function

$$\mathbb{R} \ni a \to f(a) = \frac{1}{p}|a|^p \quad \text{for } 1 < p < \infty$$

is

$$\mathbb{R} \ni b \to f^*(b) = \frac{1}{q}|b|^q \quad \text{for } 1 < q < \infty \quad \text{and} \quad \frac{1}{p} + \frac{1}{q} = 1.$$

Then the definition (13.2c) of the Legendre transform implies the Young's inequality

$$|ab| \leq \frac{1}{p}|q|^p + \frac{1}{q}|b|^q \quad \text{for all } a, b \in \mathbb{R}. \tag{13.3c}$$

The inequality continues to holds for the limiting case $p = 1$ and $q = \infty$. For a different proof see Proposition 2.1 of Chap. 6.

14c Jensen's Inequality

Proposition 14.1c (Hölder [75]) *Let $\{\alpha_i\}$ be a sequence of nonnegative numbers such that $\sum \alpha_i = 1$, and let $\{\xi_i\}$ be a sequence in $\mathbb{R}$. Then*

$$\exp\left(\sum \alpha_i \xi_i\right) \leq \sum \alpha_i \exp(\xi_i). \tag{14.1c}$$

Proof Apply (14.1) to e^x with $\eta = \xi_j$ and $\alpha = \sum \alpha_i \xi_i$, to get

$$\exp\left(\sum \alpha_i \xi_i\right) + m_j \left(\xi_j - \sum \alpha_i \xi_i\right) \leq \exp(\xi_j).$$

Multiply by α_j and add over j. ∎

Corollary 14.1c *Let $\{\alpha_i\}$ be a sequence of nonnegative numbers such that $\sum \alpha_i = 1$, and let $\{\xi_i\}$ be a sequence of positive numbers. Then*

$$\prod \xi_i^{\alpha_i} \leq \sum \alpha_i \xi_i. \tag{14.2c}$$

14.1c The Inequality of the Geometric and Arithmetic Mean

In the case where $\alpha_i = 0$ for $i > n$ and $\alpha_i = 1/n$ for $i = 1, 2, \ldots, n$, inequality (14.2c) reduces to the inequality between the geometric and arithmetic mean of n positive numbers[6]

$$(\xi_1 \xi_2 \cdots \xi_n)^{1/n} \leq \frac{\xi_1 + \xi_2 + \cdots + \xi_n}{n}. \tag{14.3c}$$

14.2c Integrals and Their Reciprocals

Proposition 14.2c *Let E be a measurable set of finite measure and let $f : E \to \mathbb{R}^+$ be measurable. Then*

$$\frac{1}{\left(\dfrac{1}{\mu(E)} \displaystyle\int_E f d\mu\right)^p} \leq \frac{1}{\mu(E)} \int_E \frac{1}{f^p} d\mu, \quad \text{for all } p > 0. \tag{14.4c}$$

Proof Assume first that f is integrable and that $f \geq \varepsilon$, and apply Jensen's inequality with $\varphi(t) = t^{-p}$. ∎

---

[6][70], Chap. II, § 5 contains an alternate proof of this inequality that does not use Jensen's inequality.

15c Extending Continuous Functions

15.1. Let f be convex in a closed interval $[a, b]$ and assume that $D_+ f(a)$ and $D_- f(b)$ are both finite. There exists a *convex* function $\tilde{f}$ defined in $\mathbb{R}$ such that $f = \tilde{f}$ in $[a, b]$.

16c The Weierstrass Approximation Theorem

16.1. Let $E \subset \mathbb{R}^N$ be open, bounded and with smooth boundary ∂E. Let $f \in C^1(\bar{E})$ vanish on ∂E, and let P_j denote the jth Stieltjes polynomial relative to f. Then

$$\lim_{j \to \infty} \frac{\partial P_j}{\partial x_i} = \frac{\partial f}{\partial x_i} \quad j = 1, \ldots, N \quad \text{in } E.$$

16.2. Let $E \subset \mathbb{R}^N$ be bounded and open, and fet $f : \bar{E} \to \mathbb{R}$ and be Lipschitz continuous in $\bar{E}$. with Lipschitz constant L. Then the Stieltjes polynomials P_j relative to f are equi-Lipschitz continuous in E, with the same constant L.

16.3. A continuous function $f : [0, 1] \to \mathbb{R}$, can be approximated by the Bernstein polynomials B_j, relative to f

$$B_j(x) = \sum_{i=1}^{j} \binom{j}{i} f\left(\frac{i}{j}\right) x^i (1 - x)^{j-i}.$$

State and prove a N-dimensional version of such an approximation ([98]).

16.4. Let f be uniformly continuous on a bounded, open set $E \subset \mathbb{R}^N$ and denote by $\mathcal{P}_n$ the set of all polynomials of degree n in the coordinate variables. Then

$$\int_E f p_n dx = 0 \quad \text{for all } p_n \in \mathcal{P}_n \text{ and all } n \in \mathbb{N} \implies f = 0.$$

17c The Stone-Weierstrass Theorem

17.1. The Stone-Weierstrass theorem fails for complex valued functions.

Let D be the closed, unit disc in the complex plane $\mathbb{C}$ and denote by $C(D; \mathbb{C})$ the linear space of all the continuous complex valued functions defined in D endowed with the topology generated by the metric in (17.1).

Consider also the subset $\mathcal{H}(D)$ of $C(D; \mathbb{C})$, consisting of all holomorphic functions defined in D. One verifies that $\mathcal{H}(D)$ is an algebra. Moreover uniform limits of holomorphic functions in D are holomorphic ([24], Chap. V,

Théoréme 1, page 145). Thus $\mathcal{H}(D)$ is closed under the metric in (17.1). The algebra $\mathcal{H}(D)$ is called the *disc algebra*.

Such an algebra separates points since it contains the holomorphic function $f(z) = z$. Moreover $\mathcal{H}(D)$ contains the constants. However $\mathcal{H}(D) \neq C(D; \mathbb{C})$. Indeed the function $f(z) = \bar{z}$ is continuous but not holomorphic in D.

17.2. Let $f : \mathbb{R} \to \mathbb{R}$ be continuous and 2π-periodic. For every $\varepsilon > 0$ there exists a function of the type

$$\varphi(x) = a_o + \sum_{n=1}^{m} (b_n \cos nx + c_n \sin nx)$$

such that $\sup_{\mathbb{R}} |f - \varphi| \leq \varepsilon$ (**Hint:** Use Stone's Theorem).

19c   A General Version of the Ascoli-Arzelà Theorem

The proof of Theorem 19.1 uses only the separability of $\mathbb{R}^N$ and the metric structure of $\mathbb{R}$. Thus it can be extended into any abstract framework with these two properties.

Let $\{f_n\}$ be a countable collection of continuous functions from a separable topological space $\{X; \mathcal{U}\}$ into a metric space $\{Y; d_Y\}$. The functions f_n are equibounded at x if the closure in $\{Y; d_Y\}$ of the set $\{f_n(x)\}$ is compact.

The functions f_n are equi-continuous at a point $x \in X$ if for every $\varepsilon > 0$, there exists an open set $\mathcal{O} \in \mathcal{U}$ containing x and such that

$$d_Y(f_n(x), f_n(y)) \leq \varepsilon \quad \text{for all } y \in \mathcal{O} \quad \text{and all } n \in \mathbb{N}.$$

Theorem 19.1c *Let $\{f_n\}$ be a sequence of continuous functions from a separable space $\{X; \mathcal{U}\}$ into a metric space $\{Y; d_Y\}$. Assume that the functions f_n are equibounded and equi-continuous at each $x \in X$. Then, there exists a subsequence $\{f_{n'}\} \subset \{f_n\}$ and a continuous function $f : X \to Y$ such that $\{f_{n'}\} \to f$ pointwise in X. Moreover the convergence is uniform on compact subsets of X.*

State and prove an analog of Proposition 19.1.

Chapter 6
The L^p Spaces

1 Functions in $L^p(E)$ and Their Norm

Let $\{X, \mathcal{A}, \mu\}$ be a measure space and let $E \in \mathcal{A}$. A measurable function $f : E \to \mathbb{R}^*$ is said to be in $L^p(E)$, for $1 \leq p < \infty$, if $|f|^p$ is integrable on E, i.e., if

$$\|f\|_p \overset{def}{=} \left(\int_E |f|^p d\mu \right)^{1/p} < \infty. \tag{1.1}$$

Equivalently, the collection of all such functions is denoted by $L^p(E)$. The quantity $\|f\|_p$ is the *norm* of f in $L^p(E)$. It follows from the definition that $\|f\|_p \geq 0$ for all $f \in L^p(E)$, and $\|f\|_p = 0$ if and only if $f = 0$ a.e. in E. Let f and g be in $L^p(E)$ and let $\alpha, \beta \in \mathbb{R}$. Then (2.2c of the Complements)

$$|\alpha f + \beta g|^p \leq 2^{p-1} \left(|\alpha|^p |f|^p + |\beta|^p |g|^p \right) \quad p \geq 1, \quad \text{a.e. in } E.$$

Therefore $L^p(E)$ is a linear space. A measurable function $f : E \to \mathbb{R}^*$ is in $L^\infty(E)$ if $|f| \leq M$ a.e. in E for some $M > 0$. Equivalently, $L^\infty(E)$ is the linear space of all such functions. Set

$$
\begin{aligned}
\operatorname*{ess\,sup}_E f &= \inf \left\{ k \mid \mu([f > k]) = 0 \right\} \\
\operatorname*{ess\,inf}_E f &= \sup \left\{ k \mid \mu([f < k]) = 0 \right\}.
\end{aligned}
\tag{1.2}
$$

A norm $\| \cdot \|_\infty$ in $L^\infty(E)$ is defined by

$$\|f\|_\infty \overset{def}{=} \operatorname*{ess\,sup}_E |f|. \tag{1.3}$$

© Springer Science+Business Media New York 2016
E. DiBenedetto, *Real Analysis*, Birkhäuser Advanced
Texts Basler Lehrbücher, DOI 10.1007/978-1-4939-4005-9_6

It follows from the definition that for $\varepsilon > 0$ arbitrarily small

$$\mu(|f| \geq \|f\|_\infty + \varepsilon) = 0 \quad \text{and} \quad \mu(|f| \geq \|f\|_\infty - \varepsilon) > 0. \tag{1.4}$$

Remark 1.1 The μ-measurability of f is essential in the definition of the spaces $L^p(E)$. For example, let $E \subset [0, 1]$ be the Vitali non-Lebesgue measurable set constructed in § 13 of Chap. 3. The function

$$f(x) = \begin{cases} 1 & \text{for } x \in E \\ -1 & \text{for } x \in [0, 1] - E \end{cases}$$

is not in $L^2[0, 1]$, although f^2 is Lebesgue integrable in $[0, 1]$.

2   The Hölder and Minkowski Inequalities

Two elements p and q in the extended real numbers $\mathbb{R}^*$ are said to be *conjugate*, if $p, q \geq 1$ and

$$\frac{1}{p} + \frac{1}{q} = 1. \tag{2.1}$$

Since $p, q \in \mathbb{R}^*$, if $p = 1$ then $q = \infty$. Likewise if $q = 1$ then $p = \infty$.

Proposition 2.1 (Young's Inequality)[1] *Let $1 \leq p, q \leq \infty$ be conjugate. Then for all $a, b \in \mathbb{R}$*

$$|a\,b| \leq \frac{1}{p}|a|^p + \frac{1}{q}|b|^q \tag{2.2}$$

and equality holds only if $|a|^p = |b|^q$.

Proof The inequality is obvious if either a or b is zero. Thus assume $|a| > 0$ and $|b| > 0$. The inequality is also obvious if either $p = 1$ or $q = 1$. Thus assume $1 < p, q < \infty$. The function

$$s \longrightarrow \left(\frac{s^p}{p} + \frac{1}{q} - s\right) \quad s \geq 0$$

has an absolute minimum at $s = 1$. Therefore for all $s > 0$

$$s \leq \frac{s^p}{p} + \frac{1}{q}$$

---

[1] When $p = q = 2$, this is the Cauchy–Schwarz inequality. An alternative proof of (2.2) is in § 13.16c of the Complements of Chap. 5. It can also be established by using Proposition 14.1c of Chap. 5. See also [70] pp. 132–133.

and equality holds only if $s = 1$. Choosing $s = |a||b|^{-q/p}$ yields

$$\frac{|a|}{|b|^{q/p}} \le \frac{1}{p}\frac{|a|^p}{|b|^q} + \frac{1}{q}.$$

∎

Proposition 2.2 (Hölder's Inequality) *Let $f \in L^p(E)$ and $g \in L^q(E)$, where $1 \le p, q \le \infty$ satisfy (2.1). Then $fg \in L^1(E)$ and*

$$\int_E |fg|d\mu \le \|f\|_p \|g\|_q. \tag{2.3}$$

Moreover if $1 < p, q < \infty$, equality holds only if $|f|^p = c|g|^q$ a.e. in E, for some constant $c \ge 0$.

Proof May assume that f and g are nonnegative and neither is zero a.e. in E. Also (2.3) is obvious if either $p = 1$ or $q = 1$. If $p, q > 1$, in (2.2) take

$$a = \frac{f}{\|f\|_p} \quad \text{and} \quad b = \frac{g}{\|g\|_q}$$

to obtain

$$\frac{fg}{\|f\|_p \|g\|_q} \le \frac{1}{p}\frac{f^p}{\|f\|_p^p} + \frac{1}{q}\frac{g^q}{\|g\|_q^q}.$$

Integrating over E

$$\frac{\int_E fg d\mu}{\|f\|_p \|g\|_q} \le \frac{1}{p} + \frac{1}{q} = 1.$$

For the indicated choice of a and b in (2.2), equality holds only if

$$f^p(x) = \frac{\|f\|_p^p}{\|g\|_q^q} g^q(x) \quad \text{for a.e. } x \in E.$$

∎

Proposition 2.3 (Minkowski Inequality) *Let $f, g \in L^p(E)$ for some $1 \le p \le \infty$. Then*

$$\|f + g\|_p \le \|f\|_p + \|g\|_p. \tag{2.4}$$

Moreover if $1 < p < \infty$, equality holds only if $f = Cg$ a.e. in E, for some constant C.

Proof The inequality is obvious if $p = 1$ and $p = \infty$. If $1 < p < \infty$

$$\|f + g\|_p^p = \int_E |f + g|^p d\mu = \int_E |f + g|^{p-1}|f + g|d\mu$$

$$\le \int_E |f + g|^{p-1}|f|d\mu + \int_E |f + g|^{p-1}|g|d\mu.$$

The integrals on the right-hand side are majorized by Hölder's inequality

$$\|f + g\|_p^p \le \|f + g\|_p^{p-1} \left(\|f\|_p + \|g\|_p \right).$$ ∎

3   More on the Spaces L^p and Their Norm

3.1   Characterizing the Norm $\|f\|_p$ for $1 \le p < \infty$

Proposition 3.1 *Let $f \in L^p(E)$ for some $1 \le p < \infty$. Then*

$$\|f\|_p = \left(\int_E |f|^p d\mu \right)^{1/p} = \sup_{\substack{g \in L^q(E) \\ \|g\|_q = 1}} \int_E f g d\mu$$

where $1 \le p < \infty$ and $1 < q \le \infty$ are conjugate.

Proof May assume that $f \not\equiv 0$. By Hölder's inequality

$$\sup_{\substack{g \in L^q(E) \\ \|g\|_q = 1}} \int_E f g d\mu \le \|f\|_p.$$

If $1 < p < \infty$ one verifies that

$$g_* = \frac{|f|^{p-2}f}{\|f\|_p^{p/q}} \in L^q(E) \quad \text{and} \quad \|g_*\|_q = 1.$$

Then

$$\sup_{\substack{g \in L^q(E) \\ \|g\|_q = 1}} \int_E f g d\mu \ge \int_E f g_* d\mu = \|f\|_p.$$

If $p = 1$ the proof is similar for the choice $g_* = \text{sign} f \in L^\infty(E)$. ∎

3.2   The Norm $\| \cdot \|_\infty$ for E of Finite Measure

Assume that $\mu(E) < \infty$. If $f \in L^q(E)$, then for all $1 \le p < q$, by the Hölder inequality, applied to the pair of functions f and $g \equiv 1$

$$\|f\|_p \le \mu(E)^{\frac{q-p}{qp}} \|f\|_q.$$

Therefore $f \in L^p(E)$ for all $1 \leq p \leq q$. In particular if $f \in L^\infty(E)$ then $f \in L^p(E)$ for all $p \geq 1$.

Proposition 3.2 *Let $\mu(E) < \infty$ and $f \in L^\infty(E)$. Then*

$$\lim_{p\to\infty} \|f\|_p = \|f\|_\infty.$$

Proof Since E is of finite measure

$$\limsup_{p\to\infty} \|f\|_p \leq \|f\|_\infty \limsup_{p\to\infty} \mu(E)^{1/p} = \|f\|_\infty.$$

Next, for any $\varepsilon > 0$

$$\int_E |f|^p d\mu \geq \int_{[|f|>\|f\|_\infty-\varepsilon]} |f|^p d\mu \geq (\|f\|_\infty - \varepsilon)^p \, \mu\left[|f| > \|f\|_\infty - \varepsilon\right].$$

From the second of (1.2) the last term is positive. Therefore taking the $(1/p)$-power and letting $p \to \infty$

$$\liminf_{p\to\infty} \|f\|_p \geq \|f\|_\infty - \varepsilon. \qquad \blacksquare$$

3.3 The Continuous Version of the Minkowski Inequality

Proposition 3.3 *Let $\{X, \mathcal{A}, \mu\}$ and $\{Y, \mathcal{B}, \nu\}$ be two complete measure spaces and assume in addition that $\{Y, \mathcal{B}, \nu\}$ is σ-finite. Then for every nonnegative $f \in L^p(X \times Y)$ for some $1 \leq p < \infty$*

$$\left(\int_X \left|\int_Y f(x, y)d\nu\right|^p d\mu\right)^{1/p} \leq \int_Y \|f(\cdot, y)\|_{p,X} d\nu.$$

Proof Assume first that $\nu(Y) < \infty$. Then for every $g \in L^q(X)$ one has $fg \in L^1(X \times Y)$. Setting

$$F = \int_Y f(\cdot, y)d\nu,$$

the left hand side is $\|F\|_{p,X}$. Then

$$\|F\|_{p,X} = \sup_{\|g\|_{q,x}=1} \int_X Fg d\mu = \sup_{\|g\|_{q,x}=1} \int_X \left(\int_Y f(x,y)d\nu\right)g(x)d\mu$$

$$= \sup_{\|g\|_{q,x}=1} \int_Y \left(\int_X f(x,y)g(x)d\mu\right)d\nu$$

$$\leq \int_Y \left(\sup_{\|g\|_{q,x}=1} \int_X f(x,y)g(x)d\mu\right)d\nu = \int_Y \|f(\cdot,y)\|_{p,X}d\nu.$$

If $\{Y, \mathcal{B}, \nu\}$ is σ-finite the proof is concluded by a limiting argument.   ∎

4 $L^p(E)$ for $1 \leq p \leq \infty$ as Normed Spaces of Equivalence Classes

Since $L^p(E)$ is a linear space it must contain a *zero element* with respect to the operations of addition and multiplication by scalars. Such an element is defined by $f + (-1)f$ for any $f \in L^p(E)$.

A *norm* in $L^p(E)$ is a function $\|\cdot\| : L^p(E) \to \mathbb{R}^+$ satisfying

$$\|f\| = 0 \iff f \text{ is the zero element of } L^p(E) \tag{4.1}$$

$$\|\alpha f\| = |\alpha|\|f\| \quad \text{for all } f \in L^p(E) \text{ and for all } \alpha \in \mathbb{R} \tag{4.2}$$

$$\|f + g\| \leq \|f\| + \|g\| \quad \text{for all } f, g \in L^p(E). \tag{4.3}$$

The norm $\|\cdot\|_p$ defined in (1.1) for $p \in [1, \infty)$ and in (1.3) for $p = \infty$, satisfies (4.2). It also satisfies (4.3) by the Minkowski inequality. Finally, it satisfies (4.1) if the zero element of $L^p(E)$ is meant in the sense

f is the zero element of $L^p(E)$ if $f(x) = 0$ for a.e. $x \in E$.

However, the norm $\|\cdot\|_p$ does not distinguish between two elements f and g in $L^p(E)$ that differ on a set of measure zero.

Motivated by this remark, we regard the elements of $L^p(E)$ as *equivalence classes*. If $\mathcal{C}_f$ is one such class and f is a representative, then

$$\mathcal{C}_f = \left\{ \begin{array}{l} \text{all measurable functions } g : E \to \mathbb{R}^* \text{ such that } |g|^p \\ \text{is integrable on } E \text{ and such that } f = g \text{ a.e. in } E \end{array} \right\}.$$

With such interpretation the function $\|\cdot\|_p : L^p(E) \to \mathbb{R}^+$ is a norm in $L^p(E)$, which then becomes a normed linear space.

4.1 $L^p(E)$ for $1 \leq P \leq \infty$ as a Metric Topological Vector Space

The norm $\|\cdot\|_p$ generates a distance in $L^p(E)$ by $d(f, g) = \|f - g\|_p$. One verifies that such a metric is translation invariant and therefore it generates a translation invariant topology in $L^p(E)$ determined by a base at the origin consisting of the balls

$$[\|f\|_p < \rho] = \{f \in L^p(E) \mid \|f\|_p < \rho\} \quad (\rho > 0).$$

Such a topology is called the *norm topology* of $L^p(E)$. By Minkowski's inequality, for $h, g \in L^p(E)$ and $t \in (0, 1)$

$$\|tg + (1 - t)h\|_p \leq t\|g\|_p + (1 - t)\|h\|_p.$$

Therefore the balls $[\|f\|_p < \rho]$ are convex, and the norm topology of $L^p(E)$ for $1 \leq p \leq \infty$, is locally convex. The unit ball $[\|f\|_p < 1]$ is *uniformly* convex if for every $\varepsilon > 0$ there exists $\delta > 0$ such that for any pair $h, g \in L^p(E)$

$$\|h\|_p = \|g\|_p = 1 \text{ and } \|h - g\|_p \geq \varepsilon \implies \left\|\frac{h + g}{2}\right\|_p \leq 1 - \delta. \qquad (4.4)$$

If this occurs the norm topology of $L^p(E)$ is said to be uniformly convex or simply that $L^p(E)$ is uniformly convex.

If $p = \infty$ one can construct examples of functions $h, g \in L^\infty(E)$ such that

$$\|h\|_\infty = \|g\|_\infty = 1 \quad \|h - g\|_\infty \geq 1 \quad \text{and} \quad \left\|\frac{h + g}{2}\right\|_\infty = 1.$$

Similar examples can be constructed in $L^1(E)$. Thus $L^\infty(E)$ and $L^1(E)$ are not uniformly convex. However $L^p(E)$ are uniformly convex for all $1 < p < \infty$ (see § 15).

5 Convergence in $L^p(E)$ and Completeness

A sequence $\{f_n\}$ of functions in $L^p(E)$ for some $1 \leq p \leq \infty$, converges in the sense of $L^p(E)$ to a function $f \in L^p(E)$ if

$$\lim \|f_n - f\|_p = 0.$$

This notion of convergence is also called convergence in the *mean* of order p, or in the *norm* $L^p(E)$ or *strong* convergence in $L^p(E)$.

The sequence $\{f_n\}$ is a Cauchy sequence in $L^p(E)$, if for every $\varepsilon > 0$ there exists a positive integer n_ε such that

$$\|f_n - f_m\|_p \le \varepsilon \quad \text{for all } n, m \ge n_\varepsilon.$$

If $\{f_n\} \to f$ in $L^p(E)$, then $\{f_n\}$ is a Cauchy sequence. Indeed if n, m are sufficiently large

$$\|f_n - f_m\|_p \le \|f_n - f\|_p + \|f_m - f\|_p \le \varepsilon.$$

The next theorem asserts the converse, i.e., if $\{f_n\}$ is a Cauchy sequence in $L^p(E)$ for some $1 \le p \le \infty$, it converges in $L^p(E)$ to some $f \in L^p(E)$. In this sense the spaces $L^p(E)$ for $1 \le p \le \infty$, are *complete*.

Theorem 5.1 (Riesz–Fischer [46, 125]) *Let $\{f_n\}$ be a Cauchy sequence in $L^p(E)$ for some $1 \le p \le \infty$. There exists $f \in L^p(E)$ such that $\{f_n\} \to f$ in $L^p(E)$.*

Proof Assume $p \in [1, \infty)$, the arguments for $L^\infty(E)$ being similar. For $j \in \mathbb{N}$ let n_j be a positive integer, such that

$$\|f_n - f_m\|_p \le \frac{1}{2^j} \qquad \text{for all } n, m \ge n_j. \tag{5.1}$$

Without loss of generality we may arrange that $n_j < n_{j+1}$ for all $j \in \mathbb{N}$. Set formally

$$f(x) = f_{n_1}(x) + \sum [f_{n_{j+1}}(x) - f_{n_j}(x)] \quad \text{for a.e. } x \in E. \tag{5.2}$$

We claim that (5.2) defines a function $f \in L^p(E)$ and that $\{f_n\} \to f$ in $L^p(E)$. For $m = 1, 2, \ldots$, set

$$g_m(x) = \sum_{j=1}^m |f_{n_{j+1}}(x) - f_{n_j}(x)| \qquad \text{for a.e. } x \in E.$$

Since $g_m \le g_{m+1}$ there exists the limit

$$\lim g_m(x) = g(x) \qquad \text{for a.e. } x \in E.$$

By Fatou's lemma, the Minkowski inequality and (5.1)

$$\left(\int_E g^p d\mu \right)^{1/p} \le \left(\liminf \int_E g_m^p d\mu \right)^{1/p} \le \sum_{j=1}^m \frac{1}{2^j} \le 1.$$

Thus $g \in L^p(E)$. The a.e. convergence of $\{g_n\}$ implies that the limit

$$\lim_{m \to \infty} \sum_{j=1}^m [f_{n_{j+1}}(x) - f_{n_j}(x)]$$

exists for a.e. $x \in E$. Therefore (5.2) defines a function f measurable in E.

From (5.2) and the definition of g

$$|f(x)| \leq |f_{n_1}(x)| + |g(x)| \qquad \text{for a.e. } x \in E.$$

Thus $f \in L^p(E)$. Next, from (5.2)–(5.1) and the Minkowski inequality, it follows that for any positive integer k

$$\|f_{n_k} - f\|_p \leq \sum_{j=k}^{\infty} \|f_{n_{j+1}} - f_{n_j}\|_p \leq \frac{1}{2^{k-1}}.$$

Therefore $\{f_{n_j}\}$ converges to f in $L^p(E)$. In particular for every $\varepsilon > 0$, there exist a positive integer j_ε such that

$$\|f_{n_j} - f\|_p \leq \tfrac{1}{2}\varepsilon \qquad \text{for all } j \geq j_\varepsilon.$$

We finally establish that the entire sequence $\{f_n\}$ converges to f in $L^p(E)$. Since $\{f_n\}$ is a Cauchy sequence, having fixed $\varepsilon > 0$, there exists a positive integer n_ε such that

$$\|f_n - f_m\|_p \leq \tfrac{1}{2}\varepsilon \qquad \text{for all } n, m \geq n_\varepsilon.$$

Therefore for $n \geq n_\varepsilon$

$$\|f_n - f\|_p \leq \|f_n - f_{n_j}\|_p + \|f_{n_j} - f\|_p \leq \varepsilon$$

provided $j \geq j_\varepsilon$ and $n_j \geq n_\varepsilon$. ∎

Remark 5.1 The spaces $L^p(E)$, for all $1 \leq p \leq \infty$, endowed with their norm topology are complete metric spaces. As such they are of second category, i.e., they are not the countable union of nowhere dense sets.

6 Separating $L^p(E)$ by Simple Functions

Proposition 6.1 *Let $f \in L^p(E)$ for some $1 \leq p \leq \infty$. For every $\varepsilon > 0$ there exists a simple function $\varphi \in L^p(E)$, such that $\|f - \varphi\|_p \leq \varepsilon$.*[2]

Proof By the decomposition $f = f^+ - f^-$, one may assume that f is nonnegative. Since f is measurable, there exists a sequence $\{\varphi_n\}$ of nonnegative, simple functions such that

$$\varphi_n \leq \varphi_{n+1} \quad \text{and} \quad \varphi_n \to f \quad \text{everywhere in } E.$$

If $1 \leq p < \infty$ the sequence $\{(f - \varphi_n)^p\}$ converges to zero a.e. in E and it is dominated by the integrable function f^p. Therefore $\|f - \varphi_n\|_p \to 0$.

---

[2]It is not claimed here that $L^p(E)$ is separable. See § 15.

If $p = \infty$ the construction of the φ_n implies that

$$\mu\left(\left[f - \varphi_n > \frac{1}{2^n}\|f\|_\infty\right]\right) = 0.$$

Thus $\|f - \varphi_n\|_\infty \le 2^{-n}\|f\|_\infty$ for all n. ∎

Proposition 6.2 *Let* $1 \le p, q \le \infty$ *be conjugate, and let* $g \in L^1(E)$ *satisfy*

$$\int_E \varphi g d\mu \le K\|\varphi\|_p \quad \text{for all simple functions } \varphi$$

for some positive constant K. *Then* $g \in L^q(E)$ *and* $\|g\|_q \le K$.

Proof If $q = 1$ it suffices to choose $\varphi = \text{sign } g \in L^\infty(E)$. Assuming $q \in (1, \infty)$, let $\{\varphi_n\}$ denote a sequence on nonnegative simple functions, such that $\varphi_n \le \varphi_{n+1}$ and $\varphi_n \to |g|^q$. Since

$$0 \le \varphi_n^{1/q} \le |g| \in L^1(E)$$

each φ_n is simple and vanishes outside a set of finite measure. Therefore the functions

$$h_n = \varphi_n^{1/p}\text{sign } g$$

are simple and in $L^p(E)$. For these choices

$$\int_E \varphi_n d\mu = \int_E \varphi_n^{1/p}\varphi_n^{1/q}d\mu \le \int_E \varphi_n^{1/p}|g|d\mu$$
$$= \int_E h_n g d\mu \le K\|h_n\|_p = K\left(\int_E \varphi_n d\mu\right)^{1/p}.$$

From this and Fatou's lemma

$$\|g\|_q \le \left(\liminf \int_E \varphi_n d\mu\right)^{1/q} \le K.$$

Consider now the case $q = \infty$. For $\varepsilon > 0$ set

$$E_\varepsilon = \{x \in E \text{ such that } |g(x)| \ge K + \varepsilon\}$$

and choose $\varphi = \chi_{E_\varepsilon}\text{sign } g$. Since $g \in L^1(E)$ the set E_ε is of finite measure and $\varphi \in L^1(E)$. Therefore

$$(K + \varepsilon)\mu(E_\varepsilon) \le \left|\int_E \varphi g d\mu\right| \le K\mu(E_\varepsilon).$$

Thus $\mu(E_\varepsilon) = 0$ for all $\varepsilon > 0$. ∎

Corollary 6.1 *Let* $1 \leq p, q \leq \infty$ *be conjugate, and let* $g \in L^1(E)$ *satisfy*

$$\int_E fgd\mu \leq K\|f\|_p \quad \text{for all} \quad f \in L^p(E) \cap L^\infty(E)$$

for some positive constant K. *Then* $g \in L^q(E)$ *and* $\|g\|_q \leq K$.

7 Weak Convergence in $L^p(E)$

Let $1 \leq p, q \leq \infty$ be conjugate. A sequence of functions $\{f_n\}$ in $L^p(E)$ for $1 \leq p \leq \infty$, converges *weakly* to a function $f \in L^p(E)$ if

$$\lim \int_E f_n gd\mu = \int_E fgd\mu \quad \text{for all} \quad g \in L^q(E).$$

If $\{f_n\}$ converges to f in $L^p(E)$ it also converges weakly to f in $L^p(E)$. Indeed by the Hölder inequality, for all $g \in L^q(E)$

$$\left| \int_E (f_n g - fg)d\mu \right| \leq \|g\|_q \|f_n - f\|_p.$$

Thus strong convergence implies weak convergence. The converse is false as indicated by the following

7.1 Counterexample

The functions $x \to \cos nx$, $n = 1, 2, \ldots,$ satisfy

$$\int_0^{2\pi} \cos^2 nxdx = \pi \quad \text{for all} \quad n \in \mathbb{N}.$$

Therefore $\{\cos nx\}$ is a sequence in $L^2[0, 2\pi]$ which does not converge to zero in $L^2[0, 2\pi]$. However it converges to zero weakly in $L^2[0, 2\pi]$. To prove it, let first $g = \chi_{(\alpha,\beta)}$, where $(\alpha, \beta) \subset [0, 2\pi]$, and compute

$$\int_0^{2\pi} \chi_{(\alpha,\beta)} \cos nxdx = \frac{1}{n}(\sin n\beta - \sin n\alpha) \to 0 \quad \text{as } n \to \infty.$$

Let now φ be a simple function of the form

$$\varphi = \sum_{i=1}^{m} g_i \chi_{(\alpha_i, \beta_i)} \tag{7.1}$$

where (α_i, β_i) are mutually disjoint subintervals of $[0, 2\pi]$ and g_i are real numbers. For any such function

$$\lim \int_0^{2\pi} \varphi \cos nx \, dx = 0.$$

Simple functions of the form (7.1) are dense in $L^2[0, 2\pi]$ (Corollary 6.1c of the Complements). Thus

$$\lim \int_0^{2\pi} g \cos nx \, dx = 0 \quad \text{for all } g \in L^2[0, 2\pi].$$

8 Weak Lower Semi-continuity of the Norm in $L^p(E)$

Proposition 8.1 *Let $\{f_n\}$ be a sequence of functions in $L^p(E)$ for some $1 \le p < \infty$, converging weakly to some $f \in L^p(E)$. Then*

$$\liminf \|f_n\|_p \ge \|f\|_p. \tag{8.1}$$

If $p = \infty$ the same conclusion holds if $\{X, \mathcal{A}, \mu\}$ is σ-finite.

Proof Assume first that $1 \le p < \infty$. The function $g = |f|^{p/q} \operatorname{sign} f$ belongs to $L^q(E)$ and

$$\lim \int_E f_n g \, d\mu = \int_E f g \, d\mu = \|f\|_p^p.$$

On the other hand, by Hölder's inequality

$$\left| \int_E f_n g \, d\mu \right| \le \|f_n\|_p \|g\|_q = \|f_n\|_p \|f\|_p^{p/q}.$$

Therefore

$$\liminf \|f_n\|_p \|f\|_p^{p/q} \ge \|f\|_p^p.$$

Assume next that $p = \infty$ and $\mu(E) < \infty$. Fix $\varepsilon > 0$ and set

$$E_\varepsilon = [|f| \ge \|f\|_\infty - \varepsilon] \quad \text{and} \quad g = \chi_{E_\varepsilon} \operatorname{sign} f.$$

For such choices

$$\lim \int_E f_n g \, d\mu = \int_E f g \, d\mu \ge (\|f\|_\infty - \varepsilon) \mu(E_\varepsilon).$$

By Hölder's inequality

$$\left| \int_E f_n g d\mu \right| \leq \|f_n\|_\infty \mu(E_\varepsilon).$$

Since $\mu(E_\varepsilon) > 0$, this implies

$$\liminf \|f_n\|_\infty \geq \|f\|_\infty - \varepsilon \quad \text{for all } \varepsilon > 0.$$

If $p = \infty$ and $\{X, \mathcal{A}, \mu\}$ is σ-finite let $A_j \subset A_{j+1}$ be a sequence of measurable sets, of finite measure whose union is X. Setting $E_j = E \cap A_j$, the previous remarks give

$$\liminf \|f_n\|_{\infty, E} \geq \|f\|_{\infty, E_j} \quad \text{for all } j \in \mathbb{N}. \qquad \blacksquare$$

Remark 8.1 The counterexample of § 7.1 shows that the inequality in (8.1) might be strict.

Corollary 8.1 *Let $p \in [1, \infty)$. The function $\|\cdot\|_p : L^p(E) \to \mathbb{R}^+$, is weakly, sequentially, lower semi-continuous. If $p = \infty$ the same conclusion holds if $\{X, \mathcal{A}, \mu\}$ is σ-finite.*

9   Weak Convergence and Norm Convergence

Weak convergence does not imply norm convergence, nor the latter implies weak convergence. The sequence in § 7.1 provides a counterexample to both statements. The next proposition relates these two notions of convergence.

Proposition 9.1 (Radon [121]) *Let $p \in (1, \infty)$ and let $\{f_n\}$ be a sequence of functions in $L^p(E)$ converging weakly to some $f \in L^p(E)$. If also $\|f_n\|_p \to \|f\|_p$, then $\{f_n\}$ converges to f strongly in $L^p(E)$.*

The counterexample of § 7.1 shows that weak convergence and norm convergence does not imply strong convergence. For this to occur the norm of the weak limit is required to coincide with the norm-limit.

The proposition is false for $p = \infty$. In $(0, 1)$ with the Lebesgue measure set

$$f_n(x) = \begin{cases} 0 & \text{for } 0 \leq x \leq \frac{1}{n} \\ 1 & \text{for } \frac{1}{n} < x \leq 1 \end{cases} \quad \text{for } n \in \mathbb{N}.$$

One verifies that $\{f_n\} \to 1$ weakly in $L^\infty(0, 1)$, and $\|f_n\|_\infty \to 1$. However $\|f_n - 1\|_\infty = 1$ for all $n \in \mathbb{N}$.

The proposition is false for $p = 1$. In $(0, 2\pi)$ with the Lebesgue measure, set

$$f_n(x) = 4 + \sin nx \quad \text{for } n \in \mathbb{N}.$$

One verifies that $\{f_n\} \to 4$ weakly in $L^1(0, 2\pi)$ and $\|f_n\|_1 \to \|4\|_1$. However $\|f_n - 4\|_1 = 4$ for all $n \in \mathbb{N}$.[3]

The proof of Proposition 9.1 rests on the following inequalities.

Lemma 9.1 *Let $p \geq 2$. There exists a constant $c \in (0, 1]$, such that for all $t \in \mathbb{R}$*

$$|1 + t|^p \geq 1 + pt + c|t|^p. \tag{9.1}$$

Lemma 9.2 *Let $1 < p < 2$. There exists a constant $c \in (0, 1)$ such that for all $t \in \mathbb{R}$*

$$|1 + t|^p \geq \begin{cases} 1 + pt + c|t|^p & \text{if } |t| \geq 1 \\ 1 + pt + c|t|^2 & \text{if } |t| \leq 1. \end{cases} \tag{9.2}$$

The proof of these lemmas is given in § 9.1c of the Complements.

9.1   Proof of Proposition 9.1 for $p \geq 2$

In (9.1) put

$$t = \frac{f_n(x) - f(x)}{f(x)} \quad \text{for } f(x) \neq 0. \tag{9.3}$$

Multiplying the inequality so obtained by $|f(x)|^p$ gives

$$|f_n|^p \geq |f|^p + p|f|^{p-2}f(f_n - f) + c|f_n - f|^p.$$

One verifies that such an inequality continues to hold also if $f(x) = 0$. Integrating it over E and taking the limit as $n \to \infty$ yields

$$c \limsup \int_E |f_n - f|^p d\mu \leq \lim \left(\|f_n\|_p^p - \|f\|_p^p \right)$$

$$-p \lim \int_E |f|^{p-2}f(f_n - f)d\mu = 0. \qquad \blacksquare$$

---

[3]This example was suggested by J. Manfredi.

9.2 *Proof of Proposition 9.1 for* $1 < p < 2$

For $n \in \mathbb{N}$, introduce the sets

$$E_n = \{ x \in E \mid |f_n(x) - f(x)| \ge |f(x)| \}.$$

In (9.2) choose t as in (9.3) and multiply by $|f(x)|^p$ to obtain

$$|f_n|^p \ge |f|^p + p|f|^{p-2}f(f_n - f) + c|f_n - f|^p \qquad \text{in } E_n$$

$$|f_n|^p \ge |f|^p + p|f|^{p-2}f(f_n - f) + c(f_n - f)^2|f|^{p-2} \quad \text{in } E - E_n.$$

Integrate the first over E_n and the second over $E - E_n$, add the resulting inequalities and let $n \to \infty$ to obtain

$$c \limsup \left\{ \int_{E_n} |f_n - f|^p d\mu + \int_{E - E_n} (f_n - f)^2 |f|^{p-2} d\mu \right\}$$

$$\le \lim \left\{ \|f_n\|_p^p - \|f\|_p^p \right\} - p \lim \int_E |f|^{p-2} f(f_n - f) d\mu = 0.$$

This implies

$$\lim \int_{E_n} |f_n - f|^p d\mu = 0$$

and

$$\lim \int_{E - E_n} (f_n - f)^2 |f|^{p-2} d\mu = 0.$$

From this, the definition of E_n and Hölder's inequality

$$\int_E |f_n - f|^p d\mu \le \int_{E_n} |f_n - f|^p d\mu + \int_{E - E_n} |f|^{p-1} |f_n - f| d\mu$$

$$\le \int_{E_n} |f_n - f|^p d\mu + \left(\int_E |f|^p d\mu \right)^{1/2} \left(\int_{E - E_n} |f|^{p-2} |f_n - f|^2 d\mu \right)^{1/2}. \qquad \blacksquare$$

10 Linear Functionals in $L^p(E)$

A map $\mathcal{F} : L^p(E) \to \mathbb{R}$ is a linear functional in $L^p(E)$ if for all f, $g \in L^p(E)$ and $\alpha, \beta \in \mathbb{R}$

$$\mathcal{F}(\alpha f + \beta g) = \alpha \mathcal{F}(f) + \beta \mathcal{F}(g).$$

The functional $\mathcal{F}$ is bounded if there exists a constant K such that

$$|\mathcal{F}(f)| \leq K\|f\|_p \quad \text{for all } f \in L^p(E).$$

The norm of $\mathcal{F}$ is the smallest of such constants K. Therefore

$$\|\mathcal{F}\| = \sup_{\|f\|_p \neq 0} \frac{|\mathcal{F}(f)|}{\|f\|_p} = \sup_{\|f\|_p=1} |\mathcal{F}(f)|. \tag{10.1}$$

A bounded linear functional in $L^p(E)$ is continuous.

Proposition 10.1 *Let $1 < p \leq \infty$ and $1 \leq q < \infty$ be conjugate. Every $g \in L^q(E)$ generates a bounded linear functional in $L^p(E)$ by the formula*

$$\mathcal{F}_g(f) = \int_E fg d\mu \quad \text{for all } f \in L^p(E). \tag{10.2}$$

Moreover $\|\mathcal{F}_g\| = \|g\|_q$. If $p = 1$ and $q = \infty$ the same conclusion holds if $\{X, \mathcal{A}, \mu\}$ is σ-finite.

Proof The map $\mathcal{F}_g$ is linear. By Hölder's inequality it is also bounded. If $1 \leq q < \infty$, Proposition 3.1 identifies the norm $\|\mathcal{F}_g\|$ as the norm $\|g\|_q$.

Let now $q = \infty$ and assume momentarily that E is of finite measure. For $\varepsilon > 0$ set

$$E_\varepsilon = \{x \in E \mid |g(x)| \geq \|g\|_{\infty,E} - \varepsilon\} \tag{10.3}$$

and in (10.2) choose $f = \chi_{E_\varepsilon} \operatorname{sign} g \in L^1(E)$. This gives

$$\mathcal{F}_g(f) = \int_{E_\varepsilon} |g| dx \geq \|f\|_{1,E}(\|g\|_{\infty,E} - \varepsilon) \quad \text{for all } \varepsilon > 0.$$

Therefore

$$\|g\|_{\infty,E} - \varepsilon \leq \|\mathcal{F}_g\| \leq \|g\|_{\infty,E}.$$

If $\{X, \mathcal{A}, \mu\}$ is σ-finite, let $A_j \subset A_{j+1}$ be a countable collection of measurable sets of finite measure, whose union is X. Set $E_j = E \cap A_j$ and define $E_{j,\varepsilon}$ as in (10.3) with E replaced by E_j. Choosing $f = \chi_{E_{j,\varepsilon}} \operatorname{sign} g \in L^1(E)$ in (10.2) gives

$$\|g\|_{\infty,E_j} - \varepsilon \leq \|\mathcal{F}_g\| \leq \|g\|_{\infty,E}. \qquad \blacksquare$$

Remark 10.1 Let $p \in (1, \infty)$. The proof shows that if g is not the zero equivalence class of $L^q(E)$, the norm $\|\mathcal{F}_g\|$ is achieved by computing $\mathcal{F}_g$ at the element

$$g^* = \frac{|g|^{q-1} \operatorname{sign} g}{\|g\|_q^{q/p}} \in L^p(E). \tag{10.4}$$

Remark 10.2 If $p = 1$ and $\{X, \mathcal{A}, \mu\}$ is not σ-finite, formula (10.2), for a given $g \in L^\infty(E)$, still defines a bounded linear functional in $L^1(E)$. However the identification $\|\mathcal{F}_g\| = \|g\|_\infty$, might fail. A counterexample can be constructed using the measure space $\{X, \mathcal{A}, \mu\}$ in **3.2** of the Complements of Chap. 3.

The Riesz representation theorem asserts that if $1 \le p < \infty$, the functionals in (10.2) are the only bounded linear functionals in $L^p(E)$.

11   The Riesz Representation Theorem

Theorem 11.1 *Let $1 < p, q < \infty$ be conjugate. For every bounded, linear functional $\mathcal{F}$ in $L^p(E)$, there exists a unique function $g \in L^q(E)$ such that $\mathcal{F}$ is represented by the formula (10.2). Moreover $\|\mathcal{F}\| = \|g\|_q$.*
If $p = 1$ and $q = \infty$ the same conclusion holds if $\{X, \mathcal{A}, \mu\}$ is σ-finite.

Remark 11.1 The conclusion is false for $p = 1$ if $\{X, \mathcal{A}, \mu\}$ is not σ-finite. A counterexample can be constructed using the measure space in **3.2** of the Complements of Chap. 3.

Remark 11.2 The theorem is false for $p = \infty$. A counterexample is in § 9.2c of the Complements of Chap. 7.

11.1   *Proof of Theorem 11.1: The Case of $\{X, \mathcal{A}, \mu\}$ Finite*

Assume first that $\mu(X) < \infty$ and that $E = X$. For every μ-measurable set $A \subset X$ the function χ_A is in $L^p(E)$. The functional $\mathcal{F}$ induces a set function ν defined on the σ-algebra $\mathcal{A}$ by the formula

$$\mathcal{A} \ni A \longrightarrow \nu(A) = \mathcal{F}(\chi_A).$$

The set function $\nu(\cdot)$ is finite for all $A \in \mathcal{A}$, it vanishes on the empty set and is countably additive. To establish the last claim, let $\{A_n\}$ be a countable collection of mutually disjoint, measurable sets in $\mathcal{A}$. Since $\mu(\cup A_n) < \infty$, for every $\varepsilon > 0$ there exists $n_\varepsilon \in \mathbb{N}$ such that $\sum_{j>n_\varepsilon} \mu(A_j) < \varepsilon$. By linearity

$$\mathcal{F}\big(\chi_{\cup A_n}\big) = \mathcal{F}\big(\chi_{\cup_{j=1}^{n_\varepsilon} A_j} + \chi_{\cup_{j>n_\varepsilon} A_j}\big)$$
$$= \mathcal{F}\Big(\sum_{j=1}^{n_\varepsilon} \chi_{A_j} + \chi_{\cup_{j>n_\varepsilon} A_j}\Big) = \sum_{j=1}^{n_\varepsilon} \mathcal{F}(\chi_{A_j}) + \mathcal{F}\big(\chi_{\cup_{j>n_\varepsilon} A_j}\big).$$

Since the A_j are disjoint and $p \in [1, \infty)$

$$\left|\mathcal{F}(\chi_{\bigcup A_n}) - \sum_{j=1}^{n_\varepsilon} \mathcal{F}(\chi_{A_j})\right| \leq \|\mathcal{F}\| \left\|\chi_{\bigcup_{j>n_\varepsilon} A_j}\right\|_p$$

$$= \|\mathcal{F}\| \left(\sum_{j>n_\varepsilon} \mu(A_j)\right)^{1/p} \leq \|\mathcal{F}\| \varepsilon^{1/p}.$$

Since ε is arbitrary, this implies

$$\nu\left(\bigcup A_n\right) = \mathcal{F}(\chi_{\bigcup A_n}) = \sum \mathcal{F}(\chi_{A_n}) = \sum \nu(A_n).$$

Therefore ν is countably additive and defines a signed measure on $\mathcal{A}$. Since $|\nu(A)| = 0$ whenever $\mu(A) = 0$, the signed measure ν is absolutely continuous with respect to μ. By the Radon–Nikodým theorem there exists a μ-measurable function $g : X \to \mathbb{R}^*$ such that

$$\mathcal{A} \ni A \longrightarrow \nu(A) = \int_E g\chi_A d\mu.$$

For every simple function $\varphi = \sum_{i=1}^n \alpha_i \chi_{A_i}$, by the linearity of $\mathcal{F}$

$$\mathcal{F}(\varphi) = \sum_{i=1}^n \alpha_i \nu(A_i) = \sum_{i=1}^n \alpha_i \int_E g\chi_{A_i} d\mu = \int_E g\varphi d\mu.$$

For all the simple functions

$$\left|\int_E g\varphi d\mu\right| \leq \|\mathcal{F}\| \|\varphi\|_{p,X}.$$

Moreover $g \in L^1(E)$, since E is of finite measure. Therefore by Proposition 6.2, $g \in L^q(E)$ and $\|g\|_q \leq \|\mathcal{F}\|$. Since the simple functions are dense in $L^p(E)$

$$\mathcal{F}(f) = \int_E fg d\mu \quad \text{for all } f \in L^p(E) \text{ and } \|\mathcal{F}\| = \|g\|_q.$$

If $g' \in L^q(E)$ identifies the same functional $\mathcal{F}$, then

$$\int_E f(g - g')d\mu = 0 \quad \text{for all } f \in L^p(E).$$

Thus $g = g'$ a.e. in E. ∎

11.2 Proof of Theorem 11.1: The Case of $\{X, \mathcal{A}, \mu\}$ σ-Finite

Let $A_j \subset A_{j+1}$ be a countable collection of sets of finite measure exhausting X, and set $E_j = E \cap A_j$. For each $f \in L^p(E)$ and $j \in \mathbb{N}$ set

$$f_j = \begin{cases} f & \text{on } E_j \\ 0 & \text{on } E - E_j. \end{cases}$$

For each $j \in \mathbb{N}$ there exists $g_j \in L^q(E_j)$ such that

$$\mathcal{F}(f_j) = \int_{E_j} f g_j d\mu \quad \text{for all } f \in L^p(E).$$

We regard g_j as defined in the whole E by setting them to be equal to zero outside E_j. If $f \in L^p(E)$ vanishes outside E_j, then

$$\mathcal{F}(f) = \int_{E_j} f g_j d\mu = \int_{E_{j+1}} f g_{j+1} d\mu.$$

Therefore

$$\int_{E_j} f(g_j - g_{j+1}) d\mu = 0 \quad \text{for all } f \in L^p(E_j)$$

and g_j coincides with g_{j+1} on E_j. The sequence $\{g_j\}$ converges a.e. on E to a measurable function g. The sequence $\{|g_j|\}$ is nondecreasing, and by monotone convergence

$$\|g\|_q = \lim \|g_j\|_q \leq \|\mathcal{F}\|, \quad 1 < q \leq \infty.$$

Thus $g \in L^q(E)$. Given now any $f \in L^p(E)$, the sequence $\{f_j g\}$ converges to fg a.e. on E and $|f_j g| \leq |fg| \in L^1(E)$. Therefore by dominated convergence

$$\int_E f g d\mu = \lim \int_E f_j g d\mu = \lim \mathcal{F}(f_j) = \mathcal{F}(f).$$

The characterization of $\|\mathcal{F}\|$ follows from Proposition 10.1.   ∎

11.3 Proof of Theorem 11.1: The Case $1 < p < \infty$

We assume $1 < p < \infty$ and place no restrictions on the measure space $\{X, \mathcal{A}, \mu\}$. If $A \subset E$ is of σ-finite measure, there exists a unique $g_A \in L^q(E)$, and vanishing on $E - A$, such that[4]

$$\mathcal{F}(f|_A) = \int_E f g_A d\mu \quad \text{for all } f \in L^p(E).$$

Moreover if $B \subset A$ is of σ-finite measure, then $g_B = g_A$ a.e. on B. The set function $A \to \|g_A\|_q$ defined on the subsets of E of σ-finite measure, is uniformly bounded since

---

[4]$f|_A$ is the restriction of f to A, defined in the whole E by setting it to be zero outside A.

$$\|g_A\|_q \leq \|\mathcal{F}\| \quad \text{for all sets } A \text{ of } \sigma - \text{ finite measure.}$$

Denote by M the supremum of $\|g_A\|_q$ as A ranges over such sets and let $\{A_n\}$ be a sequence of sets of σ-finite measure such that

$$\|g_{A_n}\|_q \leq \|g_{A_{n+1}}\|_q   \quad \text{and} \quad   \lim \|g_{A_n}\|_q = M.$$

The set $A_* = \cup A_n$ is of σ-finite measure and $\|g_{A_*}\|_q = M$. Thus the supremum of $\|g_A\|$ is actually achieved at A_*. We regard g_{A_*} as defined in the whole E by setting it to be zero outside A_*. In such a way $g_{A_*} \in L^q(E)$. Such a function g_{A_*} is the one claimed by the Riesz representation theorem.

If B is a set of σ-finite measure containing A_*, then $g_{A_*} = g_B$ a.e. on A_*. By maximality

$$\|g_{A_*}\|_q \leq \|g_B\|_q \leq \|g_{A_*}\|_q.$$

Therefore $g_B = 0$ a.e. on $B - A_*$, since $1 < q < \infty$.

Given $f \in L^p(E)$, the set $[|f| > 0]$ is of σ-finite measure. Since also the set $B = [|f| > 0] \cup A_*$ is of σ-finite measure

$$\mathcal{F}(f) = \int_E f g_B d\mu = \int_E f g_{A_*} d\mu. \qquad \blacksquare$$

12   The Hanner and Clarkson Inequalities

Proposition 12.1 (Hanner's Inequalities [64]) *Let f and g be in $L^p(E)$ for some $1 \leq p < \infty$. Then*

$$\|f+g\|_p^p + \|f-g\|_p^p \leq (\|f\|_p + \|g\|_p)^p + |\|f\|_p - \|g\|_p|^p$$
$$\text{for } p \geq 2 \qquad (12.1)$$

$$\|f+g\|_p^p + \|f-g\|_p^p \geq (\|f\|_p + \|g\|_p)^p + |\|f\|_p - \|g\|_p|^p$$
$$\text{for } p \in [1,2] \qquad (12.2)$$

$$(\|f+g\|_p + \|f-g\|_p)^p + |\|f+g\|_p - \|f-g\|_p|^p \geq 2^p(\|f\|_p^p + \|g\|_p^p)$$
$$\text{for } p \geq 2 \qquad (12.3)$$

$$(\|f+g\|_p + \|f-g\|_p)^p + |\|f+g\|_p - \|f-g\|_p|^p \leq 2^p(\|f\|_p^p + \|g\|_p^p)$$
$$\text{for } p \in [1,2] \qquad (12.4)$$

Proposition 12.2 (Clarkson's Inequalities [28]) *Let $1 < p, q < \infty$ be conjugate, and let $f, g \in L^p(E)$. Then*

$$\left\|\frac{f+g}{2}\right\|_p^p + \left\|\frac{f-g}{2}\right\|_p^p \le \frac{\|f\|_p^p + \|g\|_p^p}{2} \qquad \text{for } p \ge 2 \qquad (12.5)$$

$$\left\|\frac{f+g}{2}\right\|_p^p + \left\|\frac{f-g}{2}\right\|_p^p \ge \frac{\|f\|_p^p + \|g\|_p^p}{2} \qquad \text{for } p \in (1, 2] \qquad (12.6)$$

$$\left\|\frac{f+g}{2}\right\|_p^q + \left\|\frac{f-g}{2}\right\|_p^q \ge \left(\frac{\|f\|_p^p + \|g\|_p^p}{2}\right)^{q-1} \qquad \text{for } p \ge 2 \qquad (12.7)$$

$$\left\|\frac{f+g}{2}\right\|_p^q + \left\|\frac{f-g}{2}\right\|_p^q \le \left(\frac{\|f\|_p^p + \|g\|_p^p}{2}\right)^{q-1} \qquad \text{for } p \in (1, 2] \qquad (12.8)$$

Remark 12.1 If $p = 2$, both Hanner's and Clarkson's inequalities reduce to the standard parallelogram identity, and for $p = 1$ they coincide with the triangle inequality.

For $p > 1$ and $p \ne 2$ set

$$\varphi(s; t) = h(s) + k(s)t^p$$

where, for $s \in (0, 1]$ and $t > 0$

$$h(s) = (1 + s)^{p-1} + (1 - s)^{p-1}$$
$$k(s) = \left[(1 + s)^{p-1} - (1 - s)^{p-1}\right]s^{1-p}.$$

Lemma 12.1 *Let $1 < p, q < \infty$ be conjugate. For every fixed $t > 0$ there holds*

$$\varphi(s; t) \le |1 + t|^p + |1 - t|^p \text{ for } p \in (1, 2]$$
$$\varphi(s; t) \ge |1 + t|^p + |1 - t|^p \text{ for } p \ge 2. \qquad (12.9)$$

Moreover for all $t \in [0, 1]$

$$\left|\frac{1+t}{2}\right|^q + \left|\frac{1-t}{2}\right|^q \le \left(\frac{1+t^p}{2}\right)^{q-1} \quad \text{for } q \ge 2$$

$$\left|\frac{1+t}{2}\right|^q + \left|\frac{1-t}{2}\right|^q \ge \left(\frac{1+t^p}{2}\right)^{q-1} \quad \text{for } q \in (1, 2]. \qquad (12.10)$$

Proof Assume first $t \in (0, 1)$. By direct calculation

$$\frac{1}{p-1}\frac{d\varphi(s; t)}{ds} = \left[(1 + s)^{p-2} - (1 - s)^{p-2}\right]\frac{s^p - t^p}{s^p}.$$

Therefore if $p \in (1, 2)$ the function $s \to \varphi(s; t)$ increases for $s \in (0, t)$, decreases for $s \in (t, 1]$ and takes its maximum at $s = t$. Analogously, if $p > 2$ the function $s \to \varphi(s; t)$ takes its minimum for $s = t$. Therefore if $t \in (0, 1)$

$$\varphi(s; t) \leq \varphi(t; t) = |1 + t|^p + |1 - t|^p \quad \text{for } p \in (1, 2]$$
$$\varphi(s; t) \geq \varphi(t; t) = |1 + t|^p + |1 - t|^p \quad \text{for } p \geq 2.$$

By continuity these continue to hold for also for $t = 1$.

Assume now that $t > 1$. If $p \in (1, 2)$ then $k(s) \leq h(s)$. Indeed the function $s \to \{k(s) - h(s)\}$ vanishes for $s = 1$ and is increasing for $s \in (0, 1)$. Therefore

$$\varphi(s; t) = h(s) + k(s)t^p \leq h(s)t^p + k(s)$$
$$= t^p \left(h(s) + k(s)\frac{1}{t^p} \right) = t^p \varphi\left(s; \frac{1}{t}\right)$$
$$\leq t^p \varphi\left(\frac{1}{t}; \frac{1}{t}\right) = |1 + t|^p + |1 - t|^p.$$

If $p > 2$ and $t > 1$ the argument is similar, starting from the inequality $k(s) \geq h(s)$ for $p > 2$. Inequalities (12.10) are obvious for $t = 0$ and $t = 1$. To prove the first of (12.10) for $t \in (0, 1)$, write the second of (12.9) with q replacing p and in the resulting inequality take $s = t^p$. Such a choice is admissible since $t \in (0, 1)$. The second of (12.10) is proved analogously. ∎

12.1 Proof of Hanner's Inequalities

Having fixed f and g in $L^p(E)$ may assume $\|f\|_p \geq \|g\|_p > 0$. Let $p \in (1, 2)$ and in the first of (12.9) take $t = |g|/|f|$ provided $|f| \neq 0$. Multiplying the inequality so obtained by $|f|$ gives

$$h(s)|f|^p + k(s)|g|^p \leq |f + g|^p + |f - g|^p$$

and this inequality continues to hold if $|f| = 0$. Integrating over E

$$h(s)\|f\|_p^p + k(s)\|g\|_p^p \leq \|f + g\|_p^p + \|f - g\|_p^p \tag{12.11}$$

for all $s \in (0, 1]$. Taking $s = \|g\|_p/\|f\|_p$ proves (12.2). Inequality (12.4) follows from (12.2) by replacing f with $(f + g)$ and g with $(f - g)$. The proof of (12.1) and (12.3) is analogous starting from the second of (12.9). ∎

12.2 Proof of Clarkson's Inequalities

Since (12.11) holds for all $s \in (0, 1]$, by taking $s = 1$ proves (12.6). If $p \geq 2$ inequality (12.11) holds with the sign reversed and still for all $s \in (0, 1]$. By taking $s = 1$ proves (12.5). To establish (12.7) and (12.8), observe first that

$$\left\|\frac{f \pm g}{2}\right\|_p^q = \left(\int_E \left|\frac{f \pm g}{2}\right|^p d\mu\right)^{\frac{1}{p-1}}$$

$$= \left(\int_E \left|\frac{f \pm g}{2}\right|^{\frac{p}{p-1}(p-1)} d\mu\right)^{\frac{1}{p-1}} = \left\|\left|\frac{f \pm g}{2}\right|^q\right\|_{p-1}.$$

To prove (12.7), since $q \in (1, 2)$ the second of (12.10) implies the pointwise inequality

$$\left|\frac{f + g}{2}\right|^q + \left|\frac{f - g}{2}\right|^q \geq \left(\frac{|f|^p + |g|^p}{2}\right)^{q-1}. \tag{12.12}$$

By the Minkowski inequality

$$\left\|\frac{f + g}{2}\right\|_p^q + \left\|\frac{f - g}{2}\right\|_p^q = \left\|\left|\frac{f + g}{2}\right|^q\right\|_{p-1} + \left\|\left|\frac{f - g}{2}\right|^q\right\|_{p-1}$$

$$\geq \left(\int_E \left(\left|\frac{f + g}{2}\right|^q + \left|\frac{f - g}{2}\right|^q\right)^{p-1} d\mu\right)^{\frac{1}{p-1}}$$

$$\geq \left(\int_E \frac{|f|^p + |g|^p}{2} d\mu\right)^{\frac{1}{p-1}} = \left(\frac{\|f\|_p^p + \|g\|_p^p}{2}\right)^{q-1}.$$

Inequality (12.8) is established the same way, by making use of the reverse Minkowski inequality. Since $q > 2$, inequality (12.12) is reversed. Therefore, since $(p - 1) \in (0, 1)$

$$\left\|\frac{f + g}{2}\right\|_p^q + \left\|\frac{f - g}{2}\right\|_p^q = \left\|\left|\frac{f + g}{2}\right|^q\right\|_{p-1} + \left\|\left|\frac{f - g}{2}\right|^q\right\|_{p-1}$$

$$\leq \left(\int_E \left(\left|\frac{f + g}{2}\right|^q + \left|\frac{f - g}{2}\right|^q\right)^{p-1} d\mu\right)^{\frac{1}{p-1}}$$

$$\leq \left(\int_E \frac{|f|^p + |g|^p}{2} d\mu\right)^{\frac{1}{p-1}} = \left(\frac{\|f\|_p^p + \|g\|_p^p}{2}\right)^{\frac{1}{p-1}}. \qquad \blacksquare$$

13 Uniform Convexity of $L^p(E)$ for $1 < p < \infty$

Proposition 13.1 *The spaces $L^p(E)$ for $1 < p < \infty$ are uniformly convex.*

Proof It suffices to verify (4.4). Let $f, g \in L^p(E)$ satisfy $\|f\|_p = \|g\|_p = 1$, and $\|f - g\|_p \geq \varepsilon > 0$. By the Clarkson inequalities

$$\left\|\frac{f + g}{2}\right\|_p^p \leq 1 - \frac{\varepsilon^p}{2^p} \qquad \text{if } p \geq 2$$

$$\left\|\frac{f + g}{2}\right\|_p^q \leq 1 - \frac{\varepsilon^q}{2^q} \qquad \text{if } p \in (1, 2]. \qquad \blacksquare$$

A remarkable fact is that the Riesz representation theorem depends only on the uniform convexity of $L^p(E)$ and, in particular, is independent of the Radon–Nikodým theorem. The starting point is the following consequence of the Clarkson's inequalities.

Proposition 13.2 *Let* $1 < p, q < \infty$ *be conjugate. For a nonzero* $g \in L^q(E)$ *let* g^* *be defined by (10.4).*

If $\mathcal{F}_1$ *and* $\mathcal{F}_2$ *are two bounded linear functionals in* $L^p(E)$ *satisfying*

$$\mathcal{F}_i(g^*) = \|\mathcal{F}_i\| = 1 \qquad i = 1, 2$$

for some fixed $g \in L^q(E)$, *then* $\mathcal{F}_1 = \mathcal{F}_2$.

Proof If $\mathcal{F}_1 \neq \mathcal{F}_2$, there exists $f \in L^p(E)$ such that $\mathcal{F}_1(f) \neq \mathcal{F}_2(f)$. Set

$$\varphi = \frac{2f}{\mathcal{F}_1(f) - \mathcal{F}_2(f)} - \frac{\mathcal{F}_1(f) + \mathcal{F}_2(f)}{\mathcal{F}_1(f) - \mathcal{F}_2(f)} g^* \in L^p(E).$$

One verifies that $\mathcal{F}_1(\varphi) = 1$ and $\mathcal{F}_2(\varphi) = -1$. Let $t \in (0, 1)$ and compute

$$1 + t = \mathcal{F}_1(g^* + t\varphi) \le \|\mathcal{F}_1\| \|g^* + t\varphi\|_p = \|g^* + t\varphi\|_p$$
$$1 + t = \mathcal{F}_2(g^* - t\varphi) \le \|\mathcal{F}_2\| \|g^* - t\varphi\|_p = \|g^* - t\varphi\|_p.$$

Assume first that $p \ge 2$. Then from these inequalities and Clarkson's inequality (12.7)

$$(1+t)^q \le \left(\frac{\|g^* + t\varphi\|_p^p + \|g^* - t\varphi\|_p^p}{2} \right)^{q-1}$$
$$\le \left\| \frac{(g^* + t\varphi) + (g^* - t\varphi)}{2} \right\|_p^q + \left\| \frac{(g^* + t\varphi) - (g^* - t\varphi)}{2} \right\|_p^q$$
$$= \|g^*\|_p^q + t^q \|\varphi\|_p^q = 1 + t^q \|\varphi\|_p^q$$

since $\|g^*\|_p = 1$. Similarly if $1 < p \le 2$ using Clarkson's inequality (12.6)

$$(1+t)^p \le \frac{\|g^* + t\varphi\|_p^p + \|g^* - t\varphi\|_p^p}{2}$$
$$\le \left\| \frac{(g^* + t\varphi) + (g^* - t\varphi)}{2} \right\|_p^p + \left\| \frac{(g^* + t\varphi) - (g^* - t\varphi)}{2} \right\|_p^p$$
$$= \|g^*\|_p^p + t^p \|\varphi\|_p^p = 1 + t^p \|\varphi\|_p^p.$$

Consider the last of these. Expanding the left hand side with respect to t about $t = 0$, gives

$$pt + O(t^2) \le t^p \|\varphi\|_p^p.$$

Dividing by t and letting $t \to 0$ gives a contradiction. A similar contradiction occurs if $p \geq 2$. ∎

14 The Riesz Representation Theorem By Uniform Convexity

Theorem 14.1 *Let $1 < p, q < \infty$ be conjugate. To every bounded, linear functional $\mathcal{F}$ in $L^p(E)$, there corresponds a unique function $g \in L^q(E)$ such that $\mathcal{F}$ is represented by the formula (10.2). Moreover $\|\mathcal{F}\| = \|g\|_q$.*
  If $p = 1$ and $q = \infty$ the same conclusion holds if $\{X, \mathcal{A}, \mu\}$ is σ-finite.

14.1 Proof of Theorem 14.1. The Case $1 < p < \infty$

Without loss of generality we may assume $\|\mathcal{F}\| = 1$. By the definition (10.1) of $\|\mathcal{F}\|$, there exists a sequence $\{f_n\}$ of functions in $L^p(E)$, such that

$$\|f_n\|_p = 1 \quad |\mathcal{F}(f_n)| \geq \tfrac{1}{2} \quad \text{and} \quad \lim |\mathcal{F}(f_n)| = 1.$$

By possibly replacing f_n with $-f_n$ we may assume, without loss of generality, that $\mathcal{F}(f_n) > 0$ for all $n \in \mathbb{N}$.

We claim that $\{f_n\}$ is a Cauchy sequence in $L^p(E)$. Proceeding by contradiction, if not, there exists some $\varepsilon > 0$ such that $\|f_m - f_n\|_p \geq \varepsilon$ for infinitely many indices m and n. The uniform convexity of $L^p(E)$, then implies that there exists $\delta = \delta(\varepsilon) \in (0, 1)$, such that

$$\left\| \frac{f_m + f_n}{2} \right\|_p \leq 1 - \delta$$

for infinitely many indices m and n. Letting $m, n \to \infty$ along such indices

$$2 = \lim \{\mathcal{F}(f_m) + \mathcal{F}(f_n)\} = \lim \mathcal{F}(f_m + f_n)$$
$$\leq \lim \|f_m + f_n\|_p \leq 2(1 - \delta).$$

The contradiction proves that $\{f_n\}$ is a Cauchy sequence in $L^p(E)$ and we let f denote its limit. By construction $\|f\|_p = 1$. Set

$$g = |f|^{p/q}\operatorname{sign} f \quad \text{and} \quad g^* = |g|^{q-1}\operatorname{sign} g = f.$$

By construction $g \in L^q(E)$ and $g^* \in L^p(E)$, and

$$\|f\|_p = \|g\|_q^{q/p} = \|g^*\|_p = 1.$$

Let $\mathcal{F}_g$ be the bounded linear functional in $L^p(E)$ generated by such a $g \in L^q(E)$, by the formula (10.2). By construction the two functionals $\mathcal{F}$ and $\mathcal{F}_g$ satisfy

$$\mathcal{F}(g^*) = \mathcal{F}_g(g^*) = \|\mathcal{F}\| = \|\mathcal{F}_g\| = 1.$$

Therefore $\mathcal{F} = \mathcal{F}_g$, by Proposition 13.2.   ∎

14.2   The Case $p = 1$ and E of Finite Measure

Without loss of generality we may assume $\|\mathcal{F}\| = 1$. If $\mu(E) < \infty$, then $L^p(E) \subset L^1(E)$ for all $p \geq 1$. In particular for all $f \in L^p(E)$

$$|\mathcal{F}(f)| \leq \|f\|_1 \leq \mu(E)^{1/q}\|f\|_p.$$

Therefore for each fixed $p \in (1, \infty)$, the map $\mathcal{F}$ may be identified with a bounded linear functional in $L^p(E)$. By Theorem 14.1 for any such p, there exists a unique $g_p \in L^q(E)$ such that

$$\mathcal{F}(f) = \mathcal{F}_{g_p}(f) = \int_E f g_p d\mu \quad \text{for all } f \in L^p(E).$$

Moreover

$$\|g_p\|_q = \|\mathcal{F}_{g_p}\| = \sup_{\substack{f \in L^p(E) \\ \|f\|_p = 1}} |\mathcal{F}_{g_p}(f)| = \sup_{\substack{f \in L^p(E) \\ \|f\|_p = 1}} |\mathcal{F}(f)|$$

$$\leq \sup_{\substack{f \in L^p(E) \\ \|f\|_p = 1}} \|\mathcal{F}\|\|f\|_1 \leq \sup_{\substack{f \in L^p(E) \\ \|f\|_p = 1}} \|f\|_p \mu(E)^{1/q} = \mu(E)^{1/q}.$$

Now let $1 < p_1 < p_2 < \infty$ and let

$$g_{p_i} \in L^{q_i}(E) \quad i = 1, 2 \quad \frac{1}{p_i} + \frac{1}{q_i} = 1$$

be the two functions that identify $\mathcal{F}_{g_{p_1}}$ and $\mathcal{F}_{g_{p_2}}$. If φ is a simple function

$$\mathcal{F}(\varphi) = \mathcal{F}_{g_{p_i}}(\varphi) = \int_E \varphi g_{p_i} d\mu \quad i = 1, 2.$$

From these, by difference $(g_{p_1} - g_{p_2}) \in L^1(E)$, and

$$\int_E (g_{p_1} - g_{p_2})\varphi d\mu = 0 \quad \text{for all simple functions } \varphi.$$

Since the simple functions are dense in $L^1(E)$ this implies that $g_{p_1} = g_{p_2}$. Therefore there exists a function $g \in L^q(E)$ for all $q \in [1, \infty)$, such that

$$\mathcal{F}(f) = \int_E fg d\mu \quad \text{for all} \quad f \in L^p(E).$$

For such a function g

$$\left| \int_E fg d\mu \right| = |\mathcal{F}(f)| \le \|f\|_1 \quad \text{for all } f \in L^1(E) \cap L^\infty(E).$$

By Corollary 6.1 this implies that $g \in L^\infty(E)$ and $\|g\|_\infty \le 1$. Therefore, by density, (10.2) gives a representation of $\mathcal{F}$ for all $f \in L^1(E)$. Also

$$1 = \|\mathcal{F}\| = \sup_{\substack{f \in L^1(E) \\ \|f\|_1 = 1}} \int_E fg d\mu \le \|g\|_\infty \le 1. \qquad \blacksquare$$

14.3 The Case $p = 1$ and $\{X, \mathcal{A}, \mu\}$ σ-Finite

Let $A_n \subset A_{n+1}$ be a countable collection of sets of finite measure whose union is X. Set

$$E_n = E \cap (A_{n+1} - A_n)$$

and let $\mathcal{F}_n$ be the restriction of $\mathcal{F}$ to $L^1(E_n)$. A function $\varphi \in L^1(E_n)$ may be regarded as an element of $L^1(E)$, by possibly defining it to be zero in $E - E_n$. In this sense $L^1(E_n) \subset L^1(E)$, and $\mathcal{F} = \mathcal{F}_n$ within $L^1(E_n)$. Since E_n has finite measure, there exists $g_n \in L^\infty(E_n)$, such that

$$\mathcal{F}_n(\varphi) = \int_{E_n} g_n \varphi d\mu \quad \text{for all} \quad \varphi \in L^1(E_n) \quad \text{and} \quad \|g_n\|_{\infty, E_n} = 1.$$

By extending g_n to be zero in $(E - E_n)$ we regard it as an element of $L^\infty(E)$ and set $g = \sum g_n \chi_{E_n}$. Then, for all $f \in L^1(E)$

$$\mathcal{F}(f) = \mathcal{F}\left(\sum f \chi_{E_n}\right) = \sum \mathcal{F}\left(f \chi_{E_n}\right) = \sum \int_{E_n} g_n f d\mu = \int_E gf d\mu. \qquad \blacksquare$$

15 If $E \subset \mathbb{R}^N$ and $p \in [1, \infty)$, then $L^p(E)$ Is Separable

Let dx denote the Lebesgue measure in $\mathbb{R}^N$ and let $E \subset \mathbb{R}^N$ be Lebesgue measurable. The $\mathbb{R}^N$ structure of E affords to $L^p(E)$ further defining properties. For example by Proposition 6.1, the collection of simple functions is dense in $L^p(E)$, for E a measurable subset of any measure space $\{X, \mathcal{A}, \mu\}$. When $E \subset \mathbb{R}^N$ is Lebesgue measurable and $1 \leq p < \infty$, the space $L^p(E)$ contains a dense *countable* collection of step functions.

Theorem 15.1 *Let $E \subset \mathbb{R}^N$ be measurable. Then $L^p(E)$ for $1 \leq p < \infty$ is separable, i.e., it contains a dense countable set.*

Proof Let $\{Q_n\}$ denote the collection of closed dyadic cubes in $\mathbb{R}^N$. For a fixed positive integer n, let $\mathcal{S}_n$ denote the family of step functions defined on E and taking constant, rational values on the first n cubes, i.e.,

$$\mathcal{S}_n = \left\{ \begin{array}{c} \varphi : E \to \mathbb{R} \text{ of the form } \sum_{i=1}^{n} f_i \chi_{Q_i \cap E} \\ \text{where the numbers } f_i \text{ are rational} \end{array} \right\}.$$

Each $\mathcal{S}_n$ is countable and the union $\mathcal{S} = \cup \mathcal{S}_n$ is a countable family of simple functions. We claim that $\mathcal{S}$ is dense in $L^p(E)$ for $1 \leq p < \infty$, i.e., for every $f \in L^p(E)$ and every $\varepsilon > 0$ there exists $\varphi \in \mathcal{S}$ such that $\|f - \varphi\|_p \leq \varepsilon$.

Assume first that E is bounded and that $f \in L^p(E) \cap L^\infty(E)$. By Lusin's theorem f is quasi-continuous. Therefore having fixed $\varepsilon > 0$ there exists a closed set $E_\varepsilon \subset E$ such that

$$\mu(E - E_\varepsilon) \leq \frac{\varepsilon^p}{4^p \|f\|_\infty^p}$$

and f is uniformly continuous in E_ε. In particular there exists $\delta > 0$, such that

$$|f(x) - f(y)| \leq \frac{\varepsilon}{4\mu(E)^{1/p}}$$

for all $x, y \in E_\varepsilon$ such that $|x - y| < \delta$. Since E_ε is bounded, having determined $\delta > 0$, there exist a finite number of closed dyadic cubes, with pairwise disjoint interior and with diameter less than δ, whose union covers E_ε, say for example $\{Q_1, \ldots, Q_{n_\delta}\}$. Select $x_i \in Q_i \cap E_\varepsilon$ and then choose a rational number f_i such that

$$|f_i - f(x_i)| \leq \frac{\varepsilon}{4\mu(E)^{1/p}}.$$

Set

$$\varphi = \sum_{i=1}^{n_\delta} f_i \chi_{Q_i \cap E}$$

and compute

$$\int_E |f - \varphi|^p dx = \int_{E_\varepsilon} |f - \varphi|^p dx + \int_{E - E_\varepsilon} |f - \varphi|^p dx$$

$$= \sum_{i=1}^{n_\delta} \int_{Q_i \cap E_\varepsilon} |(f - f(x_i)) + (f(x_i) - f_i)|^p dx$$

$$+ \int_{E - E_\varepsilon} |f - \varphi|^p dx$$

$$\leq \frac{1}{2^p} \varepsilon^p + 2^p \|f\|_\infty^p \mu(E - E_\varepsilon) \leq \varepsilon^p.$$

If E is unbounded, since $f \in L^p(E)$

$$\int_E |f|^p dx = \sum \int_{E \cap \{n < |x| \leq n+1\}} |f|^p dx < \infty.$$

Therefore, having fixed $\varepsilon > 0$, there exists an index n_ε so large that

$$\int_{E \cap \{|x| \geq n_\varepsilon\}} |f|^p dx \leq \frac{1}{4^p} \varepsilon^p.$$

Also

$$n^p \mu([|f| \geq n]) \leq \int_E |f|^p dx = \|f\|_p^p.$$

Therefore, for every $\delta > 0$ there exists a positive integer n_δ such that

$$\mu([|f| \geq n]) \leq \delta \qquad \text{for all } n \geq n_\delta.$$

By the Vitali theorem on the absolute continuity of the integral, having fixed $\varepsilon > 0$, there exists $\delta > 0$ such that

$$\int_{[|f| \geq n]} |f|^p dx \leq \frac{1}{4^p} \varepsilon^p \qquad \text{for all } n \geq n_\delta.$$

Setting

$$E_n = E \cap \{[|x| \geq n] \cup [|f| \geq n]\}$$

one estimates

$$\|f\|_{p, E_n} \leq \tfrac{1}{2} \varepsilon \qquad \text{for all } n \geq \max\{n_\varepsilon; n_\delta\}.$$

The set $E - E_n$ is bounded and the restriction of f to such a set is bounded. Therefore there exists a simple function $\varphi \in S$, vanishing outside $E - E_n$ such that $\|f - \varphi\|_{p, E - E_n} \leq \tfrac{1}{2} \varepsilon$. ∎

276 6 The L^p Spaces

Remark 15.1 Let E be a Lebesgue measurable subset of $\mathbb{R}^N$. Then the spaces $L^p(E)$ for all $1 \leq p < \infty$ satisfy the second axiom of countability (Proposition 13.2 of Chap. 2).

15.1 $L^\infty(E)$ Is Not Separable

Proposition 15.1 *Let $E \subset \mathbb{R}^N$ be Lebesgue measurable and of positive measure. Then $L^\infty(E)$ is not separable, i.e., it does not contain a dense sequence $\{f_n\}$.*

Proof Denote by $\{B_s(y)\}$ the collection of open balls of radius s centered at y and such that $\mu(E \cap B_s(y)) > 0$. For each $s > 0$ fixed there exists uncountably many r such that

$$\left\| \chi_{E \cap B_s(y)} - \chi_{E \cap B_r(y)} \right\|_{\infty, E} = 1.$$

The collection

$$\left\{ \chi_{E \cap B_s(y)} \mid s > 0, \ y \in E \right\} \subset L^\infty(E)$$

is uncountable and it cannot be separated by a countable sequence $\{f_n\}$ of functions in $L^\infty(E)$. ∎

Corollary 15.1 *$L^\infty(E)$ satisfies the first but not the second axiom of countability.*

16 Selecting Weakly Convergent Subsequences

We continue to assume that E is a Lebesgue measurable subset of $\mathbb{R}^N$ and dx is the Lebesgue measure in $\mathbb{R}^N$. A sequence $\{f_n\}$ of functions in $L^p(E)$ is *bounded* in $L^p(E)$ if there exists a constant K such that $\|f_n\|_p \leq K$ for all n.

Proposition 16.1 *Let $1 < p \leq \infty$. Every sequence $\{f_n\}$ bounded in $L^p(E)$, contains a subsequence $\{f_{n'}\} \subset \{f_n\}$ weakly convergent in $L^p(E)$.*

Proof Let $q \in [1, \infty)$ be the conjugate of $p \in (1, \infty]$. The corresponding $L^q(E)$ is separable and we let $\{g_n\}$ be a countable collection of simple functions dense in $L^q(E)$. For any such simple function g_j set

$$\mathcal{F}_n(g_j) = \int_E g_j f_n \, dx.$$

By Hölder's inequality and the assumption of equiboundedness

$$|\mathcal{F}_n(g_j)| \leq K \|g_j\|_q.$$

Therefore the sequence of numbers $\{\mathcal{F}_n(g_1)\}$ is bounded and we extract a convergent subsequence $\{\mathcal{F}_{n,1}(g_1)\}$. Analogously the sequence $\{\mathcal{F}_{n,1}(g_2)\}$ is bounded and we extract a convergent subsequence $\{\mathcal{F}_{n,2}(g_2)\}$. Proceeding in this fashion, for each $m \in \mathbb{N}$, we may select a sequence $\{\mathcal{F}_{n,m}\}$. such that

$$\lim_{n \to \infty} \mathcal{F}_{n,m}(g_j) \quad \text{exists for all } j = 1, 2, \ldots, m.$$

The diagonal sequence $\mathcal{F}_{n'} = \mathcal{F}_{n,n}$ is such that

$$\lim_{n' \to \infty} \mathcal{F}_{n'}(g_j) \quad \text{exists for all simple functions } g_j \in \{g_n\}.$$

Fix $g \in L^q(E)$ and $\varepsilon > 0$, let $g_j \in \{g_n\}$ be such that $\|g - g_j\|_q < \varepsilon$. Since $\{F_{n'}(g_j)\}$ is a Cauchy sequence, there exists an index n_ε such that

$$|\mathcal{F}_{n'}(g_j) - \mathcal{F}_{m'}(g_j)| \le \varepsilon \quad \text{for all } n', m' \ge n_\varepsilon.$$

From this

$$|\mathcal{F}_{n'}(g) - \mathcal{F}_{m'}(g)| \le |\mathcal{F}_{n'}(g - g_j)| + |\mathcal{F}_{m'}(g - g_j)| + |\mathcal{F}_{n'}(g_j) - \mathcal{F}_{m'}(g_j)|$$
$$\le 2K\|g - g_j\|_q + \varepsilon \le (2K + 1)\varepsilon.$$

Thus for all $g \in L^q(E)$, the sequence $\{\mathcal{F}_{n'}(g)\}$ is a Cauchy sequence and hence convergent. Setting

$$\mathcal{F}(g) = \lim \mathcal{F}_{n'}(g) \quad \text{for all } g \in L^q(E)$$

defines a bounded, linear functional in $L^q(E)$. Since $q \in [1, \infty)$, by the Riesz representation theorem there exists $f \in L^p(E)$ such that

$$\mathcal{F}(g) = \int_E gf dx \quad \text{for all } g \in L^q(E).$$

Therefore

$$\lim \int_E gf_{n'} dx = \int_E gf dx \quad \text{for all } g \in L^q(E). \qquad \blacksquare$$

Remark 16.1 The conclusion of Proposition 16.1 is false for $p = 1$. A counterexample is provided by the sequence in **10.13** of the Complements of Chap. 4.

17 Continuity of the Translation in $L^p(E)$ for $1 \le p < \infty$

Let $E \subset \mathbb{R}^N$ be Lebesgue measurable. For $f \in L^p(E)$ and $h \in \mathbb{R}^N$, let $T_h f$ denote the translated of f, that is

$$T_h f(x) = \begin{cases} f(x+h) & \text{if } x+h \in E \\ 0 & \text{if } x+h \notin E. \end{cases}$$

The following proposition asserts that the translation operation is continuous in the norm topology of $L^p(E)$ for all $p \in [1, \infty)$.

Proposition 17.1 *Let E be a Lebesgue measurable set in $\mathbb{R}^N$ and let $f \in L^p(E)$ for some $1 \leq p < \infty$. For every $\varepsilon > 0$ there exists $\delta = \delta(\varepsilon)$, such that*

$$\sup_{|h| \leq \delta} \|T_h f - f\|_p \leq \varepsilon.$$

Proof Assume first that E is bounded, i.e., contained in a ball B_R centered at the origin and radius R, for some $R > 0$ sufficiently large. Indeed, without loss of generality, we may assume that $E = B_R$, by defining f to be zero outside E. For a subset $\mathcal{E} \subset E$ and a vector $\eta \in \mathbb{R}^N$, set

$$\mathcal{E} - \eta = \{x \in \mathbb{R}^N \mid x + \eta \in \mathcal{E}\}.$$

Having fixed $\varepsilon > 0$, let δ_p be the number claimed by Vitali's theorem on the absolute continuity of the integral, and for which

$$\int_{\mathcal{E}} |f|^p dx \leq \frac{1}{2^{p+1}} \varepsilon^p$$

whenever $\mathcal{E} \subset E$ is measurable and $\mu(\mathcal{E}) \leq \delta_p$. Since the Lebesgue measure is translation invariant, for any vector $\eta \in \mathbb{R}^N$ and any such set $\mathcal{E}$

$$\mu[(\mathcal{E} - \eta) \cap E] \leq \delta_p.$$

Therefore, for any such set $\mathcal{E}$

$$\int_{\mathcal{E}} |T_\eta f - f|^p dx \leq 2^{p-1} \left(\int_{\mathcal{E}} |f|^p dx + \int_{(\mathcal{E}-\eta) \cap E} |f|^p dx \right) \leq \frac{1}{2} \varepsilon^p.$$

Since f is measurable, by Lusin's theorem it is quasi-continuous. Therefore having fixed the positive number

$$\sigma = \frac{\delta_p}{2 + \mu(E)}$$

there exists a closed set $E_\sigma \subset E$ such that $\mu(E - E_\sigma) \leq \sigma$ and f is uniformly continuous in E_σ. In particular, there exists $\delta_\sigma > 0$, such that

$$|T_h f(x) - f(x)|^p \leq \frac{1}{2\mu(E)} \varepsilon^p \qquad \text{for all } |h| < \delta_\sigma$$

provided both, x and $x + h$ belong to E_σ. For any such vector h

$$\int_{E_\sigma \cap (E_\sigma - h)} |T_h f - f|^p dx \le \frac{1}{2} \varepsilon^p.$$

For any vector η of length $|\eta| < \sigma$, estimate[5]

$$\mu[E - (E_\sigma - \eta)] = \mu((E + \eta) - E_\sigma)$$
$$\le \mu(E - E_\sigma) + \mu((E + \eta) - E) \le \sigma + \sigma\mu(E).$$

From this

$$\mu\{E - [E_\sigma \cap (E_\sigma - \eta)]\} \le \mu(E - E_\sigma) + \mu[E - (E_\sigma - \eta)]$$
$$\le \sigma[2 + \mu(E)] = \delta_p.$$

If $|h| \le \delta = \min\{\sigma; \delta_\sigma\}$, estimate

$$\int_E |T_h f - f|^p dx \le \int_{E_\sigma \cap (E_\sigma - h)} |T_h f - f|^p dx$$
$$+ \int_{E - [E_\sigma \cap (E_\sigma - h)]} |T_h f - f|^p dx$$
$$= \int_{E_\sigma \cap (E_\sigma - h)} |T_h f - f|^p dx + \frac{1}{2}\varepsilon^p \le \varepsilon^p.$$

If E is unbounded, having fixed $\varepsilon > 0$, there exists R so large that, for all $|h| < 1$

$$\int_{E \cap \{|x| > 2R\}} |T_h f - f|^p dx \le 2^p \int_{E \cap \{|x| > R\}} |f|^p dx \le \frac{1}{2^p}\varepsilon^p.$$

For such a fixed R, there exists $\delta = \delta(\varepsilon)$, such that

$$\sup_{|h| < \delta} \|T_h f - f\|_{p, E \cap \{|x| < 2R\}} \le \frac{1}{2}\varepsilon.$$

Therefore

$$\sup_{|h| < \delta} \|T_h f - f\|_{p, E} \le \sup_{|h| < \delta} \|T_h f - f\|_{p, E \cap \{|x| \le 2R\}}$$
$$+ 2\|f\|_{p, E \cap \{|x| > R\}} \le \varepsilon. \qquad \blacksquare$$

Remark 17.1 The proposition is false for $p = \infty$. Counterexamples can be constructed as in § 15.1.

---

[5]Since $E = B_R$, we may estimate $\mu[(E + \eta) - E] \le \sigma\mu(B_R)$.

17.1   Continuity of the Convolution

Corollary 17.1 *Let $f \in L^p(E)$ and $g \in L^q(E)$ where p and q are Hólder conjugate and $1 \leq p < \infty$. Regard f and g as defined in the whole $\mathbb{R}^N$ by extending them to be zero outside E. Then the convolution $\mathbb{R}^N \ni x \to (f * g)(x)$ is a continuous function of x.*

18   Approximating Functions in $L^p(E)$ with Functions in $C^\infty(E)$

Let f be a real valued function defined in $\mathbb{R}^N$. The support of f is the closure of the set $[|f| > 0]$ and we write

$$\text{supp } \{f\} = \overline{[|f| > 0]}.$$

Let E be an open subset of $\mathbb{R}^N$. We regard a real valued function f defined in E, as defined in the whole $\mathbb{R}^N$, by extending it to be zero in $\mathbb{R}^N - E$. A function $f : E \to \mathbb{R}$ is of *compact support* in E if supp $\{f\}$ is compact and contained in E. Set

$$C^\infty(E) = \left\{ \begin{array}{c} \text{the collection of all infinitely differentiable} \\ \text{functions } f : E \to \mathbb{R} \end{array} \right\}$$

$$C_o^\infty(E) = \left\{ \begin{array}{c} \text{the collection of all infinitely differentiable} \\ \text{functions } f : E \to \mathbb{R} \text{ of compact support in } E \end{array} \right\}.$$

A function $f \in L^p(E)$ for $1 \leq p < \infty$, can approximated in its norm topology, by functions in $C^\infty(E)$, by means of the Friedrichs *mollifying* kernels [51]

$$J(x) = \begin{cases} k \exp \left\{ \dfrac{-1}{1 - |x|^2} \right\} & \text{if } |x| < 1 \\ 0 & \text{if } |x| \geq 1 \end{cases}$$

where k is a positive constant chosen so that $\|J(x)\|_{1,\mathbb{R}^N} = 1$. For $\varepsilon > 0$ let

$$J_\varepsilon(x) = \frac{1}{\varepsilon^N} J\left(\frac{x}{\varepsilon}\right).$$

The kernels J and J_ε are in $C_o^\infty(\mathbb{R}^N)$. Moreover, for all $\varepsilon > 0$

$$\int_{\mathbb{R}^N} J_\varepsilon(x)dx = 1 \quad \text{and} \quad J_\varepsilon(x) = 0 \quad \text{for } |x| \geq \varepsilon.$$

The convolution of the mollifying kernels J_ε with a function $f \in L^1(E)$ is

$$x \to (J_\varepsilon * f)(x) = \int_{\mathbb{R}^N} J_\varepsilon(x - y)f(y)dy. \tag{18.1}$$

Since $J_\varepsilon \in C_o^\infty(\mathbb{R}^N)$, the convolution $(J_\varepsilon * f)$ is in $\in C^\infty(\mathbb{R}^N)$. In this sense $(J_\varepsilon * f)$ is a regularization or mollification of f.

For fixed $x \in \mathbb{R}^N$, the domain of integration in (18.1) is the ball centered at x and radius ε. If $f \in L^p(E)$ and $x \in \partial E$, then (18.1) is well defined since f is extended to be zero outside E. If the support of f is contained in E and has positive distance from ∂E, then

$$(J_\varepsilon * f) \in C_o^\infty(E) \quad \text{provided } \varepsilon < \text{dist}\{\text{supp}\{f\}; \partial E\}.$$

Proposition 18.1 *Let $f \in L^p(E)$ for some $1 \le p < \infty$. Then*

$$(J_\varepsilon * f) \in L^p(E) \quad \text{and} \quad \|(J_\varepsilon * f)\|_p \le \|f\|_p$$

and

$$\lim_{\varepsilon \to 0} \|(J_\varepsilon * f) - f\|_p = 0. \tag{18.2}$$

If $f \in C(E)$, then for every compact subset $\mathcal{K} \subset E$

$$\lim_{\varepsilon \to 0}(J_\varepsilon * f)(x) = f(x) \quad \text{uniformly in } \mathcal{K}. \tag{18.3}$$

Proof By Hölder's inequality

$$
\begin{aligned}
|(J_\varepsilon * f)(x)| &= \left| \int_{\mathbb{R}^N} J_\varepsilon(x - y)f(y)dy \right| \\
&\le \left(\int_{\mathbb{R}^N} J_\varepsilon(x - y)dy \right)^{1/q} \left(\int_{\mathbb{R}^N} J_\varepsilon(x - y)|f|^p(y)dy \right)^{1/p} \\
&= \left(\int_{\mathbb{R}^N} J_\varepsilon(x - y)|f|^p(y)dy \right)^{1/p}.
\end{aligned}
$$

Taking the p-power of both sides and integrating over $\mathbb{R}^N$ with respect to the x-variables, gives

$$\int_{\mathbb{R}^N} |J_\varepsilon * f|^p dx \le \int_{\mathbb{R}^N} |f|^p dy.$$

To establish (18.2) write

$$|(J_\varepsilon * f)(x) - f(x)| = \left| \int_{\mathbb{R}^N} J_\varepsilon(x - y)[f(y) - f(x)]dy \right|$$

$$\leq \int_{|\eta| < \varepsilon} J_\varepsilon(\eta)|f(x + \eta) - f(x)|d\eta$$

$$\leq \left(\int_{\mathbb{R}^N} J_\varepsilon(\eta)d\eta \right)^{1/q} \left(\int_{\mathbb{R}^N} J_\varepsilon(\eta)|f(x + \eta) - f(x)|^p d\eta \right)^{1/p}.$$

Taking the p-power and integrating $\mathbb{R}^N$ with respect to the x-variables, gives

$$\|(J_\varepsilon * f) - f\|_{p,\mathbb{R}^N} \leq \sup_{|\eta| < \varepsilon} \|T_\eta f - f\|_{p,\mathbb{R}^N}.$$

Now (18.2) follows from Proposition 17.1. To prove (18.3) fix a compact subset $\mathcal{K} \subset E$ and, for $x \in \mathcal{K}$, write

$$|(J_\varepsilon * f)(x) - f(x)| = \left| \int_{\mathbb{R}^N} J_\varepsilon(x - y)[f(y) - f(x)]dy \right|$$

$$\leq \sup_{|x-y| < \varepsilon} |f(x) - f(y)|. \qquad \blacksquare$$

Proposition 18.2 *Let E be an open subset of $\mathbb{R}^N$. Then $C_o^\infty(E)$ is dense in $L^p(E)$ for $p \in [1, \infty)$.*

Proof Having fixed $\varepsilon > 0$, let K_ε be a compact subset of E such that

$$\int_{E-K_\varepsilon} |f|^p dy \leq \frac{1}{2^p} \varepsilon^p.$$

Let $2\delta_\varepsilon = \text{dist } \{K_\varepsilon; \partial E\}$ and set

$$f_\varepsilon = \begin{cases} f(x) & \text{for } x \in K_\varepsilon \\ 0 & \text{for } x \in E - K_\varepsilon. \end{cases}$$

If $\delta < \delta_\varepsilon$ the functions $f_\delta = f_\varepsilon * J_\delta$ have compact support in E. By Proposition 18.1 there exists δ_ε' such that

$$\|f_\varepsilon - f_\delta\|_p \leq \frac{1}{2}\varepsilon \quad \text{for all } \delta < \delta_\varepsilon'.$$

Therefore for all such δ

$$\|f - f_\delta\|_p \leq \|f_\varepsilon - f_\delta\|_p + \|f\|_{p,E-K_\varepsilon} < \varepsilon. \qquad \blacksquare$$

19 Characterizing Pre-compact Sets in $L^p(E)$

Let $E \subset \mathbb{R}^N$ be open and let dx denote the Lebesgue measure in $\mathbb{R}^N$. Pre-compact subsets of $L^p(E)$ for $1 \leq p < \infty$ are characterized in terms of total boundedness (Proposition 17.3 of Chap. 2). For $\delta > 0$ set

$$E_\delta = \left\{ x \in E \mid \text{dist} \{x; \partial E\} > \delta \text{ and } |x| < \frac{1}{\delta} \right\}$$

and assume that δ is so small that E_δ is not empty.

Theorem 19.1 (Kolmogorov [84] and Riesz [132]) *A bounded subset K of $L^p(E)$ for $1 \leq p < \infty$, is pre-compact in $L^p(E)$ if and only if for every $\varepsilon > 0$ there exists $\delta > 0$ such that for all vectors $h \in \mathbb{R}^N$ of length $|h| < \delta$ and for all $u \in K$*

$$\|T_h u - u\|_p < \varepsilon \quad \text{and} \quad \|u\|_{p.E-E_\delta} < \varepsilon. \tag{19.1}$$

Proof (Sufficient Condition) For all $\nu > 0$ we will construct a ν-net for K. For $\varepsilon > 0$, choose δ so that (19.1) is satisfied and let J_η be the η-mollifying kernel, for $\eta \leq \delta$. For a.e. $x \in E_\delta$

$$|(J_\eta * u)(x) - u(x)| = \left| \int_{|y| < \eta} J_\eta(y)[u(x + y) - u(x)]dy \right|$$

$$\leq \int_{|y| < \eta} J_\eta(y)|(T_{-y}u - u)(x)|dy.$$

From this and the condition of the theorem

$$\|J_\eta * u - u\|_{p.E_\delta} \leq \sup_{|h| < \eta} \|T_h u - u\|_p \leq \varepsilon.$$

The family $\{J_\eta * u \mid u \in K\}$ for a fixed $\eta \in (0, \delta)$, is equibounded and equicontinuous in $\bar{E}_\delta$. Indeed

$$|(J_\eta * u)(x)| \leq \|J_\eta\|_q \|u\|_p \quad \text{for all } x \in \bar{E}_\delta.$$

Moreover, for every $\xi \in \mathbb{R}^N$

$$|J_\eta * u(x + \xi) - J_\eta * u(x)| \leq \|J_\eta\|_q \|T_\xi u - u\|_p.$$

Therefore having fixed an arbitrary $\sigma > 0$, there exists $\tau > 0$, such that for all vectors $|\xi| < \tau$

$$|J_\eta * u(x + \xi) - J_\eta * u(x)| \leq \|J_\eta\|_q \sigma \quad \text{for all } u \in K.$$

Thus by the Ascoli–Arzelá theorem, the set $\{J_\eta * u \mid u \in \mathcal{K}\}$ is pre-compact in $C(\bar{E}_\delta)$ and for all fixed $\epsilon > 0$, it admits a finite ϵ-net $\{\varphi_1, \ldots, \varphi_m\}$ of functions in $C(E_\delta)$ (§ 19.1 of Chap. 5). Precisely, for every $u \in K$, there exists some φ_j such that

$$|\varphi_j(x) - (J_\eta * u)(x)| < \epsilon \qquad \text{for all } x \in \bar{E}_\delta.$$

Define

$$\bar{\varphi}_j = \begin{cases} \varphi_j(x) & \text{for } x \in \bar{E}_\delta \\ 0 & \text{otherwise.} \end{cases}$$

Having fixed $\nu > 0$, the numbers ϵ, δ, and ε, can be chosen so that the collection $\{\bar{\varphi}_1, \ldots, \bar{\varphi}_m\}$ is a finite ν-net for K in $L^p(E)$. Indeed, if $u \in K$

$$\|u - \bar{\varphi}_j\|_p \leq \|u\|_{p,E-E_\delta} + \|J_\eta * u - u\|_{p,E_\delta} + \|J_\eta * u - \varphi_j\|_{p,E_\delta}.$$

The proof is concluded by choosing ϵ, δ, and ε so that the right hand side does not exceed ν. ∎

Proof (Necessary Condition) The existence of ε-nets for all $\varepsilon > 0$ implies that for all $\varepsilon \in (0, 1)$ there exists $\delta > 0$ such that

$$\|u\|_{p,E-E_\delta} \leq \varepsilon \qquad \text{for all } u \in K.$$

To prove this, let $\{\varphi_1, \ldots, \varphi_n\}$ be a finite $\frac{1}{2}\varepsilon$-net. Then for every $u \in K$ there exists some φ_j such that for all $\delta > 0$

$$\|u\|_{p,E-E_\delta} \leq \|\varphi_j\|_{p,E-E_\delta} + \frac{1}{2}\varepsilon.$$

Now we may choose δ so that

$$\|\varphi_j\|_{p,E-E_\delta} \leq \frac{1}{2}\varepsilon \qquad \text{for all } j = 1, 2, \ldots, n.$$

Fix $\varepsilon > 0$ and $\delta > 0$ so small that

$$\|u\|_p \leq \|u\|_{p,E_\delta} + \frac{1}{2}\varepsilon \quad \text{for all } u \in K.$$

Next, for all $u \in K$

$$\|T_h u - u\|_p \leq 2\varepsilon + \|T_h u - u\|_{p,E_\delta}$$

and

$$\begin{aligned} \|T_h u - u\|_{p,E_\delta} &\leq \|T_h u - T_h \varphi_j\|_{p,E_\delta} + \|T_h \varphi_j - \varphi_j\|_{p,E_\delta} + \|u - \varphi_j\|_{p,E_\delta} \\ &\leq 2\varepsilon + \|T_h \varphi_j - \varphi_j\|_{p,E_\delta}. \end{aligned}$$ ∎

Problems and Complements

1c Functions in $L^p(E)$ and Their Norm

1.1c The Spaces L^p for $0 < p < 1$

A measurable function $f : E \to \mathbb{R}^*$ is in $L^p(E)$ for $0 < p < 1$ if $|f|^p \in L^1(E)$. A norm-like function $f \to \|f\|_p$ might be defined as in (1.1). The collection $L^p(E)$ for $0 < p < 1$ is a linear space. This is a consequence of the following lemma.

Lemma 1.1c *Let $0 < p < 1$. Then for nonnegative x, y*

$$(x + y)^p \le x^p + y^p.$$

Proof If either x or y is zero, the inequality is trivial. Otherwise, letting $t = x/y$, the inequality is equivalent to

$$f(t) = (1 + t)^p - (1 + t^p) \le 0 \quad \text{for } t \ge 0 \quad \text{and } 0 < p < 1 \qquad \blacksquare$$

1.2c The Spaces L^q for $q < 0$

If $0 < p < 1$, its Hölder conjugate q is negative. A measurable function $f : E \to \mathbb{R}^*$ is in $L^q(E)$ for $q < 0$ if

$$0 < \int_E |f|^q d\mu = \int_E \frac{1}{|f|^{\frac{p}{1-p}}} d\mu < \infty, \qquad \frac{1}{p} + \frac{1}{q} = 1.$$

A norm-like function $f \to \|f\|_q$ might be defined as in (1.1). If $f \in L^q(E)$ for $q < 0$, then $f \ne 0$ a.e. in E and $|f| \ne \infty$. If $q < 0$ the set $L^q(E)$ is not a linear space.

1.3c The Spaces ℓ_p for $1 \le P \le \infty$

Let $\mathbf{a} = \{a_n\}$ be a sequence of real numbers and set

$$\begin{aligned}
\|\mathbf{a}\|_p &= \left(\sum |a_n|^p \right)^{1/p} \quad \text{if } 1 \le p < \infty \\
\|\mathbf{a}\|_\infty &= \sup |a_n| \qquad\quad \text{if } p = \infty.
\end{aligned} \tag{1.1c}$$

Denote by ℓ_p the set of all sequences $\mathbf{a}$ such that $\|\mathbf{a}\|_p < \infty$. One verifies that ℓ_p is a linear space for all $1 \le p \le \infty$ and that (1.1c) is a norm. Moreover ℓ_∞ satisfies analogues of (1.2) and (1.4).

2c The Inequalities of Hölder and Minkowski

Corollary 2.1c *Let $p, q \in (1, \infty)$ be conjugate. For $a, b \in \mathbb{R}$ and $\varepsilon > 0$*

$$|ab| \le \frac{\varepsilon^p}{p}|a|^p + \frac{1}{\varepsilon^q q}|b|^q.$$

Corollary 2.2c *Let $\alpha_i \in (0, 1)$ for $i = 1, \dots, n$, and $\sum_{i=1}^{n} \alpha_i = 1$. Then for any n-tuple of real numbers $\xi_1, \dots, \xi_n$*

$$\prod_{i=1}^{n} |\xi_i|^{\alpha_i} \le \sum_{i=1}^{n} \alpha_i |\xi_i|.$$

State and prove a variant of these corollaries when p is permitted to be one, or some of the α_i is permitted to be one.

2.1c Variants of the Hölder and Minkowski Inequalities

Corollary 2.3c *Let $f_i \in L^{p_i}(E)$, for $1 \le p_i \le \infty$ and $i = 1, \dots, n$. Then*

$$\int_E \prod_{i=1}^{n} |f_i| d\mu \le \prod_{i=1}^{n} \|f_i\|_{p_i} \quad \text{whenever} \quad \sum_{i=1}^{n} \frac{1}{p_i} = 1.$$

Prove the following convolution inequality.

Corollary 2.4c *Let $f \in L^p(\mathbb{R}^N)$, $g \in L^q(\mathbb{R}^N)$ and $h \in L^r(\mathbb{R}^N)$, where $p, q, r \ge 1$ satisfy*

$$\frac{1}{p} + \frac{1}{q} + \frac{1}{r} = 2.$$

Then

$$\int_{\mathbb{R}^N} f(g * h) dx \le \|f\|_p \|g\|_q \|h\|_r. \tag{2.1c}$$

Proof Write

$$f(x)g(x - y)h(y) = [f^p(x)g^q(x - y)]^{\frac{1}{r'}}[g^q(x - y)h^r(y)]^{\frac{1}{p'}}[f^p(x)h^r(y)]^{\frac{1}{q'}}.$$

where p', q' and r' are the Hölder conjugate of p, q and r. ∎

Corollary 2.5c *Let $1 \le p, q \le \infty$ be conjugate. Then for $\mathbf{a} \in \ell_p$ and $\mathbf{b} \in \ell_q$*

$$|\mathbf{a} \cdot \mathbf{b}| = \sum |a_i b_i| \le \|\mathbf{a}\|_p \|\mathbf{b}\|_q.$$

For **a**, **b** $\in \ell_p$ *and* $1 \le p \le \infty$ *[109]*

$$\|\mathbf{a} + \mathbf{b}\|_p \le \|\mathbf{a}\|_p + \|\mathbf{b}\|_p.$$

2.2c   Some Auxiliary Inequalities

Lemma 2.1c *Let x and y be any two positive numbers. Then for $1 \le p < \infty$*

$$|x - y|^p \le |x^p - y^p|; \tag{2.2c}$$

$$(x + y)^p \ge x^p + y^p;$$

$$(x + y)^p \le 2^{p-1}(x^p + y^p). \tag{2.3c}$$

Proof Assuming $x \ge y$ and $p > 1$

$$x^p - y^p = \int_0^1 \frac{d}{ds}[sx + (1-s)y]^p ds = p(x-y)\int_0^1 [sx + (1-s)y]^{p-1} ds$$

$$= p(x-y)\int_0^1 [s(x-y) + y]^{p-1} ds$$

$$\ge p(x-y)\int_0^1 s^{p-1}(x-y)^{p-1} ds = (x-y)^p.$$

This proves the first of (2.2c). The second follows from the first, since

$$(x + y)^p - y^p \ge (x + y - y)^p.$$

To prove (2.3c) we may assume that both x and y are nonzero and that $p > 1$. Setting $t = x/y$, the inequality is equivalent to

$$f(t) = \frac{(1+t)^p}{1+t^p} \le 2^{p-1} \quad \text{for all } t > 0. \qquad \blacksquare$$

2.3c   An Application to Convolution Integrals

Proposition 2.1c *Let $f \in L^p(\mathbb{R}^N)$ and $g \in L^q(\mathbb{R}^N)$ for some $1 \le p \le \infty$, with p and q Hölder conjugate. Then*

$$f * g \in L^\infty(\mathbb{R}^N) \quad \text{and} \quad \|f * g\|_\infty \le \|f\|_p \|g\|_q.$$

Proposition 2.2c *Let* $f \in L^1(\mathbb{R}^N)$ *and* $g \in L^p(\mathbb{R}^N)$ *for some* $1 \leq p \leq \infty$. *Then*

$$f * g \in L^p(\mathbb{R}^N) \quad \text{and} \quad \|f * g\|_p \leq \|f\|_1 \|g\|_p.$$

Hint: Assume $\|f\|_1 = 1$ and observe that $|t|^p$ is a convex function of t in $\mathbb{R}$.

2.4c   The Reverse Hölder and Minkowski Inequalities

Proposition 2.3c (Reverse Hölder Inequality) *Let* $p \in (0, 1)$ *and* $q < 0$ *satisfy* (2.1). *Then for every* $f \in L^p(E)$ *and* $g \in L^q(E)$

$$\int_E |fg| d\mu \geq \|f\|_p \|g\|_q, \qquad \frac{1}{p} + \frac{1}{q} = 1.$$

Proof We may assume that $fg \in L^1(E)$. Apply Hölder's inequality with

$$\overline{p} = \frac{1}{p} > 1 \qquad \overline{q} = \frac{1}{1-p} > 1 \qquad \frac{1}{\overline{p}} + \frac{1}{\overline{q}} = 1. \qquad \blacksquare$$

Proposition 2.4c (Reverse Minkowski Inequality) *Let* $p \in (0, 1)$. *Then for all* f, $g \in$ $L^p(E)$

$$\| |f| + |g| \|_p \geq \|f\|_p + \|g\|_p.$$

2.5c $L^p(E)$-Norms and Their Reciprocals

Proposition 2.5c *Let* $E \subset \mathbb{R}^N$ *be measurable with* $\mu(E) = 1$. *Then for every nonnegative measurable function* f *defined in* E, *and all* $p > 0$

$$\|f\|_1 \left\| \frac{1}{f} \right\|_p \geq 1. \tag{2.4c}$$

Proof Apply inequality (14.4c) of Chap. 5. $\blacksquare$

3c   More on the Spaces L^p and Their Norm

3.1.   Let $\{f_n\}$ be a sequence in $L^p(E)$ for some $p \geq 1$. Then

$$\left(\int_E \left|\sum f_n\right|^p d\mu\right)^{\frac{1}{p}} \le \sum \|f_n\|_p.$$

$$\left(\sum \left|\int_E f_n d\mu\right|^p\right)^{\frac{1}{p}} \le \int_E \left(\sum |f_n|^p\right)^{\frac{1}{p}} d\mu.$$

3.2. The spaces ℓ_p satisfy analogues of Propositions 3.1 and 3.2.

3.3. Let $f \in L^p(E)$ for all $1 \le p < \infty$. Assume that $\|f\|_p \le K$ for all $1 \le p < \infty$, for some constant K, and that $\mu(E) < \infty$. Then $f \in L^\infty(E)$ and $\|f\|_\infty \le K$.

3.4. Give an example of $f \in L^p(E)$ for all $p \in [1, \infty)$, and $f \notin L^\infty(E)$.

3.5. Let $E \subset$ be open set. Give an example of $f \in L^p(E)$ for all $p \in [1, \infty)$, and $f \notin L^\infty(E')$ for any open subset $E' \subset E$. **Hint:** Properly modify the function in **17.9.** of the Complements of Chap. 4.

3.4c   A Metric Topology for $L^p(E)$ when $0 < p < 1$

If $0 < p < 1$, the norm-like function $f \to \|f\|_p$ defined as in (1.1), is not a norm. Indeed (4.3) is violated in view of the reverse Minkowski inequality. By the same token $(f, g) \to \|f - g\|_p$ is not a metric in $L^p(E)$. A topology could be generated by the balls

$$B_\rho(g) = \left\{f \in L^p(E) \mid \|f - g\|_p < \rho\right\}, \qquad p \in (0, 1).$$

where $g \in L^p(E)$ is fixed. By the reverse Minkowski inequality these balls are not convex. A distance in $L^p(E)$ for $0 < p < 1$ is introduced by setting

$$d(f, g) = \|f - g\|_p^p = \int_E |f - g|^p d\mu. \tag{3.1c}$$

One verifies that $d(\cdot, \cdot)$ satisfies the requirements (i)–(iii) of § 13 of Chap. 2, to be a metric. The triangle inequality (iv) follows from Lemma 1.1c.

3.5c   Open Convex Subsets of $L^p(E)$ for $0 < p < 1$

Let $\{X, \mathcal{A}, \mu\}$ be $\mathbb{R}^N$ with the Lebesgue measure.

Proposition 3.1c (Day [31]) *Let $L^p(E)$ for $0 < p < 1$ be equipped with the topology generated by the metric in (3.1c). Then the only open, convex subsets of $L^p(E)$ are $\emptyset$ and $L^p(E)$ itself.*

Proof Let $\mathcal{O}$ be a nonempty, open, convex neighborhood of the origin of $L^p(E)$ and let f be an arbitrary element of $L^p(E)$. Since $\mathcal{O}$ is open it contains some ball B_ρ centered at the origin. Let n be a positive integer such that

$$\frac{1}{n^{1-p}} \int_E |f|^p dx \leq \rho \qquad \text{i.e.,} \quad nf \in B_{n\rho}.$$

Partition E into exactly n disjoint, measurable subsets $\{E_1, \ldots, E_n\}$ such that

$$\int_{E_j} |f|^p dx = \frac{1}{n} \int_E |f|^p dx \quad j = 1, \ldots, n.$$

Such a partition can be carried out in view of the absolute continuity of the Lebesgue integral. Set

$$h_j = \begin{cases} nf & \text{in } E_j \\ 0 & \text{in } E - E_j \end{cases}$$

and compute

$$\int_E |h_j|^p dx = n^p \int_{E_j} |f|^p dx \leq \rho.$$

Thus $h_j \in B_\rho \subset \mathcal{O}$ for all $j = 1, \ldots, n$. Since $\mathcal{O}$ is convex

$$f = \frac{h_1 + h_2 + \cdots + h_n}{n} \in \mathcal{O}.$$

Since $f \in L^p(E)$ is arbitrary $\mathcal{O} \equiv L^p(E)$.   ∎

Corollary 3.1c *The topology generated by the metric (3.1c) in $L^p(E)$ for $0 < p < 1$ is not locally convex.*

5c   Convergence in $L^p(E)$ and Completeness

5.1.   A sequence $\{f_n\} \subset L^\infty(E)$ converges in $L^\infty(E)$ to some f, if and only if there exists a set $\mathcal{E} \subset E$ of measure zero, such that $\{f_n\} \to f$ uniformly in $E - \mathcal{E}$.

5.2.   A sequence $\{f_n\} \subset L^p(E)$ converges to some f in $L^p(E)$ if and only if every subsequence $\{f_{n'}\} \subset \{f_n\}$, contains in turn a subsequence $\{f_{n''}\}$ converging to f in $L^p(E)$.

5.3. ℓ_p is complete for all $1 \leq p \leq \infty$.

5.4.   Let $\{f_n\}$ be a sequence of functions in $L^p(E)$ for $p \in (1, \infty)$, converging a.e. in E to a function $f \in L^p(E)$. Then $\{f_n\}$ converges to f in $L^p(E)$ if and only if $\lim \|f_n\|_p = \|f\|_p$.

5.5.   Let $L^p(E)$ for $0 < p < 1$ be endowed with the metric topology generated by the metric in (3.1c). With respect to such a topology $L^p(E)$ is a complete metric space. In particular $L^p(E)$ is of second category.

5.1c   The Measure Space $\{X, \mathcal{A}, \mu\}$ and the Metric Space $\{\mathcal{A}; d\}$

Let $\{X, \mathcal{A}, \mu\}$ be a measure space and let $d(E; F)$ be the distance of any two measurable sets E and F, as introduced in **3.7** of the Complements of Chap. 3. One verifies that

$$d(E; F) = \int_X |\chi_E - \chi_F| d\mu.$$

Two measurable sets E and E' are equivalent if $d(E; E') = 0$. This identifies equivalence classes in $\mathcal{A}$. Continue to denote by $\mathcal{A}$ the collection of such equivalence classes. By the Riesz-Fisher theorem, $L^1(X)$ is complete (Theorem 5.1). Therefore $\{\mathcal{A}; d\}$ is a complete metric space and, as such, is of second category.

5.1.1c   Continuous Functions on $\{\mathcal{A}; d\}$

Lemma 5.1c *Let $\{X, \mathcal{A}, \mu\}$ be a measure space, and let λ be a signed measure on $\mathcal{A}$, absolutely continuous with respect to μ, and of finite total variation $|\lambda|$. Then the function $\lambda : \{\mathcal{A}; d\} \to \mathbb{R}$, is continuous with respect to the metric topology of $\{\mathcal{A}; d\}$.*

Proof Let $\{E_n\}$ be a sequence of equivalence classes in $\mathcal{A}$, converging to $E \in \mathcal{A}$, in the sense that $d(E; E_n) \to 0$. Equivalently,

$$\lim \mu(E - E \cap E_n) = \lim \mu(E_n - E \cap E_n) = 0.$$

Since $|\lambda|$ is finite and $\lambda \ll \mu$, by *(c)* and *(d)* of Chap. 3, this implies

$$\lim |\lambda|(E - E \cap E_n) = \lim |\lambda|(E_n - E \cap E_n) = 0.$$

Hence $\lim \lambda(E_n) = \lambda(E)$.   ∎

6c   Separating $L^p(E)$ by Simple Functions

For a measure space $\{X, \mathcal{A}, \mu\}$ and $E \in \mathcal{A}$, Proposition 6.1 asserts that the simple functions, is dense in $L^p(E)$ for all $1 \leq p \leq \infty$.

When $E \subset \mathbb{R}^N$ is Lebesgue measurable, the simple functions dense in $L^p(E)$ can be given specific forms, provided $1 \leq p < \infty$. Let $\mathcal{S}_o$ denote the collection of simple functions of the form

$$\varphi_o = \sum_{j=1}^{n} \varphi_j \chi_{E_{j,o}} \tag{6.1c}$$

where $\varphi_j \in \mathbb{R}$ and $E_{j,o}$ are open for $j = 1, \ldots, n$.

Proposition 6.1c *Let $E \subset \mathbb{R}^N$ be Lebesgue measurable. Then $\mathcal{S}_o$ is dense in $L^p(E)$ for $1 \le p < \infty$.*

Proof Fix $f \in L^p(E)$ and $\varepsilon > 0$. There exists a simple function

$$\varphi = \sum_{j=1}^{n} \varphi_j \chi_{E_j} \quad \text{such that} \quad \|f - \varphi\|_{p,E} < \tfrac{1}{2}\varepsilon.$$

Each E_j is of finite measure. Therefore, by Proposition 12.2 of Chap. 3, there exist open sets $E_{j,o} \supset E_j$ such that

$$\mu(E_{j,o} - E_j) < \frac{1}{(2n)^p \max_{j \in \{1,\ldots,n\}} |\varphi_j|^p} \, \varepsilon^p. \qquad \blacksquare$$

If $N = 1$, the sets $E_{j,o}$ can be taken to be open intervals (a_j, b_j).

Corollary 6.1c *Let $E \subset \mathbb{R}$ be Lebesgue measurable. Then the collection of simple functions of the form*

$$\varphi = \sum_{j=1}^{n} \varphi_j \chi_{(a_j, b_j)} \qquad (6.2c)$$

is dense in $L^p(E)$ for $1 \le p < \infty$.

Proof Fix $f \in L^p(E)$ and $\varepsilon > 0$, and let φ_o of the form (6.1c) be such that

$$\|f - \varphi\|_p < \frac{1}{2}\varepsilon.$$

Each $E_{j,o}$ is the countable union of disjoint open intervals $\{(a_{j,i}, b_{j,i})\}_{i \in \mathbb{N}}$. Since each $E_{j,o}$ is of finite measure, for each $j \in \{1, \ldots, n\}$, there exists an index m_j, such that

$$\sum_{i=m_{j+1}}^{\infty} (b_{j,i} - a_{j,i}) \le \frac{1}{(2n)^p \max_{j \in \{1,\ldots,n\}} |\varphi_j|^p} \, \varepsilon^p. \qquad \blacksquare$$

7c   Weak Convergence in $L^p(E)$

Throughout this section $\{X, \mathcal{A}, \mu\}$ is $\mathbb{R}^N$ with the Lebesgue measure and $E \subset \mathbb{R}^N$ is Lebesgue measurable and of finite measure.

7.1. Let $\{f_n\}$ be a sequence of functions in $L^p(E)$ converging weakly in $L^p(E)$ to some $f \in L^p(E)$, and converging a.e. in E to some g. Then $f = g$, a.e. in E.

7.2. Let $\{f_n\} \to f$ weakly in $L^p(E)$ and $\{f_n\} \to g$ in measure. Then $f = g$, a.e. in E.

7.3c   *Comparing the Various Notions of Convergence*

Denote by {a.e.} the set of all sequences $\{f_n\}$ of functions from E into $\mathbb{R}^*$, convergent a.e. in E. Analogously denote by {meas}, $\{L^p(E)\}$ and $\{\text{w-}L^p(E)\}$ the set of all sequences $\{f_n\}$ of measurable functions defined in E and convergent respectively, in measure, strongly and weakly in $L^p(E)$.

By the remarks in § 7 and the counterexample of § 7.1

$$\{L^p(E)\} \subset \{\text{w-}L^p(E)\} \quad \text{and} \quad \{\text{w-}L^p(E)\} \not\subset \{L^p(E)\}.$$

By Proposition 4.2 of Chap. 4 and the counterexample (4.1)

$$\{\text{a.e.}\} \subset \{\text{meas}\} \quad \text{and} \quad \{\text{meas}\} \not\subset \{\text{a.e.}\}.$$

Weak convergence in $L^p(E)$ does not imply convergence in measure, nor does convergence in measure imply weak convergence in $L^p(E)$, i.e.,

$$\{\text{w-}L^p(E)\} \not\subset \{\text{meas}\} \quad \text{and} \quad \{\text{meas}\} \not\subset \{\text{w-}L^p(E)\}.$$

The sequence $\{\cos nx\}$ for $x \in [0, 2\pi]$, converges weakly to zero, but not in measure. Indeed

$$\mu\left(x \in [0, 2\pi] \mid |\cos nx| \geq \tfrac{1}{2}\right) = \tfrac{4}{3}\pi.$$

This proves the first statement. For the second consider the sequence

$$f_n(x) = \begin{cases} n & \text{for } x \in [0, \tfrac{1}{n}] \\ 0 & \text{for } x \in (\tfrac{1}{n}, 1]. \end{cases}$$

Such a sequence converges to zero in measure and not weakly in $L^p[0, 1]$. Almost everywhere convergence does not imply weak convergence in $L^p(E)$. Strong convergence in $L^p(E)$ does not imply a.e. convergence, i.e.,

$$\{\text{a.e.}\} \not\subset \{\text{w-}L^p(E)\} \quad \text{and} \quad \{L^p(E)\} \not\subset \{\text{a.e.}\}.$$

Indeed the sequence $\{f_n\}$ above, converges a.e. to zero and does not converge to zero weakly in $L^p(E)$. For the second statement consider the sequence $\{\varphi_{nm}\}$ introduced in (4.1) of Chap. 4.

The relationships between these notions of convergence can be organized in the diagram below in form of inclusion of sets where each square represents convergent sequences (Fig. 1c).[6]

---

[6]This discussion on the various notions of convergence and the picture have been taken from the 1974 lectures on Real Analysis by C. Pucci at the Univ. of Florence, Italy.

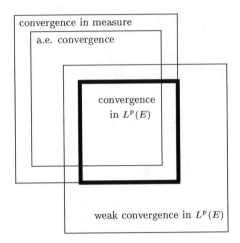

Fig. 1c Various notions of convergence

7.4.   Exhibit a sequence $\{f_n\}$ of measurable functions in $[0, 1]$ convergent to zero in measure and weakly in $L^2[0, 1]$, but not convergent almost everywhere in $[0, 1]$.

7.5c   *Weak Convergence in ℓ_p*

Let $1 \leq p, q \leq \infty$ be conjugate. Let $\{\mathbf{a}_n\}$ be a sequence of elements in ℓ_p and let $\mathbf{b} \in \ell_q$. The sequence $\{\mathbf{a}_n\}$ converges weakly in ℓ_p to some $\mathbf{a} \in \ell_p$ if

$$\lim \sum a_{j,n} b_j = \sum a_j b_j \qquad \text{for all } \mathbf{b} \in \ell_q.$$

Strong convergence implies weak convergence. For $1 < p \leq \infty$, the converse is false. For example, let

$$a_{j,n} = \begin{cases} n^{-\frac{1}{p}} & \text{for } 1 \leq j \leq n \\ 0 & \text{for } j > n \end{cases} \qquad \text{for } 1 < p < \infty;$$

$$b_{j,n} = \begin{cases} 0 \text{ for } 1 \leq j \leq n \\ 1 \text{ for } j > n \end{cases} \qquad \text{for } p = \infty.$$

Then $\|\mathbf{a}_n\|_p = 1$ for all n and $\{\mathbf{a}_n\} \to 0$ weakly in ℓ_p. Likewise $\|\mathbf{b}_n\|_\infty = 1$ for all n and $\{\mathbf{b}_n\} \to 0$ weakly in ℓ_∞. Discuss and compare the various notions of convergence in ℓ_p, for $1 < p \leq \infty$, and construct examples. For $p = 1$ weak and strong convergence are equivalent (Corollary 22.1c and § 23c).

9c Weak Convergence and Norm Convergence

When $p = 2$, the proof of Proposition 9.1 is particularly simple and elegant. Indeed

$$\|f_n - f\|_2^2 = \|f_n\|_2^2 + \|f\|_2^2 - 2\int_E f_n f d\mu.$$

9.1c Proof of Lemmas 9.1 and 9.2

For $t \neq 0$ and $x = t^{-1}$, consider the function

$$f(t) = \frac{|1 + t|^p - 1 - pt}{|t|^p} = |1 + x|^p - |x|^p - p|x|^{p-2}x = \varphi(x).$$

It suffices to prove that $\varphi(x) \geq c$ for all $x \in \mathbb{R}$ for some $c > 0$. If $x \in (-1, 0]$ we estimate directly

$$\varphi(x) \geq (1 - |x|)^p + (p - 1)|x|^{p-1} \geq \frac{1}{2^p} \min\{1; (p - 1)\}.$$

If $x \in (0, 1]$

$$
\begin{aligned}
(1 + x)^p - x^p &= \int_0^1 \frac{d}{ds}(s + x)^p ds = p \int_0^1 (s + x)^{p-1} ds \\
&= -p \int_0^1 (s + x)^{p-1} \frac{d}{ds}(1 - s) ds \\
&= px^{p-1} + p(p - 1) \int_0^1 (1 - s)(s + x)^{p-2} ds \\
&\geq px^{p-1} + \min\left\{1; \frac{p(p - 1)}{4}\right\}.
\end{aligned}
$$

Therefore for $|x| \leq 1$

$$\varphi(x) \geq c_p(p - 1) \quad \text{for some positive constant } c_p.$$

The Case $p \geq 2$ and $|x| > 1$: By direct calculation

$$|1 + x|^p - |x|^p = \int_0^1 \frac{d}{ds}|s + x|^p ds = p \int_0^1 |s + x|^{p-1} \text{sign}(s + x) ds.$$

From this, for $x \geq 1$ and $p \geq 2$, making use of the second of (2.2c)

$$|1 + x|^p - |x|^p \geq p \int_0^1 \left(x^{p\pm 1} + s^{p-1} \right) ds \geq p|x|^{p-2}x + 1.$$

For $x \leq -1$ and $p \geq 2$, making use of the first (2.2c)

$$|1 + x|^p - |x|^p = -p \int_0^1 (|x| - s)^{p-1} ds$$

$$\geq -p \int_0^1 \left(|x|^{p-1} - s^{p-1} \right) ds = p|x|^{p-2}x + 1.$$

The Case $1 < p < 2$ **and** $|x| > 1$: Assume first $t \in (0, 1)$. Then by repeated integration by parts

$$(1 + t)^p - 1 = - \int_0^t \frac{d}{ds}[1 + (t - s)]^p ds = p \int_0^t [1 + (t - s)]^{p-1} ds$$

$$= pt + p(p - 1) \int_0^t s[1 + (t - s)]^{p-2} ds$$

$$\geq pt + \frac{p(p - 1)}{4} t^2.$$

Therefore

$$\frac{(1 + t)^p - 1 - pt}{t^2} \geq \frac{p(p - 1)}{2} \qquad \text{for all } t \in (0, 1).$$

A similar calculation holds for $t \in (-1, 0)$ with the same bound below.

11c   The Riesz Representation Theorem

11.1c   *Weakly Cauchy Sequences in $L^p(X)$ for $1 < p \leq \infty$*

Let $\{X, \mathcal{A}, \mu\}$ be a measure space. A sequence of functions $\{f_n\} : E \to \mathbb{R}^*$ is *weakly Cauchy* in $L^p(X)$ if it is bounded in $L^p(X)$, and if, for all measurable subsets E, of finite measure, the sequence

$$\left\{ \int_E f_n d\mu \right\} \quad \text{is a Cauchy sequence in } \mathbb{R} \tag{11.1c}$$

In such a case, for all such sets, there exists the limit

$$\nu(E) = \lim \int_E f_n d\mu. \tag{11.2c}$$

Prove that for $1 < p < \infty$ a sequence $\{f_n\} \subset L^p(X)$ is weakly convergent to some $f \in L^p(E)$, if and only if it is weakly Cauchy in $L^p(X)$. Thus for $1 < p < \infty$, the space $L^p(X)$ is weakly complete. Prove that the same conclusion holds for $p = \infty$ if $\{X, \mathcal{A}, \mu\}$ is σ-finite.

11.2c Weakly Cauchy Sequences in $L^p(X)$ for $p = 1$

If $p = 1$, the notion of weakly Cauchy sequence is modified by requiring that (11.1c) holds for all measurable sets E. Prove that if $\{X, \mathcal{A}, \mu\}$, is σ-finite, then $L^1(X)$ is weakly complete. **Hint:** Use the indicated modified notion of weakly Cauchy sequence and the Radon–Nikodým theorem.

11.3c The Riesz Representation Theorem in ℓ_p

Let $\mathbf{a} \in \ell_p$ and $\mathbf{b} \in \ell_q$, where $1 \le p, q \le \infty$ are conjugate. Every element $\mathbf{b} \in \ell_q$ induces a bounded linear functional on ℓ_p by the formula

$$T(\mathbf{a}) = \mathbf{a} \cdot \mathbf{b} = \sum a_i b_i.$$

Theorem 11.1c Let $1 \le p < \infty$. For every bounded, linear functional $\mathcal{F}$ in ℓ_p, there exists a unique element $\mathbf{b} \in \ell_q$ such that $\mathcal{F}(\mathbf{a}) = \mathbf{a} \cdot \mathbf{b}$ for all $\mathbf{a} \in \ell_p$.

14c The Riesz Representation Theorem By Uniform Convexity

14.1c Bounded Linear Functional in $L^p(E)$ for $0 < p < 1$

Let $\{X, \mathcal{A}, \mu\}$ be $\mathbb{R}^N$ with the Lebesgue measure and let $E \subset \mathbb{R}^N$ be Lebesgue measurable. The topology generated in $L^p(E)$ for $0 < p < 1$ by the metric (3.1c) is not locally convex. A consequence is that there are no linear, bounded maps $T : L^p(E) \to \mathbb{R}$ except the identically zero map.

Proposition 14.1c (Day [31]) *Let E be a Lebesgue measurable subset of $\mathbb{R}^N$ and let $L^p(E)$ for $0 < p < 1$ be equipped with the topology generated by the metric (3.1c). Then the only bounded, linear functional on $L^p(E)$ is the identically zero functional.*

Proof Let T be a bounded linear functional in $L^p(E)$ for some $0 < p < 1$. Since T is continuous and linear, the pre-image of any open interval must be open and convex in $L^p(E)$. However by Proposition 3.1c, $L^p(E)$ for $0 < p < 1$, with the indicated topology, does not have any open convex sets except the empty set and $L^p(E)$ itself. Let $(-\alpha, \alpha)$ for some $\alpha > 0$ be an interval about the origin of $\mathbb{R}$. Then $T^{-1}(-\alpha, \alpha) = L^p(E)$ for all $\alpha > 0$. Thus $T \equiv 0$. ∎

14.2c An Alternate Proof of Proposition 14.1c

The alternate proof below is independent of the lack of open, convex neighborhoods of the origin in the metric topology of $L^p(E)$.[7]

Let T be a continuous, linear functional in $L^p(E)$ and let $f \in L^p(E)$ be such that $T(f) \neq 0$. There exists $A \in E \cap \mathcal{A}$ such that

$$\int_E \chi_A |f|^p d\mu = \frac{1}{2} \|f\|_p^p.$$

Set $f = f_o$ and $f_{o,1} = f_o \chi_A$ and $f_{o,2} = f_o \chi_{E-A}$. Therefore $f_o = f_{o,1} + f_{o,2}$, and

$$\|f_o\|_p^p = \|f_{o,1}\|_p^p + \|f_{o,2}\|_p^p, \qquad \|f_{o,1}\|_p^p = \|f_{o,2}\|_p^p = \tfrac{1}{2}\|f_o\|_p^p.$$

Since T is linear, $T(f_o) = T(f_{o,1}) + T(f_{o,2})$, and

$$|T(f_o)| \leq |T(f_{o,1})| + |T(f_{o,2})|.$$

Therefore either

$$|T(f_{o,1})| \geq \tfrac{1}{2}|T(f_o)| \qquad \text{or} \qquad |T(f_{o,2})| \geq \tfrac{1}{2}|T(f_o)|.$$

Assume the first holds true and set $f_1 = 2f_{o,1}$. For this choice

$$|T(f_1)| \geq |T(f_o)| \quad \text{and} \quad \|f_1\|_p^p = 2^p \|f_{o,1}\|_p^p = 2^{p-1}\|f_o\|_p^p.$$

Now repeat this construction with f_o replaced by f_1 and generate a function f_2 such that

$$|T(f_2)| \geq |T(f_o)| \quad \text{and} \quad \|f_2\|_p^p = 2^{p-1}\|f_1\|_p^p = 2^{2(p-1)}\|f_o\|_p^p.$$

By iteration generate a sequence of functions $\{f_n\}$ in $L^p(E)$ such that

$$|T(f_n)| \geq |T(f_o)| \qquad \text{and} \qquad \|f_n\|_p^p = 2^{n(p-1)}\|f_o\|_p^p.$$

---

[7]This proof was provided by A.E. Nussbaum.

Since $p \in (0, 1)$ the sequence $\{f_n\}$ converges to zero in the metric topology of $L^p(E)$. However $T(f_n)$ does not converge to zero. ∎

15c   If $E \subset \mathbb{R}^N$ and $p \in [1, \infty)$, then $L^p(E)$ Is Separable

15.1.   The spaces ℓ_p for $1 \le p < \infty$ are separable, whereas ℓ_∞ is not separable.

15.2. $BV[a, b]$ is not separable (§ 1.2c of the Complements of Chap. 5).

15.3.   Let E be a measurable subset in $\mathbb{R}^N$ and let $L^p(E)$ for $0 < p < 1$ be endowed with the metric topology generated by the metric in (3.1c). With respect to such a topology $L^p(E)$ is separable. In particular it satisfies the second axiom of countability.

18c   Approximating Functions in $L^p(E)$ with Functions in $C^\infty(E)$

A function $f \in L^p(\mathbb{R}^N)$ can approximated by smooth functions by forming the convolution with kernels other than the Friedrichs mollifying kernels J_ε.

We mention here two such kernels. Their advantage with respect to the Friedrichs kernels is that they satisfy specific Partial Differential Equations and therefore they are more suitable in applications related to such equations. Their disadvantage is that they are not compactly supported. Therefore even if f is of compact support in $\mathbb{R}^N$, its approximations will not be.

18.1c   Caloric Extensions of Functions in $L^p(\mathbb{R}^N)$

For $x \in \mathbb{R}^N$ and $t > 0$ set

$$\Gamma(x - y; t) = \frac{1}{(4\pi t)^{N/2}} e^{-\frac{|x-y|^2}{4t}}.$$

Set formally

$$\Delta_x = \operatorname{div} \nabla_x = \sum_{j=1}^{N} \frac{\partial^2}{\partial x_j^2}$$

and define Δ_y similarly. Verify by direct calculation that for all $x, y \in \mathbb{R}^N$ and $t > 0$

$$\Gamma_t - \Delta_x \Gamma = 0 \qquad \text{and} \qquad \Gamma_t - \Delta_y \Gamma = 0.$$

This partial differential equation is called the *heat equation*. The variables x are referred to as the *space variables* and t is referred to as the *time*.

A function $(x, t) \rightarrow u(x, t)$ that satisfies the heat equation in a space-time open set $E \subset \mathbb{R}^N \times \mathbb{R}$, is said to be *caloric* in E. For example $(x, t) \rightarrow \Gamma(x - y; t)$ is caloric in $\mathbb{R}^N \times \mathbb{R}^+$ for all $y \in \mathbb{R}^N$.

Verify that

$$\int_{\mathbb{R}^N} \Gamma(x - y; t)dy = \int_{\mathbb{R}^N} \Gamma(x - y; t)dx = 1$$

for all $x, y \in \mathbb{R}^N$ and all $t > 0$. **Hint:** Introduce the change of variables

$$\frac{x - y}{2\sqrt{t}} = \eta \qquad \text{for a fixed} \quad y \in \mathbb{R}^N \quad \text{and} \quad t > 0$$

and use **14.2** of the Complements of Chap. 4.

Let E be an open set in $\mathbb{R}^N$ and regard functions in $L^p(E)$ as functions in $L^p(\mathbb{R}^N)$ by extending them to be zero in $\mathbb{R}^N - E$. For $f \in L^p(E)$ and $t > 0$ set

$$f_t(x) = (\Gamma * f)(x) = \int_{\mathbb{R}^N} \Gamma(x - y; t)f(y)dy.$$

The function $(x, t) \rightarrow f_t(x)$ is caloric in $\mathbb{R}^N \times \mathbb{R}^+$ and is called the *caloric extension* of f in the upper half space $\mathbb{R}^N \times \mathbb{R}^+$. Such an extension is a mollification of f since $x \rightarrow f_t(x) \in C^\infty(\mathbb{R}^N)$.

Proposition 18.1c *Let $f \in L^p(E)$ for some $1 \leq p < \infty$. Then $\Gamma * f \in L^p(E)$ and $\|\Gamma * f\|_p \leq \|f\|_p$. The mollifications $\Gamma * f$ approximate f in the sense*

$$\lim_{t \to 0} \|f_t - f\|_p = 0.$$

Moreover if $f \in C(E)$ and f is bounded in $\mathbb{R}^N$ then for every compact subset $\mathcal{K} \subset E$

$$\lim_{t \to 0} f_t(x) = f(x) \quad \text{uniformly in } \mathcal{K}.$$

Proof The first two statements are proved as in Proposition 18.1. To prove the last, fix a compact set $\mathcal{K} \subset E$ and a positive number ε_o. If ε_o is sufficiently small, there exists a compact set $\mathcal{K}_{\varepsilon_o}$ such that $\mathcal{K} \subset \mathcal{K}_{\varepsilon_o} \subset E$, and dist $\{\mathcal{K}; \mathcal{K}_{\varepsilon_o}\} \geq \varepsilon_o$. For all $x \in \mathcal{K}$ and all $\varepsilon \in (0, \varepsilon_o)$, write

$$|f_t(x) - f(x)| = \left| \int_{\mathbb{R}^N} \Gamma(x - y; t)[f(y) - f(x)]dy \right|$$

$$\leq \left| \int_{|x-y|\leq\varepsilon} \Gamma(x - y; t)[f(y) - f(x)]dy \right|$$

$$+ \left| \int_{|x-y|>\varepsilon} \Gamma(x - y; t)[f(y) - f(x)]dy \right|$$

$$\leq \sup_{|x-y|\leq\varepsilon; x\in\mathcal{K}} |f(y) - f(x)| \int_{\mathbb{R}^N} \Gamma(x - y; t)dy$$

$$+ 2\|f\|_\infty \int_{|x-y|>\varepsilon} \Gamma(x - y; t)dy$$

$$\leq \omega(\varepsilon) + 2\|f\|_\infty \int_{|x-y|>\varepsilon} \Gamma(x - y; t)dy$$

where $\omega_o(\cdot)$ is the uniform modulus of continuity of f in $\mathcal{K}_{\varepsilon_o}$. The last integral is transformed into

$$\int_{|x-y|>\varepsilon} \Gamma(x - y; t)dy = \frac{1}{\pi^{N/2}} \int_{|\eta|>\frac{\varepsilon}{2\sqrt{t}}} e^{-|\eta|^2} d\eta.$$

From this, for $\varepsilon \in (0, \varepsilon_o)$ fixed

$$\lim_{t\to 0} \int_{|x-y|>\varepsilon} \Gamma(x - y; t)dy = 0.$$

Letting now $t \to 0$ in the previous inequality gives

$$\lim_{t\to 0} |f_t(x) - f(x)| = \omega_o(\varepsilon) \qquad \text{for all} \quad \varepsilon \in (0, \varepsilon_o). \qquad \blacksquare$$

Remark 18.1c The assumption that f be bounded in $\mathbb{R}^N$ can be removed. Indeed a similar approximation would hold if f grows as $|x| \to \infty$ not faster than $e^{\gamma|x|^2}$, for a positive constant γ ([34] Chap. V).

18.2c   *Harmonic Extensions of Functions in $L^p(\mathbb{R}^N)$*

For $x, y \in \mathbb{R}^N$ and $t > 0$ set

$$H(x - y; t) = \frac{1}{\omega_{N+1}} \frac{2t}{[|x - y|^2 + t^2]^{\frac{N+1}{2}}}.$$

Set formally

$$\Delta_{(x,t)} = \Delta_x + \frac{\partial^2}{\partial t^2}$$

and define $\Delta_{(y,t)}$ similarly. Verify by direct calculation that for all $x, y \in \mathbb{R}^N$ and all $t > 0$

$$\Delta_{(x,t)}H = \Delta_{(y,t)}H = 0.$$

This is the *Laplace equation* in the variables (x, t). A function that satisfies the Laplace equation in an open set $E \subset \mathbb{R}^{N+1}$ is called *harmonic* in E. As an example, $(x, t) \to H(x - y; t)$ is harmonic in $\mathbb{R}^N \times \mathbb{R}^+$ for all $y \in \mathbb{R}^N$.

Verify that for all $x, y \in \mathbb{R}^N$ and all $t > 0$

$$\int_{\mathbb{R}^N} H(x - y; t)dy = \int_{\mathbb{R}^N} H(x - y; t)dx = 1.$$

Hint: The change of variables $(x - y) = t\eta$ transforms these integrals in

$$\frac{2}{\omega_{N+1}} \int_{\mathbb{R}^N} \frac{d\eta}{\left(1 + |\eta|^2\right)^{\frac{N+1}{2}}} = 2\frac{\omega_N}{\omega_{N+1}} \int_0^\infty \frac{\rho^{N-1}d\rho}{\left(1 + \rho^2\right)^{\frac{N+1}{2}}} = 1.$$

Use also **14.1** of the Complements of Chap. 4.

Let E be an open set in $\mathbb{R}^N$ and regard functions in $L^p(E)$ as functions in $L^p(\mathbb{R}^N)$ by extending them to be zero in $\mathbb{R}^N - E$. For $f \in L^p(E)$ and $t > 0$ set

$$f_t(x) = (H * f)(x) = \int_{\mathbb{R}^N} H(x - y; t)f(y)dy.$$

The function $(x, t) \to f_t(x)$ is harmonic in $\mathbb{R}^N \times \mathbb{R}^+$ and is called the *harmonic extension* of f in the upper half space $\mathbb{R}^N \times \mathbb{R}^+$. Such and extension is a mollification of f since $x \to f_t(x) \in C^\infty(\mathbb{R}^N)$.

The integral defining $f_t(\cdot)$ is called the *Poisson Integral* of f ([34] Chap. II).

Proposition 18.2c *Let $f \in L^p(E)$ for some $1 \leq p < \infty$. Then $H * f \in L^p(E)$ and $\|H * f\|_p \leq \|f\|_p$. The mollifications $H * f$ approximate f in the sense,*

$$\lim_{t \to 0} \|f_t - f\|_p = 0.$$

Moreover if $f \in C(E)$ and f is bounded in $\mathbb{R}^N$ then for every compact subset $\mathcal{K} \subset E$

$$\lim_{t \to 0} f_t(x) = f(x) \quad \text{uniformly in } \mathcal{K}.$$

If $p = \infty$ neither $C_o^\infty(E)$ nor $C(\bar{E})$ is dense in $L^\infty(E)$.

18.3c Characterizing Hölder Continuous Functions

Let $E \subset \mathbb{R}^N$ be open and let $C^\alpha(E)$ be the space of Hölder continuous functions in E, endowed with the topology generated by the distance $d(\cdot, \cdot)$ introduced in (15.6) of Chap. 2. With respect to such a topology a function $u \in C^\alpha(E)$ cannot be approximated, in general, by smooth functions (**15.5–15.7** of § 15.1c of the Complements of Chap. 2).

However u can be approximated in the topology of the uniform convergence by its mollifications $\{u_\varepsilon\}$. Indeed the rate of convergence of $\{u_\varepsilon\}$ to u in $L^\infty(E)$ characterizes $C^\alpha(E)$.

Proposition 18.3c *Let $u \in C^\alpha(E)$ for some $\alpha \in (0, 1)$. Then*

$$\|u - u_\varepsilon\|_\infty \le [u]_\alpha \varepsilon^\alpha, \quad \text{and} \quad \|u_{\varepsilon,x_j}\|_\infty \le \|J_{x_j}\|_1 [u]_\alpha \varepsilon^{\alpha-1} \tag{18.1c}$$

for $j = 1, \dots, N$.

Proof Without loss of generality we may assume $E = \mathbb{R}^N$. Indeed, since u is uniformly continuous in E, with concave modulus of continuity, it can be extended to a Hölder continuous function defined in $\mathbb{R}^N$ with the same upper and lower bounds and with the same Hölder exponent α (Theorem 15.1 of Chap. 5). Then

$$|u(x) - u_\varepsilon(x)| \le \int_{|x-y|<\varepsilon} J_\varepsilon(x - y)|u(x) - u(y)|dy \le [u]_\alpha \varepsilon^\alpha.$$

Also by the properties of J_ε

$$|u_{\varepsilon,x_j}| = \left| \int_{\mathbb{R}^N} J_{\varepsilon,x_j} u(y)dy - \left(\int_{\mathbb{R}^N} J_\varepsilon dy \right)_{x_j} u(x) \right|$$

$$\le \int_{|x-y|<\varepsilon} |J_{\varepsilon,x_j}||u(x) - u(y)|dy$$

$$\le \|J_{x_j}\|_1 [u]_\alpha \varepsilon^{\alpha-1}. \qquad \blacksquare$$

This rate of convergence characterizes $C^\alpha(E)$ in the following sense.

Proposition 18.4c *Let u be a continuous function defined in $\mathbb{R}^N$ and assume that for some fixed $\alpha \in (0, 1)$, for all $\varepsilon > 0$ there exists $v_\varepsilon \in C^1(\mathbb{R}^N)$ such that*

$$\|u - v_\varepsilon\|_\infty \le \gamma \varepsilon^\alpha, \quad \text{and} \quad \|v_{\varepsilon,x_j}\|_\infty \le \gamma \varepsilon^{\alpha-1} \tag{18.2c}$$

for some fixed constant $\gamma > 0$. Then $u \in C^\alpha(\mathbb{R}^N)$ and $[u]_\alpha \le 3\gamma$.

Proof For any pair $x, y \in \mathbb{R}^N$ with $|x - y| < \varepsilon$

$$|u(x) - u(y)| \le |u(x) - v_\varepsilon(x)| + |u(y) - v_\varepsilon(y)| + |v_\varepsilon(x) - v_\varepsilon(y)|. \qquad \blacksquare$$

19c   Characterizing Pre-compact Sets in $L^p(E)$

19.1.   A closed, bounded subset C of ℓ_p for $1 \le p < \infty$ is compact if and only if for every $\varepsilon > 0$, there exists an index n_ε such that $\sum_{n > n_\varepsilon} |a_n|^p \le \varepsilon$ for all $\mathbf{a} \in C$.

19.2.   The closed unit ball of $L^p(E)$ is not compact since is not sequentially compact. The same conclusion holds for the unit ball of ℓ_p and $C(\bar{E})$.

19.1c   The Helly's Selection Principle

When $E = (a, b)$ is an open interval and $\{f_n\} \subset BV[a, b]$ then $L^p(E)$-convergence of a subsequence can be replaced with *everywhere pointwise* convergence, provided $\{f_n\}$ is uniformly bounded. Continue to denote by $V_f[a, b]$ the variation of f in $[a, b]$.

Proposition 19.1c   (Helly [73]) *Let (a, b) be an open interval of $\mathbb{R}$ and let $\{f_n\}$ be a sequence of real valued functions defined in $[a, b]$, such that*

$$\sup_{[a,b]} |f_n| \le M, \quad \text{and} \quad V_{f_n} \le M \tag{19.1c}$$

for some constant M independent of n. Then, there exists a function $f \in BV[a, b]$, with $V_f[a, b] \le M$ and a subsequence $\{f_{n'}\} \subset \{f_n\}$ such that $\{f_{n'}\} \to f$ everywhere in (a, b).

Proof   By the Jordan's decomposition we may assume that each of the f_n are nondecreasing. Define f_n in the whole $\mathbb{R}$ by extending them to be zero in $\mathbb{R}^N - [a, b]$. Then for $h > 0$ however small, compute

$$\int_{\mathbb{R}} |f_n(x + h) - f_n(x)| dx = \sum_{j \in \mathbb{Z}} \int_0^h |f_n(x + (j + 1)h) - f_n(x + jh)| dx$$

$$= \int_0^h \sum_{j \in \mathbb{Z}} |f_n(x + (j + 1)h) - f_n(x + jh)| dx$$

$$\le h V_{f_n}[a, b] \le h M \quad \text{for all } n \in \mathbb{N}.$$

Hence $\{f_n\}$ satisfy (19.1) uniformly in n. Therefore there exists $f \in L^1[a, b]$ and a subsequence $\{f_{n_1}\} \subset \{f_n\}$ such that $\{f_{n_1}\} \to f$ in $L^1[a, b]$. From this a further subsequence $\{f_{n_2}\} \subset \{f_{n_1}\}$ can be selected converging to f a.e. in $[a, b]$. Since $\{f_n\}$ is equibounded in $[a, b]$ a further subsequence $\{f_{n_3}\} \subset \{f_{n_2}\}$ can be selected converging to f at all rationals of $[a, b]$. Here f is properly redefined on a set of measure zero. Since the f_n are all nondecreasing, the function f is nondecreasing at the points of convergence. Also by properly redefining f on a set of measure zero we may assume that f is nondecreasing in $[a, b]$. One also verifies that $\{f_{n_3}\} \to f$ at all points of continuity of f. Thus $\{f_{n_3}\}$ might fail to converge to f only at the points of discontinuity of f. However f being nondecreasing, it has at most countably many points of discontinuity. Hence a further selection of $\{f_{n'}\} \subset \{f_{n_3}\}$ can be effected such that $\{f_{n'}\} \to f$ everywhere in $[a, b]$. ∎

20c   The Vitali-Saks-Hahn Theorem [59, 138, 170]

Theorem 20.1c (Vitali-Saks-Hahn [59, 138, 170]) *Let* $\{X, \mathcal{A}, \mu\}$ *be a measure space and let* $\{\lambda_n\}$ *be a sequence of signed measures on* $\mathcal{A}$, *absolutely continuous with respect to* μ, *each of finite variation* $|\lambda_n|$, *and such that*

$$\lim_n \lambda_n(E) \quad \text{exists for all } E \in \mathcal{A}.$$

Then

$$\lim_{\mu(E) \to 0} \lambda_n(E) = 0 \quad \text{uniformly in } n.$$

Proof Continue to denote by $\{\mathcal{A}; d\}$ the collection of equivalence classes of measurable sets at zero mutual distance, endowed with the metric topology, generated by $d(\cdot; \cdot)$. Having fixed $\varepsilon > 0$, consider the collection of equivalence classes

$$E_{m,n} = \{E \in \mathcal{A} \mid |\lambda_n(E) - \lambda_m(E)| \le \varepsilon\}, \quad \text{for } m, n = 1, 2 \dots.$$

By Lemma 5.1c, λ_m and λ_n are continuous functions in $\{\mathcal{A}; d\}$. Therefore the sets $E_{m,n}$ are closed in the metric topology of $\{\mathcal{A}; d\}$. Set

$$E_k = \bigcap_{m,n \ge k} E_{m,n}.$$

The assumption implies that every $E \in \mathcal{A}$ belongs to some E_k, and hence $\mathcal{A} = \bigcup E_k$. Since $\{\mathcal{A}; d\}$ is a complete metric space, by the Baire category theorem (Theorem 16.1 of Chap. 2), at least one of the E_k has nonempty interior. Thus there exists $k \in \mathbb{N}$, a positive number r, and an equivalence class $A \in E_k$ such that all equivalence classes F in the ball $B_r(A)$ of radius r centered at A

$$B_r(A) = \{F \in \mathcal{A} \mid d(A; F) = \mu(A \Delta F) < r\}$$

belong to E_k. Equivalently

$$|\lambda_n(F) - \lambda_n(F)| \le \varepsilon, \quad \text{for all } m, n \ge k, \quad \text{and for all } F \in B_r(A).$$

Determine $0 < \delta < r$ such that for any $E \in \mathcal{A}$ of μ-measure $\mu(E) < \delta$, there holds

$$|\lambda_n(E)| \le \varepsilon, \quad \text{for } n = 1, \dots, k.$$

Such choice is possible since λ_n are absolutely continuous with respect to μ, and k is finite. For any such set E, one verifies that both $A \cup E$, and $A - E$, belong to the ball $B_r(A)$. Then compute

$$\begin{aligned} \lambda_n(E) &= \lambda_k(E)r + \lambda_n(E) - \lambda_k(E) \\ &= \lambda_k(E) + \left[\lambda_n(A \cup E) - \lambda_k(A \cup E)\right] \\ &\quad - \left[\lambda_n(A - E) - \lambda_k(A - E)\right] \end{aligned}$$

Hence $|\lambda_n(E)| \le 3\varepsilon$, for all $n \in \mathbb{N}$. ∎

Remark 20.1c A consequence of the Vitali-Saks-Hahn theorem is that there exists a Lebesgue measurable set $E \subset [0, 1]$ such that

$$\lim \int_E n\chi_{[0,\frac{1}{n}]} dx \quad \text{does not exist.}$$

Likewise there exists a Lebesgue measurable set $E \subset \mathbb{R}^N$ such that

$$\lim \left(\frac{n}{4\pi}\right)^{\frac{N}{2}} \int_E e^{-n\frac{|y|^2}{4}} dy \quad \text{does not exist.}$$

Construct such sets explicitly.

Corollary 20.1c *Let $\{X, \mathcal{A}, \mu\}$ be a finite measure space and let $\{\lambda_n\}$ be a sequence of finite measures on $\mathcal{A}$, absolutely continuous with respect to μ, and such that*

$$\lambda(E) \overset{def}{=} \lim_n \lambda_n(E) \quad \text{exists for all } E \in \mathcal{A}.$$

Then $\lambda(\cdot)$ is a measure on $\mathcal{A}$.

Proof There is only to prove that λ is countably additive. By the Vitali-Saks-Hahn theorem, for all $\varepsilon > 0$ there exists $\delta = \delta(\varepsilon)$ independent of n such that

$$\lambda_n(E) < \varepsilon \quad \text{for all } E \in \mathcal{A} \text{ such that } \mu(E) < \delta, \quad \text{for all } n \in \mathbb{N}.$$

Let $\{E_j\}$ be a countable collection of disjoint, measurable sets. Since μ is finite, for all $\delta > 0$ there exists an index $m = m(\delta)$ such that

$$\mu\left(\bigcup_{j>m} E_j\right) < \delta.$$

Fix $\varepsilon > 0$, determine δ as claimed by the Vitali-Saks-Hahn theorem, and choose $m = m(\delta)$ accordingly. Then

$$\lambda\Big(\bigcup E_j\Big) = \lim_n \lambda_n\Big(\bigcup_{j=1}^m E_j\Big) + \lim_n \Big(\bigcup_{j>m} E_j\Big)$$

$$= \sum_{j=1}^m \lambda(E_j) + \lim_n \Big(\bigcup_{j>m} E_j\Big).$$

Thus

$$\Big|\lambda\Big(\bigcup E_j\Big) - \sum_{j=1}^m \lambda(E_j)\Big| < \varepsilon. \qquad\blacksquare$$

Corollary 20.2c (Nikodým [117]) *Let $\mathcal{A}$ be a σ-algebra on a set X, and let $\{\lambda_n\}$ be a sequence signed measures on $\mathcal{A}$, each with finite variation $|\lambda_n|$, and such that*

$$\lambda(E) \overset{def}{=} \lim_n \lambda_n(E) \quad \text{exists for all } E \in \mathcal{A}.$$

Then $\lambda(\cdot)$ is countably additive on $\mathcal{A}$.

Proof For all $E \in \mathcal{A}$ and all $n \in \mathbb{N}$ set

$$\mu_n(E) = \frac{|\lambda_n|(E)}{|\lambda_n|(X)} \quad \text{and} \quad \mu(E) = \sum \frac{1}{2^n} \mu_n(E).$$

One verifies that μ is a finite measure on $\mathcal{A}$ and $\lambda_n \ll \mu$ for all n. Thus the conclusion follows from the Vitali-Saks-Hahn theorem and Corollary 20.1c. $\blacksquare$

21c   Uniformly Integrable Sequences of Functions

The next assertions are a direct consequence of the Vitali-Saks-Hahn Theorem 20.1c. As a consequence give conditions on a sequence of functions $\{f_n\}$ to be *uniformly integrable* in X, in the sense of § 11c of Chap. 4.

Proposition 21.1c *Let $\{X, \mathcal{A}, \mu\}$ be a measure space and let $\{\lambda_n\}$ be a sequence of signed measures on $\mathcal{A}$, absolutely continuous with respect to μ, each of finite variation $|\lambda_n|$, and such that*

$$\lim_n \lambda_n(E) \quad \text{exists for all } E \in \mathcal{A}.$$

Then for all $\varepsilon > 0$ there exists δ such that

$$\mu(E) < \delta \quad \text{implies} \quad |\lambda_n|(E) < \varepsilon \quad \text{uniformly in } n.$$

Proof If the conclusion does not hold, there exists $\varepsilon > 0$ and a sequence $\{E_m\}$ of measurable subsets of X, such that

$$\mu(E_m) < \frac{1}{m} \quad \text{and} \quad |\lambda_m(E_m)| > \varepsilon.$$

For each m fixed

$$\varepsilon < |\lambda_m|(E_m) = \lambda_m^+(E_m) + \lambda_m^-(E_m).$$

Therefore either

$$\lambda_m^+(E_m) > \tfrac{1}{2}\varepsilon, \quad \text{or} \quad \lambda_m^-(E_m)\tfrac{1}{2}\varepsilon.$$

Let $X_m^+ \cup X_m^-$ be the Hanh's decomposition of X induced by λ. By replacing E_m with either $E_m \cap X_m^+$ or $E_m \cap X_m^-$, the sets $\{E_m\}$ can be chosen so that

$$\mu(E_m) < \frac{1}{m} \quad \text{and} \quad \lambda_m(E_m) > \tfrac{1}{2}\varepsilon.$$

This contradicts the conclusion of the Vitali-Hahn-Saks theorem and establishes the proposition. ∎

Corollary 21.1c *Let $\{X, \mathcal{A}, \mu\}$ be a measure space and $\{f_n\}$ be a sequence of integrable functions in X such that*

$$\left\{ \int_E f_n d\mu \right\} \quad \text{is a Cauchy sequence in } \mathbb{R}$$

for all $E \in \mathcal{A}$. Then $\{f_n\}$ is uniformly integrable in X, in the sense of § 11c of Chap. 4.

Proof Since $f_n \in L^1(X)$, setting

$$\mathcal{A} \ni E \longrightarrow \lambda_n(E) = \int_E f_n d\mu.$$

defines signed measures λ_n on $\mathcal{A}$, absolutely continuous with respect to μ, and each of finite variation $|\lambda_n|$. By the assumptions $\{\lambda_n(E)\}$ has a limit for all $E \in \mathcal{A}$. Therefore by the Vitali-Saks-Hahn theorem

$$\lim_{\mu(E)\to 0} \left| \int_E f_n d\mu \right| = 0 \quad \text{uniformly in } n.$$

If $\{f_n\}$ is not uniformly integrable, there exists $\varepsilon > 0$ and a sequence $\{E_m\}$ of measurable subsets of X, such that

$$\mu(E_m) < \frac{1}{m} \quad \text{and} \quad \int_{E_m} |f_m| d\mu > \varepsilon.$$

For each m fixed

$$\varepsilon < \int_{E_m} f_m^+ d\mu + \int_{E_m} f^- d\mu.$$

Therefore either

$$\int_{E_m} f_m^+ d\mu > \tfrac{1}{2}\varepsilon, \quad \text{or} \quad \int_{E_m} f_m^- d\mu > \tfrac{1}{2}\varepsilon.$$

Thus by replacing E_m with either $E_m \cap [f_m > 0]$ or $E_m \cap [f_m \le 0]$, the sets $\{E_m\}$ can be chosen so that

$$\mu(E_m) < \frac{1}{m} \quad \text{and} \quad \left|\int_{E_m} f_m d\mu\right| > \tfrac{1}{2}\varepsilon. \qquad \blacksquare$$

Corollary 21.2c *Let $\{X, \mathcal{A}, \mu\}$ be a measure space and $\{f_n\}$ be a sequence of integrable functions in X weakly convergent in $L^1(X)$ to some $f \in L^1(X)$. Then $\{f_n\}$ is uniformly integrable in X, in the sense of § 11c of Chap. 4.*

22c Relating Weak and Strong Convergence and Convergence in Measure

The previous statements permit one to give necessary and sufficient conditions for weak convergence to imply strong convergence.

Proposition 22.1c *Let $\{X, \mathcal{A}, \mu\}$ be a finite measure space and let $\{f_n\}$ be a sequence of integrable functions in X. Then $\{f_n\}$ converges strongly in $L^1(X)$ if and only if $\{f_n\} \to f$ weakly and in measure.*

Proof Strong convergence implies weak convergence and convergence in measure. To prove the converse, fix $\varepsilon > 0$ and determine $\delta > 0$ such that

$$\mu(E) < \delta \quad \text{implies} \quad \int_E |f_n - f| d\mu < \varepsilon \quad \text{uniformly in } n.$$

This is possible, since by Corollary 21.2c, $\{f_n\}$ is uniformly integrable. Since $\{f_n\} \to f$ in measure, there exists n_ε such that

$$\mu([|f_n - f| > \varepsilon]) < \varepsilon \qquad \text{for all} \quad n > n_\varepsilon.$$

Then since μ is finite

$$\int_X |f_n - f| d\mu = \int_{|f_n - f| > \varepsilon} |f_n - f| d\mu$$

$$+ \int_{|f_n - f| \le \varepsilon} |f_n - f| d\mu < \varepsilon(1 + \mu(X)). \qquad \blacksquare$$

The next proposition extends Proposition 22.1c to σ-finite measure spaces.

Proposition 22.2c *Let* $\{X, \mathcal{A}, \mu\}$ *be a σ-finite measure space and let* $\{f_n\}$ *be a sequence of integrable functions in X converging weakly in* $L^1(X)$ *to some* $f \in L^1(X)$. *Then* $\{f_n\}$ *converges strongly in* $L^1(X)$ *if and only if* $\{f_n\} \to f$ *in measure, on every subset of X of finite measure.*

Proof By replacing f_n with $f_n - f$, one may assume that $f = 0$. For a measurable set $E \subset X$ and $n \in \mathbb{N}$ set

$$\lambda_n(E) = \int_E f_n d\mu; \quad \nu_n(E) = \frac{\int_E |f_n| d\mu}{\int_X |f_n| d\mu}; \quad \nu(E) = \sum \frac{1}{2^n} \nu_n(E).$$

One verifies that ν is a finite measure on $\mathcal{A}$ and that $\lambda_n \ll \nu$ for all n. Moreover, since $\{f_n\} \to 0$ weakly in $L^1(X)$, the $\lim \lambda_n(E)$ exists for all $E \in \mathcal{A}$. Thus, by Proposition 22.1c, for all $\varepsilon > 0$ there exists δ such that

$$\nu(E) < \delta \quad \text{implies} \quad \left| \int_E |f_n| d\mu \right| < \varepsilon \quad \text{uniformly in } n.$$

Having fixed ε and the corresponding δ, there is $n = n_\delta$ such that

$$\sum_{j > n_\delta} \frac{1}{2^j} \nu_j(E) < \frac{1}{2}\delta \quad \text{for all } E \in \mathcal{A}.$$

Since $\{X, \mathcal{A}, \mu\}$ is σ-finite, there exists a countable collection of measurable sets $\{E_m\}$ of finite μ-measure, such that $E_m \subset E_{m+1}$, and $X = \bigcup E_m$. Since $f_j \in L^1(X)$, there exists $m = m(n_\delta)$ such that

$$\int_{X - E_{m(n_\delta)}} |f_j| d\mu < \frac{1}{2}\delta \quad \text{for } j = 1, \dots, n_\delta.$$

Therefore, by the definition of ν, the set $X - E_{m(n_\delta)}$ satisfies

$$\nu(X - E_{m(n_\delta)}) < \delta$$

and hence,

$$\int_{X - E_{m(n_\delta)}} |f_n| d\mu < \varepsilon \quad \text{uniformly in } n.$$

Then

$$\lim \int_X |f_n| d\mu = \lim \int_{X - E_{m(n_\delta)}} |f_n| d\mu + \lim \int_{E_{m(n_\delta)}} |f_n| d\mu$$

$$< \varepsilon + \lim \int_{E_{m(n_\delta)}} |f_n| d\mu.$$

Since $E_{m(n_\delta)}$ is of finite μ-measure, the last limit is zero (Proposition 22.1c). ∎

Corollary 22.1c *In ℓ_1 weak and strong convergence coincide.*

23c An Independent Proof of Corollary 22.1c

Proposition 23.1c *A sequence $\{x_n\} \subset \ell_1$ converges weakly to some $x \in \ell_1$ if and only if $\|x_n - x\|_1 \to 0$ as $n \to \infty$.*

Proof Strong convergence implies weak convergence. To show the converse assume $x = 0$. Thus the assumption is

$$\lim \langle x_n, y \rangle = \lim \sum x_{j,n} y_j = 0 \quad \text{for all } y \in \ell_\infty. \tag{23.1c}$$

Choosing y as the base elements of ℓ_∞, gives

$$\lim_n x_{j,n} = 0 \quad \text{for all } j \in \mathbb{N}. \tag{23.2c}$$

If $\limsup \|x_n\|_1 > 0$, there exists $\varepsilon > 0$ and a subsequence $\{x_{n'}\} \subset \{x_n\}$ such that

$$\|x_{n'}\|_1 > \varepsilon \quad \text{and} \quad \lim \langle x_{n''}, y \rangle = 0 \tag{23.3c}$$

for all subsequences $\{x_{n''}\} \subset \{x_{n'}\}$ and all $y \in \ell_\infty$. The proof consists of extracting a subsequence $\{x_{n''}\} \subset \{x_{n'}\}$ and an element $y \in \ell_\infty$ fow which that last statement fails to hold.

Fix the index $m_1 = n_1'$ and consider the sequence $x_{m_1} = \{x_{j,m_1}\}$. Since $x_{m_1} \in \ell_1$, there exists an index j_{m_1} such that

$$\sum_{j > j_{m_1}} |x_{j,m_1}| < \tfrac{1}{4}\varepsilon.$$

Then choose

$$y_j = \text{sign}\{x_{j,m_1}\}, \quad \text{for } j = 1, \dots, j_{m_1}.$$

For such a choice, and in view of the first of (23.3c),

$$\left| \sum_{j=1}^{j_{m_1}} y_j x_{j,m_1} \right| > \|x_{m_1}\|_1 - \sum_{j > j_{m_1}} |x_{j,m_1}| > \tfrac{3}{4}\varepsilon.$$

The index j_{m_1} being fixed, by the pointwise convergence in (23.2c), there exists an integer $m_1 < m_2 \in \{n'\}$ such that

$$\left| \sum_{j=1}^{j_{m_1}} y_j x_{j,m_2} \right| < \tfrac{1}{8}\varepsilon.$$

Since $\mathbf{x}_{m_2} \in \ell_1$, there is an index j_{m_2}, such that

$$\sum_{j>j_{m_2}} |x_{j,m_2}| < \tfrac{1}{8}\varepsilon.$$

Without loss of generality we may take $j_{m_2} > j_{m_1} + 1$ and set

$$y_j = \text{sign}\,\{x_{j,m_2}\}, \quad \text{for } j = j_{m_1}+1, \ldots, j_{m_2}.$$

Taking into account the first of (23.3c) the element $\mathbf{x}_{m_2}$ satisfies

$$\left|\sum_{j=1}^{j_{m_1}} y_j x_{j,m_2}\right| < \tfrac{1}{8}\varepsilon; \quad \sum_{j_{m_1}+1}^{j_{m_2}} y_j x_{j,m_2} > \tfrac{3}{4}\varepsilon; \quad \sum_{j>j_{m_2}} |x_{j,m_2}| < \tfrac{1}{8}\varepsilon. \tag{23.4c}$$

Proceeding by induction, assume that for a positive integer $s \geq 2$, an element $\mathbf{x}_{m_s} \in \{\mathbf{x}_{n'}\}$, an index j_{m_s} and numbers y_j for $j = 1, \ldots j_{m_s}$, have been selected satisfying

$$\left|\sum_{j=1}^{j_{m_{s-1}}} y_j x_{j,m_s}\right| < \tfrac{1}{8}\varepsilon; \quad \sum_{j_{m_{s-1}}+1}^{j_{m_s}} y_j x_{j,m_s} > \tfrac{3}{4}\varepsilon; \quad \sum_{j>j_{m_s}} |x_{j,m_s}| < \tfrac{1}{8}\varepsilon. \tag{23.4c$_s$}$$

An element $\mathbf{x}_{m_{s+1}} \in \{\mathbf{x}_{n'}\}$, an index $j_{m_{s+1}}$ and numbers y_j for $j = j_{m_s}+1, \ldots, j_{m_{s+1}}$ are constructed by first choosing $m_{s+1} \in \{n'\}$ so that

$$\left|\sum_{j=1}^{j_{m_s}} y_j x_{j,m_{s+1}}\right| < \tfrac{1}{8}\varepsilon.$$

Such a choice is possible in view of the pointwise convergence in (23.2c). Then choose $j_{m_{s+1}}$ so that

$$\sum_{j>j_{m_{s+1}}} |x_{j,m_{s+1}}| < \tfrac{1}{8}\varepsilon.$$

Such a choice is possible since $\mathbf{x}_{m_{s+1}} \in \ell_1$. Then choose

$$y_j = \text{sign}\,\{x_{j,m_{s+1}}\}, \quad \text{for } j = j_{m_s}+1, \ldots, j_{m_{s+1}}.$$

Taking into account the first of (23.3c) one verifies that $\mathbf{x}_{m_{s+1}}$ satisfies (23.4c)$_{s+1}$. This procedure identifies a subsequence $\{\mathbf{x}_{m_s}\} \subset \{\mathbf{x}_{n'}\}$, and an element $\mathbf{y} \in \ell_\infty$, of norm 1, such that

$$\langle \mathbf{y}, \mathbf{x}_{m_s} \rangle = \sum_{j=1}^{j_{m_{s-1}}} y_j x_{j,m_s} + \sum_{j=m_{s-1}+1}^{j_{m_s}} y_j x_{j,m_s} + \sum_{j>j_{m_s}} y_j x_{j,m_s} > \tfrac{1}{2}\varepsilon. \qquad \blacksquare$$

Chapter 7
Banach Spaces

1 Normed Spaces

Let X be a vector space and let Θ be its zero element. A norm on X is a function $\|\cdot\| : X \to \mathbb{R}^+$ satisfying:

   i. $\|x\| = 0$ if and only if $x = \Theta$
  ii. $\|x + y\| \leq \|x\| + \|y\|$ for all $x, y \in X$
 iii. $\|\lambda x\| = |\lambda| \, \|x\|$ for all $\lambda \in \mathbb{R}$ and $x \in X$.

Every norm on X defines a translation invariant metric by the formula $d(x, y) = \|x - y\|$. This in turn generates a translation-invariant topology in X. We denote by $\{X; \|\cdot\|\}$ the corresponding metric space. By Proposition 14.1 of Chap. 2, the sum $+ : X \times X \to X$ and the product $\bullet : \mathbb{R} \times X \to X$ are continuous with respect to the metric topology of X. Therefore the norm $\|\cdot\|$ induces a topology on X by which $\{X; \|\cdot\|\}$ is a *topological* vector space. By the requirements (ii) and (iii), the balls in $\{X; \|\cdot\|\}$ are convex. Therefore such a topology is locally convex.

Remark 1.1 The requirements (i)–(iii) distinguish between metrics and norms. While every norm is a metric, there exist metrics that do not satisfy (iii). For example the metric d_o in (13:1) of Chap. 2 does not satisfy (iii) even if d does.

The pair $\{X; \|\cdot\|\}$ is called a *normed space* and the topology generated by $\|\cdot\|$ is the norm topology of X. If $\{X; \|\cdot\|\}$ is complete, it is called a *Banach space*. The spaces $\mathbb{R}^N$, for all $N \in \mathbb{N}$, endowed with their Euclidean norm, are Banach spaces. The spaces $L^p(E)$ for $1 \leq p \leq \infty$ are Banach spaces. The spaces ℓ_p for $1 \leq p \leq \infty$ are Banach spaces. Let E be a bounded open set in $\mathbb{R}^N$. Then $C(\bar{E})$ is a Banach space by the norm

$$C(\bar{E}) \ni f \longrightarrow \|f\| = \sup_{\bar{E}} |f|. \tag{1.1}$$

This is also called the sup-norm and generates the topology of uniform convergence (§ 15 of Chap. 2).

© Springer Science+Business Media New York 2016
E. DiBenedetto, *Real Analysis*, Birkhäuser Advanced
Texts Basler Lehrbücher, DOI 10.1007/978-1-4939-4005-9_7

Let $[a, b]$ be an interval of $\mathbb{R}$. Then the space $BV[a, b]$ of the functions of bounded variations in $[a, b]$ is a Banach space by the norm

$$BV[a, b] \ni f \longrightarrow \|f\| = |f(a)| + \mathcal{V}_f[a, b] \tag{1.2}$$

where $\mathcal{V}_f[a, b]$ is the variation of f in $[a, b]$ (§ 1.1c of the Complements of Chap. 5).

Let E be a bounded open set in $\mathbb{R}^N$ and $\alpha \in (0, 1)$. The space $C^\alpha(E)$ of the Hölder continuous functions defined in E, with Hölder exponent α is a Banach space by the norm (see § 15.2 of Chap. 2)

$$C^\alpha(E) \ni f \to \|f\| = \|f\|_\infty + [f]_\alpha. \tag{1.3}$$

If X_o is a subspace of X we denote by $\{X_o; \|\cdot\|\}$ the normed space X_o with the norm inherited from $\{X; \|\cdot\|\}$. If X_o is a closed subspace of X then also $\{X_o; \|\cdot\|\}$ is a Banach space.

The same vector space X can be endowed with different norms. The notion of equivalence of two norms $\|\cdot\|_1$ and $\|\cdot\|_2$ on the same vector space X can be inferred from the notion of equivalence of the corresponding metrics (see § 13.2 of Chap. 2 and related problems; see also **1.1** and **1.4** of the Complements).

All norms in a finite dimensional, Hausdorff, topological vector space, space is equivalent (§ 12 of Chap. 2).

Every finite dimensional subspace of a normed space is closed.

1.1 Semi-norms and Quotients

A nonnegative function $p : X \to \mathbb{R}$ is a *semi-norm* if it satisfies the requirements (ii)–(iii) of a norm. A semi-norm p is a norm on X if and only if it satisfies also the requirement (i). As an example, the function $p : C[0, 1] \to \mathbb{R}$ defined by $p(f) = |f(\frac{1}{2})|$ is a semi-norm in $C[0, 1]$, which is not a norm.

The *kernel* of a semi-norm p is defined by

$$\ker\{p\} = \{x \in X \mid p(x) = 0\}.$$

Since p is nonnegative, the triangle inequality implies that $\ker\{p\}$ is a subspace of X. If p is a norm $\ker\{p\} = \{\Theta\}$ and if $p \equiv 0$ then $\ker\{p\} = X$. As an example, let $\{X; \|\cdot\|\}$ be a normed space and let X_o be a subspace of X. The distance from an element $x \in X$ to X_o

$$d(x, X_o) = \inf_{y \in X_o} \|x - y\|$$

defines a semi-norm on X whose kernel is X_o.

A semi-norm p on X and its kernel $\ker\{p\}$, induce an equivalence relation in X, by stipulating that two elements $x, y \in X$ are equivalent if and only if $p(x - y) = 0$.

Equivalently, x is equivalent to y if and only if $p(x) = p(y)$. The quotient space $X/\ker\{p\}$ consists of the equivalence classes of elements $x' = x + \ker\{p\}$.

The operation of linear combination of any two elements x' and y' of $X/\ker\{p\}$ can be introduced by operating with representatives out of these equivalence classes, and by verifying that such an operation is independent of the choice of these representatives. This turns $X/\ker\{p\}$ into a vector space whose zero element is the equivalence class $\ker\{p\}$. Moreover, the function $p : X \to \mathbb{R}$ may be redefined as a map p' from $X/\ker\{p\}$ into $\mathbb{R}$ by setting

$$p'(x') = p'(x + \ker\{p\}) = p(x).$$

One verifies that $p' : X/\ker\{p\} \to \mathbb{R}$ is now a norm on $X/\ker\{p\}$ by which $\{X/\ker\{p\}; p'\}$ is a normed space.

2 Finite and Infinite Dimensional Normed Spaces

A vector space X is of *finite dimension* if it has a finite Hamel basis and is of *infinite dimension* if any Hamel basis is infinite (§ 9.6c of the Complements of Chap. 2).

The unit sphere of a normed space $\{X; \|\cdot\|\}$, centered at the origin of X is defined as $S_1 = \{x \in X \,|\, \|x\| = 1\}$.

Let S_1 be the unit sphere centered at the origin of $\mathbb{R}^N$ and let π be an hyperplane through the origin of $\mathbb{R}^N$. Then there exists at least one element $x_o \in S_1$ whose distance from π is 1.

The analog in an infinite dimensional normed space $\{X; \|\cdot\|\}$ is stated as follows. One fixes a closed, proper subspace $\{X_o, \|\cdot\|\}$ and seeks an element $x_o \in X$ such that

$$\|x_o\| = 1 \quad \text{and} \quad d(x_o, X_o) = \inf_{x \in X_o} \|x_o - x\| = 1. \qquad (2.1)$$

Unlike the finite dimensional case, such an element in general does not exist, as shown by the following counterexample.

2.1 A Counterexample

Let X be the subspace of $C[0, 1]$ endowed with the sup-norm, of those functions vanishing at 0. Let also $X_o \subset X$ be the subspace of X of those functions with vanishing integral average over $[0, 1]$. One verifies that X_o endowed with the sup-norm is a closed, proper subspace of X.

Proposition 2.1 *There exists no function $f \in X$ such that and*

$$\|f\| = 1 \quad \text{and} \quad \|f - g\| \geq 1 \quad \text{for all functions } g \in X_o. \qquad (2.2)$$

Remark 2.1 If (2.2) holds for some f of unit norm, is equivalent to (2.1). Indeed (2.1) implies (2.2), whereas the latter implies $d(f, X_o) \geq 1$. However, $g \equiv 0$ is in X_o and $\|f\| = 1$. Therefore $d(f, X_o) = 1$.

Proof (of Proposition 2.1) Assume such a f exists. For $h \in X - X_o$ set

$$g = f - ch \qquad \text{where} \qquad c = \frac{\int_0^1 f(t)dt}{\int_0^1 h(t)dt}.$$

Then $g \in X_o$ and
$$1 \leq \|f - (f - ch)\| = |c| \|h\|,$$

that is

$$\left| \int_0^1 h(t)dt \right| \leq \left| \int_0^1 f(t)dt \right| \sup_{t \in [0,1]} |h(t)|.$$

Choosing $h = t^{1/n} \in X - X_o$ gives

$$\frac{n}{n+1} \leq \left| \int_0^1 f(t)dt \right| \qquad \text{for all } n \in \mathbb{N}.$$

Therefore letting $n \to \infty$

$$1 \leq \int_0^1 |f(t)|dt.$$

This however is impossible since f is continuous, $\|f\| = 1$ and $f(0) = 0$. ∎

2.2 The Riesz Lemma

While in general an element $x_o \in S_1$ at distance 1 from a given subspace, does not exist, there exist elements of norm 1 and at a distance arbitrarily close to 1 from the given subspace.

Lemma 2.1 (Riesz [128]) *Let $\{X; \|\cdot\|\}$ be a normed space and let $\{X_o; \|\cdot\|\}$ be a closed, proper subspace of $\{X; \|\cdot\|\}$. For every $\varepsilon \in (0, 1)$ there exists $x_\varepsilon \in X$ such that $\|x_\varepsilon\| = 1$ and $\|x_\varepsilon - x\| \geq 1 - \varepsilon$ for all $x \in X_o$.*

Proof Fix $x_o \in X - X_o$ and let d be the distance from x_o to X_o. Since X_o is a proper, closed subspace of X we have $d > 0$. There exists an element $x_1 \in X_o$, such that

$$d \leq \|x_o - x_1\| \leq d + \frac{\varepsilon d}{1 - \varepsilon}.$$

The element claimed by the lemma is

$$x_\varepsilon = \frac{x_o - x_1}{\|x_o - x_1\|} \qquad \|x_\varepsilon\| = 1.$$

To prove the lemma for such x_ε first observe that, for every $x \in X_o$

$$\widetilde{x} = x_1 + \|x_o - x_1\| x \in X_o.$$

Then, for every $x \in X_o$

$$\|x_\varepsilon - x\| = \left\| \frac{x_o - x_1}{\|x_o - x_1\|} - x \right\|$$

$$= \frac{1}{\|x_o - x_1\|} \|x_o - \widetilde{x}\|$$

$$\geq \frac{d}{\|x_o - x_1\|} \geq 1 - \varepsilon. \qquad \blacksquare$$

2.3 Finite Dimensional Spaces

A locally compact, Hausdorff, topological vector space is of finite dimension (Proposition 12.2 of Chap. 2). In normed spaces $\{X; \|\cdot\|\}$, the Riesz lemma permits one to give an independent proof.

Proposition 2.2 *Let $\{X; \|\cdot\|\}$ be a normed space. If the unit sphere S_1 is compact, then X is finite dimensional.*

Proof If not, choose $x_1 \in S_1$ and consider the subspace X_1 spanned by x_1. It is a proper subspace of X and by Corollary 12.1 of Chap. 2, it is closed. Therefore by the Riesz lemma, there exists $x_2 \in S_1$ such that $\|x_1 - x_2\| \geq \frac{1}{2}$.

The space $X_2 = \mathrm{spn}\{x_1, x_2\}$ is a proper, closed subspace of $\{X; \|\cdot\|\}$. Therefore there exists $x_3 \in S_1$ such that $\|x_2 - x_3\| \geq \frac{1}{2}$.

Proceeding in this fashion we generate a sequence $\{x_n\}$ of elements in S_1 such that $\|x_n - x_m\| \geq \frac{1}{2}$, for $n \neq m$. Such a sequence does not contain any convergent subsequence contradicting the compactness of S_1. $\blacksquare$

Corollary 2.1 *A normed space $\{X; \|\cdot\|\}$ is of finite dimension if and only if S_1 is compact.*

3 Linear Maps and Functionals

Let $\{X; \|\cdot\|_X\}$ and $\{Y; \|\cdot\|_Y\}$ be normed spaces. A linear map $T : X \to Y$ is bounded if there exists a positive number M such that

$$\|T(x)\|_Y \le M\|x\|_X \qquad \text{for all } x \in X.$$

The norm of T is defined as the smallest of such M, i.e.,

$$\|T\| = \sup_{x \in X - \Theta} \frac{\|T(x)\|_Y}{\|x\|_X} = \sup_{\|x\|_X = 1} \|T(x)\|_Y. \tag{3.1}$$

Proposition 3.1 *Let T be a linear map from a normed space $\{X; \|\cdot\|_X\}$ into a normed space $\{Y; \|\cdot\|_Y\}$. Then*

(i). If T is bounded, it is uniformly continuous.
(ii). If T is continuous at some $x_o \in X$, it is bounded.

Proof If T is bounded, for any pair of elements $x, y \in X$

$$\|T(x) - T(y)\|_Y \le \|T\|\|x - y\|_X.$$

This proves (i). To prove (ii) we may assume that $x_o = \Theta$. There exists a ball B_ε of radius ε and centered at the origin of X such that

$$\|T(y)\|_Y < 1 \qquad \text{for all } y \in B_\varepsilon.$$

If x is an element of X not in the ball B_ε, put

$$x_\varepsilon = \frac{\varepsilon}{2} \frac{x}{\|x\|_X} \in B_\varepsilon.$$

By the linearity of T

$$\frac{\varepsilon}{2\|x\|_X}\|T(x)\|_Y = \|T(x_\varepsilon)\|_Y \le 1.$$

Therefore

$$\|T(x)\|_Y \le \frac{2}{\varepsilon}\|x\|_X \qquad \text{for all } x \in X. \qquad \blacksquare$$

Let $\mathcal{B}(X; Y)$ denote the collection of all the bounded linear maps from $\{X; \|\cdot\|_X\}$ into $\{Y; \|\cdot\|_Y\}$. Such a collection has the structure of a vector space, for the operations of sum and product by scalars, i.e.,

$$(T_1 + T_2)(x) = T_1(x) + T_2(x), \qquad \alpha T(x) = T(\alpha x)$$

for all $x \in X$ and all $\alpha \in \mathbb{R}$. It has also the structure of a normed space by the norm in (3.1). Therefore a topology and the various notions of convergence in $\mathcal{B}(X; Y)$ are defined in terms of such a norm. One also verifies that the operations

$$+ : \mathcal{B}(X; Y) \times \mathcal{B}(X; Y) \to \mathcal{B}(X; Y), \qquad \bullet : \mathbb{R} \times \mathcal{B}(X; Y) \to \mathcal{B}(X; Y)$$

are continuous with respect to the corresponding product topology. Therefore $\mathcal{B}(X; Y)$ is a topological vector space by the topology generated by the norm in (3.1). The next proposition gives a sufficient condition for the normed space $\mathcal{B}(X; Y)$ to be a Banach space.

Proposition 3.2 *Let $\{Y; \|\cdot\|_Y\}$ be a Banach space. Then $\mathcal{B}(X; Y)$ endowed with the norm (3.1) is also a Banach space.*

Proof Let $\{T_n\}$ be a Cauchy sequence in $\mathcal{B}(X; Y)$, i.e., for any fixed $\varepsilon > 0$, there exist an index n_ε such that, $\|T_n - T_m\| \le \varepsilon$ for all $n, m \ge n_\varepsilon$. For each fixed $x \in X$, the sequence $\{T_n(x)\}$ is a Cauchy sequence in Y. Indeed

$$\|T_n(x) - T_m(x)\|_Y \le \|T_n - T_m\| \|x\|_X.$$

Since $\{Y; \|\cdot\|_Y\}$ is complete $\{T_n(x)\}$ has a limit in Y which is denoted by $T(x)$. For every $x \in X$ and all $n \ge n_\varepsilon$

$$\|T_n(x) - T(x)\|_Y = \lim_m \|T_n(x) - T_m(x)\|_Y \le \varepsilon \|x\|_X.$$

The map $T : X \to Y$ so constructed is linear. It is also bounded since

$$\|T\| = \sup_{\|x\|_X = 1} \lim \|T_n(x)\|_Y \le \|T_{n_\varepsilon}\| + \varepsilon.$$

This implies that $T \in \mathcal{B}(X; Y)$ and that $\{T_n\}$ converges to T in the norm of $\mathcal{B}(X; Y)$. Indeed

$$\|T_n - T\| = \sup_{\|x\|_X = 1} \|T_n(x) - T(x)\|_Y \le \varepsilon \qquad \text{for } n \ge n_\varepsilon. \qquad \blacksquare$$

If the target space $\{Y; \|\cdot\|_Y\}$ is the set of the real numbers, endowed with the Euclidean norm, then T is said to be a *functional*. The space $\mathcal{B}(X; \mathbb{R})$ of all the bounded linear functionals in X, is denoted by X^* and is called the dual space to X.

Corollary 3.1 *The dual X^* of a normed space $\{X; \|\cdot\|\}$, endowed with the norm (3.1) is a Banach space.*

4 Examples of Maps and Functionals

Let E be a bounded open set in $\mathbb{R}^N$ and let dx denote the Lebesgue measure in $\mathbb{R}^N$. Given a real valued function $K(\cdot, \cdot)$, defined and measurable in the product space $E \times E$, set formally

$$T(f)(x) = \int_E K(x, y) f(y) dy. \qquad (4.1)$$

If $K(\cdot, \cdot) \in L^q(E \times E)$ for some $1 < q < \infty$ then (4.1) defines a bounded linear map $T : L^p(E) \to L^q(E)$, where p and q are conjugate. A formal example of (4.1) is the Riesz potential

$$T(f)(x) = \int_E \frac{f(y)}{|x - y|^{N-1}} dy. \qquad (4.2)$$

whose kernel satisfies (Proposition 21.1 of Chap. 9)

$$\sup_{y \in E} \int_E |K(x, y)| dx \le \gamma \quad \left(\stackrel{\text{def}}{=} N \kappa_N^{\frac{N-1}{N}} \mu(E)^{\frac{1}{N}} \right) \qquad (4.3)$$

where κ_N is the volume of the unit ball in $\mathbb{R}^N$. Therefore the Riesz potential can be regarded as a bounded linear map from $L^1(E)$ into $L^1(E)$ or from $L^\infty(E)$ into $L^\infty(E)$.

Denote by $C^1([0, 1])$ the space of all continuously differentiable functions in $[0, 1]$ with the topology inherited from the sup-norm on $C[0, 1]$. The differentiation map

$$T(f) = f' : C^1([0, 1]) \longrightarrow C([0, 1])$$

is linear and not bounded. Let $C[0, 1]$ be equipped with the sup-norm. The map

$$T(f)(t) = \int_0^t f(s) ds, \qquad t \in [0, 1]$$

from $C[0, 1]$ into itself, is linear, continuous but not onto.

4.1 Functionals

The dual of $\mathbb{R}^N$ equipped with its Euclidean norm is isometrically isomorphic to $\mathbb{R}^N$, that is, $\mathbb{R}^{N*} = \mathbb{R}^N$, up to an isomorphism. By the Riesz representation theorem, the dual of $L^p(E)$ for $1 \le p < \infty$ is isometrically isomorphic to $L^q(E)$ where p and q are conjugate, that is, $L^p(E)^* = L^q(E)$, up to an isomorphism. Similarly $\ell_p^* = \ell_q$.

4.2 Linear Functionals on $C(\bar{E})$

Let $E \subset \mathbb{R}^N$ be bounded and open. For a fixed $x_o \in E$ set

$$T(f) = f(x_o) \qquad \text{for all } f \in C(\bar{E}). \tag{4.4}$$

This is called the *evaluation* map at x_o and it identifies a bounded linear functional $T \in C(\bar{E})^*$, with norm $\|T\| = 1$. Let now μ be a σ-finite Borel measure in $\mathbb{R}^N$ and set

$$T(f) = \int_E f \, d\mu \qquad \text{for all } f \in C(\bar{E}). \tag{4.5}$$

This identifies a functional $T \in C\left(\bar{E}\right)^*$, with norm $\|T\| = \mu(E)$. It will be shown in § 2–6 of Chap. 8, that, in a sense to be made precise, these are essentially *all* the bounded linear functionals on $C(\bar{E})$. Also the evaluation map (4.4) can be represented as in (4.5) if μ is the Dirac mass δ_{x_o}.

4.3 Linear Functionals on $C^\alpha(\bar{E})$ for Some $\alpha \in (0, 1)$

Since $C^\alpha(\bar{E}) \subset C(\bar{E})$ one has the inclusion $C(\bar{E})^* \subset C^\alpha(\bar{E})^*$. The inclusion is in general strict. Take for example $E = (-1, 1)$ and consider the map

$$C^\alpha[-1, 1] \ni f \to \lim_{\varepsilon \to 0} \int_{|x| > \varepsilon} \frac{f(x)}{x} dx. \tag{4.6}$$

One verifies that this defines a bounded, linear functional in $C^\alpha[-1, 1]$ which cannot be extended to be an element of $C(\bar{E})^*$. Precisely for all $\gamma > 0$ there is $f \in C(\bar{E})$ such that $\|f\|_\infty = 1$ and $T(f) > \gamma$.

5 Kernels of Maps and Functionals

Let T be a map from $\{X; \|\cdot\|_X\}$ into $\{Y; \|\cdot\|_Y\}$. The kernel of T is

$$\ker\{T\} = \{x \in X \mid T(x) = \Theta_Y\}.$$

If T is linear $\ker\{T\}$ is a linear subspace of X. If $T \in \mathcal{B}(X; Y)$ and $\{X; \|\cdot\|_X\}$ is a Banach space, then $\ker\{T\}$ is a closed subspace of X.

Let $T : \{X; \|\cdot\|\} \to \mathbb{R}$ be a linear functional. It is not assumed that T is bounded, nor that $\{X; \|\cdot\|\}$ is a Banach space. The mere linearity of T provides information on the structure of X in terms of the kernel of T.

Proposition 5.1 *Let X_o be a closed subspace of X such that the quotient space X/X_o is 1-dimensional. Then, there exists a nontrivial, bounded, linear functional $T : X \to \mathbb{R}$ such that $X_o = \ker\{T\}$.*

Conversely, let $T : X \to \mathbb{R}$ be a not identically zero linear functional. Then the quotient space $X/\ker\{T\}$ is 1-dimensional.

Proof To prove the first statement choose $x \in X - X_o$ such that dist$\{x; X_o\} > 0$ and write $X - X_o = \bigcup\{\lambda x | \lambda \in \mathbb{R}\}$. Every element $y \in X$ can be written as $y = x_o + \lambda x$ for some $x_o \in X_o$ and some $\lambda \in \mathbb{R}$. Then set $T(y) = \lambda$.

To establish the converse statement fix $x \in X - \ker\{T\}$. Such a choice is possible since $T \neq 0$. To show that $X = \text{span}\{x; X_o\}$, pick any element $y \in X$ and compute

$$T\left(y - \frac{T(y)}{T(x)}x\right) = T(y) - \frac{T(y)}{T(x)}T(x) = 0.$$

Therefore

$$y - \frac{T(y)}{T(x)}x \in \ker\{T\}.$$

Thus y is a linear combination of x and an element of $\ker\{T\}$.   ∎

Corollary 5.1 *Let $\{X; \|\cdot\|\}$ be a normed space. For any nonzero, linear functional $T : X \to \mathbb{R}$, the normed space $\{X; \|\cdot\|\}$ can be written as the direct sum of $\ker\{T\}$ and the 1-dimensional space spanned by an element in $X - \ker\{T\}$.*

Corollary 5.2 *Let T be a not identically zero linear functional on a normed space $\{X; \|\cdot\|\}$. Then T is continuous if and only if $\ker\{T\}$ is not dense in X. As a consequence T is continuous if and only if $\ker\{T\}$ is nowhere dense in X.*

Corollary 5.3 *Let T be a not identically zero, bounded, linear functional in a normed space $\{X; \|\cdot\|\}$. Then for every $x \in X - \ker\{T\}$*

$$\|T\| = \frac{|T(x)|}{\text{dist}\{x; \ker\{T\}\}}.$$

6   Equibounded Families of Linear Maps

If $\{X; \|\cdot\|\}$ is a Banach space, it is of second category. In such a case, the uniform boundedness principle can be applied to families of bounded linear maps in $\mathcal{B}(X; Y)$ in the following form.

Proposition 6.1 *Let $\mathcal{T}$ be a family of bounded linear maps from a Banach space $\{X; \|\cdot\|_X\}$ into a normed space $\{Y; \|\cdot\|_Y\}$. Assume that the elements of $\mathcal{T}$ are pointwise equibounded in X, i.e., for every $x \in X$, there exists a positive number $F(x)$, such that*

$$\|T(x)\|_Y \le F(x) \qquad for\ all \quad T \in \mathcal{T}.$$

Then, the elements of $\mathcal{T}$ are uniformly equibounded in $\mathcal{B}(X; Y)$, i.e., there exists a positive number M, such that

$$\|T(x)\|_Y \le M\|x\|_X \quad for\ all \quad T \in \mathcal{T} \quad and\ all \quad x \in X.$$

Proof The functions $\|T(x)\|_Y : X \to \mathbb{R}$ satisfy the assumptions of the Banach–Steinhaus theorem. Therefore there exists a positive F and a ball $B_\varepsilon(x_o) \subset X$, of radius ε and centered at some $x_o \in X$, such that

$$\|T(y)\|_Y \le F \quad for\ all \quad T \in \mathcal{T} \quad and\ all \quad y \in B_\varepsilon(x_o).$$

Given any nonzero $x \in X$, the element

$$y = x_o + \frac{\varepsilon}{2\|x\|_X}x$$

belongs to $B_\varepsilon(x_o)$. Therefore

$$\|T(x)\|_Y \le 2\frac{F + \|T(x_o)\|_Y}{\varepsilon}\|x\|_X \le \frac{4F}{\varepsilon}\|x\|_X$$

for all $T \in \mathcal{T}$. $\blacksquare$

Corollary 6.1 *Let $\{X; \|\cdot\|\}$ be a Banach space. Then a pointwise bounded family of elements in X^* is equibounded in X.*

6.1 Another Proof of Proposition 6.1

The proposition can be proved without appealing to category arguments. It suffices to establish that the elements of $\mathcal{T}$ are equiuniformly bounded in some open ball $B_\varepsilon(x_o) \subset X$. The proof proceeds by contradiction, assuming that such a ball does not exists [119].

Fix any such ball $B_1(x_o)$. There exists $x_1 \in B_1(x_o)$ and $T_1 \in \mathcal{T}$ such that $\|T_1(x_1)\|_Y > 1$. By continuity, there exists a ball $B_{\varepsilon_1}(x_1) \subset X$ such that $\|T_1(x)\|_Y > 1$ for all $x \in B_{\varepsilon_1}(x_1)$. By taking ε_1 sufficiently small, we may insure that,

$$\bar{B}_{\varepsilon_1}(x_1) \subset B_\varepsilon(x_o) \quad and \quad \varepsilon_1 < 1.$$

There exists $x_2 \in B_{\varepsilon_1}(x_1)$ and $T_2 \in \mathcal{T}$, such that $\|T_2(x_2)\|_Y > 2$. By continuity, there exists a ball $B_{\varepsilon_2}(x_2) \subset X$ such that $\|T_2(x)\|_Y > 2$ for all $x \in B_{\varepsilon_2}(x_2)$. By taking ε_2 sufficiently small, we may ensure that

$$\bar{B}_{\varepsilon_2}(x_2) \subset B_{\varepsilon_1}(x_1) \quad \text{and} \quad \varepsilon_2 < \tfrac{1}{2}.$$

Proceeding in this fashion we construct a sequence $\{x_n\}$ of elements of X, a countable family of balls $B_{\varepsilon_n}(x_n)$, and a sequence $\{T_n\}$ of elements of $\mathcal{T}$, such that

$$\bar{B}_{\varepsilon_{n+1}}(x_{n+1}) \subset B_{\varepsilon_n}(x_n) \quad \varepsilon_n < \tfrac{1}{n}$$

and

$$\|T_n(x)\|_Y > n \quad \text{for all } x \in B_{\varepsilon_n}(x_n).$$

The sequence $\{x_n\}$ is a Cauchy sequence and its limit x must belong to the closure of all $B_{\varepsilon_n}(x_n)$. Therefore $\|T_n(x)\|_Y > n$ for all $n \in \mathbb{N}$. This contradicts the assumption that the maps in $\mathcal{T}$ are pointwise equibounded. ∎

7 Contraction Mappings

A map T from a normed space $\{X; \|\cdot\|\}$ into itself is a *contraction* if there exists $t \in (0, 1)$ such that

$$\|T(x) - T(y)\| \le t\|x - y\| \quad \text{for all } x, y \in X.$$

Theorem 7.1 ([Banach [13]]) *Let T be a contraction form a Banach space $\{X; \|\cdot\|\}$ into itself. Then T has a unique fixed point, i.e., there exists a unique $x_o \in X$ such that $T(x_o) = x_o$.*

Proof Starting from an arbitrary $x_1 \in X$, define the sequence $x_{n+1} = T(x_n)$. Then

$$\|T(x_{n+1}) - T(x_n)\| \le t^n \|x_2 - x_1\|.$$

Equivalently
$$\|T(x_n) - x_n\| \le t^{n-1} \|x_2 - x_1\|.$$

Therefore, the sequences $\{x_n\}$ and $\{T(x_n)\}$ are both Cauchy sequences. If x_o is the limit of $\{x_n\}$, by continuity

$$\|T(x_o) - x_o\| = \lim \|T(x_n) - x_n\| = 0.$$

Thus $T(x_o) = x_o$. If $\bar{x}$ were another fixed point

$$\|T(\bar{x}) - T(x_o)\| \le t\|\bar{x} - x_o\| = t\|T(\bar{x}) - T(x_o)\|.$$

Thus $\bar{x} = x_o$. ∎

Remark 7.1 The theorem continues to hold in a complete metric space with the proper variants.

7.1 Applications to Some Fredholm Integral Equations

Let E be an open set in $\mathbb{R}^N$ and consider formally the integral equation [49]

$$f(x) = \int_E K(x, y) f(y) dy + h(x). \tag{7.1}$$

Assume the *kernel* $K(x, y)$ satisfies (4.3) for some given constant γ. Given a function $h \in L^1(E)$ one seeks a function $f \in L^1(E)$ satisfying (7.1) for a.e. $x \in E$.

Proposition 7.1 *Assume the constant γ in (4.3) is less than* 1. *Then the integral equation (7.1) has a unique solution.*

Proof The solution would be the unique fixed point of

$$T(f) = \int_E K(x, y) f(y) dy + h$$

provided $T : L^1(E) \to L^1(E)$ is a contraction. For $f, g \in L^1(E)$

$$\|T(f - g)\|_1 \leq \gamma \|f - g\|_1. \qquad \blacksquare$$

Remark 7.2 In the case of the Riesz kernel in (4.2), the assumption is satisfied if $\mu(E)$ is sufficiently small.

Remark 7.3 If the kernel $K(x, y)$ is as in (4.2) one gives a function $h \in L^\infty(E)$ and seeks a function $f \in L^\infty(E)$ satisfying (7.1). The integral equation (7.1) could be set in $L^2(E)$, provided $K \in L^2(E \times E)$. A more complete theory is in [34, 108, 162] Chapter IV.

8 The Open Mapping Theorem

A map T from a topological space $\{X; \mathcal{U}\}$ into a topological space $\{Y; \mathcal{V}\}$ is called *open* if it maps open sets of $\mathcal{U}$ into open sets of $\mathcal{V}$. If T is one-to-one and open, T^{-1} is continuous. An open map $T : X \to Y$ which is continuous, one-to-one, and onto, is a homeomorphism between $\{X; \mathcal{U}\}$ and $\{Y; \mathcal{V}\}$.

The next theorem called the *Open Mapping Theorem* states that continuous linear maps between Banach spaces are open mappings.

Theorem 8.1 (Open Mapping Theorem) *A bounded linear map T from a Banach space $\{X; \|\cdot\|_X\}$ onto a Banach space $\{Y; \|\cdot\|_Y\}$ is an open mapping. If T is also one-to-one, it is a homeomorphism between $\{X; \|\cdot\|_X\}$ and $\{Y; \|\cdot\|_Y\}$.*

Remark 8.1 The requirement that T be linear cannot be removed. Let $T(x) = e^x \cos x : \mathbb{R} \to \mathbb{R}$, where $\mathbb{R}$ is endowed with the Euclidean norm. Then T is continuous but not open, since the image of $(-\infty, 0)$ is not open.

Let B_ρ denote the open ball in X of radius ρ and centered at the origin of X. Denote also by $\mathcal{B}_\varepsilon$ the open ball in Y of radius ε and centered at the origin of Y.

Lemma 8.1 $T(B_1)$ *contains a ball $\mathcal{B}_\varepsilon$ for some $\varepsilon > 0$.*

Proof Since T is linear and onto

$$X = \bigcup n B_{1/2} \quad \text{and} \quad Y = \bigcup n T(B_{1/2}).$$

Since $\{Y; \|\cdot\|_Y\}$ is of second category, $T(B_{1/2})$ is not nowhere dense and its closure contains an open ball $\mathcal{B}_{2\varepsilon}(\bar{y})$ in Y centered at some $\bar{y}$. The inclusions

$$\overline{T(B_{1/2})} - \bar{y} \subset \overline{T(B_{1/2})} - \overline{T(B_{1/2})} \subset 2\overline{T(B_{1/2})} \subset \overline{T(B_1)}$$

imply that $\mathcal{B}_{2\varepsilon} \subset \overline{T(B_1)}$. The linearity of T also implies

$$\mathcal{B}_{2\varepsilon/2^n} \subset \overline{T\left(B_{1/2^n}\right)} \quad \text{for all } n \in \mathbb{N}. \tag{8.1}$$

We next show that $\mathcal{B}_\varepsilon \subset T(B_1)$. Fix $y \in \mathcal{B}_\varepsilon$. Since $y \in \overline{T(B_{1/2})}$ there exist $x_1 \in B_{1/2}$ such that

$$\|y - T(x_1)\|_Y < \tfrac{1}{2}\varepsilon.$$

In particular the element $y - T(x_1)$ belongs to $\mathcal{B}_{\varepsilon/2}$. Therefore, by (8.1) for $n = 2$, there exist $x_2 \in B_{1/4}$ such that

$$\|y - T(x_1) - T(x_2)\|_Y < \tfrac{1}{4}\varepsilon.$$

Proceeding in this fashion we find a sequence $\{x_n\}$ of elements of X such that $x_n \in B_{1/2^n}$ and

$$\left\| y - \sum_{j=1}^n T(x_j) \right\|_Y \le \frac{1}{2^n}\varepsilon.$$

Since $\|x_n\|_X < 2^{-n}$, the series $\sum x_n$ is absolutely convergent and identifies an element $x = \sum x_n$ in the ball B_1. Since T is linear and continuous

$$T(x) = T\left(\sum x_n\right) = \sum T(x_n) = y.$$

Thus $y \in T(B_1)$. Since y is an arbitrary element of $\mathcal{B}_\varepsilon$ this implies that $\mathcal{B}_\varepsilon \subset T(B_1)$. ∎

Proof (*of the Open Mapping Theorem*) Let $\mathcal{O} \subset X$ be open. To establish that $T(\mathcal{O})$ is open in Y, for every fixed $\xi \in \mathcal{O}$ and $\eta = T(\xi)$ we exhibit a ball $\mathcal{B}_\varepsilon(\eta)$ contained in $T(\mathcal{O})$. Since $\mathcal{O}$ is open there exists an open ball $B_\sigma(\xi) \subset \mathcal{O}$. Then by Lemma 8.1, $T(B_\sigma(\xi))$ contains a ball $\mathcal{B}_\varepsilon(\eta)$. $\blacksquare$

8.1 Some Applications

The open mapping theorem may be applied to finding conditions for two different norms on the same vector space, to generate the same topology.

Proposition 8.1 *Let* $\| \cdot \|_1$ *and* $\| \cdot \|_2$ *be two norms on the same vector space* X, *by which* $\{X; \| \cdot \|_1\}$ *and* $\{X; \| \cdot \|_2\}$ *are both Banach spaces. Assume that there exists a positive constant* C_1 *such that*

$$\|x\|_2 \leq C_1 \|x\|_1 \quad \textit{for all} \quad x \in X. \tag{8.2}$$

Then, there exists a positive constant C_2 *such that*

$$\|x\|_1 \leq C_2 \|x\|_2 \quad \textit{for all} \quad x \in X. \tag{8.3}$$

Proof The identity map $T(x) = x$, from $\{X; \| \cdot \|_1\}$ onto $\{X; \| \cdot \|_2\}$ is linear and one-to-one. By (8.2) is also continuous. Therefore it is a homeomorphism. In particular the inverse T^{-1} is linear and continuous, and hence bounded. $\blacksquare$

8.2 The Closed Graph Theorem

Let T be a linear map from a Banach space $\{X; \| \cdot \|_X\}$ into a Banach space $\{Y; \| \cdot \|_Y\}$. The graph $\mathcal{G}_T$ of T is a subset of $X \times Y$ defined by

$$\mathcal{G}_T = \bigcup\{(x, T(x)) \mid x \in X\}.$$

The closure of $\mathcal{G}_T$ is meant in the product topology on $X \times Y$. In particular the graph $\mathcal{G}_T$ is closed if and only if whenever $\{x_n\}$ is a Cauchy sequence in $\{X; \| \cdot \|_X\}$ and $\{T(x_n)\}$ is a Cauchy sequence in $\{Y; \| \cdot \|_Y\}$, then

$$\lim T(x_n) = T(\lim x_n). \tag{8.4}$$

This would hold true if T were continuous. The next theorem, called *Closed Graph theorem* states the converse, i.e., (8.4) implies that T is continuous.

Theorem 8.2 (Closed Graph Theorem) *Let T be a linear map from a Banach space $\{X; \|\cdot\|_X\}$ into a Banach space $\{Y; \|\cdot\|_Y\}$. If $\mathcal{G}_T$ is closed, then T is continuous.*

Proof On X introduce a new norm $\|x\| = \|x\|_X + \|T(x)\|_Y$. One verifies this is a norm on X and if $\mathcal{G}_T$ is closed, $\{X; \|\cdot\|\}$ is complete. Therefore if $\mathcal{G}_T$ is closed, $\{X; \|\cdot\|\}$ is a Banach space. Since $\|x\|_X \leq \|x\|$, by Proposition 8.1 there exists a positive constant C such that

$$\|x\|_X + \|T(x)\|_Y \leq C\|x\|_X \quad \text{for all } x \in X.$$

This implies that $\|T(x)\|_Y \leq C\|x\|_X$. Thus T is bounded and hence continuous. ∎

Remark 8.2 The assumption that both $\{X; \|\cdot\|_X\}$ and $\{Y; \|\cdot\|_Y\}$ be Banach spaces is essential for the Closed Graph theorem to hold. Indeed, without such a completeness requirement, there is no relationship between the continuity of a linear map $T :$ $X \to Y$ and the closedness of its graph $\mathcal{G}_T$, as illustrated by the following two counterexamples.

Let $C[0, 1]$ be endowed with the sup-norm. Let also $C^1[0, 1] \subset C[0, 1]$ be equipped with the topology inherited from the norm-topology of $C[0, 1]$. The map $T = \frac{d}{dt}$ from $C^1[0, 1]$ into $C[0, 1]$, is linear, has closed graph, but is discontinuous.

Regard $C[0, 1]$ as a topological subspace of $L^2[0, 1]$. The identity map from $C[0, 1]$ into $L^2[0, 1]$ is linear but its graph is not closed.

9   The Hahn–Banach Theorem

Let X be a linear vector space over the reals. A sub-linear, homogeneous, real-valued map from X into $\mathbb{R}$, is a function $p : X \to \mathbb{R}$, satisfying

$$p(x + y) \leq p(x) + p(y) \quad \text{for all } x, y \in X$$
$$p(\lambda x) = \lambda p(x) \qquad\qquad \text{for all } x \in X \text{ and all } \lambda > 0.$$

The *Dominated Extension Theorem* states that a real-valued, linear functional T_o defined on a linear vector subspace X_o of X, and dominated by a sub-linear map p, can be extended into a linear map $T : X \to \mathbb{R}$, in such a way that the extended map is dominated by p in the whole X.

While the main applications of this extension procedure are in Banach spaces, it is worth noting that such an extension is algebraic in nature and topology-independent. In particular while X is required to have a vector structure, it is not required to be a topological vector space. Likewise no topological assumptions, such as continuity, are placed on the sub-linear map p nor on the linear functional T_o defined on X_o.

Theorem 9.1 (Hahn–Banach [60, 11]) *Let X be a real vector space and let p : $X \to \mathbb{R}$ be sub-linear and homogeneous. Then, every linear functional $T_o : X_o \to \mathbb{R}$ defined on a subspace X_o of X and satisfying*

$$T_o(x) \le p(x) \quad \textit{for all } x \in X_o$$

admits a linear extension $T : X \to \mathbb{R}$ such that

$$T(x) \le p(x) \quad \textit{for all } x \in X$$

Proof If $X_o \neq X$ choose $\eta \in X - X_o$ and let X_η be the linear span of X_o and η. First we extend T_o to a linear functional T_η defined in X_η, coinciding with T_o on X_o, and dominated by p on X_η. If T_η is such an extension, then by linearity

$$T_\eta(x + \lambda\eta) = T_o(x) + \lambda T_\eta(\eta)$$

for all $\lambda \in \mathbb{R}$ and all $x \in X_o$. Therefore, to construct T_η it suffices to specify its value at η. Since T_o is dominated by the sub-linear map p on X_o

$$T_o(x) + T_o(y) = T_o(x + y) \le p(x + y) \le p(x - \eta) + p(y + \eta)$$

for all $x, y \in X_o$. Therefore

$$\sup_{x \in X_o} \{T_o(x) - p(x - \eta)\} \le \inf_{y \in X_o} \{p(y + \eta) - T_o(y)\}.$$

Define $T_\eta(\eta) = \alpha$ where α is any number satisfying

$$\sup_{x \in X_o} \{T_o(x) - p(x - \eta)\} \le \alpha \le \inf_{y \in X_o} \{p(y + \eta) - T_o(y)\}. \tag{9.1}$$

By construction $T_\eta : X_\eta \to \mathbb{R}$ is linear and it coincides with T_o on X_o. It remains to show that such an extended functional is dominated by p in X_η, that is

$$T_\eta(x + \lambda\eta) = T_o(x) + \lambda\alpha \le p(x + \lambda\eta)$$

for all $x \in X_o$ and all $\lambda \in \mathbb{R}$. If $\lambda > 0$, by the upper inequality in (9.1)

$$\lambda\alpha + T_o(x) = \lambda\left[\alpha + T_o\left(\frac{x}{\lambda}\right)\right]$$
$$\le \lambda\left[p\left(\frac{x}{\lambda} + \eta\right) - T_o\left(\frac{x}{\lambda}\right) + T_o\left(\frac{x}{\lambda}\right)\right] \le p(x + \lambda\eta).$$

If $\lambda < 0$ the same conclusion holds by using the lower inequality in (9.1).

If $X_\eta \neq X$, the construction can be repeated by extending T_η to a larger subspace of X. The extension of T_o to the whole X can now be concluded by induction.

Introduce the set $\mathcal{E}$ of all the dominated extensions of T_o, i.e., the set of pairs $\{T_\eta; X_\eta\}$ where T_η is a linear functional defined on X_η such that $T_\eta(x) \le p(x)$ for all $x \in X_\eta$ and $T_\eta(x) = T_o(x)$ for all $x \in X_o$. On the set $\mathcal{E}$ introduce an ordering relation by stipulating that

$$\{T_\xi; X_\xi\} \le \{T_\eta; X_\eta\} \quad \text{if and only if} \quad X_\xi \subset X_\eta \quad \text{and} \quad T_\eta = T_\xi \text{ on } X_\xi.$$

One verifies that such a relation is a partial ordering on $\mathcal{E}$.

Every linearly ordered subset $\mathcal{E}' \subset \mathcal{E}$ has an upper bound. Indeed, denoting by $\{T_\sigma; X_\sigma\}$ the elements of $\mathcal{E}'$, setting $X' = \bigcup X_\sigma$ and $T' = T_\sigma$ on X_σ, provides an upper bound for $\mathcal{E}'$. It follows by Zorn's lemma that $\mathcal{E}$ has a maximal element $\{T; \widetilde{X}\}$. For such a maximal element, $\widetilde{X} = X$. Indeed otherwise $X - \widetilde{X}$ would be not empty and the extension process could be repeated, contradicting the maximality of $\{T; \widetilde{X}\}$. ∎

10 Some Consequences of the Hahn–Banach Theorem

The main applications of the Hahn–Banach theorem occur in normed spaces $\{X; \|\cdot\|\}$ and for a suitable choice of the dominating sub-linear function p. For example p could be a norm or a semi-norm in X. If X_o is a subspace of $\{X; \|\cdot\|\}$, let $\{X_o; \|\cdot\|\}$ denote the corresponding normed space and let X_o^* be its dual. Typically, given $T_o \in X_o^*$ dominated by some sub-linear function p, one seeks to extend it to an element $T \in X^*$.

Proposition 10.1 *Let $\{X; \|\cdot\|\}$ be a normed space and let X_o be a subspace of X. Every $T_o \in X_o^*$ admits an extension $T \in X^*$ such that $\|T\| = \|T_o\|$.*

Proof Apply the Hahn–Banach theorem with $p(x) = \|T_o\|\|x\|$. This gives an extension T defined in X and satisfying

$$|T(x)| \le \|T_o\|\|x\| \quad \text{for all } x \in X.$$

Therefore $\|T\| \le \|T_o\|$. On the other hand $\|T\| \ge \|T_o\|$, since T is an extension of T_o. ∎

Proposition 10.2 *Let $\{X; \|\cdot\|\}$ be a normed space. For every $x_o \in X$, and $x_o \ne \Theta$, there exists $T \in X^*$ such that $\|T\| = 1$ and $T(x_o) = \|x_o\|$.*

Proof Having fixed $x_o \in X$ let $X_o = \{\lambda x_o | \lambda \in \mathbb{R}\}$ be the span of x_o. On X_o consider the functional $T_o(\lambda x_o) = \lambda\|x_o\|$ and as p take the norm $\|\cdot\|$. By the Hahn–Banach theorem T_o can be extended into a linear functional T defined in the whole X and satisfying

$$T(x) \le \|x\| \text{ for all } x \in X \quad \text{and} \quad T(\lambda x_o) = \lambda\|x_o\| \text{ for all } \lambda \in \mathbb{R}.$$

The first implies $\|T\| \le 1$. The second $\|T\| = 1$. ∎

Corollary 10.1 *Let $\{X; \|\cdot\|\}$ be a normed space. Then X^* separates the points of X, i.e., for any pair x, y of distinct points of X, there exists $T \in X^*$ such that $T(x) \ne T(y)$.*

Proof Apply Proposition 10.2 to the element $x - y$. ∎

Corollary 10.2 *Let $\{X; \|\cdot\|\}$ be a normed space. Then, for every $x \in X$*

$$\|x\| = \sup_{T \in X^*; T \ne 0} \frac{|T(x)|}{\|T\|} = \sup_{T \in X^*; \|T\|=1} |T(x)|.$$

Proposition 10.3 *Let $\{X; \|\cdot\|\}$ be a normed space and let X_o be a linear subspace of X. Assume that there exists an element $\eta \in X - X_o$ that has positive distance from X_o, i.e.,*

$$\inf_{x \in X_o} \|x - \eta\| \ge \delta > 0.$$

There exists $T \in X^*$ such that

$$\|T\| \le 1 \quad T(\eta) = \delta \quad \text{and} \quad T(x) = 0 \quad \text{for all } x \in X_o.$$

Proof Let X_η be the span of X_o and η. On X_η define the linear functional $T_\eta(\lambda\eta + x) = \lambda\delta$ for all $x \in X_o$ and all $\lambda \in \mathbb{R}$. From the definition of δ for $\lambda \ne 0$ and $x \in X_o$

$$\lambda\delta \le |\lambda|\left\|\eta + \frac{x}{\lambda}\right\| = \|\lambda\eta + x\|.$$

Therefore, denoting by $y = \lambda\eta + x$ the generic element of X_η

$$T_\eta(y) \le \|y\| \quad \text{for all } y \in X_\eta.$$

By the Hahn–Banach theorem, T_η has an extension T defined in the whole X and satisfying $T(x) \le \|x\|$ for all $x \in X$. Therefore $\|T\| \le 1$. Moreover $T(x) = 0$ for all $x \in X_o$ and $T(\eta) = \delta$. ∎

Remark 10.1 The assumptions of Proposition 10.3 are verified if X_o is a proper, closed linear subspace of X.

Corollary 10.3 *Let $\{X; \|\cdot\|\}$ be a normed space and let X_o be a linear subspace of X. If X_o is not dense in X there exists a nonzero functional $T \in X^*$ such that $T(x) = 0$ for all $x \in X_o$.*

Proposition 10.4 *Let $\{X; \|\cdot\|\}$ be a normed space. Then if X^* is separable also $\{X; \|\cdot\|\}$ is separable.*

Proof Let $\{T_n\}$ be a sequence dense in X^*. For each T_n choose $x_n \in X$ such that $\|x_n\| = 1$ and $|T_n(x_n)| \geq \frac{1}{2}\|T_n\|$. Let now X_o be the set of all finite linear combinations of elements of $\{x_n\}$ with rational coefficients. The set X_o is countable and we claim it is dense in X. If not, the closure $\bar{X}_o$ is a linear, closed, proper subspace of X. By Corollary 10.3, there exists a nonzero functional $T \in X^*$ vanishing on X_o. Since $\{T_n\}$ is dense in X^*, there exists a subsequence $\{T_{n_j}\}$ convergent to T. For such a sequence

$$\|T - T_{n_j}\| \geq |(T - T_{n_j})(x_{n_j})| \geq \frac{1}{2}\|T_{n_j}\|.$$

Thus $\{T_{n_j}\} \to 0$, contradicting that T is nonzero.   ∎

Remark 10.2 The converse of Proposition 10.4 is in general false, i.e., X separable does not imply that X^* is separable. As an example consider $L^1(E)$ where E is an open set in $\mathbb{R}^N$ with the Lebesgue measure. By the Riesz representation theorem, $L^1(E)^* = L^\infty(E)$, up to an isometric isomorphism. However $L^1(E)$ is separable and $L^\infty(E)$ is not.

10.1 Tangent Planes

Let $\{X; \|\cdot\|\}$ be a normed space. For a fixed $T \in X^*$ and $\alpha \in \mathbb{R}$, set

$$[T = \alpha] = \{x \in X \mid T(x) = \alpha\}$$

and introduce analogously the sets $[T > \alpha]$ and $[T < \alpha]$. In analogy with linear functionals in Euclidean spaces, $[T = \alpha]$ is called an *hyperplane* in X, and divides X into two disjoint *half-spaces* $[T > \alpha]$ and $[T < \alpha]$.

Let now C be a set in X and let $x_o \in C$. An hyperplane $[T = \alpha]$ is *tangent* to C at x_o, if $x_o \in [T = \alpha]$, and $T(x) \leq T(x_o)$ for all $x \in C$.

With this terminology, Proposition 10.2 can be rephrased in the following geometric form.

Proposition 10.5 *The unit ball of X has a tangent plane at any one of its boundary points.*

11 Separating Convex Subsets of a Hausdorff, Topological Vector Space $\{X; \mathcal{U}\}$

As indicated earlier, the Hahn–Banach theorem, holds in linear vector spaces, with no particular topological structure, and hinges upon specifying a suitable, dominating, homogeneous, sub-linear function $p : X \to \mathbb{R}$. For topological vector spaces $\{X; \mathcal{U}\}$, the Minkowski functional $\mu_C(\cdot)$ defined below, is one such a function.

Let $\{X;\mathcal{U}\}$ be a Hausdorff, linear, topological vector space and let C be a non-empty, convex, open neighborhood of the origin of X. Then for every $x \in X$ there exists some positive number t such that $x \in tC$. Define

$$\mu_C(x) = \inf\{t > 0 \mid x \in tC\}.$$

If C is the unit ball of a normed space $\{X; \|\cdot\|\}$, then $\mu_{B_1}(x) = \|x\|$ for all $x \in X$. An element x is in C if and only if $\mu_C(x) < 1$. If C is unbounded, then $\mu_C(x)$ vanishes for $x = 0$ and for infinitely many nonzero elements of X. It follows from the definition that $\lambda\mu_C(x) = \mu_C(\lambda x)$, for all $\lambda > 0$.

Proposition 11.1 *The map* $x \to \mu_C(x)$ *is sub-linear in* X.

Proof Fix $x, y \in X$ and let t and s be positive numbers such that $\mu_C(x) < t$ and $\mu_C(y) < s$. For such choices, $t^{-1}x \in C$ and $s^{-1}y \in C$. Then, since C is convex

$$\frac{1}{s+t}(x+y) = \frac{t}{s+t}t^{-1}x + \frac{s}{s+t}s^{-1}y \in C.$$

Therefore $\mu(x+y) \leq s + t$. ∎

By Corollary 10.1 the collection X^* of the bounded linear functionals on a normed space $\{X; \|\cdot\|\}$ separates points. The next more general proposition asserts that any two nonempty, convex subsets of a topological vector space $\{X;\mathcal{U}\}$ can be separated by a nontrivial, linear, continuous functional T, provided one of them is open.

Proposition 11.2 *Let C_1 and C_2 be two disjoint, convex subsets of a Hausdorff, linear, topological vector space $\{X;\mathcal{U}\}$, and assume C_1 is open. There exists a non-trivial, linear, continuous functional T on X, and $\alpha \in \mathbb{R}$ such that*

$$T(x) < \alpha \leq T(y) \quad \text{for all } x \in C_1 \quad \text{and all } y \in C_2. \tag{11.1}$$

Proof Fix some $x_1 \in C_1$ and $x_2 \in C_2$ and set

$$C = C_1 - C_2 + x_o \qquad \text{where } x_o = x_2 - x_1.$$

The set C is open, convex and it contains the origin. Since C_1 and C_2 are disjoint, $x_o \notin C$. On the one-dimensional span of x_o, define a bounded linear functional by $T_o(\lambda x_o) = \lambda$. Such a functional is pointwise bounded above, on the span of x_o, by the Minkowski functional μ_C relative to the set C. Indeed, for $\lambda \geq 0$

$$T_o(\lambda x_o) = \lambda \leq \lambda\mu_C(x_o) = \mu_C(\lambda x_o).$$

If $\lambda < 0$ then $T_o(\lambda x_o) = \lambda \leq \mu_C(\lambda x_o)$. By the Hahn–Banach theorem, there exists a linear functional T on X that coincides with T_o on the span of x_o and such that $T(x) \leq \mu_C(x)$ for all $x \in X$. Using the definition of $\mu_C(\cdot)$ and the linearity of T, one verifies that $T(x) \leq 1$ for all $x \in C$ and $T(x) \geq -1$ for all $x \in -C$. Therefore

$|T(x)| \leq 1$ for all $x \in C \cap (-C)$. Thus T is bounded in a neighborhood of the origin and hence is continuous (Proposition 11.1 of Chap. 2). For $x \in C_1$ and $y \in C_2$, the element $x - y + x_o$ is in C. Therefore $\mu_C(x - y + x_o) < 1$ since C is open. Using that $T(x_o) = 1$, compute

$$T(x - y + x_o) = T(x) - T(y) + 1 \leq \mu_C(x - y + x_o) < 1.$$

Thus $T(x) < T(y)$. The existence of α satisfying (11.1) follows since $T(C_1)$ and $T(C_2)$ are convex subsets of $\mathbb{R}$ and $T(C_1)$ is open. ∎

Remark 11.1 The topological structure of C_1 and C_2 is essential for the separation statement of Proposition 11.2 to hold, as shown by the following counterexample.

Let C_1 and C_2 be the two convex, subsets of $L^2[0, 1]$ defined by

$$C_1 = \{f \in C[0, 1] \text{ nonnegative and vanishing only at } t = 0\}$$
$$C_2 = \{f \in C[0, 1] \text{ nonnegative and vanishing only at } t = 1\}.$$

One verifies that $C_1 - C_2$ is dense in $L^2[0, 1]$ and hence C_1 and C_2 cannot be separated by any bounded linear functional.

11.1 Separation in Locally Convex, Hausdorff, Topological Vector Spaces $\{X; \mathcal{U}\}$

A linear functional T on a topological vector space $\{X; \mathcal{U}\}$ *strictly* separates two disjoint, convex sets C_1 and C_2 if there exist real numbers $\alpha < \beta$ such that

$$T(x) < \alpha < \beta \leq T(y) \quad \text{for all } x \in C_1 \quad \text{and all } y \in C_2. \tag{11.2}$$

The previous proposition gives sufficient conditions of separations but not strict separation. In general strict separation under the assumptions of Proposition 11.2 is not expected to hold. For example in $\mathbb{R}^2$, the two sets $[x < 0]$ and $[x \geq 0]$ are convex and disjoint, and one of them is open. Yet they are not strictly separated.

If the topology of $\{X; \mathcal{U}\}$ is locally convex, the next proposition gives some more stringent conditions on C_2 for strict separation to occur.

Proposition 11.3 *Let C_1 and C_2 be two nonempty, disjoint, convex subsets of a locally convex, Hausdorff, topological vector space $\{X; \mathcal{U}\}$. Assume that C_1 is compact and C_2 is closed. There exists a bounded linear functional T on X and real numbers $\alpha < \beta$ such that (11.2) holds.*

Proof We claim that there exists a convex, open neighborhood $\mathcal{O}$ of the origin such that $C_1 + \mathcal{O}$ is open, convex and does not intersect C_2. Whence the claim is established, Proposition 11.3 follows from Proposition 11.2 applied to the pair of sets $C_1 + \mathcal{O}$ and C_2.

To establish the claim, observe that, since C_2 is closed and $C_1 \cap C_2 = \emptyset$, for every $x \in C_1$ there exists a convex neighborhood of the origin $\mathcal{O}_x$ such that $x + \mathcal{O}_x$ does not intersect C_2. By possibly replacing $\mathcal{O}_x$ with $\mathcal{O}_x \cap (-\mathcal{O}_x)$ we may assume that $\mathcal{O}_x$ is symmetric. By the continuity of the sum and product in $\{X; \mathcal{U}\}$ the sets $\frac{1}{2}\mathcal{O}_x$ are open. The collection $\{x + \frac{1}{2}\mathcal{O}_x\}$ for $x \in C_1$ is an open covering of C_1 from which we extract a finite one

$$\left\{ x_1 + \tfrac{1}{2}\mathcal{O}_{x_1}, \dots, x_n + \tfrac{1}{2}\mathcal{O}_{x_n} \right\}.$$

Setting

$$\mathcal{O} = \bigcap_{j=1}^{n} \tfrac{1}{2}\mathcal{O}_{x_j}$$

the set $C_1 + \mathcal{O}$ is open, convex and does not intersect C_2, since

$$C_1 + \mathcal{O} \subset \bigcup_{j=1}^{n} \left(x_j + \tfrac{1}{2}\mathcal{O}_{x_j} + \mathcal{O} \right) \subset \bigcup_{j=1}^{n} \left(x_j + \tfrac{1}{2}\mathcal{O}_{x_j} + \tfrac{1}{2}\mathcal{O}_{x_j} \right)$$

and none of the sets of this union intersects C_2. ∎

Corollary 11.1 *Let C be a closed, convex subset of a locally convex, Hausdorff, topological vector space $\{X; \mathcal{U}\}$. Then C is the intersection of all the closed half-spaces $[T \geq \alpha]$ that contain C.*

Corollary 11.2 *The collection of bounded, linear functionals on a locally convex, Hausdorff, topological vector space $\{X; \mathcal{U}\}$, separates the points of X.*

12 Weak Topologies

Let $\{X; \|\cdot\|\}$ be a normed space and let X^* be its dual. The topology generated on X by its norm $\|\cdot\|$ is called the *strong* topology of X.

Since any $T \in X^*$ is a continuous linear map from X into $\mathbb{R}$, the inverse image of any open set in $\mathbb{R}$ is open in the strong topology of X. Set

$$\mathcal{B} = \left\{ \begin{array}{c} \text{the collection of the finite intersections of the inverse images} \\ T^{-1}(O) \text{ where } T \in X^* \text{ and } O \text{ are open subsets of } \mathbb{R} \end{array} \right\}.$$

The collection $\mathcal{B}$ is a base for a topology $\mathcal{W}$ on X called the *weak topology* of $\{X; \|\cdot\|\}$. The collection $\mathcal{B}$ satisfies the requirements (i)–(ii) of § 4 of Chap. 2, to be a base for a topology. The corresponding topology is constructed by the procedure of Proposition 4.1 of Chap. 2.

The weak topology $\mathcal{W}$ on X is the weakest topology for which all the functionals $T \in X^*$ are continuous. The procedure is similar to the construction of a product topology in the Cartesian product of topological spaces, as the weakest topology for which all the projections are continuous (§ 4.1 of Chap. 2).

One verifies that the operations of sum $+ : X \times X \to X$ and multiplication by scalars $\bullet : \mathbb{R} \times X \to X$, are continuous with respect to such a topology (see Proposition 12.1c of the Complements). Thus $\{X; \mathcal{W}\}$ is a topological vector space.

The topology of $\mathcal{W}$ is translation invariant and is determined by a local base at the origin of X. A neighborhood of the origin, in the topology $\mathcal{W}$, contains an element of $\mathcal{B}$ of the form

$$\mathcal{O} = \bigcap_{j=1}^{n} T_j^{-1}(-\alpha_j, \alpha_j), \quad \text{where } T_j \in X^*, \tag{12.1}$$

for some finite n and $\alpha_j > 0$. Equivalently

$$\mathcal{O} = \{x \in X \mid |T_j(x)| < \alpha_j \quad \text{for } j = 1, \dots, n\}. \tag{12.2}$$

The collection of open sets of the form (12.1) forms a local base $\mathcal{B}_o$ at the origin, for the weak topology $\mathcal{W}$. These open sets are convex, since T_j are linear. Thus $\mathcal{W}$ is a locally convex topology. A sequence $\{x_n\}$ of elements of X converges weakly to 0 if and only if every weak neighborhood of the origin contains all but finitely many elements of $\{x_n\}$. From (12.2) it follows that $\{x_n\}$ converges weakly to 0 if and only if $\{T(x_n)\} \to 0$ for all $T \in X^*$. More generally, $\{x_n\}$ converges weakly to some $x_o \in X$ if and only if $\{T(x_n)\} \to T(x_o)$, for all $T \in X^*$. Strong convergence implies weak convergence, but the converse is false (see § 9 of Chap. 6).

12.1 Weak Boundedness

The weak topology $\mathcal{W}$ contains, roughly speaking, fewer open sets than the strong topology. The two topologies are markedly different also in terms of local boundedness.

A set $E \subset X$ is weakly bounded if and only if for every $\mathcal{O} \in \mathcal{B}_o$ there exists some positive number t, depending upon $\mathcal{O}$, such that $E \subset t\mathcal{O}$.

In view of the structure (12.1)–(12.2) of the open sets of $\mathcal{B}_o$, a set $E \subset X$ is bounded in the weak topology of X if and only if for every $T \in X^*$, there exists a positive number γ_T such that $|T(x)| \le \gamma_T$ for all $x \in E$.

Proposition 12.1 Let X be infinite dimensional. Then every weak neighborhood of the origin, contains an infinite dimensional subspace X_o.

Proof Having fixed a weak neighborhood of the origin, we may assume is of the form (12.1)–(12.2) and set

$$X_o = \bigcap_{j=1}^{n} \ker\{T_j\}.$$

This is a subspace of X and $X_o \subset \mathcal{O}$. To prove that X_o is infinite dimensional, consider the map $F : X \to \mathbb{R}^n$ defined by

$$X \ni x \to F(x) = (T_1(x), \dots, T_n(x)) \in \mathbb{R}^n.$$

Such a map is linear, continuous and its kernel is X_o. It is also a one-to-one map between the quotient space X/X_o and $\mathbb{R}^n$. Thus $\dim\{X\} \leq n + \dim\{X_o\}$. ■

Corollary 12.1 *Let X be infinite dimensional. Then every weak neighborhood of the origin, is unbounded.*

In particular a ball B_ρ, open in the strong topology of $\{X; \|\cdot\|\}$ cannot contain any weak neighborhood of the origin.

Corollary 12.2 *The weak topology of an infinite dimensional normed space does not satisfy the first axiom of countability.*

Proof Assume by contradiction that $\{\mathcal{O}_n\}$ is a countable base for the weak topology at the origin. By Proposition 12.1 and Corollary 12.1 one may pick

$$x_n \in \bigcap_{j=1}^{n} \mathcal{O}_j \quad \text{such that} \quad \|x_n\| \geq n.$$

The sequence $\{x_n\}$ is not strongly bounded and hence not weakly bounded. Thus there exists a weak neighborhood of the origin $\mathcal{O}$ such that $\lambda\mathcal{O}$ does not contain $\{x_n\}$ for all $\lambda\mathbb{R}$. On the other hand $\mathcal{O}_n \subset \mathcal{O}$ for some n, since $\{\mathcal{O}_n\}$ is a base for the weak topology at the origin. Therefore $\mathcal{O}$ contains all but finitely many elements of $\{x_n\}$. ■

Corollary 12.3 *The weak topology of an infinite dimensional normed space is not metrizable, i.e., there is no metric $d(\cdot, \cdot)$ on $X \times X$ that generates its weak topology.*

Proof If the weak topology of $\{X; \|\cdot\|\}$ were metrizable, it would satisfy the first axiom of countability.   ■

12.2 Weakly and Strongly Closed Convex Sets

Sets that are closed in the weak topology are also closed in the strong topology. The converse, while false in general, holds for convex sets.

Proposition 12.2 (Mazur [105]) *Let E be a convex subset of a normed space $\{X; \|\cdot\|\}$. Then, the weak closure of E coincides with its strong closure.*

Proof Denote by $\bar{E}_w$ and $\bar{E}_s$ the closure of E in the weak and, respectively, strong topology. Since $\mathcal{W}$ is weaker than the strong topology, $\bar{E}_s \subset \bar{E}_w$. For the converse inclusion it suffices to show that $X - \bar{E}_s$ is a weakly open set. This in turn would follow if every point $x_o \in X - \bar{E}_s$ admits a weakly open neighborhood not intersecting $\bar{E}_s$.

Having fixed $x_o \in X - \bar{E}_s$, consider the two disjoint, closed, convex sets x_o and $\bar{E}_s$. Since x_o is compact, by Proposition 11.3, there exists $T \in X^*$ and $\alpha \in \mathbb{R}$, such that

$$T(x_o) < \alpha < T(x) \qquad \text{for all } x \in \bar{E}_s.$$

Therefore the half-space $[T < \alpha]$ is a weakly open neighborhood of x_o that does not intersect $\bar{E}_s$. ∎

Corollary 12.4 *Let $\{X; \|\cdot\|\}$ be a normed space. Then any weakly closed subspace $X_o \subset X$, is also strongly closed.*

Corollary 12.5 *Let $\{X; \|\cdot\|\}$ be a normed space and let $\{x_n\}$ be a sequence of elements of X converging weakly to some $x \in X$. Then there exists a sequence $\{y_m\}$ of elements of X, such that each y_m is the convex combination of finitely many x_n, that is,*

$$y_m = \sum_{j=1}^{n_m} \alpha_j x_{n_j} \quad \text{where } \alpha_j > 0 \quad \text{and} \quad \sum_{j=1}^{n_m} \alpha_j = 1$$

and $\{y_m\} \to x$ strongly.

Proof Let $c(\{x_n\})$ be convex hull of $\{x_n\}$ and denote by $\overline{c(\{x_n\})}_w$ its weak closure. By assumption the weak limit x belongs to $\overline{c(\{x_n\})}_w$. The conclusion follows since weak and strong closure coincide. ∎

13 Reflexive Banach Spaces

Let $\{X; \|\cdot\|\}$ be a normed space and let X^* be its dual. The collection of all bounded linear functionals $f : X^* \to \mathbb{R}$ is denoted by X^{**} and is called the *double dual* of the *second dual* of X. It is a Banach space by the norm

$$\|f\| = \sup_{T \in X^*; T \neq 0} \frac{f(T)}{\|T\|} = \sup_{T \in X^*; \|T\|=1} f(T). \tag{13.1}$$

Every element $x \in X$ identifies an element $f_x \in X^{**}$ by the formula

$$X^* \ni T \longrightarrow f_x(T) = T(x). \tag{13.2}$$

Let X_{**} denote the collection of all such functionals, that is

$$X_{**} = \left\{ \begin{array}{l} \text{the collection of all functionals } f_x \in X^{**} \\ \text{of the form (13.2) as } x \text{ ranges over } X \end{array} \right\}. \tag{13.3}$$

From Corollary 10.2 and (13.1) it follows that $\|f_x\| = \|x\|$. Therefore the injection map

$$X \ni x \longrightarrow f_x \in X_* \subset X^{**} \tag{13.4}$$

is an isometric isomorphism between X and X_{**}. In general, not all the bounded linear functionals $f : X^* \to \mathbb{R}$ are derived from the injection map (13.4); otherwise said, the inclusion $X_{**} \subset X^{**}$ is in general strict.

A Banach space $\{X; \|\cdot\|\}$ is *reflexive* if $X_{**} = X^{**}$, i.e., if all the bounded linear functionals $f \in X^{**}$ are derived from the injection map (13.4). In such a case $X = X^{**}$ up to the isometric isomorphism in (13.4). By the Riesz representation theorem, the spaces $L^p(E)$ are reflexive for all $1 < p < \infty$. The spaces $L^1(E)$ and $L^\infty(E)$ are not reflexive since the dual of $L^\infty(E)$ is strictly larger than $L^1(E)$ (§ 9.2c of the Complements). Also the spaces ℓ_p are reflexive for all $1 < p < \infty$. The spaces ℓ_1 and ℓ_∞ are not reflexive.

Proposition 13.1 *Let X_o be a closed, linear, proper subspace of a reflexive Banach space $\{X; \|\cdot\|\}$. Then X_o is reflexive.*

Proof By Proposition 10.1, every $x_o^* \in X_o^*$ can be regarded as the restriction to X_o of some $x^* \in X^*$, such that $\|x^*\|_{X^*} = \|x_o^*\|_{X_o^*}$. Now fix $f_o \in X_o^{**}$ and set

$$f'(x^*) = f_o(x^*|_{X_o}) \quad \text{for all } x^* \in X^*.$$

One verifies that this is a bounded linear functional in X^*. Since X is reflexive, by the injection map (13.4), there exists $x_o \in X$ such that $f' = f_{x_o}$. To establish the proposition, it suffices to show that $x_o \in X_o$. If not, there exists $T \in X^*$ such that $T(x) = 0$ for all $x \in X_o$ and $T(x_o) \neq 0$. Therefore such a T, when restricted to X_o is the zero element of X_o^*. Thus

$$0 \neq T(x_o) = f_{x_o}(T) = f'(T) = f_o(T|_{X_o}) = 0. \qquad \blacksquare$$

Remark 13.1 The assumption that X_o be closed is essential. Indeed $L^\infty(E)$ for a bounded, Lebesgue measurable set $E \subset \mathbb{R}^N$, is a nonreflexive linear subspace of $L^p(E)$ for all $1 < p < \infty$.

The following statements are a consequence of the definitions modulo isometric isomorphisms.

Proposition 13.2 *A reflexive normed space is weakly complete.*
If $\{X; \|\cdot\|_X\}$ and $\{Y; \|\cdot\|_Y\}$ are isometrically isomorphic Banach spaces, X is reflexive if and only if Y is reflexive.
A Banach space $\{X; \|\cdot\|\}$ is reflexive if and only if X^* is reflexive.

14 Weak Compactness

Let $\{X; \| \cdot \|\}$ be a normed space and let X^* be its dual. A subset $E \subset X$ is weakly closed if $X - E$ is weakly open. The weak sequential closure of a set $E \subset X$ need not coincide with its weak closure (§ 12.3c of the Complements).

A set $E \subset X$ is weakly bounded if and only if for every $T \in X^*$, there exists a constant γ_T, such that

$$|T(x)| \leq \gamma_T \qquad \text{for all } x \in E. \tag{14.1}$$

Proposition 14.1 *Let $\{X; \| \cdot \|\}$ be a normed space. A set $E \subset X$ is weakly bounded if and only if is strongly bounded.*

Proof If E is strongly bounded, there exists a constant R such that $\|x\| \leq R$ for all $x \in E$. Then for all $T \in X^*$

$$|T(x)| \leq \|T\|\|x\| \leq \|T\|R.$$

Thus E is weakly bounded. Assume now E is weakly bounded so that (14.1) holds. Let f_x be the injection map (13.4). Then, for each fixed $T \in X^*$, (14.1) takes the form $|f_x(T)| \leq \gamma_T$ for all $x \in E$. The family $\{f_x\}$ for $x \in E$ is a collection of bounded linear maps from the Banach space X^* into $\mathbb{R}$, which are pointwise uniformly bounded in X^*. Therefore, by Proposition 6.1 they are equiuniformly bounded, i.e., there is a positive constant C such that $|T(x)| \leq C\|T\|$ for all $x \in E$ and all $T \in X^*$. Thus $\|x\| \leq C$ for all $x \in E$. ∎

Corollary 14.1 *Let E be a weakly, countably compact subset of a normed space $\{X; \| \cdot \|\}$. Then E is strongly bounded.*

Proof Let $\mathcal{O}$ be a weakly open neighborhood of the origin. Then the collection $\{n\mathcal{O}\}_{n \in \mathbb{N}}$ is a countable, weakly open covering for E, since $X = \bigcup n\mathcal{O}$. Therefore $E \subset t\mathcal{O}$ for some $t > 0$. Thus E is weakly bounded and hence strongly bounded. ∎

14.1 Weak Sequential Compactness

Corollary 14.2 *Let $\{X; \| \cdot \|\}$ be a normed space and let $\{x_n\}$ be a sequence of elements of X weakly convergent to some $x \in X$. There exists a positive constant C such that $\|x_n\| \leq C$ for all n. Moreover*

$$\|x\| \leq \liminf \|x_n\|. \tag{14.2}$$

Therefore in a normed linear space, the norm $\|\cdot\| : X \to \mathbb{R}$ is a weakly lower semi-continuous function.[1]

Proof The uniform upper bound of $\|x_n\|$ follows from Proposition 14.1. For all $T \in X^*$, by definition of weak limit

$$|T(x)| \le \liminf |T(x_n)| \le \|T\| \liminf \|x_n\|.$$

The conclusion now follows from Corollary 10.2.   ∎

Proposition 14.2 *Let $\{X; \|\cdot\|\}$ be a reflexive Banach space. Then every bounded sequence $\{x_n\}$ of elements of X contains a weakly convergent subsequence $\{x_{n'}\}$.*[2]

Proof Let X_o be the closed linear span of $\{x_n\}$. Such a subspace is separable since the finite linear combinations of elements of $\{x_n\}$ with rational coefficients is a countable dense subset.

Since X_o is reflexive, its double-dual X_o^{**}, being isometrically isomorphic to X_o is also separable. Then by Proposition 10.4, also X_o^* is separable.

Let $\{T_n\}$ be a countable, dense subset of X_o^*. The sequence $\{T_1(x_n)\}$ is bounded in $\mathbb{R}$ and we may extract a convergent subsequence $\{T_1(x_{n_1})\}$. The sequence $\{T_2(x_{n_1})\}$ is bounded in $\mathbb{R}$ and we may extract a convergent subsequence $\{T_2(x_{n_2})\}$. Proceeding in this fashion at the kth step we extract a subsequence $\{x_{n_k}\}$ such that

$$\{T_j(x_{n_k})\} \quad \text{is convergent for all} \quad j = 1, 2, \ldots k.$$

The diagonal sequence $\{x_{n'}\} = \{x_{n_n}\}$ is such that $\{T_j(x_{n'})\}$ is convergent for all $j \in \mathbb{N}$. Since $\{T_n\}$ is dense in X_o^*, the sequences $\{T(x_{n'})\}$ are convergent for all $T \in X_o^*$. Every $T \in X_o^*$ can be regarded as the restriction to X_o of some element of X^*. Therefore $\{T(x_{n'})\}$ is convergent for all $T \in X^*$. In particular for all $T \in X^*$ there exists $\alpha_T \in \mathbb{R}$, such that

$$\lim T(x_{n'}) = \alpha_T.$$

By the identification map (13.4), each $x_{n'}$ identifies a functional $f_{x_{n'}} \in X^{**}$. Therefore the previous limit can be rewritten as

$$\lim f_{x_{n'}}(T) = \alpha_T \quad \text{for all} \quad T \in X^*.$$

This process identifies an element $h \in X^{**}$ by the formula

$$h(T) = \lim f_{x_{n'}}(T) \quad \text{for all} \quad T \in X^*.$$

---

[1]Compare with Proposition 10.1 of Chap. 6.
[2]Compare with Proposition 16.1 of Chap. 6.

Since X is reflexive, there exists $x \in X$ such that $h = f_x$. Now we claim that $\{x_{n'}\} \to x$ weakly in X. Indeed for any fixed $T \in X^*$

$$\lim T(x_{n'}) = \lim f_{x_{n'}}(T) = f_x(T) = T(x).$$ ∎

Corollary 14.3 *Let* $\{X; \|\cdot\|\}$ *be a reflexive Banach space. A subset* $C \subset X$ *is weakly sequentially compact if and only if is both weakly bounded and weakly sequentially closed.*

Proof Sequential compactness implies countable compactness (Proposition 5.2 of Chap. 2). ∎

As an example consider the space $L^p(E)$ where E is a Lebesgue measurable subset of $\mathbb{R}^N$ and $1 < p < \infty$. By Proposition 14.2 and Corollary 14.2, the unit ball $\{\|f\|_p \leq 1\}$ is weakly sequentially compact. However, the unit sphere $\{\|f\|_p = 1\}$ is not weakly sequentially compact. For example, the unit sphere of $L^2[0, 2\pi]$ is bounded, but not weakly closed and therefore is not sequentially compact (§ 9.1 of Chap. 6).

The unit ball of $L^p(E)$ for $1 < p < \infty$ is not sequentially compact in the strong topology of $L^p(E)$, since it does not satisfy the necessary and sufficient conditions for compactness given in § 19 of Chap. 6. Compare also with Proposition 2.2.

15   The Weak* Topology of X^*

The dual X^* of a normed space $\{X; \|\cdot\|\}$ is a Banach space and as such can be endowed with, the corresponding weak topology, that is, the weakest topology for which all the elements of X^{**} are continuous, with respect to the norm topology of X^*.

The weak* topology on X^* is the weakest topology which renders continuous all the functionals $f_x \in X^{**}$ of the form (13.4), that is, those that are in a natural one-to-one correspondence with the elements of X. If $\{X; \|\cdot\|\}$ is a reflexive Banach space then $X = X^{**}$ up to an isometric isomorphism, and the weak topology of X^* coincides with its weak* topology.

The collection $\mathcal{W}^*$ of weak* open sets in X^* is constructed starting from the base

$$\mathcal{B}^* = \left\{ \begin{array}{l} \text{the collection of finite intersections of the inverse images} \\ f_x^{-1}(O) \text{ where } x \in X \text{ and } O \text{ are open subsets of } \mathbb{R} \end{array} \right\}.$$

One verifies that the operations of sum and multiplication by scalars

$$+ : X^* \times X^* \longrightarrow X^* \quad \bullet : \mathbb{R} \times X^* \longrightarrow X^*$$

are continuous with respect to the topology of $\mathcal{W}^*$. Thus $\{X^*; \mathcal{W}^*\}$ is a topological vector space. The topology of $\mathcal{W}^*$ is translation invariant and it is determined by a local base at the origin of X^*. A weak* open neighborhood of the origin contains an element of $\mathcal{B}^*$ of the form

$$\mathcal{O}^* = \bigcap_{j=1}^{n} f_{x_j}^{-1}(-\alpha_j, \alpha_j) \tag{15.1}$$

for some finite n and $\alpha_j > 0$. Equivalently,

$$\mathcal{O}^* = \{T \in X^* \mid |T(x_j)| < \alpha_j \quad \text{for } j = 1, \ldots, n\}. \tag{15.1'}$$

Since these open sets are convex, $\mathcal{W}^*$ is a locally convex topology. A sequence $\{T_n\}$ of elements of X^* converges weakly* to 0 if and only if every weak* open neighborhood of the origin contains all but finitely many elements of $\{T_n\}$. From (15.1)' it follows that $\{T_n\} \to 0$ if and only if $\{T_n(x)\} \to 0$ for all $x \in X$. More generally, $\{T_n\} \to T_o$ if and only if $\{T_n(x)\} \to T_o(x)$, for all $x \in X$. Weak convergence implies weak* convergence. The converse is false. Thus in general, the weak* topology $\mathcal{W}^*$ contains, roughly speaking, fewer open sets than those of the weak topology generated by X^{**}.

16 The Alaoglu Theorem

Theorem 16.1 (Alaoglu [2]) *Let* $\{X; \|\cdot\|\}$ *be a normed space and let* X^* *be its dual. The closed unit ball* $B^* = \{T \in X^* \mid \|T\| \le 1\}$ *in* X^*, *is weak* compact.*

Proof If $T \in B^*$ then $T(x) \in [-\|x\|, \|x\|]$ for all $x \in X$. Consider now the Cartesian product

$$P = \prod_{x \in X} \big[-\|x\|, \|x\| \big].$$

A point in P is a function $f : X \to \mathbb{R}$ such that $f(x) \in [-\|x\|, \|x\|]$ and P is the collection of all such functions. The set B^* is a subset of P and as such inherits the product topology of P. On the other hand, as a subset of X^* it also inherits the weak* topology of X^*.

Lemma 16.1 *These two topologies coincide on* B^*.

Proof Every weak* open neighborhood of a point $T_o \in X^*$ contains an open set of the form

$$\mathcal{O} = \left\{ \begin{array}{l} T \in X^* \mid |T(x_j) - T_o(x_j)| < \delta \text{ for some } \delta > 0 \\ \quad \text{for finitely many } x_j, \ j = 1, \ldots n. \end{array} \right\}.$$

Likewise, every neighborhood of a point $T_o \in P$, open in the product topology of P, contains an open set of the form

$$\mathcal{V} = \left\{ \begin{array}{l} f \in P \mid |f(x_j) - T_o(x_j)| < \delta \text{ for some } \delta > 0 \\ \text{for finitely many } x_j, \ j = 1, \ldots n. \end{array} \right\}.$$

These collections of open sets form a base for the corresponding topologies. Since $B^* = P \cap X^*$

$$\mathcal{O} \cap B^* = \mathcal{V} \cap B^*.$$

These intersections form a base for the corresponding relative topologies inherited by B^*. Therefore, the weak* topology and the relative product topology coincide on B^*.   ∎

Lemma 16.2 B^* *is closed in its relative product topology.*

Proof Let f_o be in the closure of B^* in the relative product topology. Fix $x, y \in X$ and $\alpha, \beta \in \mathbb{R}$ and consider the three points

$$x_1 = x \qquad x_2 = y \qquad x_3 = \alpha x + \beta y.$$

For $\varepsilon > 0$, the sets

$$V_\varepsilon = \{ f \in P \mid |f(x_j) - f_o(x_j)| < \varepsilon \text{ for } j = 1, 2, 3. \}$$

are open neighborhoods of f_o. Since they intersect B^*, there exists $T \in B^*$ such that

$$|f_o(x) - T(x)| < \varepsilon \qquad |f_o(y) - T(y)| < \varepsilon$$

and, since T is linear

$$|f_o(\alpha x + \beta y) - \alpha T(x) - \beta T(y)| < \varepsilon.$$

From this

$$|f_o(\alpha x + \beta y) - \alpha f_o(x) - \beta f_o(y)| < (1 + |\alpha| + |\beta|)\varepsilon$$

for all $\varepsilon > 0$. Thus f_o is linear and it belongs to B^*.   ∎

Proof (of Theorem 16.1 concluded) Each $[-\|x\|, \|x\|]$ as a bounded, closed interval in $\mathbb{R}$, equipped with its Euclidean topology, is compact. Therefore, by Tychonov's theorem P is compact in its product topology.

Then B^*, as a closed subset of a compact space, is compact in its relative product topology and hence in its relative weak* topology.   ∎

Corollary 16.1 *Let* $\{X; \|\cdot\|\}$ *be a reflexive Banach space. Then its unit ball is weakly compact.*

Corollary 16.2 *Let* $\{X; \|\cdot\|\}$ *be a reflexive Banach space. Then a subset of the unit ball of X is weakly compact if and only if it is weakly closed.*

Remark 16.1 The weak* compactness of the unit ball B^* of X^*, does not imply B^* is compact in the norm topology of X^*. As an example let $X = L^2(E)$. Then $L^2(E) = L^2(E)^* = L^2(E)^{**}$ up to isometric isomorphisms. The unit ball of $L^2(E)^*$ is weak* compact but not compact in the strong norm topology.

Remark 16.2 The weak* compactness of the unit ball $\|x\| \leq 1$ of a normed space, does not imply that the unit sphere $\|x\| = 1$ is weak* compact. For example the unit sphere of $L^2[0, 2\pi]$ is bounded but not weak* closed and therefore it is not weak* compact.

17 Hilbert Spaces

Let X be a vector space over $\mathbb{R}$. A *scalar* or *inner* product on X over $\mathbb{R}$ is a function, $\langle \cdot, \cdot \rangle : X \times X \to \mathbb{R}$, satisfying

  (i) $\langle x, y \rangle = \langle y, x \rangle$ for all $x, y \in X$
 (ii) $\langle x_1 + x_2, y \rangle = \langle x_1, y \rangle + \langle x_2, y \rangle$ for all $x_1, x_2, y \in X$
(iii) $\langle \lambda x, y \rangle = \lambda \langle x, y \rangle$ for all $x, y \in X$ and all $\lambda \in \mathbb{R}$
(iv) $\langle x, x \rangle \geq 0$ for all $x \in X$ and
 (v) $\langle x, x \rangle = 0$ if and only if $x = \Theta$.

A vector space X equipped with a scalar product $\langle \cdot, \cdot \rangle$ is called a *pre-Hilbert* space. Set

$$\langle x, x \rangle = \|x\|^2. \tag{17.1}$$

17.1 The Schwarz Inequality

Let X be a pre-Hilbert space for a scalar product $\langle \cdot, \cdot \rangle$. Then for all $x, y \in X$

$$\langle x, y \rangle \leq \|x\| \|y\| \tag{17.2}$$

and equality holds if and only if $x = \lambda y$ for some $\lambda \in \mathbb{R}$. Indeed for all $x, y \in X$ and $\lambda \in \mathbb{R}$

$$0 \leq \|x - \lambda y\|^2 = \langle x - \lambda y, x - \lambda y \rangle = \|x\|^2 - 2\lambda \langle x, y \rangle + \lambda^2 \|y\|^2.$$

From this, $2\lambda \langle x, y \rangle \leq \|x\|^2 + \lambda^2 \|y\|^2$. Inequality (17.2) is trivial if either $y = \Theta$ or $x = \Theta$. Otherwise choose $\lambda = \|x\|/\|y\|$.

17.2   The Parallelogram Identity

Let X be a pre-Hilbert space for a scalar product $\langle \cdot, \cdot \rangle$. Then for all $x, y \in X$

$$\|x + y\|^2 + \|x - y\|^2 = 2(\|x\|^2 + \|y\|^2). \qquad (17.3)$$

From the properties of a scalar product

$$\|x + y\|^2 = \|x\|^2 + 2\langle x, y \rangle + \|y\|^2$$
$$\|x - y\|^2 = \|x\|^2 - 2\langle x, y \rangle + \|y\|^2.$$

Adding these identities yields (17.3).

Using the properties (i)–(v) of a scalar product, and the Schwarz inequality (17.2), one verifies that the function $\|\cdot\| : X \to \mathbb{R}$, introduced in (17.1), defines a norm in X. Therefore a pre-Hilbert space is a normed space $\{X; \|\cdot\|\}$ by the norm in (17.1).

A Hilbert space is a pre-Hilbert space which is complete with respect to the topology generated by the norm (17.1). Equivalently, a Hilbert space is a Banach space whose norm is generated by an inner product $\langle \cdot, \cdot \rangle$. The N-dimensional Euclidean spaces are Hilbert spaces for the Euclidean scalar product. Let $\{X, \mathcal{A}, \mu\}$ be a measure space and let $E \in \mathcal{A}$. Then $L^2(E)$ is a Hilbert space for the scalar product

$$\langle f, g \rangle = \int_E f g \, d\mu \qquad \text{for all } f, g \in L^2(E).$$

In particular ℓ_2 is a Hilbert space for the inner product

$$\langle \mathbf{a}, \mathbf{b} \rangle = \sum a_i b_i, \qquad \text{for } \mathbf{a}, \mathbf{b} \in \ell_2.$$

We will denote by H a Hilbert space for the inner product $\langle \cdot, \cdot \rangle$.

18   Orthogonal Sets, Representations and Functionals

Two elements $x, y \in H$ are *orthogonal* if $\langle x, y \rangle = 0$ and in such a case we write $x \perp y$. In $L^2(0, 2\pi)$ the two elements $t \to \sin t, \cos t$ are orthogonal.

An element $x \in H$ is said to be orthogonal to a set $H_o \subset H$ if $x \perp y$ for all $y \in H_o$, and in such a case we write $x \perp H_o$.

Proposition 18.1 (Riesz [133]) *Let H_o be a closed, convex, proper subset of H. Then for every $x \in H - H_o$ there exists a unique $x_o \in H_o$, such that*

$$\inf_{y \in H_o} \|x - y\| = \|x_o - x\|. \qquad (18.1)$$

Proof Let $\{y_n\}$ be a sequence in H_o such that

$$\delta \stackrel{\text{def}}{=} \inf_{y \in H_o} \|x - y\| = \lim \|y_n - x\|. \qquad (18.2)$$

Since H_o is convex

$$\frac{y_n + y_m}{2} \in H_o \quad \text{for any two elements } y_n, y_m \in \{y_n\}.$$

Therefore by the parallelogram identity

$$\begin{aligned}
\|y_n - y_m\|^2 &= \|(y_n - x) + (x - y_m)\|^2 \\
&= 2\|y_n - x\|^2 + 2\|y_m - x\|^2 - \|y_n + y_m - 2x\|^2 \\
&= 2\|y_n - x\|^2 + 2\|y_m - x\|^2 - 4\left\|\frac{y_n + y_m}{2} - x\right\|^2 \\
&\leq 2\|y_n - x\|^2 + 2\|y_m - x\|^2 - 4\delta^2.
\end{aligned}$$

Therefore $\{y_n\}$ is a Cauchy sequence and, since H_o is closed, it converges to some $x_o \in H_o$ which satisfies (18.1). If x_o and x'_o both satisfy (18.1), then by the parallelogram identity

$$\|x'_o - x_o\|^2 = 2\|x'_o - x\|^2 + 2\|x_o - x\|^2 - 4\left\|\frac{x_o + x'_o}{2} - x\right\|^2 \leq 0. \qquad \blacksquare$$

Let H_o be a subset of H. The orthogonal complement $H_o^\perp$ of H_o is defined as the collection of all $x \in H$, such $x \perp H_o$.

Proposition 18.2 *Let H_o be a closed, proper subspace of H. Then $H = H_o \oplus H_o^\perp$, i.e., every $x \in H$ can be represented in a unique way as*

$$x = x_o + \eta \quad \text{for some } x_o \in H_o \quad \text{and} \quad \eta \in H_o^\perp. \qquad (18.3)$$

Proof If $x \in H_o$ it suffices to take $x_o = x$ and $y = \Theta$. If $x \in H - H_o$ let x_o be the unique element claimed by Proposition 18.1 and let δ be defined as in (18.2). Set now $x = x_o + \eta$ where $\eta = x - x_o$. To prove that $\eta \perp H_o$, fix $y \in H_o$ and consider the function

$$\mathbb{R} \ni t \to h(t) \stackrel{\text{def}}{=} \|\eta + ty\|^2 = \|\eta\|^2 + 2t\langle y, \eta \rangle + t^2\|y\|^2.$$

Such a function takes its minimum for $t = 0$. Indeed $h(0) = \delta^2$ and

$$\inf_{t \in \mathbb{R}} \|\eta - ty\|^2 = \inf_{t \in \mathbb{R}} \|x - (x_o + ty)\|^2 \geq \inf_{y \in H_o} \|x - y\|^2 = \delta^2.$$

Therefore $h'(0) = 0$, i.e., $\langle \eta, y \rangle = 0$ for all $y \in H_o$.

If the representation (18.3) were not unique there would exist $x_o' \neq x_o$ and $\eta' \neq \eta$ such that

$$x = x_o' + \eta' \qquad x_o' \in H_o \quad \text{and} \quad \eta' \in H_o^\perp. \tag{18.3}'$$

Then, by difference

$$x_o - x_o' \in H_o \qquad \eta - \eta' \in H_o^\perp \qquad \text{and} \qquad x_o - x_o' = \eta' - \eta.$$

Thus $x_o - x_o'$ and $\eta - \eta'$ are perpendicular to themselves and therefore must both be equal to Θ. ■

18.1   Bounded Linear Functionals on H

Every $y \in H$ identifies a bounded linear functional $T_y \in H^*$ by the formula

$$T_y(x) = \langle y, x \rangle \qquad \text{for all } x \in H. \tag{18.4}$$

Moreover $\|T_y\| = \|y\|$. The next proposition asserts that these are the only bounded, linear functionals on H.

Proposition 18.3 *For every $T \in H^*$ there exists a unique $y \in H$, such that T can be represented as in (18.4).*

Proof The conclusion is trivial if $T \equiv 0$. If $T \not\equiv 0$ its kernel H_o is a closed, proper subspace of H. Select a nontrivial element $\eta \in H_o^\perp$ and observe that, for all $x \in H$

$$T(\eta)x - T(x)\eta \in H_o \quad \text{i.e.,} \quad T(x)\eta = T(\eta)x + \eta_o \quad \text{for some } \eta_o \in H_o.$$

By taking the inner product of both sides by η

$$T(x) = \langle y, x \rangle \quad \text{where} \quad y = T(\eta)\frac{\eta}{\|\eta\|^2}.$$

Since $x \in H$ is arbitrary, this implies also $\|T\| = \|y\|$.

If y_1 and y_2 were to identify the same functional T, then $\langle y_1 - y_2, x \rangle = 0$ for all $x \in H$. Thus $\|y_1 - y_2\| = 0$. ■

Roughly speaking T is identified by the unique "direction" orthogonal to its kernel. Compare with Proposition 5.1.

19 Orthonormal Systems

A set S of elements of H is said to be orthogonal, if any two distinct elements x and y of S are orthogonal. The set S is *orthonormal* if it is orthogonal and all its elements have norm 1. In such a case S is called an orthonormal system. In $\mathbb{R}^N$ with its Euclidean norm, an orthonormal system is given by any n-tuple of mutually orthogonal unit vectors. In $L^2(0, 2\pi)$ an orthonormal system is given by

$$\frac{1}{\sqrt{2\pi}}, \frac{1}{\sqrt{\pi}}\cos t, \frac{1}{\sqrt{\pi}}\cos 2t, \frac{1}{\sqrt{\pi}}\cos 3t, \ldots$$

$$\frac{1}{\sqrt{\pi}}\sin t, \frac{1}{\sqrt{\pi}}\sin 2t, \frac{1}{\sqrt{\pi}}\sin 3t, \ldots \tag{19.1}$$

In ℓ_2 an orthonormal system is given by,

$$\mathbf{e}_1 = \{1, 0, 0, \ldots, 0_m, 0, \ldots\}$$
$$\mathbf{e}_2 = \{0, 1, 0, \ldots, 0_m, 0, \ldots\}$$
$$\cdots$$
$$\mathbf{e}_m = \{0, 0, 0, \ldots, 1_m, 0, \ldots\} \tag{19.2}$$
$$\cdots = \cdots$$

Lemma 19.1 *Let S be an orthonormal system in H. Any two distinct elements x and y in S are at mutual distance $\sqrt{2}$.*

Proof For any $x, y \in S$ and $x \neq y$, compute

$$\|x - y\|^2 = \langle x - y, x - y \rangle = \|x\|^2 - 2\langle x, y \rangle + \|y\|^2.$$

The conclusion follows since $\|x\| = \|y\| = 1$. ∎

19.1 The Bessel Inequality

Proposition 19.1 *Let H be a Hilbert space and let S be an orthonormal system in H. Then for any n-tuple $\{\mathbf{u}_1, \ldots, \mathbf{u}_n\}$ of elements of S[3]*

$$\sum_{i=1}^{n} \langle \mathbf{u}_i, x \rangle^2 \leq \|x\|^2 \quad \text{for all } x \in H. \tag{19.3}$$

---

[3] S need not be countable.

Moreover, for any fixed $x \in H$ *the inner product* $\langle \mathbf{u}, x \rangle$ *vanishes except for at most countably many* $\mathbf{u} \in \mathcal{S}$, *and*

$$\sum_{\mathbf{u} \in \mathcal{S}} \langle \mathbf{u}, x \rangle^2 \leq \|x\|^2 \qquad \text{for all } x \in H. \tag{19.4}$$

Proof For any such n-tuple and any $x \in H$

$$0 \leq \left\| x - \sum_{i=1}^n \langle \mathbf{u}_i, x \rangle \mathbf{u}_i \right\|^2 = \left(x - \sum_{i=1}^n \langle \mathbf{u}_i, x \rangle \mathbf{u}_i, x - \sum_{i=1}^n \langle \mathbf{u}_i, x \rangle \mathbf{u}_i \right)$$

$$= \|x\|^2 - \sum_{i=1}^n \langle \mathbf{u}_i, x \rangle^2.$$

This establishes (19.3). To prove (19.4), fix $x \in H$ and observe that for any $m \in \mathbb{N}$ the set

$$\left\{ \mathbf{u} \in \mathcal{S} \quad \text{such that} \quad \frac{1}{2^{m+1}} \|x\|^2 \leq \langle \mathbf{u}, x \rangle^2 < \frac{1}{2^m} \|x\|^2 \right\}$$

contains at most finitely many elements. Therefore the collection of those $\mathbf{u} \in \mathcal{S}$ such that $\langle \mathbf{u}, x \rangle \neq 0$ is countable and (19.4) holds. ∎

19.2   Separable Hilbert Spaces

Proposition 19.2 *Let H be a separable Hilbert space. Then any orthonormal system $\mathcal{S}$ in H is countable.*

Proof Let H_o be a countable subset of H dense in H and let $\mathcal{S}$ be an orthonormal system in H. For every $\mathbf{u} \in \mathcal{S}$ there exists $x(\mathbf{u}) \in H_o$ such that

$$\|\mathbf{u} - x(\mathbf{u})\| < \frac{\sqrt{2}}{3}. \tag{19.5}$$

If $\mathbf{u}_1$ and $\mathbf{u}_2$ are distinct elements of $\mathcal{S}$ then any two elements $x(\mathbf{u}_1)$ and $x(\mathbf{u}_2)$ in H_o for which (19.5) holds are distinct. Indeed

$$\sqrt{2} = \|\mathbf{u}_1 - \mathbf{u}_2\| \leq \|\mathbf{u}_1 - x(\mathbf{u}_1)\| + \|\mathbf{u}_2 - x(\mathbf{u}_2)\| + \|x(\mathbf{u}_1) - x(\mathbf{u}_2)\|$$

$$\leq 2\frac{\sqrt{2}}{3} + \|x(\mathbf{u}_1) - x(\mathbf{u}_2)\|.$$

Thus $\mathcal{S}$ can be put in a one-to-one correspondence with a subset of H_o. ∎

20 Complete Orthonormal Systems

An orthonormal system S in H is said to be *complete* if

$$\langle x, \mathbf{u} \rangle = 0 \quad \text{for all } \mathbf{u} \in S \quad \text{implies } x = \Theta. \tag{20.1}$$

The orthonormal system in (19.2) is complete in ℓ_2. It can be shown that the orthonormal system in (19.1) is complete in $L^2(0, 2\pi)$.

Proposition 20.1 *Let S be a complete orthonormal system in H. Then for every $x \in H$*

$$x = \sum_{\mathbf{u} \in S} \langle x, \mathbf{u} \rangle \mathbf{u} \quad \text{(representation of } x\text{)}. \tag{20.2}$$

Moreover

$$\|x\|^2 = \sum_{\mathbf{u} \in S} |\langle x, \mathbf{u} \rangle|^2 \quad \text{(Parseval's identity)}. \tag{20.3}$$

Proof As $\mathbf{u}$ ranges over S, only countably many of the numbers $\langle x, \mathbf{u} \rangle$ are not zero and we order them in some fashion $\{\langle x, \mathbf{u}_n \rangle\}$. By the Bessel inequality the series $\sum \langle x, \mathbf{u}_n \rangle^2$ converges. Therefore for any two positive integers $n < m$

$$\left\| \sum_{i=n}^{m} \langle x, \mathbf{u}_i \rangle \mathbf{u}_i \right\|^2 = \sum_{i=n}^{m} \langle x, \mathbf{u}_i \rangle^2 \longrightarrow 0 \quad \text{as } m, n \to \infty.$$

This implies that the sequence $\{\sum_{i=1}^{n} \langle x, \mathbf{u}_i \rangle \mathbf{u}_i\}$ is a Cauchy sequence in H and has a limit

$$y = \sum \langle x, \mathbf{u}_i \rangle \mathbf{u}_i = \sum_{\mathbf{u} \in S} \langle x, \mathbf{u} \rangle \mathbf{u}.$$

For any $\mathbf{u} \in S$

$$\langle x - y, \mathbf{u} \rangle = \lim \left(x - \sum_{i=1}^{m} \langle x, \mathbf{u}_i \rangle \mathbf{u}_i, \mathbf{u} \right) = 0.$$

Thus $\langle x - y, \mathbf{u} \rangle = 0$ for all $\mathbf{u} \in S$ and since S is complete $x = y$.
From (20.2), by taking the inner product with respect to x

$$\|x\|^2 = \lim \left(x, \sum_{i=1}^{n} \langle x, \mathbf{u}_i \rangle \mathbf{u}_i \right) = \lim \sum_{i=1}^{n} \langle x, \mathbf{u}_i \rangle^2. \qquad \blacksquare$$

20.1 Equivalent Notions of Complete Systems

Let S be an orthonormal system in H. If (20.3) holds for all $x \in H$ then S is complete. Indeed if not there would be an element $x \in H$ such that $\langle x, \mathbf{u} \rangle = 0$ for all $\mathbf{u} \in S$ and $x \neq \Theta$. However if (20.3) holds $x = \Theta$.

The proof of Proposition 20.1 shows that the notion (20.1) of complete system implies (20.2) and this in turn implies (20.3). We have just observed that (20.3) implies the notion (20.1) of complete system. Thus (20.1)–(20.3) are equivalent and each could be taken as a definition of complete system.

20.2 Maximal and Complete Orthonormal Systems

An orthonormal system S in H is *maximal* if it is not properly contained in any other orthonormal system of H. From the definitions it follows that an orthonormal system S in H is complete if and only if it is maximal.

The family Σ of all orthonormal systems in H is partially ordered by set inclusion. Moreover every linearly ordered subset $\Sigma' \subset \Sigma$ has an upper bound given by the union of all orthonormal systems in Σ'. Therefore by Zorn's lemma H has a maximal orthonormal system.

Zorn's lemma provides an abstract notion of existence of a maximal orthonormal system in H. Of greater interest is the actual construction of a complete system.

20.3 The Gram–Schmidt Orthonormalization Process ([142])

Let $\{x_n\}$ be a countable collection of linearly independent elements of H and set

$$\mathbf{u}_1 = \frac{x_1}{\|x_1\|} \qquad \mathbf{u}_{n+1} = \frac{x_{n+1} - \sum_{i=1}^{n} \langle x_{n+1}, \mathbf{u}_i \rangle \mathbf{u}_i}{\left\| x_{n+1} - \sum_{i=1}^{n} \langle x_{n+1}, \mathbf{u}_i \rangle \mathbf{u}_i \right\|} \qquad \text{for } n = 1, 2, \ldots.$$

These are well defined since $\{x_n\}$ are linearly independent. One verifies that $\{\mathbf{u}_n\}$ forms an orthonormal system and $\{\mathbf{u}_n\} = \{x_n\}$.

If H is separable this procedure can be used to generate a maximal orthonormal system S in H, independent of Zorn's lemma.

20.4 On the Dimension of a Separable Hilbert Space

If H is separable any complete orthonormal system $\mathcal{S}$, is either finite or infinite-countable.

Assume first $\mathcal{S}$ is infinite-countable and index its elements as $\{\mathbf{u_n}\}$. By Parseval's identity, any element $x \in H$ generates an element of ℓ_2 by the formula $\{a_n\} = \{\langle x, \mathbf{u}_n \rangle\}$. Vice versa any element $\mathbf{a} \in \ell_2$ generates a unique element $x \in H$ by the formula $x = \sum a_i \mathbf{u}_i$. Let x and y be two elements in H and let $\mathbf{a}$ and $\mathbf{b}$ be their corresponding elements in ℓ_2. By same reasoning $\langle x, y \rangle_H = \langle \mathbf{a}, \mathbf{b} \rangle_{\ell_2}$. This implies that the isomorphism between H and ℓ_2 is an isometry. Thus any separable Hilbert space with an infinite-countable orthonormal system $\mathcal{S}$ is isometrically isomorphic to ℓ_2. Equivalently, any complete, infinite-countable, orthonormal system $\mathcal{S}$ of a Hilbert space H, can be put in one-to-one correspondence with the system (19.2) which forms a complete orthonormal system of ℓ_2.

We say that the dimension of a separable Hilbert space with an infinite-countable orthonormal system, is $\aleph_o$, i.e., the cardinality of $\{e_n\}$.

If $\mathcal{S}$ is finite, say for example $\{\mathbf{u}_1, \ldots, \mathbf{u}_N\}$ then by the same procedure H is isometrically isomorphic to $\mathbb{R}^N$ and its dimension is N.

Problems and Complements

1c Normed Spaces

1.1. Let E be a bounded, open subset of $\mathbb{R}^N$. The space $C(\bar{E})$ endowed with the norm of $L^p(E)$ is not a Banach space.

1.2. Every normed space is homeomorphic to its open unit ball.

1.3. A normed space $\{X; \|\cdot\|\}$ is complete if and only if the intersection of a countable family of nested, closed balls is nonempty.

1.4. The L^2-norm and the sup-norm on $C[0, 1]$ are not equivalent. In particular $C[0, 1]$ is not complete in $L^p[0, 1]$ for all $1 \leq p < \infty$. The norms of $L^p(E)$ and $L^q(E)$ for $1 \leq q < p < \infty$, are not equivalent.

The next proposition provides a criterion for a normed space to be a Banach space.

Proposition 1.1c *A normed space $\{X; \|\cdot\|\}$ is complete if and only if every absolutely convergent series converges to an element of X.*

Proof Let $\{X; \|\cdot\|\}$ be complete and let $\sum x_n$ be absolutely convergent in $\{X; \|\cdot\|\}$, so that $\sum \|x_n\| \leq M$ for some $M > 0$. Then $\{\sum_{j=1}^n x_j\}$ is Cauchy and hence convergent to some $x \in X$.

Conversely, let $\{x_n\}$ be Cauchy in $\{X; \|\cdot\|\}$, so that for each $j \in \mathbb{N}$ there exists n_j such that

$$\|x_n - x_m\| \le 2^{-j} \quad \text{for } n, m \ge n_j.$$

Set

$$y_j = x_{n_{j+1}} - x_{n_j}.$$

The series $\sum y_j$ is absolutely convergent and we let x denote its limit. Thus the subsequence $\{x_{n_j}\} \subset \{x_n\}$ converges to x. Since $\{x_n\}$ is Cauchy, the whole sequence converges to x.   ∎

In the context of $L^p(E)$ the criterion has been used in the proof of Theorem 5.1 of Chap. 6.

Proposition 1.2c *Every norm p on $\mathbb{R}^N$ is equivalent to the Euclidean norm $\| \cdot \|$.*

Remark 1.1c The statement is a particular case of Proposition 12.1 of Chap. 2. The proof below is more direct using that the topology of $\mathbb{R}^N$ is generated by a norm $p(\cdot)$.

Proof (of Proposition 1.2c) Let $\{\mathbf{e}_1, \dots, \mathbf{e}_N\}$ be a basis of $\mathbb{R}^N$. Then

$$x = \sum_{j=1}^{N} c_j \mathbf{e}_j \quad \Longrightarrow \quad p(x) \le \sum_{j=1}^{N} |c_j| p(\mathbf{e}_j) \le C\|x\|$$

for a constant C independent of x. This implies that

$$|p(x) - p(y)| \le p(x - y) \le C\|x - y\| \quad \text{for all } x, y \in \mathbb{R}^N.$$

Thus $p(\cdot)$ is continuous in the topology of the Euclidean norm. Its restriction to the Euclidean unit sphere S_1 is a continuous function pointwise strictly bounded below in S_1. Since the S_1 is compact in $\mathbb{R}^N$ with its Euclidean topology, $p(\cdot)$ attains its minimum $c > 0$ there. From this

$$p(x) = p\left(\|x\| \frac{x}{\|x\|}\right) = \|x\| p\left(\frac{x}{\|x\|}\right) \ge c\|x\| \quad \text{for all } x \in \mathbb{R}^N - \{0\}. \qquad ∎$$

1.1c Semi-Norms and Quotients

1.5. The quantities $V_f[a, b]$ in (1.2), and $[f]_\alpha$ in (1.3) are semi-norms in their respective spaces.

1.6. Let $\{p_\alpha\}$ be a collection of semi-norms on X such that

$$p_\infty(x) = \sup_\alpha p_\alpha(x) < \infty \quad \text{for all } x \in X.$$

Then $p_\infty(\cdot)$ defines a semi-norm on X.

1.7. Let $\{p_j\}$ for $j = 1, \ldots, n$ be a finite collection of semi-norms on X. Then

$$q_1(x) = \sum_{j=1}^{n} p_j(x); \quad q_2(x) = \sqrt{\sum_{j=1}^{n} p_j^2(x)}; \quad q_\infty(x) = \max_{1 \leq j \leq n} p_j(x),$$

are also semi-norms.

1.8. Prove that a semi-norm p on X is convex. Conversely a convex function $f :$ $X \to \mathbb{R}^+$ which is homogeneous of order 1 in X, defines a semi-norm in X.
Hint: Homogeneous of order $\alpha \in \mathbb{R}$ means that $f(tx) = |t|^\alpha f(x)$ for all $t \in \mathbb{R}$.

2c Finite and Infinite Dimensional Normed Spaces

2.1. An infinite dimensional Banach space $\{X; \|\cdot\|\}$ cannot have a countable Hamel basis (**16.9** of the Complements of Chap. 2).
2.2. Let ℓ_o be the collection of all sequences of real numbers $\{c_n\}$ with only finitely many nonzero elements. There is no norm on ℓ_o by which ℓ_o would be a Banach space. See also § 9.6c of the Complements of Chap. 2.
2.3. $L^p(E)$ and ℓ_p are of infinite dimension for all $1 \leq p \leq \infty$ and their dimension is larger than $\aleph_o$.
2.4. Let $C^1[0, 1]$ be the collections of all continuously differentiable functions in $[0, 1]$ with norm

$$C^1[0, 1] \ni f \to \|f\|_{C^1[0,1]} = \sup_{[0,1]} |f| + \sup_{[0,1]} |f'|.$$

Let X_o be a closed subspace of $C^1[0, 1]$ which is also closed in $L^2[0, 1]$. Prove that

   i. The norms $\|\cdot\|_{C^1[0,1]}$ and $\|\cdot\|_{L^2[0,1]}$ are equivalent on X_o, i.e., there exists two positive constants $c \leq C$ such that

$$c\|f\|_{C^1[0,1]} \leq \|f\|_{L^2[0,1]} \leq C\|f\|_{C^1[0,1]}.$$

   ii. X_o is a finite dimensional subspace of $L^2[0, 1]$.

2.5. The dimension of $C[0, 1]$ and $BV[0, 1]$ is uncountably infinite. The collection $\{x^n\}$ is an infinite set of linearly independent elements of $C[0, 1]$ and $BV[0, 1]$.
2.6. An infinite dimensional Banach space $\{X; \|\cdot\|\}$ has an infinite dimensional non-closed subspace. Fix a Hamel basis $\{x_\alpha\}$, for X, where α is an uncountable index, select an infinite, countable collection $\{x_n\} \subset \{x_\alpha\}$ of linearly independent elements and set $Y = \{x_n\}$. Then Y is non-closed, for otherwise it would be an infinite dimensional Banach space with a countable Hamel basis.

3c Linear Maps and Functionals

3.1. A linear map T from $\{X; \| \cdot \|_X\}$ into $\{Y; \| \cdot \|_Y\}$ is continuous if and only if it maps sequences $\{x_n\}$ converging to Θ_X into bounded sequences of $\{Y; \| \cdot \|_Y\}$.

3.2. Two normed spaces $\{X; \|\cdot\|_1\}$ and $\{X; \|\cdot\|_2\}$ are homeomorphic if and only if there exist positive constants $0 < c_o \leq 1 \leq c_1$ such that

$$c_o\|x\|_1 \leq \|x\|_2 \leq c_1\|x\|_1 \quad \text{for all } x \in X.$$

3.3. A linear functional on a finite dimensional normed space is continuous.

3.4. Let E be a bounded, open subset of $\mathbb{R}^N$. A linear map $T : C(\bar{E}) \to \mathbb{R}$ is a positive functional, if $T(f) \geq 0$ whenever $f \geq 0$. A positive linear functional on $C(\bar{E})$ is bounded. Thus positivity implies continuity. Moreover any two of the conditions

$$\text{(i)} \quad \|T\| = 1 \qquad \text{(ii)} \quad T(1) = 1 \qquad \text{(iii)} \quad T \geq 0$$

implies the remaining one.

3.5. Let $\{X; \| \cdot \|\}$ be an infinite dimensional Banach space. There exists a discontinuous, linear map $T : X \to X$.

Having fixed a Hamel basis $\{x_\alpha\}$ for X, after a possible renormalization, we may assume that $\|x_\alpha\| = 1$ for all α. Every element $x \in X$ can be represented in a unique way, as the *finite* linear combination of elements of $\{x_\alpha\}$, i.e., for every $x \in X$ there exists a unique m-tuple of real numbers $\{c_1, \ldots, c_m\}$ such that

$$x = \sum_{j=1}^{m} c_j x_{\alpha_j}. \tag{3.1c}$$

Since X is of infinite dimension, the index α ranges over some set A such that $(A) \geq (\mathbb{N})$. Out of $\{x_\alpha\}$ select a countable collection $\{x_n\} \subset \{x_\alpha\}$. Then set

$$T(x_n) = nx_n \quad \text{and} \quad T(x_\alpha) = \Theta \quad \text{if } \alpha \notin \mathbb{N}.$$

For $x \in X$, having determined its representation (3.1c), set also

$$T(x) = \sum_{j=1}^{m} c_j T(x_{\alpha_j}).$$

In view of the uniqueness of the representation (3.1c), this defines a linear map from X into X. Such a map, however, is discontinuous since $\|T(x_n)\| = n\|x_n\| \to \infty$ as $n \to \infty$.

3.6. Let $\{X; \| \cdot \|\}$ be an infinite dimensional Banach space. There exists a discontinuous, linear functional $T : X \to \mathbb{R}$.

3.7. Are the constructions of **3.5** and **3.6** possible if $\{X; \| \cdot \|\}$ is an infinite dimensional normed space?

Remark 3.1c The conclusion of **3.6** is in general false for metric spaces. For example if X is a set, the discrete metric, generates the discrete topology on X. With respect to such a topology there exists no discontinuous maps $T : X \to \mathbb{R}$.

However there exist metric, not normed, spaces that admit discontinuous linear functionals. As an example consider $L^p(E)$ for $0 < p < 1$ endowed with the metric (3.1c) of § 3.4c of the Complements of Chap. 6. If E is a Lebesgue measurable subset of $\mathbb{R}^N$ and μ is the Lebesgue measure, then every nontrivial, linear functional on $L^p(E)$ is discontinuous.[4]

3.8. Let the unit ball in $\mathbb{R}^N$ in some norm $\| \cdot \|$, be $\prod_{j=1}^{N}[-1, 1]$. Compute the unit ball of the dual space.

3.9. Let $X = \bigoplus_{j=1}^{k} X_j$ where X_j for $j = 1, \ldots, k$ are Banach spaces. Then $X^* = \bigoplus_{j=1}^{k} X_j^*$. In particular $(L^p(E)^k)^* = L^q(E)^k$ where $1 \leq p < \infty$ and p and q are Hölder conjugate.

6c   Equibounded Families of Linear Maps

Let $\{X; \| \cdot \|_X\}$ and $\{Y; \| \cdot \|_Y\}$ be Banach spaces.

6.1. Let $T(x, y) : X \times Y \to \mathbb{R}$ be a functional linear and continuous in each of the two variables. Then T is linear and bounded with respect to both variables.

6.2. Let $\{T_n\}$ be a sequence in $\mathcal{B}(X; Y)$, such that the limit of $\{T_n(x)\}$ exists for all $x \in X$. Then $T(x) = \lim T_n(x)$ defines a bounded, linear map from $\{X; \| \cdot \|_X\}$ into $\{Y; \| \cdot \|_Y\}$.

6.3. Let $\{T_n\}$ be a sequence in $\mathcal{B}(X; Y)$, such that $\|T_n\| \leq C$ for some positive constant C and all $n \in \mathbb{N}$. Let X_o be the set of $x \in X$ for which $\{T_n(x)\}$ converges. Then X_o is a closed subspace of $\{X; \| \cdot \|_X\}$.

6.4. Let T_1 and T_2 be elements of $\mathcal{B}(X; X)$ and let $T_1 T_2(\cdot) = T_1(T_2(\cdot))$ be the composition map. Then $T_1 T_2 \in \mathcal{B}(X; X)$ and $\|T_1 T_2\| \leq \|T_1\| \|T_2\|$.

Let $T \in \mathcal{B}(X; X)$ satisfy $\|T\| < 1$. Then $(I + T)^{-1}$ exists as an element of $\mathcal{B}(X; X)$ and

$$(I + T)^{-1} = I + \sum (-1)^n T^n.$$

Hint: Since $\|T\| < 1$, the series $\sum \|T^n\| \leq \sum \|T\|^n$ converges. Since $\mathcal{B}(X; X)$ is a Banach space, $\sum(-T)^n$ converges to an element $\mathcal{B}(X; X)$.

6.5. Let E be a bounded open set in $\mathbb{R}^N$. Fix $h \in L^\infty(E)$ and find $f \in L^\infty(E)$ such that

---

[4] These remarks on existence and nonexistence of unbounded linear functionals in metric spaces were suggested by Allen Devinatz[†].

$$h = (I + T)f \qquad \text{i.e., formally} \qquad f = (I + T)^{-1}h \qquad (6.1c)$$

where $T(f)$ is the Riesz potential introduced in (4.2). Verify that T is a bounded linear map from $L^\infty(E)$ into itself. Give conditions on E so that such a formal solution formula is actually justified and exhibit explicitly the solution f.

6.6. Let E be a Lebesgue measurable subset of $\mathbb{R}^N$ of finite measure and let $1 \leq p, q \leq \infty$ be conjugate. Then if $q > p$ the space $L^q(E)$ is of first category in $L^p(E)$. **Hint:** $L^q(E)$ is the union of $\left[\|g\|_q \leq n\right]$.

8c   The Open Mapping Theorem

$\{X; \|\cdot\|_X\}$ and $\{Y; \|\cdot\|_Y\}$ are Banach spaces:

8.1. $T \in \mathcal{B}(X; Y)$ is a homeomorphism if and only if it is onto, and there exist positive constants $c_1 \leq c_2$ such that

$$c_1\|x\|_X \leq \|T(x)\|_Y \leq c_2\|x\|_X \qquad \text{for all } x \in X.$$

8.2. A bounded linear map T from a subspace $X_o \subset X$ into Y, has closed graph if and only if its domain is closed.

8.3. The sup-norm on $C[0, 1]$ generates a strictly stronger topology than the L^2-norm.

8.4. Let $T \in \mathcal{B}(X; Y)$. If $T(X)$ is of second category in Y, then T is onto.

9c   The Hahn–Banach Theorem

Let X be a vector space over the complex field $\mathbb{C}$. The norm $\|\cdot\|$ is defined as in (i)–(iii) of § 1, except that $\lambda \in \mathbb{C}$. In such a case $|\lambda|$ is the modulus of λ as element of $\mathbb{C}$.

Denote by $X_\mathbb{R}$ the vector space X when multiplication is restricted to scalars in $\mathbb{R}$. A linear functional $T : X \to \mathbb{C}$ is separated into its real and imaginary part by

$$T(x) = T_\mathbb{R}(x) + iT_i(x) \qquad (9.1c)$$

where the maps $T_\mathbb{R}$ and T_i are functionals from $X_\mathbb{R}$ into $\mathbb{R}$. Since $T : X \to \mathbb{C}$ is linear, $T(ix) = iT(x)$ for all $x \in X$. From this compute

$$T(ix) = T_\mathbb{R}(ix) + iT_i(ix) \qquad iT(x) = iT_\mathbb{R}(x) - T_i(x).$$

This implies

$$T_i(x) = -T_\mathbb{R}(ix) \qquad \text{for all } x \in X. \qquad (9.2c)$$

Thus $T : X \to \mathbb{C}$ is identified by its real part $T_{\mathbb{R}}$ regarded as a linear functional from $X_{\mathbb{R}}$ into $\mathbb{R}$. Vice versa any such real-valued functional $T_{\mathbb{R}} : X_{\mathbb{R}} \to \mathbb{R}$ identifies a linear functional $T : X \to \mathbb{C}$ by the formulae (9.1c)–(9.2c).

9.1c The Complex Hahn–Banach Theorem

Theorem 9.1c ([17, 151]) *Let X be a complex vector space and let $p : X \to \mathbb{R}$ be a semi-norm on X. Then, every linear functional $T_o : X_o \to \mathbb{C}$ defined on a subspace X_o of X and satisfying $|T_o(x)| \leq p(x)$ for all $x \in X_o$, admits an extension $T : X \to \mathbb{C}$ such that*

$$|T(x)| \leq p(x) \quad \text{for all } x \in X \quad \text{and} \quad T(x) = T_o(x) \quad \text{for } x \in X_o.$$

Proof Denote by $X_{o,\mathbb{R}}$ the real subspace of X_o and by $T_{o,\mathbb{R}} : X_{o,\mathbb{R}} \to \mathbb{R}$ the real part of T. By the representation (9.1c)–(9.2c) it suffices to extend $T_{o,\mathbb{R}}$ into a linear map $T_{\mathbb{R}} : X_{\mathbb{R}} \to \mathbb{R}$. This follows from the Hahn–Banach theorem, since

$$T_{o,\mathbb{R}}(x) \leq |T(x)| \leq p(x) \qquad \text{for all } x \in X. \qquad \blacksquare$$

9.2c Linear Functionals in $L^\infty(E)$

The Riesz representation theorem for the bounded linear functionals in $L^p(E)$, fails for $p = \infty$. A counterexample can be constructed as follows.

On $C[-1, 1]$ let $T_o(f) = f(0)$. This is bounded, linear functional on $C[-1, 1]$. The boundedness of T_o is meant in the sense of $L^\infty[-1, 1]$, i.e.,

$$\|T_o\| = \sup_{\substack{\varphi \in C[-1,1] \\ \|\varphi\|_\infty = 1}} |T_o(\varphi)|.$$

Then, by the Hahn–Banach theorem, T_o can be extended to a bounded linear functional T in $L^\infty[-1, 1]$ coinciding with T_o on $C[-1, 1]$ and such that $\|T\| = \|T_o\|$. For such an extension there exists no function $g \in L^1[-1, 1]$ such that

$$T(f) = \int_{-1}^{1} fg\,dx \qquad \text{for all} \quad f \in L^\infty[-1, 1].$$

11c   Separating Convex Subsets of X

11.1c   A Counterexample of Tukey [164]

The nonempty interior assumption is essential. We produce two disjoint, closed, convex sets C_1 and C_2 in ℓ_2, such that $C_1 - C_2$ is dense in ℓ_2. This implies that $C_1 - C_2$ cannot be separated from Θ and hence C_1 and C_2 cannot be separated. Identify $x \in \ell_2$ with its corresponding sequence $\{x_n\}$ and define

$$C_1 = \left\{ x \in \ell_2 \mid x_1 > \left| n^2 \left(x_n - \tfrac{1}{n} \right) \right|, \text{ for all } n > 1 \right\}$$

$$C_2 = \left\{ x \in \ell_2 \mid x_n = 0 \text{ for all } n > 1 \right\}.$$

One verifies that C_1 and C_2 are convex and closed. They are also disjoint. Indeed if $y \in C_1 \cap C_2$

$$y_1 \geq \left| n^2 \left(0 - \tfrac{1}{n} \right) \right| = n \quad \text{for all } n \geq 2.$$

Thus $y_1 = \infty$ and hence $y \notin \ell_2$. To show that $C_1 - C_2$ is dense in ℓ_2, fix $z \in \ell_2$ and $\varepsilon > 0$ and pick n_ε so large that

$$\sum_{n \geq n_\varepsilon} n^{-2} + \sum_{n \geq n_\varepsilon} z_n^2 < \frac{\varepsilon^2}{4}.$$

Choose a number

$$x_1 > \left| n^2 \left(z_n - \tfrac{1}{n} \right) \right| \qquad \text{for } 2 \leq n_\varepsilon < n$$

and let $x = \{x_n\} \in C_1$ and $y = \{y_n\} \in C_2$ be defined by

$$x_n = \begin{cases} x_1 & \text{for } n = 1; \\ z_n & \text{for } 2 \leq n < n_\varepsilon; \\ \tfrac{1}{n} & \text{for } n \geq n_\varepsilon; \end{cases} \qquad y_n = \begin{cases} z_1 + x_1 & \text{for } n = 1; \\ 0 & \text{for } n > 1. \end{cases}$$

By the triangle inequality $\| z - (x - y) \| < \varepsilon$.

11.1.1c   A Variant of Tukey's Counterexample

Corollary 11.1c *Every infinite dimensional Banach space X contains a 1-dimensional subspace C_1 and a closed, convex set C_2 that cannot be separated by a bounded linear functional.*

Proof Let $\{e_n\}$ be a sequence of linearly independent elements on the unit sphere of X and set $C_2 = \{e_1\}$ and

$$C_1 = \left\{ \begin{array}{l} x \in X \text{ of the form } x = \sum x_n e_n \text{ with} \\ x_1 \geq n^3 |x_n - n^{-2}| \text{ for all } n \geq 2 \end{array} \right\} \qquad \blacksquare$$

11.2c   A Counterexample of Goffman and Pedrick [56]

Let X be a linear space with a countably infinite Hamel basis $\{x_n\}$. Let C_1 be the collections of all elements $x \in X$ whose last nonzero coefficient of its representations in terms of $\{x_n\}$ is positive. The set C_1 is convex and $\Theta \notin C_1$. Any functional T that separates $\{\Theta\}$ from C_1 must be either non-positive or nonnegative on C_1. Let then T be a linear functional on X, which is nonnegative on C_1. For every $x_j \in \{x_n\}$ and all $\alpha \in \mathbb{R}$ the element $\alpha x_j + x_{j+1} \in C_1$. Therefore

$$T(\alpha x_j + x_{j+1}) = \alpha T(x_j) + T(x_{j+1}) \geq 0 \quad \text{for all } \alpha \in \mathbb{R}.$$

Thus $T(x_j) = 0$. Since $x_j \in \{x_n\}$ is arbitrary, T vanishes on all basis elements of X and therefore is the zero functional.

11.3c   Extreme Points of a Convex Set

Let C be a convex set in a linear, normed space. A point $x \in C$ is an *extreme point* of C if there do not exist distinct points $u, v \in C$ and $t \in (0, 1)$, such that $x = tu + (1 - t)v$, that is, if no line segment in C has x as an interior point.

11.1. The extreme points of a convex, closed polyhedron in $\mathbb{R}^N$ are its vertices.

11.2. An open convex polyhedron in $\mathbb{R}^N$ has no extreme points.

11.3. A closed $\frac{1}{2}$-space in $\mathbb{R}^N$ has no extreme points.

11.4. The set of extreme points of the closed unit ball of a uniformly convex linear normed space $\{X; \| \cdot \|\}$ is the unit sphere. In particular the extreme points of $L^p(E)$ for $1 < p < \infty$ are all those functions in $L^p(E)$ such that $\|f\|_p = 1$.

11.5. The set of the extreme points of the closed unit ball in $L^\infty(E)$ are those $f \in L^\infty(E)$ such that $|f| = 1$ a.e. in E.

11.6. The set of the extreme points of the closed unit ball in $L^1(E)$ is empty.

11.7. Let E be a bounded, connected, open set in $\mathbb{R}^N$. The closed unit ball of $C(\bar{E})$ has no, non constant, extreme points. Examine the case when E has several connected components.

A set $E \subset C$ is an *extremal* subset of C if it satisfies the property:

$$\begin{cases} \text{if there exist } u, v \in C \text{ and } t \in (0, 1) \text{ such that} \\ \qquad tu + (1 - t)v \in E, \text{ then } u, v \in E. \end{cases}$$

Extreme points are extremal sets. Convex portions of a face of a closed, convex polyhedron are extremal sets of that polyhedron. Prove the following:

Lemma 11.1c *If E_1 is an extremal subset of C and E_2 is an extremal subset of E_1, then E_2 is an extremal subset of C.*

Lemma 11.2c *A nonempty, convex, compact subset C of a linear, normed space X, has extreme points.*

Proof Consider the collection $\mathcal{E}$ of all closed, extremal subsets of C, partially ordered by inclusion. Such a collection is not empty since $C \in \mathcal{E}$. Any linearly ordered subcollection $\mathcal{E}_1 \subset \mathcal{E}$ has the finite intersection property (§ 5 of Chap. 2) and therefore

$$E_o = \bigcap \{E \mid E \in \mathcal{E}_1\} \quad \text{is not empty, closed and compact.}$$

We claim that E_o is a singleton which is extreme of C. If E_o contains more than one point, pick $p, q \in E_o$ with $p \neq q$, let $T \in X^*$ be such that $T(p) > T(q)$ (Corollary 10.1), and set

$$E_o' = \left\{ x \in E_o \mid T(x) = \inf_{y \in E_o} T(y) \right\}.$$

The set E_o' is well defined, since T is continuous and E_o is compact, and is a proper subset of E_o, since $T(p) > T(q)$. The proof consists in verifying that E_o' is a convex, closed, extremal subset of E_o. Thus by Lemma 11.1c it would be a closed, extremal subset of C, properly contained in E_o, thereby contradicting the definition of E_o.

If there exists $u, v \in C$ and $t \in (0, 1)$ such that $tu + (1 - t)v \in E_o' \subset E_o$, then $u, v \in E_o$ and

$$tT(u) + (1 - t)T(v) = T(tu + (1 - t)v) = \inf_{y \in E_o} T(y).$$

Now

$$T(u) > \inf_{y \in E_o} T(y) \quad \text{implies} \quad T(v) < \inf_{y \in E_o} T(y)$$

contradicting that $v \in E_o$. Thus

$$T(u) = T(v) = \inf_{y \in E_o} T(y) \quad \text{which implies } u, v \in E_o'. \qquad \blacksquare$$

Theorem 11.1c (Krein–Milman [86]) *A compact, convex set C in a normed linear space is the closed convex hull of its extreme points.*

Proof Let E be the set of the extreme points of C. By the definition of convex hull, $c(E) \subset C$. To prove the converse inclusion, proceed by contradiction, assuming that there exists $x_o \in C$ such that $x_o \notin \overline{c(E)}$. By Proposition 11.3 there exists $T \in X^*$ such that $T(x_o) < T(\overline{c(E)})$. Set

$$C_1 = \left\{ x \in C \mid T(x) = \inf_{y \in C} T(y) \right\}.$$

Proceeding as in the proof of Lemma 11.2c one verifies that C_1 is a nonempty, convex, closed, extremal subset of C. Moreover $C_1 \cap E = \emptyset$. Since C_1 is convex and compact it has at least one extreme point y_o, which by Lemma 11.1c is also an extreme point of C. ∎

11.8. Give an example to show that the compactness assumption on C is essential.

11.4c A General Version of the Krein–Milman Theorem

Prove that Theorem 11.1c continues to hold, with essentially the same proof if X is a Hausdorff topological vector space on which X^* separates points. Compactness of C is referred to the topology of X.

12c Weak Topologies

Proposition 12.1c *Let $\{X; \|\cdot\|\}$ be a normed space and let $\mathcal{W}$ denote its weak topology. Denote also by $\mathcal{O}$ a weak neighborhood of the origin of X. Then:*

(i) *For every $V \in \mathcal{W}$ and $x_o \in V$, there exists $\mathcal{O}$ such that $x_o + \mathcal{O} \subset V$.*
(ii) *For every $V \in \mathcal{W}$ and $x_o \in V$, there exists $\mathcal{O}$ such that $\mathcal{O} + \mathcal{O} + x_o \subset V$.*
(iii) *For every $\mathcal{O}$ there exists $\mathcal{O}'$ such that $\lambda \mathcal{O}' \subset \mathcal{O}$, for all $|\lambda| \leq 1$.*

Proof (of (i)) An open weak neighborhood V of $x_o \in X$ contains an open set of the form

$$V_o = \bigcap_{j=1}^{m} T_j^{-1}(T_j(x_o) - \alpha_j, T_j(x_o) + \alpha_j)$$

for some finite m and $\alpha_j > 0$. Equivalently

$$V_o = \{ x \in X \mid |T_j(x - x_o)| < \alpha_j \quad \text{for } j = 1, \dots, n \}.$$

The neighborhood of the origin

$$\mathcal{O} = \{ x \in X \mid |T_j(x)| < \alpha_j \quad \text{for } j = 1, \dots, n \}$$

is such that $x_o + \mathcal{O} \subset V_o$.   ∎

The remaining statements are proved similarly.

12.1. In a finite dimensional normed linear space, the notions of weak and strong convergence coincide.

12.2. Construct examples and counterexamples for the following statements:

    i. A weakly closed subset of a normed linear space is also strongly closed. The converse is false.

    ii. A strongly sequentially compact subset of a normed linear space is also weakly sequentially compact. The converse is false.

12.3. A normed linear space is weakly complete if and only if it is complete in its strong topology. A weakly dense set in $\{X; \|\cdot\|\}$ is also strongly dense.

12.4. The weak topology on a normed linear space is Hausdorff (Proposition 11.3).

12.1c   Infinite Dimensional Normed Spaces

Let $\{X; \|\cdot\|\}$ be an infinite dimensional Banach space. There exists a countable collection $\{X_n\}$ of infinite dimensional, closed subspaces of X such that $X_{n+1} \subset X_n$ with strict inclusion. For example one might take a nonzero functional $T_1 \in X^*$ and set $X_1 = \ker\{T_1\}$. Such a subspace is infinite dimensional (**Hint**: Proposition 5.1). Then select a nonzero functional $T_2 \in X_1^*$, set $X_2 = \ker\{T_2\}$, and proceed by induction.

For each n, select an element $x_n \in X_{n+1} - X_n$ so that $\|x_n\| = 2^{-n}$. This generates a sequence $\{x_n\}$ of linearly independent elements of X whose span X_o is isomorphic to ℓ_∞ by the representation

$$\ell_\infty \ni \{c_n\} \longrightarrow \sum c_n x_n \in X$$

Since the dimension of ℓ_∞ is not less than the cardinality of $\mathbb{R}$ the dimension of an infinite dimensional Banach space is at least the cardinality of $\mathbb{R}$. Compare with **2.1**.

12.5. A Banach space $\{X; \|\cdot\|\}$ is finite dimensional if and only if every linear subspace is closed (**2.6**). However an infinite dimensional Banach space X contains infinitely many, infinite dimensional, closed subspaces.

12.6. The weak topology of an infinite dimensional normed space X is not normable, i.e., there exists no norm $\|\cdot\|_w$ on X that generates the weak topology. This follows from Corollary 12.3. Give an alternate proof. **Hint:** If such $\|\cdot\|_w$ exists, the open unit ball, with respect to such a norm, would be an open neighborhood of the origin.

12.2c   About Corollary 12.5

In the context of $L^p(E)$ spaces the corollary had been established in [14]. When $p = 2$ the coefficients α_j can be given an elegant form as established by the following proposition.

Proposition 12.2c *Let* $\{f_n\}$ *be a sequence of functions in* $L^2(E)$ *weakly convergent to some* $f \in L^2(E)$. *There exists a subsequence* $\{f_{n_j}\}$ *such that setting*

$$\varphi_m = \frac{f_{n_1} + f_{n_2} + \cdots + f_{n_m}}{m}$$

the sequence $\{\varphi_m\}$ *converges to* f *strongly in* $L^2(E)$.

Proof By possibly replacing f_n with $f_n - f$, we may assume that $f = 0$. Fix $n_1 = 1$. Since $\{f_n\} \to 0$ weakly in $L^2(E)$, there exists an index n_2 such that

$$\left| \int_E f_{n_1} f_{n_2} d\mu \right| \leq \frac{1}{2}.$$

Then there is an index n_3 such that

$$\left| \int_E f_{n_1} f_{n_3} d\mu \right| \leq \frac{1}{3} \quad \text{and} \quad \left| \int_E f_{n_2} f_{n_3} d\mu \right| \leq \frac{1}{3}.$$

Proceeding in this fashion we extract, out of $\{f_n\}$, a subsequence $\{f_{n_j}\}$, such that, for all $k \geq 2$

$$\left| \int_E f_{n_\ell} f_{n_k} d\mu \right| \leq \frac{1}{k} \quad \text{for all } \ell = 1, \dots, k-1.$$

Denoting by M the upper bound of $\|f_n\|_2$, compute

$$\int_E \varphi_m^2 d\mu = \frac{1}{m^2} \int_E (f_{n_1} + f_{n_2} + \cdots f_{n_m})^2 d\mu$$

$$\leq \frac{1}{m^2} \left(mM^2 + 2 + 4\frac{1}{2} + \cdots + 2m\frac{1}{m} \right)$$

$$\leq \frac{M^2 + 2}{m} \longrightarrow 0 \quad \text{as } m \to \infty. \qquad \blacksquare$$

12.3c   Weak Closure and Weak Sequential Closure

For a subset E of a normed space $\{X; \|\cdot\|\}$ denote by $\bar{E}_{w-\sigma}$ its is weak sequential closure, that is the set of all limit points of weakly convergent sequences in E. Equivalently, by $\bar{E}_{w-\sigma}$ is the set of all points $x \in X$, for which there exists a sequence

$\{x_n\} \subset E$ weakly convergent to x. By the definition $\bar{E}_{w-\sigma} \subset \bar{E}_w$. The inclusion is in general strict, as shown by the following examples.

12.7. Let $E = \{\sqrt{n}\mathbf{e}_n\} \subset \ell_2$, where $\mathbf{e}_n$ is the infinite sequence consisting of zeroes except the nth entry which is one. Denoting by $\mathbf{0}$ the zero element of ℓ_2

$$\mathbf{0} \in \overline{\{\sqrt{n}\mathbf{e}_n\}}_w \qquad \text{but} \qquad \mathbf{0} \notin \overline{\{\sqrt{n}\mathbf{e}_n\}}_{w-\sigma}. \tag{12.1c}$$

To prove the first of these, pick $T \in \ell_2^*$ and for a fixed $\alpha > 0$, consider the weak, open neighborhood of the origin

$$\mathcal{O}_{T,\alpha} = \{x \in \ell_2 \mid |T(x)| < \alpha\}.$$

By the Riesz representation theorem any such T is identified by some $\mathbf{t} \in \ell_2$ acting on elements $\mathbf{a} \in \ell_2$ by the formula

$$T(\mathbf{a}) = \mathbf{t} \cdot \mathbf{a} = \sum t_n a_n.$$

Since $\mathbf{t} \in \ell_2$, for all $M > 0$, there exists $n > M$ such that $|t_n| \le \alpha/\sqrt{n}$. Indeed otherwise

$$\|\mathbf{t}\|^2 \ge \sum_{n>M} t_n^2 > \alpha \sum \frac{1}{n}.$$

For such an index, $|T(\sqrt{n}\mathbf{e}_n)| < \alpha$, and hence $\sqrt{n}\mathbf{e}_n \in \mathcal{O}_{T,\alpha}$. Let now

$$\mathcal{O} = \bigcap_{j=1}^k T_j^{-1}(-\alpha_j, \alpha_j), \quad \text{for } \alpha_j > 0 \quad \text{for some } k \in \mathbb{N}. \tag{12.2c}$$

Prove that n can be chosen so large that $\sqrt{n}\mathbf{e}_n \in \mathcal{O}$. Therefore any weak, open neighborhood of the origin intersects $\{\sqrt{n}\mathbf{e}_n\}$.

To prove the second of (12.1c) we establish that there exists no subsequence $\{\sqrt{m}\mathbf{e}_m\} \subset \{\sqrt{n}\mathbf{e}_n\}$, weakly convergent to $\mathbf{0}$. Having picked such a subsequence assume first that some fixed index $j \in \mathbb{N}$ occurs infinitely many times in $\{\sqrt{m}\mathbf{e}_m\}$. Pick $T_j \in \ell_2^*$ corresponding to $\mathbf{e}_j \in \ell_2$. For such a choice the sequence $\{T_j(\sqrt{m}\mathbf{e}_m)\}$ contains the value $\sqrt{j}$ infinitely many times, and thus it cannot converge to zero. If no index $j \in \mathbb{N}$ occurs infinitely many times in $\{\sqrt{m}\mathbf{e}_m\}$ pick a subsequence

$$\{\sqrt{m_j}\mathbf{e}_{m_j}\} \subset \{\sqrt{m}\mathbf{e}_m\} \quad \text{so that} \quad m_j \ge 2^j$$

and set

$$\mathbf{t} = \sum_{j \in \mathbb{N}} \frac{1}{\sqrt{m_j}} \mathbf{e}_{m_j} \quad \text{satisfying} \quad \|\mathbf{t}\|^2 = \sum_{j \in \mathbb{N}} \frac{1}{m_j} < \infty.$$

Therefore $\mathbf{t} \in \ell_2$ and we let $T \in \ell_2^*$ be its corresponding functional. For such a functional $T(\sqrt{m_j} \mathbf{e}_{m_j}) = 1$. Therefore the sequence $\{T(\sqrt{m} \mathbf{e}_m)\}$ contains 1 infinitely many times and thus it cannot converge to zero.

Remark 12.1c The set E is countable, so that countability alone is not sufficient to identifty weak closure with weak sequential closure.

Remark 12.2c The set $\{\sqrt{n} \mathbf{e}_n\}$ is unbounded in ℓ_2. However the strict inclusion $\bar{E}_{w-\sigma} \subset \bar{E}_w$ continues to hold even for bounded sets, as shown by the next example.

12.8. Consider the set $E \subset \ell_1$

$$E = \bigcup_{n,m \in \mathbb{N}; m \neq n} \{\mathbf{e}_m - \mathbf{e}_n\}.$$

We claim that E is bounded in ℓ_1, its weak closure contains $\mathbf{0}$ but its weak sequential closure does not contain $\mathbf{0}$.

To establish the first claim let $\mathcal{O}$ be a weakly open set of the form (12.2c). Every $T_j \in \ell_1^*$ is identified with an element $\mathbf{t}_j \in \ell_\infty$ acting on elements $\mathbf{a} \in \ell_1$ by the formula

$$T_j(\mathbf{a}) = \mathbf{t}_j \cdot \mathbf{a} = \sum t_{j,n} a_n.$$

Since $\mathbf{t}_j \in \ell_\infty$ for $j = 1, \ldots, k$, as n ranges over $\mathbb{N}$, the k-tuple $(t_{1,n}, \ldots, t_{k,n})$ ranges over a bounded set $K \subset \mathbb{R}^k$, which, without loss of generality we may assume to be a closed cube with faces parallel to the coordinate planes. Pick $\varepsilon = \min\{\alpha_1, \ldots, \alpha_k\}$ and subdivide K into no less than ε^{-k} homotetic closed sub-cubes K_ε of edge not exceeding ε. At least one of these sub-cubes must contain infinitely many k-tuples $(t_{1,n}, \ldots, t_{k,n})$. Select one such cube and relabel the corresponding k-tuples as $\{(t_{1,n_i}, \ldots, t_{k,n_i})\}_{i \in \mathbb{N}}$. For the element $\mathbf{e}_{m_i} - \mathbf{e}_{n_i} \in \{\mathbf{e}_m - \mathbf{e}_n\}$ compute

$$|T_j(\mathbf{e}_{m_i} - \mathbf{e}_{n_i})| = |t_{j,m_i} - t_{j,n_i}| \leq \varepsilon \leq \alpha_j \quad \text{for} \quad j = 1, \ldots k.$$

Therefore $\mathbf{e}_{m_i} - \mathbf{e}_{n_i} \in \mathcal{O}$. Thus every weak open neighborhood of $\mathbf{0}$ intersects $\{\mathbf{e}_m - \mathbf{e}_n\}$ and hence $\mathbf{0} \in \overline{\{\mathbf{e}_m - \mathbf{e}_n\}}_w$.

To show that no subsequence $\{\mathbf{e}_{m_j} - \mathbf{e}_{n_j}\} \subset \{\mathbf{e}_m - \mathbf{e}_n\}$ converges weakly to $\mathbf{0}$ it suffices for any such subsequence, to construct a functional $T \in \ell_1^*$ such that $\{T(\mathbf{e}_{m_j} - \mathbf{e}_{n_j})\}$ does not converge to zero. Construct one such T by examining separately the following cases

    i. Some pair (m_j, n_j) occurs infinitely many times;
   ii. Some index m_j occurs infinitely many times and $n_j \to \infty$, or vice versa.
  iii. Both $m_j, n_j \to \infty$.

12.9. Let $E = \ell_2$ be defined by (von Neuman [113])

$$E = \{\mathbf{e}_n + n\mathbf{e}_m\} \quad \text{for } 0 \le n < m.$$

Prove that E is strongly closed (**Hint:** all the sequences in E, Cauchy in ℓ_2 are constant).

Prove that the origin of ℓ_2 is in the weak closure but not in the weak sequential closure of E.

12.10. Prove that ℓ_2 and ℓ_1 equipped with their weak topology do not satisfy the first axiom of countability.

14c Weak Compactness

14.1. A weakly compact subset of a normed linear space is weakly closed. See also Proposition 5.1 of Chap. 2.

14.2. A weakly compact, convex subset of a normed linear space is strongly closed.

14.1c Linear Functionals on Subspaces of $C(\bar{E})$

Let E be a bounded, open set in $\mathbb{R}^N$ and let $C(\bar{E})$ denote the space of the continuous functions in $\bar{E}$ equipped with the sup-norm.

    Let X_o be a subspace of $C(\bar{E})$ closed in the topology of $L^2(E)$. Prove that there exist positive constants $C_o \le C_1$, such that

$$C_o \|f\|_\infty \le \|f\|_2 \le C_1 \|f\|_\infty.$$

Let $T_{o,y} \in X_o^*$ be the evaluation map at y

$$T_{o,y}(f) = f(y) \qquad \text{for all } f \in X_o.$$

By the Hahn–Banach theorem there exists a functional $T_y \in L^2(E)^*$ such that $T_y = T_{o,y}$ on X_o. By the Riesz representation of the bounded linear functionals in $L^2(E)$, there exists a function $K(\cdot, y) \in L^2(E)$, such that

$$f(y) = \int_E K(x, y) f(x) d\mu \qquad \text{for all } f \in X_o. \tag{14.1c}$$

Proposition 14.1c *The unit ball of X_o is compact.*

Proof The closed unit ball $\bar{B}_{o,1}$ of X_o is also weakly closed. Since $L^2(E)$ is reflexive, also X_o is reflexive. Therefore $\bar{B}_{o,1}$ is bounded and weakly closed and hence sequentially compact. In particular every sequence $\{f_n\} \subset \bar{B}_{o,1}$ contains in turn a subsequence $\{f_{n'}\}$ weakly convergent to some $f \in \bar{B}_{o,1}$. By (14.1c) such a sequence converges pointwise to f and

$$\|f_n\|_\infty \leq (\text{const})\|f_n\|_2 \leq (\text{const})'$$

for a constant independent of n. Therefore, by the Lebesgue dominated convergence theorem, $\{f_{n'}\} \to f$ strongly in $L^2(E)$.

Thus every sequence $\{f_n\} \subset \bar{B}_{o,1}$ contains in turn a strongly convergent subsequence. This implies that $\bar{B}_{o,1}$ is compact in the strong topology inherited form $L^2(E)$. ∎

Corollary 14.1c *Every subspace $X_o \subset C(\bar{E})$, closed in $L^2(E)$ is finite dimensional.*

14.2c Weak Compactness and Boundedness

14.3. Give a different proof of Corollary 14.1 by means of the uniform boundedness principle. **Hint**: For all $T \in X^*$, the image $T(E)$ is a compact and hence bounded subset of $\mathbb{R}$.

15c The Weak* Topology of X^*

15.1c Total Sets of X

The linear span X_* of a collection of linear functionals on a linear vector space X is a total set of X, if it separates its points, that is, if for any two distinct elements $x, y \in X$, there exists $T \in X_*$ such that $T(x) \neq T(y)$. The dual X^* of a normed space $\{X; \|\cdot\|\}$ separates the points of X and therefore is a total set of X. However the notion of total set does not require a topology being placed on X nor the continuity of the linear maps in X_*.

If X_* is total for X, one might endow X with the weakest topology $\mathcal{W}_*$, for which all the elements of X_* are continuous. A $\mathcal{W}_*$-open base neighborhood of the origin is of the form

$$\mathcal{O}_* = \left\{ \begin{array}{l} x \in X \mid |T_j(x)| < \alpha_j \text{ for } T_j \in X_* \text{ and } \alpha_j \in \mathbb{R}^+ \\ \text{for } j = 1, \ldots, n \text{ for some finite } n. \end{array} \right\} \tag{15.1c}$$

The construction is in all similar to the construction of the weak topology of a normed space $\{X; \|\cdot\|\}$. Since X_* is total, the $\mathcal{W}_*$ topology on X is Hausdorff and one verifies that the operations of sum and product by scalars are continuous in the indicated topology. Thus $\{X; \mathcal{W}_*\}$ is a Hausdorff, topological vector space. By construction the elements of X_* are continuous from $\{X; \mathcal{W}_*\}$ into $\mathbb{R}$. The next proposition asserts that these are the only bounded linear functional on $\{X; \mathcal{W}_*\}$.

Proposition 15.1c *Let $T : \{X; \mathcal{W}_*\} \to \mathbb{R}$ be nonidentically zero, linear and continuous. Then $T \in X_*$.*

Proof By the continuity of T, there exists a $\mathcal{W}_*$-open set of the form (15.1c) such that $\mathcal{O}_* \subset T^{-1}(-1, 1)$. Let $T_j \in X_*$ for $j = 1, \ldots, n$ be the functionals in X_* that identify $\mathcal{O}_*$. If $x \in \bigcap_{j=1}^n \ker\{T_j\}$ then $x \in T^{-1}(-\varepsilon, \varepsilon)$ for all $\varepsilon > 0$. Therefore T vanishes on $\bigcap_{j=1}^n \ker\{T_j\}$, and by Proposition 11.2 of Chap. 2, T is a linear combination of the T_j. $\blacksquare$

15.1. Let $\{X; \|\cdot\|\}$ be a normed space. Then X_{**} as defined in (13.3) separates the points in X^* and is total for X^*. The elements of X_{**} are the only bounded linear functionals on X^* equipped with its weak* topology. In particular every $T_* \in X^{**} - X_{**}$ is discontinuous with respect to the weak* topology of X^*,

15.2. Let $\{X; \|\cdot\|\}$ be a normed space. Then X^* equipped with its weak* topology is a topological vector space whose dual is isometrically isomorphic to X.

15.3. Let $\{X; \|\cdot\|\}$ be a normed space. Then every convex, weak* compact subset of X^* is the weak* closed convex hull of its extreme points (§ 11.4c).

15.2c *Metrization Properties of Weak* Compact Subsets of X^**

Proposition 15.2c *Let $\{X; \|\cdot\|\}$ be a Banach space. Then the weak* topology of a weak* compact set $K \subset X^*$ is metric if and only if $\{X; \|\cdot\|\}$ is separable.*

Proof (sufficient condition) Assume first that X is separable and let $\{x_n\} \subset X$ be a sequence dense in X. For $T_1, T_2 \in X^*$ set

$$d(T_1, T_2) = \sum \frac{1}{2^n} \frac{|(T_1 - T_2)(x_n)|}{1 + |(T_1 - T_2)(x_n)|}.$$

Keeping in mind that X as a subset of X^{**} separates the points of X^*, one verifies that

$$d : X^* \times X^* \to \mathbb{R}^+$$

is a metric which generates a metric topology on X^* (§ 13.11c of the Complements of Chap. 2). One also verifies that every ball $B_\varepsilon(T) \subset X^*$ of center $T \in X^*$ and radius $\varepsilon > 0$ in the metric $d(\cdot, \cdot)$ contains a weak* open neighborhood of T. Therefore the identity map from X^* equipped with its weak* topology onto X^* equipped with the metric topology of $d(\cdot, \cdot)$, is continuous. Let $\{K; \text{weak}^*\}$ and $\{K; d\}$ be the topological spaces formed by K equipped with the topologies inherited respectively from the weak* and metric topologies of X^*. The identity map from the compact space $\{K; \text{weak}^*\}$ onto $\{K; d\}$ is continuous and one-to-one. To show that it is a homeomorphism we appeal to Proposition 5.1 of Chap. 2. A weak*-closed set $E \subset \{K; \text{weak}^*\}$ is weak* compact. Since the identity map from $\{K; \text{weak}^*\}$ onto $\{K; d\}$ is continuous, the image of E in $\{K; d\}$ is compact and hence closed in the metric topology of $\{K; d\}$, since the latter is Hausdorff. ∎

Remark 15.1c The identity map from $\{K; \text{weak}^*\}$ onto $\{K; d\}$, being a homeomorphism, identifies the two topologies. For this is essential that $\{K; \text{weak}^*\}$ be weak* compact. In particular, the proposition does not imply that the weak* topology of the whole X^* is metrizable.

Proof (necessary condition) Assume conversely that the weak* topology of $K \subset X^*$ is metric. Up to a translation, may assume that $\Theta \in K$. There exists a countable collection $\{\mathcal{O}_n^*\}$ of weak* open neighborhoods of Θ^* such that $\bigcap \mathcal{O}_n^* = \Theta^*$. Without loss of generality the $\mathcal{O}_n^*$ are of the form

$$\mathcal{O}_n^* = \left\{ \begin{array}{c} T \in X^* \mid |x_{j_n}(T)| < \alpha_{j_n} \text{ for } \alpha_{j_n} > 0 \text{ and } x_{j_n} \in X \\ \text{for } j_n \in \{j_{n,1}, \ldots, j_{n,k}\} \subset \mathbb{N} \end{array} \right\}.$$

The collection of $\{x_{j_n}\}$ is countable and its closed linear span is separable. We claim that $\overline{\{x_{j_n}\}} = X$. If not, there exists a nontrivial $T \in X^*$ vanishing on $\overline{\{x_{j_n}\}}$ (Corollary 10.3). However if $T(x_{j_n}) = 0$ for all x_{j_n} then $T \in \mathcal{O}_n^*$ for all n and thus $T = \Theta^*$. ∎

16c The Alaoglu Theorem

16.1. Let E be a bounded open set in $\mathbb{R}^N$. Prove that there is no linear normed space $\{X; \|\cdot\|\}$ whose dual is $L^1(E)$ with respect to the Lebesgue measure. **Hint**: Problems **11.6**, and **15.3**.

16.2. Let E be a bounded open set in $\mathbb{R}^N$. Prove that $C(\bar{E})$ is not the dual of any linear normed space. **Hint**: Problems **11.7**, and **15.3**.

16.3. The weak* topology of the closed unit ball of $L^\infty(E)$ and ℓ_∞ are metrizable.

16.4. A bounded and weakly closed subset E of a Banach space X, need not be weakly compact. In particular the closed unit balls of $L^1(E)$, $L^\infty(E)$, ℓ_1 and ℓ_∞ are weakly closed but not weakly compact.

16.5. Let $\{X; \|\cdot\|\}$ be a normed space and let

$$B_1 = \{x \in X \mid \|x\| \le 1\}, \qquad S_1 = \{x \in X \mid \|x\| = 1\}$$

be respectively the unit ball and the unit sphere in X. Prove or disprove by a counterexample the following statements:

i. B_1 is strongly and weakly bounded.
ii. B_1 is weakly sequentially closed.
iii. S_1 is weakly sequentially closed.
iv. If $\{X; \|\cdot\|\}$ is reflexive then S_1 is weakly sequentially closed.
v. If $\{X; \|\cdot\|\}$ is reflexive then B_1 is weakly sequentially compact.
vi. If $\{X; \|\cdot\|\}$ is reflexive then it is Banach.
vii. If $\{X; \|\cdot\|\}$ is Banach then it is reflexive.
viii. The weak topology of $\{X; \|\cdot\|\}$ is Hausdorff.

16.1c   The Weak* Topology of X^{**}

Denote by X^{***} the dual of X^{**}, that is the collection of all bounded linear functionals $f : X^{**} \to \mathbb{R}$. It is a Banach space by the norm

$$\|f\| = \sup_{x^{**} \in X^{**}; x^{**} \neq 0} \frac{f(x^{**})}{\|x^{**}\|} = \sup_{x^{**} \in X^{**}; \|x^{**}\|=1} f(x^{**}). \tag{16.1c}$$

Every element $x^* \in X^*$ identifies an element $f_{x^*} \in X^{***}$ by the formula

$$X^{**} \ni x^{**} \longrightarrow f_{x^*}(x^{**}) = x^{**}(x^*). \tag{16.2c}$$

Let X_{***} denote the collection of all such functionals, that is

$$X_{***} = \left\{ \begin{array}{c} \text{the collection of all functionals } f_{x^*} \in X^{***} \\ \text{of the form (13.2) as } x^* \text{ ranges over } X^* \end{array} \right\}.$$

From Corollary 10.2 and (16.1c) it follows that $\|f_{x^*}\| = \|x^*\|$. Therefore the injection map

$$X^* \ni x^* \longrightarrow f_{x^*} \in X_{***} \subset X^{***} \tag{16.3c}$$

is an isometric isomorphism between X^* and X_{***}. In general, not all the bounded linear functionals $f : X^{**} \to \mathbb{R}$ are derived from the injection map (16.3c); otherwise said, the inclusion $X_{***} \subset X^{***}$ is in general strict. The set X_{***} is total for X^{**} and it generates a topology $\mathcal{W}_{**}$ on X^{**} which is the weak* topology generated by X^* on X^{**} (§ 15.1c). By this topology, $\{X^{**}; \mathcal{W}_{**}\}$ turns into a Hausdorff, linear, topological vector space (§ 15.1c), and for such a space the separation Proposition 11.3 holds. In particular $\mathcal{W}_{**}$-closed sets are separated from points by a bounded linear functional on $\{X^{**}; \mathcal{W}_{**}\}$. By Proposition 15.1c, any such a functional is of the form (16.2c). Let

B the unit ball of X closed in the norm of $\{X; \|\cdot\|_X\}$

B^{**} the unit ball of X^{**} closed in the norm of $\{X^{**}; \|\cdot\|_{**}\}$.

Since the natural injection $X \ni x \to f_x \in X^{**}$ defined by (14.2) is an isometric isomorphism, the ball B when regarded as a subset of B^{**} is a closed subset of B^{**} and the inclusion is proper unless X is reflexive. If B^{**} is given the $\mathcal{W}_{**}$ topology, by the Alaoglu Theorem is weak* compact, and since $\mathcal{W}_{**}$ is Hausdorff, B^{**} is both $\|\cdot\|_{**}$-closed and $\mathcal{W}_{**}$-closed (Proposition 5.1-(iii) of Chap. 2). Set

$$B_{**} = \left\{ \begin{array}{c} \text{the } \mathcal{W}_{**}\text{-closure of norm-closed unit ball of } X \\ \text{regarded as a subset of } \{X^{**}; \mathcal{W}_{**}\} \end{array} \right\}$$

The next proposition asserts that while B with its norm topology is a subset B^{**}, with in general strict inclusion, when it is given the $\mathcal{W}_{**}$ topology, it is actually dense in B^{**}.

Proposition 16.1c ([57]) $B_{**} = B^{**}$.

Proof B_{**} is a closed and convex subset of B^{**} (**10.1**-(iii) of the Complements of Chap. 2). If $B_{**} \subset B^{**}$ with proper inclusion, select and fix $x^{**} \in B^{**} - B_{**}$. By Proposition 11.3 there exists a nontrivial, bounded, linear functional f on $\{X^{**}; \mathcal{W}_{**}\}$ and real numbers $\alpha < \beta$ such that

$$f(y) \le \alpha < \beta \le f(x^{**}) \quad \text{for all } y \in B_{**}.$$

By Proposition 15.1c, such a functional f is of the form (16.2c). Therefore, there exists $x^* \in X^*$ such that

$$x^*(y) \le \alpha < \beta \le x^{**}(x^*) \quad \text{for all } y \in B. \tag{16.4c}$$

Now $y \in B$ implies $-y \in B$. Therefore

$$\|x^*\| = \sup_{\|y\|=1} x^*(y) \le \alpha.$$

On the other hand, since $x^{**} \in B^{**}$

$$\alpha < \beta \le x^{**}(x^*) \le \|x^{**}\| \, \|x^*\| \le \alpha.$$

This contradicts (16.4c) and proves the proposition.   ∎

Corollary 16.1c *The Banach space* $\{X; \|\cdot\|\}$ *when regarded as a subspace of* X^{**} *and equipped with the weak* topology of* X^{**}, *is dense in* X^{**}.

16.2c   Characterizing Reflexive Banach Spaces

The next statements follow from the previous proposition, upon observing that the W_{**} topology of X, as a subset of X^{**}, is precisely the weak topology generated on X by X^*.

Corollary 16.2c *A Banach space $\{X; \|\cdot\|\}$ is reflexive if and only if its closed unit ball is weakly compact.*

Corollary 16.3c *A Banach space $\{X; \|\cdot\|\}$ is reflexive if and only if a bounded and weakly closed set is weakly compact.*

Thus in some sense, reflexive Banach spaces are those for which a version of the Heine–Borel Theorem holds (Proposition 6.4 of Chap. 2).

16.6. Some of the following statements contain fallacies. Identify and disprove them, by an argument or a counterexample.

  i. The closed unit ball B of a normed space $\{X; \|\cdot\|\}$, is convex and hence weakly closed (Proposition 12.2).

  ii. Regarding B as a subset of B^{**} its W_{**} topology coincides precisely with its weak topology.

  iii. Therefore B as a subset of B^{**} is W_{**} closed and hence W_{**}-compact, since it is a closed subset of a compact set.

  iv. Since the W_{**} topology on $B \subset B^{**}$ coincides with its weak topology, B is weakly compact.

16.3c   Metrization Properties of the Weak Topology
of the Closed Unit Ball of a Banach Space

Proposition 16.2c *Let $\{X; \|\cdot\|\}$ be a Banach space. Then the weak topology of the closed unit ball $B \subset X$ is metric, if and only if the dual X^* is separable.*

Proof If X^* is separable, the weak* topology of $B^{**} \subset X^{**}$ is metric. Identify X as a subset of X^{**} by the natural injection map $X \ni x \to f_x \in X^{**}$ and observe that the weak* topology inherited by X as a subset of X^{**} is precisely the weak topology of X.

Assume next that the weak topology of the closed unit ball $B \subset X$ is metric. There exists a countable collection $\{\mathcal{O}_n\}$ of weakly open neighborhoods of the origin such that every weak neighborhood of the origin contains some $\mathcal{O}_n$. Without loss of generality the $\mathcal{O}_n$ can be taken of the form

$$\mathcal{O}_n = \left\{ \begin{array}{l} x \in X \mid |T_{j_n}(x)| < \varepsilon_n \text{ for } \varepsilon_n > 0 \text{ and } T_{j_n} \in X^* \\ \text{for } j_n \in \{j_{n,1}, \ldots, j_{n,k}\} \subset \mathbb{N} \end{array} \right\}$$

The collection of $\{T_{j_n}\}$ is countable and its closed linear span is separable. We claim that $\overline{\{T_{j_n}\}} = X^*$. If not pick $\eta^* \in X^* - \overline{\{T_{j_n}\}}$ at distance $\delta \in (0, 1)$ from $\overline{\{T_{j_n}\}}$ and determine a nontrivial, bounded linear functional $x^{**} \in X^{**}$, of norm $\|x^{**}\| \leq 1$, vanishing on $\overline{\{T_{j_n}\}}$ and such that $x^{**}(\eta^*) = \delta$ (Proposition 10.3). The set

$$\mathcal{O}_\delta = \left\{ x \in X \mid |\eta^*(x)| < \tfrac{1}{2}\delta \right\}$$

is a weak neighborhood of the origin of X and hence $\mathcal{O}_n \subset \mathcal{O}_\delta$ for some n. Since $x^{**} \in B^{**}$, by the density Proposition 16.1c, there exists $x \in B$ such that

$$|x^{**}(T_{j_n}) - T_{j_n}(x)| < \varepsilon_n \text{ for all } j_n \in \{j_{n,1}, \ldots, j_{n,k}\}$$

and simultaneously

$$|\delta - \eta^*(x)| = |x^{**}(\eta^*) - \eta^*(x)| < \varepsilon_n.$$

By picking ε_n sufficiently small we may insure that $\varepsilon_n < \tfrac{1}{2}\delta$. Then, since x^{**} vanishes on $\overline{\{T_{j_n}\}}$, the first of these gives,

$$|T_{j_n}(x)| < \varepsilon_n \quad \text{for all } j_n \in \{j_{n,1}, \ldots, j_{n,k}\}$$

which implies that $x \in \mathcal{O}_n$. However the second of these implies $|\eta^*(x)| > \tfrac{1}{2}\delta$. That is $x \notin \mathcal{O}_\delta \subset \mathcal{O}_n$. ∎

16.7. The weak topology of the closed unit ball of $L^1(E)$ and ℓ_1 are not metrizable.

16.4c   Separating Closed Sets in a Reflexive Banach Space

Proposition 16.3c *Let C_1 and C_2 be two nonempty, disjoint, closed subsets of a reflexive Banach space $\{X; \|\cdot\|\}$. Assume that at least one of them is bounded. There exists a bounded linear functional T on X and real numbers $0 < \alpha < \beta$ such that (11.2) holds. Equivalently, any two closed subsets of a reflexive Banach space, can be strictly separated, provided at least one of them is bounded.*

Proof If C_1 is closed and bounded it is weak* compact. The statement then follows from Proposition 11.3. ∎

Remark 16.1c The requirement that at least one of the two closed sets C_1 and C_2 be bounded is essential, in view of the counterexample in § 11.1c

17c Hilbert Spaces

17.1c On the Parallelogram Identity

The parallelogram identity is equivalent to the existence of an inner product on a vector space X in the following sense.

A scalar product $\langle \cdot, \cdot \rangle$ on a vector space X generates a norm $\|\cdot\|$ on X that satisfies the parallelogram identity. Vice versa, let $\{X; \|\cdot\|\}$ be a normed space whose norm $\|\cdot\|$ satisfies the parallelogram identity. Then setting

$$4\langle x, y \rangle = \|x + y\|^2 - \|x - y\|^2$$

defines a scalar product in X.

17.2. If $p \neq 2$ then $L^p(E)$ is not a Hilbert space.

17.3. Let $\{x_n\}$ and $\{y_n\}$ be Cauchy sequences in a Hilbert space H. Then $\{\langle x_n, y_n \rangle\}$ is a Cauchy sequence in $\mathbb{R}$.

18c Orthogonal Sets, Representations and Functionals

18.1. For an n-tuple $\{x_1, x_2, \ldots, x_n\}$ of orthogonal elements in a Hilbert space H

$$\left\| \sum_{i=1}^{n} x_i \right\|^2 = \sum_{i=1}^{n} \|x_i\|^2 \qquad \text{(Pythagora's theorem)}$$

18.2. Let E be a subset of H. Then $E^\perp$ is a linear subspace of H and $(E^\perp)^\perp$ is the smallest, closed, linear subspace of H containing E.

18.3. Every closed convex set of H has a unique element of least norm. More generally, let C be a weakly closed subset of a reflexive Banach space. Then the functional $h(x) = \|x\|$ takes its minimum on C, i.e., there exists $x_o \in C$ such that $\inf_{x \in C} \|x\| = \|x_o\|$.

18.4. Let E be a bounded open set in $\mathbb{R}^N$ and let $f \in C(\bar{E})$. Denote by $\mathcal{P}_n$ the collections of polynomials of degree at most n in the coordinate variables. There exists a unique $P_o \in \mathcal{P}_n$ such that

$$\int_E |f - P|^2 dx \geq \int_E |f - P_o|^2 dx \qquad \text{for all } P \in \mathcal{P}_n.$$

19c   Orthonormal Systems

19.1. Let H be a Hilbert space and let S be an orthonormal system in H. Then for any pair x, y of elements in H

$$\sum_{\mathbf{u} \in S} |\langle \mathbf{u}, x \rangle||\langle \mathbf{u}, y \rangle| \le \|x\| \|y\|.$$

19.2. Let S be an orthonormal system in H and denote by H_o the closure of the linear span of S. The projection of an element $x \in H$ into H_o is defined by

$$x_o = \sum_{\mathbf{u} \in S} \langle x, \mathbf{u} \rangle \mathbf{u}.$$

Such a formula defines x_o uniquely. Moreover $x_o \in H_o$ and $(x - x_o) \perp H_o$.

19.3. A Non Separable Hilbert Space: In $L^2_{\text{loc}}(\mathbb{R})$ with respect to the Lebesgue measure, define

$$L^2_{\text{loc}}(\mathbb{R}) \ni f, g \to \langle f, g \rangle \overset{\text{def}}{=} \lim_{\rho \to \infty} \frac{1}{\rho} \int_{-\rho}^{\rho} fg \, dx.$$

Set

$$H_o = \{ f \in L^2_{\text{loc}}(\mathbb{R}) \mid \|f\| = 0 \}$$
$$H_1 = \{ f \in L^2_{\text{loc}}(\mathbb{R}) \mid \|f\| < \infty \}.$$

Since H_o contains nonzero elements, $\| \cdot \|$ is not a norm. Introduce the quotient space

$$H = \frac{H_1}{H_o} \quad \text{of equivalence classes } f + H_o \quad \text{for } f \in H_1.$$

Verify that $\langle \cdot, \cdot \rangle$ is an inner product and $\| \cdot \|$ is a norm on H. The system

$$\left\{ \sin \alpha x + H_o \right\}_{\alpha \in \mathbb{R}}$$

is orthonormal in H, and uncountable. Therefore H is non separable, by Proposition 19.2.

Chapter 8
Spaces of Continuous Functions, Distributions, and Weak Derivatives

1 Bounded Linear Functionals on $C_o(\mathbb{R}^N)$

Let $C_o(\mathbb{R}^N)$ denote the space of continuous functions of compact support in $\mathbb{R}^N$, equipped with the sup-norm. Continuity of functionals $T \in C_o(\mathbb{R}^N)^*$ is meant with respect to such a norm. A finite Radon measures μ in $\mathbb{R}^N$, generates a bounded linear functional in $C_o(\mathbb{R}^N)$, by the formula

$$C_o(\mathbb{R}^N) \ni f \to T(f) = \int_{\mathbb{R}^N} f \, d\mu. \tag{1.1}$$

One verifies that $\|T\| = \mu(\mathbb{R}^N)$. Given two finite Radon measures μ_1 and μ_2 in $\mathbb{R}^N$, the signed measure $\mu = \mu_1 - \mu_2$ generates a bounded linear functional $T \in C_o(\mathbb{R}^N)^*$ by the formula

$$C_o(\mathbb{R}^N) \ni f \to T(f) = \int_{\mathbb{R}^N} f \, d\mu_1 - \int_{\mathbb{R}^N} f \, d\mu_2. \tag{1.2}$$

One verifies that $\|T_\mu\| = |\mu|(\mathbb{R}^N)$, where $|\mu|$ is the total variation of μ. More generally given a Radon measure μ in $\mathbb{R}^N$ and a μ-integrable function w the formula

$$C_o(\mathbb{R}^N) \ni f \to T(f) = \int_{\mathbb{R}^N} f w \, d\mu. \tag{1.3}$$

identifies a bounded linear functional on $C_o(\mathbb{R}^N)$ with $\|T\| = \|w\|_{1,\mathbb{R}^N}$. One also checks that (1.3) is of the same form as (1.2).

A linear map $T : C_o(\mathbb{R}^N) \to \mathbb{R}$ is *locally bounded* if for every compact set $K \subset \mathbb{R}^N$ there exists a positive constant γ_K such that

$$|T(f)| \leq \gamma_K \|f\| \quad \text{for all } f \in C_o(\mathbb{R}^N) \text{ with } \operatorname{supp}\{f\} \subset K. \tag{1.4}$$

© Springer Science+Business Media New York 2016
E. DiBenedetto, *Real Analysis*, Birkhäuser Advanced
Texts Basler Lehrbücher, DOI 10.1007/978-1-4939-4005-9_8

Elements $T \in C_o(\mathbb{R}^N)^*$ are locally bounded; the converse is false. If in (1.3) $w \in L^1_{\text{loc}}(\mathbb{R}^N)$, the corresponding functional is locally bounded. A relevant class of locally bounded linear functionals is that of positive functionals.

1.1 Positive Linear Functionals on $C_o(\mathbb{R}^N)$

A linear map $T : C_o(\mathbb{R}^N) \to \mathbb{R}$ is *positive* if $T(f) \geq 0$ whenever $f \geq 0$. Since T is linear $f \geq g$ implies $T(f) \geq T(g)$. The functional in (1.1) is positive even if μ is not finite. Thus positive functionals need not be bounded; however they are locally bounded.

Proposition 1.1 *A positive, linear functional T on $C_o(\mathbb{R}^N)$ is locally bounded.*

Proof For a compact set $K \subset \mathbb{R}^N$ choose $\varphi \in C_o(\mathbb{R}^N)$ such that $0 \leq \varphi \leq 1$, and $\varphi = 1$ on K. Given $f \in C_o(\mathbb{R}^N)$ with $\text{supp}\{f\} \subset K$, the two functions $\|f\|\varphi \pm f$ are both nonnegative. Therefore $\pm T(f) \leq \|f\| T(\varphi)$. ∎

1.2 The Riesz Representation Theorem

For the integrals in (1.1)–(1.3) to be well defined, f has to be μ-measurable, that is the sets $[f > c]$ must be μ-measurable for all $c \in \mathbb{R}$. Since these sets are open it suffices that μ be defined only on the Borel sets. For a Radon measure μ denoted by $\mu|_\mathcal{B}$ its restriction to the Borel σ-algebra. The Riesz representation theorem asserts that all locally bounded linear functionals in $C_o(\mathbb{R}^N)$ are of the form (1.3) for some *Borel measure* $\mu|_\mathcal{B}$ and some locally μ-integrable function w.

Theorem 1.1 *Let $T : C_o(\mathbb{R}^N) \to \mathbb{R}$ be linear and locally bounded. There exists a Radon measure μ in $\mathbb{R}^N$, and a μ-measurable real valued function w, with $|w| = 1$, μ-a.e. in $\mathbb{R}^N$, such that T can be represented as in (1.3). Moreover the pair $\{\mu|_\mathcal{B}, w\}$ is unique.*

Corollary 1.1 (Riesz Representation) *Every $T \in C_o(\mathbb{R}^N)^*$ has a unique representation of the form (1.3) for a finite Radon measure $\mu|_\mathcal{B}$ and a μ-measurable function w, with $|w| = 1$, μ-a.e. in $\mathbb{R}^N$.*

The first form of this theorem for $C_o[0, 1]$ is in [127]. Through various extensions it is known to hold for $C_o(X)$ where X is a locally compact Hausdorff topological space ([74], vol. I, § 11).

2 Partition of Unity

Let E be a subset of $\mathbb{R}^N$ and let $\mathcal{U}$ be an open covering of E. A countable collection $\{\varphi_n\} \subset C_o^\infty(\mathbb{R}^N)$ is a locally finite *partition of the unity* for E, subordinate to $\mathcal{U}$ if

(i)   for any compact set $K \subset E$ all but finitely many of the functions φ_n are identically zero on K

(ii) $0 \leq \varphi_j \leq 1$ in E

(iii) for any φ_j, there exists $\mathcal{O} \in \mathcal{U}$ such that $\text{supp}\{\varphi_j\} \subset \mathcal{O}$

(iv) $\sum \varphi_n(x) = 1$ for all $x \in E$.

Proposition 2.1 *For every open covering $\mathcal{U}$ of E, there exists a partition of unity for E, subordinate to $\mathcal{U}$.*

Proof Consider the collection of balls $B_{r_i}(x_j)$ centered at points $x_j \in E$ of rational coordinates and rational radii r_i, and contained in some $\mathcal{O} \in \mathcal{U}$. The union of $B_{\frac{1}{2}r_i}(x_j)$ covers E. For each such ball construct

$$\psi_{ij} \in C_o^\infty\big(B_{r_i}(x_j)\big) \qquad 0 \leq \psi_{ij} \leq 1 \qquad \psi_{ij} = 1 \text{ on } B_{\frac{1}{2}r_i}(x_j).$$

The ψ_{ij} can be constructed for example by mollifying the characteristic functions of $B_{\frac{2}{3}r_i}(x_j)$. Order the $\{\psi_{ij}\}$ in some fashion, say $\{\psi_n\}$, and construct a partition of unity $\{\varphi_n\}$ by setting

$$\varphi_1 = \psi_1 \quad \text{and} \quad \varphi_{j+1} = \psi_{j+1} \prod_{i=1}^{j}(1 - \psi_i) \quad \text{for } j = 1, 2, \ldots$$

The collection $\{\varphi_n\}$ satisfies (i)–(iii). Moreover, for all $m = 1, 2, \ldots$

$$\sum_{j=1}^{m} \varphi_j = 1 - \prod_{i=1}^{m}(1 - \psi_i).$$

This holds true for $m = 1$ and is verified by induction for all $m \in \mathbb{N}$, by making use of the definition of the φ_n. To verify (iv) observe that for each $x \in E$ there exists some ψ_{ij} such that $\psi_{ij}(x) = 1$. ∎

3 Proof of Theorem 1.1. Constructing μ

For a non-void open set $\mathcal{O} \subset \mathbb{R}^N$, introduce the class of functions

$$\Gamma_{\mathcal{O}} = \big\{ f \in C_o(\mathcal{O}) \mid |f| \leq 1 \big\}. \tag{3.1}$$

The collection $\mathcal{Q}$ of all non-void, open sets in $\mathbb{R}^N$, complemented with the empty set $\emptyset$, forms a sequential covering for $\mathbb{R}^N$. On $\mathcal{Q}$ define a nonnegative set function λ, by setting $\lambda(\emptyset) = 0$ and

$$\lambda(\mathcal{O}) = \sup_{f \in \Gamma_{\mathcal{O}}} T(f) \quad \text{for all, non empty, open sets } \mathcal{O} \subset \mathbb{R}^N.$$

This generates an outer measure μ_e by

$$\mathbb{R}^N \supset E \to \mu_e(E) = \inf \left\{ \sum \lambda(\mathcal{O}_n) \mid \mathcal{O}_n \in \mathcal{Q} \text{ and } E \subset \bigcup \mathcal{O}_n \right\}.$$

Lemma 3.1 *The set function $\lambda : \mathcal{Q} \to \mathbb{R}^*$ is monotone, countably sub-additive, and it coincides with μ_e on the open sets.*

Proof The monotonicity follows from the definition. Let $\{\mathcal{O}_n\}$ be a countable collection of open sets in $\mathbb{R}^N$ and let $\mathcal{O} = \cup \mathcal{O}_n$. For $f \in \Gamma_{\mathcal{O}}$, the collection $\{\mathcal{O}_n\}$ is an open covering for supp$\{f\}$. Construct a partition of unity for supp$\{f\}$ subordinate to $\{\mathcal{O}_n\}$ and let $\{\varphi_n\}$ be the sum of those elements of the partition supported in $\mathcal{O}_n$. Then $f\varphi_n \in \Gamma_{\mathcal{O}_n}$, and $f = \sum f\varphi_n$. Since supp$\{f\}$ is compact, this sum involves only finitely many, non identically zero terms, and by the linearity of T

$$T(f) = \sum T(f\varphi_n) \le \sum \lambda(\mathcal{O}_n).$$

Since $f \in \Gamma_{\mathcal{O}}$ is arbitrary, by the definition of $\lambda(\mathcal{O})$

$$\lambda(\mathcal{O}) = \lambda(\bigcup \mathcal{O}_n) = \sup_{f \in \Gamma_{\mathcal{O}}} T(f) \le \sum \lambda(\mathcal{O}_n).$$

By construction $\mu_e(\mathcal{O}) \le \lambda(\mathcal{O})$. On the other hand since λ is countably sub-additive and monotone

$$\mu_e(\mathcal{O}) \ge \inf\{\lambda(\bigcup \mathcal{O}_n) \mid \mathcal{O}_n \text{ open and } \mathcal{O} \subset \bigcup \mathcal{O}_n\} \ge \lambda(\mathcal{O}). \qquad \blacksquare$$

The outer measure μ_e generates in turn a measure μ in $\mathbb{R}^N$ defined on the σ-algebra $\mathcal{A}$ of all sets E satisfying the Carathéodory measurability condition (6.2) of Chap. 3.

Proposition 3.1 *The open sets are μ-measurable.*

Proof An open set $\mathcal{O}$ is μ-measurable if it satisfies the Carathéodory condition, for all sets $A \subset \mathbb{R}^N$ of finite outer measure. Assume first that A is itself open. Then $A \cap \mathcal{O}$ is open and from the definition of $\lambda(\cdot)$, for any $\varepsilon > 0$ there exists $f \in \Gamma_{A \cap \mathcal{O}}$ such that

$$T(f) \ge \lambda(A \cap \mathcal{O}) - \varepsilon.$$

The set $A - \text{supp}\{f\}$ is open and there exists $g \in \Gamma_{A-\text{supp}\{f\}}$ such that

$$T(g) \geq \lambda(A - \text{supp}\{f\}) - \varepsilon.$$

Then $f + g \in \Gamma_A$ and by the linearity of T

$$\begin{aligned}
\mu_e(A) = \lambda(A) &\geq T(f + g) = T(f) + T(g) \\
&\geq \lambda(A \cap \mathcal{O}) + \lambda(A - \text{supp}\{f\}) - 2\varepsilon \\
&\geq \mu_e(A \cap \mathcal{O}) + \mu_e(A - \mathcal{O}) - 2\varepsilon
\end{aligned}$$

since λ and μ_e coincide on open sets. If A is any subset of $\mathbb{R}^N$ of finite outer measure, having fixed $\varepsilon > 0$ there exists an open set A_ε containing A, and such that $\mu_e(A) \geq \mu_e(A_\varepsilon) - \varepsilon$. From this

$$\begin{aligned}
\mu_e(A) &\geq \mu_e(A_\varepsilon) - \varepsilon \geq \mu_e(A_\varepsilon \cap \mathcal{O}) + \mu_e(A_\varepsilon - \mathcal{O}) - \varepsilon \\
&\geq \mu_e(A \cap \mathcal{O}) + \mu_e(A - \mathcal{O}) - \varepsilon. \qquad \blacksquare
\end{aligned}$$

Thus the σ-algebra $\mathcal{A}$ contains the Borel σ-algebra $\mathcal{B}$. By construction

$$\mu = \mu_e \big|_{\mathcal{A}}, \quad \text{and} \quad \mu(\mathcal{O}) = \mu_e(\mathcal{O}) = \lambda(\mathcal{O})$$

for all open sets $\mathcal{O}$. The process by which μ is constructed from μ_e, generates a σ-algebra $\mathcal{A}$ which might be strictly larger than the Borel σ-algebra. We restrict μ to $\mathcal{B}$.

4 An Auxiliary Positive Linear Functional on $C_o(\mathbb{R}^N)^+$

Denote by $C_o(\mathbb{R}^N)^+$ the collection of all nonnegative $f \in C_o(\mathbb{R}^N)$. Given a locally bounded linear functional $T : C_o(\mathbb{R}^N) \to \mathbb{R}$, set

$$C_o(\mathbb{R}^N)^+ \ni f \to T_+(f) = \sup\{|T(h)| \mid h \in C_o(\mathbb{R}^N), \quad |h| \leq f\}. \qquad (4.1)$$

Note that T is assumed to be locally bounded but not to be positive.

Proposition 4.1 *The functional $T_+ : C_o(\mathbb{R}^N)^+ \to \mathbb{R}^+$ is positive and linear.*

Proof One verifies that $T_+(\alpha f) = \alpha T_+(f)$, for all $f \in C_o(\mathbb{R}^N)^+$ and $\alpha > 0$. To show that T_+ is linear, fix two nonnegative functions f_1 and f_2 in $C_o(\mathbb{R}^N)$ and select two functions h_1 and h_2 in $C_o(\mathbb{R}^N)$ such that $|h_i| \leq f_i$. Then

$$|h_1 + h_2| \leq |h_1| + |h_2| \leq f_1 + f_2.$$

Without loss of generality we may assume that $T(h_i) \geq 0$ for $i = 1, 2$. Then

$$T_+(f_1 + f_2) \geq |T(h_1 + h_2)| = |T(h_1)| + |T(h_2)|.$$

Since $|h_i| \leq f_i$ are arbitrary this gives

$$T_+(f_1 + f_2) \geq T_+(f_1) + T_+(f_2).$$

To prove the reverse inequality, select $h \in C_o(\mathbb{R}^N)$ such that $|h| \leq f_1 + f_2$, and set

$$h_j = \begin{cases} \dfrac{f_j \, h}{f_1 + f_2} & \text{if } f_1 + f_2 > 0 \\ 0 & \text{otherwise} \end{cases} \qquad \text{for } j = 1, 2.$$

One verifies that $|h_j| \leq f_j$ and therefore

$$|T(h)| \leq |T(h_1)| + |T(h_2)| \leq T_+(f_1) + T_+(f_2).$$

Since the function h is arbitrary, this implies

$$T_+(f_1 + f_2) \leq T_+(f_1) + T_+(f_2). \qquad \blacksquare$$

4.1 Measuring Compact Sets by T_+

For a compact set $K \subset \mathbb{R}^N$, introduce the class of functions

$$\Gamma_K = \{ f \in C_o(\mathbb{R}^N)^+ \mid f \geq 1 \text{ on } K \}. \tag{4.2}$$

Proposition 4.2 *Let T be a locally bounded linear functional on $C_o(\mathbb{R}^N)$ and let μ be its corresponding, previously constructed Radon measure. Then for every compact set $K \subset \mathbb{R}^N$*

$$\mu(K) = \inf_{f \in \Gamma_K} T_+(f). \tag{4.3}$$

Proof Since K is a Borel set, $\mu_e(K) = \mu(K)$ and

$$\mu(K) \leq \inf\{\lambda(\mathcal{O}) \mid \mathcal{O} \text{ open and } K \subset \mathcal{O}\}.$$

Fix $f \in \Gamma_K$ and $\varepsilon \in (0, 1)$, and consider the open set $[f > 1 - \varepsilon]$. By definition of $\Gamma_{\mathcal{O}}$

$$h \in \Gamma_{[f > 1 - \varepsilon]} \implies |h| \leq \frac{1}{1 - \varepsilon} f.$$

Therefore

$$\mu(K) \leq \lambda([f > 1 - \varepsilon]) = \sup_{h \in \Gamma_{[f>1-\varepsilon]}} T(h) \leq \frac{1}{1 - \varepsilon} T_+(f).$$

Since $f \in \Gamma_K$ is arbitrary

$$\mu(K) \leq \frac{1}{1 - \varepsilon} \inf_{f \in \Gamma_K} T_+(f), \qquad \text{for all } \varepsilon \in (0, 1).$$

From the construction of $\mu_e(K)$, for any $\varepsilon > 0$, there exists an open set $\mathcal{O}$ containing K, and such that $\mu(K) \geq \lambda(\mathcal{O}) - \varepsilon$. There exists $f \in \Gamma_{\mathcal{O}}$ such that $f = 1$ on K. From this and the definition of $\lambda(\cdot)$

$$\mu(K) \geq \sup_{h \in \Gamma_{\mathcal{O}}} T(h) - \varepsilon \geq T_+(f_K) - \varepsilon \geq \inf_{f \in \Gamma_K} T_+(f) - \varepsilon.$$

∎

5 Representing T_+ on $C_o(\mathbb{R}^N)^+$ as in (1.1) for a Unique $\mu_{\mathcal{B}}$

Having fixed $f \in C_o(\mathbb{R}^N)^+$, we may assume that $\|f\| = 1$ and for $n \in \mathbb{N}$ set

$$K_o = \text{supp}\{f\}$$

$$K_j = \left[f \geq \frac{j}{n}\right]$$

$$f_j = \begin{cases} \dfrac{1}{n} & \text{in } K_j \\ f(x) - \dfrac{j-1}{n} & \text{in } K_{j-1} - K_j \\ 0 & \text{in } \mathbb{R}^N - K_{j-1} \end{cases}$$

for $j = 1, \ldots, n$. One verifies that $f_j \in C_o(\mathbb{R}^N)^+$, and

$$\frac{1}{n}\chi_{K_j} \leq f_j \leq \frac{1}{n}\chi_{K_{j-1}}.$$

Let now μ be the Radon measure constructed in § 3. From such a construction and Proposition 4.2

$$\frac{1}{n}\mu(K_j) \leq \int_{\mathbb{R}^N} f_j d\mu \leq \frac{1}{n}\mu(K_{j-1})$$

$$\frac{1}{n}\mu(K_j) \leq T_+(f_j) \leq \frac{1}{n}\mu(K_{j-1}).$$

By construction $f = \sum_{j=1}^{n} f_j$. Therefore

$$\frac{1}{n} \sum_{j=1}^{n} \mu(K_j) \le \int_{\mathbb{R}^N} f d\mu \le \frac{1}{n} \sum_{j=1}^{n} \mu(K_{j-1})$$

$$\frac{1}{n} \sum_{j=1}^{n} \mu(K_j) \le T_+(f) \le \frac{1}{n} \sum_{j=1}^{n} \mu(K_{j-1}).$$

From these, by difference

$$\left| T_+(f) - \int_{\mathbb{R}^N} f d\mu \right| \le \frac{1}{n} \big[\mu(K_o) - \mu(K_n) \big].$$

Since T is locally bounded $\mu(K_o) < \infty$. Therefore, letting $n \to \infty$ gives

$$C_o(\mathbb{R}^N)^+ \ni f \to T_+(f) = \int_{\mathbb{R}^N} f d\mu.$$

If μ_1 and μ_2 are two Radon measures identifying the same T_+, by formula (1.1), they must coincide on open and compact subsets of $\mathbb{R}^N$. Thus, by Theorem 11.1 of Chap. 3, they coincide on the Borel sets.   ∎

6 Proof of Theorem 1.1. Representing T on $C_o(\mathbb{R}^N)$ as in (1.3) for a Unique μ-Measurable w

The functional $T : C_o(\mathbb{R}^N) \to \mathbb{R}$ can be bounded above as

$$|T(f)| \le \sup \big\{ |T(h)| \ \big| \ |h| \le |f| \big\}$$
$$= T_+(|f|) = \int_{\mathbb{R}^N} |f| d\mu = \|f\|_{1,\mathbb{R}^N}.$$

Therefore $T(f)$ on $C_o(\mathbb{R}^N)$ is dominated by the L^1-norm of f, where the integrals are meant with respect to the Radon measure μ. By the Hahn-Banach dominated extension theorem (§ 9 of Chap. 7) T can be extended to a linear functional $\tilde{T}$ from $L^1(\mathbb{R}^N)$ into $\mathbb{R}$, still dominated by the L^1-norm. Then, by the Riesz representation theorem in L^1 (§ 11 of Chap. 6) there exists a unique μ-measurable function $w \in L^\infty(\mathbb{R}^N)$ such that

$$\tilde{T}(f) = \int_{\mathbb{R}^N} f w \, d\mu \quad \text{for all } f \in L^1(\mathbb{R}^N).$$

Since $\tilde{T} = T$ on $C_o(\mathbb{R}^N)$, the representation (1.3) holds for T. If such a representation is realized by two μ-measurable functions w_1 and w_2 in $L^\infty(\mathbb{R}^N)$, then

$$C_o(\mathbb{R}^N) \ni f \to \int_{\mathbb{R}^N} f(w_1 - w_2) d\mu = 0.$$

This implies that $w_1 = w_2$, μ-a.e. in $\mathbb{R}^N$. Having identified w, fix an open set $\mathcal{O} \subset \mathbb{R}^N$ and let $\{f_n\} \subset C_o(\mathcal{O})$ be such that $|f_n| \leq 1$ and $\{f_n\} \to \text{sign}\{w\}$, μ-a.e. in $\mathcal{O}$. From the construction of μ one has $\mu(\mathcal{O}) = \mu_e(\mathcal{O}) = \lambda(\mathcal{O})$, and by dominated convergence

$$
\begin{aligned}
\mu(\mathcal{O}) &= \sup_{f \in \Gamma_\mathcal{O}} T(f) \\
&= \sup \left\{ \int_{\mathbb{R}^N} f \, w \, d\mu \mid f \in C_o(\mathcal{O}), \text{ and } |f| \leq 1 \right\} \\
&\geq \lim \int_{\mathbb{R}^N} f_n \, w \, d\mu = \int_\mathcal{O} |w| d\mu.
\end{aligned}
$$

Also by the same construction

$$\mu(\mathcal{O}) \leq \int_\mathcal{O} |w| d\mu.$$

Thus $|w| = 1$ μ-a.e. in $\mathcal{O}$. ∎

Corollary 6.1 *Let $T : C_o(\mathbb{R}^N) \to \mathbb{R}$ be linear and locally bounded. There exist two Radon measures μ_1 and μ_2 in $\mathbb{R}^N$, such that*

$$T(f) = \int_{\mathbb{R}^N} f \, d\mu_1 - \int_{\mathbb{R}^N} f \, d\mu_2 \quad \text{for all } f \in C_o(\mathbb{R}^N).$$

Moreover the restrictions of μ_i to the Borel σ-algebra $\mathcal{B}$ is unique.

The two Radon measures need not be finite. However they are finite on bounded sets and the representation formula is well defined since $f \in C_o(\mathbb{R}^N)$. If T is linear and bounded, then the two measures μ_1 and μ_2 are both finite.

7 A Topology for $C_o^\infty(E)$ for an Open Set $E \subset \mathbb{R}^N$

A N-dimensional multi-index α of size $|\alpha|$ is a N-tuple of nonnegative integers, whose sum is $|\alpha|$, that is $\alpha = (\alpha_1, \dots, \alpha_N)$, with $|\alpha| = \sum_{j=1}^N \alpha_j$. If all the components of α are zero, α is the null multi-index. For $f \in C^{|\alpha|}(E)$ set

$$D^\alpha f = \frac{\partial^{\alpha_1 + \dots + \alpha_N} f}{\partial x_1^{\alpha_1} \dots \partial x_N^{\alpha_N}}.$$

If some of the components of α are zero, say for example if $\alpha_j = 0$, then $\partial^{\alpha_j} f / \partial x_j^{\alpha_j} = f$. If α is the null multi-index, $D^\alpha f = f$. Set

$$p_j(f) = \max_{x \in E}\{|D^\alpha f(x)|; \ |\alpha| \le j\} \quad j = 0, 1, \ldots .$$

These are norms in $C_o^\infty(E)$ satisfying $p_j(f) \le p_{j+1}(f)$ for all $f \in C_o^\infty(E)$. Introduce the neighborhoods of the origin of $C_o^\infty(E)$

$$\mathcal{O}_j = \left\{f \in C_o^\infty(E) \ \Big| \ p_j(f) < \frac{1}{j+1}\right\} \quad j = 0, 1, \ldots$$

and the neighborhoods $B_{\varphi,j} = \varphi + \mathcal{O}_j$ of a given $\varphi \in C_o^\infty(E)$. By construction $\mathcal{O}_{j+1} \subset \mathcal{O}_j$ and $B_{\varphi,j+1} \subset B_{\varphi,j}$.

Lemma 7.1 *For any $f \in \mathcal{O}_j$, there exists an index ℓ_j such that $f + \mathcal{O}_\ell \subset \mathcal{O}_j$ for all $\ell \ge \ell_j$.*

Proof Having fixed $f \in \mathcal{O}_j$, there exists $\varepsilon > 0$ such that

$$p_j(f) \le \frac{1}{j+1} - \varepsilon.$$

Let ℓ_j be a positive integer satisfying $\ell_j \ge \max\{j+1; \varepsilon^{-1}\}$. Then for every $g \in \mathcal{O}_\ell$, for $\ell \ge \ell_j$

$$p_j(f+g) \le p_j(f) + p_j(g) < \frac{1}{j+1} - \varepsilon + \frac{1}{\ell+1} \le \frac{1}{j+1}.$$

Thus $f + g \in \mathcal{O}_j$.   ∎

Proposition 7.1 *The collection $\mathcal{B} = \{B_{\varphi,j}\}$ as φ ranges over $C_o^\infty(E)$ and $j = 0, 1, \ldots$, forms a base for a topology $\mathcal{U}$ of $C_o^\infty(E)$. The topology $\mathcal{U}$ satisfies the first axiom of countability and is translation invariant. For all $\delta \in (0, 1]$, the sets $\delta\mathcal{O}_j \in \mathcal{U}$. Finally $\delta\mathcal{O}_j$ for all $\delta \ne 0$, are open, convex, symmetric neighborhoods of the origin of $C_o^\infty(E)$.*

Proof We verify that $\mathcal{B}$ satisfies the requirements (i)–(ii) of § 4 of Chap. 2 to be a base for a topology. Let $B_{\varphi,i}$ and $B_{\psi,j}$ be out of $\mathcal{B}$, and with non-empty intersection. Fix $\eta \in B_{\varphi,i} \cap B_{\psi,j}$, so that $\eta - \varphi \in \mathcal{O}_i$ and $\eta - \psi \in \mathcal{O}_j$. There exist positive integers ℓ_i and ℓ_j such that

$$\eta - \varphi + \mathcal{O}_\ell \subset \mathcal{O}_i \quad \text{and} \quad \eta - \psi + \mathcal{O}_\ell \subset \mathcal{O}_j \quad \text{for all } \ell \ge \max\{\ell_i; \ell_j\}.$$

For all such ℓ, $\eta + \mathcal{O}_\ell \subset \varphi + \mathcal{O}_i$ and $\eta + \mathcal{O}_\ell \subset \psi + \mathcal{O}_j$.

Let $f \in \delta\mathcal{O}_j$. There exists an index $\ell(\delta, j)$ such that $f + \mathcal{O}_\ell \subset \delta\mathcal{O}_j$, for all $\ell > \ell(\delta, j)$. Therefore, by the construction procedure of the topology $\mathcal{U}$ from the base $\mathcal{B}$, the set $\delta\mathcal{O}_j$ is open.   ∎

The space $C_o^\infty(E)$ endowed with the topology $\mathcal{U}$ is denoted by $D(E)$.

Proposition 7.2 *$D(E)$ is a topological vector space.*

Proof Let $\varphi_1, \varphi_2 \in D(E)$ and let U be an open set containing $\varphi_1 + \varphi_2$. There exists an open neighborhood of the origin $\mathcal{O}_j$ such that $\varphi_1 + \varphi_2 + \mathcal{O}_j \subset U$. Since $\mathcal{O}_j$ is convex

$$(\varphi_1 + \tfrac{1}{2}\mathcal{O}_j) + (\varphi_2 + \tfrac{1}{2}\mathcal{O}_j) \subset \varphi_1 + \varphi_2 + \mathcal{O}_j \subset U.$$

This implies that the sum $+ : D(E) \times D(E) \to D(E)$ is continuous. Fix $\varphi_o \in D(E)$, a real number λ_o and an open neighborhood of the origin $\mathcal{O}_j$. To establish that the product $\bullet : \mathbb{R} \times D(E) \to D(E)$, is continuous, one needs to show that for all $\varepsilon > 0$ there exists a positive number $\delta = \delta(\mathcal{O}_j, \varphi_o, \lambda_o, \varepsilon)$ such that

$$\lambda\varphi \in \lambda_o\varphi_o + \varepsilon\mathcal{O}_j \quad \text{for all } |\lambda - \lambda_o| < \delta$$

and for all $\varphi \in \varphi_o + \delta\mathcal{O}_j$. The element φ_o belongs to $\sigma\mathcal{O}_j$ for some $\sigma > 0$. Having fixed $\varepsilon > 0$, the number δ is chosen from

$$\lambda\varphi - \lambda_o\varphi_o = \lambda(\varphi - \varphi_o) + (\lambda - \lambda_o)\varphi_o \subset \lambda\delta\mathcal{O}_j + \delta\sigma\mathcal{O}_j \subset \varepsilon\mathcal{O}_j$$

for a suitable choice of δ. ∎

8 A Metric Topology for $C_o^\infty(E)$

As an alternative construction of a topology for $C_o^\infty(E)$, introduce the metric (§ 15 of Chap. 2)

$$d(f, g) = \sum \frac{1}{2^j} \frac{p_j(f - g)}{1 + p_j(f - g)}.$$

Since each of the p_j is translation invariant, $d(\cdot, \cdot)$ is also translation invariant, and generates a translation invariant topology in $C_o^\infty(E)$. The sum and the multiplications by scalars, are continuous with respect to such a topology. The continuity of the sum follows from Proposition 14.1 of Chap. 2. The continuity of the product follows from the definition of p_j and $d(\cdot, \cdot)$. Thus $C_o^\infty(E)$ equipped with the topology generated by $d(\cdot, \cdot)$ is a metric, topological, vector space.

8.1 *Equivalence of These Topologies*

A base for the metric topology of $C_o^\infty(E)$ is the collection of open balls

$$B_\rho(\varphi) = \{f \in C_o^\infty(E) \mid d(f, \varphi) < \rho\}$$

for $\varphi \in C_o^\infty(E)$ and rational $\rho \in (0, 1)$. For fixed $j \in \mathbb{N}$, the ball B_ρ centered at the origin of $C_o^\infty(E)$ and radius

$$\rho = \frac{1}{2^{j+1}(j + 1)}$$

is contained in $\mathcal{O}_j$. Indeed for every $f \in B_\rho$

$$\sum \frac{1}{2^i} \frac{p_i(f)}{1 + p_i(f)} < \frac{1}{2^{j+1}(j + 1)}.$$

From this

$$\frac{p_j(f)}{1 + p_j(f)} < \frac{1}{2(j + 1)} \qquad \text{i.e.,} \quad p_j(f) < \frac{1}{j + 1}.$$

Thus $f \in \mathcal{O}_j$. Vice versa, every ball B_ρ about the origin, contains an open neighborhood of the origin $\mathcal{O}_j$ for some $j \in \mathbb{N}$. Indeed, let ℓ be a nonnegative integer such that $\rho > 4(\ell + 1)^{-1}$. Then, for every $f \in \mathcal{O}_\ell$

$$d(f, 0) = \sum \frac{1}{2^j} \frac{p_j(f)}{1 + p_j(f)} \leq \sum_{j=1}^{\ell} \frac{1}{2^j} \frac{p_j(f)}{1 + p_j(f)} + \frac{1}{2^\ell} \leq \frac{2}{\ell + 1} + \frac{1}{2^\ell} < \rho$$

provided ℓ is sufficiently large. Since the topologies generated by the base $\mathcal{B}$ and the one generated by the metric $d(\cdot, \cdot)$ are both translation invariant, every open set $\varphi + \mathcal{O}_j \in \mathcal{B}$ contains a ball $B_\rho(\varphi)$ and viceversa.

8.2 $D(E)$ Is Not Complete

Cauchy sequences in $D(E)$ need not converge to an element of $C_o^\infty(E)$. As an example let $E = \mathbb{R}$. Having fixed some $f \in C_o^\infty(0, 1)$ consider the sequence

$$f_n(x) = \sum_{j=1}^{n} \frac{1}{j} f(x - j).$$

One verifies that $f_n \in C_o^\infty(\mathbb{R})$ and that $\{f_n\}$ is a Cauchy sequence in $D(\mathbb{R})$. However $\{f_n\}$ does not converge to a function in $C_o^\infty(\mathbb{R})$.

For an example in bounded domains, let $E = B_1$ be the open unit ball centered at the origin of $\mathbb{R}^N$. The functions

$$f_n(x) = \begin{cases} \exp\left\{\dfrac{n^2}{|nx|^2 - (n - 1)^2}\right\} & \text{for } |x| < \dfrac{n - 1}{n} \\ 0 & \text{for } |x| \geq \dfrac{n - 1}{n} \end{cases}$$

are in $C_o^\infty(B_1)$ and form a Cauchy sequence in $D(B_1)$. However their limit is not in $C_o^\infty(B_1)$. An indirect proof of the non completeness of $D(E)$ will be given in § 9.2 by a category argument.

9   A Topology for $C_o^\infty(K)$ for a Compact Set $K \subset E$

For a compact subset $K \subset E$ denote by $C_o^\infty(K)$ the space of all $f \in C_o^\infty(E)$ whose support is contained in K. On $C_o^\infty(K)$ introduce the norms

$$p_{K;j}(f) = \max_{x \in K}\{|D^\alpha f(x)| \mid |\alpha| \le j\}, \quad j = 0, 1, 2, \ldots$$

the neighborhoods of the origin of $C_o^\infty(K)$

$$\mathcal{O}_{K;j} = \left\{f \in C_o^\infty(K) \mid p_{K;j}(f) < \frac{1}{j+1}\right\} \quad j = 0, 1, 2, \ldots$$

and the neighborhoods $B_{K;\varphi,j} = \varphi + \mathcal{O}_{K;j}$, of a given $\varphi \in C_o^\infty(K)$. Proceeding exactly as in the previous sections, the collection $\mathcal{B}_K = \{B_{K;\varphi,j}\}$ as φ ranges over $C_o^\infty(K)$ and j ranges over $\{0, 1, \ldots\}$, forms a base for a translation invariant, first countable topology $\mathcal{U}_K$ of $C_o^\infty(K)$. Moreover for all $\delta \ne 0$ the sets $\delta\mathcal{O}_{K;j}$ are open. Finally $C_o^\infty(K)$ equipped with the topology $\mathcal{U}_K$ is a topological vector space, and is denoted by $\mathcal{D}(K)$. A topology in $C_o^\infty(K)$ can also be constructed by the the metric

$$d_K(f, g) = \sum \frac{1}{2^j} \frac{p_{K;j}(f-g)}{1 + p_{K;j}(f-g)}.$$

The equivalence of $\mathcal{U}_K$ with the metric topology generated by d_K can be established as § 8.1.

9.1 $\mathcal{D}(K)$ Is Complete

The notion of convergence in $\mathcal{D}(K)$ can be given in terms of the metric $d_K(\cdot, \cdot)$, that is a sequence $\{f_n\}$ of functions in $\mathcal{D}(K)$ converges to some $f \in \mathcal{D}(K)$ if and only if $\{D^\alpha f_n\} \to D^\alpha f$ uniformly in K, for every N-dimensional multi-index α. With respect to such a notion of convergence $\mathcal{D}(K)$ is complete. Indeed for every Cauchy sequence $\{f_n\}$ in $\mathcal{D}(K)$ the sequences $\{D^\alpha f_n\}$ are Cauchy in $C(K')$ for every compact subset $K' \subset E$ containing K, for all multi-indices α. Thus $\{f_n\} \to f$ and $\{D^\alpha f_n\} \to f_\alpha$ in the uniform topology of $C(K')$, for functions $f, f_\alpha \in C(K')$, vanishing outside K. By working with difference quotients, one identifies $f_\alpha = D^\alpha f$.

9.2   Relating the Topology of $D(E)$ to the Topology of $\mathcal{D}(K)$

The construction of $\mathcal{U}$ and $\mathcal{U}_K$, by means of the open sets $\mathcal{O}_j$ in $D(E)$ and $\mathcal{O}_{K;j}$ in $\mathcal{D}(K)$, implies that the topology $\mathcal{U}_K$ is the restriction of $\mathcal{U}$ to $\mathcal{D}(K)$. Equivalently $\mathcal{U}_K = \mathcal{U} \cap \mathcal{D}(K)$. As K ranges over the compact subsets of E, each $\mathcal{D}(K)$ is a closed subspace of $D(E)$. Moreover $\mathcal{D}(K)$ does not contain any open set of $D(E)$. Therefore $\mathcal{D}(K)$ is nowhere dense in $D(E)$. By construction, $D(E) = \cup \mathcal{D}(K_n)$ for a countable collection $\{K_n\}$ of nested, compact subsets of E, exhausting E. If $D(E)$ were complete, it would be the countable union of nowhere dense sets, against the Baire Category Theorem.

10   The Schwartz Topology of $\mathcal{D}(E)$

The topology of $D(E)$ allows, roughly speaking, for too many sequences $\{f_n\}$ in $C_o^\infty(E)$ to be Cauchy. Equivalently, it does not contain sufficiently many open sets to restrict the Cauchy sequences to the ones with limit in $C_o^\infty(E)$. This accounts for its lack of completeness. A topology $\mathcal{W}$ by which $C_o^\infty(E)$ would be complete, would have to be stronger than $\mathcal{U}$, that is $\mathcal{U} \subset \mathcal{W}$. However since the topology of each $\mathcal{D}(K)$ is completely determined by $\mathcal{U}$, such a stronger topology $\mathcal{W}$, would have to preserve such a property. Specifically, when restricted to $\mathcal{D}(K)$, it should not generate new open sets in $\mathcal{D}(K)$.

To construct $\mathcal{W}$, define first the collection $\mathcal{V}_o$ of open, base-neighborhoods of the origin. A set V containing the origin, is in $\mathcal{V}_o$ if and only if

i. V is convex and $\delta V \subset V$ for all $\delta \in (0, 1]$
ii. $V \cap \mathcal{D}(K) \in \mathcal{U}_K$ for all compact subsets $K \subset E$.

The collection $\mathcal{V}_o$ is not empty, since $\mathcal{O}_j \in \mathcal{V}_o$ for all $j \in \mathbb{N} \cup \{0\}$. The base-neighborhoods of a given $\varphi \in C_o^\infty(E)$ are of the form $\varphi + V$ for some $V \in \mathcal{V}_o$.

Proposition 10.1 (Schwartz [141]) *The collection $\mathcal{V} = \{\varphi + V\}$ as φ ranges over $C_o^\infty(E)$ and V ranges over $\mathcal{V}_o$, forms a base for a topology $\mathcal{W}$ on $C_o^\infty(E)$.*

Proof Let $\varphi_1 + V_1$ and $\varphi_2 + V_2$ be any two elements of $\mathcal{V}$ with non empty intersection. Choose

$$\eta \in (\varphi_1 + V_1) \cap (\varphi_2 + V_2)$$

and let K be a compact subset of E such that

$$\text{supp}\{\varphi_1\} \cup \text{supp}\{\varphi_2\} \cup \text{supp}\{\eta\} \subset K.$$

Then $\eta - \varphi_i \in \mathcal{D}(K)$, and $\eta - \varphi_i \in V_i \cap \mathcal{D}(K)$ for $i = 1, 2$. Since $V_i \cap \mathcal{D}(K)$ are open in $\mathcal{D}(K)$, there exists $\varepsilon > 0$ such that

$$\eta - \varphi_i \in (1 - \varepsilon) V_i \cap \mathcal{D}(K) \subset (1 - \varepsilon) V_i.$$

Since the sets V_i are convex

$$\eta - \varphi_i + \varepsilon V_i \subset (1 - \varepsilon) V_i + \varepsilon V_i.$$

From this, $\eta + \varepsilon V_i \subset \varphi_i + V_i$, and

$$\eta + \varepsilon (V_1 \cap V_2) \subset (\varphi_1 + V_1) \cap (\varphi_2 + V_2).$$

Since $\mathcal{V}_o$ is closed under intersection, $V_1 \cap V_2 \in \mathcal{V}_o$. ∎

The continuity of the sum and multiplication by scalars, with respect to such a topology, is proved as in § 7. Thus $C_o^\infty(E)$ equipped with the topology $\mathcal{W}$ is a topological vector space and is denoted by $\mathcal{D}(E)$.

Proposition 10.2 *The restriction of $\mathcal{W}$ to $\mathcal{D}(K)$ is $\mathcal{U}_K$.*

Proof The inclusion $\mathcal{U} \subset \mathcal{W}$, implies the inclusion $\mathcal{U}_K \subset \mathcal{W} \cap \mathcal{D}(K)$. For the converse inclusion, let $W \in \mathcal{W}$ and select $\varphi \in W \cap \mathcal{D}(K)$. There exists $V \in \mathcal{V}_o$ such that $(\varphi + V) \subset W$, and for such V

$$(\varphi + V) \cap \mathcal{D}(K) = \varphi + V \cap \mathcal{D}(K).$$

Therefore, $(\varphi + V) \cap \mathcal{D}(K) \in \mathcal{U}_K$, since $V \cap \mathcal{D}(K)$ is open in $\mathcal{D}(K)$. Thus $W \cap \mathcal{D}(K) \in \mathcal{U}_K$. ∎

By construction $\mathcal{U} \subset \mathcal{W}$. In the next section it will be shown that the inclusion is strict.

11 $\mathcal{D}(E)$ Is Complete

A set $B \subset \mathcal{D}(E)$ is bounded if and only if, for every $W \in \mathcal{W}$, containing the origin, $B \subset \delta W$ for some $\delta > 0$.

Proposition 11.1 *Let B be a bounded subset of $\mathcal{D}(E)$. There exists a compact subset $K \subset E$, such that $B \subset \mathcal{D}(K)$.*

Proof Assuming that such a K does not exists, there exists be a countable collection $\{K_n\}$ of compact, nested subsets of E, exhausting E, a sequence of functions $\{f_n\} \subset B$, and a sequence of points $\{x_n\}$, such that $x_n \in K_{n+1} - K_n$, and $|f_n(x_n)| > 0$. By construction the sequence $\{x_n\} \subset E$ has no limit in E.

Let V be the collection of functions $f \in \mathcal{D}(E)$ satisfying

$$|f(x_n)| < \frac{1}{n}|f_n(x_n)| \quad \text{for all } n \in \mathbb{N}.$$

Such a set is convex, contains the origin and $\delta V \subset V$ for all $\delta \in (0, 1]$. Let $K \subset E$ be compact. Since finitely many points $\{x_n\}$ are in K, there exists $\delta_K > 0$ such that $V \cap \mathcal{D}(K) = \delta_K \mathcal{O}_{K;0}$. Therefore $V \cap \mathcal{D}(K) \in \mathcal{U}_K$. Since K is arbitrary, $V \in \mathcal{V}_o$. On the other hand B is not contained in δV for any $\delta > 0$. ∎

11.1   Cauchy Sequences in $\mathcal{D}(E)$ and Completeness

A sequence $\{f_n\} \subset C_o^\infty(E)$ is a Cauchy sequence in $\mathcal{D}(E)$ if for every base-neighborhood of the origin $V \in \mathcal{V}_o$, there exists $n_V \in \mathbb{N}$, such that $f_n - f_m \in V$ for all $n, m \geq n_V$. Therefore a Cauchy sequence in $\mathcal{D}(E)$ is a bounded set in $\mathcal{D}(E)$. In this sense the enlargement of the topology of $C_o^\infty(E)$ from $\mathcal{U}$ to $\mathcal{W}$, places stringent restrictions on a sequence $\{f_n\}$ to be Cauchy. Indeed if $\{f_n\}$ is a Cauchy sequence, then by the previous proposition, $\{f_n\} \subset \mathcal{D}(K)$ for some compact subset $K \subset E$. Thus if $\{f_n\}$ is a Cauchy sequence in $\mathcal{D}(E)$, the notion of convergence coincides with that of the metric topology of $\mathcal{D}(K)$, since the restriction of $\mathcal{W}$ to $\mathcal{D}(K)$ is precisely the metric topology $\mathcal{U}_K$.

Corollary 11.1 $\mathcal{D}(E)$ is complete.

Corollary 11.2   A sequence $\{f_n\} \subset \mathcal{D}(E)$ is Cauchy, if and only if there exists a compact subset $K \subset E$, and a function $f \in \mathcal{D}(K)$, such that $\{D^\alpha f_n\} \to D^\alpha f$ uniformly in K for all multi-indices α.

11.2   The Topology of $\mathcal{D}(E)$ Is Not Metrizable

There exists no metric $d(\cdot, \cdot)$ on $C_o^\infty(E)$ that generates the same topology $\mathcal{W}$. If such a metric were to exist, then $\mathcal{D}(E)$ would be a complete metric space and hence of second category. Let $\{K_n\}$ be a countable collection of compact, nested subsets of E, exhausting E. Then $\mathcal{D}(E) = \bigcup \mathcal{D}(K_n)$ would be the countable union of nowhere dense, closed subsets, thereby violating Baire's Category Theorem.

Corollary 11.3   The inclusion $\mathcal{U} \subset \mathcal{W}$ is strict.

Proof If $\mathcal{W} = \mathcal{U}$, the topology $\mathcal{W}$ would be metric. ∎

12 Continuous Maps and Functionals

A set $B \subset \mathcal{D}(K)$ is bounded if and only if for every neighborhood of the origin $\mathcal{O}_K$ there exists a positive number δ such that $B \subset \delta \mathcal{O}_K$. This in turn implies that there exists positive numbers λ_j such that

$$p_{K;j}(f) < \lambda_j \quad \text{for all} \quad f \in B. \tag{12.1}$$

12.1 Distributions in E

A continuous linear functional $T : \mathcal{D}(E) \to \mathbb{R}$ maps bounded neighborhoods of the origin of $\mathcal{D}(E)$ into bounded intervals about the origin of $\mathbb{R}$ (Proposition 10.3 of Chap. 2). A bounded neighborhood of the origin of $\mathcal{D}(E)$ is a bounded neighborhood of the origin of $\mathcal{D}(K)$ for some compact subset $K \subset E$. Therefore T is continuous if and only if its restriction to every $\mathcal{D}(K)$ is a continuous functional $T_K : \mathcal{D}(K) \to \mathbb{R}$. Since $\mathcal{D}(K)$ is a metric space the continuity of T can be characterized in terms of sequences.

Proposition 12.1 *A linear functional T in $\mathcal{D}(E)$ is continuous if and only if for every sequence $\{f_n\}$ converging to zero in the sense of $\mathcal{D}(E)$, the sequence $\{T(f_n)\} \to 0$ in $\mathbb{R}$.*

Corollary 12.1 *A linear functional T in $\mathcal{D}(E)$ is continuous if and only if for every compact subset $K \subset E$ and for every sequence $\{f_n\} \subset \mathcal{D}(K)$ converging to zero in $\mathcal{D}(K)$, the sequence $\{T(f_n)\} \to 0$ in $\mathbb{R}$.*

A distribution on E is a continuous linear functional $T : \mathcal{D}(E) \to \mathbb{R}$. Its action on $\varphi \in \mathcal{D}(E)$ is denoted by either symbol $T[\varphi] = \langle T, \varphi \rangle$. The latter is also referred to as the distribution *pairing* of T and φ. The linear space of all the continuous, linear functionals T on $\mathcal{D}(E)$ is denoted by $\mathcal{D}'(E)$. The topology on $\mathcal{D}'(E)$ is the weak* topology induced by $\mathcal{D}(E)$. Each element $\varphi \in \mathcal{D}(E)$ can be identified with a linear functional on $\mathcal{D}'(E)$ by the distribution pairing. The weak* topology of $\mathcal{D}'(E)$ is the weakest topology on $\mathcal{D}'(E)$ by which all elements of $\mathcal{D}(E)$ define a continuous linear functional on $\mathcal{D}'(E)$ by the distribution pairing. With respect to such a topology, sets of the type

$$\mathcal{O} = \left\{ \begin{array}{c} \text{collection of } T \in \mathcal{D}'(E) \text{such that } \langle T, \varphi \rangle \in (\alpha, \beta) \\ \text{for some } \alpha, \beta \in \mathbb{R} \text{ and some } \varphi \in \mathcal{D}(E) \end{array} \right\}$$

are open. A base for the weak* topology of $\mathcal{D}'(E)$ is the collection $\mathcal{B}$ of finite intersections of these open sets. This induces a notion of convergence in $\mathcal{D}'(E)$, by which $\{T_n\} \to T \in \mathcal{D}'(E)$ if and only if $\lim\langle T_n - T, \varphi \rangle = 0$ for all $\varphi \in \mathcal{D}(E)$. Given Radon measures μ and ν in $\mathbb{R}^N$, the formula

$$\int_E \varphi d(\mu - \nu) = \langle \mu - \nu, \varphi \rangle \quad \varphi \in \mathcal{D}(E)$$

identifies an element of $\mathcal{D}'(E)$. As an example, for $\nu = 0$ and $\mu = \delta_{x_o}$, for a fixed $x_o \in E$, the evaluation map $\langle \delta_{x_o}, \varphi \rangle = \varphi(x_o)$ is a distribution. If ν is the Lebesgue measure and $f \in L^1(E)$ with respect to ν, then $\mu = f d\nu$ is the difference of two signed Radon measures and identifies an element of $\mathcal{D}'(E)$. Thus $L^1(E) \hookrightarrow \mathcal{D}'(E)$ up to an identification.

12.2   Continuous Linear Maps $T : \mathcal{D}(E) \to \mathcal{D}(E)$

Proposition 12.2 *A linear map $T : \mathcal{D}(E) \to \mathcal{D}(E)$ is continuous if and only if is bounded.*

The statement holds true for maps T between metric spaces. The proof consists in establishing that if T is bounded, it is restricted to some $\mathcal{D}(K)$, which is a metric space.

Proof Continuity of T implies T is bounded. To prove the converse, if $B \subset \mathcal{D}(E)$ is bounded, it is a bounded subset of $\mathcal{D}(K)$ for some compact subset $K \subset E$. Since $T(B)$ is bounded in $\mathcal{D}(E)$, it is a bounded subset of $\mathcal{D}(K')$, for some compact subset $K' \subset E$. Since both $\mathcal{D}(K)$ and $\mathcal{D}(K')$ are metric spaces, the restriction of T to $\mathcal{D}(K)$ is continuous (Propositions 10.3 and 14.2 of Chap. 2).   ∎

Corollary 12.2 *The differentiation maps $D^\alpha : \mathcal{D}(E) \to \mathcal{D}(E)$ are continuous for every multi-index α.*

Proof Let $B \subset \mathcal{D}(E)$ be bounded and let λ_j be the positive numbers claimed by the characterization (12.1) of the bounded subsets of $\mathcal{D}(E)$. Then

$$p_{K;j}(D^\alpha f) < \lambda_{j+|\alpha|}.$$

Thus $D^\alpha(B) \subset B'$ for some bounded set $B' \subset \mathcal{D}(E)$.   ∎

13   Distributional Derivatives

Let α be a multi-index and let $f \in C^{|\alpha|}(E)$. By integration by parts

$$\int_E D^\alpha f \varphi dx = (-1)^{|\alpha|} \int_E f D^\alpha \varphi dx \quad \text{for all } \varphi \in \mathcal{D}(E).$$

This motivates the following definition of derivative of a distribution. The derivative $D^\alpha T$ of a distribution T, is a distribution acting on $\varphi \in \mathcal{D}(E)$ as

$$\langle D^\alpha T, \varphi \rangle = (-1)^{|\alpha|} \langle T, D^\alpha \varphi \rangle \quad \text{for all } \varphi \in \mathcal{D}(E).$$

Such a definition coincides with the classical one, when $T \in C^{|\alpha|}(E)$. One also verifies the formula $D^\alpha(D^\beta T) = D^\beta(D^\alpha T)$, valid for any pair of multi-indices α and β. For $f \in L^1_{\text{loc}}(E)$ the derivative $D^\alpha f$ is that distribution acting on $\varphi \in \mathcal{D}(E)$, as

$$\langle D^\alpha f, \varphi \rangle = (-1)^{|\alpha|} \int_E f D^\alpha \varphi \, dx.$$

The distributional derivative of $f(x) = |x|$ for $x \in \mathbb{R}$, is the Heaviside graph

$$H(x) = \begin{cases} 1 & \text{if } x > 0 \\ [0, 1] & \text{if } x = 0 \\ -1 & \text{if } x < 0. \end{cases}$$

Taking now the distributional derivative of H, gives

$$H'[\varphi] = -\int_{\mathbb{R}} H(s)\varphi'(s)ds = \int_{-\infty}^0 \varphi'ds - \int_0^\infty \varphi'ds = 2\varphi(0).$$

Therefore $|x|'' = 2\delta_o$ in the sense of $\mathcal{D}'(\mathbb{R})$.

Two distributions T_1 and T_2 in $\mathcal{D}'(E)$ are equal if and only if $\langle T_1, \varphi \rangle = \langle T_2, \varphi \rangle$ for all $\varphi \in \mathcal{D}(E)$. If $T \in \mathcal{D}'(E)$ coincides with a function in $C^k(E)$ for some positive integer k, then the distributional derivatives $D^\alpha T$, for all multi-indices $|\alpha| \le k$, coincide with the classical derivatives of T.

The product of two distributions is, in general, not defined. However, if $\psi \in \mathcal{D}(E)$ and $T \in \mathcal{D}'(E)$ the product ψT is defined as that distribution acting on $\varphi \in \mathcal{D}(E)$, as

$$\langle \psi T, \varphi \rangle = \langle T, \psi\varphi \rangle \quad \text{for all } \varphi \in \mathcal{D}(E).$$

Let J_ε be the Friedrichs mollifying kernel introduced in § 18 of Chap. 6. For $T \in \mathcal{D}'(\mathbb{R}^N)$, the convolution $T_\varepsilon = T * J_\varepsilon$ is defined as that distribution acting on $\varphi \in \mathcal{D}(\mathbb{R}^N)$ as

$$\langle T * J_\varepsilon(\cdot - x), \varphi \rangle = \langle T, \varphi * J_\varepsilon(x - \cdot) \rangle.$$

To justify such a formula, assume first that $T \in L^1_{\text{loc}}(\mathbb{R}^N)$. Then for every $\varphi \in \mathcal{D}(\mathbb{R}^N)$

$$\langle T, \varphi * J_\varepsilon(x - \cdot) \rangle = \int_{\mathbb{R}^N} T(x) \int_{\mathbb{R}^N} \varphi(y) J_\varepsilon(x - y) dy \, dx$$

$$= \int_{\mathbb{R}^N} \left(\int_{\mathbb{R}^N} T(x) J_\varepsilon(x - y) dx \right) \varphi(y) dy$$

$$= \langle T * J_\varepsilon(\cdot - x), \varphi \rangle.$$

More generally, the convolution of a distribution T with a kernel $K \in L^1_{\mathrm{loc}}(\mathbb{R}^N)$ is defined by the formula

$$\langle T * K(\cdot - x), \varphi \rangle = \langle T, \varphi * K(x - \cdot) \rangle \quad \text{for all } \varphi \in \mathcal{D}(\mathbb{R}^N)$$

provided proper assumptions are made on K to insure that $K * \varphi \in \mathcal{D}(\mathbb{R}^N)$. For example $T * D^\alpha J_\varepsilon(\cdot - x)$ is well defined with $K = D^\alpha J_\varepsilon$. One verifies that $x \to T * J_\varepsilon(\cdot - x) \in C^\infty(\mathbb{R}^N)$, by the formula

$$D^\alpha(T * J_\varepsilon(\cdot - x)) = T * D^\alpha J_\varepsilon(\cdot - x).$$

Moreover $T_\varepsilon \to T$ in $\mathcal{D}'(\mathbb{R}^N)$. Thus a distribution T can be approximated, in the weak* topology, by functions in $C^\infty(\mathbb{R}^N)$.

14 Fundamental Solutions

For a multi-index α let a_α denote an N-tuple of real numbers labeled with the entries of α, as in $a_\alpha = (a_{\alpha_1}, \ldots, a_{\alpha_N})$. A linear differential operator $\mathcal{L}$ of order m, with constant coefficients, and its adjoint $\mathcal{L}^*$ are formally defined by

$$\mathcal{L} = \sum_{|\alpha| \le m} a_\alpha D^\alpha \qquad \mathcal{L}^* = \sum_{|\alpha| \le m} (-1)^{|\alpha|} a_\alpha D^\alpha.$$

Let $f \in \mathcal{D}'(E)$ be fixed and consider formally the partial differential equation

$$\mathcal{L}(u) = f \quad \text{in} \quad E.$$

A distribution $u \in \mathcal{D}'(E)$ is a solution of this equation, if

$$u[\mathcal{L}^*(\varphi)] = f[\varphi] \quad \text{for all } \varphi \in \mathcal{D}(E).$$

If $f = \delta_x$ the corresponding solution is called a *fundamental solution* of the operator $\mathcal{L}$ with pole at x. When regarding x as varying over E a fundamental solution of $\mathcal{L}$ is a family of distributions, parameterized with $x \in E$, satisfying

$$\mathcal{L}_y u = \delta_x \quad \text{in the sense} \quad u[\mathcal{L}_y^*(\varphi)] = \varphi(x) \quad \text{for all } \varphi \in \mathcal{D}(E).$$

Here $\mathcal{L}_y$ and $\mathcal{L}_y^*$ are the differential operator $\mathcal{L}$ and $\mathcal{L}^*$ where the derivatives are taken with respect to the variables y.

A linear differential operator of any fixed order m, with constant coefficients admits a fundamental solution [40, 100].

Here, as an example, we compute the fundamental solution of two particular linear differential operators.

14.1   The Fundamental Solution of the Wave Operator in $\mathbb{R}^2$

Let $E = \mathbb{R}^2$, fix $(\xi, \eta) \in \mathbb{R}^2$ and set

$$u(x, y) = \begin{cases} 1 & \text{if } x > \xi \text{ and } y > \eta \\ 0 & \text{otherwise.} \end{cases}$$

Compute in $\mathcal{D}'(\mathbb{R}^2)$

$$\frac{\partial^2 u}{\partial x \partial y}[\varphi] = \iint_{\mathbb{R}^2} u \varphi_{xy} dx dy = \int_\eta^\infty \int_\xi^\infty \varphi_{xy} dx dy = \varphi(\xi, \eta).$$

Therefore

$$\frac{\partial^2 u}{\partial x \partial y} = \delta_{(\xi, \eta)} \qquad \text{in } \mathcal{D}'(\mathbb{R}^2).$$

For fixed $(\xi, \eta) \in \mathbb{R}^2$ consider now the function

$$u(x, y) = \begin{cases} \frac{1}{2} & \text{if } |x - \xi| < y - \eta \\ 0 & \text{otherwise.} \end{cases}$$

This is the characteristic function of the sector $S_{(\xi, \eta)}$ delimited by the two half lines originating at (ξ, η)

$$\ell^+_{(\xi, \eta)} = \{x - y = \xi - \eta\} \cap \{x \geq \xi\}$$
$$\ell^-_{(\xi, \eta)} = \{x + y = \xi + \eta\} \cap \{x \leq \xi\}.$$

The exterior normal to such a sector is

$$\frac{(1, -1)}{\sqrt{2}} \quad \text{on} \quad \ell^+_{(\xi, \eta)} \qquad \text{and} \qquad \frac{(-1, -1)}{\sqrt{2}} \quad \text{on} \quad \ell^-_{(\xi, \eta)}.$$

Compute in $\mathcal{D}'(E)$

$$\left(\frac{\partial^2 u}{\partial y^2} - \frac{\partial^2 u}{\partial x^2} \right)[\varphi] = \iint_{\mathbb{R}^2} u(\varphi_{yy} - \varphi_{xx}) dx dy = \frac{1}{2} \iint_{S_{(\xi, \eta)}} (\varphi_{yy} - \varphi_{xx}) dx dy$$

$$= -\frac{1}{2\sqrt{2}} \int_{\ell^+_{(\xi, \eta)}} (\varphi_x + \varphi_y) ds + \frac{1}{2\sqrt{2}} \int_{\ell^-_{(\xi, \eta)}} (\varphi_x - \varphi_y) ds$$

where s is the abscissa along $\ell^{\pm}_{(\xi,\eta)}$ and ds is the corresponding measure. On $\ell^{\pm}_{(\xi,\eta)}$ one computes $\varphi_x \pm \varphi_y = \sqrt{2}\varphi'(s)$. Therefore

$$\frac{\partial^2 u}{\partial y^2} - \frac{\partial^2 u}{\partial x^2} = \delta_{(\xi,\eta)}, \quad \text{in } \mathcal{D}'(E).$$

14.2   The Fundamental Solution of the Laplace Operator

For $x, y \in \mathbb{R}^N$ and $x \neq y$, set

$$F(x; y) = \begin{cases} \dfrac{1}{(N-2)\omega_N} \dfrac{1}{|x-y|^{N-2}} & \text{if } N \geq 3 \\[3mm] \dfrac{-1}{2\pi} \ln|x-y| & \text{if } N = 2 \end{cases} \tag{14.1}$$

where ω_N is the area of the unit sphere in $\mathbb{R}^N$. Compute

$$\nabla_y F(x; y) = \frac{1}{\omega_N} \frac{x-y}{|x-y|^N} \quad \text{for } x \neq y, \quad N \geq 2. \tag{14.2}$$

From this

$$\Delta_y F(x; y) = \operatorname{div}_y \nabla_y F(x; y) = 0 \quad \text{for } x \neq y, \quad N \geq 2. \tag{14.3}$$

Proposition 14.1 (Stokes Formula)[1] *For all $\varphi \in C_o^\infty(\mathbb{R}^N)$ and all $x \in \mathbb{R}^N$*

$$\varphi(x) = - \int_{\mathbb{R}^N} F(x; y)\Delta\varphi dy. \tag{14.4}$$

Proof For a fixed $x \in \mathbb{R}^N$, the function $y \to F(x; y)$ is integrable about x. Therefore

$$\int_{\mathbb{R}^N} F(x; y)\Delta\varphi dy = \lim_{\varepsilon \to 0} \int_{\mathbb{R}^N - B_\varepsilon(x)} F(x; y)\Delta\varphi dy$$

where $B_\varepsilon(x)$ is the ball centered at x and of radius ε. The last integral is transformed by applying recursively the Gauss–Green Theorem

---

[1]This is a particular case of a more general Stokes representation formula when φ is not required to vanish on ∂E ([34], Chap. II).

$$\int_{\mathbb{R}^N - B_\varepsilon(x)} F(x; y)\Delta\varphi dy = \int_{|x-y|=\varepsilon} F(x; y)\nabla\varphi \cdot \frac{(x-y)}{\varepsilon} dy$$

$$- \int_{\mathbb{R}^N - B_\varepsilon(x)} \nabla_y F(x; y) \cdot \nabla\varphi dy$$

$$= \int_{|x-y|=\varepsilon} F(x; y)\nabla\varphi \cdot \frac{(x-y)}{\varepsilon} dy$$

$$- \int_{|x-y|=\varepsilon} \varphi\nabla_y F(x; y) \cdot \frac{(x-y)}{\varepsilon} dy$$

$$+ \int_{\mathbb{R}^N - B_\varepsilon(x)} \varphi\Delta_y F(x; y) dy$$

$$= I_{1,\varepsilon} + I_{2,\varepsilon} + I_{3,\varepsilon}.$$

The last integral is zero for all $\varepsilon > 0$ in view of (14.3), since the domain of integration excludes the singular point $y = x$. Using (14.1) for $|x - y| = \varepsilon$ one computes $\lim_{\varepsilon \to 0} I_{1,\varepsilon} = 0$. The second integral is computed with the aid of (14.2) for $|x - y| = \varepsilon$, and gives

$$I_{2,\varepsilon} = -\frac{1}{\omega_N \varepsilon^{N-1}} \int_{|x-y|=\varepsilon} \varphi dy$$

$$= \frac{-1}{\omega_N \varepsilon^{N-1}} \int_{|x-y|=\varepsilon} [\varphi(y) - \varphi(x)] dy - \frac{1}{\omega_N \varepsilon^{N-1}} \int_{|y-x|=\varepsilon} \varphi(x) dy.$$

The last integral equals $\varphi(x)$ for all $\varepsilon > 0$, whereas the first integral tends to zero as $\varepsilon \to 0$, since its modulus is majorized by $\sup_{|x-y|=\varepsilon} |\varphi(y) - \varphi(x)|$. ∎

The Stokes formula can be rewritten in terms of distributions as

$$-\Delta_y F(x; y) = \delta_x.$$

Therefore $F(\cdot; \cdot)$ given by (14.1) is the fundamental solution of the Laplace operator.

15 Weak Derivatives and Main Properties

Let $u \in L^1_{loc}(E)$ and let α be a multi-index. If the distribution $D^\alpha u$ coincides a.e., with a function $w \in L^1_{loc}(E)$, then w is called the *weak D^α-derivative* of u and

$$\int_E u D^\alpha \varphi dx = (-1)^{|\alpha|} \int_E w\varphi dx \quad \text{for all } \varphi \in \mathcal{D}(E).$$

If $u \in C^{|\alpha|}_{loc}(E)$ then w coincides with the classical D^α derivative of u. Let $1 \le p \le \infty$ and let m be a nonnegative integer. A function $u \in L^p(E)$ is said to be in $W^{m,p}(E)$, if all its weak derivatives $D^\alpha u$ for all $|\alpha| \le m$ are in $L^p(E)$. Equivalently ([148])

$$W^{m,p}(E) = \left\{ \begin{array}{c} \text{the collection of all } u \in L^p(E) \text{such that} \\ D^\alpha u \in L^p(E) \ \text{for all } |\alpha| \le m \end{array} \right\}.$$

A norm in $W^{m,p}(E)$ is

$$\|u\|_{m,p} = \sum_{|\alpha| \le m} \|D^\alpha u\|_p.$$

If $m = 0$, then $W^{m,p}(E) = L^p(E)$ and $\|\cdot\|_{0,p} = \|\cdot\|_p$. Define also

$$H^{m,p}(E) = \left\{\text{the closure of } C^\infty(E) \text{ with respect to } \|\cdot\|_{m,p}\right\}$$
$$W^{m,p}_o(E) = \left\{\text{the closure of } C^\infty_o(E) \text{ with respect to } \|\cdot\|_{m,p}\right\}.$$

Proposition 15.1 $W^{m,p}(E)$ *is a Banach space.*

Proof If $\{u_n\}$ is a Cauchy sequence in $W^{m,p}(E)$, the sequences $\{D^\alpha u_n\}$ are Cauchy in $L^p(E)$ for all multi-indices $0 \le |\alpha| \le m$. By the completeness of $L^p(E)$, there exist $u \in L^p(E)$ and functions $u_\alpha \in L^p(E)$, such that $\{u_n\} \to u$ and $\{D^\alpha u_n\} \to u_\alpha$ in $L^p(E)$. For all $\varphi \in \mathcal{D}(E)$

$$\int_E u_\alpha \varphi dx = \lim \int_E D^\alpha u_n \varphi dx$$
$$= \lim (-1)^{|\alpha|} \int_E u_n D^\alpha \varphi dx = (-1)^{|\alpha|} \int_E u D^\alpha \varphi dx.$$

Therefore u_α is the weak D^α-derivative of u. ∎

Corollary 15.1 $H^{m,p}(E) \subset W^{m,p}(E)$.

Theorem 15.1 (Meyers–Serrin [107]) *Let* $1 \le p < \infty$. *Then* $C^\infty(E)$ *is dense in* $W^{m,p}(E)$, *and as a consequence* $H^{m,p}(E) = W^{m,p}(E)$.

Proof Having chosen $u \in W^{m,p}(E)$ and $\varepsilon \in (0,1)$, we exhibit a function $\varphi \in C^\infty(E)$ such that $\|u - \varphi\|_{m,p} < \varepsilon$. For $j = 1, 2, \ldots$, set

$$E_j = \left\{x \in E \mid \text{dist}\{x, \partial E\} > \frac{1}{j} \ \text{and} \ |x| < j\right\}, \qquad E_o = E_{-1} = \emptyset.$$

Set also $\mathcal{O}_j = E_{j+1} \cap \bar{E}^c_{j-1}$. The set $\mathcal{O}_j$ for $j \ge 2$ is the set of points of E such that

$$\frac{1}{j+1} < \text{dist}\{x, \partial E\} < \frac{1}{j-1} \qquad \text{and} \qquad j - 1 < |x| < j + 1.$$

The sets $\mathcal{O}_j$ are open and their collection $\mathcal{U}$, forms an open covering of E. Let Φ be a partition of unity subordinate to $\mathcal{U}$ and let ψ_j be the sum of the finitely many $\varphi \in \Phi$, whose support is contained in $\mathcal{O}_j$. Then $\psi_j \in C_o^\infty(\mathcal{O}_j)$, and $\sum \psi_j(x) = 1$, for all $x \in E$. If ε_j are positive numbers satisfying

$$0 < \varepsilon_j < \frac{1}{(j+1)(j+2)}$$

the mollification $J_{\varepsilon_j} * (\psi_j u)$ has support in $E_{j+2} \cap \bar{E}_{j-2}^c = \mathcal{E}_j$. Since $\psi_j u \in W^{m,p}(E)$, one may choose ε_j so small that

$$\|J_{\varepsilon_j} * (\psi_j u) - \psi_j u\|_{m,p;\mathcal{E}_j} < \frac{1}{2^j}\varepsilon.$$

Set $\varphi = \sum J_{\varepsilon_j} * (\psi_j u)$, and observe that within any compact subset of E, all the terms in the sum vanish except at most finitely many. Therefore $\varphi \in C^\infty(E)$. For $x \in E_j$

$$\varphi(x) = \sum_{i=1}^{j+2} J_{\varepsilon_i} * (\psi_i u)(x) \quad \text{and} \quad u(x) = \sum_{i=1}^{j+2} \psi_i(x) u(x).$$

Therefore, for all $j = 1, 2, \ldots$

$$\|u - \varphi\|_{m,p;E_j} \le \sum_{i=1}^{j+2} \|J_{\varepsilon_i} * (\psi_i u) - \psi_i u\|_{m,p;E} \le \varepsilon \sum \frac{1}{2^i}.$$

From this, by monotone convergence, $\|u - \varphi\|_{m,p} < \varepsilon$. ∎

16 Domains and Their Boundaries

Let E denote an open set in $\mathbb{R}^N$ and let ∂E denote its boundary. We list here some structural assumptions that might be necessary to impose on ∂E. A countable collection of open balls $\{B_\rho(x_j)\}$ centered at points $x_j \in \partial E$ and radius ρ is an open, *locally finite* covering of ∂E if $\partial E \subset \bigcup B_\rho(x_j)$ and there exists a positive integer k such that any $(k+1)$ distinct elements of $\{B_\rho(x_j)\}$ have empty intersection.

16.1 ∂E of Class C¹

The boundary ∂E is said to be of class C^1 if there exist a positive ρ and an open, locally finite covering $\{B_\rho(x_j)\}$ of ∂E such that for all $x_j \in \partial E$ the portion of ∂E within the ball $B_\rho(x_j)$ can be represented, in a local system of coordinates with the origin at x_j, as the graph of function f_j of class class C^1 in a neighborhood of the origin of the new local coordinates and such that $f_j(0) = 0$ and $Df_j(0) = 0$. Denote by K_j the $(N-1)$-dimensional domain where f_j is defined and set

$$\||\partial E\||_1 = \sup_j \max_{K_j} |Df_j|. \tag{16.1}$$

This quantity depends upon the choice of the covering $\{B_\rho(x_j)\}$. However, having fixed one such covering, it is invariant under homotetic transformations of the coordinates. In particular it does not depend upon the size of E.

16.2   Positive Geometric Density and ∂E Piecewise Smooth

The boundary ∂E satisfies the property of *positive geometric density* at some $x_o \in \partial E$, with respect to the Lebesgue measure μ in $\mathbb{R}^N$, if there exists $\theta \in (0, 1)$ and $\rho_o > 0$ such that for every ball $B_\rho(x_o)$ centered at x_o and radius $\rho \le \rho_o$

$$\mu\big(E \cap B_\rho(x_o)\big) \le (1 - \theta)\mu\big(B_\rho(x_o)\big). \tag{16.2}$$

The boundary ∂E satisfies the property of uniform geometric density, if such inequality is satisfied for all $x_o \in \partial E$ for the same value of ρ_o and θ.

16.3   The Segment Property

The boundary ∂E has the *segment property* if there exists a locally finite, open covering of ∂E with balls $\{B_t(x_j)\}$, a corresponding sequence of unit vectors $\mathbf{n}_j$ and a number $t^* \in (0, 1)$, such that

$$x \in \bar{E} \cap B_t(x_j) \implies x + t\mathbf{n}_j \in E \text{ for all } t \in (0, t^*). \tag{16.3}$$

Such a requirement forces, in some sense, the domain E to lie, locally on one side of its boundary. For example, the unit disc from which a radius is removed, does not satisfy the segment property. For $x \in \mathbb{R}$ set

$$h(x) = \begin{cases} \sqrt{|x|} \sin^2 \dfrac{1}{x} & \text{for } |x| > 0 \\ 0 & \text{for } x = 0, \end{cases} \quad E = (-1, 1) \times \{y > h(x)\}. \tag{16.4}$$

The bidimensional set E satisfies the segment property.

16.4 The Cone Property

Let $\mathcal{C}_o$ denote a closed, circular, spherical cone of solid angle ω, height h and vertex at the origin. Such a cone has volume

$$\mu(\mathcal{C}_o) = \frac{\omega}{N} h^N. \tag{16.5}$$

A domain E has the *cone property* if there exist some $\mathcal{C}_o$, such that for all $x \in \bar{E}$, there exists a circular, spherical cone $\mathcal{C}_x$ with vertex at x and congruent to $\mathcal{C}_o$, all contained in $\bar{E}$.

16.5 On the Various Properties of ∂E

The cone property does not imply the segment property. For example the unit disc from which a radius is removed, satisfies the cone property and does not satisfy the segment property.

The segment property does not imply the cone property. For example the set in (16.4), does not satisfy the cone property.

The cone property does not imply the property of positive geometric density. For example the unit disc from which a radius is removed, satisfies the cone property and does not satisfy the property of positive geometric density.

The property of positive geometric density does not imply the cone property. For example the boundary of the cusp-like domain

$$E = \left\{ (x_1, x_2) \in \mathbb{R}^2 \mid 0 < x_1 < 1;\ 0 < x_2 < x_1^\alpha \right\} \tag{16.6}$$

for some $\alpha > 1$, satisfies the property of positive geometric density, but not the cone property.

The segment property does not imply that ∂E is of class C^1 as indicated by the domain in (16.4). Conversely, ∂E of class C^1 does not imply the segment property. For example let

$$E = \{x^2 + y^2 < 1\} - \{x^2 + y^2 = \tfrac{1}{4}\}. \tag{16.7}$$

The boundary of E is regular but ∂E does not satisfy the segment property.

17 More on Smooth Approximations

The approximations constructed in the proof of the Meyers–Serrin's Theorem, might deteriorate near ∂E and it is natural to ask whether a function in $W^{m,p}(E)$ can be approximated, in the sense of $W^{m,p}(E)$ by functions that are smooth up to $\bar{E}$. This in general is not the case as indicated by the following example. Let E as in (16.7) and set

$$u(x, y) = \begin{cases} 1 & \text{for } \frac{1}{4} < x^2 + y^2 < 1 \\ 0 & \text{for } \quad\quad x^2 + y^2 < \frac{1}{4}. \end{cases}$$

The function u is in $W^{m,p}(E)$ but there is no smooth function up to ∂E that approximates u in the norm of $W^{m,p}(E)$. This example shows that such an approximation property is in general false for domains that do not satisfy the segment property.

Proposition 17.1 *Let E be a bounded domain in $\mathbb{R}^N$ with boundary ∂E satisfying the segment property. Then $C_o^\infty(\mathbb{R}^N)$ is dense in $W^{m,p}(E)$ for $1 \le p < \infty$.*

Proof Since ∂E is bounded, the open covering claimed by the segment property is finite, say for example

$$\{B_t(x_1), \ldots, B_t(x_n)\} \text{ for some } n \in \mathbb{N} \quad \text{and some } t > 0. \tag{17.1}$$

Denote by $\mathbf{n}_j$ the corresponding unit vectors pointing inside E and that realize the segment property. By reducing t if necessary we may assume that for all $j = 1, \ldots, n$

$$\text{for all } x \in \partial E \cap B_{8t}(x_j) \quad x + \tau \mathbf{n}_j \in E \quad \text{for all } \tau \in (0, 8t).$$

Construct an open covering $\mathcal{U}$ of E, by

$$\mathcal{U} = \{B_o, B_{2t}(x_1), \ldots, B_{2t}(x_n)\}, \quad B_o = E - \bigcup_{j=1}^n \bar{B}_t(x_j). \tag{17.2}$$

Let Φ be a partition of unity subordinate to $\mathcal{U}$, and for $j = 1, \ldots, n$, let ψ_j be the sum of the finitely many $\varphi \in \Phi$ supported in $B_{2t}(x_j)$. For $j = 0$ define ψ_o analogously by replacing $B_{2t}(x_j)$ with B_o. Set

$$u_j(x) = \begin{cases} (u\psi_j)(x) & \text{for } x \in E \\ 0 & \text{otherwise} \end{cases} \quad j = 0, 1, \ldots, n.$$

Let Γ_j be the portion of ∂E within the ball $B_{4t}(x_j)$. By definition of weak derivative $u_j \in W^{m,p}(\mathbb{R}^N - \Gamma_j)$. To prove the Proposition, having fixed $\varepsilon > 0$, it suffices to find functions $\varphi_j \in C_o^\infty(\mathbb{R}^N)$ such that

$$\|u_j - \varphi_j\|_{m,p} < \frac{\varepsilon}{n+1} \quad \text{for all } j = 0, 1, \ldots, n.$$

Indeed putting $\varphi = \sum_{j=n}^n \varphi_j$ it would give

$$\|u - \varphi\|_{m,p} \le \sum_{j=0}^n \|u_j - \varphi_j\|_{m,p} < \varepsilon.$$

For $j = 0$ such a φ_o can be constructed by a standard mollification since u_o is compactly supported in E. To construct φ_j, for $j \in \{1, 2, \ldots, n\}$, move Γ_j, towards the outside of E by setting

$$\Gamma_{j,\tau} = \Gamma_j - \tau \mathbf{n}(x_j) \quad \text{for} \quad \tau \in (0, 8t).$$

Then define

$$u_{j,\tau}(x) = u_j(x + \tau \mathbf{n}(x_j)) \quad \text{for all } x \in \mathbb{R}^N - \Gamma_{j,\tau}.$$

By definition of weak derivative $u_{j,\tau} \in W^{m,p}(\mathbb{R}^N - \Gamma_{j,\tau})$ and

$$D^\alpha u_{j,\tau}(x) = D^\alpha u_j(x + \tau \mathbf{n}(x_j)) \quad \text{for all } x \in \mathbb{R}^N - \Gamma_{j,\tau}.$$

Since the translation operation is continuous in $L^p(E)$, there exists $\tau_\varepsilon \in (0, 4t)$, such that

$$\|u_{j,\tau} - u_j\|_{m,p;E} \leq \frac{\varepsilon}{2(n+1)}.$$

The function $\varphi_j \in C_o^\infty(\mathbb{R}^N)$ is constructed by the mollification $\varphi_j = J_\delta * u_{j,\tau}$, where for a fixed $\tau \in (0, \tau_\varepsilon)$ the positive number δ is chosen so small that

$$\|J_\delta * u_{j,\tau} - u_{j,\tau}\|_{m,p;E} \leq \frac{\varepsilon}{2(n+1)}. \qquad \blacksquare$$

18 Extensions into $\mathbb{R}^N$

Proposition 18.1 ([93]) *Let E be a bounded domain in $\mathbb{R}^N$ with boundary ∂E satisfying the segment property and of class C^1, and let $\{B_t(x_j)\}_{j=1}^n$ be a finite open covering of ∂E with balls of radius t centered at points $x_j \in \partial E$ as in (17.1). A function $u \in W^{1,p}(E)$, for some $1 \leq p < \infty$, admits an extension $w \in W_o^{1,p}(\mathbb{R}^N)$ such that*

$$\|w\|_{p,\mathbb{R}^N} \leq \gamma(1 + \|\|\partial E\|\|_1)\|u\|_{p,E}$$

$$\|Dw\|_{p;\mathbb{R}^N} \leq \gamma(1 + \|\|\partial E\|\|_1)\left(\|Du\|_{p,E} + \frac{1}{t}\|u\|_{p;E}\right) \tag{18.1}$$

where γ depends only upon N, p, n and the number of local overlaps of the covering $\{B_t(x_j)\}_{j=1}^n$.

Proof By density we may assume $u \in C^1(\bar{E})$. Consider first the following special case. For $\bar{x} = (x_1, \ldots, x_{N-1})$ and $R > 0$ let $\mathcal{B}_R = [|\bar{x}| < R]$ be the $(N-1)$-dimensional ball of radius R centered at the origin. Set also

$$Q_R^+ = \mathcal{B}_R \times [0, R), \quad Q_R^- = \mathcal{B}_R \times (-R, 0], \quad Q_R = Q_R^+ \cup Q_R^-.$$

Assume that, as a function of the first $(N - 1)$ variables, $\bar{x} \to u(\bar{x}, x_N)$ is compactly supported in $\mathcal{B}_R$, and that $u(\bar{x}, R) = 0$. Thus u vanishes on the top and on the lateral boundary of Q_R^+ and it is of class C^1 up to $x_N = 0$. The claimed extension, in such a case is

$$\tilde{u}(\bar{x}, x_N) = -3u(\bar{x}, -x_N) + 4u(\bar{x}, -\tfrac{1}{2}x_N) \quad x_N \leq 0. \tag{18.2}$$

The function w defined as u within Q_R^+ and as $\tilde{u}$ within Q_R^- satisfies the indicated requirements. The general case is proved by a local *flattening* of ∂E. Referring back to the proof of Proposition 17.1, consider the finite covering (17.2), and construct the functions ψ_j. These can be chosen to satisfy

$$\sup_{B_{2t}(x_j)} |D\psi_j| \leq \frac{\gamma}{t} \quad \text{for a positive constant } \gamma.$$

Represent $\partial E \cap B_t(x_j)$, in a local system of coordinates as the graph of $x_N = f_j(\bar{x})$, for $\bar{x} = (x_1, \ldots, x_{N-1})$, where f_j is of class C^1 within the $(N - 1)$-dimensional ball $\mathcal{B}_t$. For each j fixed, flatten $\partial E \cap B_t(x_j)$ by introducing the system of coordinates

$$(y_1, \ldots, y_{N-1}, y_N) = (\bar{x}, x_N - f_j(\bar{x})).$$

This maps $\partial E \cap B_t(x_j)$ into $\mathcal{B}_t$ and, by taking t even smaller if necessary, $E \cap B_t(x_j)$ is mapped into $Q_t^+ = \mathcal{B}_t \times [0, t)$. The functions $\widetilde{u\psi_j}$ obtained from the $u\psi_j$ with these transformations, are of class C^1 in $\bar{Q}_t^+$, and $\bar{x} \to \widetilde{u\psi_j}(\bar{x}, y_N)$ has compact support in $\mathcal{B}_t$. Next extend $\widetilde{u\psi_j}$ with a function $\tilde{w}_j$ of class C^1 in the whole cylinder $Q_t = \mathcal{B}_t \times (-t, t)$, by the procedure of (18.2). Let w_j denote the function obtained from $\tilde{w}_j$ by the change of variables that maps Q_t^+ back to $E \cap B_t(x_j)$. The extension claimed by the Proposition can be constructed by setting $w = u\psi_o + \sum_{j=1}^n w_j$. Indeed for each $j = 1, \ldots, n$

$$\|\tilde{w}_j\|_{p;Q_t} \leq \gamma(1 + \|\partial E\|_1) \|u\|_{p;E\cap B_t(x_j)}$$

$$\|D\tilde{w}_j\|_{p;Q_t} \leq \gamma(1 + \|\partial E\|_1)\left(\|Du\|_{p;E\cap B_t(x_j)} + \frac{1}{t}\|u\|_{p;E\cap B_t(x_j)}\right) \tag{18.3}$$

and each $B_t(x_j)$ overlaps at most finitely many balls $B_t(x_i)$.   ∎

Remark 18.1 The boundedness of E is not needed. It suffices to require that ∂E admits a *locally finite*, open covering. In such a case however one has to assume $u \in W^{1,p}(E)$.

Remark 18.2 If E is the ball B_R, the balls $B_t(x_j)$ can be chosen so that, for example, $t = \frac{1}{8}R$ and the number of their mutual overlap can be estimates by an absolute number depending only on the dimension and independent of R. Then the extension $w \in W^{1,p}(\mathbb{R}^N)$ of a function $u \in W^{1,p}(B_R)$ can be constructed as to satisfy

$$\|w\|_{p;\mathbb{R}^N} \leq \gamma \|u\|_{p;B_R}$$

$$\|Dw\|_{p;\mathbb{R}^N} \leq \gamma \left(\|Du\|_{p;B_R} + \frac{1}{R} \|u\|_{p;B_R} \right) \tag{18.4}$$

where γ is an absolute constant depending only on N and independent of R.

19 The Chain Rule

Proposition 19.1 *Let* $u \in W^{1,p}(E)$ *for some* $1 \leq p < \infty$, *and let* $f \in C^1(\mathbb{R})$ *satisfy* $\sup|f'| \leq M$ *for some positive constant* M. *Then the composition* $f(u)$ *belongs to* $W^{1,p}(E)$ *and* $Df(u) = f'(u)Du$.

Proof Let $C^\infty(E) \supset \{u_n\} \to u$ in $W^{1,p}(E)$. By possibly passing to a subsequence we may assume that $\{u_n\} \to u$ a.e. in E. Then $\{f(u_n)\} \to f(u)$ and

$$\lim \|Df(u_n) - f'(u)Du\|_p = \lim \|f'(u_n)Du_n - f'(u)Du\|_p$$
$$\leq \lim M\|Du_n - Du\|_p + \lim \left\||f'(u_n) - f'(u)||Du|\right\|_p.$$

The sequence $\{(f'(u_n) - f'(u))^p |Du|^p\}$ tends to zero a.e. in E. Moreover it is dominated by the integrable function $(2M)^p |Du|^p$. Therefore $Df(u_n) \to f'(u)Du$ in $L^p(E)$. Also, for all $\varphi \in C_o^\infty(E)$

$$\lim \int_E Df(u_n)\varphi dx = -\int_E f(u)D\varphi dx.$$

Thus $Df(u) = f'(u)Du$ in $\mathcal{D}'(E)$. ∎

Proposition 19.2 *Let* $u \in W^{1,p}(E)$ *for some* $1 \leq p < \infty$. *Then* u^+, u^- *and* $|u|$ *belong to* $W^{1,p}(E)$ *and*

$$Du^+ = \begin{cases} Du & a.e. \ [u > 0] \\ 0 & a.e. \ [u \leq 0] \end{cases} \qquad Du^- = \begin{cases} -Du & a.e. \ [u < 0] \\ 0 & a.e. \ [u \geq 0] \end{cases}$$

$$D|u| = \begin{cases} Du & a.e. \ [u > 0] \\ 0 & a.e. \ [u = 0] \\ -Du & a.e. \ [u < 0]. \end{cases}$$

Proof For $\varepsilon > 0$ let

$$f_\varepsilon(u) = \begin{cases} \sqrt{u^2 + \varepsilon^2} - \varepsilon & \text{if } u > 0 \\ 0 & \text{if } u \leq 0. \end{cases}$$

Then $f_\varepsilon \in C^1(\mathbb{R})$ and $|f'_\varepsilon| \leq 1$. Therefore for all $\varphi \in C_o^\infty(E)$

$$\int_E f_\varepsilon(u) D\varphi \, dx = -\int_E Df_\varepsilon(u)\varphi \, dx = -\int_{[u>0]} \frac{u \, Du}{\sqrt{u^2 + \varepsilon^2}} \varphi \, dx.$$

Letting $\varepsilon \to 0$

$$\int_E u^+ D\varphi \, dx = -\int_{[u>0]} Du \varphi \, dx.$$

Thus the conclusion holds for u^+. The remaining statements follow from $u^- = (-u)^+$ and $|u| = u^+ + u^-$. ∎

Corollary 19.1 *Let $u \in W^{1,p}(E)$. Then $Du = 0$ a.e. on any level set of u.*

Corollary 19.2 *Let $f, g \in W^{1,p}(E)$. Then $\max\{f; g\}$ and $\min\{f; g\}$ are in $W^{1,p}(E)$ and*

$$D\max\{f; g\} = \begin{cases} Df & \text{a.e. } [f > g] \\ Dg & \text{a.e. } [f < g] \\ 0 & \text{a.e. } [f = g]. \end{cases}$$

A similar formula holds for $\min\{f; g\}$.

Proof It follows from Proposition 19.2 and the formulae

$$2\max\{f; g\} = (f + g) + |f - g|$$
$$2\min\{f; g\} = (f + g) - |f - g|.$$ ∎

20   Steklov Averagings

Regard $u \in L^p(E)$ as defined in the whole $\mathbb{R}^N$ by setting it to be zero outside E. For $h \neq 0$ set

$$u_{h,i}(x) = \frac{1}{h} \int_{x_i}^{x_i+h} u(x_1, \ldots, \xi_i, \ldots x_N) d\xi_i.$$

These are the Steklov averages of u with respect to the variable x_i.

Proposition 20.1 *Let* $u \in L^p(E)$ *for some* $p \in [1, \infty)$. *Then* $u_{h,i} \to u$ *in* $L^p(E)$ *as* $h \to 0$.

Proof For almost all $x \in E$

$$|u_{h,i}(x) - u(x)| = \left| \frac{1}{h} \int_{x_i}^{x_i+h} [u(\ldots, \xi_i, \ldots) - u(x)] d\xi_i \right|$$

$$= \left| \frac{1}{h} \int_0^h [u(\ldots, x_i + \sigma, \ldots) - u(x)] d\sigma \right|$$

$$\leq \frac{1}{h} \int_0^h |T_{\sigma,i} u(x) - u(x)| d\sigma$$

$$\leq \frac{1}{|h|^{1/p}} \left(\int_0^{|h|} |T_{\sigma,i} u(x) - u(x)|^p \, d\sigma \right)^{1/p}$$

where $T_{\sigma,i}$ is the translation operator in $L^p(E)$ with respect to x_i. Taking the p-power and integrating in dx over E, gives

$$\|u_{h,i} - u\|_p \leq \sup_{|\sigma| \leq |h|} \|T_{\sigma,i} u - u\|_p.$$ ∎

For $\delta > 0$ let $E_\delta = \{x \in E \mid \text{dist}\{x; \partial E\} > \delta\}$ and assume that δ is so small that $E_\delta \neq \emptyset$. Denote by $\mathbf{h} = (h_1, \ldots, h_N) \in \mathbb{R}^N$ a vector of length $|\mathbf{h}| < \delta$, and set

$$w_{h_i,i}(x) = \frac{u(\ldots, x_i + h_i \ldots) - u(\ldots, x_i, \ldots)}{h_i} = \frac{\partial u_{h_i,i}}{\partial x_i}$$

a.e. in E_δ. If $h_i = 0$ set $w_{0,i} = 0$ and denote by $\mathbf{w_h}$ the vector of components $w_{h_i,i}$.

Proposition 20.2 *Let* $u \in L^p(E)$ *for some* $1 < p < \infty$ *and assume that there exists positive constant* C_p *and* δ_o, *such that*

$$\|\mathbf{w_h}\|_{p,E_\delta} \leq C_p \quad \text{for all } \delta \leq \delta_o \text{ and all } |\mathbf{h}| < \delta.$$

Then $u \in W^{1,p}(E)$ *and* $\|Du\|_p \leq C_p$. *If* E *is of finite measure, the conclusion continues to hold for* $p = \infty$.

Proof Fix $\delta \in (0, \delta_o)$ and $\varphi \in C_o^\infty(E_\delta)$. For fixed $i \in \{1, \ldots, N\}$

$$\lim_{h_i \to 0} \int_E w_{h_i,i} \varphi dx = -\lim_{h_i \to 0} \int_E u_{h_i,i} D_i \varphi dx = -\int_E u D_i \varphi dx.$$

The family $\{w_{h_i,i}\}$ for $|h_i| \in (0, \delta)$, is uniformly bounded in $L^p(E_\delta)$. Therefore for a subsequence, relabelled with h, $\{w_{h_i,i}\} \to w_i$, weakly in $L^p(E_\delta)$. For such a subsequence

$$\lim_{h_i \to 0} \int_E w_{h_i,i}\varphi dx = \int_E w_i\varphi dx \qquad \text{for all} \;\; \varphi \in C_o^\infty(E_\delta).$$

Thus $w_i = D_i u$ in E_δ. This identifies w_i as the distributional D_i-derivative of u in E_δ. Once the limit has been identified, the selection of subsequences is unnecessary and the entire family $\{w_{h_i,i}\}$ converges to $D_i u$ weakly in $L^p(E_\delta)$. By weak lower semi-continuity of the L^p-norm

$$\|Du\|_p^p \le \liminf_{\delta \to 0} \int_{E_\delta} |Du|^p dx \le \liminf_{\delta \to 0} \liminf_{h \to 0} \int_{E_\delta} |w_h|^p dx \le C^p.$$

If $p = \infty$ and E is of finite measure, the same arguments give

$$\|Du\|_p \le C_\infty \mu(E)^{1/p} \qquad \text{for all} \;\; p > 1. \qquad \blacksquare$$

20.1 Characterizing $W^{1,p}(E)$ for $1 < p < \infty$

Proposition 20.3 *A function u belongs to $W^{1,p}(E)$ for some $1 < p < \infty$, if and only if there exists a positive number δ_o and a vector valued function $\mathbf{w} \in L^p(E)$ such that*

$$\|u(\cdot + \mathbf{h}) - u - \mathbf{h} \cdot \mathbf{w}\|_{p,E_\delta} = o(|\mathbf{h}|) \quad as \quad |\mathbf{h}| \to 0 \qquad (20.1)$$

for every $0 < \delta \le \delta_o$ and every $|\mathbf{h}| < \delta$. In such a case $\mathbf{w} = Du$ in $\mathcal{D}'(E)$.

Proof (Sufficient Condition) Let $u \in L^p(E)$ satisfy (20.1) and for $|h| > 0$ let $\mathbf{w_h}$ denote its discrete gradient. Fix $\delta > 0$ and choose $\mathbf{h} = (0, \ldots, h_i, \ldots 0)$ with $0 < |h_i| < \delta$. For such a choice

$$\|\mathbf{w_h}\|_{p,E_\delta} = \frac{1}{|\mathbf{h}|}\|u(\cdot + \mathbf{h}) - u\|_{p,E_\delta}$$

$$\le \frac{1}{|\mathbf{h}|}\|u(\cdot + \mathbf{h}) - u - \mathbf{h} \cdot \mathbf{w}\|_{p,E_\delta} + \frac{1}{|\mathbf{h}|}\|\mathbf{h} \cdot \mathbf{w}\|_{p,E_\delta}$$

$$\le 1 + \|\mathbf{w}\|_p.$$

Therefore $u \in W^{1,p}(E)$ and $\mathbf{w} = Du$ by Proposition 20.2. $\qquad \blacksquare$

Proof (Necessary Condition) Let $u \in W^{1,p}(E)$, fix $\delta > 0$ and compute

$$u(\ldots, x_i + h_i, \ldots) - u(\ldots, x_i, \ldots) = \int_0^{h_i} u_{x_i}(\ldots, x_i + \sigma, \ldots)d\sigma$$

$$= h_i (u_{x_i})_{h_i} \qquad \text{a.e. in } E_\delta$$

where $(u_{x_i})_{h_i}$ is the Steklov average of u_{x_i}. From this

$$\|u(\cdot + \mathbf{h}) - u - \mathbf{h} \cdot Du\|_{p,E_\delta} \le |\mathbf{h}| \|(Du)_\mathbf{h} - Du\|_{p,E_\delta} = o(\mathbf{h}).$$ ∎

20.2 Remarks on $W^{1,\infty}(E)$

Assume (20.1) holds for $p = \infty$ for some $\mathbf{w} \in L^\infty(E)$. Then

$$\frac{\|u(\cdot + \mathbf{h}) - u\|_{\infty,E_\delta}}{|\mathbf{h}|} \le (1 + \|\mathbf{w}\|_{\infty,E}) \quad \text{for all } |\mathbf{h}| < \delta.$$

If E is of finite measure, this implies $u \in W^{1,\infty}(E)$ and $Du = \mathbf{w}$. Thus if (20.1) holds for $p = \infty$, then u is Lipschitz continuous in E_δ. Conversely, if u is Lipschitz continuous in some domain E it also is in $W^{1,\infty}(E')$ for every subdomain $E' \subset E$ of finite measure. Indeed the discrete gradient $\mathbf{w}_h$ of u is pointwise bounded above by the Lipschitz constant of u. It remains to investigate whether a Taylor formula of the type of (20.1) would hold for such functions. This is the content of the Rademacher Theorem.

21 The Rademacher's Theorem

A continuous function $f : \mathbb{R}^N \to \mathbb{R}$ is differentiable at $x \in \mathbb{R}^N$ if there exists a vector $Df(x) \in \mathbb{R}^N$ such that

$$f(y) = f(x) + Df(x) \cdot (y - x) + o(|x - y|) \quad \text{as } y \to x. \tag{21.1}$$

If such a vector $Df(x)$ exists, then $Df = (f_{x_1}, \dots, f_{x_N}) = \nabla f$ at x. However the existence of $\nabla f(x)$ does not imply f is differentiable at x. For a unit vector $\mathbf{u}$ and $x \in \mathbb{R}^N$ set

$$D'_\mathbf{u} f(x) = \liminf_{\tau \to 0} \frac{f(x + \tau \mathbf{u}) - f(x)}{\tau}$$

$$D''_\mathbf{u} f(x) = \limsup_{\tau \to 0} \frac{f(x + \tau \mathbf{u}) - f(x)}{\tau}.$$

Since f is continuous, the limit can be taken along τ rational. Therefore $D'_\mathbf{u} f$ and $D''_\mathbf{u} f$ are measurable. Set also

$$D_\mathbf{u} f(x) = \lim_{\tau \to 0} \frac{f(x + \tau \mathbf{u}) - f(x)}{\tau}$$

provided the limit exists.

Proposition 21.1 *Let $f : \mathbb{R}^N \to \mathbb{R}$ be locally Lipschitz continuous. Then $D_{\mathbf{u}} f$ exists a.e. in $\mathbb{R}^N$. In particular ∇f exists a.e. in $\mathbb{R}^N$, and $D_{\mathbf{u}} f = \mathbf{u} \cdot \nabla f$, a.e. in $\mathbb{R}^N$.*

Proof Let $E_{\mathbf{u}} = [D'_{\mathbf{u}} f < D''_{\mathbf{u}} f]$, be the set where $D_{\mathbf{u}} f$ does not exist. Since $D'_{\mathbf{u}} f$ and $D''_{\mathbf{u}} f$ are measurable $E_{\mathbf{u}}$ is measurable. For $x \in \mathbb{R}^N$ fixed, the function of one variable $t \to f(x + t\mathbf{u})$ is absolutely continuous in every sub-interval of $\mathbb{R}$, and hence a.e. differentiable in $\mathbb{R}$. Therefore the intersection of $E_{\mathbf{u}}$ with any line parallel to $\mathbf{u}$ has 1-dimensional Lebesgue measure zero. Thus by Fubini's Theorem, $\mu(E_{\mathbf{u}}) = 0$. Let $\{\tau_n\}$ be the rationals in $(-1, 1)$ and let $\mathbf{e}_j$ be the unit vector along the jth coordinate axis of $\mathbb{R}^N$. For all $\zeta \in C_o^\infty(\mathbb{R}^N)$

$$\left| \frac{f(x + \tau_n \mathbf{u}) - f(x)}{\tau_n} \zeta \right| \leq L_\zeta |\zeta|$$

where L_ζ is the Lipschitz constant of f over a sufficiently large ball containing the support of ζ. By dominated convergence and change of variables

$$\int_{\mathbb{R}^N} D_{\mathbf{u}} f \zeta dx = \lim_{\tau_n \to 0} \int_{\mathbb{R}^N} \frac{f(\cdot + \tau_n \mathbf{u}) - f}{\tau_n} \zeta dx$$

$$= -\lim_{\tau_n \to 0} \int_{\mathbb{R}^N} f \frac{\zeta(\cdot + \tau_n \mathbf{u}) - \zeta}{\tau_n} dx = -u_j \int_{\mathbb{R}^N} f \zeta_{x_j} dx$$

$$= -u_j \lim_{\tau_n \to 0} \int_{\mathbb{R}^N} f \frac{\zeta(\cdot + \tau_n \mathbf{e}_j) - \zeta}{\tau_n} dx$$

$$= u_j \lim_{\tau_n \to 0} \int_{\mathbb{R}^N} \frac{f(\cdot + \tau_n \mathbf{e}_j) - f}{\tau_n} \zeta dx$$

$$= \int_{\mathbb{R}^N} \mathbf{u} \cdot \nabla f \zeta dx.$$

∎

Theorem 21.1 (Rademacher [120]) *Let $f : \mathbb{R}^N \to \mathbb{R}$ be locally Lipschitz continuous. Then f is a.e. differentiable in $\mathbb{R}^N$.*

Proof Let $\{\mathbf{u}_n\}$ be a countable dense subset of the unit sphere in $\mathbb{R}^N$, and let $E_{\mathbf{u}_n} = [D'_{\mathbf{u}_n} f < D''_{\mathbf{u}_n} f]$, that is the set where the directional derivative along $\mathbf{u}_n$ does not exist. By the previous proposition, $\mu(\bigcup E_{\mathbf{u}_n}) = 0$. To prove the theorem we establish that f is differentiable in $\mathbb{R}^N - \bigcup E_{\mathbf{u}_n}$. Fix $x \in \mathbb{R}^N - \bigcup E_{\mathbf{u}_n}$ and $y \in B_1(x)$ with $y \neq x$ and set

$$\mathbf{u} = \frac{y - x}{|x - y|}, \qquad t = |x - y|, \qquad y = x + t\mathbf{u}.$$

Let also L_R be the Lipschitz constant of f in the ball B_R centered at the origin and radius $R = 2\max\{|x|; 1\}$. Then, for $\mathbf{u}_j \in \{\mathbf{u}_n\}$

$$
\begin{aligned}
|f(y) - f(x) - \nabla f(x) \cdot (y - x)| &\leq |f(x + t\mathbf{u}) - f(x) - t\mathbf{u} \cdot \nabla f(x)| \\
&= |f(x + t\mathbf{u}_j) - f(x) - t\mathbf{u}_j \cdot \nabla f(x)| \\
&\quad + |f(x + t\mathbf{u}) - f(x + t\mathbf{u}_j)| + t|(\mathbf{u} - \mathbf{u}_j) \cdot \nabla f(x)| \\
&\leq t\left|\frac{f(x + t\mathbf{u}_j) - f(x)}{t} - \mathbf{u}_j \cdot \nabla f(x)\right| + t 2 L_R |\mathbf{u} - \mathbf{u}_j|.
\end{aligned}
$$

Having fixed $\varepsilon > 0$ fix $y \in B_\delta(x)$, where $\delta > 0$ is to be chosen. Then for $\mathbf{u}$ fixed, choose $\mathbf{u}_j$ such that $2L_R|\mathbf{u} - \mathbf{u}_j| \leq \frac{1}{2}\varepsilon$. Such a choice is independent of δ. For $\mathbf{u}_j$ fixed, there exists $\delta > 0$ such that

$$
\left|\frac{f(x + t\mathbf{u}_j) - f(x)}{t} - \mathbf{u}_j \cdot \nabla f(x)\right| \leq \frac{1}{2}\varepsilon \qquad \text{for all } 0 < t < \delta.
$$

Therefore for all $\varepsilon > 0$, there exists $\delta > 0$ such that

$$
|f(y) - f(x) - \nabla f(x) \cdot (y - x)| \leq |x - y|\varepsilon
$$

provided $|x - y| \leq \delta$.   ∎

Remark 21.1 Rademacher's Theorem continues to hold for a function f, Lipschitz continuous on a subset $E \subset \mathbb{R}^N$, modulo a preliminary application of the extension Theorem 15.1 of Chap. 5.

Problems and Complements

1c   Bounded Linear Functionals on $C_o(\mathbb{R}^N; \mathbb{R}^m)$

For a positive integer m denote by $C_o(\mathbb{R}^N; \mathbb{R}^m)$ the collection of all vector valued functions $\mathbf{f} = (f_1, \dots, f_m)$ with $f_j \in C_o(\mathbb{R}^N)$ for all $j = 1, \dots, m$, equipped with the norm

$$
\|\mathbf{f}\| = \sup_{\mathbb{R}^N} |\mathbf{f}|
$$

where $|\cdot|$ is the Euclidean length. Continuity of functionals $T \in C_o(\mathbb{R}^N; \mathbb{R}^m)^*$ is meant with respect to such a norm. Given a finite Radon measure μ in $\mathbb{R}^N$ and a μ-integrable vector valued function $\mathbf{e} = (e_1, \dots, e_m)$, the formula

$$
C_o(\mathbb{R}^N; \mathbb{R}^m) \ni \mathbf{f} \to T(\mathbf{f}) = \int_{\mathbb{R}^N} \mathbf{f} \cdot \mathbf{e}\, d\mu \tag{1.1c}
$$

generates a bounded linear functional in $C_o(\mathbb{R}^N; \mathbb{R}^m)$ with norm $\|T\| = \|\mathbf{e}\|_1$, where the integral is meant with respect to the measure μ.

However (1.1c) continues to be well defined, if μ is a Radon measure (not necessarily finite) and if $\mathbf{e}$ is locally μ-integrable in $\mathbb{R}^N$.

A linear functional T on $C_o(\mathbb{R}^N; \mathbb{R}^m)$ is *locally bounded* if for every compact set $K \subset \mathbb{R}^N$ there exists a constant γ_K such that

$$|T(\mathbf{f})| \leq \gamma_K \|f\| \quad \text{for all } \mathbf{f} \in C_o(\mathbb{R}^N; \mathbb{R}^m) \text{ with } \mathrm{supp}\{\mathbf{f}\} \subset K. \tag{1.2c}$$

The vector valued version of the Riesz representation theorem asserts the elements of $C_o(\mathbb{R}^N; \mathbb{R}^m)^*$ are of the form (1.1c) for some finite Radon measure μ and a μ-integrable function $\mathbf{e}$ The theorem is more general as it applies to *locally bounded* linear functionals on $C_o(\mathbb{R}^N; \mathbb{R}^m)$.

Theorem 1.2c *Let T be a linear, locally bounded functional in $C_o(\mathbb{R}^N; \mathbb{R}^m)$. There exists a Radon measure μ and a μ-measurable function $\mathbf{e} : \mathbb{R}^N \to \mathbb{R}^m$ such that $|\mathbf{e}| = 1$, μ-a.e. in $\mathbb{R}^N$ and $T(\mathbf{f})$ has the form (1.1c).*

The proof follows the main ideas as for the scalar case $m = 1$ as presented in § 1– § 6 with minor changes. The measure μ is constructed as in § 3 by using vector valued functions. The linear functional T_+ in § 4 is constructed still on *scalar* nonnegative functions f but using vector valued $\mathbf{h}$. The remainder of the proof follows the same steps with minor changes.

2c Convergence of Measures

Endow $C_o(\mathbb{R}^N)^*$ with its weak* topology. Then a sequence of measures $\{\mu_n\} \subset C_o(\mathbb{R}^N)^*$ converges weak* to some $\mu \in C_o(\mathbb{R}^N)^*$ if and only if (§ 15 of Chap. 7)

$$\lim \int_{\mathbb{R}^N} f d\mu_n = \int_{\mathbb{R}^N} f d\mu \quad \text{for all } f \in C_o(\mathbb{R}^N). \tag{2.1c}$$

The next proposition provides alternative ways of characterizing such a convergence.

Proposition 2.1c *A sequence of measures $\{\mu_n\} \subset C_o(\mathbb{R}^N)^*$ converges weak* to a measure $\mu \in C_o(\mathbb{R}^N)*$ if and only if either*

$$\begin{aligned} &\limsup \mu_n(K) \leq \mu(K) \quad \textit{for all compact sets } K \subset \mathbb{R}^N, \textit{ and} \\ &\liminf \mu_n(\mathcal{O}) \leq \mu(\mathcal{O}) \quad \textit{for all open sets } \mathcal{O} \subset \mathbb{R}^N \end{aligned} \tag{2.2c}$$

or

$$\lim \mu_n(E) = \mu(E) \quad \begin{array}{l} \textit{for all bounded Borel sets } E \subset \mathbb{R}^N \\ \textit{such that } \mu(\partial E) = 0. \end{array} \tag{2.3c}$$

Thus the notions (2.1c)–(2.3c) are equivalent and each of them can be taken as a notion of weak* convergence of measures. To prove the proposition we establish that $(2.1c) \Longrightarrow (2.2c) \Longrightarrow (2.3c) \Longrightarrow (2.1c)$.

Proof $(2.1c) \Longrightarrow (2.2c)$ Having fixed a compact set $K \subset \mathbb{R}^N$ let $\mathcal{O}$ be an open set cotaining K and construct a nonnegative function $f \in C_o(\mathcal{O})$ such that $f = 1$ on K. Then

$$\mu(K) \le \int_{\mathbb{R}^N} f d\mu = \lim \int_{\mathbb{R}^N} f d\mu_n \le \liminf \mu_n(\mathcal{O}).$$

This proves the first of (2.1c). For the second having fixed $\mathcal{O}$ take any compact set $K \subset \mathcal{O}$ and construct a similar f. Then

$$\mu(\mathcal{O}) \ge \int_{\mathbb{R}^N} f d\mu = \lim \int_{\mathbb{R}^N} f d\mu_n \ge \limsup \mu_n(K). \qquad \blacksquare$$

Proof $(2.2c) \Longrightarrow (2.3c)$ Let $E \subset \mathbb{R}^N$ be a bounded Borel set such that $\mu(\partial E) = 0$. Then $\mu(E) < \infty$ and denoting by $\overset{\circ}{E}$ its interior

$$\mu(E) = \mu\left(\overset{\circ}{E}\right) \le \liminf \mu_n\left(\overset{\circ}{E}\right) \le \limsup \mu_n\left(\bar{E}\right) \le \mu\left(\bar{E}\right) = \mu(E). \qquad \blacksquare$$

Proof $(2.3c) \Longrightarrow (2.1c)$ Let $f \in C_o(\mathbb{R}^N)$ be nonnegative and supported in some ball B_ρ centered at the origin and radius ρ. Having fixed $\varepsilon > 0$ select a finite sequence

$$\begin{array}{l} 0 = s_o < s_1 < \cdots < s_{k-1} < s_k = \sup f + 1, \quad \text{such that} \\ s_j - s_{j-1} < \varepsilon \quad \text{and} \quad \mu[f = s_j] = 0 \quad \text{for} \quad j = 1, \dots, k. \end{array} \tag{2.4c}$$

Setting

$$E_j = [s_{j-1} < f \le s_j] \quad \text{for} \quad j = 1, \dots, k$$

estimate for all n

$$\sum_{j=2}^{k} s_{j-1} \mu_n(E_j) \le \int_{\mathbb{R}^N} f d\mu_n \le \sum_{j=2}^{k} s_j \mu_n(E_j) + s_1 \mu_n(B_\rho)$$

and

$$\sum_{j=2}^{k} s_{j-1} \mu(E_j) \le \int_{\mathbb{R}^N} f d\mu \le \sum_{j=2}^{k} s_j \mu(E_j) + s_1 \mu(B_\rho).$$

By the construction of $\{s_j\}$ the sets E_j are Borel sets and $\mu(\partial E_j) = 0$. Then letting $n \to \infty$ and using (2.3c) these inequalities yield

$$\left| \int_{\mathbb{R}^N} f\, d\mu_n - \int_{\mathbb{R}^N} f\, d\mu \right| \leq 2\varepsilon \mu(B_\rho).$$
∎

2.1. Prove that $C_o(\mathbb{R}^N)$ is separable.

2.2. Prove that $C_o(\mathbb{R}^N)$ is not reflexive.

2.3. Prove that the construction in (2.4c) can be effected.

2.4. **Weak* Sequential Compactness:** The space $C_o(\mathbb{R}^N)$, and hence $C_o(\mathbb{R}^N)^*$, is not reflexive. Therefore Proposition 14.2 of Chap. 7 does not apply. Nevertheless a similar statement continues to hold for sequences of measures $\{\mu_n\} \subset C_o(\mathbb{R}^N)^*$.

Proposition 2.2c *Let* $\{\mu_n\} \subset C_o(\mathbb{R}^N)^*$ *be a sequence of measures equibounded on compact sets, i.e., for every compact set* $K \subset \mathbb{R}^N$ *there exists a constant* C_K *such that*

$$\mu_n(K) \leq C_K \quad \text{for all } n \in \mathbb{N}. \tag{2.5c}$$

Then, there exists a subsequence $\{\mu_{n'}\} \subset \{\mu_n\}$ *and measure* $\mu \in C_o(\mathbb{R}^N)^*$ *such that* $\{\mu_{n'}\} \to \mu$ *in the weak* topology.*

Proof Assume first the $\{\mu_n\}$ is uniformly bounded in $\mathbb{R}^N$, i.e., that (2.5c) holds with K replaced by $\mathbb{R}^N$. Then mimic the proof of Proposition 14.2 of Chap. 7 or Proposition 16.1 of Chap. 6. ∎

3c Calculus with Distributions[2]

3.1. If $u \in C^1(\mathbb{R})$ there holds the differentiation formula

$$(u^3)' = 3u^2 u' \quad \text{in } \mathbb{R}.$$

This formula is no longer valid if u is a distribution in $\mathbb{R}$ or even a function in $L^1_{\text{loc}}(\mathbb{R})$, even if both sides of this formula are well defined. The left-hand side is well defined in $\mathcal{D}'(\mathbb{R})$ if $u \in L^3_{\text{loc}}(\mathbb{R})$ and the right-hand side is well defined if $u^2 \in C^\infty(\mathbb{R})$. Even under these more restrictive condition the indicated differentiation formula is false in $\mathcal{D}'(\mathbb{R})$. Consider for example

$$u = \text{sign}\, x, \quad \text{so that} \quad u^2 = 1 \in C^\infty(\mathbb{R}), \quad \text{and} \quad u^3 = u.$$

---

[2]Most of the problems in Sections 3c-6c, were provided by U. Gianazza and V. Vespri.

Then compute in $\mathcal{D}'(\mathbb{R})$

$$(u^3)' = 2\delta_o, \quad \text{and} \quad 3u^2 u' = 6\delta_0.$$

3.2. The function $u(x) = x^{-1}$ is measurable but not locally integrable. Prove that the limit

$$\lim_{\varepsilon \to 0} \frac{1}{x} \chi_{|x| \geq \varepsilon} = PV\left(\frac{1}{x}\right) \text{ in } \mathcal{D}'(\mathbb{R})$$

defines a distribution in $\mathbb{R}$ called the Cauchy Principal Value of x^{-1}. Show that

$$PV\left(\frac{1}{x}\right) = (\ln |x|)' \quad \text{in } \mathcal{D}'(\mathbb{R}).$$

3.3. Compute the limit

$$\lim_{\varepsilon \to 0} \left(\frac{1}{x} \chi_{[\varepsilon,\infty)} + (\ln \varepsilon)\delta_o\right) = \left(H(x) \ln |x|\right)' \quad \text{in } \mathcal{D}'(\mathbb{R}).$$

3.4. The function $u(x) = x^{-2}$ is measurable but not locally integrable. Prove that the limit

$$\lim_{\varepsilon \to 0} \left(\frac{1}{x^2} \chi_{|x| \geq \varepsilon} - \frac{2}{\varepsilon}\delta_o\right) = FP\left(\frac{1}{x^2}\right) \text{ in } \mathcal{D}'(\mathbb{R})$$

defines a distribution in $\mathbb{R}$ called the Finite Part of x^{-2}. Show that

$$FP\left(\frac{1}{x^2}\right) = -PV\left(\frac{1}{x}\right)' = -(\ln |x|)'' \quad \text{in } \mathcal{D}'(\mathbb{R}).$$

Verify that

$$x PV\left(\frac{1}{x}\right) = x^2 FP\left(\frac{1}{x^2}\right) = 1 \quad \text{in } \mathcal{D}'(\mathbb{R}).$$

3.5. Prove that

$$\lim_{\varepsilon \to 0} \frac{1}{x \pm i\varepsilon} = PV\left(\frac{1}{x}\right) \mp i\frac{\pi}{2}\delta_o \quad \text{in } \mathcal{D}'(\mathbb{R}).$$

Hint:

$$\left(\frac{1}{x \pm i\varepsilon}, \varphi\right) = \int_{\mathbb{R}} \frac{x \mp i\varepsilon}{x^2 + \varepsilon^2} \varphi dx.$$

3.6. Let $\ln z$ and $\sin z$ be the holomorphic branches of the homologous maps, defined in the complex plane $\mathbb{C}$ from which the closed, negative imaginary semi-axis has been removed. Compute the limits

(a) $\lim \ln \left(x + \dfrac{i}{n}\right) = \ln |x| + i\pi H(-x);$

(b) $\lim \sin \sqrt{x + \dfrac{i}{n}} = \begin{cases} \sin x & \text{if } x > 0; \\ \sin(-ix) & \text{if } x < 0. \end{cases}$

in $\mathcal{D}'(\mathbb{R})$.

3.7. A distribution $T \in \mathcal{D}'(\mathbb{R}^N)$ is homogeneous of order $\lambda \in \mathbb{R}$ if

$$t^\lambda \langle T, \varphi \rangle = t^{-N} \left\langle T, \varphi\left(\dfrac{\cdot}{t}\right) \right\rangle \quad \text{for all } t > 0 \quad \text{and all } \varphi \in C_o^\infty(\mathbb{R}^N).$$

   i. Verify that this definition coincides with the classical one whenever $T \in C(\mathbb{R}^N)$.

  ii. Prove that the distributions $T_\pm = [\ln |x| \, H(\pm x)]'$ are homogeneous of order -1 in $\mathbb{R}$.

 iii. If $T \in \mathcal{D}'(\mathbb{R})$ is homogeneous of order λ, then its T' is homogeneous of order $\lambda - 1$. The converse is false.

3.8. Let $\{c_n\}$ be a sequence of real numbers. Prove that

$$\sum c_n \delta_{\frac{1}{n^2}} \in \mathcal{D}'(\mathbb{R}^N) \quad \Longleftrightarrow \quad \sum c_n < \infty.$$

3.9. Given an interval $(a, b) \subset \mathbb{R}$ consider a partition

$$\mathcal{P} = \{a = x_o < x_1 < \cdots < x_n = b\}.$$

Given n functions $f_j \in C^2[x_{j-1}, x_j]$ for $j = 1, \ldots, n$ introduce the piecewise continuous function

$$f = f_j \quad \text{in } [x_{j-1}, x_j) \quad \text{for } j = 1, \ldots, n.$$

Compute f' and f'' in $\mathcal{D}'(a, b)$. Give conditions on the functions f_j for $f' \in L^1_{\text{loc}}(a, b)$. Similarly give conditions on the f_j for $f'' \in L^1_{\text{loc}}(a, b)$.

3.10. Let $f \in BV[a, b]$. Compute f' in $\mathcal{D}'(a, b)$. Give conditions on f for $f' \in L^1_{\text{loc}}(a, b)$. Note that the distributional derivative of f need not coincide with the a.e. derivative of f. **Hint:** Use the function of the jumps introduced in § 1.1c and **3.4.** of Chap. 5.

3.11. Let $E \subset \mathbb{R}^2$ be bounded with boundary ∂E, locally representable by a smooth curve γ. Compute $D_x \chi_E$ and $D_y \chi_E$.

3.12. Compute

$$\Delta(1 - |x|^2)_+ \quad \text{in } \mathcal{D}'(\mathbb{R}^N).$$

3.13. Let E be a domain in $\mathbb{R}^N$ with smooth boundary ∂E, and let $f \in C^2(E) \cap C(\bar{E})$ vanish on ∂E. Regard f as defined in the whole $\mathbb{R}^N$ by extending it to be zero outside E, and compute Δf in $\mathcal{D}'(\mathbb{R}^N)$.

4c   Limits in $\mathcal{D}'$

4.1. Prove that
$$\lim \frac{\sin nx}{\pi x} = \delta_o \quad \text{in } \mathcal{D}'(\mathbb{R}).$$

4.2. Prove that
$$\lim \frac{2}{\pi} \arctan \frac{n}{x} = \operatorname{sign} x \quad \text{in } \mathcal{D}'(\mathbb{R})$$

and indeed also in $L^1_{\text{loc}}(\mathbb{R})$.

4.3. Prove that in $\mathcal{D}'(\mathbb{R}^N)$,
$$\lim \frac{n^\alpha}{\pi^{N/2}} \exp\{-n^2|x|^2\} = \begin{cases} 0 \text{ if } \alpha < N; \\[2mm] \delta_o \text{ if } \alpha = N. \end{cases}$$

For $\alpha > N$, the distributional limit does not exists, whereas the a.e. limit exists and is zero, and the $L^1(\mathbb{R}^N)$-limit exists and is ∞.

4.4. Prove that in $\mathcal{D}'(\mathbb{R})$,
$$\lim n^\alpha \exp\{-n^2|y|\} = \begin{cases} 0 & \text{if } \alpha < 2; \\[2mm] 2\delta_o & \text{if } \alpha = 2. \end{cases}$$

For $\alpha > 2$ the distributional limit does not exist.

4.5. Prove that
$$\lim \frac{n}{1 + n^2 x^2} = \pi \delta_o \quad \text{in } \mathcal{D}'(\mathbb{R}).$$

4.6. Let $f_n : (0, 1) \to \mathbb{R}$, for $n = 2, 3, \ldots$, be defined by ([8], p. 661)
$$f_n(x) = \begin{cases} \dfrac{n^2}{2} & \text{for } x \in \bigcup_{j=1}^{n} \left(\dfrac{j}{n+1} - \dfrac{1}{n^3}, \dfrac{j}{n+1} + \dfrac{1}{n^3} \right) \\ 0 & \text{otherwise.} \end{cases}$$

Verify that:

i. The intervals where $f_n > 0$ do not overlap;

ii. $\|f_n\|_1 = 1$ for all $n = 2, 3, \ldots$;

iii. The measure of the set $[f_n > 0]$ is $2/n^2$.

Therefore $\{f_n\} \to 0$ in measure but not in $L^1(0, 1)$. However $\{f_n\} \to 1$ in $\mathcal{D}'(0, 1)$. Indeed for $\varphi \in C_o^\infty(0, 1)$ and the mean value theorem,

$$\int_0^1 f_n \varphi dx = \sum_{j=1}^n \int_{\frac{j}{n+1} - \frac{1}{n^3}}^{\frac{j}{n+1} + \frac{1}{n^3}} \frac{n^2}{2} \varphi(x) dx$$

$$= \frac{n^2}{2} \sum_{j=1}^n \frac{2}{n^3} \varphi(y_{j,n}) = \frac{1}{n} \sum_{j=1}^n \varphi(y_{j,n})$$

for some points

$$y_{j,n} \in \left(\frac{j}{n+1} - \frac{1}{n^3}, \frac{j}{n+1} + \frac{1}{n^3} \right).$$

Now as $n \to \infty$

$$\lim \int_0^1 f_n \varphi dx = \lim \frac{1}{n} \sum_{j=1}^n \varphi(y_{j,n}) = \int_0^1 \varphi dx.$$

Prove that $\{f_n\} \to 0$ a.e. in $(0, 1)$. Thus distributional limits in general do not coincide with a.e. limits nor limits in measure. Compare with the example in Remark 4.2 of Chap. 4, and **10.13** of the Complements of Chap. 4.

4.7. Consider the sequence

$$u_n = \begin{cases} \dfrac{n}{2} + \dfrac{n^2}{2}x & \text{for} & -\dfrac{1}{n} < x < 0; \\[2mm] \dfrac{n}{2} - \dfrac{n^2}{2}x & \text{for} & 0 \le x < \dfrac{1}{n}; \\[2mm] 0 & \text{otherwise} \end{cases}$$

Prove that $\|u\|_1 = \frac{1}{2}$ uniformly in n and that

$$\lim u_n = \tfrac{1}{2}\delta_o \quad \text{in } \mathcal{D}'(\mathbb{R}).$$

4.8. Consider the sequence

$$v_n = \begin{cases} \dfrac{n^2}{2} & \text{for} & -\dfrac{1}{n} < x < 0; \\[2mm] -\dfrac{n^2}{2} & \text{for} & 0 \le x < \dfrac{1}{n}. \\[2mm] 0 & \text{otherwise} \end{cases}$$

Prove that

$$\lim v_n = \tfrac{1}{2}\delta_o' \quad \text{in } \mathcal{D}'(\mathbb{R}).$$

Verify that $v_n = u'_n$ in $\mathcal{D}'(\mathbb{R})$ and that

$$\lim u'_n = (\lim u_n)' = \lim v_n, \quad \text{in } \mathcal{D}'(\mathbb{R}).$$

4.9. Prove that if $\{T_n\} \to T$ in $\mathcal{D}'(E)$ then also $\{D^\alpha T_n\} \to D^\alpha T$ in $\mathcal{D}'(E)$ for every multi-index α.

4.10. Compute in $\mathcal{D}'(\mathbb{R})$

$$\lim \left(2n^3 x e^{-(nx)^2} + \sin n^2 x\right) = -\sqrt{\pi}\delta'_o.$$

Hint: $2n^3 x e^{-(nx)^2} = -[n e^{-(nx)^2}]'$.

4.11. Compute the distributional limits

$$\lim \arctan x^{2n} = \frac{\pi}{2} \begin{cases} 0 & \text{for } |x| < 1; \\ \frac{1}{2} & \text{for } x = \pm 1; \\ 1 & \text{for } |x| > 1. \end{cases}$$

$$\lim \frac{x^{2n+1}}{1 + |x|^{2n+1}} = \begin{cases} 0 & \text{for } |x| < 1; \\ \pm\frac{1}{2} & \text{for } |x| < 1; \\ \operatorname{sign} x & \text{for } |x| > 1. \end{cases}$$

Compute the pointwise limits and show that they are different from the distributional limits.

4.12. Prove that sequence

$$\left\{\frac{n^2}{1 + n^2 x^2}\right\} \quad \text{has no limit in } \mathcal{D}'(\mathbb{R}).$$

4.13. Let $f \in C(\mathbb{R}) \cap L^1(\mathbb{R})$ be nonnegative, and not identically zero, and set

$$f_n(x) = n^\alpha f(n^\beta x) \quad \text{for parameters } \alpha, \beta \in \mathbb{R}.$$

Verify the distributional limit

$$\lim f_n = \begin{cases} 0 & \text{if } \alpha < \beta; \\ 0 & \text{if } \alpha = \beta < 0; \\ f & \text{if } \alpha = \beta = 0; \\ \|f\|_1 \delta_o & \text{if } \alpha = \beta > 0. \end{cases} \quad \text{in } \mathcal{D}'(\mathbb{R}).$$

The distributional limit does not exists if $\alpha > \beta$.

4.14. Prove that

$$\lim n^2 \chi_{\left(-\frac{1}{n}, \frac{1}{n}\right)} \sin 2n\pi x = -\frac{8}{\pi}\delta'_o \quad \text{in } \mathcal{D}'(\mathbb{R}).$$

4.15. Prove that
$$\lim n(4nx - n^2x^2 - 3)_+ = \tfrac{4}{3}\delta_o \quad \text{in } \mathcal{D}'(\mathbb{R}).$$

4.16. Compute the limit

$$\lim \frac{2nx^{2n-1}}{1 + x^{4n}} = \tfrac{1}{2}\pi\big(\delta_1 - \delta_{-1}\big) \quad \text{in } \mathcal{D}'(\mathbb{R}).$$

Hint: The expression is the derivative of $\arctan x^{2n}$. See **4.11**.

4.17. Prove that for every positive integer k

$$\lim n^k \sin nx = \lim n^k \cos nx = 0, \quad \text{in } \mathcal{D}'(\mathbb{R}).$$

Hint:
$$\cos nx = \left(\frac{\sin nx}{n}\right)', \qquad \sin nx = -\left(\frac{\cos nx}{n}\right)'.$$

4.18. Compute the limits in $\mathcal{D}'(\mathbb{R})$ of the sequences

$$\{\sin^2 nx\}, \quad \{\cos^2 nx\}, \quad \{\sin^k nx\}, \quad \{\cos^k nx\}$$

for a positive integer k.

4.19. Compute the limit

$$\lim \frac{n^2 x}{1 + n^2 x^2} = (\ln |x|)' \quad \text{in } \mathcal{D}'(\mathbb{R}).$$

5c Algebraic Equations in $\mathcal{D}'$

5.1. Find all the solutions of the "algebraic" equation

$$xu - u = \sin 3(x - 1) \quad \text{in } \mathcal{D}'(\mathbb{R}).$$

The associated homogeneous equation is $(x - 1)u = 0$, whose solutions are $\gamma\delta_1$ for an arbitrary contant γ. The non-homogeneous equation is then

$$v = \frac{\sin 3(x - 1)}{x - 1} \in L^1_{\text{loc}}(\mathbb{R}).$$

Thus all solutions are
$$u = \gamma\delta_1 + \frac{\sin 3(x - 1)}{x - 1}.$$

5.2. Find all distributional solutions of

$$(x^2 - 1)u = \delta_o \quad \text{in } \mathcal{D}'(\mathbb{R}).$$

The associated homogeneous equation is solved by a linear combination of $\delta_{\pm 1}$. The full equation is then solved by

$$u = \gamma_1 \delta_1 + \gamma_{-1} \delta_{-1} - \delta_o.$$

5.3. Find all distributional solutions of

$$xu = \delta_o'' \quad \text{in } \mathcal{D}'(\mathbb{R}).$$

The homogeneous equation is solved by $\gamma \delta_o$ for an arbitrary constant γ. To solve the non-homogeneous equation observe that $x\delta_o$ is the zero distribution. From this by repeated distributional differentiation

$$(j + 1)\delta_o^j + x\delta^{j+1} = 0 \quad \text{in } \mathcal{D}'(\mathbb{R}).$$

For $j = 2$ this permits one to find all solutions

$$u = \gamma \delta_o - \tfrac{1}{3}\delta_o'''.$$

5.4. Find all distributional solutions of

$$x^2 u = x \quad \text{in } \mathcal{D}'(\mathbb{R}).$$

The homogeneous equation is solved

$$u_o = \gamma_o \delta_o + \gamma_1 \delta_o'$$

for arbitrary constants γ_o and γ_1. The non-homogeneous equation is solved by

$$u = \gamma_o \delta_o + \gamma_1 \delta_o' + PV\left(\frac{1}{x}\right) \quad \text{in } \mathcal{D}'(\mathbb{R}).$$

5.5. Solve the "algebraic" distributional equation

$$\sin x^3 u = \delta_o \quad \text{in } \mathcal{D}'(\mathbb{R}).$$

The associated homogeneous equation has solution

$$u_o = \gamma_o \delta_o + \gamma_1 \delta_o' + \gamma_2 \delta_o'' + \sum_{j \in \mathbb{N}} c_j^{\pm} \delta_{\pm \sqrt[3]{j\pi}}$$

for constants γ_i for $i = 0, 1, 2$ and $c_j^{\pm}$. Prove that a particular solution of the non-homogenous equation is

$$v = -\tfrac{1}{6}\delta_o''' \quad \text{in } \mathcal{D}'(\mathbb{R})$$

5.6. Solve the "algebraic" distributional equation

$$(\cos x)\, u = \sin \frac{1}{|x|} + \frac{\pi}{2}\left(1 - \chi_{(-1,1)}\right).$$

Setting $x_j = \frac{\pi}{2} + j\pi$, the solution is

$$u = \sum_{j\in\mathbb{Z}} c_j\delta_{x_j} + PV\left(\frac{1}{\cos x}\sin\frac{1}{x}\right) + \frac{\pi}{2}PV\left(\frac{1}{\cos x}\left(1 - \chi_{(-1,1)}\right)\right).$$

5.7. Compute the limit

$$\lim \frac{1 - \cos nx}{x} = PV\left(\frac{1}{x}\right) + \gamma\delta_o, \quad \text{in } \mathcal{D}'(\mathbb{R})$$

for an arbitrary constant γ. **Hint:** Naming u_n the argument of the limit, observe that $\{xu_n\} \to 1$ in $\mathcal{D}'(\mathbb{R})$. Then solve $xu = 1$ in $\mathcal{D}'(\mathbb{R})$.

6c   Differential Equations in $\mathcal{D}'$

6.1. Solve the differential equation in $\mathcal{D}'(\mathbb{R})$

$$(x^2 u'')' = \pi.$$

A first integration in $\mathcal{D}'(\mathbb{R})$ gives

$$x^2 u'' = \pi x + \gamma_o$$

for an arbitrary constant γ_o. Proceeding as before, i.e., considering this first as an "algebraic" equation in $\mathcal{D}'(\mathbb{R})$ one has

$$u'' = \pi PV\left(\frac{1}{x}\right) + +\gamma_o FP\left(\frac{1}{x^2}\right) + \gamma_o\delta_o + \gamma_1\delta_o'$$

for arbitrary constants γ_o, and γ_1. From this

$$u' = \pi \ln |x| - \gamma_o PV\left(\frac{1}{x}\right) + \gamma_o H(x) + \gamma_1 \delta_o + \gamma_2;$$

$$u = \pi x \ln |x| - \gamma_o \ln |x| + \gamma_o x H(x) + \gamma_1 H(x) + \gamma_2 x + \gamma_3,$$

for arbitrary constants γ_i, for $i = 0, 1, 2, 3$.

6.2. Solve the differential equation

$$(x^2 - 4)u'' = \sum_{j \in \mathbb{Z}} \delta'_j \quad \text{in } \mathcal{D}'(\mathbb{R}).$$

setting $u'' = v$ solve first the associated "algebraic" equation in v. The corresponding homogeneous equation for v is solved by a linear combination of $\delta_{\pm 2}$. A particular solution of the non-homogeneous equation for v is the sum of v_j where

$$(x^2 - 4)v_j = \delta'_j \quad \text{in } \mathcal{D}'(\mathbb{R}).$$

For $j \neq \pm 2$ one computes

$$\langle v_j, \varphi \rangle = \left\langle \frac{1}{x^2 - 4} \delta'_j, \varphi \right\rangle = \left\langle \frac{2j}{(j^2 - 4)^2} \delta_j + \frac{1}{j^2 - 4} \delta'_j, \varphi \right\rangle.$$

For $j = 2$ the non-homogeneous equation of v_2 reduces to

$$(x - 2)v_2 = \frac{1}{x + 2} \delta'_2 = -\frac{1}{16} \delta_2 + \frac{1}{4} \delta'_2 \quad \text{in } \mathcal{D}'(\mathbb{R}).$$

From $(x - 2)\delta_2 = 0$ compute by double differentiation

$$(x - 2)\delta'_2 + \delta_2 = 0;$$
$$(x - 2)\delta''_2 + 2\delta'_2 = 0.$$

From this compute

$$-\tfrac{1}{16}\delta_2 + \tfrac{1}{4}\delta'_2 = (x - 2)\left(\tfrac{1}{16}\delta'_2 - \tfrac{1}{8}\delta''_2\right).$$

Hence

$$v_2 = \tfrac{1}{16}\delta'_2 - \tfrac{1}{8}\delta''_2.$$

Similarly one computes

$$v_{-2} = \tfrac{1}{8}\delta''_{-2} + \tfrac{1}{16}\delta'_{-2}.$$

Combining these remarks gives

$$v = \gamma_2 \delta_2 + \gamma_{-2}\delta_{-2} + \tfrac{1}{16}\delta_2' - \tfrac{1}{8}\delta_2'' + \tfrac{1}{16}\delta_{-2}' + \tfrac{1}{8}\delta_{-2}''$$
$$+ \sum_{\substack{j\in\mathbb{Z}\\j\neq\pm 2}} \left(\frac{2j}{(j^2-4)^2}\delta_j + \frac{1}{j^2-4}\delta_j' \right).$$

Finally, by double distributional integration

$$u = \gamma_o + \gamma_1 x + \gamma_2(x-2)H(x-2) + \gamma_{-2}(x+2)H(x+2)$$
$$- \tfrac{1}{8}(\delta_2 - \delta_{-2}) + \tfrac{1}{16}\left(H(x-2) + H(x+2)\right)$$
$$+ \sum_{\substack{j\in\mathbb{Z}\\j\neq\pm 2}} \left(\frac{2j}{(j^2-4)^2}(x-j)H(x-j) + \frac{1}{j^2-4}H(x-j) \right).$$

6.3. Solve the differential equations in $\mathcal{D}'(\mathbb{R})$

(**i**) $u'' + u = \delta_o$:   (**ii**) $u'' + u = \delta_o'$;   (**iii**) $u''' + u = \delta_o''$.

Seek solutions of the form

(**i**) $u = v + \gamma_1|x|$:   (**ii**) $u = v + \gamma_2 H(x)$;   (**iii**) $u = v + \gamma_3 H(x)$

for constants γ_1, γ_2 and γ_3 to be chosen. This recasts the problems in terms of v solutions of

(**i**) $v'' + v = -\tfrac{1}{2}\operatorname{sign} x$;   (**ii**) $v'' + v = -\tfrac{1}{2}|x|$;   (**iii**) $v''' + v = -H(x)$.

These can be solved by classical methods since the right-hand sides are bounded functions.

6.4. Solve $(xu')' = 0$ in $\mathcal{D}'(\mathbb{R})$. Establish first that the differential equation implies

$$u' = \gamma_1 PV\left(\frac{1}{x}\right) + \gamma_2\delta_o \quad \text{in } \mathcal{D}'(\mathbb{R}).$$

7c   Miscellaneous Problems

7.1. Characterize $C(\bar{E})^*$
7.2. Positive distributions on E are identified by Radon measures.
7.3. Let $AC(a, b)$ denote the space of absolutely continuous functions in (a, b). A function $u \in W^{1,p}(a, b)$ if and only if $u \in L^p(a, b) \cap AC(a, b)$ and $u' \in L^p(a, b)$.

7.4. Introduce the notion of ∂E of class C^m for some positive integer m. The extension Proposition 18.1 is a special case of

Proposition 7.1c *Let E be a bounded domain in $\mathbb{R}^N$ with boundary ∂E of class C^m for some positive integer m. Every function $u \in C^m(\bar{E})$ admits an extension $w \in C_o^m(\mathbb{R}^N)$ such that $w = u$ in $\bar{E}$ and*

$$\|w\|_{m,p;\mathbb{R}^N} \le C \|u\|_{m,p;E}$$

where C is a constant depending only upon N, m, p and $\|\|\partial E\|\|_m$.

Proof If $E = \{x_N > 0\}$, extend u in $\{x_N \le 0\}$ by

$$\tilde{u}(\bar{x}, x_N) = \sum_{j=1}^{m+1} c_j u\left(\bar{x}, -\frac{1}{j}x_N\right) \qquad \text{where} \qquad \sum_{j=1}^{m+1} (-1)^h \frac{c_j}{j^h} = 1. \qquad \blacksquare$$

7.5. Let E be the unit disc in $\mathbb{R}^2$ from which a diameter has been removed. There exists $u \in W^{1,2}(E)$ that cannot be approximated, in $W^{1,2}(E)$ by functions in $C^1(\bar{E})$.

7.6. Approximating Characteristic Functions of Some Sets by Functions in C_o^∞: Let $J : \mathbb{R} \to \mathbb{R}^+$ be defined by

$$J(x) = \begin{cases} e^{-\frac{1}{x}} & \text{if } x > 0; \\ 0 & \text{otherwise}; \end{cases}$$

The function $J \in C^\infty(\mathbb{R})$ and $J(x) \to 1$ as $x \to \infty$. Consider now the sets

$$(a, b) \subset \mathbb{R}; \qquad R = \prod_{i=1}^N (a_i, b_i); \qquad D = [|x| < R]$$

and the sequences

$$J_{(a,b);n}(x) = J\left[n(b - x)(a - x)\right] \in C_o^\infty(\mathbb{R});$$

$$J_{R;n}(x) = \prod_{i=1}^N J\left[n(b_i - x_i)(a_i - x_i)\right] \in C_o^\infty(\mathbb{R}^N);$$

$$J_{D;n}(x) = J\left[n(R^2 - |x|^2)\right] \in C_o^\infty(\mathbb{R}^N).$$

Prove that as $n \to \infty$ they tend pointwise to the characteristic function of the interval (a, b), the rectangle R and the disc D respectively.

Chapter 9
Topics on Integrable Functions
of Real Variables

1 A Vitali-Type Covering

Let μ be the Lebesgue measure in $\mathbb{R}^N$ and refer the notions of measurability and integrability to such a measure. Let $E \subset \mathbb{R}^N$ be measurable and of finite measure, and let $\mathcal{F}$ be a collection of cubes in $\mathbb{R}^N$, with faces parallel to the coordinate planes whose union covers E. Such a covering is a Vitali-type covering. The cubes making up $\mathcal{F}$ are not required to be open or closed.

The next theorem asserts that E can be covered, in a measure theoretical sense, by a countable collections of pairwise disjoint cubes in $\mathcal{F}$. The key feature of this theorem is that, unlike the Vitali, or the Besicovitch measure theoretical covering Theorems, the covering $\mathcal{F}$ is not required to be fine (see § 17 and 18 of Chap. 3).

This limited information on the covering $\mathcal{F}$ results in a weaker covering, that is, the measure theoretical covering is realized through an *estimation*, rather than equality of the measure of the set E in terms of the measure of the selected cubes.

Theorem 1.1 (Wiener [175]) *Let $E \subset \mathbb{R}^N$ be of finite measure and let $\mathcal{F}$ be a Vitali-type covering for E. There exists a countable collection $\{Q_n\}$ of pairwise disjoint cubes in $\mathcal{F}$, such that*

$$\mu(E) \leq 5^N \sum \mu(Q_n). \tag{1.1}$$

Proof Label $\mathcal{F}$ by $\mathcal{F}_1$ and set

$$2\rho_1 = \{\text{the supremum of the edges of cubes in } \mathcal{F}_1\}.$$

If $\rho_1 = \infty$, we select a cube Q of edge so large that

$$\mu(E) \leq 5^N \mu(Q).$$

© Springer Science+Business Media New York 2016
E. DiBenedetto, *Real Analysis*, Birkhäuser Advanced
Texts Basler Lehrbücher, DOI 10.1007/978-1-4939-4005-9_9

If $\rho_1 < \infty$ select a cube $Q_1 \in \mathcal{F}_1$ of edge $\ell_1 > \rho_1$, and subdivide $\mathcal{F}_1$ into two subcollections $\mathcal{F}_2$ and $\mathcal{F}_2'$, by setting

$$\mathcal{F}_2 = \{\text{the collection of cubes } Q \in \mathcal{F}_1 \text{ that do not intersect } Q_1\};$$
$$\mathcal{F}_2' = \{\text{the collection of cubes } Q \in \mathcal{F}_1 \text{ that intersect } Q_1\}.$$

Denote by Q_1' the cube with the same center as Q_1 and edge $5\ell_1$. Then by construction

$$\bigcup \{Q \mid Q \in \mathcal{F}_2'\} \subset Q_1'.$$

If $\mathcal{F}_2$ is empty then

$$E \subset Q_1' \qquad \text{and} \qquad \mu(E) \leq 5^N \mu(Q_1).$$

If $\mathcal{F}_2$ is not empty, set

$$2\rho_2 = \{\text{the supremum of the edges of the cubes in } \mathcal{F}_2\}$$

and select a cube $Q_2 \in \mathcal{F}_2$ of edge $\ell_2 > \rho_2$. Then subdivide $\mathcal{F}_2$ into the two subcollections

$$\mathcal{F}_3 = \{\text{the collection of cubes } Q \in \mathcal{F}_2 \text{ that do not intersect } Q_2\};$$
$$\mathcal{F}_3' = \{\text{the collection of cubes } Q \in \mathcal{F}_2 \text{ that intersect } Q_2\}.$$

Denote by Q_2' the cube with the same center as Q_2 and edge $5\ell_2$. Then by construction

$$\bigcup \{Q \mid Q \in \mathcal{F}_3'\} \subset Q_2'.$$

If $\mathcal{F}_3$ is empty

$$E \subset Q_1' \cup Q_2' \qquad \text{and} \qquad \mu(E) \leq 5^N \big(\mu(Q_1) + \mu(Q_2)\big).$$

If $\mathcal{F}_3$ is not empty, we repeat the process to define inductively subfamilies of cubes $\{\mathcal{F}_n\}$, positive numbers $\{\rho_n\}$ and $\{\ell_n\}$, and cubes $\{Q_n\}$ and $\{Q_n'\}$, by the procedure,

$$\mathcal{F}_n = \{\text{collection of cubes in } \mathcal{F}_{n-1} \text{ that do not intersect } Q_{n-1}\};$$
$$2\rho_n = \{\text{the supremum of the edges of the cubes in } \mathcal{F}_n\};$$
$$Q_n = \{\text{a cube selected out of } \mathcal{F}_n \text{ of edge } \ell_n > \rho_n\};$$
$$Q_n' = \{\text{a cube with the same center as } Q_n \text{ and edge } 5\ell_n\}.$$

If $\mathcal{F}_{n+1}$ is empty for some $n \in \mathbb{N}$, then

$$E \subset Q_1' \cup \cdots \cup Q_n' \qquad \text{and} \qquad \mu(E) \leq 5^N \sum_{j=1}^{n} \mu(Q_j).$$

If $\mathcal{F}_n$ is not empty for all $n \in \mathbb{N}$, consider series $\sum \mu(Q_n)$. If the series diverges then (1.1) is trivial. If the series converges then $\rho_n \to 0$ as $n \to \infty$. In such a case we claim that every cube $Q \in \mathcal{F}$ belongs to some Q'_n. Indeed if not, Q must belong to all $\mathcal{F}_n$ and therefore the length of its edge is zero. Thus

$$ E \subset \bigcup Q'_n \quad \text{and} \quad \mu(E) \leq \sum \mu(Q'_n) = 5^N \sum \mu(Q_n). \quad \blacksquare $$

Remark 1.1 The theorem is more general in that the set E is not required to be measurable. The same conclusion continues to hold, by the same proof, with μ replaced by the Lebesgue outer measure μ_e.

Remark 1.2 The set E is not required to be bounded. It is not claimed here that the union of the disjoint cubes Q_n, satisfying (1.1), covers E.

Corollary 1.1 *Let $E \subset \mathbb{R}^N$ be measurable and of finite measure, and let $\mathcal{F}$ be a collection of cubes in $\mathbb{R}^N$, with faces parallel to the coordinate planes and covering E. For every $\varepsilon > 0$, there exists a finite collection $\{Q_1, \ldots, Q_m\}$ of disjoint cubes in $\mathcal{F}$, such that*

$$ \mu(E) - \varepsilon \leq 5^N \bigcup_{j=1}^{m} \mu(Q_j). \tag{1.2} $$

2 The Maximal Function (Hardy–Littlewood [69] and Wiener [175])

Let Q denote a cube centered at the origin and with faces parallel to the coordinate planes. For $x \in \mathbb{R}^N$, we let $Q(x)$ denote the cube centered at x and congruent to Q. The maximal function $M(f)$ of a function $f \in L^1_{\text{loc}}(\mathbb{R}^N)$ is defined by

$$ M(f)(x) = \sup_Q \frac{1}{\mu(Q)} \int_{Q(x)} |f(y)| dy. $$

From the definition it follows that $M(f)$ is nonnegative and sub-additive with respect to the argument f, i.e.,

$$ M(f + g) \leq M(f) + M(g). $$

Moreover $M(\alpha f) = |\alpha| M(f)$, for all $\alpha \in \mathbb{R}$.

Proposition 2.1 *$M(f)$ is measurable and lower semi-continuous. Moreover, if f is the characteristic function of a bounded measurable set E, there exists positive constants C_o, C_1, and γ, depending only upon E, such that*

$$\frac{C_o}{|x|^N} \le M(\chi_E)(x) \le \frac{C_1}{|x|^N} \quad \text{for all } |x| > \gamma. \tag{2.1}$$

Proof If $|f| = 0$, also $M(f) = 0$. Otherwise $M(f)(x) > 0$ for all $x \in \mathbb{R}^N$. Let $c > 0$ and $x \in [M(f) > c]$. There exist $\varepsilon > 0$ and a cube Q, such that

$$\frac{1}{\mu(Q)} \int_{Q(x)} |f| dx \ge M(f)(x) - \varepsilon > c.$$

By the absolute continuity of the integral, there exists $\delta > 0$ such that

$$M(f)(y) \ge \frac{1}{\mu(Q)} \int_{Q(y)} |f| dx > c \quad \text{for all } |y - x| < \delta.$$

Thus $[M(f) > c]$ is open and hence $M(f)$ is lower semi-continuous and measurable. To prove (2.1) observe that

$$M(\chi_E)(x) = \sup_Q \frac{\mu(E \cap Q(x))}{\mu(Q)}.$$

Since E is bounded is included in some cube Q_o. Let d be the maximum distance of points in Q_o from the origin. Fix $x \in \mathbb{R}^N$ such that $|x| \ge 2d/\sqrt{N}$ and let $Q(x)$ be the smallest cube centered at x and containing E. Then

$$M(\chi_E)(x) \ge \frac{\mu(E \cap Q(x))}{\mu(Q)} = \frac{\mu(E)}{\mu(Q)} = \frac{C_o}{|x|^N}.$$

The bound above is estimated analogously. ∎

Corollary 2.1 *Let $f \in L^1(\mathbb{R}^N)$ be of compact support in $\mathbb{R}^N$ and not identically zero. There exists positive constants C_o, C_1 and γ, depending only upon f, such that*

$$\frac{C_o}{|x|^N} \le M(f)(x) \le \frac{C_1}{|x|^N}, \quad \text{for all } |x| > \gamma. \tag{2.2}$$

It follows from (2.1) and (2.2) that $M(f)$ is not in $L^1(\mathbb{R}^N)$ even if f is bounded and compactly supported, unless $f = 0$.

Proposition 2.2 *Let $f \in L^1(\mathbb{R}^N)$. Then for all $t > 0$*

$$\mu([M(f) > t]) \le \frac{5^N}{t} \int_{\mathbb{R}^N} |f| dx. \tag{2.3}$$

Proof Assume first that f is of compact support. Then (2.2) implies that $[M(f) > t]$ is of finite measure. For every $x \in [M(f) > t]$, there exists a cube $Q(x)$ centered at x and with faces parallel to the coordinate planes such that

$$\mu(Q) \le \frac{1}{t} \int_{Q(x)} |f|\,dy. \qquad (2.4)$$

The collection $\mathcal{F}$ of all such cubes is a covering of $[M(f) > t]$. Out of $\mathcal{F}$ we may extract a countable collection $\{Q_n\}$ of disjoint cubes such that

$$\mu([M(f) > t]) \le 5^N \sum \mu(Q_n).$$

From this and (2.4),

$$\mu([M(f) > t]) \le 5^N \sum \mu(Q_n)$$
$$\le \frac{5^N}{t} \sum \int_{Q_n} |f|\,dx \le \frac{5^N}{t} \int_{\mathbb{R}^N} |f|\,dx.$$

If $f \in L^1(\mathbb{R}^N)$, we may assume that $f \ge 0$. Let $\{f_n\}$ be a nondecreasing sequence of compactly supported, nonnegative functions in $L^1(\mathbb{R}^N)$, converging to f a.e. in $\mathbb{R}^N$. Then $\{M(f_n)\}$ is a nondecreasing sequence of measurable functions, converging to $M(f)$ a.e. in $\mathbb{R}^N$. Therefore, by monotone convergence

$$\mu([M(f) > t]) = \lim \mu([M(f_n) > t])$$
$$\le \frac{5^N}{t} \lim \int_{\mathbb{R}^N} f_n\,dx \le \frac{5^N}{t} \int_{\mathbb{R}^N} f\,dx. \qquad \blacksquare$$

3 Strong L^p Estimates for the Maximal Function

Proposition 3.1 *Let $f \in L^p(\mathbb{R}^N)$ for some $p \in (1, \infty]$. Then the maximal function $M(f)$ is in $L^p(\mathbb{R}^N)$ and*

$$\|M(f)\|_p \le \gamma_p \|f\|_p \qquad \text{where } \gamma_p^p = \frac{2^p \, p 5^N}{p - 1}. \qquad (3.1)$$

Proof The estimate is obvious for $p = \infty$. Assuming then $p \in (1, \infty)$ fix $t > 0$ and set

$$g(x) = \begin{cases} f(x) & \text{if } |f(x)| \ge \frac{1}{2}t; \\ 0 & \text{if } |f(x)| < \frac{1}{2}t. \end{cases}$$

Such a function g is in $L^1(\mathbb{R}^N)$. Indeed,

$$\int_{\mathbb{R}^N} |g|\,dx = \int_{[|f| \ge \frac{1}{2}t]} |f|\,dx \le \left(\frac{2}{t}\right)^{p-1} \int_{\mathbb{R}^N} |f|^p\,dx.$$

Since $|f| \leq |g| + \frac{1}{2}t$

$$M(f)(x) \leq \sup \frac{1}{\mu(Q)} \int_{Q(x)} |g| dy + \frac{1}{2}t = M(g)(x) + \frac{1}{2}t.$$

Therefore

$$[M(f) > t] \subset \left[M(g) > \tfrac{1}{2}t \right].$$

By Proposition 2.2, applied to g and $M(g)$

$$\mu([M(f) > t]) \leq \mu\left(\left[M(g) > \tfrac{1}{2}t \right] \right) \leq \frac{2 \cdot 5^N}{t} \int_{\mathbb{R}^N} |g| dx$$

$$= \frac{2 \cdot 5^N}{t} \int_{[|f| \geq \frac{1}{2}t]} |f| dx.$$

Next express the integral of $M(f)^p$ in terms of the distribution function of $M(f)$ (§ 15.1 of Chap. 4), and in the integral so obtained interchange the order of integration by means of Fubini's theorem. This gives

$$\int_{\mathbb{R}^N} M(f)^p dx = p \int_0^\infty t^{p-1} \mu([M(f) > t]) dt$$

$$\leq 2 p 5^N \int_0^\infty t^{p-2} \left(\int_{[|f| \geq \frac{1}{2}t]} |f| dx \right) dt$$

$$= 2 p 5^N \int_{\mathbb{R}^N} |f| \left(\int_0^{2|f|} t^{p-2} dt \right) dx$$

$$= \frac{2^p p 5^N}{p - 1} \int_{\mathbb{R}^N} |f|^p dx. \qquad \blacksquare$$

3.1   Estimates of Weak and Strong Type

Let $E \subset \mathbb{R}^N$ be measurable. A measurable function $g : E \to \mathbb{R}$ is in the space weak-$L^1(E)$, denoted by $L_w^1(E)$, if there exists a constant C, depending only upon g, such that

$$\mu([|g| > t]) \leq \frac{C}{t} \qquad \text{for all } t > 0.$$

If $g \in L^1(E)$

$$\mu([|g| > t]) \leq \frac{1}{t} \int_E |g| dx.$$

Therefore $L^1(E) \subset L_w^1(E)$. The converse inclusion is false. The function

$$g(x) = \begin{cases} |x|^{-N} & \text{for } |x| > 0 \\ 0 & \text{for } |x| = 0 \end{cases}$$

is in weak-$L^1(\mathbb{R}^N)$ and not to $L^1(\mathbb{R}^N)$. By Proposition 2.2 the maximal function $M(f)$ of a function $f \in L^1(\mathbb{R}^N)$ is in weak-$L^1(\mathbb{R}^N)$ and, in general, not in $L^1(\mathbb{R}^N)$.

Let T be a map acting on $L^1(E)$ and such that $T(f)$ is a real-valued, measurable function defined on E. Examples include the maximal function $T(f) = M(f)$ and the convolution $T(f) = J_\varepsilon * f$ for a mollifying kernel J_ε.

A map T is of weak type in $L^1(E)$ if $T(f)$ is in weak-$L^1(E)$, for every $f \in L^1(E)$. By Proposition 2.2, the map $T(f) = M(f)$ is of weak type in $L^1(\mathbb{R}^N)$.

A map T is of strong type in $L^p(E)$ for some $1 \le p \le \infty$, if

$$f \in L^p(E) \implies T(f) \in L^p(E).$$

The convolution $T(f) = J_\varepsilon * f$ is of strong type in $L^p(\mathbb{R}^N)$ for all $p \in [1, \infty)$ (§. 18 and §18c of Chap. 6). By Proposition 3.1, the maximal function $T(f) = M(f)$ is of strong type in $L^p(E)$ for $p \in (1, \infty)$.

4 The Calderón–Zygmund Decomposition Theorem [20]

Theorem 4.1 *Let f be a nonnegative function in $L^1(\mathbb{R}^N)$. Then, for any fixed $\alpha > 0$, $\mathbb{R}^N$ can be partitioned into two disjoint sets E and F, such that*

(i) $f \le \alpha$ *a.e. in E;*

(ii) *F is the countable union of closed cubes Q_n, with faces parallel to the coordinate planes and with pairwise disjoint interior. For each of these cubes,*

$$\alpha < \frac{1}{\mu(Q_n)} \int_{Q_n} f(y)dy \le 2^N \alpha. \tag{4.1}$$

Proof Let $\alpha > 0$ be fixed and partition $\mathbb{R}^N$ into closed cubes with pairwise disjoint interior, with faces parallel to the coordinate planes, and of equal edge. Since $f \in L^1(\mathbb{R}^N)$, such a partition can be realized so that for every cube Q' of such a partition,

$$\frac{1}{\mu(Q')} \int_{Q'} f(y)dy \le \alpha.$$

Having fixed one such cube Q', we partition it into 2^N equal cubes, by bisecting Q' with hyperplanes parallel to the coordinate planes. Let Q'' be any one of these new cubes. Then either

$$\frac{1}{\mu(Q'')}\int_{Q''} f(y)dy \le \alpha \qquad\qquad \textbf{(a)}$$

or

$$\frac{1}{\mu(Q'')}\int_{Q''} f(y)dy > \alpha. \qquad\qquad \textbf{(b)}$$

If the second case occurs, then Q'' is not further subdivided and is taken as one of the cubes of the collection $\{Q_n\}$ claimed by the theorem. Indeed for such a cube

$$\alpha < \frac{1}{\mu(Q'')}\int_{Q''} f(y)dy \le \frac{2^N}{\mu(Q')}\int_{Q'} f(y)dy \le 2^N\alpha.$$

If (a) occurs, we subdivide further Q'' into 2^N sub-cubes and on each of then repeat the same alternative.

For each of the cubes Q' of the initial partition of $\mathbb{R}^N$, we carry on this recursive partitioning process. The process terminates only if case (b) occurs. Otherwise it is continued recursively.

Let $F = \bigcup Q_n$, where Q_n are cubes for which (b) occurs. By construction these are cubes with faces parallel to the coordinate planes, and with pairwise disjoint interior. Moreover (4.1) holds for all of them.

Setting $E = \mathbb{R}^N - F$, it remains to prove (i). Let x be a Lebesgue point of f in E. There exists a sequence of cubes $\tilde{Q}_j$ with faces parallel to the coordinate planes and containing x, resulting from the recursive partition such that

$$\lim_{j\to\infty} \operatorname{diam}\{\tilde{Q}_j\} = 0 \qquad \text{and} \qquad \frac{1}{\mu(\tilde{Q}_j)}\int_{\tilde{Q}_j} f(y)dy \le \alpha.$$

The collection of cubes $\{\tilde{Q}_j\}$ forms a regular family $\mathcal{F}_x$ at x. Therefore (§ 12 and Proposition 12.1 of Chap. 5)

$$f(x) = \lim_{j\to\infty} \frac{1}{\mu(\tilde{Q}_j)}\int_{\tilde{Q}_j} f(y)dy \le \alpha. \qquad\qquad \blacksquare$$

5 Functions of Bounded Mean Oscillation

Let Q_o be a cube in $\mathbb{R}^N$ centered at the origin and with faces parallel to the coordinate planes. For a function $f \in L^1_{\text{loc}}(Q_o)$ and a cube $Q \subset Q_o$ with faces parallel to the coordinate planes, let f_Q denote the integral average of f in Q

$$f_Q = \frac{1}{\mu(Q)}\int_Q f(y)dy.$$

A function $f \in L^1(Q_o)$ is of *bounded mean oscillation* if

$$|f|_o = \sup_{Q \subset Q_o} \frac{1}{\mu(Q)} \int_Q |f - f_Q| dy < \infty. \tag{5.1}$$

The collection of all $f \in L^1(Q_o)$ of bounded mean oscillation is denoted by $BMO(Q_o)$. One verifies that $BMO(Q_o)$ is a linear space and that

$$\|f\|_o = \|f\|_1 + |f|_o$$

defines a norm on $BMO(Q_o)$. Moreover, from the definition, and the completeness of $L^1(Q_o)$, it follows that $BMO(Q_o)$ is complete.

Theorem 5.1 (John-Nirenberg [77]) *There exist two positive constants C_1, C_2, depending only on N such that, for every $f \in BMO(Q_o)$, for all cubes $Q \in Q_o$, and all $t \geq 0$*

$$\mu([|f - f_Q| > t] \cap Q) \leq C_1 \exp\left\{-\frac{C_2 t}{|f|_o}\right\} \mu(Q). \tag{5.2}$$

5.1 Some Consequences of the John–Nirenberg Theorem

Proposition 5.1 *If $f \in BMO(Q_o)$ then for all cubes $Q \subset Q_o$*

$$f - f_Q \in L^p(Q) \quad and \quad f \in L^p(Q) \quad for\ all\ 1 \leq p < \infty.$$

Proof From (5.2)

$$\int_Q |f - f_Q|^p dy = p \int_0^\infty t^{p-1} \mu([|f - f_Q| > t] \cap Q) dt$$

$$\leq pC_1 \mu(Q) \int_0^\infty t^{p-1} \exp\left\{-\frac{C_2 t}{|f|_o}\right\} dt$$

$$\leq pC_1 \mu(Q) \left(\frac{|f|_o}{C_2}\right)^p \int_0^\infty t^{p-1} e^{-t} dt$$

$$\leq \gamma(N, p) |f|_o^p \mu(Q)$$

for a constant $\gamma(N, p)$ depending only upon N and p. This in turn implies that $f \in L^p(Q)$, since

$$\int_Q |f|^p dy \leq \gamma(p) \left(\int_Q |f - f_Q|^p dy + |f_Q|^p \mu(Q)\right). \qquad \blacksquare$$

Proposition 5.2 *Let $f \in L^1(Q_o)$ and assume that for every sub-cube $Q \subset Q_o$ there is a constant γ_Q such that*

$$\mu([|f - \gamma_Q| > t] \cap Q) \le \gamma_1 \exp\{-\gamma_2 t\}\mu(Q) \quad \text{for all } t > 0 \qquad (5.3)$$

for two given constants γ_1 and γ_2 independent of Q and t. Then f is of bounded mean oscillation in Q_o and $|f|_o \le 2\gamma_1/\gamma_2$.

Proof Assume first that $\gamma_Q = f_Q$. Then

$$\int_Q |f - f_Q| dy = \int_0^\infty \mu([|f - f_Q| > t] \cap Q) dt$$

$$\le \gamma_1 \mu(Q) \int_0^\infty e^{-\gamma_2 t} dt \le \frac{\gamma_1}{\gamma_2} \mu(Q)$$

for all sub-cubes $Q \subset Q_o$. Hence $f \in BMO(Q_o)$ and $|f|_o \le \gamma_1/\gamma_2$. For general γ_Q, by a similar argument

$$\sup_{Q \in Q_o} \frac{1}{\mu(Q)} \int_Q |f - \gamma_Q| dy \le \frac{\gamma_1}{\gamma_2}.$$

From this, for every $Q \subset Q_o$

$$\frac{1}{\mu(Q)} \int_Q |f - f_Q| dy \le \frac{2}{\mu(Q)} \int_Q |f - \gamma_Q| dy. \qquad \blacksquare$$

Corollary 5.1 *The inequalities (5.2) and (5.3) are necessary and sufficient for a function $f \in L^1(Q_o)$ to be of bounded mean oscillation in Q_o.*

Proposition 5.3 *The function $x \to \ln |x|$ is of bounded mean oscillation in the unit cube Q_o, centered at the origin of $\mathbb{R}^N$.*

Proof Having fixed $Q \in Q_o$, let ξ be the element of largest Euclidean length in Q and set $\gamma_Q = \ln |\xi|$. Then

$$\Sigma_t = \{x \in Q \mid |\ln|x| - \gamma_Q| > t\}$$

$$= \left\{x \in Q \mid \ln \frac{|\xi|}{|x|} > t\right\}$$

$$= \{x \in Q \mid |x| < |\xi|e^{-t}\}.$$

Let h be the edge of Q and denote by η the element of least Euclidean length in Q. If Σ_t is not empty, it must contain η. Hence

$$|\xi|e^{-t} \ge |\eta| \ge |\xi| - |\xi - \eta| \ge |\xi| - \sqrt{N}h$$

which implies

$$|\xi| \le \frac{\sqrt{N}h}{1 - e^{-t}}.$$

Therefore Σ_t is contained in the ball

$$|x| \le \frac{\sqrt{N}h}{e^t - 1}.$$

This implies

$$\mu(\Sigma_t) \le \frac{\omega_N}{N} \left(\frac{\sqrt{N}}{e^t - 1}\right)^N \mu(Q).$$

If $t \ge 1$ this gives (5.3) for $\gamma_2 = N$ and a suitable constant γ_1. If $0 < t < 1$, the inequality (5.3) is still satisfied by possibly modifying γ_1. ∎

Remark 5.1 A function $f \in L^\infty(Q_o)$ is in $BMO(Q_o)$. The converse is false as the function $x \to \ln|x|$ is in of bounded mean oscillation in the unit cube Q_o centered at the origin but is not bounded in Q_o.

Remark 5.2 The converse to Proposition 5.1 is false. Indeed the function $(-1, 1) \ni x \to f(x) = (\ln|x|)^2$ is in $L^p(-1, 1)$ for all $1 \le p < \infty$, but is not in $BMO[-1, 1]$.

6 Proof of the John–Nirenberg Theorem 5.1

Having fixed some cube $Q \subset Q_o$ we may assume, without loss of generality, that $Q = Q_o$. Also, by possibly replacing f with $f/\|f\|_o$ we may assume that $\|f\|_o = 1$. Set

$$f_o = \begin{cases} |f - f_{Q_o}| & \text{in } Q_o; \\ 0 & \text{otherwise.} \end{cases}$$

Since $f_o \in L^1(\mathbb{R}^N)$, having fixed some $\alpha > 1$, by the Calderón–Zygmund decomposition theorem there exists a countable collection of closed cubes $\{Q_n^1\}$, with faces parallel to the coordinate planes and with pairwise disjoint interior, such that

$$\alpha < \frac{1}{\mu(Q_n^1)} \int_{Q_n^1} |f - f_{Q_o}| dy \le 2^N \alpha \quad \text{and}$$

$$|f - f_{Q_o}| \le \alpha \quad \text{a.e. in } Q_o - \bigcup Q_n^1. \tag{6.1}$$

It follows from the first of (6.1) and $\|f\|_o = 1$, that

$$\sum \mu(Q_n^1) \le \frac{1}{\alpha} \int_{Q_o} |f - f_{Q_o}| dy \le \frac{1}{\alpha} \mu(Q_o). \tag{6.2}$$

Also, using again that $\|f\|_o = 1$

$$|f_{Q_n^1} - f_{Q_o}| \le \frac{1}{\mu(Q_n^1)} \int_{Q_n^1} |f - f_{Q_o}| dy \le 2^N \alpha. \tag{6.3}$$

Finally since $\alpha > 1$ and $\|f\|_o = 1$

$$\frac{1}{\mu(Q_n^1)} \int_{Q_n^1} |f - f_{Q_n^1}| dy < \alpha \quad \text{for all cubes } Q_n^1. \tag{6.4}$$

For $n \in \mathbb{N}$ fixed, set

$$f_{1,n}(x) = \begin{cases} |f - f_{Q_n^1}| & \text{in } Q_n^1; \\ 0 & \text{otherwise.} \end{cases}$$

Apply again the Calderón–Zygmund decomposition, for the same $\alpha > 1$ to the function $f_{1,n}$, starting from the cube Q_n^1 and using (6.4). This generates a countable collection of cubes $\{Q_{n,m}^2\}$, with faces parallel to the coordinate planes and with pairwise disjoint interior, such that

$$\alpha < \frac{1}{\mu(Q_{n,m}^2)} \int_{Q_{n,m}^2} |f - f_{Q_n^1}| dy \le 2^N \alpha \quad \text{and}$$

$$|f - f_{Q_n^1}| \le \alpha \quad \text{a.e. in} \quad Q_n^1 - \bigcup_m Q_{n,m}^2. \tag{6.5}$$

For such a collection, also the analog of (6.2) is satisfied, i.e.,

$$\sum_m \mu(Q_{n,m}^2) \le \frac{1}{\alpha} \sum_m \int_{Q_{n,m}^2} |f - f_{Q_n^1}| dy$$

$$\le \frac{1}{\alpha} \int_{Q_n^1} |f - f_{Q_n^1}| dy \le \frac{1}{\alpha} \mu(Q_n^1) \tag{6.6}$$

where we have used that $\|f\|_o = 1$. Next we claim that

$$|f - f_{Q_o}| \le 2 \cdot 2^N \alpha \quad \text{a.e. in } Q_o - \bigcup_{n,m} Q_{n,m}^2.$$

If $x \in Q_o - \bigcup Q_n^1$, this follows from the second of (6.1). If $x \in Q_n^1 - \bigcup_m Q_{n,m}^2$ then by (6.3) and the second of (6.5)

$$|f(x) - f_{Q_o}| \le |f(x) - f_{Q_n^1}| + |f_{Q_n^1} - f_{Q_o}| \le 2 \cdot 2^N \alpha.$$

Adding (6.6) with respect to n and taking into account (6.2), gives

$$\sum_{n,m} \mu(Q^2_{n,m}) \leq \frac{1}{\alpha^2} \mu(Q_o).$$

Also, using the first of (6.5) and (6.3)

$$\begin{aligned}
|f_{Q^2_{n,m}} - f_{Q_o}| &\leq |f_{Q^2_{n,m}} - f_{Q^1_n}| + |f_{Q^1_n} - f_{Q_o}| \\
&\leq \frac{1}{\mu(Q^2_{n,m})} \int_{Q^2_{n,m}} |f - f_{Q^1_n}| dy + |f_{Q^1_n} - f_{Q_o}| \\
&\leq 2 \cdot 2^N \alpha.
\end{aligned}$$

We now relabel $\{Q^2_{n,m}\}$ to obtain a countable collection $\{Q^2_n\}$ of closed cubes, with faces parallel to the coordinate planes, with pairwise disjoint interior and such that

$$|f - f_{Q_o}| \leq 2 \cdot 2^N \alpha \quad \text{a.e. in } Q_o - \bigcup Q^2_n;$$

$$\sum \mu(Q^2_n) \leq \frac{1}{\alpha^2} \mu(Q_o); \tag{$6.7)_2$}$$

$$|f_{Q^2_n} - f_{Q_o}| \leq 2 \cdot 2^N \alpha.$$

Repeating the process k times generates a countable collection $\{Q^k_n\}$ of closed sub-cubes of Q_o, with faces parallel to the coordinate planes, with pairwise disjoint interior, and such that

$$|f - f_{Q_o}| \leq k \cdot 2^N \alpha \quad \text{a.e. in } Q_o - \bigcup_n Q^k_n;$$

$$\sum_n \mu(Q^k_n) \leq \frac{1}{\alpha^k} \mu(Q_o); \tag{$6.7)_k$}$$

$$|f_{Q^k_n} - f_{Q_o}| \leq k \cdot 2^N \alpha.$$

From this, for a fixed positive integer k,

$$\mu([|f - f_{Q_o}| > k2^N \alpha] \cap Q_o) \leq \sum_n \mu(Q^k_n) \leq \frac{1}{\alpha^k} \mu(Q_o).$$

This inequality continues to hold for $k = 0$. Fix now $t > 0$ and let $k \geq 0$ be such that

$$k2^N \alpha < t \leq (k+1)2^N \alpha.$$

Then

$$\begin{aligned}
\mu([|f - f_{Q_o}| > t] \cap Q_o) &\leq \mu([|f - f_{Q_o}| > k2^N \alpha] \cap Q_o) \\
&\leq \frac{1}{\alpha^k} \mu(Q_o) \leq \alpha e^{-\gamma t} \mu(Q_o)
\end{aligned}$$

where $\gamma = (\ln \alpha)/2^N \alpha$. $\blacksquare$

7   The Sharp Maximal Function

Continue to denote by Q_o and Q closed cubes in $\mathbb{R}^N$, with faces parallel to the coordinate planes. Given a measurable function f defined in Q_o, we regard it as defined in the whole $\mathbb{R}^N$ by setting it to be zero outside Q_o.

For a cube Q such that $\mu(Q \cap Q_o) > 0$, let $f_{Q \cap Q_o}$ denote the integral average of f over $Q \cap Q_o$, i.e.,

$$f_{Q \cap Q_o} = \frac{1}{\mu(Q \cap Q_o)} \int_{Q \cap Q_o} f \, dy.$$

Set also,

$$|f|_{Q_o} = \frac{1}{\mu(Q_o)} \int_{Q_o} |f| \, dy.$$

The *sharp* maximal function $Q_o \ni x \to f^{\#}(x)$, is defined by

$$f^{\#}(x) = \sup_{Q \ni x} \frac{1}{\mu(Q \cap Q_o)} \int_{Q \cap Q_o} |f - f_{Q \cap Q_o}| \, dy \tag{7.1}$$

where the supremum is taken over all cubes Q containing x. This is also called the function of *maximal mean oscillation*.

It follows from the definition that if $f^{\#} \in L^{\infty}(Q_o)$ then $f \in BMO(Q_o)$. Also, from (7.1) and the definition of maximal function $M(f)$

$$f^{\#}(x) \le 2^{N+1} M(f)(x) \qquad \text{for a.e. } x \in Q_o.$$

By Proposition 2.2, this implies that if $f \in L^1(Q_o)$ then

$$\mu([f^{\#} > t]) \le \frac{2 \cdot 10^N}{t} \int_{Q_o} |f| \, dy \qquad \text{for all } t > 0.$$

Hence $f^{\#} \in L^1_w(Q_o)$. If $f \in L^p(Q_o)$ for $p \in (1, \infty)$, by Proposition 3.1

$$\|f^{\#}\|_p \le 2^{N+1} \gamma_p \|f\|_p \quad \text{where } \gamma_p^p = \frac{2^p p 5^N}{p-1}.$$

The next theorem asserts the converse, i.e., if $f^{\#} \in L^p(Q_o)$ then also $f \in L^p(Q_o)$.

Theorem 7.1 (Fefferman–Stein [45]) *Let $f \in L^1(Q_o)$ and assume the corresponding sharp maximal function $f^{\#}$ is in $L^p(Q_o)$. Then $f \in L^p(Q_o)$ and there exists a positive constant $\gamma = \gamma(N, p)$ depending only upon N and p, such that*

$$\|f\|_p \le \gamma(N, p) \big(\|f^{\#}\|_p + \mu(Q_o)|f|_{Q_o} \big). \tag{7.2}$$

8 Proof of the Fefferman–Stein Theorem

Fix $t > |f|_{Q_o}$ and apply the Calderón–Zygmund decomposition to the function $|f|$, for $\alpha = t$. This generates a countable collection $\{Q_n^t\}$ of closed cubes with faces parallel to the coordinate planes, with pairwise disjoint interior and such that

$$t < \frac{1}{\mu(Q_n^t)} \int_{Q_n^t} |f| dy \leq 2^N t \quad \text{and}$$

$$|f| \leq t \quad \text{a.e. in} \quad Q_o - \bigcup Q_n^t. \tag{8.1}_t$$

Without loss of generality, we may arrange that Q_o is part of the initial partition of $\mathbb{R}^N$ in the Calderón–Zygmund process. Therefore, the cubes Q_n^t result from repeated bisections starting from the parent cube Q_o.

Let $t > \tau > |f|_{Q_o}$, and $\{Q_j^\tau\}$ be the corresponding Calderón–Zygmund decomposition, for $\alpha = \tau$, satisfying the analog of $(8.1)_t$, i.e.,

$$\tau < \frac{1}{\mu(Q_n^\tau)} \int_{Q_n^\tau} |f| dy \leq 2^N \tau \quad \text{and}$$

$$|f| \leq \tau \quad \text{a.e. in} \quad Q_o - \bigcup Q_n^\tau. \tag{8.1}_\tau$$

Moreover the cubes Q_j^τ result from a repeated bisection of the parent cube Q_o. By the Calderón–Zygmund recursive bisection process, and since $t > \tau$, each of the cubes Q_n^t is a sub-cube of some Q_j^τ. Therefore

$$m(t) = \sum \mu(Q_n^t) \leq \sum \mu(Q_j^\tau) = m(\tau).$$

Also for any $t > \tau > |f|_{Q_o}$

$$\mu([|f| > t] \cap Q_o) \leq m(t). \tag{8.2}$$

Lemma 8.1 *Let* $t > 2^{N+1} |f|_{Q_o}$. *Then*

$$m(t) \leq \mu([f^\# > t\delta] \cap Q_o) + 2\delta m(t 2^{-(N+1)})$$

where δ *is an arbitrary positive constant.*

Proof Set $\tau = t 2^{-(N+1)}$ and determine the two countable families of cubes $\{Q_n^t\}$ and $\{Q_j^\tau\}$ satisfying $(8.1)_t$ and $(8.1)_\tau$, respectively. Fix one of the cubes Q_j^τ and consider those cubes Q_n^t out of $\{Q_n^t\}$ that are contained in Q_j^τ. For such a cube Q_j^τ, either

$$Q_j^\tau \subset [f^\# > t\delta], \quad \text{or} \quad Q_j^\tau \not\subset [f^\# > t\delta].$$

If the first alternative occurs

$$\sum_{Q_n^t \subset Q_j^\tau} \mu(Q_n^t) \le \mu([f^\# > t\delta] \cap Q_j^\tau). \tag{8.3}$$

If the second alternative occurs, then there exists some $x \in Q_j^\tau$ such that $f^\#(x) \le t\delta$. From the definition of $f^\#$, for such a cube

$$\frac{1}{\mu(Q_j^\tau)} \int_{Q_j^\tau} |f - f_{Q_j^\tau}| dy \le t\delta.$$

By the lower bound in the first of $(8.1)_t$ and the upper bound in the first of $(8.1)_\tau$,

$$|f|_{Q_n^t} > t \quad\text{and}\quad |f|_{Q_j^\tau} \le 2^N \tau.$$

From this, for each of the cubes Q_n^t contained in the fixed cube Q_j^τ

$$\int_{Q_n^t} |f - f_{Q_j^\tau}| dy \ge (|f|_{Q_n^t} - |f|_{Q_j^\tau})\mu(Q_n^t)$$
$$\ge (t - 2^N \tau)\mu(Q_n^t) = \tfrac{1}{2} t\mu(Q_n^t).$$

Adding over all the cubes Q_n^t contained in Q_j^τ gives

$$\sum_{Q_n^t \subset Q_j^\tau} \mu(Q_n^t) \le \frac{2}{t} \sum_{Q_n^t \subset Q_j^\tau} \int_{Q_n^t} |f - f_{Q_j^\tau}| dy$$
$$\le \frac{2}{t} \int_{Q_j^\tau} |f - f_{Q_j^\tau}| dy \le 2\delta\mu(Q_j^\tau). \tag{8.4}$$

Combining the first alternative, leading to (8.3) and the second alternative, leading to (8.4) gives

$$\sum_{Q_n^t \subset Q_j^\tau} \mu(Q_n^t) \le \mu([f^\# > t\delta] \cap Q_j^\tau + 2\delta\mu(Q_j^\tau).$$

Adding over j proves the lemma.   ∎

Taking into account (8.2), the estimate (7.2) of the theorem will be derived from the limiting process

$$\int_{Q_o} |f|^p dy = \lim_{s \to \infty} p \int_0^s t^{p-1} \mu([|f| > t] \cap Q_o) dt$$
$$\le \limsup_{s \to \infty} p \int_0^s t^{p-1} m(t) dt \tag{8.5}$$

provided the last limit is finite. To estimate such a limit fix $s > 2^{N+1}|f|_{Q_o}$ and use Lemma 8.1 to compute

$$
\begin{aligned}
p\int_0^s t^{p-1}m(t)dt = p\int_0^{2^{N+1}|f|_{Q_o}} t^{p-1}m(t)dt &+ p\int_{2^{N+1}|f|_{Q_o}}^s t^{p-1}m(t)dt \\
\leq (2^{N+1}|f|_{Q_o})^p\mu(Q_o) &+ p\int_0^\infty t^{p-1}\mu([f^\# > t\delta]\cap Q_o)dt \\
&+ 2p\delta\int_0^s t^{p-1}m(t2^{-(N+1)})dt \\
= (2^{N+1}|f|_{Q_o})^p\mu(Q_o) &+ \delta^{-p}\|f^\#\|_p^p \\
&+ 2\delta 2^{(N+1)p}p\int_0^s t^{p-1}m(t)dt.
\end{aligned}
$$

Choosing $\delta^{-1} = 4\cdot 2^{(N+1)p}$ gives

$$
p\int_0^s t^{p-1}m(t)dt \leq 2(2^{N+1}|f|_{Q_o})^p\mu(Q_o) + 2\delta^{-p}\|f^\#\|_p^p.
$$

Putting this in (8.5) proves the theorem. ∎

9 The Marcinkiewicz Interpolation Theorem

Let E be a measurable subset of $\mathbb{R}^N$ and let $1 \leq p < \infty$. A measurable function $f : E \to \mathbb{R}$ is in weak-$L^p(E)$, denoted by $L_w^p(E)$, if there is a positive constant F such that

$$
\mu([|f| > t]) \leq \frac{F^p}{t^p} \qquad \text{for all } t > 0. \tag{9.1}
$$

Set

$$
\|f\|_{p,w} = \inf\{F \text{ for which (9.1) holds}\}.
$$

Let $f, g \in L_w^p(E)$ and let α and β be nonzero real numbers. Then for all $t > 0$

$$
[|\alpha f + \beta g| > t] \subset \left[|f| > \frac{t}{2|\alpha|}\right] \cup \left[|g| > \frac{t}{2|\beta|}\right].
$$

Thus $L_w^p(E)$ is a linear space. However $\|\cdot\|_{p,w}$ is not a norm on $L_w^p(E)$. If $f \in L^p(E)$, then for all $t > 0$

$$
\mu([|f| > t]) \leq \frac{1}{t^p}\|f\|_p^p.
$$

Therefore $f \in L_w^p(E)$, and $\|f\|_p \ge \|f\|_{p,w}$. However there exist functions $f \in L_w^p(E)$ that are not in $L^p(E)$ (examples can be constructed as in § 3.1).

The space $L_w^\infty(E)$ is defined as the collection of measurable functions for which (9.1) holds, for some constant F, for all $p \ge 1$ and all $t > 0$.

If $t > F$ then $\mu([|f| > t]) = 0$. Thus if $f \in L_w^\infty(E)$, then $f \in L^\infty(E)$ and $\|f\|_\infty \le \|f\|_{\infty,w}$. On the other hand if $f \in L^\infty(E)$ then (9.1) holds for $F = \|f\|_\infty$. Thus $L_w^\infty(E) = L^\infty(E)$ and $\|\cdot\|_{\infty,w} = \|\cdot\|_\infty$.

9.1 Quasi-linear Maps and Interpolation

Let T be a map T defined in $L^p(E)$ and such that $T(f)$ is measurable for all $f \in L^p(E)$. The map T is *quasi-linear* if there exists a positive constant C such that for all f and g in $L^p(E)$

$$|T(f+g)| \le C(|T(f)| + |T(g)|) \quad \text{a.e. in } E.$$

If T is quasi-linear, then for all $t > 0$

$$[|T(f+g)| > t] \subset \left[|T(f)| > \frac{t}{2C}\right] \cup \left[|T(g)| > \frac{t}{2C}\right].$$

A quasi-linear map $T : L^p(E) \to L^q(E)$, for some pair $p, q \ge 1$, is of *strong type* (p, q) if there exists a positive constant $M_{p,q}$ such that

$$\|T(f)\|_q \le M_{p,q} \|f\|_p \quad \text{for all } f \in L^p(E).$$

A quasi-linear map T defined in $L^p(E)$ and such that $T(f)$ is measurable for all $f \in L^p(E)$, is of *weak type* (p, q) if there exists a positive constant $N_{p,q}$ such that

$$\|T(f)\|_{q,w} \le N_{p,q} \|f\|_p \quad \text{for all } f \in L^p(E).$$

When $p = q$ we set $M_{p,p} = M_p$ and $N_{p,p} = N_p$. Examples of maps of strong and weak type are in § 3.1. Further example will arise from the Riesz potentials in § 24.

Theorem 9.1 (Marcinkiewicz [103]) *Let T be a quasi-linear map defined both in $L^p(E)$ and $L^q(E)$ for some pair $1 \le p < q \le \infty$. Assume that T is both of weak type (p, p) and of weak type (q, q), i.e., there exist positive constants N_p and N_q such that*

$$\begin{aligned} \|T(f)\|_{p,w} &\le N_p \|f\|_p \quad &\text{for all } f \in L^p(E); \\ \|T(f)\|_{q,w} &\le N_q \|f\|_q \quad &\text{for all } f \in L^q(E). \end{aligned} \tag{9.2}$$

Then T is of strong type (r, r) for every $p < r < q$ and

$$\|T(f)\|_r \leq \gamma N_p^\delta N_q^{1-\delta} \|f\|_r \quad \text{for all } f \in L^r(E) \tag{9.3}$$

where

$$\delta = \begin{cases} \dfrac{p(q-r)}{r(q-p)} & \text{if } q < \infty; \\ \dfrac{p}{r} & \text{if } q = \infty; \end{cases} \tag{9.4}$$

and

$$\gamma = 2C \begin{cases} \left(\dfrac{r(q-p)}{(r-p)(q-r)}\right)^{1/r} & \text{if } q < \infty; \\ \left(\dfrac{r}{r-p}\right)^{1/r} & \text{if } q = \infty; \end{cases} \tag{9.5}$$

where C is the constant appearing in the definition of semi-linear map.

10   Proof of the Marcinkiewicz Theorem

Having fixed $r \in (p, q)$, and some $f \in L^r(E)$, decompose it as $f = f_1 + f_2$, where

$$f_1 = \begin{cases} f & \text{for } |f| > \lambda t; \\ 0 & \text{for } |f| \leq \lambda t; \end{cases} \qquad f_2 = \begin{cases} 0 & \text{for } |f| \geq \lambda t; \\ f & \text{for } |f| < \lambda t \end{cases}$$

where $t > 0$ and λ is a positive constant to be chosen later.

We claim that $f_1 \in L^p(E)$ and $f_2 \in L^q(E)$. Since $f \in L^r(E)$, the set $[|f| > \lambda t]$ has finite measure. Therefore by Hölder's inequality

$$\int_E |f_1|^p dy \leq \|f\|_r^p \left(\mu([|f| > \lambda t])\right)^{1-\frac{p}{r}}.$$

Thus $f_1 \in L^p(E)$. Moreover

$$\int_E |f_2|^q dy \leq \int_E |f_2|^{q-r} |f|^r dy \leq (\lambda t)^{q-r} \int_E |f|^r dy.$$

Assume first $1 \leq p < q < \infty$. Then, using the quasi-linear structure of T, and the assumptions (9.2)

$$\mu([|T(f)| > t]) \leq \mu\left(\left[|T(f_1)| > \frac{t}{2C}\right]\right) + \mu\left(\left[|T(f_2)| > \frac{t}{2C}\right]\right)$$

$$\leq \frac{(2CN_p)^p}{t^p}\|f_1\|_p^p + \frac{(2CN_q)^q}{t^q}\|f_2\|_q^q \qquad (10.1)$$

$$= \frac{(2CN_p)^p}{t^p}\int_{|f|>\lambda t} |f|^p dy + \frac{(2CN_q)^q}{t^q}\int_{|f|\leq\lambda t} |f|^q dy.$$

From this

$$\int_E |T(f)|^r dy = r\int_0^\infty t^{r-1}\mu([|T(f)| > t])dt$$

$$\leq r(2CN_p)^p \int_0^\infty t^{r-p-1}\int_{|f|>\lambda t} |f|^p dy$$

$$+ r(2CN_q)^q \int_0^\infty t^{r-q-1}\int_{|f|\leq\lambda t} |f|^q dy.$$

The integrals on the right-hand side are transformed by means of Fubini's Theorem and give

$$\int_0^\infty t^{r-p-1}\int_{|f|>\lambda t} |f|^p dy = \int_E |f|^p\left(\int_0^{|f|/\lambda} t^{r-p-1}dt\right)dy$$

$$= \frac{1}{r-p}\frac{1}{\lambda^{r-p}}\int_E |f|^r dy.$$

$$\int_0^\infty t^{r-q-1}\int_{|f|\leq\lambda t} |f|^q dy = \int_E |f|^q\left(\int_{|f|/\lambda}^\infty t^{r-q-1}dt\right)dy$$

$$= \frac{1}{q-r}\lambda^{q-r}\int_E |f|^r dy.$$

Combining these estimates

$$\|T(f)\|_r^r \leq r\left\{\frac{(2C)^p}{r-p}\frac{N_p^p}{\lambda^{r-p}} + \frac{(2C)^q}{q-r}N_q^q\lambda^{q-r}\right\}\|f\|_r^r. \qquad (10.2)$$

Minimizing the right-hand side with respect to λ proves (9.3) with the value of $\lambda = \delta$ given by (9.4) and (9.5).

Turning now to the case $q = \infty$, we begin by choosing the parameter λ so large that

$$\mu\left(\left[|T(f_2)| > \frac{t}{2C}\right]\right) = 0. \qquad (10.3)$$

If this is violated, then $\|T(f_2)\|_\infty > t/2C$. From this and the second of (9.2) with $q = \infty$

$$\lambda t \geq \|f_2\|_\infty \geq \frac{1}{N_\infty}\|T(f_2)\|_\infty \geq \frac{1}{N_\infty}\frac{t}{2C}.$$

Therefore choosing

$$\lambda = \frac{1}{2CN_\infty}, \quad \text{one has} \quad \|T(f_2)\|_\infty = \frac{t}{2C},$$

and (10.3) holds. We proceed now as before, starting from (10.1) where the terms involving f_2 are discarded. This gives an analog of (10.2) without the terms involving q, i.e.,

$$\|T(f)\|_r^r \le r \frac{(2C)^p}{r-p} \frac{N_p^p}{\lambda^{r-p}} \|f\|_r^r, \quad \lambda = \frac{1}{2CN_\infty}. \qquad \blacksquare$$

11 Rearranging the Values of a Function

Let $E \subset \mathbb{R}^N$ be measurable and of finite Lebesgue measure $\mu(E)$. The set E is symmetrically rearranged about the origin of $\mathbb{R}^N$ into an open ball $B_R = E^*$, of equal measure as E. Thus

$$\kappa_N R^N = \mu(E)$$

where κ_N is the volume of the unit ball in $\mathbb{R}^N$. The symmetric rearrangement of the characteristic function of E is the characteristic function of E^*, i.e.,

$$\chi_E^* = \chi_{E^*}.$$

Next, let f be a nonnegative, simple function taking n distinct positive values $f_1 < \cdots < f_n$, on mutually disjoint sets $\{E_1, \ldots, E_n\}$, each of finite measure. Rewrite f as

$$f = f_1 \chi_{E_1 \cup \cdots \cup E_n} + (f_2 - f_1)\chi_{E_2 \cup \cdots \cup E_n} + \cdots (f_n - f_{n-1})\chi_{E_n} \qquad (11.1)$$

and define the rearrangement of f as

$$f^* = f_1 \chi_{(E_1 \cup \cdots \cup E_n)^*} + (f_2 - f_1)\chi_{(E_2 \cup \cdots \cup E_n)^*} + \cdots + (f_n - f_{n-1})\chi_{E_n^*}.$$

By construction

$$\begin{aligned}
E_1 \cup \cdots \cup E_n &= [f > t] &\quad \text{for all } t \in [0, f_1); \\
E_2 \cup \cdots \cup E_n &= [f > t] &\quad \text{for all } t \in [f_1, f_2); \\
\cdots\cdots &= \cdots\cdots &\quad \cdots\cdots \\
E_{n-1} \cup E_n &= [f > t] &\quad \text{for all } t \in [f_{n-2}, f_{n-1}); \\
E_n &= [f > t] &\quad \text{for all } t \in [f_{n-1}, f_n).
\end{aligned}$$

Likewise

$$(E_1 \cup \cdots \cup E_n)^* = [f^* > t] \quad \text{for all } t \in [0, f_1);$$
$$(E_2 \cup \cdots \cup E_n)^* = [f^* > t] \quad \text{for all } t \in [f_1, f_2);$$
$$\cdots \cdots \;\; = \;\; \cdots \cdots \qquad \cdots \cdots$$
$$(E_{n-1} \cup E_n)^* = [f^* > t] \quad \text{for all } t \in [f_{n-2}, f_{n-1});$$
$$E_n^* = [f^* > t] \quad \text{for all } t \in [f_{n-1}, f_n).$$

Hence, picking some $x \in (E_1 \cup \cdots \cup E_n)^*$ the value of f^* at x is determined by

$$f^*(x) = \sup\{t \mid \mu([f > t]) > \kappa_N |x|^N\}. \tag{11.2}$$

Also by this construction

$$\mu([f > t]) = \mu([f^* > t]) \quad \text{for all } t \geq 0. \tag{11.3}$$

One also verifies that if f and g are any two such nonnegative, simple functions, then $f \leq g$ implies $f^* \leq g^*$. Thus the operation of symmetric decreasing rearrangement of such simple functions preserves the ordering.

Let now f be a real-valued, measurable, nonnegative function defined in $\mathbb{R}^N$ and such that

$$\mu([f > t]) < \infty \quad \text{for all } t > 0. \tag{11.4}$$

There exists a sequence of nonnegative, measurable, simple functions $\{f_n\} \to f$ pointwise in $\mathbb{R}^N$, and $f_n \leq f_{n+1}$. The assumption (11.4) and the construction of the f_n in § 3 of Chap. 4, implies that each f_n takes finitely many, distinct values on distinct, measurable sets of finite measure. Hence, f_n^* is well defined for each n. Moreover $f_n^* \leq f_{n+1}^*$ and the limit of $\{f_n^*\}$ exists. Define

$$f^*(x) = \lim f_n^*(x) \quad \text{pointwise in } \mathbb{R}^N$$
$$= \sup_n \sup\{t \mid \mu([f_n > t]) > \kappa_N |x|^N\} \tag{11.5}$$
$$= \sup\{t \mid \mu([f > t]) > \kappa_N |x|^N\}.$$

Hence (11.2) can be taken as the definition of the symmetric, decreasing rearrangement of a real-valued, nonnegative, measurable function f defined in $\mathbb{R}^N$ and satisfying (11.4).

Proposition 11.1 *Let f and g be real-valued, nonnegative, measurable functions satisfying (11.4). Then*

(i) *f^* is nonnegative, radially symmetric and nonincreasing;*
(ii) *$f \leq g$ implies $f^* \leq g^*$;*
(iii) *Let $F : \mathbb{R}^+ \to \mathbb{R}^+$ be, nondecreasing. Then $F(f)^* = F(f^*)$. In particular $(f - t)_+^* = (f^* - t)_+$ for any constant t;*

(iv) $[f^* > t]$ are open and hence f^* is measurable.
(v) f and f^* are equi-measurable, in the sense that (11.3) holds.
(vi) For $s \leq t$, $\mu([s < f \leq t]) = \mu([s < f^* \leq t])$;
(vii) If $f \in L^p(\mathbb{R}^N)$ for some $1 \leq p \leq \infty$, then $f^* \in L^p(\mathbb{R}^N)$ and moreover $\|f\|_p = \|f^*\|_p$.

Proof The statements **(i)**–**(vi)** follow from the construction of f^* leading to (11.5). As for **(vii)**, the statement is obvious if $p = \infty$. If $1 \leq p < \infty$, since f and f^* are equi-measurable

$$\|f\|_p^p = \int_0^\infty t^{p-1}\mu([f > t])dt = \int_0^\infty t^{p-1}\mu([f^* > t])dt = \|f^*\|_p^p. \qquad \blacksquare$$

12 Some Integral Inequalities for Rearrangements

Proposition 12.1 *Let f and g be real-valued, nonnegative, measurable functions in $\mathbb{R}^N$ satisfying (11.4). Then*

$$\int_{\mathbb{R}^N} fg\,d\mu \leq \int_{\mathbb{R}^N} f^*g^*\,d\mu. \qquad (12.1)$$

Proof Assume first that f and g are simple and both take only the values 0 and 1. Set

$$E = [f = 1], \ E^* = [f^* = 1];$$
$$G = [g = 1], \ G^* = [g^* = 1].$$

Now compute and estimate

$$\int_{\mathbb{R}^N} fg\,d\mu = \mu(E \cap G)$$

$$\leq \min\{\mu(E); \mu(G)\}$$
$$= \min\{\mu(E^*); \mu(G^*)\}$$
$$= \mu(E^* \cap G^*).$$

By linear combinations and iterations the statement holds for nonnegative simple functions satisfying (11.4). By approximation and a limiting process, it continues to hold for nonnegative, measurable functions satisfying (11.4).

Remark 12.1 Neither f, g or fg are required to be integrable. If fg is not integrable (12.1) is meant in the sense that if the left-hand side is infinite so is the right-hand side.

Corollary 12.1 *Let f and g be real-valued, nonnegative, measurable functions in $\mathbb{R}^N$ satisfying (11.4). Then*

$$\int_{\mathbb{R}^N} f \chi_{[g \leq t]} d\mu \geq \int_{\mathbb{R}^N} f^* \chi_{[g^* \leq t]} d\mu. \tag{12.2}$$

Proof Assume first $f \in L^1(\mathbb{R}^N)$, and apply (12.1) to the pair of functions f and $\chi_{[g>t]}$. Writing the latter as $1 - \chi_{[g \leq t]}$ gives

$$\int_{\mathbb{R}^N} f(1 - \chi_{[g \leq t]}) d\mu \leq \int_{\mathbb{R}^N} f^* \chi_{[g^* > t]} d\mu$$

$$= \int_{\mathbb{R}^N} f^* (1 - \chi_{[g^* \leq t]}) d\mu.$$

The general case follows from this by a limiting process, understanding that if the right-hand side of (12.2) is infinite, so is the left-hand side. ∎

12.1 Contracting Properties of Symmetric Rearrangements

Theorem 12.1 (Chiti [27]; also in [29]) *Let f and g be nonnegative functions in $L^p(\mathbb{R}^N)$ for $1 \leq p \leq \infty$. If $p = \infty$ assume also that f and g satisfy (11.4). Then*

$$\| f^* - g^* \|_p \leq \| f - g \|_p. \tag{12.3}$$

Proof The inequality is obvious for $p = \infty$. Let $1 \leq p < \infty$ and assume first that $f \geq g$. Compute

$$(f - g)^p = -\int_g^f \frac{d}{dt}(f - t)^p dt$$

$$= p \int_g^f (f - t)^{p-1} dt$$

$$= p \int_0^\infty (f - t)_+^{p-1} \chi_{[g \leq t]} dt.$$

Integrate over $\mathbb{R}^N$ and interchange the order of integration by means of Fubini's theorem. In the integral so obtained use (12.2). These operation yield

$$\|f - g\|_p^p = p \int_0^\infty \int_{\mathbb{R}^N} (f - t)_+^{p-1} \chi_{[g \le t]} d\mu dt$$

$$\ge p \int_0^\infty \int_{\mathbb{R}^N} (f^* - t)_+^{p-1} \chi_{[g^* \le t]} d\mu dt$$

$$= \int_{\mathbb{R}^N} p \int_0^\infty (f^* - t)_+^{p-1} \chi_{[g^* \le t]} dt d\mu$$

$$= \int_{\mathbb{R}^N} \int_{g^*}^{f^*} p(f^* - t)^{p-1} dt d\mu$$

$$= \int_{\mathbb{R}^N} \int_{g^*}^{f^*} \frac{d}{dt}(f^* - t)^p dt d\mu$$

$$= \int_{\mathbb{R}^N} (f^* - g^*)^p d\mu = \|f^* - g^*\|_p^p.$$

In general,

$$\|f - g\|_p = \|f \vee g - f \wedge g\|_p \ge \|(f \vee g)^* - (f \wedge g)^*\|_p \ge \|f^* - g^*\|_p. \quad \blacksquare$$

12.2 Testing for Measurable Sets E Such that $E = E^*$ a.e. in $\mathbb{R}^N$

Proposition 12.2 *Let $E \subset \mathbb{R}^N$ be measurable and of finite measure and let f be nonnegative, radially symmetric, and strictly decreasing in $|x|$. If*

$$\int_{\mathbb{R}^N} f \chi_E dx = \int_{\mathbb{R}^N} f^* \chi_{E^*} dx \tag{12.4}$$

then $E = E^*$ a.e. in $\mathbb{R}^N$.

Proof The assumptions on f imply that $f = f^*$ and $f > 0$ in $\mathbb{R}^N$. The sets $[f > t]$ are balls of radius $\rho(t)$ about the origin, for some positive function $\rho(\cdot)$. While f may exhibit jump discontinuities, the assumptions on f imply that the function

$$\mathbb{R}^+ \ni t \to \mu([f > t]) = \kappa_N \rho^N(t)$$

is continuous and ρ ranges over $(0, \infty)$. By (12.1) for all $t > 0$,

$$\int_{\mathbb{R}^N} \chi_{[f > t]} \chi_E dx \le \int_{\mathbb{R}^N} \chi_{[f > t]} \chi_{E^*} dx.$$

Integrating in dt over $\mathbb{R}^+$

$$\int_{\mathbb{R}^N} f\chi_E dx = \int_0^\infty \left(\int_{\mathbb{R}^N} \chi_{[f>t]}\chi_E dx \right) dt$$

$$\le \int_0^\infty \left(\int_{\mathbb{R}^N} \chi_{[f>t]}\chi_{E^*} dx \right) dt$$

$$= \int_{\mathbb{R}^N} f\chi_{E^*} dx = \int_{\mathbb{R}^N} f\chi_E dx$$

where we have used the assumption (12.4). From this

$$\int_{\mathbb{R}^N} \chi_{[f>t]}\chi_E dx = \int_{\mathbb{R}^N} \chi_{[f>t]}\chi_{E^*} dx \quad \text{for a.e. } t > 0.$$

From the continuity of $\rho(\cdot)$ this implies

$$\mu\big([f > t] \cap E\big) = \mu\big([f > t] \cap E^*\big) \quad \text{for all } t > 0.$$

This equality is possible if for all fixed $\rho > 0$, either E and E^* are both contained in B_ρ, except for a set of measure zero, or if E and E^* both contain B_ρ, except for a set of i measure zero.   ∎

Corollary 12.2 *Let g be a real-valued, nonnegative, measurable function in $\mathbb{R}^N$ satisfying (11.4), and let f be nonnegative, radially symmetric, and strictly decreasing in $|x|$. Then (12.1) holds with equality if and only if $g = g^*$.*

Proof For all $t > 0$ by (12.1)

$$\int_{\mathbb{R}^N} f\chi_{[g>t]} dx \le \int_{\mathbb{R}^N} f\chi_{[g^*>t]} dx.$$

Integrating in dt over $\mathbb{R}^+$,

$$\int_{\mathbb{R}^N} fg\,dx = \int_0^\infty \left(\int_{\mathbb{R}^N} f\chi_{[g>t]} dx \right) dt$$

$$\le \int_0^\infty \left(\int_{\mathbb{R}^N} f\chi_{[g^*>t]} dx \right) dt$$

$$= \int_{\mathbb{R}^N} fg^* dx = \int_{\mathbb{R}^N} fg\,dx.$$

Thus

$$\int_{\mathbb{R}^N} f\chi_{[g>t]} dx = \int_{\mathbb{R}^N} f\chi_{[g^*>t]} dx \quad \text{for a.e. } t > 0.$$

Apply now Proposition 12.2 with $E = [g > t]$.   ∎

13 The Riesz Rearrangement Inequality

Theorem 13.1 (Riesz [131]; also Zygmund [178]) *Let f, g and h be real-valued, nonnegative, measurable functions in $\mathbb{R}^N$ satisfying (11.4). Then*

$$
\begin{aligned}
\mathcal{I} = \int_{\mathbb{R}^N} \int_{\mathbb{R}^N} f(x)g(y)h(y-x)dxdy \\
\leq \int_{\mathbb{R}^N} \int_{\mathbb{R}^N} f^*(x)g^*(y)h^*(y-x)dxdy = \mathcal{I}^*.
\end{aligned}
\tag{13.1}
$$

Since f, g and h are measurable and nonnegative, the integrals in (13.1), finite or infinite are well defined, and the inequality holds with the understanding that if $\mathcal{I} = \infty$ then also $\mathcal{I}^* = \infty$.

In the next sections we will prove this inequality for $N = 1$. The proof, albeit lengthy, is rather elementary, being based on examining the various occurrences and overlaps of the support of f, g, and h. While 1-dimensional, it suffices to establish the main potential estimates, in any number of dimensions (§ 18–§ 22), needed for the Sobolev embedding theorems of Chap. 10. The proof for $N = 2$ and $N > 2$ will be given in § 25–§ 26. It is more intricate and based on different ways of symmetrizing sets and functions (Steiner symmetrization in § 23). In all cases, the starting point is the reduction of the proof to the case when f, g, and h are characteristic functions of measurable, bounded sets.

13.1 Reduction to Characteristic Functions of Bounded Sets

It suffices to prove the theorem for characteristic functions of measurable sets. Indeed f, g, and h are the pointwise limit of nondecreasing sequences of simple functions $\{f_n\}, \{g_n\}$, and $\{h_n\}$, each having the representation

$$
f_n = \sum_{j=1}^{n} \varphi_j \chi_{F_j}, \quad F_j \supset F_{j+1}, \ j = 1, \ldots, n;
$$

$$
g_m = \sum_{s=1}^{m} \gamma_s \chi_{G_s}, \quad G_s \supset G_{s+1}, \ s = 1, \ldots, m;
$$

$$
h_k = \sum_{\ell=1}^{k} \theta_\ell \chi_{H_\ell}, \quad H_\ell \supset H_{\ell+1}, \ \ell = 1, \ldots, k,
$$

where φ_j, γ_s, and θ_ℓ are positive constants, and F_j, G_s, and H_ℓ are measurable subsets of $\mathbb{R}^N$, of finite measure. Their symmetric rearrangements are

$$
f_n^* = \sum_{j=1}^{n} \varphi_j \chi_{F_j^*}, \quad g_m^* = \sum_{s=1}^{m} \gamma_s \chi_{G_s^*}, \quad h_k^* = \sum_{\ell=1}^{k} \theta_\ell \chi_{H_\ell^*}.
$$

Assuming (13.1) holds true for characteristic functions of measurable sets, by monotone convergence,

$$\begin{aligned}
\mathcal{I} &= \int_{\mathbb{R}^N} \int_{\mathbb{R}^N} f(x)g(y)h(y-x)dxdy \\
&= \lim_{n,m,k\to\infty} \int_{\mathbb{R}^N} \int_{\mathbb{R}^N} f_n(x)g_m(y)h_k(y-x)dxdy \\
&= \lim_{n,m,k\to\infty} \sum_{j,s,\ell} \varphi_j \gamma_s \theta_\ell \int_{\mathbb{R}^N} \int_{\mathbb{R}^N} \chi_{F_j}(x)\chi_{G_s}(y)\chi_{H_\ell}(y-x)dxdy \\
&\le \lim_{n,m,k\to\infty} \sum_{j,s,\ell} \varphi_j \gamma_s \theta_\ell \int_{\mathbb{R}^N} \int_{\mathbb{R}^N} \chi_{F_j}^*(x)\chi_{G_s}^*(y)\chi_{H_\ell}^*(y-x)dxdy \\
&\le \lim_{n,m,k\to\infty} \int_{\mathbb{R}^N} \int_{\mathbb{R}^N} f_n^*(x)g_m^*(y)h_k^*(y-x)dxdy \\
&= \int_{\mathbb{R}^N} \int_{\mathbb{R}^N} f^*(x)g^*(y)h^*(y-x)dxdy.
\end{aligned}$$

Thus in what follows we may assume that

$$f = \chi_F, \qquad g = \chi_G, \qquad h = \chi_H \tag{13.2}$$

where F, G, and H are measurable subsets of $\mathbb{R}^N$ of finite measure.

For a positive integer n let B_n denote the ball of radius n centered at the origin. Assume that (13.1) holds for measurable, bounded sets F, G, and H. Then by monotone convergence

$$\begin{aligned}
\mathcal{I} &= \lim \int_{\mathbb{R}^N} \int_{\mathbb{R}^N} \chi_{F\cap B_n}(x)\chi_{G\cap B_n}(y-x)\chi_{H\cap B_n}(y)dxdy \\
&\le \lim \int_{\mathbb{R}^N} \int_{\mathbb{R}^N} \chi_{F\cap B_n}^*(x)\chi_{G\cap B_n}^*(y-x)\chi_{H\cap B_n}^*(y)dxdy = \mathcal{I}^*.
\end{aligned}$$

Thus the proof of the Riesz rearrangement inequality (13.1) reduces to the case when f, g, and h are of the form (13.2) where F, G, and H are measurable and *bounded* sets in $\mathbb{R}^N$.

14 Proof of (13.1) for $N = 1$

14.1 *Reduction to Finite Union of Intervals*

Since F is measurable and of finite measure, for every $\varepsilon > 0$ there exists an open set $F_{o,\varepsilon}$ containing F and such that (Proposition 16.2 of Chap. 3),

$$\mu(F_{o,\varepsilon} - F) < \tfrac{1}{2}\varepsilon.$$

Such an open set $F_{o,\varepsilon}$ is the countable union of mutually disjoint open intervals $\{I_n\}$ (Proposition 1.1 of Chap. 3). Since $F_{o,\varepsilon}$ is of finite measure, there exists a positive integer n_ε such that

$$\sum_{j > n_\varepsilon} \mu(I_j) < \tfrac{1}{2}\varepsilon.$$

Setting

$$F_\varepsilon = \bigcup_{j=1}^{n_\varepsilon} I_j, \qquad F_{1,\varepsilon} = F \cap \bigcup_{j > n_\varepsilon} I_j, \qquad F_{2,\varepsilon} = F_\varepsilon - F$$

the set F can be represented as

$$F = F_\varepsilon \bigcup F_{1,\varepsilon} - F_{2,\varepsilon} \quad \text{with} \quad \mu(F_{1,\varepsilon} \cup F_{2,\varepsilon}) < \varepsilon$$

where F_ε is the finite union of open disjoint intervals. Moreover,

$$\chi_F = \chi_{F_\varepsilon} + \chi_{F_{1,\varepsilon}} - \chi_{F_{2,\varepsilon}}.$$

Similar decompositions hold for G and H. It is apparent that sets of arbitrarily small measure give arbitrarily small contributions in the integrals $\mathcal{I}$ and $\mathcal{I}^*$. Therefore, in proving (13.1) for $N = 1$ one may assume that f, g, and h are characteristic functions of sets F, G, and H, each finite union of disjoint, open intervals. Moreover, by changing ε if necessary, we may assume that the end points of the intervals making up F and respectively G and H are rational.

In such a case, in the integral $\mathcal{I}$ we may introduce a change of variables by rescaling x and y of a multiple equal to the minimum, common denominator of the end points of the intervals making up F, G, and H. This reduces the proof of Theorem 13.1 for $N = 1$, to the case when each of the sets F, G, and H is the finite union of intervals of the type $(j, j + 1)$ for integral j.

Finally, we may assume that the number of intervals making up each of the sets F, G, and H is even. This can be realized by bisecting each of these intervals and by effecting a further change of variables.

Thus in proving Theorem 13.1 for $N = 1$, we may assume that f, g, and h are characteristic functions of sets F, G, and H of the form,

$$F = \bigcup_{i=1}^{2R} (m_i, m_i + 1) \quad \begin{array}{l} \text{for positive integers } m_i \\ \text{and some positive integer } R; \end{array}$$

$$G = \bigcup_{j=1}^{2S} (n_j, n_j + 1) \quad \begin{array}{l} \text{for positive integers } n_j \\ \text{and some positive integer } S; \end{array}$$

$$H = \bigcup_{\ell=1}^{2T} (k_\ell, k_\ell + 1) \quad \begin{array}{l} \text{for positive integers } k_\ell \\ \text{and some positive integer } T. \end{array}$$

From this

$$f^* = \chi_{F^*}, \qquad g^* = \chi_{G^*}, \qquad h^* = \chi_{H^*},$$

where,
$$F^* = (-R, R), \qquad G^* = (-S, S), \qquad H^* = (-T, T).$$

From the definition of symmetric, decreasing rearrangement, it follows that for all $x \in \mathbb{R}$,
$$h^*(\cdot - x) = \chi_{H_x^*} \qquad \text{where} \qquad H_x^* = (x - T, x + T).$$

With this notation we rewrite $\mathcal{I}$ and $\mathcal{I}^*$ as

$$\mathcal{I} = \int_{\mathbb{R}} f(x)\Gamma(x)dx, \qquad \text{where} \quad \Gamma(x) = \int_{\mathbb{R}} g(y)h(y - x)dy;$$
$$\mathcal{I}^* = \int_{\mathbb{R}} f^*(x)\Gamma^*(x)dx, \quad \text{where} \quad \Gamma^*(x) = \int_{\mathbb{R}} g^*(y)h^*(y - x)dy.$$

Moreover

$$\mathcal{I}^* = \int_{-R}^{R} \Gamma^*(x)dx \quad \text{and} \quad \Gamma^*(x) = \int_{-S}^{S} \chi_{(x-T,x+T)}(y)dy.$$

14.2   Proof of (13.1) for $N = 1$. The Case $T + S \leq R$

Without loss of generality we may assume that

$$\mu(H) \leq \mu(G) \qquad \text{i.e.,} \qquad T \leq S.$$

Indeed we may always reduce to such a case, by interchanging the role of g and h and effecting a suitable change of variables in the integral $\mathcal{I}$. Estimate and compute

$$\mathcal{I} \leq \int_{\mathbb{R}} \int_{\mathbb{R}} g(y)h(y - x)dydx$$
$$= \int_{\mathbb{R}} g(y)dy \int_{\mathbb{R}} h(\eta)d\eta$$
$$= \mu(G)\mu(H) = 4ST.$$

Next we show that $\mathcal{I}^* = 4ST$. From the definition of Γ^*

$$\Gamma^*(x) = \mu\big((-S, S) \cap (x - T, x + T)\big)$$
$$= \begin{cases} 0 & \text{for} & x \leq -(S + T); \\ x + (S + T) & \text{for} & -(S + T) \leq x \leq -(S - T); \\ 2T & \text{for} & -(S - T) \leq x \leq (S - T); \\ (S + T) - x & \text{for} & (S - T) \leq x \leq (S + T); \\ 0 & \text{for} & (S + T) \leq x. \end{cases} \qquad (14.1)$$

Assuming now that $(S + T) \le R$, compute

$$\mathcal{I}^* = \int_{-R}^{R} \Gamma^*(x)dx = \int_{-(S+T)}^{(S+T)} \Gamma^*(x)dx = 4ST.$$

So far no use has been made of the structure of the sets F, G, and H. Such a structure will be employed in examining the case $S + T > R$.

14.3 Proof of (13.1) for $N = 1$. The Case $S + T > R$

Since S, T and R are positive integers, the difference $(S + T) - R$ is a positive integer, i.e.,

$$\tfrac{1}{2}\mu(G) + \tfrac{1}{2}\mu(H) - \tfrac{1}{2}\mu(F) = S + T - R = n$$

for some positive integer n. The arguments of the previous section show that the theorem holds for $n \le 0$. We show by induction that if it does hold for some integer $(n - 1) \ge 0$, then it continues to hold for n. Set

$$G_1 = \bigcup_{j=1}^{2S-1} (n_j, n_j + 1) \quad \text{i.e., the set } G \text{ from which the last}$$
interval on the right has been removed;

$$H_1 = \bigcup_{\ell=1}^{2T-1} (k_\ell, k_\ell + 1) \quad \text{i.e., the set } H \text{ from which the last}$$
interval on the right has been removed.

By construction

$$\tfrac{1}{2}\mu(G_1) + \tfrac{1}{2}\mu(H_1) - \tfrac{1}{2}\mu(F) = (S - \tfrac{1}{2}) + (T - \tfrac{1}{2}) - R = n - 1 \ge 0. \quad (14.2)$$

Set also $g_1 = \chi_{G_1}$ and $h_1 = \chi_{H_1}$. From the definitions it follows that

$$g_1^* = \chi_{G_1^*} \quad \text{where} \quad G_1^* = (-S + \tfrac{1}{2}, S - \tfrac{1}{2});$$
$$h_1^* = \chi_{H_1^*} \quad \text{where} \quad H_1^* = (-T + \tfrac{1}{2}, T - \tfrac{1}{2}).$$

Moreover,

$$h_1^*(\cdot - x) = \chi_{H_{1,x}^*} \quad \text{where} \quad H_{1,x}^* = (x - T + \tfrac{1}{2}, x + T - \tfrac{1}{2}).$$

Taking into account (14.2), the induction hypothesis is that

$$\mathcal{I}_1 = \int_{\mathbb{R}} f(x) \int_{\mathbb{R}} g_1(y) h_1(y - x) dy dx$$

$$= \int_{\mathbb{R}} f(x) \Gamma_1(x) dx$$

$$\leq \int_{\mathbb{R}} f^*(x) \int_{\mathbb{R}} g_1^*(y) h_1^*(y - x) dy dx$$

$$= \int_{\mathbb{R}} f^*(x) \Gamma_1^*(x) dx = \mathcal{I}_1^*.$$

Next observe that Γ_1^* is defined by (14.1) with S and T replaced, respectively, by $S - \frac{1}{2}$ and $T - \frac{1}{2}$, i.e.,

$$\Gamma_1^*(x) = \begin{cases} 0 & \text{for} & x \leq -(S + T - 1); \\ x + (S + T - 1) & \text{for} & -(S + T - 1) \leq x \leq -(S - T); \\ 2T - 1 & \text{for} & -(S - T) \leq x \leq (S - T); \\ (S + T - 1) - x & \text{for} & (S - T) \leq x \leq (S + T - 1); \\ 0 & \text{for} & (S + T - 1) \leq x. \end{cases}$$

From this and (14.1) one verifies that

$$\Gamma^*(x) - \Gamma_1^*(x) = 1 \qquad \text{for all } |x| \leq (S + T - 1).$$

In particular this holds true for all $|x| \leq R$, since $R \leq (S + T - 1)$. Using these remarks, compute

$$\mathcal{I}^* - \mathcal{I}_1^* = \int_{\mathbb{R}} f^*(x) \Gamma^*(x) dx - \int_{\mathbb{R}} f^*(x) \Gamma_1^*(x) dx$$

$$= \int_{-R}^{R} (\Gamma^*(x) - \Gamma_1^*(x)) dx = 2R.$$

Next we examine the structure of the function

$$x \to \Gamma(x) - \Gamma_1(x) = \mu(\{G \cap H_x\}) - \mu(G_1 \cap H_{1,x}).$$

Here by H_x and $H_{1,x}$ we have denoted the sets H and H_1 shifted by x.

Lemma 14.1 $0 \leq \Gamma(x) - \Gamma_1(x) \leq 1$, *for all* $x \in \mathbb{R}$.

Assuming the lemma for the moment, compute

$$\mathcal{I} - \mathcal{I}_1 = \int_{\mathbb{R}} f(x) \Gamma(x) dx - \int_{\mathbb{R}} f(x) \Gamma_1(x) dx$$

$$= \int_{\mathbb{R}} f(x)(\Gamma(x) - \Gamma_1(x)) dx$$

$$\leq \int_{\mathbb{R}} \chi_F dx = 2R.$$

From this

$$\mathcal{I} - \mathcal{I}_1 \leq \mathcal{I}^* - \mathcal{I}_1^*.$$

This implies the theorem since, by the induction hypothesis, $\mathcal{I}_1 \leq \mathcal{I}_1^*$. ∎

14.4 Proof of the Lemma 14.1

It is apparent that such a function is affine within any interval of the form $(n, n+1)$ for integral n. Therefore it must take its extrema for some integral value of x. If x is an integer, the set H_x is the finite union of unit intervals whose end points are integers. Now, still for integral x, the set $H_{1,x}$ is precisely H_x from which the last interval on the right has been removed. Set

$$I_G = \{\text{the rightmost interval of } G\};$$
$$I_{H_x} = \{\text{the rightmost interval of } H_x\}.$$

If I_G coincides with I_{H_x}, then removing them both, amounts to removing a single interval of unit length out of $G \cap H_x$. Therefore

$$\mu(G \cap H_x) - \mu(G_1 \cap H_{1,x}) = 1.$$

If I_{H_x} is on the right with respect to I_G, then removing it, has no effect on the intersection $G \cap H_x$, i.e.,

$$G \cap H_x = G \cap H_{1,x}.$$

Now, by removing I_G out of G, the two sets $G \cap H_x$ and $G_1 \cap H_{1,x}$, differ at most by one interval of unit length. Thus

$$\mu(G \cap H_x) - \mu(G_1 \cap H_{1,x}) \leq 1.$$

Finally, if I_{H_x} is on the left with respect to I_G, we arrive at the same conclusion by interchanging the role of G and H_x. ∎

15 The Hardy's Inequality

Proposition 15.1 (Hardy [66]) *Let $f \in L^p(\mathbb{R}^+)$ for some $p > 1$, be nonnegative. Then*

$$\int_0^\infty \frac{1}{x^p} \left(\int_0^x f(t)dt \right)^p dx \leq \left(\frac{p}{p-1} \right)^p \int_0^\infty f^p dx. \tag{15.1}$$

Proof Fix $0 < \xi < \eta < \infty$. Then, by integration by parts

$$\int_\xi^\eta \frac{1}{x^p}\Big(\int_0^x f(t)dt\Big)^p dx = \frac{-1}{p-1}\int_\xi^\eta \Big(\int_0^x f(t)dt\Big)^p \frac{d}{dx}x^{1-p}dx$$

$$= \frac{\xi^{1-p}}{p-1}\Big(\int_0^\xi f(t)dt\Big)^p - \frac{\eta^{1-p}}{p-1}\Big(\int_0^\eta f(t)dt\Big)^p$$

$$+ \frac{p}{p-1}\int_\xi^\eta x^{1-p}f(x)\Big(\int_0^x f(t)dt\Big)^{p-1} dx.$$

The second term on the right-hand side is nonpositive and it is discarded. The first term tends to zero as $\xi \to 0$. Indeed by Hölder's inequality

$$\xi^{1-p}\Big(\int_0^\xi f(t)dt\Big)^p \le \int_0^\xi f^p(t)dt.$$

Therefore letting $\xi \to 0$ and applying Hölder's inequality in the resulting inequality, gives

$$\int_0^\eta \frac{1}{x^p}\Big(\int_0^x f(t)dt\Big)^p dx$$

$$\le \frac{p}{p-1}\int_0^\eta x^{1-p}f(x)\Big(\int_0^x f(t)dt\Big)^{p-1} dx$$

$$\le \frac{p}{p-1}\Big[\int_0^\eta \frac{1}{x^p}\Big(\int_0^\eta f(t)dt\Big)^p dx\Big]^{\frac{p-1}{p}}\Big(\int_0^\eta f^p dx\Big)^{\frac{1}{p}}. \qquad \blacksquare$$

The constant on the right-hand side of (15.1) is the best possible as it can be tested for the family of functions

$$f_\varepsilon(x) = \begin{cases} x^{-\frac{1}{p}-\varepsilon} & \text{for } x \ge 1; \\ 0 & \text{for } 0 \le x < 1, \end{cases}$$

for $\varepsilon > 0$. Assume (15.1) were to hold for a smaller constant, say for example

$$\Big(\frac{p}{p-1}\Big)^p(1-\delta)^p \qquad \text{for some } \delta \in (0,1).$$

If (15.1) were applied to f_ε it would give

$$\int_1^\infty \frac{1}{x^p}\Big(\int_1^x t^{-\frac{1}{p}-\varepsilon}dt\Big)^p dx \le \Big(\frac{p}{p-1}\Big)^p(1-\delta)^p\frac{1}{p\varepsilon}. \qquad (15.2)$$

To estimate below the left-hand side, set

$$A_\varepsilon = \left(\frac{p}{p-1-\varepsilon p}\right)^p, \qquad B_{\varepsilon,\rho} = \left(1 - \frac{1}{(1+\rho)^{1-\frac{1}{p}-\varepsilon}}\right)^p$$

where $\varepsilon > 0$ is so small that $(p - 1 - \varepsilon p) > 0$ and ρ is an arbitrary positive number. Then

$$\int_1^\infty \frac{1}{x^p}\left(\int_1^x t^{-\frac{1}{p}-\varepsilon}dt\right)^p dx = A_\varepsilon \int_1^\infty \frac{1}{x^p}\left(x^{1-\frac{1}{p}-\varepsilon} - 1\right)^p dx$$

$$\geq A_\varepsilon B_\varepsilon \int_{1+\rho}^\infty x^{-1-p\varepsilon}dx$$

$$= A_\varepsilon B_\varepsilon \frac{1}{p\varepsilon}\frac{1}{(1+\rho)^{\varepsilon p}}.$$

Putting this in (15.2), multiplying by $p\varepsilon$ and letting $\varepsilon \to 0$ gives

$$1 - \frac{1}{(1+\rho)^{\frac{p-1}{p}}} \leq 1 - \delta.$$

Since $\rho > 0$ is arbitrary this is a contradiction.

16   The Hardy–Littlewood–Sobolev Inequality for $N = 1$

Theorem 16.1 ([67, 68]) *Let f and g be nonnegative measurable functions in $\mathbb{R}$ and let $p, q > 1$ and $\sigma \in (0, 1)$ be linked by*

$$\frac{1}{p} + \frac{1}{q} + \sigma = 2. \tag{16.1}$$

There exists a constant C depending only upon p, q, and σ, such that

$$\int_\mathbb{R} \int_\mathbb{R} \frac{f(x)g(y)}{|x-y|^\sigma}dxdy \leq C\|f\|_p\|g\|_q. \tag{16.2}$$

Remark 16.1 The constant $C(p, q, \sigma)$ can be computed explicitly as

$$C(p, q, \sigma) = \frac{4}{1-\sigma}\left\{\left(\frac{p}{p-1}\right)^{p\frac{q-1}{q}} + \left(\frac{q}{q-1}\right)^{q\frac{p-1}{p}}\right\}. \tag{16.3}$$

Thus $C(p, q, \sigma)$ tends to infinity as either $\sigma \to 1$ or $p \to 1$. Also $q = 1$ is not permitted in (16.2) and (16.3).

16.1   Some Reductions

Assume that (16.2) holds true for nonnegative and symmetrically decreasing functions. Then, for general nonnegative functions f and g, by the Riesz rearrangement inequality of Theorem 13.1 applied to f, g and $h(x) = |x|^{-\sigma}$

$$\int_{\mathbb{R}} \int_{\mathbb{R}} \frac{f(x)g(y)}{|x-y|^\sigma} dxdy \le \int_{\mathbb{R}} \int_{\mathbb{R}} \frac{f^*(x)g^*(y)}{|x-y|^\sigma} dxdy$$
$$\le C\|f^*\|_p\|g^*\|_q = C\|f\|_p\|g\|_q.$$

Thus it suffices to prove (16.2) for nonnegative and symmetrically decreasing functions f and g. Next, it suffices to prove the theorem in the seemingly weaker form

$$\int_0^\infty \int_0^\infty \frac{f(x)g(y)}{|x-y|^\sigma} dxdy \le C_o\|f\|_p\|g\|_q \tag{16.4}$$

for a constant C_o depending only upon p and q. For this divide the domain of integration in (16.2), into the four coordinate quadrants. By changing the sign of both variables one verifies that the contributions of the first and third quadrant to the integral in (16.2) are equal. The contribution of the second quadrant is majorized by the contribution of the first quadrant. Indeed, by changing x into $-x$,

$$\int_0^\infty \int_{-\infty}^0 \frac{f(x)g(y)}{|x-y|^\sigma} dxdy = \int_0^\infty \int_0^\infty \frac{f(x)g(y)}{|x+y|^\sigma} dxdy$$
$$\le \int_0^\infty \int_0^\infty \frac{f(x)g(y)}{|x-y|^\sigma} dxdy.$$

Similarly, the contribution of the fourth quadrant is majorized by the contribution of the first quadrant. We conclude that

$$\int_{\mathbb{R}} \int_{\mathbb{R}} \frac{f(x)g(y)}{|x-y|^\sigma} dxdy \le 4 \int_0^\infty \int_0^\infty \frac{f(x)g(y)}{|x-y|^\sigma} dxdy.$$

17   Proof of Theorem 16.1

Divide further the first quadrant into the two octants $[x \ge y]$ and $[y > x]$ and write

$$\int_0^\infty \int_0^\infty \frac{f(x)g(y)}{|x-y|^\sigma} dxdy = \int_0^\infty f(x) \left(\int_0^x \frac{g(y)}{(x-y)^\sigma} dy \right) dx$$
$$+ \int_0^\infty g(y) \left(\int_0^y \frac{f(x)}{(y-x)^\sigma} dx \right) dy$$
$$= J_1 + J_2.$$

We estimate the first of these integrals in terms of the right-hand side of (16.2), the estimation of the second being similar.

Lemma 17.1 *Let* $t \to u(t)$, $v(t)$ *be nonnegative and measurable on a measurable subset* $E \subset \mathbb{R}$ *of finite measure. Assume in addition that* u *is nondecreasing and* v *is nonincreasing. Then*

$$\int_E uv\,dt \le \frac{1}{\mu(E)} \left(\int_E u\,dt \right) \left(\int_E v\,dt \right).$$

Proof By the stated monotonicity of u and v,

$$\int_E \int_E \big(u(x) - u(y) \big) \big(v(y) - v(x) \big) dx\,dy \ge 0.$$

From this

$$\int_E \int_E u(x)v(y)dx\,dy + \int_E \int_E u(y)v(x)dx\,dy$$
$$\ge \int_E \int_E u(x)v(x)dx\,dy + \int_E \int_E u(y)v(y)dx\,dy. \qquad \blacksquare$$

Applying the Lemma with

$$E = (0, x), \qquad u(t) = (x - t)^{-\sigma}, \qquad v(t) = g(t)$$

gives

$$\int_0^x \frac{g(y)}{(x - y)^\sigma} dy \le \frac{1}{(1 - \sigma)x^\sigma} \int_0^x g(y)dy. \qquad (17.1)$$

Using Hölder's inequality, estimate

$$G(x) = \int_0^x g(y)dy \le \left(\int_0^x g^q(y)dy \right)^{\frac{1}{q}} x^{\frac{q-1}{q}}.$$

Return now to J_1. Using (17.1), the expression of $G(x)$ and Hölder's inequality

$$J_1 \le \frac{1}{1 - \sigma} \int_0^\infty f(x)x^{-\sigma}G(x)dx$$
$$\le \frac{1}{1 - \sigma} \|f\|_p \left(\int_0^\infty x^{-\sigma \frac{p}{p-1}} G(x)^{\frac{p}{p-1}} dx \right)^{\frac{p-1}{p}}$$
$$= \frac{1}{1 - \sigma} \|f\|_p \left(\int_0^\infty x^{-\sigma \frac{p}{p-1}} G(x)^q G(x)^{\frac{p}{p-1}-q} dx \right)^{\frac{p-1}{p}}$$

$$\leq \frac{1}{1-\sigma}\|f\|_p\|g\|_q^{1-q\frac{p-1}{p}}\left(\int_0^\infty x^{-\sigma\frac{p}{p-1}}x^{\frac{q-1}{q}(\frac{p}{p-1}-q)}G(x)^q dx\right)^{\frac{p-1}{p}}$$

$$\leq \frac{1}{1-\sigma}\|f\|_p\|g\|_q^{1-q\frac{p}{p-1}}\left(\int_0^\infty \frac{1}{x^q}\left(\int_0^x g(y)dy\right)^q dx\right)^{\frac{p-1}{p}}.$$

The last integral is estimated by means of Hardy's inequality and gives

$$\left(\int_0^\infty \frac{1}{x^q}\left(\int_0^x g(y)dy\right)^q dx\right)^{\frac{p-1}{p}} \leq \left(\frac{q}{q-1}\right)^{q\frac{p-1}{p}}\|g\|_q^{q\frac{p-1}{p}}.$$

Putting this in the previous inequality proves the theorem. ∎

18　The Hardy–Littlewood–Sobolev Inequality for $N \geq 1$

Theorem 18.1 ([67, 68, 148]) *Let f and g be nonnegative measurable functions in $\mathbb{R}^N$ and let $p, q > 1$ and $\sigma \in (0, N)$ be linked by*

$$\frac{1}{p}+\frac{1}{q}+\frac{\sigma}{N}=2. \tag{18.1}$$

There exists a constant $C(p, q, \sigma, N)$ depending only upon $p, q, \sigma,$ and N, such that

$$\int_{\mathbb{R}^N}\int_{\mathbb{R}^N}\frac{f(x)g(y)}{|x-y|^\sigma}dxdy \leq C(p, q, \sigma, N)\|f\|_p\|g\|_q. \tag{18.2}$$

Remark 18.1 The constant $C(p, q, N, \sigma)$ can be computed explicitly as

$$C(p, q, \sigma, N) = \frac{1}{N^{\sigma/2}}\left(\frac{4N}{N-\sigma}\right)^N\left\{\left(\frac{p}{p-1}\right)^{p\frac{q-1}{q}}+\left(\frac{q}{q-1}\right)^{q\frac{p-1}{p}}\right\}^N. \tag{18.3}$$

Thus $C(p, q, N, \sigma)$ tends to infinity as either $\sigma \to N$ or $p \to 1$. Also $q = 1$ is not permitted in (18.2).

18.1　Proof of Theorem 18.1

The arithmetic mean of N positive numbers is more than the corresponding geometric mean (§ 14.1c of Chap. 5). Therefore

$$|x-y| = \left(\sum_{i=1}^N (x_i-y_i)^2\right)^{\frac{1}{2}} \geq \sqrt{N}\prod_{i=1}^N |x_i-y_i|^{\frac{1}{N}}.$$

Set

$$\bar{x} = (x_1, \ldots, x_{N-1}) \quad \text{and} \quad \bar{y} = (y_1, \ldots, y_{N-1}).$$

From this and Fubini's Theorem

$$\int_{\mathbb{R}^N} \int_{\mathbb{R}^N} \frac{f(x)g(y)}{|x-y|^\sigma} dxdy \leq \frac{1}{N^{\frac{\sigma}{2}}} \int_{\mathbb{R}^{N-1}} \int_{\mathbb{R}^{N-1}} \frac{1}{\prod_{i=1}^{N-1} |x_i - y_i|^{\frac{\sigma}{N}}}$$

$$\times \left(\int_{\mathbb{R}} \int_{\mathbb{R}} \frac{f(\bar{x}, x_N)g(\bar{y}, y_N)}{|x_N - y_N|^{\frac{\sigma}{N}}} dx_N dy_N \right) d\bar{x} d\bar{y}$$

$$\leq \frac{C}{N^{\frac{\sigma}{2}}} \int_{\mathbb{R}^{N-1}} \int_{\mathbb{R}^{N-1}} \frac{\|f(\bar{x}, \cdot)\|_p \|g(\bar{y}, \cdot)\|_q}{\prod_{i=1}^{N-1} |x_i - y_i|^{\frac{\sigma}{N}}} d\bar{x} d\bar{y}$$

where C is the constant appearing in (16.3) with σ replaced by σ/N. Repeated application of this procedure proves the theorem. ∎

Remark 18.2 The structure of the constant C_N in (18.3) shows that neither $p = 1$, nor $q = 1$, nor $\sigma = N$ are permitted in (18.2).

19 Potential Estimates

Let f be a nonnegative function in $L^p(\mathbb{R}^N)$ for some $p \in (1, \infty)$ and set,

$$h(y) = \int_{\mathbb{R}^N} \frac{f(x)}{|x-y|^\sigma} dx \quad \text{for some } \sigma \in (0, N).$$

This is the *formal potential* of f of order σ, and it is natural to ask for what values of p and σ such a potential is well defined as an integrable function.

Let $p > 1$, $q > 1$ and $\sigma \in (0, N)$ satisfy (18.1), which we rewrite as

$$\frac{1}{p} + \frac{\sigma}{N} = 1 + \frac{1}{p^*}$$

where $p^* = q'$ is the Hölder conjugate of q. Such a number is also called the Sobolev conjugate of p. The next proposition asserts that $h \in L^{p^*}(\mathbb{R}^N)$.

Theorem 19.1 *There exists a constant C_N depending only upon σ, N and p, such that*

$$\|h\|_{p^*} \leq C_N \|f\|_p \quad \text{where} \quad \frac{1}{p^*} = \frac{1}{p} + \frac{\sigma}{N} - 1. \quad (19.1)$$

The constant C_N is the same as the one appearing in (18.3).

Proof Since $1 < p^* < \infty$ the norm $\|h\|_{p^*}$ is characterized by (§ 3.1 of Chap. 6)

$$\|h\|_{p^*} = \sup_{\substack{g \in L^q(\mathbb{R}^N) \\ \|g\|_q = 1}} \int_{\mathbb{R}^N} hg \, dy \tag{19.2}$$

$$\leq C_N \|f\|_p \|g\|_q$$

where we have applied Theorem 18.1.   ∎

Remark 19.1 The previous argument shows that Theorem 18.1 implies Theorem 19.1. On the other hand, assuming (19.1) holds true, (19.2) implies Theorem 18.1. Thus these two theorems are equivalent.

Remark 19.2 The constant C_N in (19.1) is the same as the constant $C(p, q, \sigma, N)$ in (18.3) with $p^* = \frac{q}{q-1}$. From the explicit form (18.3) it follows that the values $p = 1$ and $p^* = \infty$ are not permitted in (19.1).

20 L^p Estimates of Riesz Potentials

Let E be a Lebesgue measurable subset of $\mathbb{R}^N$ and let $f \in L^p(E)$ for some $1 \leq p \leq \infty$. The Riesz potential generated by f in $\mathbb{R}^N$, is defined by ([129, 130]),

$$V_f(x) = \int_E \frac{f(y)}{|x - y|^{N-1}}.$$

The definition is formal and it is natural to ask whether such a potential is well defined as an integrable function. When $p \in (1, N)$ an answer in this direction is provided by Theorem 19.1.

One may regard f as a function in $L^p(\mathbb{R}^N)$ by extending it to be zero outside E. By such an extension, one might regard the domain of integration in (20.1) as the whole $\mathbb{R}^N$. By Theorem 19.1 with $\sigma = N - 1$

Theorem 20.1 *Let* $f \in L^p(E)$ *for* $1 < p < N$. *There exists a constant* $C(N, p)$ *depending only upon* N *and* p, *such that*

$$\|V_f\|_{p^*} \leq C(N, p) \|f\|_p \quad \text{where} \quad p^* = \frac{Np}{N - p}. \tag{20.1}$$

Remark 20.1 The constant $C(N, p)$ in (20.2) is the same as the one in (18.3) with $\sigma = N - 1$. As such it tends to infinity as either $p \to 1$ or $p \to N$.

Remark 20.2 The value $p = 1$ is not permitted in Theorem 20.1. To construct a counterexample, let J_ε be the Friedrichs mollifying kernels introduced in § 18 of Chap. 6. If (20.2) where to hold for $p = 1$, then

$$\int_{\mathbb{R}^N} \left(\int_{\mathbb{R}^N} \frac{J_\varepsilon(y)}{|x-y|^{N-1}} dy \right)^{\frac{N}{N-1}} dx \le C$$

for a constant C independent of ε. Letting $varep \to 0$, gives the contradiction

$$\int_{\mathbb{R}^N} \frac{1}{|x|^N} dx \le C.$$

Remark 20.3 The values $p = N$ and $p^* = \infty$, are not permitted as indicated by the following counterexample. For $0 < \varepsilon \ll 1$ and $N \ge 2$, set

$$f(x) = \begin{cases} \dfrac{1}{|x|} \dfrac{1}{\big| \ln |x| \big|^{\frac{1+\varepsilon}{N}}} & \text{for } |x| \le \tfrac{1}{2}; \\ 0 & \text{for } |x| > \tfrac{1}{2}. \end{cases}$$

One verifies that $f \in L^N(\mathbb{R}^N)$ and that the corresponding potential $V_f(x)$ is unbounded near the origin.

20.1 Motivating L^p Estimates of Riesz Potentials as Embeddings

Consider the Stokes formula (14.1) of Chap. 8 written for a function $\varphi \in C_o^\infty(E)$. After an integration by parts

$$\varphi(x) = \int_E \nabla_y F(x; y) \cdot \nabla \varphi dy.$$

Using (14.2) of Chap. 8

$$|\varphi(x)| \le \frac{1}{\omega_N} \int_E \frac{|\nabla \varphi|}{|x-y|^{N-1}} dy. \tag{20.2}$$

Given now a function $u \in W_o^{1,p}(E)$ there is a sequence $\{\varphi_n\}$ of functions in $C_o^\infty(E)$ such that $\varphi_n \to u$ in the norm of $W_o^{1,p}(E)$. Thus up to a limiting process (20.2) continues to hold for functions $\varphi \in W_o^{1,p}(E)$.

Given that $|\nabla \varphi| \in L^p(E)$ it is natural to ask what is the order of integrability of φ. This motivates Theorem 20.1. Statements of this kind are called *embedding theorems* and are systematically treated in Chap. 10.

Remark 20.4 A remarkable feature of Theorem 20.1 is that the constant $C(N, p)$ is independent of E and hence it continues to hold for E of infinite measure. However $p = 1$ and $p \ge N$ are not allowed. If one permits the constant C to depend on N, p

and the measure of E, then estimates of the type of (20.2) continue to hold for $p = 1$ and $p \geq N$ as presented in the next sections.

21 L^p Estimates of Riesz Potentials for $p = 1$ and $p > N$

Similar estimates for the limiting cases $p = 1$ and $p \geq N$, require a preliminary estimation of the potential generated by a function f, constant on a set E of finite measure.

Proposition 21.1 *Let E be of finite measure. For every $r \in [1, \frac{N}{N-1})$,*

$$\sup_{x \in E} \int_E \frac{dy}{|x - y|^{(N-1)r}} \leq \frac{\kappa_N^{\frac{N-1}{N}r}}{1 - \frac{N-1}{N}r} \mu(E)^{1-\frac{N-1}{N}r} \tag{21.1}$$

where κ_N is the volume of the unit ball in $\mathbb{R}^N$.

Proof Fix $x \in E$. The symmetric rearrangement $(E - x)^*$ of $(E - x)$ is a ball about the origin of radius $\rho > 0$ such that $\mu(E) = \mu(B_\rho(x))$. Then by Proposition 12.1,

$$\begin{aligned}
\int_E \frac{dy}{|x - y|^{(N-1)r}} &= \int_{\mathbb{R}^N} \frac{\chi_{(E-x)}}{|y|^{(N-1)r}} dy \\
&\leq \int_{\mathbb{R}^N} \frac{\chi_{(E-x)^*}}{|y|^{(N-1)r}} dy \\
&= \int_{B_\rho} \frac{dy}{|y|^{(N-1)r}} = \frac{N\kappa_N}{N - (N-1)r} \rho^{N-(N-1)r}. \quad \blacksquare
\end{aligned}$$

Proposition 21.2 *Let E be of finite measure, and let $f \in L^1(E)$. Then $V_f \in L^q(E)$ for all $q \in [1, \frac{N}{N-1})$, and*

$$\|V_f\|_q \leq \frac{\kappa_N^{\frac{N-1}{N}}}{\left(1 - \frac{N-1}{N}q\right)^{\frac{1}{q}}} \mu(E)^{\frac{1}{q} - \frac{N-1}{N}} \|f\|_1. \tag{21.2}$$

Proof Fix $q \in (1, \frac{N}{N-1})$ and write

$$\frac{|f(y)|}{|x - y|^{N-1}} = |f(y)|^{1-\frac{1}{q}} \frac{|f(y)|^{\frac{1}{q}}}{|x - y|^{N-1}}.$$

Then by Hölder's inequality,

$$V_f(x) \leq \|f\|_1^{1-\frac{1}{q}} \left(\int_E \frac{|f(y)|}{|x - y|^{(N-1)q}} dy \right)^{\frac{1}{q}}.$$

Take the q-power and integrate in dx over E, to obtain

$$\|V_f\|_q^q \leq \|f\|_1^{q-1} \int_E |f(y)| \left(\int_E \frac{dx}{|x - y|^{(N-1)q}} \right) dy$$

$$\leq \|f\|_1^q \sup_{y \in E} \int_E \frac{dx}{|x - y|^{(N-1)q}}$$

$$\leq \frac{\kappa_N^{\frac{N-1}{N}q}}{1 - \frac{N-1}{N}q} \mu(E)^{1 - \frac{N-1}{N}q} \|f\|_1^q.$$ ∎

Remark 21.1 The limiting integrability $q = \frac{N}{N-1}$ is not permitted in (21.2).

Proposition 21.3 *Let E be of finite measure, and let $f \in L^p(E)$ for some $p > N$. Then $V_f \in L^\infty(E)$ and*

$$\|V_f\|_\infty \leq C(N, p)\mu(E)^{\frac{p-N}{Np}} \|f\|_p \tag{21.3}$$

where

$$C(N, p) = \kappa_N^{\frac{N-1}{N}} \left[\frac{N(p-1)}{p - N} \right]^{\frac{p-1}{p}}. \tag{21.4}$$

Proof By Hölder's inequality, and the definition of V_f,

$$\|V_f\|_\infty \leq \|f\|_p \left(\sup_{x \in E} \int_E \frac{dy}{|x - y|^{(N-1)\frac{p}{p-1}}} \right)^{\frac{p-1}{p}}.$$

The estimate (21.3) and the form (21.4) of the constant $C(N, p)$ now follow from Proposition 21.1. ∎

22 The Limiting Case $p = N$

The value $p = N$ is not permitted neither in Theorem 20.1c nor in Proposition 21.3c. The next theorem indicates that the potential V_f of a function $f \in L^N(E)$ belongs to some intermediate space lying, roughly speaking between every $L^q(E)$ and $L^\infty(E)$.

Theorem 22.1 (Trudinger [163]) *Let E be of finite measure, and let $f \in L^N(E)$. There exist constants C_1 and C_2 depending only upon N, such that*

$$\int_E \exp\left(\frac{|V_f|}{C_1\|f\|_N}\right)^{\frac{N}{N-1}} dx \le C_2\mu(E). \tag{22.1}$$

Proof For any $q > N$ and $1 < r < \frac{N}{N-1}$ satisfying

$$\frac{1}{r} = 1 + \frac{1}{q} - \frac{1}{N},$$

write formally

$$\frac{|f(y)|}{|x-y|^{N-1}} = \frac{|f(y)|^{\frac{N}{q}}}{|x-y|^{(N-1)\frac{r}{q}}} \frac{|f(y)|^{1-\frac{N}{q}}}{|x-y|^{(N-1)\frac{r(N-1)}{N}}}.$$

Then by Hölder's inequality applied with the conjugate exponents

$$\frac{q-N}{Nq} + \frac{1}{q} + \frac{N-1}{N} = 1,$$

we obtain, at least formally,

$$|V_f(x)| \le \|f\|_N^{1-\frac{N}{q}} \left(\int_E \frac{|f(y)|^N}{|x-y|^{(N-1)r}} dy\right)^{\frac{1}{q}} \left(\int_E \frac{dy}{|x-y|^{(N-1)r}}\right)^{\frac{N-1}{N}}.$$

Take the q power of both sides and integrate in dx over E to obtain

$$\|V_f\|_q \le \|f\|_N^{1-\frac{N}{q}} \left\{ \int_E |f(y)|^N \left(\int_E \frac{dx}{|x-y|^{(N-1)r}}\right) dy \right\}^{\frac{1}{q}}$$

$$\times \sup_{x\in E} \left(\int_E \frac{dy}{|x-y|^{(N-1)r}}\right)^{\frac{N-1}{N}}$$

$$\le \|f\|_N \sup_{x\in E} \left(\int_E \frac{dy}{|x-y|^{(N-1)r}}\right)^{\frac{1}{r}}.$$

These formal calculations become rigorous provided the last term is finite. By Proposition 21.1 this occurs if $r < \frac{N}{N-1}$ and we estimate

$$\sup_{x\in E}\left(\int_E \frac{dy}{|x-y|^{(N-1)r}}\right)^{\frac{1}{r}} \le \frac{\kappa_N^{\frac{N-1}{N}}}{\left(1-\frac{N-1}{N}r\right)^{\frac{1}{r}}}\mu(E)^{\frac{1}{q}}$$

$$= \kappa_N^{\frac{N-1}{N}}\left(\frac{q}{r}\right)^{1+\frac{1}{q}-\frac{1}{N}}\mu(E)^{\frac{1}{q}}$$

$$\le \kappa_N^{\frac{N-1}{N}} q^{\frac{N-1}{N}+\frac{1}{q}}\mu(E)^{\frac{1}{q}}$$

for all $q > N$. Therefore for all such q,

$$\|V_f\|_q \le \kappa_N^{\frac{N-1}{N}} q^{\frac{N-1}{N}+\frac{1}{q}} \mu(E)^{\frac{1}{q}} \|f\|_N. \tag{22.2}$$

Set $q = \frac{N}{N-1}s$ and let s range over the positive integers larger than $N - 1$. Then from (22.2), after we take the q power, we derive

$$\int_E \left[\left(\frac{|V_f|}{\|f\|_N}\right)^{\frac{N}{N-1}}\right]^s dx \le \frac{N}{N-1}\mu(E)\left(\frac{N\kappa_N}{N-1}\right)^s s^{s+1}.$$

Let C be a constant to be chosen. Divide both sides by $C^{\frac{Ns}{N-1}}s!$ and add for all integer $s = N, N+1, \ldots, k$. This gives

$$\int_E \sum_{s=N}^{k} \frac{1}{s!}\left[\left(\frac{|V_f|}{C\|f\|_N}\right)^{\frac{N}{N-1}}\right]^s dx$$
$$\le \frac{N}{N-1}\mu(E) \sum_{s=0}^{\infty} \left(\frac{N\kappa_N}{(N-1)C^{\frac{N}{N-1}}}\right)^s \frac{s^s}{(s-1)!}$$

for all $k \ge N$. The right-hand side is convergent provided we choose C so large that

$$\frac{N\kappa_N}{(N-1)C^{\frac{N}{N-1}}} < \frac{1}{e}.$$

Making use of (22.2), it is readily seen that the sum on the left-hand side can be taken for $s = 0, 1, \ldots$, by possibly modifying the various constants on the right-hand side. Letting $k \to \infty$ and using the monotone convergence theorem proves (22.1). ∎

23 Steiner Symmetrization of a Set $E \subset \mathbb{R}^N$

For a unit vector in $\mathbf{u} \in \mathbb{R}^N$ let

$$\pi_{\mathbf{u}} = \{P \in \mathbb{R}^N \mid P \cdot \mathbf{u} = 0\}$$

denote the plane through the origin normal to $\mathbf{u}$. Also, for $P \in \mathbb{R}^N$ let $\ell_{P;\mathbf{u}}$ be the line through P and directed as $\mathbf{u}$, i.e., in parametric form

$$\ell_{P;\mathbf{u}} = \bigcup_{t \in \mathbb{R}}\{P + t\mathbf{u}\}.$$

The Steiner symmetrization of a set $E \subset \mathbb{R}^N$ with respect to $\pi_{\mathbf{u}}$ is defined by

$$E_{\mathbf{u}}^* = \bigcup_{P \in \pi_{\mathbf{u}} \,:\, E \cap \ell_{P;\mathbf{u}} \ne \emptyset} \{P + t\mathbf{u} \mid |t| \le \tfrac{1}{2}\mathcal{H}_1(E \cap \ell_{P;\mathbf{u}})\} \tag{23.1}$$

where $\mathcal{H}_1(\cdot)$ is the 1-dimensional Hausdorff *outer measure* on $\mathbb{R}$ as defined in § 5 of Chap. 3. As such (23.1) is well defined for all $E \subset \mathbb{R}^N$.

Roughly speaking, from each $P \in \pi_{\mathbf{u}}$ we draw a line normal to $\pi_{\mathbf{u}}$ and look at the 1-dimensional set of intersection of $\ell_{P;\mathbf{u}}$ with E. If such intersection is nonempty, we take its 1-dimensional Hausdorff outer measure, and construct a segment of length $\mathcal{H}_1(E \cap \ell_{P;\mathbf{u}})$, symmetric about P and normal to $\pi_{\mathbf{u}}$. Thus, roughly speaking, the points of E along the line $\ell_{P;\mathbf{u}}$ are "rearranged" symmetrically, in a measure theoretical sense, about the hyperplane $\pi_{\mathbf{u}}$ starting at P, and along the same line.

Lemma 23.1 diam $E_{\mathbf{u}}^* \leq$ diam E.

Proof May assume that diam $E < \infty$ and that E is closed. Having fixed $\varepsilon > 0$, let $x, y \in E_{\mathbf{u}}^*$ be such that

$$\text{diam } E_{\mathbf{u}}^* \leq |x - y| + \varepsilon$$

and set

$$P = x - (x \cdot \mathbf{u})\mathbf{u}, \qquad Q = y - (y \cdot \mathbf{u})\mathbf{u}.$$

By the definitions $P, Q \in \pi_{\mathbf{u}}$. Set

$$\alpha = \sup\{t \mid P + t\mathbf{u} \in E\}, \qquad \beta = \inf\{t \mid P + t\mathbf{u} \in E\};$$
$$\gamma = \sup\{t \mid Q + t\mathbf{u} \in E\}, \qquad \delta = \inf\{t \mid Q + t\mathbf{u} \in E\}.$$

Without loss of generality may assume $\gamma - \beta \geq \alpha - \delta$. Then

$$\gamma - \beta \geq \tfrac{1}{2}(\gamma - \beta) + \tfrac{1}{2}(\alpha - \delta)$$
$$= \tfrac{1}{2}(\alpha - \beta) + \tfrac{1}{2}(\gamma - \delta)$$
$$\geq \tfrac{1}{2}\mathcal{H}_1(E \cap \ell_{P;\mathbf{u}}) + \tfrac{1}{2}\mathcal{H}_1(E \cap \ell_{Q;\mathbf{u}}).$$

On the other hand

$$|x \cdot \mathbf{u}| \leq \tfrac{1}{2}\mathcal{H}_1(E \cap \ell_{P;\mathbf{u}}), \qquad |y \cdot \mathbf{u}| \leq \tfrac{1}{2}\mathcal{H}_1(E \cap \ell_{Q;\mathbf{u}}).$$

Therefore

$$|x \cdot \mathbf{u} - y \cdot \mathbf{u}| \leq \gamma - \beta.$$

From this

$$(\text{diam } E_{\mathbf{u}}^* - \varepsilon)^2 \leq |x - y|^2$$
$$= |P - Q|^2 + |x \cdot \mathbf{u} - y \cdot \mathbf{u}|^2$$
$$\leq |P - Q|^2 + (\gamma - \beta)^2$$
$$= |(P + \beta\mathbf{u}) - (Q + \gamma\mathbf{u})|^2$$
$$\leq (\text{diam } E)^2. \qquad \blacksquare$$

The Steiner symmetrization does not require that E be measurable. However on measurable sets the symmetrization is volume preserving.

Lemma 23.2 *Let E be Lebesgue measurable. Then $E_{\mathbf{u}}^*$ is Lebesgue measurable and $\mu(E_{\mathbf{u}}^*) = \mu(E)$.*

Proof By Proposition 5.2 of Chap. 3, the 1-dimensional Hausdorff outer measure coincides with the 1-dimensional Lebesgue outer measure and the latter coincides with the 1-dimensional Lebesgue measure μ_1 on measurable sets. Therefore

$$\mathcal{H}_1(E \cap \ell_{P;\mathbf{u}}) = \mu_1(E \cap \ell_{P;\mathbf{u}})$$

whenever $E \cap \ell_{P;\mathbf{u}}$ is μ_1-measurable. Since the Lebesgue measure is rotation invariant, having fixed $\mathbf{u}$ on the unit sphere of $\mathbb{R}^N$, we may assume $\mathbf{u} = \mathbf{e}_N$, so that $\pi_{\mathbf{u}} = \mathbb{R}^{N-1}$. Denote by μ_{N-1} the Lebesgue measure on $\mathbb{R}^{N-1}$.

By the Tonelli version of Fubini's Theorem (§ 14.1 of Chap. 4) the sets

$$E \cap \ell(P; \mathbf{e}_N) \text{ are } \mu_1 \text{-measurable, for } \mu_{N-1} \text{-a.e. } P \in \mathbb{R}^{N-1}.$$

Moreover, the nonnegative valued function

$$\mathbb{R}^{N-1} \ni P \to f(P) \overset{\text{def}}{=} \mu_1\left(E \cap \ell_{P;\mathbf{e}_N}\right)$$

is μ_{N-1}-measurable and

$$\mu_N(E) = \int_{\mathbb{R}^{N-1}} f(P) dP.$$

It follows that the set

$$E_{\mathbf{e}_N}^* = \left\{(P, y) \mid -\tfrac{1}{2} f(P) \le y \le \tfrac{1}{2} f(P)\right\} - \left\{(P, 0) \mid E \cap \ell_{P;\mathbf{e}_N} = \emptyset\right\},$$

is Lebesgue measurable in $\mathbb{R}^N$ and (§ 15.1c of Chap. 4)

$$\mu_N(E_{\mathbf{e}_N}^*) = \int_{\mathbb{R}^{N-1}} f(P) dP. \qquad \blacksquare$$

24 Some Consequences of Steiner's Symmetrization

24.1 *Symmetrizing a Set About the Origin*

For a set $E \subset \mathbb{R}^N$ apply the Steiner symmetrization recursively with respect to all the coordinate unit vectors $(\mathbf{e}_1, \dots, \mathbf{e}_N)$, and set

$$E_1^* = E_{\mathbf{e}_1}^*, \quad E_2^* = (E_1^*)_{\mathbf{e}_2}^*, \quad \dots, \quad E_N^* = (E_{N-1}^*)_{\mathbf{e}_N}^*. \tag{24.1}$$

The set E_1^* is symmetric with respect to $\pi_{\mathbf{e}_1}$ and E_2^* is symmetric with respect to $\pi_{\mathbf{e}_2}$. We claim that E_2^* is also symmetric with respect to $\pi_{\mathbf{e}_1}$. Given $P \in \pi_{\mathbf{e}_2}$ let P' be its symmetric with respect to $\pi_{\mathbf{e}_1}$. If for some $t \in \mathbb{R}$

$$P + \mathbf{e}_2 t \in E_1^*, \quad \text{then also} \quad P' + \mathbf{e}_2 t \in E_1^*$$

since E_1^* is symmetric about $\pi_{\mathbf{e}_1}$. Therefore

$$\{t \mid P + \mathbf{e}_2 t \in E_1^*\} = \{t \mid P' + \mathbf{e}_2 t \in E_1^*\}.$$

Thus E_2^* is also symmetric with respect to the plane $\pi_{\mathbf{e}_1}$. Applying the same argument to E_3^*, as constructed by Steiner symmetrization from E_2^*, shows that E_3^* is symmetric with respect to the planes $\pi_{\mathbf{e}_j}$ for $j = 1, 2, 3$. By induction E_N^* is symmetric with respect to all the coordinate planes, and hence is symmetric about the origin.

24.2   The Isodiametric Inequality

Proposition 24.1 *For every $E \subset \mathbb{R}^N$*

$$\mu_e(E) \le \kappa_N \left(\tfrac{1}{2} \operatorname{diam} E \right)^N \tag{24.2}$$

where μ_e is the Lebesgue outer measure in $\mathbb{R}^N$ and κ_N is the measure of the unit ball in $\mathbb{R}^N$.

Proof Assuming $\operatorname{diam} E < \infty$, let E_N^* be constructed in (24.2). Since E_N^* is symmetric about the origin, if $x \in E_N^*$ then also $-x \in E_N^*$. Therefore $2|x| \le \operatorname{diam} E_N^*$ and hence E_N^* is contained in a ball of radius $\tfrac{1}{2} \operatorname{diam} E_N^*$ about the origin. From this

$$\mu_e(E_N^*) \le \kappa_N \left(\tfrac{1}{2} \operatorname{diam} E_N^* \right)^N.$$

The set $\bar{E}$ is measurable and by Lemmas 23.1 and 23.2,

$$\mu(\bar{E}_N^*) = \mu(\bar{E}), \quad \text{and} \quad \operatorname{diam} \bar{E}_N^* \le \operatorname{diam} \bar{E}.$$

From this

$$\mu_e(E) \le \mu(\bar{E}) = \mu(\bar{E}_N^*) \le \kappa_N \left(\tfrac{1}{2} \operatorname{diam} \bar{E}_N^* \right)^N$$
$$\le \kappa_N \left(\tfrac{1}{2} \operatorname{diam} \bar{E} \right)^N = \kappa_N \left(\tfrac{1}{2} \operatorname{diam} E \right)^N. \qquad \blacksquare$$

24.3 Steiner Rearrangement of a Function

Let $E \subset \mathbb{R}^N$ be Lebesgue measurable and of finite measure. The Steiner rearrangement of χ_E with respect to a unit vector $\mathbf{u}$ is the characteristic function of $E_\mathbf{u}^*$, i.e.,

$$(\chi_E)_\mathbf{u}^* = \chi_{E_\mathbf{u}^*}.$$

Next, let f be a nonnegative, simple function taking n distinct positive values $f_1 < \cdots < f_n$, on mutually disjoint sets $\{E_1, \ldots, E_n\}$, each of finite measure. Rewrite f as in (11.1), and define the Steiner rearrangement of f with respect to $\mathbf{u}$ as

$$f_\mathbf{u}^* = f_1 \chi_{(E_1 \cup \cdots \cup E_n)_\mathbf{u}^*} + (f_2 - f_1)\chi_{(E_2 \cup \cdots \cup E_n)_\mathbf{u}^*} + \cdots + (f_n - f_{n-1})\chi_{(E_n)_\mathbf{u}^*}.$$

By construction

$$[f > t]_\mathbf{u}^* = [f_\mathbf{u}^* > t], \quad \text{for all } t \geq 0. \tag{24.3}$$

Let now f be a real-valued, measurable, nonnegative function defined in $\mathbb{R}^N$ and satisfying (11.4). There exists a sequence of nonnegative, measurable, simple functions $\{f_n\} \to f$ pointwise in $\mathbb{R}^N$, and $f_n \leq f_{n+1}$. Moreover each f_n takes finitely many, distinct values on distinct, measurable sets of finite measure. Hence, $(f_n)_\mathbf{u}^*$ is well defined for each n. Moreover $(f_n)_\mathbf{u}^* \leq (f_{n+1})_\mathbf{u}^*$ and the pointwise limit of $\{(f_n)_\mathbf{u}^*\}$ exists. Define

$$f_\mathbf{u}^* = \lim (f_n)_\mathbf{u}^* = \int_0^\infty \chi_{[f>t]_\mathbf{u}^*} dt. \tag{24.4}$$

One verifies that (24.3) continues to hold in the limit and that statements analogous to those in Proposition 11.1 are in force.

25 Proof of the Riesz Rearrangement Inequality for $N = 2$

Let M_θ be the counterclockwise rotation matrix in $\mathbb{R}^2$ of an angle θ and denote by F_θ, G_θ and H_θ the measurable bounded sets obtained from F, G and H by a counterclockwise rotation of an angle θ, i.e., for example

$$F_\theta \ni (x, y) \quad \Longleftrightarrow \quad M_\theta^{-1}(x, y) \in F.$$

Denote also by S_x and S_y the operations of Steiner symmetrization of a set, about the x- and y-axes, respectively. Then, by repeated application of Fubini's theorem and the 1-dimensional version of the Riesz rearrangement inequality (13.1)

$$\mathcal{I}(F, G, H) = \int_{\mathbb{R}^2}\int_{\mathbb{R}^2} \chi_F(x)\chi_G(x - y)\chi_H(y)dxdy$$

$$= \int_{\mathbb{R}^2}\int_{\mathbb{R}^2} \chi_{F_\theta}(x)\chi_{G_\theta}(x - y)\chi_{H_\theta}(y)dxdy$$

$$\leq \int_{\mathbb{R}^2}\int_{\mathbb{R}^2} S_x\chi_{F_\theta}(x)S_x\chi_{G_\theta}(x - y)S_x\chi_{H_\theta}(y)dxdy$$

$$\leq \int_{\mathbb{R}^2}\int_{\mathbb{R}^2} S_yS_x\chi_{F_\theta}(x)S_yS_x\chi_{G_\theta}(x - y)S_yS_x\chi_{H_\theta}(y)dxdy$$

$$= \int_{\mathbb{R}^2}\int_{\mathbb{R}^2} \chi_{F_1}(x)\chi_{G_1}(x - y)\chi_{H_1}(y)dxdy = \mathcal{I}(F_1, G_1, H_1)$$

where we have set

$$F_1 = (S_yS_xM_\theta)F, \quad G_1 = (S_yS_xM_\theta)G, \quad H_1 = (S_yS_xM_\theta)H.$$

Set

$$T_\theta^1 = (S_yS_xM_\theta) \quad \text{and} \quad T_\theta^n = (S_yS_xM_\theta)T_\theta^{n-1} \quad \text{for } n = 2, 3, \ldots.$$

Repeating this process n times and setting

$$F_n = T_\theta^n F, \qquad G_n = T_\theta^n G, \qquad H_n = T_\theta^n H$$

one has

$$\mathcal{I}(F, G, H) \leq \mathcal{I}(F_n, G_n, H_n) \quad \text{for all } n \in \mathbb{N}.$$

Since the sets F, G, and H are bounded, they are contained in some disc D about the origin of $\mathbb{R}^2$. Then, by the definition of T_θ^n, the sets F_n, G_n, and H_n are all contained in the same disc D. In particular the sequences $\{\chi_{F_n}\}$, $\{\chi_{G_n}\}$, and $\{\chi_{H_n}\}$ are equi-uniformly bounded.

By the 1-dimensional version of the Riesz rearrangement inequality (13.1) the sequence $\{\mathcal{I}(F_n, G_n, H_n)\}$ is nondecreasing and hence it has a limit. Therefore, the proof of the Riesz rearrangement inequality for $N = 2$ reduces to showing that

$$\lim \chi_{F_n}, \chi_{G_n}, \chi_{H_n} = \chi_{F^*}, \chi_{G^*}, \chi_{H^*} \quad \text{a.e. in } \mathbb{R}^2$$

where F^*, G^*, and H^* are the symmetric, decreasing rearrangements of F, G, and H about the origin of $\mathbb{R}^2$. Indeed, if this is established, by dominated convergence

$$\lim \mathcal{I}(F_n, G_n, H_n) = \mathcal{I}(F^*, G^*, H^*).$$

We will establish (25.1) for $\{F_n\}$, the proof for the remaining sets being analogous.

25.1 The Limit of $\{F_n\}$

Proposition 25.1 *There exists a measurable set $F_* \subset D$, and a subsequence $\{F_{n'}\} \subset \{F_n\}$, such that*

$$\lim \chi_{F_{n'}} = \chi_{F_*} \quad \text{a.e. in } \mathbb{R}^2. \tag{25.1}$$

Moreover

$$S_y S_x M_\theta F_* = S_x M_\theta F_* = S_y M_\theta F_* = M_\theta F_*. \tag{25.2}$$

Proof Consider the portion of F_n in the right, upper, quarter plane, i.e.,

$$F_n^+ = F_n \cap [x > 0] \cap [y > 0].$$

The least upper bound of F_n^+ in $[x > 0] \cap [y > 0]$ is the graph of a function f_n, which by the symmetrizations S_x and S_y is nonincreasing. The family $\{f_n\}$ is uniformly bounded and uniformly of bounded variation in some common interval $(0, b)$ for some $b > 0$. Hence by the Helly's selection principle (§ 19.1c of the Complements of Chap. 6), there exists a nonincreasing function f defined in $(0, b)$ and a subsequence $\{f_{n'}\} \subset \{f_n\}$ such that $\{f_{n'}\} \to f$ pointwise *everywhere* in $(0, b)$. Define

$$F_*^+ = \bigcup \{(x, y) \mid 0 < y \le f(x), \text{ for } x > 0\}.$$

Since f is measurable, the set F_*^+ is measurable, and

$$\mu(F_*^+) = \int_0^\infty f \, dx = \int_0^\infty \int_0^\infty \chi_{F_*^+} \, dx \, dy$$

By dominated convergence

$$\lim \|\chi_{F_{n'}^+} - \chi_{F_*^+}\|_2 = 0.$$

Since the sets $F_{n'}$ are symmetric with respect to the x- and y-axes, there exists a set F_* symmetric with respect to the x- and y-axes, such that

$$\lim \chi_{F_{n'}} = \chi_{F_*} \quad \text{a.e. in } \mathbb{R}^2.$$

To establish (25.2) we first observe that for any two sets E_1 and E_2 of finite measure

$$\|S_y \chi_{E_1} - S_y \chi_{E_2}\|_2 \le \|\chi_{E_1} - \chi_{E_2}\|_2;$$
$$\|S_x \chi_{E_1} - S_x \chi_{E_2}\|_2 \le \|\chi_{E_1} - \chi_{E_2}\|_2;$$
$$\|S_x S_y \chi_{E_1} - S_x S_y \chi_{E_2}\|_2 \le \|\chi_{E_1} - \chi_{E_2}\|_2; \tag{25.3}$$
$$\|M_\theta \chi_{E_1} - M_\theta \chi_{E_2}\|_2 = \|\chi_{E_1} - \chi_{E_2}\|_2;$$
$$\|T_\theta^1 \chi_{E_1} - T_\theta^1 \chi_{E_2}\|_2 \le \|\chi_{E_1} - \chi_{E_2}\|_2.$$

The first three follow from the contracting properties of symmetric rearrangements of Theorem 12.1, applied for $N = 1$ and repeated application of Fubini's theorem. The fourth is implied by the rotational invariance of the Lebesgue measure in $\mathbb{R}^N$. The last one is a sequential application of the previous ones. Next compute

$$\lim \| T_\theta^1 \chi_{F_{n'}} - T_\theta^1 \chi_{F_*} \|_2 \leq \lim \| \chi_{F_{n'}} - \chi_{F_*} \|_2 = 0.$$

Let φ be a radially symmetric, strictly decreasing positive function. For such a φ

$$S_x \varphi = S_y \varphi = M_\theta \varphi = T_\theta^1 \varphi = \varphi.$$

By dominated convergence

$$\lim \| \varphi - \chi_{F_{n'}} \|_2 = \| \varphi - \chi_{F_*} \|_2;$$
$$\lim \| \varphi - T_\theta^1 \chi_{F_{n'}} \|_2 = \| \varphi - T_\theta^1 \chi_{F_*} \|_2.$$

Moreover by (25.3)

$$\| \varphi - T_\theta^1 \chi_{F_*} \|_2 \leq \| \varphi - \chi_{F_*} \|_2.$$

By repeated application of the contracting properties of symmetric rearrangements and (25.3), for all n', there exists an integer $k(n') \geq 1$ such that

$$\| \varphi - \chi_{F_{(n+1)'}} \|_2 = \| \varphi - \chi_{F_{n'+k(n')}} \|_2$$
$$= \| \varphi - T_\theta^{k(n')} \chi_{F_{n'}} \|_2 \leq \| \varphi - T_\theta^1 \chi_{F_{n'}} \|_2.$$

From this

$$\| \varphi - \chi_{F_*} \|_2 = \lim \| \varphi - \chi_{F_{(n+1)'}} \|_2 \leq \lim \| \varphi - T_\theta^1 \chi_{F_{n'}} \|_2$$
$$= \| \varphi - T_\theta^1 \chi_{F_*} \|_2 \leq \| \varphi - S_x M_\theta \chi_{F_*} \|_2 \leq \| \varphi - \chi_{F_*} \|_2.$$

A similar chain of inequalities holds with S_x replaced by S_y. Hence

$$\| \varphi - \chi_{F_*} \|_2 = \| \varphi - M_\theta \chi_{F_*} \|_2 = \| \varphi - T_\theta^1 \chi_{F_*} \|_2$$
$$= \| \varphi - S_x M_\theta \chi_{F_*} \|_2 = \| \varphi - S_y M_\theta \chi_{F_*} \|_2 = \| \varphi - \chi_{F_*} \|_2.$$

Expanding the L^2-norm and using the volume preserving properties of the rearrangements, implies

$$\int_{\mathbb{R}^2} \varphi \chi_{M_\theta F_*} dx = \int_{\mathbb{R}^2} \varphi S_x \chi_{M_\theta F_*} dx = \int_{\mathbb{R}^2} \varphi S_y \chi_{M_\theta F_*} dx.$$

Repeated application of Proposition 12.2, along the x- and y-axes, with the aid of Fubini's theorem yields (25.2).   ∎

25.2 The Set F_* Is the Disc F^*

The set F_* is a disc if $M_\delta F_* = F_*$ for all $\delta \in [0, 2\pi)$. The angle θ so far is arbitrary and it will be chosen shortly. By (25.2) the set $M_\theta F_*$ is symmetric with respect to both coordinate axes. Therefore $M_\theta F_* = M_{-\theta} F_*$, implying

$$M_{2\theta} F_* = F_* \quad \text{and hence} \quad M_{k2\theta} F_* = F_* \quad \text{for all } k \in \mathbb{Z}.$$

Now choose 2θ to be an irrational multiple of 2π. For such a choice, the collection of numbers of the form $\{k2\theta \bmod 2\pi\}$ for $k \in \mathbb{Z}$, is dense in $[0, 2\pi)$. Therefore the function

$$[0, 2\pi) \ni \delta \rightarrow \|\chi_{F_*} - M_\delta \chi_{F_*}\|_2$$

vanishes on a dense subset of $[0, 2\pi)$. If such a function were continuous, then $M_\delta F_* = F_*$ a.e. in $\mathbb{R}^2$, for all $\delta \in [0, 2\pi)$ and F_* is a disc. To prove the continuity of such a function of δ it suffices to show that

$$[0, 2\pi) \ni \delta \rightarrow \int_{\mathbb{R}^2} \chi_{F_*} M_\delta \chi_{F_*} dx$$

is continuous. Since $C_o^\infty(D)$ is dense in $L^2(D)$, for $\varepsilon > 0$ fixed, there exists $\psi_\varepsilon \in C_o^\infty$ such that

$$\int_{\mathbb{R}^2} |(\chi_{F_*} - \psi_\varepsilon) M_\delta \chi_{F_*}| dx \leq \|\chi_{F_*} - \psi_\varepsilon\|_2 \sqrt{\mu(F)} \leq \tfrac{1}{3}\varepsilon$$

uniformly in δ. The function ψ_ε being fixed, there exists $\eta_\varepsilon > 0$ such that

$$\int_{\mathbb{R}^2} |M_{-\delta-\eta}\psi_\varepsilon - M_{-\delta}\psi_\varepsilon| \chi_{F_*} dx \leq \tfrac{1}{3}\varepsilon \quad \text{for all } |\eta| < \eta_\varepsilon.$$

Therefore

$$\left| \int_{\mathbb{R}^2} \chi_{F_*} M_\delta \chi_{F_*} dx - \int_{\mathbb{R}^2} \chi_{F_*} M_{\delta+\eta} \chi_{F_*} dx \right|$$

$$\leq \left| \int_{\mathbb{R}^2} (\chi_{F_*} - \psi_\varepsilon) M_\delta \chi_{F_*} dx \right| + \left| \int_{\mathbb{R}^2} (\chi_{F_*} - \psi_\varepsilon) M_{\delta+\eta} \chi_{F_*} dx \right|$$

$$+ \left| \int_{\mathbb{R}^2} \psi_\varepsilon M_{\delta+\eta} \chi_{F_*} dx - \int_{\mathbb{R}^2} \psi_\varepsilon M_\delta \chi_{F_*} dx \right|$$

$$\leq \tfrac{2}{3}\varepsilon + \left| \int_{\mathbb{R}^2} (M_{-\delta-\eta}\psi_\varepsilon - M_{-\delta}\psi_\varepsilon) \chi_{F_*} dx \right| \leq \varepsilon.$$

∎

26 Proof of the Riesz Rearrangement Inequality for $N > 2$

The proof is by induction. Denote the coordinates of $\mathbb{R}^N$ by $x = (\bar{x}, x_N)$ and set

$$S_N = \{\text{the Steiner symmetrization with respect to } x_N\};$$
$$\Sigma_{\bar{x}} = \{\text{the symmetric rearrangement with respect to } \bar{x} \in \mathbb{R}^{N-1}\}.$$

Denote also by M the unitary matrix that rotates the x_N axis by $\pi/2$, interchanges the x_{N-1} and x_N axes, and keeps the remaining $(N-2)$ axes unchanged. Set also

$$T^1 = (\Sigma_{\bar{x}} S_N M) \quad \text{and} \quad T^n = (\Sigma_{\bar{x}} S_N M) T^{n-1} \quad \text{for } n = 2, 3, \ldots.$$

Proceeding as before we introduce sets

$$F_n = T^n F, \qquad G_1 = T^n G, \qquad H_1 = T^n H.$$

The proof reduces to showing that these sets tend, in some appropriate sense, to F^*, G^*, and H^*. Concentrating on $\{F_n\}$, observe that these sets are radially symmetric with respect to the first $(N-1)$ variables, and symmetric with respect to x_N. Their least upper bound in the $\frac{1}{2} -$ space $[x_N > 0]$ are graphs of functions $\{f_n\}$, defined in $\mathbb{R}^{N-1}$, radially symmetric, and nonincreasing in $|\bar{x}|$. As such, they can be regarded as functions of one variable to which the Helly's selection principle can be applied. The same procedure as before now yields a limiting set F_* satisfying

$$\Sigma_{\bar{x}} S_N M F_* = S_N M F_* = M F_* = F_* \tag{26.1}$$

in the sense of the characteristic functions of these sets. The set F_* is radially symmetric with respect to the first $(N-1)$ variables, and symmetric with respect to x_N. Moreover by the last of (26.1), the role of x_N and x_{N-1} can be interchanged. Thus χ_{F_*} depends on x, formally, as

$$\chi\left(\sqrt{\textstyle\sum_{j=1}^{N-1} x_j^2}, \pm x_N\right) = \chi\left(\sqrt{\textstyle\sum_{j=1}^{N-2} x_j^2 + x_N^2}, \pm x_{N-1}\right).$$

By setting first $x_{N-1} = 0$ and then $x_N = 0$, this implies that χ_{F_*} depends on x radially, thereby proving that F_* is a ball. The argument can be made nonformal, by approximating χ_{F_*} in $L^2(\mathbb{R}^N)$, by smooth functions that preserve the indicated symmetry. ∎

Problems and Complements

11c Rearranging the Values of a Function

Let f be a real-valued, nonnegative, measurable function, satisfying (11.4).

11.1 Prove that the definition of f^* introduced in (11.2) or (11.5) is equivalent to (compare with (15.3) of Chap. 4)

$$f^* = \int_0^\infty \chi^*_{[f>t]} dt. \tag{11.1c}$$

11.2 Give a detailed proof of **(iii)** of Proposition 11.1.

11.3 Prove the following more general version of **(vii)** of Proposition 11.1. Let $\varphi(\cdot) : \mathbb{R}^+ \to \mathbb{R}^+$ be monotone increasing. Then

$$\int_{\mathbb{R}^N} \varphi(f) d\mu = \int_{\mathbb{R}^N} \varphi(f^*) d\mu. \tag{vii}'$$

11.4 Prove that **(vii)**$'$ continues to hold for $\varphi = \varphi_1 - \varphi_2$, where each of the φ_j are monotone increasing and for at least one of them the corresponding integral in **(vii)**$'$ is finite.

11.5 For a measurable set E of finite measure redefine E^*_c as the *closed* ball about the origin of radius

$$\kappa_N R^N = \mu(E).$$

Then redefine the nondecreasing, symmetric rearrangement $f^{*'}_c$ of a nonnegative function f, satisfying (11.4), by the same procedure as in § 11 with the proper modifications. Prove that all statements in Proposition 11.1 continue to hold, except that f^*_c is upper semi-continuous.

11.6 If f does not satisfy (11.4) then the symmetric, decreasing rearrangement of f can still be defined by setting

$$\chi^*_{[f>t]} = \begin{cases} 0 & \text{if } \mu([f>t]) = 0; \\ \chi_{|x|<R} & \text{if } \mu([f>t]) < \infty; \\ & \text{where } \kappa_N R^N = \mu([f>t]); \\ 1 & \text{if } \mu([f>t]) = \infty. \end{cases}$$

The symmetric rearrangement of f is then defined by the formula (11.1c) using this new definition of $\chi^*_{[f>t]}$. Prove that all statements of Proposition 11.1 remain force.

12c   Some Integral Inequalities for Rearrangements

Let f and g be real-valued, nonnegative, measurable functions, satisfying (11.4).

12.1 Let E be a measurable set in $\mathbb{R}^N$ of finite measure. Let B_ρ be a ball of radius ρ about the origin and apply (12.1) with $f = \chi_{B_\rho}$ and $g = \chi_E$. Assume that for all balls B_ρ (12.1) holds with equality, i.e.,

$$\int_{\mathbb{R}^N} \chi_{B_\rho}\chi_E d\mu = \int_{\mathbb{R}^N} \chi_{B_\rho}^* \chi_E^* d\mu.$$

Prove that $E = E^*$.

12.2 Let $f = f^*$ be strictly decreasing. Prove that (12.1) holds with equality if and only if $g = g^*$.

12.3 Let $f = f^*$ be strictly decreasing. Prove that (12.2) holds with equality for all $t \geq 0$ if and only if $g = g^*$.

12.4 Prove the following more general version of Theorem 12.1

Theorem 12.1c *Let φ be a nonnegative convex function in $\mathbb{R}$ vanishing at the origin. Then*

$$\int_{\mathbb{R}^N} \varphi(f^* - g^*)d\mu \leq \int_{\mathbb{R}^N} \varphi(f - g)d\mu. \tag{12.1c}$$

When $\varphi(t) = |t|^p$ for $p \geq 1$ this is Theorem 12.1.

12.5 Let φ be strictly convex and let $f = f^*$ be strictly decreasing. Prove that (12.1c) holds with equality if and only if $g = g^*$.

20c L^p Estimates of Riesz Potentials

Let E be a domain in $\mathbb{R}^N$ and for $\alpha \in (0, N)$ and $f \in L^p(E)$, consider the potentials

$$V_{\alpha,f}(x) = \int_E \frac{f(y)}{|x - y|^{N-\alpha}} dy.$$

If $\alpha = 1$ these coincide with the Riesz potentials.

Theorem 20.1c *Let $f \in L^p(E)$ for $1 < p < \frac{N}{\alpha}$. There exists a constant $C(N, p, \alpha)$ depending only upon N, p and α such that*

$$\|V_{\alpha,f}\|_q \leq C(N, p, \alpha)\|f\|_p, \qquad where \quad q = \frac{Np}{N - \alpha p}. \tag{20.1c}$$

Compute explicitly the constant $C(N, p, \alpha)$ and verify that it tends to infinity as either $p \to 1$ or $p \to \frac{N}{\alpha}$.

21c $\;L^p$ Estimates of Riesz Potentials for $p = 1$ and $p > N$

Proposition 21.1c *For every $\alpha \in (0, N)$ and every $1 \le r < \frac{N}{N-\alpha}$*

$$\sup_{x \in E} \int_E \frac{dy}{|x - y|^{(N-\alpha)r}} \le \frac{\kappa_N^{\frac{N-\alpha}{N}r}}{1 - \frac{N-\alpha}{N}r} \mu(E)^{1 - \frac{N-\alpha}{N}r}. \tag{21.1c}$$

Proposition 21.2c *Let E be of finite measure, and let $f \in L^1(E)$. Then $V_{\alpha, f} \in L^q(E)$ for all $q \in [1, \frac{N}{N-\alpha})$, and*

$$\|V_{\alpha, f}\|_q \le \frac{\kappa_N^{\frac{N-\alpha}{N}}}{\left(1 - \frac{N-\alpha}{N}q\right)^{\frac{1}{q}}} \mu(E)^{\frac{1}{q} - \frac{N-\alpha}{N}} \|f\|_1. \tag{21.2c}$$

Remark 21.1c The limiting integrability $q = \frac{N}{N-\alpha}$ is not permitted in (21.2c).

Proposition 21.3c *Let E be of finite measure, and let $f \in L^p(E)$ for some $p > \frac{N}{\alpha}$. Then $V_{\alpha, f} \in L^\infty(E)$ and*

$$\|V_{\alpha, f}\|_\infty \le C(N, p, \alpha)\mu(E)^{\frac{\alpha p - N}{Np}} \|f\|_p \tag{21.3c}$$

where

$$C(N, p, \alpha) = \kappa_N^{\frac{N-\alpha}{N}} \left(\frac{N(p - 1)}{\alpha p - N}\right)^{\frac{p-1}{p}}. \tag{21.4c}$$

22c $\;$ The Limiting Case $p = \frac{N}{\alpha}$

The value $\alpha p = N$ is not permitted neither in Theorem 20.1c nor in Proposition 21.3c. Prove the following:

Theorem 22.1c *Let E be of finite measure, and let $f \in L^p(E)$ for $p = \frac{N}{\alpha}$. There exist constants C_1 and C_2 depending only upon N and α, such that*

$$\int_E \exp\left(\frac{|V_{\alpha, f}|}{C_1 \|f\|_p}\right)^{\frac{N}{N-\alpha}} dx \le C_2\mu(E). \tag{22.1c}$$

23c Some Consequences of Steiner's Symmetrization

23.1c Applications of the Isodiametric Inequality

The Hausdorff outer measure $\mathcal{H}_\alpha$, introduced in § 5 of Chap. 3, can be properly re-normalized by a factor γ_α, so that when $\alpha = N$ it coincides with the Lebesgue outer measure μ_e in $\mathbb{R}^N$. Set

$$\gamma_\alpha = \frac{\pi^{\frac{\alpha}{2}}}{2^\alpha \Gamma\left(\frac{\alpha}{2}+1\right)} \qquad \text{where} \qquad \Gamma(t) = \int_0^\infty e^{-x} x^{t-1} dx, \quad t > 0,$$

is the Euler gamma function. One verifies that for $\alpha = N$

$$\gamma_N = \frac{\kappa_N}{2^N}, \qquad \kappa_N = \{\text{volume of the unit ball in } \mathbb{R}^N\}.$$

Define the re-normalized Hausdorff outer measure as

$$H_\alpha = \gamma_\alpha \mathcal{H}_\alpha.$$

Proposition 23.1c $H_N(E) = \mu_e(E)$ *for all subsets* $E \subset \mathbb{R}^N$.

Proof We may assume that $\mu_e(E) < \infty$. Having fixed $\varepsilon > 0$, let $\{E_j\}$ be a countable collection of sets in $\mathbb{R}^N$ each of diameter not exceeding ε and such that $E \subset \cup E_j$. By the isodiametric inequality

$$\mu_e(E) \leq \sum \mu_e(E_j) \leq \sum \kappa_N (\tfrac{1}{2} \operatorname{diam} E_j)^N = \gamma_N \sum (\operatorname{diam} E_j)^N.$$

Taking the infimum of all such collections $\{E_j\}$, and then letting $\varepsilon \to 0$, gives

$$\mu_e(E) \leq \gamma_N \mathcal{H}_\alpha(E) = H_\alpha(E).$$

For $\varepsilon > 0$ fixed, let $\{Q_j\}$ be a countable collection of cubes in $\mathbb{R}^N$ with faces parallel to the coordinate planes, such that $E \subset \bigcup Q_j$ and

$$\mu_e(E) \geq \sum \mu_e(Q_j) - \varepsilon.$$

By the Besicovitch measure theoretical covering of open sets in $\mathbb{R}^N$ (Proposition 18.1c of Chap. 3) for each Q_j there exists a countable collection of disjoint, closed balls $\{B_{i,j}\}$, contained in $\overset{o}{Q}_j$, of diameter not exceeding ε such that

$$\mu_e\left(Q_j - \bigcup B_{i,j}\right) = \mu\left(Q_j - \bigcup B_{i,j}\right) = \mu\left(\overset{o}{Q}_j - \bigcup B_{i,j}\right) = 0.$$

For these residual sets, by Proposition 5.2 of Chap. 3,

$$\mathcal{H}_N\left(Q_j - \bigcup B_{i,j}\right) = 0.$$

Then compute and estimate

$$
\begin{aligned}
H_N(E) &\leq \sum H_N(Q_j) \leq \sum_j H_N\left(\bigcup_i B_{i,j}\right) \\
&\leq \sum_{i,j} \gamma_N (\operatorname{diam} B_{i,j})^N = \sum_{i,j} \mu_e(B_{i,j}) \\
&= \sum \mu_e(Q_j) \leq \mu_e(E) + \varepsilon.
\end{aligned}
$$

$\blacksquare$

Chapter 10
Embeddings of $W^{1,p}(E)$ into $L^q(E)$

1 Multiplicative Embeddings of $W_o^{1,p}(E)$

Let E be a domain in $\mathbb{R}^N$. An embedding from $W^{1,p}(E)$ into $L^q(E)$ is an estimate of the $L^q(E)$-norm of a function $u \in W^{1,p}(E)$, in terms of its $W^{1,p}(E)$-norm. The structure of such an estimate and the various constants involved should not depend neither on the particular function $u \in W^{1,p}(E)$ nor on the size of E, although they might depend on the structure of ∂E. Since typically $q > p$, an embedding estimate amounts, roughly speaking, to an improvement on the local order of integrability of u. Also if p is sufficiently large one might expect a function $u \in W^{1,p}(E)$ to possess some local regularity, beyond a higher degree of integrability.

The backbone of such embeddings is that of $W_o^{1,p}(E)$ into $L^q(E)$, in view of its relative simplicity. Functions in $W_o^{1,p}(E)$ are limits of functions in $C_o^\infty(E)$ in the norm of $W_o^{1,p}(E)$, and in this sense they vanish on ∂E. This permits embedding inequalities in a multiplicative form, such as (1.1) below. Such an inequality would be false for functions not vanishing, in some sense, on a subset of $\bar{E}$. For example a constant, nonzero function would not satisfy (1.1). The proof is only based on calculus ideas and applications of Hölder's inequality.

Theorem 1.1 (Gagliardo [54]–Nirenberg[118]) *Let E be a domain in $\mathbb{R}^N$ and let $u \in W_o^{1,p}(E) \cap L^r(E)$ for some $r \geq 1$. There exists a constant C depending upon N, p, r such that*

$$\|u\|_q \leq C\|Du\|_p^\theta \|u\|_r^{1-\theta} \tag{1.1}$$

where $\theta \in [0, 1]$ and $p, q \geq 1$ are linked by

$$\theta = \left(\frac{1}{r} - \frac{1}{q}\right)\left(\frac{1}{N} - \frac{1}{p} + \frac{1}{r}\right)^{-1} \tag{1.2}$$

© Springer Science+Business Media New York 2016
E. DiBenedetto, *Real Analysis*, Birkhäuser Advanced
Texts Basler Lehrbücher, DOI 10.1007/978-1-4939-4005-9_10

and their admissible range is

$$
\begin{cases}
\text{If } N = 1 \quad \text{then} \quad q \in [r, \infty] \quad \text{and} \\
\theta \in \left[0, \dfrac{p}{p + r(p-1)}\right], \quad C = \left(1 + \dfrac{p-1}{rp}\right)^{\theta};
\end{cases}
\tag{1.3}
$$

$$
\begin{cases}
\text{If } N > p \geq 1 \quad \text{then} \\[2mm]
q \in \left[r, \dfrac{Np}{N-p}\right] \quad \text{if} \quad r \leq \dfrac{Np}{N-p} \\[3mm]
q \in \left[\dfrac{Np}{N-p}, r\right] \quad \text{if} \quad r \geq \dfrac{Np}{N-p} \\[3mm]
\theta \in [0, 1] \quad \text{and} \quad C = \left[\dfrac{p(N-1)}{N-p}\right]^{\theta};
\end{cases}
\tag{1.4}
$$

$$
\begin{cases}
\text{If} \quad p > N > 1 \quad \text{then} \quad q \in [r, \infty] \quad \text{and} \\[2mm]
\theta \in \left[0, \dfrac{Np}{Np + r(p-N)}\right] \\[3mm]
\text{If} \quad p = N \quad \text{then } q \in [r, \infty) \text{ and } \theta \in [0, 1)
\end{cases}
\tag{1.5}
$$

moreover the constant C is given explicitly in (6.1) below.

By taking $\theta = r = 1$ in (1.4), yields the embedding

Corollary 1.1 *Let* $u \in W_o^{1,p}(E)$ *for* $1 \leq p < N$. *Then*

$$
\|u\|_{p^*} \leq \frac{p(N-1)}{N-p}\|Du\|_p \quad \text{where} \quad p^* = \frac{Np}{N-p}.
\tag{1.6}
$$

Remark 1.1 The constant $\frac{p(N-1)}{N-p}$ in (1.6) is not optimal. The best constant is computed in [157]. When $p = 1$, (1.6) with best constant takes the form

$$
\|u\|_{\frac{N}{N-1}} \leq \frac{1}{N}\left(\frac{N}{\omega_N}\right)^{1/N}\|Du\|_1.
\tag{1.7}
$$

where ω_N is the area of the unit sphere in $\mathbb{R}^N$.

1.1 Proof of Theorem 1.1

Since $C_o^\infty(E)$ is dense in $W_o^{1,p}(E)$, in proving Theorem 1.1 may assume that $u \in C_o^\infty(E)$. Also since u is compactly supported, we may assume, possibly after a translation, that its support is contained in a cube centered at the origin and faces parallel to the coordinate planes, say for example

$$Q = \left\{ x \in \mathbb{R}^N \mid \max_{1 \le i \le N} |x_i| < M \right\} \quad \text{for some} \quad M > 0.$$

Thus $u \in C_o^\infty(Q)$. The embedding constant C in (1.1) is independent of M.

2 Proof of Theorem 1.1 for $N = 1$

Assume first $p > 1$ and $q < \infty$. For $r \ge 1$, $s > 1$ and $q \ge r$, and all $x \in Q$

$$|u(x)|^q = |u(x)|^r \left(|u(x)|^s \right)^{\frac{q-r}{s}} = |u(x)|^r \left(\int_{-\infty}^x D|u(\xi)|^s d\xi \right)^{\frac{q-r}{s}}$$

$$\le |u(x)|^r \left(\int_E s|u(\xi)|^{s-1}|Du(\xi)|d\xi \right)^{\frac{q-r}{s}}.$$

Integrate in dx over E and apply Hölder's inequality to the last integral to obtain

$$\|u\|_q^q \le \|u\|_r^r \left(s \|Du\|_p \|u\|_{\frac{p(s-1)}{p-1}}^{s-1} \right)^{\frac{q-r}{s}}.$$

Choose s from

$$\frac{p(s-1)}{p-1} = r \quad \text{i.e.,} \quad s = 1 + \frac{r(p-1)}{p}$$

and set

$$\theta = \frac{q-r}{qs} = \left(\frac{1}{r} - \frac{1}{q} \right) \left(\frac{1}{N} - \frac{1}{p} + \frac{1}{r} \right)^{-1}.$$

Then

$$\|u\|_q \le C\|Du\|_p^\theta \|u\|_r^{1-\theta} \quad \text{where} \quad C = \left[1 + \frac{r(p-1)}{p} \right]^\theta.$$

The limiting cases $p = 1$ and $q = \infty$ are established similarly. ∎

3   Proof of Theorem 1.1 for $1 \le p < N$

The proof for the case (1.4) uses two auxiliary lemmas.

Lemma 3.1 $\|u\|_{\frac{N}{N-1}} \le \|Du\|_1$.

Proof It suffices to establish the stronger inequality

$$\|u\|_{\frac{N}{N-1}} \le \prod_{i=1}^{N} \|u_{x_i}\|_1^{1/N}. \tag{3.1}$$

Such inequality holds for $N = 2$. Indeed

$$
\begin{aligned}
\iint_E u^2(x_1, x_2) dx_1 dx_2 &= \iint_E u(x_1, x_2) u(x_1, x_2) dx_1 dx_2 \\
&\le \iint_E \max_{x_2} u(x_1, x_2) \max_{x_1} u(x_1, x_2) dx_1 dx_2 \\
&= \int_{\mathbb{R}} \max_{x_2} u(x_1, x_2) dx_1 \int_{\mathbb{R}} \max_{x_1} u(x_1, x_2) dx_2 \\
&\le \left(\iint_E |u_{x_1}| dx \right) \left(\iint_E |u_{x_2}| dx \right).
\end{aligned}
$$

The lemma is now proved by induction, that is if (3.1) hold for $N \ge 2$ it continues to hold for $N + 1$. Set

$$\bar{x} = (x_1, \ldots, x_N), \qquad t = x_{N+1}, \qquad x = (\bar{x}, t).$$

Then by Hölder's inequality

$$
\begin{aligned}
\|u\|_{\frac{N+1}{N}}^{\frac{N+1}{N}} &= \int_{\mathbb{R}} \int_{\mathbb{R}^N} |u(\bar{x}, t)|^{\frac{N+1}{N}} d\bar{x} dt = \int_{\mathbb{R}} \int_{\mathbb{R}^N} |u(\bar{x}, t)| |u(\bar{x}, t)|^{\frac{1}{N}} d\bar{x} dt \\
&\le \int_{\mathbb{R}} \left(\int_{\mathbb{R}^N} |u(\bar{x}, t)| d\bar{x} \right)^{\frac{1}{N}} \left(\int_{\mathbb{R}^N} |u(\bar{x}, t)|^{\frac{N}{N-1}} d\bar{x} \right)^{\frac{N-1}{N}} dt.
\end{aligned}
$$

Observe that

$$\int_{\mathbb{R}^N} |u(\bar{x}, t)| d\bar{x} \le \int_{\mathbb{R}} \int_{\mathbb{R}^N} |u_t(\bar{x}, t)| d\bar{x} dt = \int_E |u_t| dx.$$

Moreover, by the induction hypothesis

$$\left(\int_{\mathbb{R}^N} |u(\bar{x}, t)^{\frac{N}{N-1}} d\bar{x} \right)^{\frac{N-1}{N}} \le \prod_{i=1}^{N} \left(\int_{\mathbb{R}^N} |u_{x_i}(\bar{x}, t)| d\bar{x} \right)^{\frac{1}{N}}.$$

Therefore

$$\int_E |u|^{\frac{N+1}{N}}\,dx \le \int_{\mathbb{R}} \prod_{i=1}^{N} \left(\int_{\mathbb{R}^N} |u_{x_i}(\bar{x},t)|d\bar{x} \right)^{\frac{1}{N}} dt \left(\int_E |u_t|dx \right)^{\frac{1}{N}}.$$

By the generalized Hölder inequality

$$\int_{\mathbb{R}} \prod_{i=1}^{N} \left(\int_{\mathbb{R}^N} |u_{x_i}(\bar{x},t)|d\bar{x} \right)^{\frac{1}{N}} dt \le \prod_{i=1}^{N} \left(\int_{\mathbb{R}} \int_{\mathbb{R}^N} |u_{x_i}(\bar{x},t)|d\bar{x}dt \right)^{\frac{1}{N}}$$

$$= \left(\prod_{i=1}^{N} \int_E |u_{x_i}|dx \right)^{\frac{1}{N}}.$$

Combining these last two inequalities proves the lemma. ∎

Lemma 3.2 $\|u\|_{\frac{Np}{N-p}} \le \frac{p(N-1)}{N-p} \|Du\|_p$.

Proof Write

$$\|u\|_{\frac{Np}{N-p}} = \left(\int_E \left(|u|^{\frac{p(N-1)}{N-p}} \right)^{\frac{N}{N-1}} dx \right)^{\frac{N-1}{N} \frac{N-p}{p(N-1)}}$$

and apply (3.1) to the function $w = |u|^{p(N-1)/(N-p)}$. This gives

$$\|u\|_{\frac{Np}{N-p}} \le \left[\prod_{i=1}^{N} \left(\int_E \left| \left(|u|^{\frac{p(N-1)}{N-p}} \right)_{x_i} \right| dx \right)^{\frac{1}{N}} \right]^{\frac{N-p}{p(N-1)}}$$

$$= \gamma \prod_{i=1}^{N} \left(\int_E |u|^{\frac{p(N-1)}{N-p}-1} |u_{x_i}|dx \right)^{\frac{N-p}{Np(N-1)}}$$

where

$$\gamma = \left[\frac{p(N-1)}{N-p} \right]^{\frac{N-p}{p(N-1)}}.$$

Now for all $i = 1, \ldots, N$, by Hölder's inequality

$$\int_E |u|^{\frac{p(N-1)}{N-p}-1} |u_{x_i}|dx \le \left(\int_E |u_{x_i}|^p dx \right)^{\frac{1}{p}} \left(\int_E |u|^{\frac{Np}{N-p}} dx \right)^{\frac{p-1}{p}}$$

and

$$\prod_{i=1}^{N} \left(\int_E |u|^{\frac{p(N-1)}{N-p}-1} |u_{x_i}|dx \right)^{\frac{N-p}{Np(N-1)}} = \prod_{i=1}^{N} \left(\|u_{x_i}\|_p \right)^{\frac{N-p}{Np(N-1)}} \|u\|_{\frac{Np}{N-p}}^{\frac{p-1}{p(N-1)}}$$

$$\le \|Du\|_p^{\frac{N-p}{p(N-1)}} \|u\|_{\frac{Np}{N-p}}^{\frac{p-1}{p} \frac{N}{N-1}}.$$

Combining these inequalities proves the Lemma. ∎

4 Proof of Theorem 1.1 for $1 \le p < N$ Concluded

According to (1.2) the case $\theta = 0$ corresponds to $q = r$ and therefore (1.1) is trivial.
The case $\theta = 1$ corresponds to $q = \frac{Np}{N-p}$ and is contained in Lemma 3.2. Let $r \in [1, \infty)$ and $p < N$ be fixed and choose $\theta \in (0, 1)$ and q such that

$$\min \left\{ r; \frac{Np}{N-p} \right\} < q < \max \left\{ r; \frac{Np}{N-p} \right\}, \qquad \theta q < \frac{Np}{N-p}. \qquad (4.1)$$

Then by Hölder's inequality

$$\int_E |u|^q dx = \int_E |u|^{\theta q} |u|^{(1-\theta)q} dx$$

$$\le \left(\int_E |u|^{\frac{Np}{N-p}} dx \right)^{\frac{N-p}{Np} \theta q} \left(\int_E |u|^r dx \right)^{\frac{(1-\theta)q}{r}}$$

where we have set

$$(1 - \theta)q \frac{Np}{Np - (N-p)\theta q} = r. \qquad (4.2)$$

From this, by Lemma 3.2

$$\|u\|_q \le C \|Du\|_p^\theta \|u\|_r^{1-\theta}, \quad \text{where} \quad C = \left[\frac{p(N-1)}{N-p} \right]^\theta.$$

By direct calculation one verifies that (4.2) is exactly (1.1). Moreover the ranges
indicated in (1.4) correspond to the compatibility of (4.1) and (4.2).

5 Proof of Theorem 1.1 for $p \ge N > 1$

Let $F(x; y)$ be the fundamental solution of the Laplace operator introduced in (14.3)
of Chap. 8. Then by Stokes formula (Proposition 14.1 of Chap. 8), for all $u \in C_o^\infty(E)$
and all $x \in E$

$$u(x) = -\int_{\mathbb{R}^N} F(x; y) \Delta u(y) dy = \frac{1}{\omega_N} \int_{\mathbb{R}^N} Du(y) \cdot \frac{(x-y)}{|x-y|^N} dy. \qquad (5.1)$$

Fix $\rho > 0$ and rewrite this as

$$\omega_N u(x) = \int_{|x-y|<\rho} Du(y) \cdot \frac{(x-y)}{|x-y|^N} dy$$
$$+ \int_{|x-y|>\rho} Du(y) \cdot \frac{x-y}{|x-y|^N} dy. \qquad (5.2)$$

The second integral can be computed by an integration by parts, as

$$\int_{|x-y|>\rho} Du(y) \cdot \frac{x-y}{|x-y|^N} dy = \frac{1}{N-2} \int_{|x-y|>\rho} Du(y) \cdot D\frac{1}{|x-y|^{N-2}} dy$$

$$= \frac{1}{\rho^{N-1}} \int_{|x-y|=\rho} u(y) d\sigma$$

since $F(x; y)$ is harmonic in $\mathbb{R}^N - \{x\}$. Here $d\sigma$ denotes the surface measure on the sphere $\{|x-y| = \rho\}$. Put this in (5.2), multiply by $N\rho^{N-1}$ and integrate in $d\rho$ over $(0, R)$ where R is a positive number to be chosen later. This gives

$$\omega_N R^N |u(x)| \leq N \int_0^R \left(\int_{|x-y|<\rho} \frac{|Du(y)|}{|x-y|^{N-1}} dy \right) \rho^{N-1} d\rho$$

$$+ N \int_0^R \left(\int_{|x-y|=\rho} |u(y)| d\sigma \right) d\rho.$$

From this, for all $x \in E$

$$\omega_N |u(x)| \leq \int_{B_R(x)} \frac{|Du(y)|}{|x-y|^{N-1}} dy + \frac{N}{R^N} \int_{B_R(x)} |u(y)| dy \qquad (5.3)$$

$$= I_1(x, R) + N I_2(x, R)$$

where $B_R(x)$ is the ball of radius R centered at x.

5.1 Estimate of $I_1(x, R)$

Choose two positive numbers $a, b < N$ such that

$$\frac{a}{q} + b\left(1 - \frac{1}{p}\right) = N - 1. \qquad (5.4)$$

Since

$$\frac{N}{q} + N\left(1 - \frac{1}{p}\right) > N\left(1 - \frac{1}{N}\right) = N - 1$$

such a choice can be made. Now write

$$\frac{|Du(y)|}{|x-y|^{N-1}} = |Du|^{p(\frac{1}{p}-\frac{1}{q})} \frac{|Du|^{\frac{p}{q}}}{|x-y|^{\frac{a}{q}}} \frac{1}{|x-y|^{b(1-\frac{1}{p})}}$$

and apply the generalized Hölder inequality with conjugate exponents

$$\left(\frac{1}{p} - \frac{1}{q}\right) + \frac{1}{q} + \left(1 - \frac{1}{p}\right) = 1.$$

This gives

$$I_1(x, R) \leq \|Du\|_p^{1-\frac{p}{q}} \left(\int_{B_R(x)} \frac{|Du(y)|^p}{|x - y|^a} dy\right)^{\frac{1}{q}} \left(\int_{B_R(x)} \frac{1}{|x - y|^b} dy\right)^{1-\frac{1}{p}}.$$

Taking the q-power and integrating over E, gives

$$\|I_1(R)\|_q \leq \frac{\omega_N^{1-\frac{1}{p}+\frac{1}{q}} R^{N\left(\frac{1}{N} - \frac{1}{p} + \frac{1}{q}\right)}}{(N - a)^{\frac{1}{q}}(N - b)^{1-\frac{1}{p}}} \|Du\|_p.$$

5.2   Estimate of $I_2(x, R)$

$$I_2(x, R) \leq R^{-N} \left(\int_{|x-y|<R} |u(y)|^r dy\right)^{\frac{1}{r}} \left(\int_{|x-y|<R} 1 dy\right)^{1-\frac{1}{r}}$$

$$\leq \left(\frac{\omega_N}{N}\right)^{1-\frac{1}{r}} R^{-\frac{N}{r}} \|u\|_r^{1-\frac{r}{q}} \left(\int_{|\xi|<R} |u(x + \xi)|^r d\xi\right)^{\frac{1}{q}}.$$

Take the q-power and integrate in dx over $\mathbb{R}^N$ to obtain

$$\|I_2\|_q \leq \left(\frac{\omega_N}{N}\right)^{1+\frac{1}{q}-\frac{1}{r}} R^{-N(\frac{1}{r}-\frac{1}{q})} \|u\|_r.$$

6   Proof of Theorem 1.1 for $p \geq N > 1$ Concluded

Combining these estimates into (5.3) gives

$$\|u\|_q \leq \frac{\omega_N^{\frac{1}{q}-\frac{1}{p}}}{(N - a)^{\frac{1}{q}}(N - b)^{1-\frac{1}{p}}} \|Du\|_p R^{N\left(\frac{1}{N}-\frac{1}{p}+\frac{1}{q}\right)}$$

$$+ \left(\frac{\omega_N}{N}\right)^{\frac{1}{q}-\frac{1}{r}} \|u\|_r R^{-N(\frac{1}{r}-\frac{1}{q})}.$$

Setting

$$A = \frac{\omega_N^{\frac{1}{q}-\frac{1}{p}} \|Du\|_p}{(N-a)^{\frac{1}{q}}(N-b)^{1-\frac{1}{p}}}, \qquad B = \left(\frac{\omega_N}{N}\right)^{\frac{1}{q}-\frac{1}{r}} \|u\|_r$$

the previous inequality takes the form

$$\|u\|_q \leq A R^\beta + B R^{-\bar{\beta}}$$

where

$$\beta = N\left(\frac{1}{N} - \frac{1}{p} + \frac{1}{q}\right), \qquad \bar{\beta} = N\left(\frac{1}{r} - \frac{1}{q}\right).$$

Minimizing the right-hand side with respect to the parameter $R \in (0, \infty)$, we find

$$\|u\|_q \leq \left[\left(\frac{\bar{\beta}}{\beta}\right)^{\frac{\beta}{\beta+\bar{\beta}}} + \left(\frac{\beta}{\bar{\beta}}\right)^{\frac{\bar{\beta}}{\beta+\bar{\beta}}}\right] A^{\frac{\bar{\beta}}{\beta+\bar{\beta}}} B^{\frac{\beta}{\beta+\bar{\beta}}}.$$

Setting

$$\frac{\bar{\beta}}{\beta+\bar{\beta}} = \theta, \qquad \frac{\beta}{\beta+\bar{\beta}} = 1 - \theta \qquad \text{so that} \qquad \frac{\beta}{\bar{\beta}} = \frac{1-\theta}{\theta}$$

proves (1.1)–(1.2), with the constant C given explicitly by

$$C = \left[\left(\frac{\theta}{1-\theta}\right)^{1-\theta} + \left(\frac{1-\theta}{\theta}\right)^{\theta}\right]\left(\frac{\omega_N}{N}\right)^{(\frac{1}{q}-\frac{1}{r})(1-\theta)}$$
$$\times \left(\frac{\omega_N^{\frac{1}{q}-\frac{1}{p}}}{(N-a)^{\frac{1}{q}}(N-b)^{1-\frac{1}{p}}}\right)^\theta. \tag{6.1}$$

The ranges indicated in (1.5) depend on the possibility of choosing numbers a and b as in (5.4). ∎

7 On the Limiting Case $p = N$

A function $u \in W_o^{1,p}(E)$ for $p > N$ belongs to $L^\infty(E)$, by the embedding (1.1)–(1.5). However, the constant $C = C(N, p)$ in (6.1) deteriorates as $p \to N$. Indeed as $p \to N$, the number b in (5.4) tends to N and consequently $C(N, p) \to \infty$. If $p = N$ the same embedding implies that u is $L^q(E)$ for all $1 \leq q < \infty$. On the one hand such an embedding is rather precise as there exist functions in $W^{1,N}(E)$ for $N > 1$, that are not essentially bounded. For example $u(x) = \ln|\ln|x||$ is not bounded near the origin and belongs to $W^{1,N}(B_{1/e})$. On the other it does not provide the sharp embedding space for $W_o^{1,N}(E)$. A sharp embedding for $p = N$ can be derived from the limiting estimates of the Riesz potentials (§ 22 of Chap. 9).

Theorem 7.1 *Let* $u \in W_o^{1,N}(E)$. *There exist constants* C_1 *and* C_2 *depending only upon* N, *such that*

$$\int_E \exp\left\{\frac{|u|}{C_1\|Du\|_N}\right\}^{\frac{N}{N-1}} dx \le C_1\mu(E).$$

Proof We may assume $u \in C_o^\infty(E)$. From the representation formula (5.1)

$$|u(x)| \le \frac{1}{\omega_N}\int_E \frac{|Du|}{|x-y|^{N-1}}dy.$$

The embedding now follows from the limiting potential estimates. ∎

8   Embeddings of $W^{1,p}(E)$

Embedding inequalities for functions in $W^{1,p}(E)$ depend in general, on the structure of ∂E. A minimal requirement is that E satisfy the cone property (§ 16.4 of Chap. 8). The embedding constants are independent of E and its size, and depend on the structure of ∂E only through the cone property. Let κ_N denote the volume of the unit ball in $\mathbb{R}^N$ and denote by $C(N, p)$ a positive constant depending only on N and p and independent of E.

Theorem 8.1 (Sobolev–Nikol'skii [149]) *Let* E *satisfy the cone condition for a fixed circular cone* C_o *of solid angle* ω, *height* h *and vertex at the origin, and let* $u \in W^{1,p}(E)$.
  If $1 < p < N$ *then* $u \in L^{p^*}(E)$ *where* $p^* = \frac{Np}{N-p}$, *and*

$$\|u\|_{p^*} \le \frac{C(N,p)}{\omega}\left(\frac{1}{h}\|u\|_p + \|Du\|_p\right), \qquad p^* = \frac{Np}{N-p}. \qquad (8.1)$$

The constant $C(N, p)$ *in (8.1) tends to infinity as either* $p \to 1$ *or* $p \to N$. *If* $p = 1$ *and* $\mu(E) < \infty$, *then* $u \in L^q(E)$ *for all* $q \in [1, \frac{N}{N-1})$ *and*

$$\|u\|_q \le \frac{C(N,q)}{\omega}\mu(E)^{\frac{1}{q}-\frac{N-1}{N}}\left(\frac{1}{h}\|u\|_1 + \|Du\|_1\right) \qquad (8.2)$$

where

$$C(N,q) = \frac{\kappa_N^{\frac{N-1}{N}}}{\left(1 - \frac{N-1}{N}q\right)^{1/q}}. \qquad (8.2)'$$

If $p > N$, then $u \in L^\infty(E)$ and

$$\|u\|_\infty \le \frac{C(N, p, \omega)}{\mu(C_o)^{1/p}}\big[\|u\|_p + h\|Du\|_p\big] \tag{8.3}$$

where

$$C(N, p, \omega) = \frac{1}{N}\Big(\frac{\omega_N}{\omega}\Big)^{\frac{N-1}{N}}\Big(\frac{N(p-1)}{p-N}\Big)^{\frac{p-1}{p}}. \tag{8.3$'$}$$

If $p > N$ and in addition E is convex, then u has a representative, still denoted by u, which is Hölder continuous in $\bar{E}$, and for every $x, y \in \bar{E}$ such that $|x - y| < h$

$$|u(x) - u(y)| \le \frac{C(N, p)}{\omega}|x - y|^{1 - \frac{N}{p}}\|Du\|_p \tag{8.4}$$

where

$$C(N, p) = 2^{N+1}\kappa_N^{\frac{p-1}{p}}\frac{Np}{p - N}. \tag{8.4$'$}$$

Remark 8.1 The estimates exhibit an explicit dependence on the height h and the solid angle ω of the cone C_o. They deteriorate as either h or ω tend to zero. Estimate (8.4) depends on the height h of the cone C_o through the requirement that $|x - y| < h$.

Remark 8.2 The value $q = 1^* = \frac{N}{N-1}$ is not permitted in (8.2). Such a limiting value is permitted for embeddings of $W_o^{1,p}(E)$ as indicated in Corollary 1.1. The limiting case $p = N$, not permitted neither in (8.3) nor in (8.4), will be given a sharp form in § 13.

Remark 8.3 The assumption that ∂E satisfies the cone property is essential for the embedding (8.3). Consider the domain $E \subset \mathbb{R}^2$ introduced in (16.6) of Chap. 8 and the function $u(x_1, x_2) = x_1^{-\beta}$ defined in E, for some $\beta > 0$. The parameters $\beta > 0$ and $\alpha > 1$ can be chosen so that $u \in W^{1,p}(E)$ for some $p > 2$, and u is unbounded near the origin.

Remark 8.4 If E is not convex, the estimate (8.4) can be applied locally. Thus if $p > N$ a function $u \in W^{1,p}(E)$ is locally Hölder continuous in E.

9 Proof of Theorem 8.1

Proof of (8.1) and (8.2): It suffices to prove the various assertions for $u \in C^\infty(E)$. Fix $x \in E$ and let $C_x \subset \bar{E}$ be a cone congruent to C_o and claimed by the cone property. Then

$$|u(x)| = \left| \int_0^h \frac{\partial}{\partial \rho} \left(1 - \frac{\rho}{h} \right) u(x + \rho \mathbf{n}) d\rho \right|$$

$$\leq \int_0^h |Du(x + \rho \mathbf{n})| d\rho + \frac{1}{h} \int_0^h |u(x + \rho \mathbf{n})| d\rho$$

where $\mathbf{n}$ denotes an arbitrary unit vector ranging on the same solid angle as C_x. Integrating over such a solid angle

$$\omega |u(x)| \leq \int_{C_x} \frac{|Du(y)|}{|x - y|^{N-1}} dy + \frac{1}{h} \int_{C_x} \frac{|u(y)|}{|x - y|^{N-1}} dx \tag{9.1}$$

$$\leq \int_E \frac{|Du(y)|}{|x - y|^{N-1}} dy + \frac{1}{h} \int_E \frac{|u(y)|}{|x - y|^{N-1}} dx.$$

The right hand of (9.1) is the sum of two Riesz potentials. Therefore inequality (8.1) follows from the second of (9.1) and the $L^{p^*}(E)$-estimates of the Riesz potentials in § 20 of Chap. 9. The behavior of the constant $C(N, p)$ follows from Remark 20.1 of the same chapter. Analogously, (8.2) and the form of the constant $C(N, p)$ follows from the $L^q(E)$-estimates of the Riesz potentials (Proposition 21.2 of Chap. 9).

To establish (8.3), start from the first of (9.1) and estimate

$$\omega |u(x)| \leq \sup_{z \in C_x} \int_{C_x} \frac{|Du(y)|}{|z - y|^{N-1}} dy + \frac{1}{h} \sup_{z \in C_x} \int_{C_x} \frac{|u(y)|}{|z - y|^{N-1}} dx.$$

Therefore by the $L^\infty(E)$-estimates of the Riesz potentials of Proposition 21.3 of Chap. 9 with E replaced by C_x

$$\omega |u(x)| \leq C(N, p) \mu(C_o)^{\frac{p-N}{Np}} \left(\frac{1}{h} \|u\|_{p, C_x} + \|Du\|_{p, C_x} \right)$$

$$\leq \frac{C(N, p)}{h} \mu(C_o)^{\frac{p-N}{Np}} \left(\|u\|_p + h \|Du\|_p \right). \qquad \blacksquare$$

The form of the constant $C(N, p)$ is in (21.4) of Chap. 9. The form (8.3)' of the constant $C(N, p, \omega)$ is computed from this and the expression of the volume of C_o.

Proof of (8.4) (Morrey [111]): For $x \in \bar{E}$, let C_x be a circular, spherical cone congruent to C_o and all contained $\bar{E}$. Then for all $0 < \rho \leq h$, the circular, spherical cone $C_{x,\rho}$ of vertex at x, radius ρ, coaxial with C_x and with the same solid angle ω, is contained in $\bar{E}$ and its volume is

$$\mu(C_{x,\rho}) = \frac{\omega}{N} \rho^N.$$

Denote by $u_{C_{x,\rho}}$ the average of u over $C_{x,\rho}$

$$u_{C_{x,\rho}} = \frac{1}{\mu(C_{x,\rho})} \int_{C_{x,\rho}} u(\xi) d\xi.$$

Lemma 9.1 *For every pair* $x, y \in \bar{E}$ *such that* $|x - y| = \rho \leq h$

$$|u(y) - u_{\mathcal{C}_{x,\rho}}| \leq 2^{N+1-\frac{N}{p}} \frac{\kappa_N^{\frac{p-1}{p}}}{\omega} \frac{Np}{p-N} \rho^{1-\frac{N}{p}} \|Du\|_p.$$

Proof Fix $x, y \in E$ such that $|x - y| = \rho \leq h$. Since E is convex, for all $\xi \in \mathcal{C}_{x,\rho}$

$$|u(y) - u(\xi)| = \left| \int_0^1 \frac{\partial}{\partial t} u(y + t(\xi - y)) dt \right|.$$

First integrate in $d\xi$ over $\mathcal{C}_{x,\rho}$, and then in the resulting integral perform the change of variables

$$y + t(\xi - y) = \eta.$$

The Jacobian is t^{-N} and the new domain of integration is transformed in those η given by

$$|y - \eta| = t|\xi - y| \quad \text{as } \xi \text{ ranges over } \mathcal{C}_{x,\rho}.$$

Therefore such a transformed domain is contained in the ball $B_{2\rho t}(y)$, of center y and radius $2\rho t$. These operations give

$$\frac{\omega}{N} \rho^N |u(y) - u_{\mathcal{C}_{x,\rho}}| \leq \int_0^1 \left(\int_{\mathcal{C}_{x,\rho}} |\xi - y| |Du(y + t(\xi - y))| d\xi \right) dt$$

$$\leq \int_0^1 t^{-(N+1)} \int_{E \cap B_{2\rho t}(y)} |\eta - y| |Du(\eta)| d\eta dt$$

$$\leq \kappa_N^{\frac{p-1}{p}} \int_0^1 t^{-(N+1)} (2\rho t)^{N(1-\frac{1}{p})+1} \|Du\|_p dt. \qquad \blacksquare$$

To conclude the proof of (8.4) fix $x, y \in \bar{E}$ and let

$$z = \tfrac{1}{2}(x + y) \qquad \rho = |x - z| = |y - z| = \tfrac{1}{2}|x - y|.$$

Then

$$|u(x) - u(y)| \leq |u(x) - u_{\mathcal{C}_{z,\rho}}| + |u(y) - u_{\mathcal{C}_{z,\rho}}|$$

$$\leq \frac{2^{N+1}}{\omega} \kappa_N^{\frac{p-1}{p}} \frac{Np}{p-N} |x - y|^{1-\frac{N}{p}} \|Du\|_p. \qquad \blacksquare$$

10 Poincaré Inequalities

The multiplicative inequalities of Theorem 1.1 cannot hold for functions in $W^{1,p}(E)$ as it can be verified if u is a nonzero constant. In general an integral norm of u cannot be controlled in terms of some integral norm of its gradient unless one has some information on the values of the function in some subset of $\bar{E}$.

10.1   The Poincaré Inequality

Let E be a bounded domain in $\mathbb{R}^N$ and for $u \in L^1(E)$, let

$$u_E = \fint_E u dx = \frac{1}{\mu(E)} \int_E u dx$$

denote the integral average of u over E.

Theorem 10.1 *Let E be bounded and convex and let $u \in W^{1,p}(E)$ for some $1 < p < N$. There exists a constant C depending only upon N and p, such that*

$$\|u - u_E\|_{p^*} \le C \frac{(diam\ E)^N}{\mu(E)} \|Du\|_p \quad where\ p^* = \frac{Np}{N-p} \qquad (10.1)$$

$$\|u - u_E\|_1 \le C(diam\ E)^N \|Du\|_N \qquad (10.2)$$

Proof Having fixed $x, y \in E$, denote by $R(x, y)$ the distance from x to ∂E, along the direction of $(y - x)$ and write

$$|u(x) - u(y)| \le \int_0^{R(x,y)} \left| \frac{\partial}{\partial \rho} u(x + \rho \mathbf{n}) \right| d\rho \qquad \mathbf{n} = \frac{(y - x)}{|y - x|}.$$

Integrate in dy over E, to obtain

$$\mu(E)|u(x) - u_E| \le \int_E \left(\int_0^{R(x,y)} |Du(x + \rho \mathbf{n})| d\rho \right) dy.$$

The integral in dy is calculated by introducing polar coordinates with pole at x. Therefore if $\mathbf{n}$ is the angular variable spanning the sphere $|\mathbf{n}| = 1$, the right-hand side is majorized by

$$(diam\ E)^{N-1} \int_0^{diam\ E} \int_{|\mathbf{n}|=1} \int_0^{R(x,y)} \rho^{N-1} \frac{|Du(x + \rho \mathbf{n})|}{|x - y|^{N-1}} d\rho d\mathbf{n} dr$$

$$\le (diam\ E)^N \int_E \frac{|Du|}{|x - y|^{N-1}} dy.$$

Therefore

$$|u(x) - u_E| \le \frac{(diam\ E)^N}{\mu(E)} \int_E \frac{|Du(y)|}{|x - y|^{N-1}} dy. \qquad (10.3)$$

The proof of (10.1) now follows from this and the $L^{p^*}(E)$-estimates of the Riesz potentials given in Theorem 20.1 of Chap. 9. Following Remark 20.1 there, the constant C in (20.1) tends to infinity as either $p \to 1$ or $p \to N$.

Inequality (10.2) follows from (10.1) and Hölder's inequality. Indeed having fixed $1 < p < N$

$$\|u - u_E\|_1 \leq \|u - u_E\|_{p^*}\mu(E)^{1 - \frac{N-p}{Np}}$$

$$\leq C\mu(E)^{1 - \frac{N-p}{Np}} \frac{(\text{diam } E)^N}{\mu(E)} \|Du\|_p$$

$$\leq C(\text{diam } E)^N \|Du\|_N. \qquad \blacksquare$$

Remark 10.1 The estimate depends upon the structure of the convex set E through the ratio $(\text{diam } E)^N/\mu(E)$. If E is a ball, then such a ratio is $2^N/\kappa_N$, where κ_N is the volume of the unit ball in $\mathbb{R}^N$. In general if R is the radius of the smallest ball containing E and ρ is the radius of the largest ball contained in E

$$\frac{(\text{diam } E)^N}{\mu(E)} \leq \frac{2^N}{\kappa_N}\left(\frac{R}{\rho}\right)^N.$$

Remark 10.2 For a ball $B_\rho(x)$ denote by $(u)_{x,\rho}$ the integral average of u over $B_\rho(x)$, i.e.,

$$u_{B_\rho(x)} = (u)_{x,\rho} = \frac{1}{\mu(B_\rho)} \int_{B_\rho(x)} u \, dy = \fint_{B_\rho(x)} u \, dy. \qquad (10.4)$$

Then (10.1)–(10.2) imply

$$\fint_{B_\rho(x)} |u - (u)_{x,\rho}| dx \leq C'\left(\rho^p \fint_{B_\rho(x)} |Du|^p dy\right)^{\frac{1}{p}} \qquad (10.5)$$

for all $1 \leq p < N$ and for a constant C' depending only on N and p. Also

$$\fint_{B_\rho(x)} |u - (u)_{x,\rho}|^P dx \leq C'\rho^p \fint_{B_\rho(x)} |Du|^p dy \qquad (10.6)$$

for all $1 < p < N$.

10.2 Multiplicative Poincaré Inequalities

Proposition 10.1 *Let E be a bounded and convex subset of $\mathbb{R}^N$, and let $u \in W^{1,p}(E)$ for some $1 < p < N$. There exists a constant C depending only upon N and p, such that*

$$\|u - u_E\|_q \leq C^\theta \left[\frac{(\text{diam } E)^N}{\mu(E)}\right]^\theta \|Du\|_p^\theta \|u - u_E\|_r^{1-\theta} \qquad (10.7)$$

where the numbers $r > 1$, $1 < p < N$ and $\theta \in [0, 1]$ are linked by (1.2).

Proof The case $\theta = 0$ corresponds to $q = r$ and (10.7) is trivial. The case $\theta = 1$ corresponds to $q = \frac{Np}{N-p}$ and coincides with (10.1). Let $r \in (1, \infty)$ and $p \in (1, N)$ be fixed and choose $\theta \in (0, 1)$ and q satisfying (4.1). By Hölder's inequality

$$\int_E |u - u_E|^q dx = \int_E |u - u_E|^{\theta q} |u - u_E|^{(1-\theta)q} dx$$

$$\leq \left(\int_E |u - u_E|^{\frac{Np}{N-p}} dx \right)^{\frac{N-p}{Np} \theta q} \left(\int_E |u - u_E|^r dx \right)^{\frac{(1-\theta)q}{r}}$$

where r is chosen as in (4.2). The inequality follows from this and (10.1).   ∎

Remark 10.3 It follows from (10.7) that the multiplicative embedding inequality (1.1) continues to hold for functions $u \in W^{1,p}(E)$ of zero average over a bounded, convex domain E.

10.3   Extensions of $(u - u_E)$ for Convex E

Let E be a bounded domain with the segment property and ∂E of class C^1. Then ∂E admits a finite open covering with balls $\{B_t(x_j)\}_{j=1}^n$, of radius t centered at $x_j \in \partial E$. By Proposition 19.1 of Chap. 8, a function $u \in W^{1,p}(E)$ can be extended into a function $w \in W_o^{1,p}(\mathbb{R}^N)$ in such a way that $\|w\|_{1,p;\mathbb{R}^N} \leq C\|u\|_{1,p;E}$ for a constant C that depends only on the geometric structure of ∂E, the radius t of the finite open covering of ∂E, and the number of local overlaps of the balls $B_t(x_j)$. If E is convex, the Poincaré inequality of Theorem 10.1 affords an extention of $(u - u_E)$ into a function $w \in W_o^{1,p}(\mathbb{R}^N)$ in such a way that $\|Dw\|_{p;\mathbb{R}^N}$ is controlled only by $\|Du\|_{p;E}$.

Proposition 10.2 *Let E be a bounded, convex domain in $\mathbb{R}^N$ with boundary ∂E of class C^1, and let $u \in W^{1,p}(E)$ for some $1 \leq p < \infty$. Then $(u - u_E)$ admits an extension $w \in W_o^{1,p}(\mathbb{R}^N)$ such that*

$$\|Dw\|_{p;\mathbb{R}^N} \leq \gamma(1 + \|\partial E\|_1)\left(1 + \frac{1}{t} \frac{(diam\ E)^N}{\mu(E)^{\frac{N-1}{N}}}\right)\|Du\|_{p,E} \tag{10.8}$$

where t is the radius of the balls $\{B_t(x_j)\}_{j=1}^n$, making up an open covering of ∂E, and γ is a constant depending only on N, p and the number of local overlaps of the balls $B_t(x_j)$.

Proof Write down the estimate (18.1), of Chap. 8, for $(u - u_E)$ and apply the Poincaré inequality to estimate the last term.   ∎

Remark 10.4 We stress that $w \in W_o^{1,p}(\mathbb{R}^N)$ is an extension of $(u - u_E)$ and not an extension of u.

Corollary 10.1 *Let B_R be the ball of radius R centered at the origin of $\mathbb{R}^N$, let $u \in W^{1,p}(B_R)$ for some $1 \leq p < \infty$. Then $(u - u_{B_R})$ admits an extension $w \in W_o^{1,p}(\mathbb{R}^N)$ such that*

$$\|Dw\|_{p;\mathbb{R}^N} \leq \gamma \|Du\|_{p,E} \tag{10.9}$$

for an absolute constant γ depending only on N, p and independent of R.

Proof It follows from (10.8) and Remark 18.2 of Chap. 8. ∎

Corollary 10.2 *Let B_R be the ball of radius R centered at the origin of $\mathbb{R}^N$, let $u \in W^{1,p}(B_R)$ for some $1 \leq p < \infty$. Then $(u - u_{B_R})_\pm$ admit extensions $w_\pm \in W_o^{1,p}(\mathbb{R}^N)$ such that*

$$\|Dw_\pm\|_{p;\mathbb{R}^N} \leq \gamma \|Du\|_{p,E} \tag{10.10}$$

for an absolute constant γ depending only on N, p and independent of R.

Proof Extend $w_\pm$ as in Proposition 19.1 of Chap. 8, and following Remark 18.2, write down (10.4) of Chap. 8. In the last of these majorize

$$\|(u - u_{B_R})_\pm\|_{p;B_R} \leq \|u - u_{B_R}\|_{p;B_R}$$

and apply Poincaré inequality. ∎

11 Level Sets Inequalities

Theorem 11.1 (DeGiorgi [32]) *Let E be a bounded convex open set in $\mathbb{R}^N$ and let $u \in W^{1,1}(E)$. Assume that the set where u vanishes has positive measure. Then*

$$\|u\|_1 \leq \kappa_N^{\frac{N-1}{N}} \frac{(diam\ E)^N \mu(E)^{\frac{1}{N}}}{\mu([u = 0])} \|Du\|_1 \tag{11.1}$$

where κ_N is the volume of the unit ball in $\mathbb{R}^N$.

Proof Let $\mathbf{n}$ denote the unit vector ranging over the unit sphere in $\mathbb{R}^N$. For almost all $x \in E$ and almost all $y \in [u = 0]$

$$|u(x)| = \left| \int_0^{|y-x|} \frac{\partial}{\partial\rho} u(x + \mathbf{n}\rho) d\rho \right| \leq \int_0^{|y-x|} |Du(x + \mathbf{n}\rho)| d\rho.$$

Integrating in dx over E and in dy over $[u = 0]$, gives

$$\mu([u = 0]) \|u\|_1 \leq \int_E \left(\int_{[u=0]} \int_0^{|y-x|} |Du(x + \mathbf{n}\rho)| d\rho dy \right) dx.$$

The integral over $[u = 0]$ is computed by introducing polar coordinates with center at x. Denoting by $R(x, y)$ the distance from x to ∂E in the along **n**

$$\int_{[u=0]} \int_0^{|y-x|} |Du(x + \mathbf{n}\rho)| d\rho dy$$

$$\leq \int_0^{R(x,y)} s^{N-1} ds \int_{|\mathbf{n}|=1} \int_0^{R(x,y)} |Du(x + \mathbf{n}\rho)| d\rho d\mathbf{n}.$$

Combining these remarks we arrive at

$$\mu([u = 0])\|u\|_1 \leq \frac{1}{N} (\text{diam } E)^N \int_E \int_E \frac{|Du(y)|}{|x - y|^{N-1}} dy dx.$$

Inequality (11.1) follows from this since (Proposition 21.1 of Chap. 9, with $r = 1$)

$$\sup_{y \in E} \int_E \frac{dx}{|x - y|^{N-1}} \leq N \kappa_N^{\frac{N-1}{N}} \mu(E)^{\frac{1}{N}}. \qquad \blacksquare$$

If E is the ball B_R of radius R centered at the origin

$$\|u\|_1 \leq 2^N \kappa_N \frac{R^{N+1}}{\mu([|u| = 0])} \|Du\|_1. \tag{11.2}$$

For a real number ℓ and $u \in W^{1,1}(E)$, set

$$u_\ell = \begin{cases} \ell & \text{if } u > \ell, \\ u & \text{if } u \leq \ell. \end{cases}$$

For $k \in \mathbb{R}$ the function $(u_\ell - k)_+$ belongs to $W^{1,1}(E)$, by Proposition 20.2 of Chap. 8. Putting such a function in (11.1) and assuming $k < \ell$, gives

$$(\ell - k)\mu([u > \ell]) \leq \kappa_N^{\frac{N-1}{N}} \frac{(\text{diam } E)^N \mu(E)^{\frac{1}{N}}}{\mu([u < k])} \int_{[k < u < \ell]} |Du| dx. \tag{11.3}$$

This is referred to as a discrete version of the isoperimetric inequality [32].

12   Morrey Spaces [110]

A function $f \in L^1(E)$ belongs to the Morrey space $M^p(E)$, for some $p \geq 1$ if there exists a constant C_f depending upon f, such that

$$\sup_{x \in E} \int_{B_\rho(x) \cap E} |f(y)| dy \leq C_f \rho^{N(1-\frac{1}{p})} \quad \text{for all} \quad \rho > 0. \tag{12.1}$$

If $f \in M^p(E)$ set

$$\|f\|_{M^p} = \inf \left\{C_f \text{ for which (12.1) holds}\right\}. \tag{12.2}$$

It follows from the definition that $L^p(E) \subset M^p(E)$ and

$$\|f\|_p \geq \kappa_N^{\frac{1}{p}-1} \|f\|_{M^p}. \tag{12.3}$$

Moreover

$$L^1(E) = M^1(E) \qquad \text{and} \qquad \|f\|_1 = \|f\|_{M^1}$$
$$L^\infty(E) = M^\infty \qquad \text{and} \qquad \|f\|_\infty = \frac{1}{\kappa_N} \|f\|_{M^\infty}.$$

We regard f as being defined in the whole $\mathbb{R}^N$ by setting it to be zero outside E. Thus if $f \in M^p(E)$

$$\|f\|_1 \leq \|f\|_{M^p} (\text{diam } E)^{N(1-\frac{1}{p})}. \tag{12.4}$$

12.1 Embeddings for Functions in the Morrey Spaces

For a given $p > 1$ and $\alpha \in (0, \frac{N}{p})$ let $V_{\alpha,f}$ be the potential generated by some $f \in M^p(E)$, that is

$$V_{\alpha,f}(x) = \int_E \frac{f(y)}{|x-y|^{N-\alpha}} dy.$$

By Theorem 20.1c of Chap. 9 such a potential is well defined as a function in $L^q(E)$ for $q = \frac{N}{N-\alpha p}$. The next proposition gives an estimate of $\|V_{\alpha,f}\|_\infty$ in terms of $\|f\|_{M^p}$, provided $p > \frac{N}{\alpha}$ and E is bounded.

Proposition 12.1 *Let* $f \in M^p(E)$ *for some* $p > \frac{N}{\alpha}$. *Then*

$$\|V_{\alpha,f}\|_\infty \leq \frac{N(p-1)}{\alpha p - N} (\text{diam } E)^{\frac{\alpha p - N}{p}} \|f\|_{M^p}. \tag{12.5}$$

Proof For x and y in $\mathbb{R}^N$ let $\mathbf{n} = \frac{y-x}{|x-y|}$. Then by making use of polar coordinates, compute

$$\frac{d}{d\rho} \int_{B_\rho(x)} |f(y)| dy = \frac{d}{d\rho} \int_{|\mathbf{n}|=1} \int_0^\rho r^{N-1} |f(x + \mathbf{n}r)| dr d\mathbf{n}$$
$$= \int_{|\mathbf{n}|=1} \rho^{N-1} |f(x + \mathbf{n}\rho)| d\mathbf{n}.$$

Using this calculation in the definition of $V_{\alpha,f}$

$$
\begin{aligned}
|V_{\alpha,f}(x)| &\le \int_{|\mathbf{n}|=1} \int_0^{\text{diam } E} \frac{1}{\rho^{N-\alpha}} \rho^{N-1} |f(x+\mathbf{n}\rho)| d\mathbf{n} d\rho \\
&= \int_0^{\text{diam } E} \frac{1}{\rho^{N-\alpha}} \left(\frac{d}{d\rho} \int_{B_\rho(x)} |f(y)| dy \right) d\rho \\
&= \frac{\|f\|_1}{(\text{diam } E)^{N-\alpha}} - \lim_{\varepsilon \to 0} \frac{1}{\varepsilon^{N-\alpha}} \int_{B_\varepsilon(x)} |f(y)| dy \\
&\quad + (N-\alpha) \int_0^{\text{diam } E} \frac{1}{\rho^{N+1-\alpha}} \int_{B_\rho(x)} |f(y)| dy d\rho.
\end{aligned}
$$

To prove (12.5) estimate each of these terms by making use of the definition (12.1)–(12.2) of $\|f\|_{M^p}$.   ∎

The main interest of inequality (12.5) is for $\alpha = 1$, when applied to the Riesz potential of the type of the right-hand side of (10.3). By the embedding (8.4) of Theorem 8.1, if $p > N$ and if E is convex, then a function in $W^{1,p}(E)$ is Hölder continuous with Hölder exponent $\eta = \frac{p-N}{p}$. The next theorem asserts that the same embedding (8.4) continues to hold for functions whose weak gradient is in $M^p(E)$ for $p > N$.

Theorem 12.1 *Let $u \in W^{1,1}(E)$ and assume that $|Du| \in M^p(E)$ for some $p > N$. Then $u \in C^\eta(E)$ with $\eta = \frac{p-N}{p}$. Moreover for every ball $B_\rho(x) \subset E$*

$$
\operatorname*{ess\,osc}_{B_\rho(x)} u \le \frac{2^{N+1}}{\kappa_N} \frac{N(p-1)}{p-N} \|Du\|_{M^p} \rho^\eta. \tag{12.6}
$$

Proof Consider the potential inequality (10.3) written for the ball $B_\rho(x)$ replacing E. It gives

$$
|u(x) - (u)_{x,\rho}| \le \frac{2^N}{\kappa_N} \int_{B_\rho(x)} \frac{|Du(y)|}{|x-y|^{N-1}} dy
$$

where $(u)_{x,\rho}$ denotes the integral average of u over $B_\rho(x)$ as in (10.4). This implies (12.6) in view of Proposition 12.1 applied with $\alpha = 1$.   ∎

Since $L^p(E) \subset M^p(E)$ the embedding (12.6) generalizes the embedding (8.4).

13   Limiting Embedding of $W^{1,N}(E)$

Let f be in the Morrey space $M^N(E)$, and consider its Riesz potential

$$
V_f(x) = \int_E \frac{f(y)}{|x-y|^{N-1}} dy.
$$

Proposition 13.1 *Let $f \in M^N(E)$. There exist constants C_1 and C_2 depending only upon N, such that*

$$\int_E \exp\left(\frac{|V_f|}{C_1 \|f\|_{M^N}}\right) dy \leq C_2 (\text{diam } E)^N.$$

Proof Let $q > 1$ and rewrite the integrand in the potential V_f, as

$$\frac{|f(y)|^{\frac{1}{q}}}{|x - y|^{(N - \frac{1}{q})\frac{1}{q}}} \frac{|f(y)|^{1 - \frac{1}{q}}}{|x - y|^{(N - \frac{q+1}{q})(1 - \frac{1}{q})}}.$$

By Hölder's inequality

$$|V_f(x)| \leq \left(\int_E \frac{|f(y)|}{|x - y|^{N - \frac{1}{q}}} dy\right)^{\frac{1}{q}} \left(\int_E \frac{|f(y)|}{|x - y|^{N - \frac{q+1}{q}}} dy\right)^{1 - \frac{1}{q}}.$$

The second integral is estimated by Proposition 12.1 with $\alpha = \frac{q+1}{q}$ and $p = N$. This gives

$$\int_E \frac{|f(y)|}{|x - y|^{N - \frac{q+1}{q}}} dy \leq (N - 1)q (\text{diam } E)^{\frac{1}{q}} \|f\|_{M^N}.$$

Put this in the previous inequality, take the q-power of both sides and integrate over E in dx, to obtain

$$\int_E |V_f|^q dx \leq \left[(N - 1)q (\text{diam } E)^{\frac{1}{q}} \|f\|_{M^N}\right]^{q-1} \int_E |f(y)| \left(\int_E \frac{dx}{|x - y|^{N - \frac{1}{q}}}\right) dy.$$

The last double integral is estimated by using (12.4) with $p = N$, and Theorem 20.1c of Chap. 9 with $\alpha = \frac{1}{q}$, and Proposition 21.1c of the same chapter for $r = 1$. It yields

$$\int_E |f(y)| \left(\int_E \frac{dx}{|x - y|^{N - \frac{1}{q}}} dx\right) dy \leq \|f\|_1 \sup_{y \in E} \int_E \frac{dx}{|x - y|^{N - \frac{1}{q}}}$$

$$\leq \omega_N q (\text{diam } E)^{\frac{1}{q}} \|f\|_1$$

$$\leq \omega_N q (\text{diam } E)^{N - 1 + \frac{1}{q}} \|f\|_{M^N}.$$

Combining these estimates

$$\int_E |V_f|^q dx \leq \omega_N (N - 1)^{q-1} q^q (\text{diam } E)^N \|f\|_{M^N}^q.$$

Since $q > 1$ is arbitrary we write this inequality for $q = 2, 3, \ldots$. Then in a manner similar to the proof of Theorem 22.1 of Chap. 9

$$\int_E \sum_{q=0}^k \frac{1}{q!} \left(\frac{|V_f|}{C_1 \|f\|_N} \right)^q dx \le \frac{\omega_N (\text{diam } E)^N}{(N-1)} \sum_{q=0}^\infty \left(\frac{N-1}{C_1} \right)^q \frac{q^q}{q!}$$

for all $k \in \mathbb{N}$ and all $C_1 > 0$.   ∎

Let $u \in W^{1,1}(E)$ be such that $|Du| \in M^N(E)$, so that

$$\int_{B_\rho(x)} |Du| dy \le \|Du\|_{M^N} \rho^{N-1} \qquad \text{for all } B_\rho(x) \subset E. \tag{13.1}$$

Theorem 13.1 (John–Nirenberg ([77]) *Let $u \in W^{1,1}(E)$ and let (13.1) hold. There exist constants C_1 and C_2 depending only upon N, such that*

$$\int_{B_\rho(x)} \exp \left(\frac{|u - (u)_{x,\rho}|}{C_1 \|Du\|_{M^N}} \right) dy \le C_2 \mu(B_\rho)$$

where $(u)_{x,rho}$ is the integral average of u over $B_\rho(x)$ as in (10.4).

Proof It follows from Proposition 13.1, starting from the potential inequality (10.3). ∎

This estimate can be regarded as a limiting case of the Hölder estimates of Theorem 12.1 when $p \to N$.

14   Compact Embeddings

Let E be a bounded domain in $\mathbb{R}^N$ with the cone property and let K be a bounded set in $W^{1,p}(E)$, say

$$K = \{ u \in W^{1,p}(E) \mid \|u\|_{1,p} \le C \} \quad \text{for some positive constant } C.$$

If $1 \le p < N$, by the embedding Theorem 8.1, K is a bounded subset of $L^{p^*}(E)$ where $p^* = \frac{Np}{N-p}$. Since E is of finite measure, K is also a bounded set of $L^q(E)$ for all $1 \le q \le p^*$. The next theorem asserts that K is a compact subset of $L^q(E)$ for all $1 \le q < p^*$.

Theorem 14.1 (Rellich–Kondrachov ([123, 85]) *Let E be a bounded domain in $\mathbb{R}^N$ with the cone property, and let $1 \le p < N$. Then the embedding of $W^{1,p}(E)$ into $L^q(E)$ is compact for all $1 \le q < p^*$.*

Proof The proof consists of verifying that a bounded subset of $W^{1,p}(E)$ satisfies the conditions for a subset of $L^q(E)$ to be compact, given in § 19 of Chap. 6. For $\delta > 0$ let

$$E_\delta = \{x \in E \mid \text{dist}(x, \partial E) > \delta\}$$

and let δ be so small that E_δ is not empty. For $q \in [1, p^*)$ and $u \in W^{1,p}(E)$

$$\|u\|_{q, E - E_\delta} \leq \|u\|_{p^*} \, \mu(E - E_\delta)^{\frac{1}{q} - \frac{1}{p^*}}.$$

Next, for a vector $h \in \mathbb{R}^N$ of length $|h| < \delta$ compute

$$\int_{E_\delta} |u(x + h) - u(x)| dx \leq \int_{E_\delta} \left(\int_0^1 \left| \frac{d}{dt} u(x + th) \right| dt \right) dx$$

$$\leq |h| \int_0^1 \left(\int_{E_\delta} |Du(x + th)| dx \right) dt$$

$$\leq |h| \mu(E)^{\frac{p-1}{p}} \|Du\|_p.$$

Therefore $\forall \sigma \in (0, \frac{1}{q})$

$$\int_{E_\delta} |T_h u - u|^q dx = \int_{E_\delta} |T_h u - u|^{q\sigma + q(1 - \sigma)} dx$$

$$\leq \left(\int_{E_\delta} |T_h u - u| dx \right)^{q\sigma} \left(\int_{E_\delta} |T_h u - u|^{\frac{q(1-\sigma)}{1-q\sigma}} dx \right)^{1 - q\sigma}.$$

Choose σ so that

$$\frac{(1 - \sigma)q}{1 - q\sigma} = p^*, \qquad \text{that is} \qquad \sigma q = \frac{p^* - q}{p^* - 1}.$$

Such a choice is possible if $1 < q < p^*$. Applying the embedding Theorem 8.1

$$\int_{E_\delta} |T_h u - u|^q dx \leq \gamma^{(1-\sigma)q} \left(\int_{E_\delta} |T_h u - u| dx \right)^{\frac{p^* - q}{p^* - 1}} \|u\|_{1,p}^{(1-\sigma)q}$$

for a constant γ depending only upon N, p and the geometry of the cone property of E, as indicated in (8.1). Combining these estimates gives

$$\|T_h u - u\|_{q, E_\delta} \leq \gamma_1 |h|^\sigma \|u\|_{1,p}, \qquad \text{where } \gamma_1 = \gamma^{\frac{1}{q} - \sigma} \mu(E)^{\frac{p-1}{p} \sigma}.$$

The assertion of the theorem is now a consequence of the characterization of the precompact subsets in $L^q(E)$. $\blacksquare$

Corollary 14.1 *Let E be a bounded domain in $\mathbb{R}^N$ with the cone property, and let $1 \leq p < N$. Then, every sequence $\{f_n\}$ of functions equi-bounded in $W^{1,p}(E)$, has a subsequence $\{f_{n'}\}$ strongly convergent in $L^q(E)$ for all $1 \leq q < p^*$.*

15   Fractional Sobolev Spaces in $\mathbb{R}^N$

Let $s \in (0, 1)$, $p \geq 1$ and $N \geq 1$ and consider the collection of functions $u \in L^p(\mathbb{R}^N)$ with finite semi-norm

$$\|\|u\|\|_{s,p} = \left(\int_{\mathbb{R}^N} \int_{\mathbb{R}^N} \frac{|u(x) - u(y)|^p}{|x - y|^{N+sp}} dx dy \right)^{\frac{1}{p}}. \tag{15.1}$$

They form a linear subspace of $L^p(\mathbb{R}^N)$ which is a Banach space when equipped with the norm

$$\|u\|_{s,p} = \|u\|_p + \|\|u\|\|_{s,p}.$$

Such a Banach space is denoted with $W^{s,p}(\mathbb{R}^N)$ and it is called the fractional Sobolev space of order s.

Proposition 15.1 $W^{1,p}(\mathbb{R}^N)$ *is continuously embedded into* $W^{s,p}(\mathbb{R}^N)$ *for all $s \in (0, 1)$. Moreover*

$$\|\|u\|\|_{s,p} \leq \left(\frac{2\omega_N}{ps(1-s)} \right)^{\frac{1}{p}} \|Du\|_p^s \|u\|_p^{1-s}.$$

Proof A change of variable in the definition of the semi-norm in (15.1) gives

$$\|\|u\|\|_{s,p}^p = \int_{\mathbb{R}^N} |\xi|^{-(N+sp)} d\xi \int_{\mathbb{R}^N} |u(x+\xi) - u(x)|^p dx$$

$$= \int_{|\xi|<\lambda} |\xi|^{-(N+sp)} d\xi \int_{\mathbb{R}^N} |u(x+\xi) - u(x)|^p dx$$

$$+ \int_{|\xi|>\lambda} |\xi|^{-(N+sp)} d\xi \int_{\mathbb{R}^N} |u(x+\xi) - u(x)|^p dx$$

where λ is a positive parameter to be chosen later. The second integral is majorized by

$$\frac{2^p \omega_N \|u\|_p^p}{sp\lambda^{sp}}.$$

The first integral is estimated by

$$\int_{|\xi|<\lambda} |\xi|^{-(N+sp)} d\xi \int_{\mathbb{R}^N} \left| \int_0^1 \frac{\partial}{\partial t} u(x+t\xi) dt \right|^p dx$$

$$\leq \int_{|\xi|<\lambda} |\xi|^{p-(N+sp)} d\xi \int_{\mathbb{R}^N} |Du|^p dx$$

$$\leq \frac{\omega_N}{p(1-s)} \lambda^{p(1-s)} \|Du\|_p^p.$$

Combining these estimates we arrive at

$$\|u\|_{s,p}^p \leq \frac{2^p \omega_N}{sp} \frac{1}{\lambda^{sp}} \|u\|_p^p + \frac{\omega_N}{p(1-s)} \lambda^{p(1-s)} \|Du\|_p^p$$

valid for every $\lambda > 0$. To prove the proposition, minimize the right-hand side with respect to λ. ∎

Proposition 15.2 $C_o^\infty(\mathbb{R}^N)$ *is dense in* $W^{s,p}(\mathbb{R}^N)$.

Proof For $\varepsilon > 0$ let ζ_ε denote a nonnegative piecewise smooth cutoff function in $\mathbb{R}^N$, such that

$$\zeta_\varepsilon = 1 \quad \text{for} \quad |x| < \frac{1}{\varepsilon}$$

$$\text{and} \quad |D\zeta_\varepsilon| \leq \varepsilon.$$

$$\zeta_\varepsilon = 0 \quad \text{for} \quad |x| \geq \frac{2}{\varepsilon}$$

One verifies that

$$\|u\zeta_\varepsilon\|_{s,p} \leq \|u\|_{s,p} + \varepsilon^s \gamma \|u\|_p$$

for all $u \in W^{s,p}(\mathbb{R}^N)$, where γ depends only upon N and p, and that

$$\|u - u\zeta_\varepsilon\|_{s,p} \to 0, \quad \text{as} \quad \varepsilon \to 0.$$

The functions $J_\varepsilon * (u\zeta_\varepsilon)$ are in $C_o^\infty(\mathbb{R}^N)$ and as u ranges over $W^{s,p}(\mathbb{R}^N)$ and ε ranges over $(0, 1)$, they span a dense subset of $W^{s,p}(\mathbb{R}^N)$. For this we verify that

$$\|J_\varepsilon * u - u\|_{s,p} \to 0, \quad \text{as} \quad \varepsilon \to 0.$$

For almost every pair $x, y \in \mathbb{R}^N$

$$\frac{\left| [(J_\varepsilon * u)(x) - u(x)] - [(J_\varepsilon * u)(y) - u(y)] \right|^p}{|x - y|^{N+sp}}$$

$$\leq \int_{|\xi|<\varepsilon} J_\varepsilon(\xi) \left| \frac{u(x+\xi) - u(y+\xi)}{|(x+\xi) - (y+\xi)|^{\frac{N}{p}+s}} - \frac{u(x) - u(y)}{|x-y|^{\frac{N}{p}+s}} \right|^p d\xi.$$

Integrating in $dxdy$ over $\mathbb{R}^N \times \mathbb{R}^N$, gives

$$\||J_\varepsilon * u - u\||_{s,p}^p \le \sup_{|\xi| < \varepsilon} \iint_{\mathbb{R}^N \times \mathbb{R}^N} \left| \frac{u(x+\xi) - u(y+\xi)}{|(x+\xi) - (y+\xi)|^{\frac{N}{p}+s}} - \frac{u(x) - u(y)}{|x-y|^{\frac{N}{p}+s}} \right|^p dxdy.$$

If $u \in W^{s,p}(\mathbb{R}^N)$, by the definition (15.1),

$$w(x, y) = \frac{u(x) - u(y)}{|x-y|^{\frac{N}{p}+s}} \in L^p(\mathbb{R}^N \times \mathbb{R}^N).$$

Therefore, if T_ξ is the translation operator in $L^p(\mathbb{R}^N \times \mathbb{R}^N)$,

$$\||J_\varepsilon * u - u\||_{s,p}^p \le \sup_{|\xi| < \varepsilon} \|T_\xi w - w\|_{p,\mathbb{R}^N \times \mathbb{R}^N}.$$

By the continuity of the translation in $L^p(\mathbb{R}^N \times \mathbb{R}^N)$, the right-hand side tends to zero as $\varepsilon \to 0$ (§ 17 of Chap. 6). ∎

16   Traces

Let $\mathbb{R}_+^N = \mathbb{R}^{N-1} \times \mathbb{R}^+$ denote the upper half-space $x_N > 0$ whose points we denote by $(\bar{x}, x_N)$, where $\bar{x} = (x_1, \ldots, x_{N-1})$. If u is a function in $W^{1,p}(\mathbb{R}_+^N)$, continuous in $\mathbb{R}_+^N$ up to $x_N = 0$, the trace of u on the hyperplane $x_N = 0$ is defined as its restriction to $x_N = 0$, that is $tr(u) = u(\bar{x}, 0)$.

Proposition 16.1 *Let $u \in W^{1,p}(\mathbb{R}_+^N)$ be continuous in $\overline{\mathbb{R}}_+^N$. Then*

$$\|u(\cdot, 0)\|_{r,\mathbb{R}^{N-1}}^r \le r \|Du\|_{p,\mathbb{R}_+^N} \|u\|_{q,\mathbb{R}_+^N}^{r-1} \tag{16.1}$$

for all $q, r \ge 1$ such that $q(p-1) = p(r-1)$, provided $u \in L^q(\mathbb{R}_+^N)$.

Proof May assume that $u \in C_o^\infty(\mathbb{R}^N)$. For $\bar{x} \in \mathbb{R}^{N-1}$ and $r \ge 1$

$$|u(\bar{x}, 0)|^r \le r \int_0^\infty |u(\bar{x}, x_N)|^{r-1} \left| \frac{\partial}{\partial x_N} u(\bar{x}, x_N) \right| dx_N.$$

Integrate both sides in $d\bar{x}$ over $\mathbb{R}^{N-1}$ and apply Hölder's inequality to the resulting integral on the right-hand side. ∎

If $u \in W^{1,p}(\mathbb{R}_+^N)$ there exists a sequence of functions $\{u_n\}$ in $C_o^\infty(\mathbb{R}^N)$ converging to u in $W^{1,p}(\mathbb{R}_+^N)$. By (16.1) with $r = p$

$$\|u_n(\cdot, 0) - u_m(\cdot, 0)\|_{p,\mathbb{R}^{N-1}} \le \|u_n - u_m\|_{1,p,\mathbb{R}_+^N}.$$

Therefore $\{u_n\}$ is a Cauchy sequence in $L^p(\mathbb{R}^{N-1})$ converging to some function $tr(u) \in L^p(\mathbb{R}^{N-1})$. Such a function we define as the trace of $u \in W^{1,p}(\mathbb{R}_+^N)$, on the hyperplane $x_N = 0$. We will use the perhaps improper but suggestive symbolism $tr(u) = u(\cdot, 0)$.

Proposition 16.2 *Let* $u \in W^{1,p}(\mathbb{R}_+^N)$ *for some* $p \geq 1$. *Then*

$$\|u(\cdot, 0)\|_{p, \mathbb{R}^{N-1}} \leq p^{\frac{1}{p}} \|u\|_{p, \mathbb{R}_+^N}^{1-\frac{1}{p}} \|Du\|_{p, \mathbb{R}_+^N}^{\frac{1}{p}}. \tag{16.2}$$

If $1 \leq p < N$, *then* $u(\cdot, 0) \in L^{p\frac{N-1}{N-p}}(\mathbb{R}^{N-1})$, *and*

$$\|u(\cdot, 0)\|_{p\frac{N-1}{N-p}, \mathbb{R}^{N-1}} \leq \frac{p(N-1)}{N-p} \|Du\|_{p, \mathbb{R}_+^N}. \tag{16.3}$$

If $p > N$, *the equivalence class* $u(\cdot, 0)$ *has a Hölder continuous representative, which we continue to denote by* $u(\cdot, 0)$, *and there exists a constant* $\gamma(N, p)$ *depending only upon* N *and* p, *such that, for all* $\bar{x}, \bar{y} \in \mathbb{R}^{N-1}$

$$\|u(\cdot, 0)\|_{\infty, \mathbb{R}^{N-1}} \leq \gamma \|u\|_{p, \mathbb{R}_+^N}^{1-\frac{N}{p}} \|Du\|_{p, \mathbb{R}_+^N}^{\frac{N}{p}} \tag{16.4}$$

$$|u(\bar{x}, 0) - u(\bar{y}, 0)| \leq \gamma |\bar{x} - \bar{y}|^{1-\frac{N}{p}} \|Du\|_{p, \mathbb{R}_+^N}. \tag{16.5}$$

Proof Inequality (16.2) follows from (16.1) with $r = p$. The domain $\mathbb{R}_+^N$ satisfies the cone condition with cone $\mathcal{C}_o$ of solid angle $\frac{1}{2}\omega_N$ and height $h \in (0, \infty)$. Then (16.5) follows from (8.4) of Theorem 8.1, whereas (16.4) follows from (8.3) by minimizing over $h \in (0, \infty)$. To prove (16.3) let $\{u_n\}$ be a sequence of functions in $C_o^\infty(\mathbb{R}^N)$ converging to u in $W^{1,p}(\mathbb{R}_+^N)$. For these, by (1.6) of Corollary 1.1

$$\|u_n\|_{\frac{Np}{N-p}, \mathbb{R}^N} \leq \frac{p(N-1)}{N-p} \|Du_n\|_{p, \mathbb{R}^N}.$$

Then from (16.1) with $r = p\frac{N-1}{N-p}$

$$\|u_n(\cdot, 0)\|_{r, \mathbb{R}^{N-1}} \leq r^{\frac{1}{r}} \|u_n\|_{\frac{Np}{N-p}, \mathbb{R}_+^N}^{1-\frac{1}{r}} \|Du_n\|_{p, \mathbb{R}_+^N}^{\frac{1}{r}} \leq r \|Du_n\|_{p, \mathbb{R}_+^N}. \qquad \blacksquare$$

Remark 16.1 The constant $\gamma(N, p)$ can be computed explicitly from (8.3)–(8.4) of Theorem 8.1. This shows that $\gamma(N, p) \to \infty$ as $p \to N$.

17   Traces and Fractional Sobolev Spaces

Denote by (x, t) the coordinates in $\mathbb{R}_+^{N+1}$ where $x \in \mathbb{R}^N$ and $t \geq 0$. For a function $u \in W^{1,p}(\mathbb{R}_+^{N+1})$ for some $p > 1$, we describe the regularity of its trace on the hyperplane $t = 0$ in terms of the fractional Sobolev spaces

$$W^{s,p}(\mathbb{R}^N) \quad \text{where} \quad s = 1 - \frac{1}{p}.$$

Introduce the symbolism

$$D_N = \left(\frac{\partial}{\partial x_1}, \ldots, \frac{\partial}{\partial x_N} \right) \quad \text{and} \quad D = \left(D_N, \frac{\partial}{\partial t} \right).$$

Proposition 17.1 *Let $u \in W^{1,p}(\mathbb{R}_+^{N+1})$ for some $p > 1$. Then the trace of u on the hyperplane $t = 0$ belongs to the fractional Sobolev space $W^{1-\frac{1}{p},p}(\mathbb{R}^N)$. Moreover*

$$\|\|u(\cdot, 0)\|\|_{1-\frac{1}{p}, p; \mathbb{R}^N} \leq \frac{p^2 [2(p-1)]^{\frac{1}{p}}}{(p-1)^2} \|u_t\|_{p, \mathbb{R}_+^{N+1}}^{\frac{1}{p}} \|D_N u\|_{p, \mathbb{R}_+^{N+1}}^{1-\frac{1}{p}}.$$

Proof For every pair $x, y \in \mathbb{R}^N$ set $2\xi = x - y$ and consider the point $z \in \mathbb{R}_+^{N+1}$ of coordinates $z = (\frac{1}{2}(x + y), \lambda|\xi|)$, where λ is a positive parameter to be chosen. Then

$$|u(x, 0) - u(y, 0)| \leq |u(z) - u(x, 0)| + |u(z) - u(y, 0)|$$

$$\leq |\xi| \int_0^1 |D_N u(x - \rho\xi, \lambda\rho|\xi|)| d\rho + |\xi| \int_0^1 |D_N u(y + \rho\xi, \lambda\rho|\xi|)| d\rho$$

$$+ \lambda|\xi| \int_0^1 |u_t(x - \rho\xi, \lambda\rho|\xi|)| d\rho + \lambda|\xi| \int_0^1 |u_t(y + \rho\xi, \lambda\rho|\xi|)| d\rho.$$

From this

$$\frac{|u(x, 0) - u(y, 0)|^p}{|x - y|^{N+(p-1)}} \leq \frac{1}{2^p} \left(\int_0^1 \frac{|D_N u(x - \rho\xi, \lambda\rho|\xi|)|}{|x - y|^{\frac{N-1}{p}}} d\rho \right)^p$$

$$+ \frac{1}{2^p} \left(\int_0^1 \frac{|D_N u(y + \rho\xi, \lambda\rho|\xi|)|}{|x - y|^{\frac{N-1}{p}}} d\rho \right)^p$$

$$+ \frac{1}{2^p} \lambda^p \left(\int_0^1 \frac{|u_t(x - \rho\xi, \lambda\rho|\xi|)|}{|x - y|^{\frac{N-1}{p}}} d\rho \right)^p$$

$$+ \frac{1}{2^p} \lambda^p \left(\int_0^1 \frac{|u_t(y + \rho\xi, \lambda\rho|\xi|)|}{|x - y|^{\frac{N-1}{p}}} d\rho \right)^p.$$

Next integrate both sides over $\mathbb{R}^N \times \mathbb{R}^N$. In the resulting inequality take the $\frac{1}{p}$-power and estimate the various integrals on the right-hand side by the continuous version of Minkowski's inequality (§ 3.3 of Chap. 6). This gives

$$\|u(\cdot, 0)\|_{1-\frac{1}{p}, \mathbb{R}^N} \leq \int_0^1 \left(\int_{\mathbb{R}^N} \int_{\mathbb{R}^N} \frac{|D_N u(x - \rho\xi, \lambda\rho|\xi|)|^p}{|x - y|^{N-1}} dx dy \right)^{\frac{1}{p}} d\rho$$

$$+ \lambda \int_0^1 \left(\int_{\mathbb{R}^N} \int_{\mathbb{R}^N} \frac{|u_t(x - \rho\xi, \lambda\rho|\xi|)|^p}{|x - y|^{N-1}} dx dy \right)^{\frac{1}{p}} d\rho.$$

Compute the first integral by integrating first in dy and perform such integration in polar coordinates with pole at x. Denoting with $\mathbf{n}$ the unit vector spanning the unit sphere in $\mathbb{R}^N$ and recalling that $2|\xi| = |x - y|$, we obtain

$$\int_{\mathbb{R}^N} \int_{\mathbb{R}^N} \frac{|D_N u(x - \rho\xi, \lambda\rho|\xi|)|^p}{|x - y|^{N-1}} dx dy$$

$$= 2 \int_{|\mathbf{n}|=1} d\mathbf{n} \int_0^\infty d|\xi| \int_{\mathbb{R}^N} |D_N u(x + \rho\mathbf{n}|\xi|, \lambda\rho|\xi|)|^p dx$$

$$= 2 \frac{\omega_N}{\lambda\rho} \int_{\mathbb{R}_+^{N+1}} |D_N u|^p dx.$$

Compute the second integral in a similar fashion and combine them into

$$\|u(\cdot, 0)\|_{1-\frac{1}{p}, p; \mathbb{R}^N} \leq 2^{\frac{1}{p}} \lambda^{-\frac{1}{p}} \|D_N u\|_{p, \mathbb{R}_+^{N+1}} \int_0^1 \rho^{-\frac{1}{p}} d\rho$$

$$+ 2^{\frac{1}{p}} \lambda^{1-\frac{1}{p}} \|u_t\|_{p, \mathbb{R}_+^{N+1}} \int_0^1 \rho^{-\frac{1}{p}} d\rho$$

$$= 2^{\frac{1}{p}} \frac{p}{p-1} \left(\lambda^{-\frac{1}{p}} \|D_N u\|_{p, \mathbb{R}_+^{N+1}} + \lambda^{1-\frac{1}{p}} \|u_t\|_{p, \mathbb{R}_+^{N+1}} \right).$$

The proof is completed by minimizing with respect to λ. ∎

Remark 17.1 Proposition 17.1 admits a converse, that is, a function $u \in W^{1-\frac{1}{p}, p}(\mathbb{R}^N)$, is the trace on the hyperplane $t = 0$ of a function in $W^{1, p}(\mathbb{R}_+^{N+1})$ (see § 17c of the Complements). Thus a measurable function u defined in $\mathbb{R}^N$ is in $W^{1-\frac{1}{p}, p}(\mathbb{R}^N)$ if and only it is the trace on $t = 0$ of a function in $W^{1, p}(\mathbb{R}_+^{N+1})$.

18 Traces on ∂E of Functions in $W^{1, p}(E)$

Let E be a bounded domain in $\mathbb{R}^N$ with boundary ∂E of class C^1 and with the segment property. There exists a finite covering of ∂E with open balls $B_t(x_j)$ of radius $t > 0$ and center $x_j \in \partial E$, such that the portion of ∂E within $B_t(x_j)$, can be represented, in a local system of coordinates, as the graph of a function f_j of class C^1 in a neighborhood of the origin of the local coordinate system. Consider now the covering of E given by

$$\mathcal{U} = \{B_o, B_t(x_1), \ldots, B_t(x_n)\} \quad \text{where} \quad B_o = E - \bigcup_{j=1}^{n} \bar{B}_{\frac{1}{2}t}(x_j).$$

Let Φ be a partition of unity subordinate to $\mathcal{U}$ and construct functions $\psi_j \in C_o^\infty(B_t(x_j))$ satisfying $\sum_{j=1}^{n}\psi_j = 1$ in $\bar{E}$, and $|D\psi_j| \leq 2/t$. For each x_j fixed, introduce a local system of coordinates $\bar{\xi} = (\xi_1, \ldots \xi_{N-1})$, and $\xi = (\bar{\xi}, \xi_N)$, such that $E \cap B_t(x_j)$ is mapped into the cylinder $Q_t^+ = \{|\bar{\xi}| < t\} \times [0, t]$ and $\partial E \cap B_t(x_j)$, is mapped into the portion of the hyperplane $\{\xi_N = 0\} \cap \{|\bar{\xi}| < t\}$. If f is a measurable function defined in E denote by $\tilde{f}$ the transformed of f by the new coordinate system. In these new coordinates

$$\widetilde{u\psi}_j \in W^{1,p}(\mathbb{R}^{N-1} \times \mathbb{R}^+)$$

and its trace on $\{\xi_N = 0\}$ can be defined as in § 16. In particular (16.1) and Proposition 16.1 hold for it.

If $\widetilde{u\psi}_j(\cdot, 0)$ is such a trace, we define the trace of $u\psi_j$ on $\partial E \cap B_t(x_j)$ as the function obtained from $\widetilde{u\psi}_j(\cdot, 0)$ upon returning to the original coordinates. With a perhaps improper but suggestive notation, we denote it by $u\psi_j|_{\partial E}$. We then define the trace of u on ∂E as

$$tr(u) = \sum_{j=1}^{n} u\psi_j|_{\partial E}$$

and denote it by $u|_{\partial E}$. Applying (16.1) to each $\widetilde{u\psi}_j$

$$\left(\int_{|\bar{\xi}|<t} |\widetilde{u\psi}_j|^r d\bar{\xi}\right)^{\frac{1}{r}} \leq r^{\frac{1}{r}} \left(\iint_{Q_t^+} |D\widetilde{u\psi}_j|^p d\xi\right)^{\frac{1}{rp}} \left(\iint_{Q_t^+} |\widetilde{u\psi}_j|^q d\xi\right)^{\frac{1}{q}(1-\frac{1}{r})}$$

$$\leq \gamma \|D\tilde{u}\|_{p,Q_t^+}^{\frac{1}{r}} \|\tilde{u}\|_{q,Q_t^+}^{1-\frac{1}{r}} + \gamma t^{-\frac{1}{r}} \|\tilde{u}\|_{p,Q_t^+}^{\frac{1}{r}} \|\tilde{u}\|_{q,Q_t^+}^{1-\frac{1}{r}}$$

for all $q, r \geq 1$ such that $q(r - 1) = p(r - 1)$, provided that $u \in L^q(E)$.

Next return to the original coordinates and add up the resulting inequalities for $j = 1, \ldots, n$. Recalling that only finitely many $B_t(x_j)$ have nonempty mutual intersection, we deduce that

$$\|u\|_{r,\partial E} \leq \gamma \left(\|Du\|_p + \|u\|_p\right)^{\frac{1}{r}} \|u\|_q^{1-\frac{1}{r}}. \tag{18.1}$$

Remark 18.1 In deriving (18.1), the requirement that E be bounded can be eliminated. It is only necessary that the open covering of ∂E be locally finite whence we observe that the notion of trace of u on ∂E is of local nature. We conclude that the trace on ∂E of a function $u \in W^{1,p}(E)$ is well defined for any domain $E \subset \mathbb{R}^N$ with boundary of class C^1 and with the segment property. Moreover such a trace satisfies (18.1).

Proceeding as before we may derive a counterpart of the embedding Proposition 16.2. Namely:

Proposition 18.1 *Let $u \in W^{1,p}(E)$ and assume that ∂E is of class C^1 and with the segment property. There exists a constant γ that can be determined a priori only in terms of N, p and the structure of ∂E, such that $\forall p \geq 1$ and $\forall \varepsilon > 0$*

$$\|u\|_{p,\partial E} \leq \varepsilon^{p-1} \|Du\|_p + \gamma(1 + \varepsilon^{-1})\|u\|_p. \tag{18.2}$$

If $1 \leq p < N$ the trace $u|_{\partial E}$ belongs to $L^{p\frac{N-1}{N-p}}(\partial E)$ and

$$\|u\|_{p\frac{N-1}{N-p},\partial E} \leq \gamma \|u\|_{1,p}. \tag{18.3}$$

If $p > N$, the equivalence class $u \in W^{1,p}(E)$ has a representative which is Hölder continuous in $\bar{E}$, and

$$\|u\|_{\infty,\partial E} \leq \gamma \varepsilon \|Du\|_p + \gamma(1 + \varepsilon^{-1})\|u\|_p \tag{18.4}$$

$$|u(x) - u(y)| \leq \gamma|x - y|^{1-\frac{N}{p}}\|u\|_{1,p} \text{ for all } x, y \in \bar{E}. \tag{18.5}$$

18.1   Traces and Fractional Sobolev Spaces

Let E be a bounded domain in $\mathbb{R}^N$ whith boundary ∂E of class C^1 and with the segment property. Motivated by Proposition 17.1, we may introduce a notion of fractional Sobolev space $W^{s,p}(\partial E)$ for $s \in (0, 1)$ on the $(N-1)$-dimensional domain ∂E. A function $u \in L^p(\partial E)$, belongs to $W^{s,p}(\partial E)$ if the semi-norm

$$\|u\|_{s,p;\partial E} = \left(\int_{\partial E} \int_{\partial E} \frac{|u(x) - u(y)|^p}{|x - y|^{(N-1)+sp}} d\sigma(x)d\sigma(y) \right)^{\frac{1}{p}}$$

is finite. Here $d\sigma(\cdot)$ is the surface measure on ∂E. A norm in $W^{s,p}(\partial E)$ is given by

$$\|u\|_{s,p;\partial E} = \|u\|_{p,\partial E} + \|u\|_{s,p;\partial E}.$$

Statements concerning $W^{s,p}(\partial E)$ can be derived from § 17, by working with the functions $\widetilde{u\psi}_j$, returning to the original coordinates and adding over j.

Theorem 18.1 *Let $u \in W^{1,p}(E)$ for $p > 1$ and let ∂E be of class C^1 and with the segment property. Then the trace of u on ∂E belongs to $W^{s,p}(\partial E)$ where $s = 1 - 1/p$, and*

$$\|u\|_{1-\frac{1}{p},p;\partial E} \leq \gamma \|u\|_{1,p}$$

for a constant γ depending only upon N, p and the structure of ∂E.

Remark 18.2 This theorem admits a converse, that is, a function $u \in W^{1-\frac{1}{p},p}(\partial E)$ is the trace on ∂E of a function in $W^{1,p}(E)$. Thus a measurable function u defined in ∂E is in $W^{1-\frac{1}{p},p}(\partial E)$ if and only it is the trace on ∂E of a function in $W^{1,p}(E)$ (see Theorems 18.2c and 18.3c of the Complements).

19   Multiplicative Embeddings of $W^{1,p}(E)$

Multiplicative embeddings in the form of (1.1) hold for functions in $W_o^{1,p}(E)$ and are in general false for functions in $W^{1,p}(E)$. The Poincaré inequalities of § 10, recover a multiplicative form of the embedding of $W^{1,p}(E)$ for functions of zero integral average on E. The discrete form of the isoperimetric inequality (11.1) would be vacuous if u were a nonzero constant. It is meaningful only if the measure of the set $[u = 0]$ is positive. These remarks imply that a multiplicative embedding of $W^{1,p}(E)$ into $L^q(E)$ is only possible if some information is available on the values of u on some subset of $\bar{E}$. The next Theorem provides a multiplicative embedding in terms of the trace of u on some subset Γ of ∂E, provided E is convex.

Theorem 19.1 (DiBenedetto–Diller ([35, 36])) *Let E be a bounded, open, convex subset of $\mathbb{R}^N$ for some $N \geq 1$ and let $\Gamma \subset \partial E$ be open in the relative topology of ∂E. There exist constants γ and C_Γ, such that for every $u \in W^{1,p}(E)$*

$$\|u\|_{q,E} \leq \gamma C_\Gamma^{\frac{N}{q}} \left(\|u\|_{m,E}^{1-\alpha} \|u\|_{s,\Gamma}^{\alpha} + C_\Gamma^{2\theta} \|u\|_{r,E}^{1-\theta} \|\nabla u\|_{p,E}^{\theta} \right) \tag{19.1}$$

where the parameters $\{\alpha, \theta, m, s, r, p, q\}$, satisfy

$$m, r \geq 1, \quad sp > 1, \quad q \geq \max\{m; r\}, \quad \alpha, \theta \in [0, 1] \tag{19.2}$$

and in addition the two sets of parameters $\{\alpha, m, s, q, N\}$ and $\{\theta, r, p, q, N\}$ are linked by

$$\theta = \left(\frac{1}{r} - \frac{1}{q}\right)\left(\frac{1}{N} - \frac{1}{p} + \frac{1}{r}\right)^{-1}$$
$$\alpha = \left(\frac{1}{m} - \frac{1}{q}\right)\left(\frac{1}{Ns} - \frac{1}{s} + \frac{1}{m}\right)^{-1} \tag{19.3}$$

and their admissible range is restricted by

$$s \geq \max\left\{1; m\frac{N-1}{N}\right\}; \quad r \leq \frac{(s-1)p}{p-1} \leq q \tag{19.4}$$

$$q < \infty \text{ if } p \geq N, \quad \text{and} \quad q \leq \frac{Np}{N-p} \text{ if } p < N \tag{19.5}$$

The constant γ depends only on the parameters $\{\alpha, \theta, m, s, r, p, q\}$ and is independent of u. The constant C_Γ depends only on the geometry of E and Γ.

Remark 19.1 If u vanishes on Γ in the sense of the traces, then (19.1) coincides with the multiplicative embedding of Theorem 1.1, that is

$$\|u\|_q \leq \gamma C_\Gamma^{\frac{N}{q}+2\theta} \|u\|_r^{1-\theta} \|Du\|_p^\theta. \tag{19.1$'$}$$

By taking $\alpha = 0$ in the second of (19.3), and using the arbitrariness of m and s, yields for the parameters $\{\theta, r, q, p, N\}$, the same range as the one in (1.3)–(1.5). Thus the multiplicative embedding (1.1) continues to hold for functions in $W^{1,p}(E)$ vanishing only on a nontrivial portion of ∂E.

Remark 19.2 The difference between (19.1)$'$ and the embedding (1.1) is the presence of the constant C_Γ in the right-hand side of (19.1)$'$. This is because u is known to vanish only on $\Gamma \subset \partial E$ as opposed to (1.1) where u vanishes on the whole ∂E in the sense of $W_o^{1,p}(E)$.

Remark 19.3 Since $q \geq m$ it follows from (19.3) that $\alpha, \theta \in [0, 1]$. The links (19.3) arise naturally from the proof. Through a rescaling argument one can see that these are the only possible links between the two sets of parameters $\{\theta, r, p, q, N\}$ and $\{\alpha, m, s, r, N\}$. For example let E be the ball B_R of radius R centered at the origin of $\mathbb{R}^N$. By rescaling R, inequality (19.1) is independent of the radius of E only if (19.3) holds.

Remark 19.4 The constant γ in (19.1) depends only upon the indicated parameters. In particular it is independent of u and possible rotations, translations, and dilations of E and Γ.

Remark 19.5 The constant C_Γ depends on the geometry of E and Γ in the following manner. Let $\bar{x} \in \Gamma$ and for $\varepsilon > 0$ let $B_\varepsilon(\bar{x})$ be the N-dimensional ball centered at $\bar{x}$ and radius ε. The number ε is chosen as the largest radius for which $B_\varepsilon(\bar{x}) \cap \partial E \subset \Gamma$. Then choose $x_o \in B_\varepsilon(\bar{x}) \cap E$ and let $\rho > 0$ be the largest radius for which $B_\rho(x_o) \subset E$. Finally, let $R > 0$ be the smallest radius for which $E \subset B_R(x_o)$. Thus

$$B_\rho(x_o) \subset B_\varepsilon(\bar{x}) \quad \text{and} \quad E \subset B_R(x_o). \tag{19.6}$$

Then the constant C_Γ is the smallest value of the ratio R/ρ, for all the possible choices of $\bar{x} \in \Gamma$ and $x_o \in B_\varepsilon(\bar{x})$.

Remark 19.6 By choosing $\theta = \alpha = m = r = 1$ in (19.3) gives

$$\|u\|_{p^*, E} \leq \gamma C_\Gamma^{N/p^*} \left(\|u\|_{s^*, \Gamma} + C_\Gamma^2 \|\nabla u\|_{p, E} \right) \tag{19.1$''$}$$

where

$$p^* = \frac{Np}{N-p} \quad \text{and} \quad s^* = \frac{(N-1)p}{N-p}.$$

Remark 19.7 It is an open question to establish the embedding (19.1) for nonconvex domains.

20   Proof of Theorem 19.1. A Special Case

Assume first that E is the unit cube in $\mathbb{R}^N$ for $N \geq 2$ with edges on the positive coordinate semiaxes, that is $Q = \bigcap_{i=1}^{N}\{0 \leq x_i < 1\}$. As the portion Γ_o of ∂Q we take the union of the faces of the cube lying on the coordinate planes, that is $\Gamma_o = \bigcup_{i=1}^{N} Q \cap \{x_i = 0\}$. To establish (19.1) for such a domain and such a Γ_o, we may assume u is nonnegative and in $C^1(\bar{Q})$. Set

$$\bar{x}_i = (x_1, \ldots, \underbrace{0}_{\text{i-th entry}}, \ldots, x_N)$$

$$(\bar{x}_i, t) = (x_1, \ldots, \underbrace{t}_{\text{i-th entry}}, \ldots, x_N) \qquad i = 1, \ldots, N.$$

Then for all $s \geq 1$ and all $x \in Q$

$$
\begin{aligned}
u^s(x) &= u^s(\bar{x}_i) + s \int_0^{x_i} u^{s-1}(\bar{x}_i, t)\frac{\partial u(\bar{x}_i, t)}{\partial x_i}dt \\
&\leq u^s(\bar{x}_i) + s \int_0^1 u^{s-1}(\bar{x}_i, t)\left|\frac{\partial u(\bar{x}_i, t)}{\partial x_i}\right|dt \overset{\text{def}}{=} w_i(\bar{x}_i).
\end{aligned}
\tag{20.1}
$$

Choose numbers $q \geq \ell \geq 0$ and all $s \geq 1$ satisfying $(N-1)(q-\ell) < Ns$ and set

$$k = \frac{Ns\ell}{Ns - (N-1)(q-\ell)} \quad \Longleftrightarrow \quad \ell = \frac{[Ns - (N-1)q]k}{Ns - (N-1)k}. \tag{20.2}$$

One verifies that the requirement $0 \leq \ell \leq q$ is equivalent to $0 \leq k \leq q$. Having fixed $m, r \geq 1$, choose k to satisfy $\max\{r; m\} \leq k \leq q$. Then from (20.1), for all $x \in Q$

$$u^q(x) = u^\ell(x)[u^{Ns}(x)]^{\frac{q-\ell}{Ns}} = u^\ell(x) \prod_{i=1}^{N} w_i^{\frac{q-\ell}{Ns}}(\bar{x}_i).$$

Integrating this in dx_1 over $(0, 1)$ and applying Hölder's inequality

$$\int_0^1 u^q(x)dx_1 \leq w_1(\bar{x}_1) \int_0^1 u^\ell(x) \prod_{i=2}^{N} w_i^{\frac{q-\ell}{Ns}}(\bar{x}_i)dx_1$$

$$\leq w_1(\bar{x}_1)\left(\int_0^1 u^k(x)dx_1\right)^{\frac{\ell}{k}}\prod_{i=2}^N\left(\int_0^1 w_i(\bar{x}_i)dx_1\right)^{\frac{q-\ell}{Ns}}.$$

Next integrate this in dx_2 over $(0, 1)$ and apply Hölder's inequality with the same conjugate exponents. Proceeding by induction to exhaust all the variables x_j, yields

$$\int_Q u^q(x)dx \leq \left(\int_Q u^k(x)dx\right)^{\frac{\ell}{k}}\prod_{i=1}^N\left(\int_{Q_{N-1}^i} w_i(\bar{x}_i)d\bar{x}_i\right)^{\frac{q-\ell}{Ns}}$$

where $Q_{N-1}^i = Q \cap \{x_i = 0\}$ is the $(N-1)$-dimensional cube excluding the ith coordinate. By Hölder's inequality and the definition of $w_i(\bar{x}_i)$

$$\int_{Q_{N-1}^i} w_i(\bar{x}_i)d\bar{x}_i \leq \int_{\Gamma_o} u^s(x)d\sigma + s\left(\int_Q |Du|^p dx\right)^{\frac{1}{p}}\left(\int_Q u^{\frac{(s-1)p}{p-1}}dx\right)^{\frac{p-1}{p}}$$

where $d\sigma$ is the surface measure on Γ_o. Combining these estimates, we arrive at

$$\|u\|_{q,Q} \leq \gamma\|u\|_{k,Q}^{\frac{\ell}{q}}\|u\|_{s,\Gamma_o}^{\frac{q-\ell}{q}} + \gamma\|u\|_{k,Q}^{\frac{\ell}{q}}\|u\|_{\frac{(s-1)p}{p-1},Q}^{\frac{s-1}{s}\frac{q-\ell}{q}}\|Du\|_{p,Q}^{\frac{q-\ell}{qs}}. \qquad (20.3)$$

If either $m = q$ or $r = q$ then $k = q$ and (20.3) imply that $k = \ell = q$. In such a case the multiplicative inequality (19.1) follows from (20.5) and is vacuous. Thus in the definition (19.3) of θ it is stipulated that $\theta = 0$ if $r = q$. A similar stipulation holds for α.

Assume that $\max\{m; r\} < q$ and choose k satisfying $\max\{m; r\} < k < q$. By Hölder's inequality

$$\|u\|_{k,Q} \leq \|u\|_{m,Q}^{\frac{m(q-k)}{k(q-m)}}\|u\|_{q,Q}^{\frac{q(k-m)}{k(q-m)}}$$

$$\|u\|_{k,Q} \leq \|u\|_{r,Q}^{\frac{r(q-k)}{k(q-r)}}\|u\|_{q,Q}^{\frac{q(k-r)}{k(q-r)}}. \qquad (20.4)$$

Assume first that the second of (19.4) holds with strict inequalities. Then by Hölder's inequality

$$\|u\|_{\frac{(s-1)p}{(p-1)},Q} \leq \|u\|_{r,Q}^{\frac{r[q(p-1)-p(s-1)]}{p(s-1)(q-r)}}\|u\|_{q,Q}^{\frac{q[p(s-1)-r(p-1)]}{p(s-1)(q-r)}}. \qquad (20.5)$$

Assume that of the two terms on the right-hand side of (20.3), the first majorizes the second. Then using the first of (20.5)

$$\|u\|_{q,Q} \leq 2\gamma\|u\|_{q,Q}^{\frac{\ell(k-m)}{k(q-m)}}\|u\|_{m,Q}^{\frac{m(q-k)}{k(q-m)}}\|u\|_{s,\Gamma_o}^{\frac{q-\ell}{q}}.$$

By reducing the powers of $\|u\|_{q,Q}$ this gives

$$\|u\|_{q,Q} \le \gamma' \|u\|_{m,Q}^{1-\alpha} \|u\|_{s,\Gamma_o}^{\alpha}$$

where α is given by the second of (19.3), and γ' is a constant depending only on the set of parameters $\{N, p, q, s, m, r\}$. If of the two terms on the right-hand side of (20.3) the second majorizes the first, then, using the second of (20.4), and (20.5), gives

$$\|u\|_{q,Q} \le 2\gamma \|u\|_{q,Q}^{\frac{\ell(k-r)}{k(q-r)} + \frac{(q-\ell)[(s-1)p-r(p-1)]}{s(q-r)}}$$
$$\times \|u\|_{r,Q}^{\frac{\ell r(q-k)}{qk(q-r)} + \frac{r(q-\ell)[q(p-1)-p(s-1)]}{qsp(q-r)}} \|Du\|_{p,Q}^{\frac{q-\ell}{qs}}.$$

By reducing the powers of $\|u\|_{q,Q}$ this gives

$$\|u\|_{q,Q} \le \gamma'' \|u\|_{r,Q}^{1-\theta} \|Du\|_{p,Q}^{1-\theta}$$

where θ is given by the first of (19.3), and γ'' is a constant depending only on the set of parameters $\{N, p, q, s, m, r\}$. Finally, if the second of (19.4) holds with some equality, the arguments are similar and indeed simpler.

21   Constructing a Map Between E and Q. Part I

Let E be a bounded, convex subset of $\mathbb{R}^N$ and let $\Gamma \subset \partial E$ be open in the relative topology of ∂E. Having fixed $\bar{x} \in \Gamma$ construct the ball $B_\varepsilon(\bar{x})$ where ε is the largest radius so that $B_\varepsilon(\bar{x}) \cap \partial E \subset \Gamma$. Then pick $x_o \in B_\varepsilon(\bar{x}) \cap E$ and construct the balls $B_\rho(x_o)$ and $B_R(x_o)$ as in (19.6). Continue to denote by Γ_o that portion of the boundary of the cube Q consisting of the faces lying on the coordinate planes.

Proposition 21.1 *There exists a map* $\mathcal{F} : \bar{E} \to \overline{Q}$ *and positive, absolute constants* $C > c > 0$, *independent of the geometry of* ∂E *and* Γ, *such that* $Q = \mathcal{F}^{-1}(E)$, $\Gamma_o \subset \mathcal{F}^{-1}(\Gamma)$ *and*

$$c\rho|x - y| \le |\mathcal{F}(x) - \mathcal{F}(y)| \le CR\frac{R}{\rho}|x - y|. \tag{21.1}$$

Moreover the Jacobian J *of* $\mathcal{F}$, *satisfies*

$$c\rho^N \le J(x) \le CR^N \qquad \text{for all } x \in E. \tag{21.2}$$

Proof Let $\mathbf{n}$ be the unit vector in $\mathbb{R}^N$ ranging over the unit sphere S_1, and consider the map $\phi_{x_o,E} : S_1 \to \partial E$ defined by

$$\phi_{x_o,E}(\mathbf{n}) = x_o + t\mathbf{n} \tag{21.3}$$

where t is the unique positive number such that $x_o + t\mathbf{n} \in \partial E$. Such a map is well defined since E is bounded and convex.

Lemma 21.1 *There exists a constant C_o, depending on x_o and E, such that*

$$\rho|\mathbf{n}_1 - \mathbf{n}_2| \le |\phi_{x_o,E}(\mathbf{n}_1) - \phi_{x_o,E}(\mathbf{n}_2)| \le 4R\frac{R}{\rho}|\mathbf{n}_1 - \mathbf{n}_2|. \tag{21.4}$$

Proof Up to a translation we may assume that $x_o = 0$ and set $\phi_{x_o,E} = \phi$. Having fixed $\mathbf{n}_1$ and $\mathbf{n}_2$ on S_1, the bound below in (21.4) follows from the definition (21.3) since the sphere S_ρ centered at the origin is contained in the interior of E. For the bound above, by intersecting E with the hyperplane through x_o and containing $\mathbf{n}_1$ and $\mathbf{n}_2$, it suffices to assume $N = 2$. Denoting by $\mathbf{i}$ and $\mathbf{j}$ the coordinate unit vectors in $\mathbb{R}^2$, we may assume up to a possible rotation that $\mathbf{n}_1 = \mathbf{i}$. Then

$$\mathbf{n}_2 = (\cos\lambda, \sin\lambda) \overset{\text{def}}{=} \mathbf{n} \quad \text{for some } \lambda \in (-\pi, \pi).$$

Therefore the bound above in (21.4) reduces to

$$|\phi(\mathbf{n}) - \phi(\mathbf{i})| \le 4R\frac{R}{\rho}\sqrt{1 - \cos\lambda}. \tag{21.4}'$$

Let $\Sigma_\mathbf{n}$ be the line segment joining $\phi(\mathbf{i})$ and $\phi(\mathbf{n})$. Let also $L_\mathbf{n}$ be the line through $\phi(\mathbf{i})$.

21.1 Case 1. $\Sigma_\mathbf{n}$ intersects B_ρ

Since $L_\mathbf{n}$ intersects B_ρ, the point x_* on the line $L_\mathbf{n}$ which minimizes the distance from $L_\mathbf{n}$ to the origin 0, is in B_ρ and thus on the line segment $\Sigma_\mathbf{n}$. Let λ_1 and λ_2 be the angles

$$\lambda_1 = \widehat{\phi(\mathbf{i})0x_*} \qquad \lambda_2 = \widehat{x_*0\phi(\mathbf{n})}.$$

Then $\lambda_1 + \lambda_2 = \lambda$ and by elementary trigonometry

$$|\phi(\mathbf{n}) - \phi(\mathbf{i})| = |\phi(\mathbf{i})|\sin\lambda_1 + |\phi(\mathbf{n})|\sin\lambda_2 \le R(\sin\lambda_1 + \sin\lambda_2).$$

At least one of the λ_i is less than $\frac{1}{2}\lambda$, say, for example λ_1. Then, since $\frac{1}{2}\lambda \in (0, \frac{\pi}{2})$

$$\sin\lambda_1 + \sin\lambda_2 \le \sin\tfrac{1}{2}\lambda + \sin(\lambda - \lambda_1)$$
$$\le \sin\tfrac{1}{2}\lambda + \sin\lambda\cos\lambda_1 - \cos\lambda\sin\lambda_1$$
$$\le 2\sin\tfrac{1}{2}\lambda + \sin\lambda \le \tfrac{4}{\sqrt{2}}\sqrt{1 - \cos\lambda}.$$

21.2   Case 2. $\Sigma_\mathbf{n}$ does not intersect B_ρ

Without loss of generality, by possibly interchanging the role of $\mathbf{i}$ and $\mathbf{n}$, we may assume that $|\phi(\mathbf{i})| > |\phi(\mathbf{n})|$. Let $\lambda_o = \widehat{\phi(\mathbf{n})\phi(\mathbf{i})}0$. Then by the law of sines

$$\frac{\sin \lambda}{|\phi(\mathbf{n}) - \phi(\mathbf{i})|} = \frac{\sin \lambda_o}{|\phi(\mathbf{n})|}.$$

From this

$$|\phi(\mathbf{n}) - \phi(\mathbf{i})| = |\phi(\mathbf{n})|\frac{\sin \lambda}{\sin \lambda_o} \leq R\frac{\sin \lambda}{\sin \lambda_o}.$$

Since the line segment $\Sigma_\mathbf{n}$ does not intersect B_ρ, the smallest λ_o could possibly be is if the line $L_\mathbf{n}$ is tangent to B_ρ. In such a case

$$\sin \lambda_o \geq \frac{\rho}{|\phi(\mathbf{i})|} \geq \frac{\rho}{R}.$$

Combining these estimates proves the Lemma.   ∎

22   Constructing a Map Between E and Q. Part II

Extend $\phi_{x_o,E}$ to a map $\varphi_{x_o,E}$ from the whole unit ball B_1 onto E by

$$B_1 \ni x \rightarrow \varphi_{x_o,E}(x) = x_o + |x|t(\mathbf{n}_x)\mathbf{n}_x \quad \mathbf{n}_x = \begin{cases} \dfrac{x}{|x|} & \text{if } x \neq 0 \\ 0 & \text{if } x = 0 \end{cases} \qquad (22.1)$$

and $t(\mathbf{n}_x)$ is defined as in (21.3), in correspondence of the unit vector $\mathbf{n}_x$.

Lemma 22.1   *For all $x, y \in B_1$*

$$\frac{1}{4}\rho\frac{\rho}{R}|x - y| \leq |\varphi_{x_o,E}(x) - \varphi_{x_o,E}(y)| \leq 5R\frac{R}{\rho}|x - y|. \qquad (22.2)$$

Moreover, denoting by J_φ the Jacobian of $\varphi_{x_o,E}$

$$J_\varphi(x) = |t(\mathbf{n}_x)|^N \qquad \text{for all } x \in B_1. \qquad (22.3)$$

Finally from the definition of $t(\mathbf{n}_x)$ it follows that

$$\rho^N \leq J_\varphi(x) \leq R^N \qquad \text{for all } x \in B_1. \qquad (22.4)$$

Proof Assume $x_o = 0$, set $\varphi_{x_o,E} = \varphi$, and fix any two nonzero vectors $x, y \in B_1$. By intersecting E with the hyperplane through the origin and containing x and y, it suffices to consider the case $N = 2$. Assume for example that $|y| \le |x|$. Then by elementary plane geometry

$$\left| \frac{x}{|x|} - \frac{y}{|y|} \right| \le \frac{1}{|y|} |x - y|.$$

From this, making also use of the upper estimate in Lemma 21.1

$$|\varphi(x) - \varphi(y)| \le R|x - y| + 4R \frac{R}{\rho} |y| \left| \frac{x}{|x|} - \frac{y}{|y|} \right| \le 5R \frac{R}{\rho} |x - y|.$$

For the lower estimate in (22.2) assume for example $|x| \ge |y|$. Suppose first that

$$|x| - |y| \le \frac{1}{4} \frac{\rho}{R} |x - y|.$$

Then by the lower estimate of Lemma 21.1

$$
\begin{aligned}
|\varphi(x) - \varphi(y)| &= \left| |x| t(\mathbf{n}_x)\mathbf{n}_x - |y| t(\mathbf{n}_y)\mathbf{n}_y \right| \\
&\ge |x| \left| t(\mathbf{n}_x)\mathbf{n}_x - t(\mathbf{n}_y)\mathbf{n}_y \right| - R(|x| - |y|) \\
&\ge \rho|x| \left| \frac{x}{|x|} - \frac{y}{|y|} \right| - \frac{1}{4}\rho|x - y| \\
&\ge \rho|x| \left| \frac{x - y}{|x|} + \frac{y}{|x|} - \frac{y}{|y|} \right| - \frac{1}{4}\rho|x - y| \\
&\ge \rho|x - y| - \rho(|x| - |y|) - \tfrac{1}{4}\rho|x - y| \ge \tfrac{1}{2}\rho|x - y|.
\end{aligned}
$$

If on the other hand

$$|x| - |y| > \frac{1}{4} \frac{\rho}{R} |x - y|$$

then by elementary plane geometry

$$|\varphi(x) - \varphi(y)| = \left| |x| t(\mathbf{n}_x)\mathbf{n}_x - |y| t(\mathbf{n}_y)\mathbf{n}_y \right| \ge \rho(|x| - |y|) \ge \frac{1}{4}\rho \frac{\rho}{R} |x - y|.$$

To establish (22.3) express the Lebesgue measure $d\nu$ of B_1 in polar coordinates as $d\nu = r^{N-1} dr d\mathbf{n}$ for $r \in (0, 1)$ and $\mathbf{n}$ ranging over the unit sphere of $\mathbb{R}^N$. Likewise the Lebesgue measure of E in polar coordinates is $d\mu = \tau^{N-1} d\tau d\mathbf{n}$ for $\tau \in (0, t(\mathbf{n}))$ where $t(\mathbf{n})$ is the polar representation of ∂E with pole at the origin. From the definition (22.1) it follows that $\tau = rt(\mathbf{n})$. Therefore

$$d\mu = \tau^{N-1} d\tau d\mathbf{n} = |t(\mathbf{n})|^N r^{N-1} dr d\mathbf{n} = |t(\mathbf{n})|^N d\nu. \qquad \blacksquare$$

The transformation $\varphi^{-1}_{x_o,E}$ is a one-to-one Lipschitz map between E and B_1 with Lipschitz continuous inverse. The boundary of E is mapped into ∂B_1 and the portion $\Gamma \subset \partial E$ is mapped into a subset $\Gamma_1 \subset \partial B_1$, open in the relative topology of the unit sphere S_1. The Lipschitz constants and the Jacobian of such a transformation are controlled by (22.2)–(22.4). Therefore the proof of Proposition 21.1 reduces to the case where $E = B_1$ and $\Gamma_1 \subset \partial B_1$. Up to a rotation, we may assume that $(0, \dots, 0, 1) \in \Gamma_1$. Since Γ_1 is open, there is $\varepsilon > 0$ such that $[x_N > 1 - \varepsilon] \cap S_1 \subset \Gamma_1$. Set $x_\varepsilon = (0, \dots, 0, 1 - \varepsilon)$ and construct the map $\varphi_{x_\varepsilon, B_1} : B_1 \to B_1$. Such a map satisfies estimates analogous to (22.2)–(22.4) with ρ replaced by ε and $R = 2$. Moreover Γ_1 is mapped into an open portion $\Gamma_2 \in \partial B_1$ such that $\varphi_{x_\varepsilon, B_1}([x_N > 0] \cap S_1) \subset \Gamma_2$. Thus we have reduced to the case when $E = B_1$ and $\Gamma_2 = [x_N > 0] \cap S_1$. Then, after an appropriate rotation, the map $\varphi_{x_\#, Q} : B_1 \to B_1$ where $x_\# = (\frac{1}{2}, \dots, \frac{1}{2})$, maps the cube Q onto B_1 and Γ_o onto Γ_2. The map $\mathcal{F}$ claimed by Proposition 21.1 is obtained by composing the maps occurring in the various steps of the construction.

23   Proof of Theorem 19.1 Concluded

Let E be a bounded, convex subset of $\mathbb{R}^N$ and let $\Gamma \subset \partial E$ be open in the relative topology of ∂E. Having fixed $\bar{x} \in \Gamma$ construct the ball $B_\varepsilon(\bar{x})$, where ε is the largest radius for which $B_\varepsilon(\bar{x}) \cap \partial E \subset \Gamma$. Then pick $x_o \in B_\varepsilon(\bar{x}) \cap E$ and construct the balls $B_\rho(x_o)$ and $B_R(x_o)$ as in (19.6). Let also $\mathcal{F}$ be the map claimed by Proposition 21.1 for these choices of $\bar{x}$, ρ and R. The Jacobian is estimated in (21.2), whereas by (21.1) $|D\mathcal{F}| \le CR(R/\rho)$. Given $u \in W^{1,p}(E)$, by Theorem 19.1 applied for the unit cube Q

$$
\begin{aligned}
\|u\|_{q,E} = \left(\int_Q |u(\mathcal{F})|^q J\,dy \right)^{\frac{1}{q}} &\le \gamma R^{\frac{N}{q}} \|u(\mathcal{F})\|_{q,Q} \\
&\le \gamma R^{\frac{N}{q}} \left(\|u(\mathcal{F})\|^{1-\alpha}_{m,Q}\, \|u(\mathcal{F})\|^{\alpha}_{s,\Gamma_o} + \|u(\mathcal{F})\|^{1-\theta}_{r,Q}\|D_y u(\mathcal{F})\|^{\theta}_{p,Q} \right) \\
&\le \gamma R^{\frac{N}{q}} \rho^{-\frac{N(1-\alpha)}{m}} \|u\|^{1-\alpha}_{m,E} \rho^{-\frac{(N-1)\alpha}{s}} \|u\|^{\alpha}_{s,\Gamma} \qquad \blacksquare \\
&\quad + \gamma R^{\frac{N}{q}} \left(\frac{R^2}{\rho} \right)^{\theta} \rho^{-\frac{N(1-\theta)}{r}} \|u\|^{1-\theta}_{r,E} \rho^{-\frac{N\theta}{p}} \|D_x u\|^{\theta}_{p,E} \\
&\le \gamma \left(\frac{R}{\rho} \right)^{\frac{N}{q}} \left[\|u\|^{1-\alpha}_{m,E} \|u\|^{\alpha}_{s,\Gamma} + \left(\frac{R}{\rho} \right)^{2\theta} \|u\|^{1-\theta}_{r,E} \|D_x u\|^{\theta}_{p,E} \right].
\end{aligned}
$$

24   The Spaces $W^{1,p}_{p*}(E)$

The proof of Corollary 1.1 only requires that $Du \in L^p(E)$, and that u is the limit of a sequence $\{u_n\} \subset C^\infty_o(E)$, in some topology, by which

$$\limsup \|Du_n\|_p \le \|Du\|_p, \qquad \text{and} \qquad \|u\|_{p^*} \le \liminf \|u_n\|_{p^*}.$$

Then $u \in L^{p^*}(E)$. However u need not be in $L^p(E)$, unless E is of finite measure. An example is

$$u(x) = \begin{cases} 1 & \text{for } |x| \le 1 \\ \dfrac{1}{|x|^\alpha} & \text{for } |x| > 1 \end{cases} \qquad \text{for} \qquad \frac{N-p}{p} < \alpha < \frac{N}{p}.$$

This suggest introducing the linear space

$$W_{p^*}^{1,p}(E) = \{u \in L^{p^*}(E), \text{ with } Du \in L^p(E)\}$$

with norm

$$\|u\|_{1,p;p^*} = \|u\|_{p^*} + \|Du\|_p.$$

One verifies that $W_{p^*}^{1,p}(E)$ with such a norm is a Banach space. If $\mu(E) < \infty$ the norms $\|\cdot\|_{1,p}$ and $\|\cdot\|_{1,p;p^*}$ are equivalent and $W_{p^*}^{1,p}(E) = W^{1,p}(E)$, up to a bijection. In general, by virtue of (1.6), $W^{1,p}(E) \subset W_{p^*}^{1,p}(E)$. The next proposition asserts that $W_{p^*}^{1,p}(E)$ is the closure of $W^{1,p}(E)$ in the norm $\|\cdot\|_{1,p;p^*}$ and that functions in $W_{p^*}^{1,p}(E)$ are only those satisfying the embedding of Corollary 1.1.

Proposition 24.1 Let $u \in W_{p^*}^{1,p}(E)$. For all $\varepsilon > 0$ there exists $u_\varepsilon \in W^{1,p}(E)$ such that $\|u - u_\varepsilon\|_{1,p;p^*} < \varepsilon$. Moreover u and Du satisfy (1.6).

Proof May assume $E = \mathbb{R}^N$. Let ζ_n be a nonnegative, piecewise smooth cutoff function in $\mathbb{R}^N$, such that

$$\begin{aligned} \zeta_n &= 1 \ \text{ for } |x| < n \\ \zeta_n &= 0 \ \text{ for } |x| \ge 2n \end{aligned} \qquad \text{and} \qquad |D\zeta_n| \le \frac{1}{n}.$$

Set $u_n = u\zeta_n$, verify that $u_n \in W^{1,p}(\mathbb{R}^N)$ and compute

$$\int_{\mathbb{R}^N} |u - u_n|^{p^*} dx \le \int_{|x|>n} |u|^{p^*} dx.$$

Moreover

$$\begin{aligned} \int_{\mathbb{R}^N} |Du - Du_n|^p dx &\le 2^{p-1} \left(\int_{\mathbb{R}^N} (1 - \zeta_n)|Du|^p dx + \int_{\mathbb{R}^N} |u \, D\zeta_n|^p dx \right) \\ &\le 2^{p-1} \left(\int_{|x|>n} |Du|^p dx + \frac{1}{n^p} \int_{n<|x|<2n} |u|^p dx \right) \\ &\le 2^{p-1} \left[\int_{|x|>n} |Du|^p dx + \omega_N^{\frac{p}{N}} \left(\int_{|x|>n} |u|^{p^*} dy \right)^{\frac{p}{p^*}} \right]. \end{aligned}$$

The second assertion follows from (1.6) applied to u_n.   ∎

Proposition 24.2 *Let* $\{u_n\}$ *be a sequence of nonnegative functions in* $W_{p*}^{1,p}(E)$ *and set*

$$v = \sup u_n, \qquad w_j = \sup |u_{n,x_j}|, \qquad \mathbf{w} = (w_1, \dots, w_N).$$

If $\mathbf{w} \in L^p(E)$, *then* $v \in W_{p*}^{1,p}(E)$ *and* $|v_{x_j}| \le w_j$ *for* $j = 1, \dots, N$.

Proof May assume $E = \mathbb{R}^N$. The assertion is obvious if $\{u_n\} = \{u_1, u_2\}$, and by induction, if $\{u_n\}$ is a finite sequence. Otherwise set

$$v_n = \max\{u_1, \dots u_n\} \quad \text{and} \quad w_{n,j} = \max\{|u_{1,x_j}|, \dots, |u_{n,x_j}|\}$$

for $j = 1, \dots, N$. Then $v_n \in W_{p*}^{1,p}(E)$ and

$$|v_{n,x_j}| \le w_{n,j} \le w_j \quad \text{for all} \quad j = 1, \dots, N.$$

Since $\{v_n\}$ is monotone increasing, by monotone convergence

$$\|v\|_{p*} = \lim \|v_n\|_{p*} \le \frac{p(N-1)}{N-p} \liminf \|Dv_n\|_p \le \frac{p(N-1)}{N-p} \|\mathbf{w}\|_p.$$

This implies that $v \in L^{p^*}(E)$. It remains to show that the distributional derivatives v_{x_j} are real valued functions in $L^p(\mathbb{R}^N)$, and $|v_{x_j}| \le w_j$.
Let $\varphi \in C_o^\infty(\mathbb{R}^N)$ and compute

$$\int_{\mathbb{R}^N} v\varphi_{x_j} dy = \lim \int_{\mathbb{R}^N} v_n \varphi_{x_j} dy = -\lim \int_{\mathbb{R}^N} v_{n,x_j} \varphi dy \le \int_{\mathbb{R}^N} w_j |\varphi| dy.$$

The left-hand side defines a linear functional T_j on $C_o^\infty(\mathbb{R}^N)$ with upper bound, uniform in φ, given by

$$C_o^\infty(\mathbb{R}^N) \ni \varphi \to T_j(\varphi) \le \|\varphi\|_q \|w_j\|_p \quad \text{where} \quad \frac{1}{p} + \frac{1}{q} = 1.$$

Assume first $1 < p < N$, so that $\frac{N}{N-1} < q < \infty$. Endow $C_o^\infty(\mathbb{R}^N)$ with the norm $\|\cdot\|_q$ and, as such, regard it as a subspace of $L^q(\mathbb{R}^N)$. By the Hahn–Banach theorem 9.1 of Chap. 7 (dominated extension of functionals), T_j admits an extension $\tilde{T}_j$ to $L^q(\mathbb{R}^N)$ satisfying the same uniform upper bound on $L^q(\mathbb{R}^N)$. Since $C_o^\infty(\mathbb{R}^N)$ is dense in $L^q(\mathbb{R}^N)$, such an extension is unique. By the Riesz representation theorem 11.1 of Chap. 6 there exists a unique function in $L^p(\mathbb{R}^N)$, which we denote by $-v_{x_j} \in L^p(\mathbb{R}^N)$ such that

$$\tilde{T}_j(\varphi) = \int_{\mathbb{R}^N} (-v_{x_j})\varphi dy \quad \text{for all} \quad \varphi \in L^q(\mathbb{R}^N).$$

Since $\tilde{T}_j = T_j$ on $C_o^\infty(\mathbb{R}^N)$

$$-\int_{\mathbb{R}^N} v\varphi_{x_j} dy = \int_{\mathbb{R}^N} v_{x_j}\varphi \le \int_{\mathbb{R}^N} w_j|\varphi|dy \quad \text{for all} \quad \varphi \in C_o^\infty(\mathbb{R}^N).$$

Thus v_{x_j} is the weak x_j-derivative of v and $|v_{x_j}| \le w_j$.

If $p = 1$ endow $C_o^\infty(\mathbb{R}^N)$ with the sup-norm $\|\cdot\|$ and, as such, regard it as a dense subset of $C_o(\mathbb{R}^N)$. By the dominated extension theorem, T_j admits a unique extension $\tilde{T}_j \in C_o(\mathbb{R}^N)^*$ satisfying the upper bound

$$\tilde{T}_j(\varphi) \le \int_{\mathbb{R}^N} |\varphi| w_j dy \le \|\varphi\| \, \|w_j\|_1 \quad \text{for all} \quad \varphi \in C_o(\mathbb{R}^N).$$

By the Riesz representation theorem 1.1 of Chap. 8, there exists a Radon measure μ_j and a μ_j-measurable function η_j such that $|\eta_j| = 1$ and

$$\tilde{T}_j(\varphi) = \int_{\mathbb{R}^N} \varphi \eta_j d\mu_j \quad \text{for all} \quad \varphi \in C_o(\mathbb{R}^N).$$

The construction of the measure μ_j in § 3 of Chap. 8 implies that

$$\mu_j(E) \le \int_E w_j dy$$

for all Lebesgue measurable sets $E \subset \mathbb{R}^N$. Therefore μ_j is absolutely continuous with respect to the Lebesgue measure and is finite, since $w_j \in L^1(\mathbb{R}^N)$. The signed measure

$$d\tilde{\mu}_j = \eta_j d\mu_j$$

is also absolutely continuous with respect to the Lebesgue measure and its total variation

$$|\tilde{\mu}_j| = \tilde{\mu}_j^+ + \tilde{\mu}_j^-$$

is also finite. By the Radon–Nykodým theorem 18.3 of Chap. 4, there exists ia function in $L^1(\mathbb{R}^N)$, which we denote by $-v_{x_j} \in L^1(\mathbb{R}^N)$ such that $d\tilde{\mu}_j = -v_{x_j} dy$. Thus

$$\tilde{T}_j(\varphi) = \int_{\mathbb{R}^N} \varphi(-v_{x_j})dy \quad \text{for all} \quad \varphi \in C_o(\mathbb{R}^N).$$

Since $\tilde{T}_j$ coincides with T_j on $C_o^\infty(\mathbb{R}^N)$

$$-\int_{\mathbb{R}^N} v\varphi_{x_j} dy = \int_{\mathbb{R}^N} v_{x_j}\varphi dy \le \int_{\mathbb{R}^N} w_j|\varphi|dy \quad \text{for all} \quad \varphi \in C_o^\infty(\mathbb{R}^N).$$

Thus v_{x_j} is the weak x_j-derivative of v and $|v_{x_j}| \le w_j$. ∎

Problems and Complements

1c Multiplicative Embeddings of $W_o^{1,p}(E)$

1.1.   The following two Propositions are established by a minor variant of the arguments in § 2–4. Their significance is that u is not required to vanish, in some sense, on ∂E. Let $Q = \prod_{j=1}^{N}(a_j, b_j)$ be a cube in $\mathbb{R}^N$.

Proposition 1.1c *Let $u \in W^{1,p}(Q)$ for some $p \in [1, N)$. Then $u \in L^{p^*}(Q)$, and*

$$\|u\|_{p^*} \le \sum_{j=1}^{N} \left(\frac{1}{N(b_j - a_j)} \|u\|_p + \frac{p(N-1)}{N(N-p)} \|u_{x_j}\|_p \right).$$

The inequality continues to hold if $(b_j - a_j) = \infty$ for some j, provided we set $(b_j - a_j)^{-1} = 0$.

Proposition 1.2c *Let $u \in W^{1,p}(\mathbb{R}^N)$ for some $p \in [1, N)$. Then $u \in L^{p^*}(\mathbb{R}^N)$ and*

$$\|u\|_{p^*} \le \frac{p(N-1)}{N(N-p)} \sum_{j=1}^{N} \|u_{x_j}\|_p.$$

1.2.   There exists a function $u \in W_o^{1,N}(E)$, that is not essentially bounded.

1.3.   The functional $u \to \|Du\|_p$ is a semi-norm in $W^{1,p}(E)$ and a norm in $W_o^{1,p}(E)$. Such a norm is equivalent to $\|u\|_{1,p}$, that is, there exists a constant γ depending only upon N and p such that

$$\gamma^{-1} \|Du\|_p \le \|u\|_{1,p} \le \gamma \|Du\|_p \quad \text{for all } u \in W_o^{1,p}(E).$$

8c Embeddings of $W^{1,p}(E)$

8.1.   Let $W^{1,p}(E)^*$ denote the dual of $W^{1,p}(E)$ for some $1 \le p < \infty$ and let q be the Hölder conjugate of p. Prove that $L^q(E)^N \subset W^{1,p}(E)^*$. Give an example to show that inclusion is, in general strict.

8.2.   Let E satisfy the cone condition, and let $W^{1,p}(E)^*$ denote the dual of $W^{1,p}(E)$ for some $1 \le p < N$. Prove that $L^q(E)^N \subset W^{1,p}(E)^*$ where q is the Hölder conjugate of p_*. Give an example to show that inclusion is, in general strict.

8.3.   Let E be the unit ball of $\mathbb{R}^N$ and consider formally the integral

$$W^{1,p}(E) \ni f \to T(f) = \int_E |x|^{-\alpha} f \, dx.$$

Find the values of α for which this defines a bounded linear functional in $W^{1,p}(E)$. Note that the previous problem provides only a sufficient conditions. **Hint**: Compute $(x_i|x|^{-\alpha})_{x_i}$ weakly.

8.4. Let $E \subset$ be open set. Give an example of $f \in W^{1,p}(E)$ unbounded in every open subset of E. **Hint**: Properly modify the function in **17.9.** of the Complements of Chap. 4.

8.1c Differentiability of Functions in $W^{1,p}(E)$ for $p > N$

A continuous function u defined in an open set $E \subset \mathbb{R}^N$, is differentiable at $x \in E$ if it admits a Taylor expansion of the form of (21.1) of Chap. 8 in the context of the Rademacher's theorem.

Functions $u \in W^{1,p}(E)$ for $1 < p < \infty$ are characterized as admitting a Taylor type expansion in the topology of $L^p(E)$ (Proposition 20.3 of Chap. 8). The embedding Theorem 8.1 implies that for $p > N$, such an expansion holds a.e. in E.

Theorem 8.1c *Functions* $u \in W^{1,p}_{loc}(E)$ *for* $N < p \le \infty$ *are a.e. differentiable in* E.

Proof Assume first $N < p < \infty$. Since $Du \in L^p_{loc}(E)$,

$$\lim_{\rho \to 0} \rlap{\,\,\,\,\,\,\,\,\,\,\,\,-}\int_{B_\rho(x)} |Du - Du(x)|^p dz = 0 \quad \text{for almost all } x \in E.$$

For any such x fixed, set

$$v(y) = u(y) - u(x) - Du(x) \cdot (y - x) \quad \text{for } y \in B_\rho(x).$$

In particular $v(x) = 0$. Apply (8.4) to the function $v(\cdot)$, with E being the ball $B_{|y-x|}(x)$. It gives

$$|v(y)| = |v(y) - v(x)| \le C(N, p)|x - y| \left(\rlap{\,\,\,\,\,\,\,\,\,\,\,\,-}\int_{B_{|y-x|}(x)} |Dv - Dv(x)|^p dz \right)^{\frac{1}{p}}.$$

Thus for $|y - x| \ll 1$

$$\frac{\left| u(y) - u(x) - Du(x) \cdot (y - x) \right|}{|y - x|} = O(|y - x|).$$

Let now $p = \infty$. The previous arguments are local and $L^\infty_{loc}(E) \subset L^p_{loc}(E)$. Hence one can always assume $N < p < \infty$. ∎

Remark 8.1c The theorem is an extension of the Rademaker's theorem to functions in $W^{1,p}(E)$ for $N < p \le \infty$. In particular for $p = \infty$ it provides an alternative proof of the Rademaker's theorem.

14c   Compact Embeddings

14.1.   Theorem 14.1 is false if E is unbounded. To construct a counterexample, consider a sequence of balls $\{B_{\rho_j}(x_j)\}$ all contained in E such that $|x_j| \to \infty$ as $j \to \infty$. Then construct a function $\varphi \in W_o^{1,p}\big(B_{\rho_o}(x_o)\big)$ such that its translated and rescaled copies φ_j, all satisfy $\|\varphi_j\|_{1,p;B_{\rho_j}(x_j)} = 1$ for all $j \in \mathbb{N}$. The sequence $\{\varphi_j\}$ does not have a subsequence strongly convergent in any $L^q(E)$ for all $q \ge 1$.

14.2.   Theorem 14.1 is false for $q = p^*$ for all $p \in [1, N)$. Construct an example.

17c   Traces and Fractional Sobolev Spaces

17.1c   Characterizing Functions in $W^{1-\frac{1}{p},p}(\mathbb{R}^N)$ as Traces

Proposition 17.1c *Every   function $\varphi \in W^{1-\frac{1}{p},p}(\mathbb{R}^N)$ has   an   extension $u \in W^{1,p}(\mathbb{R}_+^{N+1})$, such that the trace of u on the hyperplane $t = 0$, coincides with φ and*

$$\|u(\cdot, t)\|_{p,\mathbb{R}^N} \le \|\varphi\|_{p,\mathbb{R}^N} \quad \text{for all } t > 0 \tag{17.1c}$$

$$\|Du\|_{p,\mathbb{R}_+^{N+1}} \le \gamma \|\!|\varphi\|\!|_{1-\frac{1}{p},p;\mathbb{R}^N} \tag{17.2c}$$

where γ depends only on N and p.

Proof Assume $N \ge 2$ and let

$$F(x - y; t) = \frac{1}{(N-1)\omega_{N+1}} \frac{1}{\left(|x - y|^2 + t^2\right)^{\frac{N-1}{2}}}$$

be the fundamental solution of the Laplace equation in $\mathbb{R}^{N+1}$ with pole at $(x, 0)$, introduced in § 14 of Chap. 8. Let also

$$\Phi(x, t) = \frac{2}{\omega_{N+1}} \int_{\mathbb{R}^N} \frac{t}{\left(|x - y|^2 + t^2\right)^{\frac{N+1}{2}}} \varphi(y)dy$$

$$= 2 \int_{\mathbb{R}^N} F_t(x - y; t)\varphi(y)dy \overset{\text{def}}{=} (H * \varphi)(x)$$

be the Poisson integral of φ in $\mathbb{R}_+^{N+1}$ introduced in § 18.2c of the Complements of Chap. 6. Since the kernel $H = 2F_t$ has mass one for all $t > 0$ and is harmonic in $\mathbb{R}^N \times \mathbb{R}^+$ we may regard $2F_t$ as a mollifying kernel following the parameter t. Therefore (17.1c) follows from the properties of the mollifiers and in particular

Proposition 18.2c of the Complements of Chap. 6. Again, since $H = 2F_t$ has mass one

$$\frac{\partial}{\partial t} \int_{\mathbb{R}^N} F_t(x - y; t) dy = 0 \quad \text{and} \quad \frac{\partial}{\partial x_i} \int_{\mathbb{R}^N} F_t(x - y; t) dy = 0$$

for $i = 1, \ldots, N$. Therefore denoting by η any one of the components of (x, t)

$$\frac{\partial}{\partial \eta} \Phi(x, t) = \int_{\mathbb{R}^N} \frac{\partial^2}{\partial \eta \partial t} F(x - y; t) [\varphi(y) - \varphi(x)] dy$$

and by direct calculation

$$|D\Phi(x, t)| \leq \gamma \int_{\mathbb{R}^N} \frac{|\varphi(x) - \varphi(y)|}{(|x - y| + t)^{N+1}} dy$$

for a constant γ depending only upon N. Integrate in dy by introducing polar coordinates with pole at x and radial variable ρt. If $\mathbf{n}$ denotes the unit vector spanning the unit sphere in $\mathbb{R}^N$

$$|D\Phi(x, t)| \leq \gamma \int_{|\mathbf{n}|=1} d\mathbf{n} \int_0^\infty \frac{\rho^{N-1}}{(1 + \rho)^{N+1}} \frac{|\varphi(x + \rho t \mathbf{n}) - \varphi(x)|}{t} d\rho.$$

By the continuous version of Minkowski's inequality

$$\|D\Phi\|_{p, \mathbb{R}_+^{N+1}} \leq \gamma \int_{|\mathbf{n}|=1} d\mathbf{n} \int_0^\infty \frac{\rho^{N-1}}{(1 + \rho)^{N+1}} d\rho$$
$$\times \left(\int_0^\infty \frac{\|\varphi(\cdot + \rho t \mathbf{n}) - \varphi(\cdot)\|_{p, \mathbb{R}^N}^p}{t^p} dt \right)^{\frac{1}{p}}.$$

The last integral is computed by the change of variable $r = \rho t$. This gives

$$\|D\Phi\|_{p, \mathbb{R}_+^{N+1}} \leq \gamma \left(\int_0^\infty \frac{\rho^{N-\frac{1}{p}}}{(1 + \rho)^{N+1}} d\rho \right)$$
$$\times \int_{|\mathbf{n}|=1} \left(\int_0^\infty r^{N-1} \frac{\|\varphi(\cdot + r\mathbf{n}) - \varphi(\cdot)\|_{p, \mathbb{R}^N}^p}{r^{N+p-1}} \right)^{\frac{1}{p}} d\mathbf{n}$$
$$\leq \gamma(N, p) \left(\int_{\mathbb{R}^N} \int_{\mathbb{R}^N} \frac{|\varphi(x) - \varphi(y)|^p}{|x - y|^{N+p-1}} dx dy \right)^{\frac{1}{p}}$$
$$= \gamma(N, p) \|\|\varphi\|\|_{1-\frac{1}{p}, p; \mathbb{R}^N}.$$

The extension claimed by the proposition can be taken to be

$$u(x, t) = e^{-t/p} \Phi(x, t).$$

To prove that φ is the trace of u, consider a sequence $\{\varphi_n\} \subset C_o^\infty(\mathbb{R}^N)$ that approximate φ in the norm of $W^{s,p}(\mathbb{R}^N)$ for $s = 1 - 1/p$. Such a sequence exists by virtue of Proposition 15.2. Then construct the corresponding Poisson integrals Φ_n and let $u_n = e^{-t/p}\Phi_n$. Writing

$$u_n - u = 2e^{-t/p} \int_{\mathbb{R}^N} F_t(x - y; t)[\varphi_n(y) - \varphi(y)]dy$$

and applying (17.1c)–(17.2c) proves that $u_n \to u$ in $W^{1,p}(\mathbb{R}_+^{N+1})$. By the definition of trace this implies that $u(\cdot, 0) = \varphi$.   ∎

Combining Proposition 17.1 and this extension procedure, gives the following characterization of the traces.

Theorem 17.1c *A function φ defined and measurable in $\mathbb{R}^N$ belongs to $W^{s,p}(\mathbb{R}^N)$ for some $s \in (0, 1)$ if and only if it is the trace on the hyperplane $t = 0$ of a function $u \in W^{1,p}(\mathbb{R}_+^{N+1})$ where $p = 1/(1 - s)$.*

18c   Traces on ∂E of Functions in $W^{1,p}(E)$

Theorem 18.2c *Let E be a bounded domain in $\mathbb{R}^N$ with boundary ∂E of class C^1 and with the segment property. A function $\varphi \in W^{1-\frac{1}{p},p}(\partial E)$ for some $p > 1$ admits an extension $u \in W^{1,p}(E)$ such that the trace of u on ∂E is φ.*

Theorem 18.3c *Let E be a bounded domain in $\mathbb{R}^N$ with boundary ∂E of class C^1 and with the segment property. A function φ defined and measurable on ∂E belongs to $W^{s,p}(\partial E)$ for some $s \in (0, 1)$ if and only if it is the trace on ∂E of a function $u \in W^{1,p}(E)$ where $p = 1/(1 - s)$.*

18.1c   Traces on a Sphere

The embedding inequalities of Proposition 18.1 might be simplified if E is of relatively simple geometry, such as a ball or a cube in $\mathbb{R}^N$. Let B_R the ball of radius R about the origin of $\mathbb{R}^N$ and let $S_R = \partial B_R$.

Proposition 18.1c *Let $u \in W^{1,p}(B_R)$ for some $p \in [1, \infty)$. Then*

$$\|u\|_{p,S_R}^p \leq \frac{N}{R}\|u\|_p^p + p\|u\|_p^{p-1}\left\|\frac{\partial u}{\partial|x|}\right\|_p. \tag{18.3c}$$

If $p \in [1, N)$ then

$$\|u\|_{m,S_R}^m \le N \kappa_N^{\frac{1}{N}} \|u\|_{p^*}^m + m \|u\|_{p^*}^{m-1} \left\| \frac{\partial u}{\partial |x|} \right\|_p \qquad (18.4c)$$

where

$$p^* = \frac{Np}{N-p} \quad and \quad m = \frac{N-1}{N} p^*.$$

Proof We may assume that $u \in C^1(\bar{B}_R)$. Having fixed $q \ge p \ge 1$, set

$$\theta = \max\left\{ p; 1 + q\left(1 - \frac{1}{p}\right) \right\} \quad \text{so that} \quad p \le \theta \le q.$$

Then for any unit vector **n**

$$R^N |u(R\mathbf{n})|^\theta = \int_0^R \frac{\partial}{\partial \rho}(\rho^N |u(\rho\mathbf{n})|^\theta) d\rho$$

$$= N \int_0^R \rho^{N-1} |u(\rho\mathbf{n})|^\theta d\rho + \theta \int_0^R \rho^N |u(\rho\mathbf{n})|^{\theta-1} \frac{\partial u(\rho\mathbf{n})}{\partial \rho} \text{sign } u(\rho\mathbf{n}) d\rho.$$

Integrating in $d\mathbf{n}$ over the unit sphere $|\mathbf{n}| = 1$ gives

$$R \int_{\mathbf{n}=1} |u(R\mathbf{n})|^\theta R^{N-1} d\mathbf{n} = N \int_{B_R} |u|^\theta dx + \theta \int_{B_R} |x| |u|^{\theta-1} \frac{\partial u}{\partial |x|} \text{sign } u \, dx$$

$$\le N \mu(B_R)^{1-\frac{\theta}{q}} \left(\int_{B_R} |u|^q dx \right)^{\frac{\theta}{q}} + \theta R \left(\int_{B_R} |u|^{(\theta-1)\frac{p}{p-1}} dx \right)^{\frac{p-1}{p}} \left\| \frac{\partial u}{\partial |x|} \right\|_p.$$

Inequalities (18.3c) and (18.4c) follow form this for $q = p$ and $q = p^*$. ∎

Chapter 11
Topics on Weakly Differentiable Functions

1 Sard's Lemma [140]

Let E be a bounded domain in $\mathbb{R}^N$ and let $f \in C^\infty(E)$. For a multi-index α, and a positive integer k, set

$$D_k = \left\{ x \in E \ \Big| \ \sum_{1 \le |\alpha| \le k} |D^\alpha f(x)| = 0 \right\} = \bigcap_{|\alpha|=1}^{k} [D^\alpha f = 0].$$

Denote also by μ_1 the Lebesgue measure on $\mathbb{R}$. The next lemma asserts that the image of D_N by f is a subset of measure zero in the range of f. Equivalently, for almost all level sets $[f = t]$

$$\sum_{|\alpha|=1}^{N} |D^\alpha f(x)| > 0 \quad \text{for all } x \in [f = t].$$

Lemma 1.1 $\mu_1[f(D_N)] = 0.$

Proof Fix $\epsilon \in (0, 1)$. For each $x \in D_N$ and for all $0 < \varepsilon \le \epsilon$ there exists an open ball $B_r(x)$ centered at x and radius $0 < r \le \varepsilon$ such that

$$\sup_{y \in B_r(x)} |f(y) - f(x)| \le \gamma r^{N+1}$$

for a constant depending only of f and independent of x and r. Thus

$$f[B_r(x)] \subset I_{x,\varepsilon} = \left[f(x) - \inf_{B_r(x)} (f - f(x)), \ f(x) + \sup_{B_r(x)} (f - f(x)) \right].$$

© Springer Science+Business Media New York 2016
E. DiBenedetto, *Real Analysis*, Birkhäuser Advanced
Texts Basler Lehrbücher, DOI 10.1007/978-1-4939-4005-9_11

The collection of intervals $\{I_{x,\varepsilon}$ for $x \in D_N$ and $\varepsilon \in (0, \epsilon)$, forms a fine Vitali covering for $f(D_N)$. By the Vitali measure theoretical covering theorem (Theorem 17.1 of Chap. 3), one may extract a countable subcollection of intervals $\{I_n\}$ with pairwise disjoint interior, such that

$$\mu_1\left[f(D_N) - \bigcup I_n\right] = 0.$$

To each I_n there corresponds a ball $B_{r_n}(x_n)$ such that

$$B_{r_n}(x_n) \subset f^{-1}(I_n).$$

The pre-images $f^{-1}(I_n)$ have pairwise disjoint interior, and we estimate

$$\mu_1[f(D_N)] \le \sum \mu_1(I_n) \le \gamma \sum r_n^{N+1}$$
$$\le \epsilon\gamma \frac{N}{\omega_N} \sum \frac{\omega_N}{N} r_n^N = \epsilon\gamma \frac{N}{\omega_N} \sum \mu[B_{r_n}(x_n)]$$
$$\le \epsilon\gamma \frac{N}{\omega_N} \sum \mu[f^{-1}(I_n)] \le \epsilon \frac{\gamma N}{\omega_N} \mu(E). \qquad \blacksquare$$

The next lemma asserts that $f(D_1)$ has 1-dimensional Lebesgue measure zero. Equivalently, for almost all level sets $[f = t]$

$$|Df(x)| > 0 \quad \text{for all } x \in [f = t].$$

As a consequence, by the implicit function theorem, for almost all t in the range of f, the level sets $[f = t]$ are, smooth $(N - 1)$-dimensional surfaces.

Lemma 1.2 $\mu_1[f(D_1)] = 0.$

Proof By the previous lemma it suffices to prove that

$$\mu_1\left[f(D_1) - f(D_N)\right] = \mu_1\left[f(D_1 - D_N)\right] = 0.$$

The proof is by induction on the dimension N. The lemma holds trivially for $N = 1$. We will establish that if it does hold for $(N - 1)$ it continues to hold for N. Now

$$D_1 - D_N = \bigcup_{k=1}^{N-1} (D_k - D_{k+1}) \quad \text{and}$$

$$f(D_1 - D_N) = \bigcup_{k=1}^{N-1} f(D_k - D_{k+1}).$$

Therefore, it suffices to show that

$$\mu_1\left[f(D_k - D_{k+1})\right] = 0 \quad \text{for all } k = 1, \ldots, N - 1.$$

Having fixed $x \in (D_k - D_{k+1})$, there exists a multi-index α of size $|\alpha| = k$, and an index $j \in \{1, \ldots, N\}$, such that

$$D^\alpha f(x) = 0 \quad \text{and} \quad \frac{\partial}{\partial x_j} D^\alpha f(x) \neq 0.$$

Up to a possible reordering of the coordinates, may assume without loss of generality that $j = N$. By the implicit function theorem, there exists a open neighborhood $\mathcal{O}$ of x, and a local system of coordinates,

$$y = (\bar{y}, y_N), \quad \text{where} \quad \bar{y} = (y_1, \ldots, y_{N-1})$$

such that, within $\mathcal{O}$ the set $[D^\alpha f = 0]$ can be represented as the graph of a smooth function $y_N = g(\bar{y})$. Precisely, there is an open set $\mathcal{U}$ and a homeomorphism $h : \mathcal{O} \to \mathcal{U}$ such that

$$f\big|_{\mathcal{O}} = f\big(h^{-1}\big|_{\mathcal{U}}\big) \overset{\text{def}}{=} \varphi : \mathcal{U} \to \mathbb{R}.$$

The restriction

$$\psi = \varphi\big|_{\mathcal{U} \cap [y_N = g(\bar{y})]} = f\Big(h^{-1}\big(\mathcal{U} \cap [y_N = g(\bar{y})]\big)\Big)$$

defines a smooth function of $(N - 1)$ independent variables in $\mathcal{U} \cap [y_N = g(\bar{y})]$. Set

$$D_{1,\psi} = \bigcup \big\{\bar{y} \in \mathcal{U} \cap [y_N = g(\bar{y})] \mid D\psi(\bar{y}) = 0\big\}.$$

By the induction hypothesis $\mu_1[\psi(D_{1,\psi})] = 0$. By construction, if $\bar{y} \in h\big((D_k - D_{k+1}) \cap \mathcal{O}\big)$, then $\bar{y} \in D_{1,\psi}$. Therefore,

$$h\big((D_k - D_{k+1}) \cap \mathcal{O}\big) \subset D_{1,\psi}.$$

From this

$$f\big((D_k - D_{k+1}) \cap \mathcal{O}\big) \subset \psi(D_{1,\psi})$$

Hence

$$\mu_1\big[f\big((D_k - D_{k+1}) \cap \mathcal{O}\big)\big] \leq \mu_1(\psi(D_{1,\psi})) = 0.$$

As x ranges over $(D_k - D_{k+1})$ the open sets $\mathcal{O}$ about x, form an open covering of $(D_k - D_{k+1})$, from which we may select a countable one $\{\mathcal{O}_n\}$. Such a selection is possible by virtue of Proposition 5.3 of Chap. 2. To establish the lemma it suffices to observe that

$$\mu_1\big[f(D_k - D_{k+1})\big] = \mu_1\big[f\big(\bigcup(D_k - D_{k+1}) \cap \mathcal{O}_n\big)\big]$$
$$\leq \sum \mu_1\big[f\big((D_k - D_{k+1}) \cap \mathcal{O}_n\big)\big] = 0. \qquad \blacksquare$$

2 The Co-area Formula for Smooth Functions

Proposition 2.1 *Let E be a bounded domain in $\mathbb{R}^N$ and let $u \in C^\infty(E)$. Then for all nonnegative $\varphi \in C(E)$*

$$\int_E \varphi |\nabla u| dx = \int_0^\infty \left(\int_{\partial[|u|>t]} \varphi d\sigma \right) dt \qquad (2.1)$$

where $d\sigma$ is the $(N-1)$ Hausdorff measure on $[u = t]$.

Remark 2.1 Since $u \in C^\infty(E)(E)$, by Sard's Lemma, the level sets $[u = t]$ are smooth $(N-1)$-dimensional surfaces for a.e. $t \in \mathbb{R}$. Therefore, since φ is continuous in E and nonnegative, the integrals in (2.1), finite or infinite, are well defined.

Proof Assume first $\varphi \in C_o^\infty(E)$ and, for $\varepsilon > 0$ set

$$\mathbf{h} = \varphi \frac{\nabla u}{\sqrt{|\nabla u|^2 + \varepsilon}} \in C_o^\infty(E).$$

Then, by integration by parts and formula (15.5) of Chap. 4

$$\int_E \mathbf{h} \cdot \nabla u dx = - \int_E u \, \mathrm{div} \, \mathbf{h} \, dx$$
$$= - \int_0^\infty \left(\int_{[u>t]} \mathrm{div} \, \mathbf{h} dx \right) dt + \int_{-\infty}^0 \left(\int_{[u<t]} \mathrm{div} \, \mathbf{h} dx \right) dt.$$

By Sard's Lemma, the inner unit normal to $\partial[u > t]$ is well defined for a.e. $t \in \mathbb{R}$, and it is given by $\nabla u / |\nabla u|$. Then by the divergence theorem,

$$- \int_{[u>t]} \mathrm{div} \, \mathbf{h} dx = \int_{\partial[u>t]} \mathbf{h} \cdot \frac{\nabla u}{|\nabla u|} d\sigma \quad \text{for a.e. } t > 0;$$

$$\int_{[u<t]} \mathrm{div} \, \mathbf{h} dx = \int_{\partial[u<t]} \mathbf{h} \cdot \frac{\nabla u}{|\nabla u|} d\sigma \quad \text{for a.e. } t < 0.$$

Therefore, taking into account the choice of $\mathbf{h}$

$$\int_E \varphi \frac{|\nabla u|^2}{\sqrt{|\nabla u|^2 + \varepsilon}} dx = \int_0^\infty \left(\int_{\partial[|u|>t]} \varphi \frac{|\nabla u|}{\sqrt{|\nabla u|^2 + \varepsilon}} d\sigma \right) dt.$$

Letting $\varepsilon \to 0$ and passing to the limit under integral by means of the dominated convergence theorem proves (2.1) for $\varphi \in C_o^\infty(E)$. If $\varphi \in C_o(E)$, it is the pointwise limit of functions in $C_o^\infty(E)$. Thus a limiting process by dominated convergence establishes (2.1) for all nonnegative $\varphi \in C_o(E)$. A nonnegative $\varphi \in C(E)$ is the monotone, pointwise limit of functions in $C_o(E)$. Thus, by monotone convergence, (2.1) continues to hold for nonnegative $\varphi \in C(E)$. ∎

3 The Isoperimetric Inequality for Bounded Sets E with Smooth Boundary ∂E

For a Lebesgue measurable set $E \subset \mathbb{R}^N$ with smooth boundary ∂E, denote by $\mu(E)$ its Lebesgue measure, and by $\sigma(\partial E)$ the $(N-1)$-dimensional Hausforff measure of ∂E.

Proposition 3.1 *Let E be a bounded, open set in $\mathbb{R}^N$, with smooth boundary ∂E. Then*

$$\frac{\sigma(\partial E)^{\frac{N}{N-1}}}{\mu(E)} \geq N \omega_N^{\frac{1}{N-1}}. \tag{3.1}$$

Proof Denote by $E \ni x \to \delta_E(x)$ the distance from x to ∂E and for $\delta > 0$ let E_δ be defined as

$$E_\delta = \{ x \in E \mid \delta_E(x) > \delta \} \tag{3.2}$$

where δ is so small that E_δ is not empty. Introduce the family of functions

$$u_\delta = \begin{cases} 1 & \text{if } x \in E_\delta; \\[2mm] \dfrac{1}{\delta} \delta_E(x) & \text{if } x \in E - E_\delta; \\[2mm] 0 & \text{if } x \in \mathbb{R}^N - E. \end{cases} \tag{3.3}$$

The distance function $\delta_E(\cdot)$ is a Lipschitz continuous function of x, with Lipschitz constant $L = 1$ (Lemma 15.2 of Chap. 4). Hence $\delta_E \in W^{1,\infty}(E)$ (§ 20.2 of Chap. 8). The functions u_δ can also be regarded as in $W_o^{1,1}(\mathbb{R}^N)$. From this and Corollary 1.1 of Chap. 10, for $p = 1$, applied in the form (1.7) with the best constant,

$$\mu(E) = \lim \int_E u_\delta^{\frac{N}{N-1}} dx$$

$$\leq \left(N \omega_N^{\frac{1}{N-1}} \right)^{-1} \lim \left(\int_E |Du_\delta| dx \right)^{\frac{N}{N-1}}$$

$$\leq \left(N \omega_N^{\frac{1}{N-1}} \right)^{-1} \lim \left(\frac{1}{\delta} \mu(E - E_\delta) \right)^{\frac{N}{N-1}}$$

$$= \left(N \omega_N^{\frac{1}{N-1}} \right)^{-1} \sigma(\partial E)^{\frac{N}{N-1}}.$$ ∎

Remark 3.1 In computing the last limit we have used that ∂E is smooth. If ∂E is not regular, such a limiting process could be used to define the measure of the "*perimeter*" of E, i.e., [33]

$$\lim_{\delta \to 0} \int |Du_\delta| dx \overset{\text{def}}{=} |\partial E|.$$

Remark 3.2 The quantity in (3.1) is dimensionless and it measures the *isoperimetric ratio* of ∂E relative to the volume of E. It is then natural to ask for what sets E such a ratio is the least. A theorem of DeGiorgi [33] states that the inequality in (3.1) is strict, unless E is a ball. Thus the balls are the domains of least perimeter among all those of equal volume.

3.1 Embeddings of $W_o^{1,p}(E)$ Versus the Isoperimetric Inequality

The isoperimetric inequality is a consequence of the embedding of $W_o^{1,1}(E)$ into $L^{\frac{N}{N-1}}(E)$, as stated in Corollary 1.1 of Chap. 10, in the form (1.7) with best constant. This embedding is the building block of all embeddings of $W_o^{1,p}(E)$ into $L^{p^*}(E)$ for all $1 \leq p < N$, as indicated in § 3–§ 4 of Chap. 10.

Let now $u \in C_o^\infty(E)$, so that by Sard's Theorem, the sets $[u > t]$ have smooth boundaries $[u = t]$, for a.e. $t \in \mathbb{R}$. Apply the co-area formula (2.1), with $\varphi = 1$, and the isoperimetric inequality to the sets $[u > t]$ and their boundaries $[u = t]$. This gives

$$
\begin{aligned}
\frac{1}{N}\left(\frac{N}{\omega_N}\right)^{\frac{1}{N}} \int_E |Du| dx &= \frac{1}{N}\left(\frac{N}{\omega_N}\right)^{\frac{1}{N}} \int_{\mathbb{R}} \sigma([u = t]) dt \\
&\geq \int_{\mathbb{R}} \mu([u > t])^{\frac{N-1}{N}} dt = \int_0^\infty \mu([|u| > t])^{\frac{N-1}{N}} dt \\
&\geq \int_0^\infty \frac{1}{h}\{h^{\frac{N}{N-1}}\mu([|u| > t])\}^{\frac{N-1}{N}} dt \\
&\geq \int_0^\infty \frac{1}{h}\left\{\int_E \left(\min\{|u|; t+h\} - \min\{|u|; t\}\right)^{\frac{N}{N-1}}\right\}^{\frac{N-1}{N}} dt \\
&= \int_0^\infty \frac{1}{h}\left\| \min\{|u|; t+h\} - \min\{|u|; t\}\right\|_{\frac{N}{N-1}} dt \\
&\geq \int_0^\infty \frac{1}{h}\left[\left\| \min\{|u|; t+h\}\right\|_{\frac{N}{N-1}} - \left\| \min\{|u|; t\}\right\|_{\frac{N}{N-1}}\right] dt.
\end{aligned}
$$

In these calculations $0 < h \ll 1$. Letting $h \to 0$ yields

$$\frac{1}{N}\left(\frac{N}{\omega_N}\right)^{\frac{1}{N}} \int_E |Du| dx \geq \int_0^\infty \frac{d}{dt}\left\| \min\{|u|; t\}\right\|_{\frac{N}{N-1}} dt = \|u\|_{\frac{N}{N-1}}. \tag{3.4}$$

Thus the isoperimetric inequality implies the embedding of Corollary 1.1 of Chap. 10, for $p = 1$, in its form (1.7) with best constant.

Remark 3.3 The isoperimetric inequality (3.1) is applied to the sets $[u > t]$ and their boundaries $[u = t]$, which are smooth surfaces for a.e. t. The resulting Gagliardo embedding (3.4) holds for domains E with boundary ∂E not necessarily smooth.

4 The p-Capacity of a Compact Set $K \subset \mathbb{R}^N$, for $1 \le p < N$

For a non-void, compact set $K \subset \mathbb{R}^N$ introduce the classes of functions

$$\Gamma(K) = \left\{ u \in C_o^\infty(\mathbb{R}^N) \text{ with } u \ge 1 \text{ on } K \right\};$$

$$\Gamma_o(K) = \left\{ \begin{array}{l} u \in C_o^\infty(\mathbb{R}^N) \text{ with } 0 \le u \le 1 \text{ in } \mathbb{R}^N \\ u \ge 1 \text{ in an open neighborhood of } K \end{array} \right\}; \qquad (4.1)$$

$$\Gamma_1(K) = \left\{ \begin{array}{l} u \in C_o^\infty(\mathbb{R}^N) \text{ with } 0 \le u \le 1 \text{ in } \mathbb{R}^N \\ u = 1 \text{ in an open neighborhood of } K \end{array} \right\}.$$

For $1 \le p < N$ the p-capacity of K is equivalently defined as

$$c_p(K) = \inf \left\{ \int_{\mathbb{R}^N} |Du|^p dx \mid u \in \Gamma(K) \right\};$$

$$c_p^o(K) = \inf \left\{ \int_{\mathbb{R}^N} |Du|^p dx \mid u \in \Gamma_o(K) \right\}; \qquad (4.2)$$

$$c_p^1(K) = \inf \left\{ \int_{\mathbb{R}^N} |Du|^p dx \mid u \in \Gamma_1(K) \right\}.$$

If $K = \emptyset$, set $c_p(K) = c_p^o(K) = c_p^1(K) = 0$.

Proposition 4.1 $c_p(K) = c_p^o(K) = c_p^1(K)$.

Proof By construction $c_p(K) \le c_p^o(K) \le c_p^1(K)$. To establish the converse inequalities, may assume that $c_p(K) < \infty$. For $\varepsilon > 0$ there exists $u_\varepsilon \in C_o^\infty(\mathbb{R}^N)$ with $u_\varepsilon \ge 1$ on K such that

$$\int_{\mathbb{R}^N} |Du_\varepsilon|^p dx - \varepsilon \le c_p(K).$$

For $0 < \varepsilon < \frac{1}{4}$ introduce the function

$$\mathbb{R} \ni t \to \zeta_\varepsilon(t) = \begin{cases} 0 & \text{for} \quad -\infty < t \le \varepsilon; \\[2mm] \dfrac{t-\varepsilon}{1-2\varepsilon} & \text{for} \quad \varepsilon \le t \le 1-\varepsilon; \\[3mm] 1 & \text{for} \quad 1-\varepsilon \le t < \infty. \end{cases}$$

One verifies that ζ_ε is Lipschitz continuous and $0 \le \zeta_\varepsilon \le 1$. Moreover

$$0 \le \zeta_\varepsilon' \le 1 + 2\varepsilon \quad \text{a.e. in } \mathbb{R}. \tag{4.3}$$

The mollifications $\zeta_{\varepsilon,\nu}$ for $0 < \nu \ll \varepsilon$ share the same properties of ζ_ε. The function $v_\varepsilon = \zeta_{\varepsilon,\nu}(u_\varepsilon)$ is in the class $\Gamma_1(K)$. Then, by the definition of $c_p^o(K)$ and $c_p^1(K)$,

$$c_p^o(K) \le c_p^1(K) \le \int_{\mathbb{R}^N} |Dv_\varepsilon|^p dx = \int_{\mathbb{R}^N} \zeta_{\varepsilon,\nu}'^p |Du_\varepsilon|^p dx$$

$$\le (1+2\varepsilon)^p \int_{\mathbb{R}^N} |Du_\varepsilon|^p dx \le (c_p(K)+\varepsilon)(1+2\varepsilon)^p. \qquad \blacksquare$$

4.1 Enlarging the Class of Competing Functions

Let now $W_{p*}^{1,p}(\mathbb{R}^N)$ be the space, introduced in § 24 of Chap. 10, for $E = \mathbb{R}^N$, and set

$$\Gamma_*(K) = \left\{ u \in W_{p*}^{1,p}(\mathbb{R}^N) \cap C(\mathbb{R}^N) \text{ with } u \ge 1 \text{ on } K \right\}$$
$$c_p^*(K) = \inf \left\{ \int_{\mathbb{R}^N} |Du|^p dx \mid u \in \Gamma_*(K) \right\}. \tag{4.4}$$

Proposition 4.2 $c_p^*(K) = c_p(K)$.

Proof By construction $c_p^*(K) \le c_p(K)$. For $\varepsilon > 0$ there exists $u_\varepsilon \in \Gamma_*(K)$ such that

$$\int_{\mathbb{R}^N} |Du_\varepsilon|^p dx - \varepsilon \le c_p^*(K). \tag{4.5}$$

Arguing as in the proof of Proposition 4.1, there exists $\zeta_\varepsilon(u_\varepsilon) \in \Gamma_*(K)$ such that

$$0 \le \zeta_\varepsilon(u_\varepsilon) \le 1 \quad \text{and} \quad \zeta_\varepsilon(u_\varepsilon) \ge 1 \text{ in an open neighborhood of } K.$$

Since ζ_ε' satisfies (4.3), inequality (4.5) yields

$$\int_{\mathbb{R}^N} |D\zeta_\varepsilon(u_\varepsilon)|^p dx \le (1 + \gamma_p \varepsilon) \int_{\mathbb{R}^N} |Du_\varepsilon|^p dx$$
$$\le c_p^*(K) + \varepsilon \bar{\gamma}_p c_p(K) \tag{4.6}$$

for positive constants γ_p and $\bar{\gamma}_p$ depending only on p. By Proposition 24.1 of Chap. 10, and its proof, there exists $w_\varepsilon \in W_o^{1,p}(\mathbb{R}^N)$, such that

$$\|D\zeta_\varepsilon(u_\varepsilon) - Dw_\varepsilon\|_p \le \varepsilon^{\frac{1}{p}}.$$

The construction of w_ε insures that w_ε is compactly supported in $\mathbb{R}^N$ and

$$0 \le w_\varepsilon \le 1 \quad \text{and} \quad w_\varepsilon \ge 1 \text{ in an open neighborhood of } K.$$

A proper Friedrichs mollification of w_ε generates $w_{\varepsilon,o} \in C_o^\infty(\mathbb{R}^N)$, such that

$$0 \le w_{\varepsilon,o} \le 1 \quad \text{and} \quad w_{\varepsilon,o} \ge 1 \text{ in an open neighborhood of } K$$

and

$$\|Dw_{\varepsilon,o} - Dw_\varepsilon\|_p \le \varepsilon^{\frac{1}{p}}.$$

From this and (4.6)

$$\int_{\mathbb{R}^N} |Dw_{\varepsilon,o}|^p dx - \varepsilon \le c_p^*(K) + \bar{\gamma}_p \varepsilon c_p(K).$$

The function $w_{\varepsilon,o}$ is in $\Gamma_o(K)$. Therefore taking the infimum of the left hand side over all such functions yields

$$c_p(K) \le c_p^* + \varepsilon\big(1 + \bar{\gamma}_p \varepsilon c_p(K)\big). \qquad \blacksquare$$

Corollary 4.1 *Let $K \subset \mathbb{R}^N$ be compact. Then for all $\varepsilon > 0$ there exists an open set $\mathcal{O} \supset K$ such that*

$$c_p(K) \le c_p(K') + \varepsilon \quad \text{for all compact sets } K \subset K' \subset \mathcal{O}. \tag{4.7}$$

This property is also called *right continuity* of the set function $c_p(\cdot)$ on compact sets. The definition of p-capacity implies that if H and K are compact subsets of $\mathbb{R}^N$, then

$$H \subset K \implies c_p(H) \le c_p(K) \tag{4.8}$$

that is $c_p(\cdot)$ is a monotone set function on compact sets. The definition implies also that for all $t > 0$ and all rototranslations $\mathcal{R}$ of the coordinate axes

$$c_p(tK) = t^{N-p} c_p(K); \qquad c_p(\mathcal{R}K) = c_p(K). \tag{4.9}$$

5   A Characterization of the p-Capacity of a Compact Set $K \subset \mathbb{R}^N$, for $1 \leq p < N$

For a given compact set $K \subset \mathbb{R}^N$ continue to denote by $\Gamma(K)$ the class of competing functions introduced in (4.1). Set also

$$\mathcal{E}_K = \left\{ \begin{array}{c} \text{the collection of open sets } E \text{ containing } K \\ \text{such that } \partial E \text{ is a smooth surface} \end{array} \right\} \tag{5.1}$$

and denote by $\sigma(\partial E)$ the $(N-1)$ Hausdorff measure of ∂E.

Theorem 5.1 *Let K be a compact subset of $\mathbb{R}^N$. Then for $1 < p < N$*

$$c_p(K) = \inf_{u \in \Gamma(K)} \left[\int_0^1 \left(\int_{\partial[u>t]} |Du|^{p-1} d\sigma \right)^{\frac{1}{1-p}} \right]^{1-p}. \tag{5.2}$$

If $p = 1$, then

$$c_1(K) = \inf_{E \in \mathcal{E}_K} \sigma(\partial E). \tag{5.3}$$

Proof (The Case $1 < p < N$. The Lower Estimate in (5.2)) Let u be a nonnegative function in $C_o^\infty(\mathbb{R}^N)$. By the co-area formula (2.1), for smooth functions

$$\int_{\mathbb{R}^N} |Du|^p dx \geq \int_0^1 f(t) dt$$

where

$$f(t) = \int_{\partial[u>t]} |Du|^{p-1} d\sigma \tag{5.4}$$

where $d\sigma$ is the $(N-1)$ Hausdorff measure on $\partial[u>t]$. By Sard's Lemma, the set $\partial[u>t]$ is a smooth $(N-1)$-dimensional surface for a.e. $t > 0$, so that f is well defined, nonnegative and a.e. finite as a measurable function of $t > 0$. As such the integrals of f and f^{-1}, finite or infinite, are well defined.

Let now f be any such function and assume in addition that

$$f \quad \text{and} \quad f^{-\frac{1}{p-1}} \quad \text{are integrable in } (0, 1).$$

Then by Hölder's inequality

$$1 = \int_0^1 f^{\frac{1}{p}} f^{-\frac{1}{p}} dt \leq \left(\int_0^1 f dt \right)^{\frac{1}{p}} \left(\int_0^1 f^{\frac{1}{1-p}} dt \right)^{\frac{p-1}{p}}.$$

Therefore

$$\left(\int_0^1 f^{\frac{1}{1-p}} dt \right)^{1-p} \leq \int_0^1 f dt. \tag{5.5}$$

One verifies that this inequality continues to hold if either f or $f^{-\frac{1}{p-1}}$ or both, fail to be integrable. Returning to the definition (4.2) of $c_p(K)$, for all $\varepsilon > 0$ there exists $u \in \Gamma_1(K)$ such that

$$c_p(K) + \varepsilon \ge \int_{\mathbb{R}^N} |Du|^p dx = \int_0^1 f(t)dt$$

with f defined by (5.4). Thus, taking into account (5.5)

$$c_p(K) \ge \inf_{u \in \Gamma_1(K)} \left[\int_0^1 \left(\int_{\partial[u>t]} |Du|^{p-1} d\sigma \right)^{\frac{1}{1-p}} \right]^{1-p}. \qquad \blacksquare$$

Proof (The Case $1 < p < N$. The Upper Estimate in (5.2)) Let $\zeta \in C^\infty(\mathbb{R})$ be nondecreasing and satisfying

$$\zeta = 0 \text{ in } (-\infty, 0]; \quad \zeta = 1 \text{ in } [1, \infty); \quad \zeta \in [0, 1] \text{ in } [0, 1];$$
$$0 \le \zeta'(t) \le \gamma \text{ for some } \gamma > 0 \text{ for all } t \in (0, 1).$$

For any such function ζ and any $u \in \Gamma(K)$,

$$c_p(K) \le \int_{\mathbb{R}^N} \zeta'^p(u)|Du|^p dx = \int_0^1 \zeta'^p(t) f(t)dt$$

where f is introduced in (5.4). Assume momentarily that f and $f^{-\frac{1}{p-1}}$ are integrable in $(0, 1)$ and choose

$$\zeta(t) = \begin{cases} 0 & \text{for } -\infty < t \le 0; \\[2mm] \dfrac{\int_0^t (f + \varepsilon)^{-\frac{1}{p-1}} d\tau}{\int_0^1 (f + \varepsilon)^{-\frac{1}{p-1}} dt} & \text{for } 0 \le t \le 1; \\[2mm] 1 & \text{for } 1 \le t < \infty, \end{cases}$$

modulo a mollification process. For such a choice,

$$c_p(K) \le \left(\int_0^1 (f + \varepsilon)^{\frac{1}{1-p}} dt \right)^{1-p}$$

From this, by monotone convergence and the definition (5.4)

$$c_p(K) \le \left(\int_0^1 f^{\frac{1}{1-p}} dt \right)^{1-p}$$
$$= \left[\int_0^1 \left(\int_{\partial[u>t]} |Du|^{p-1} d\sigma \right)^{\frac{1}{1-p}} \right]^{1-p}$$

for all $u \in \Gamma(K)$. In this inequality we may assume that f is integrable for some $u \in \Gamma(K)$, otherwise $c_p(K) = \infty$. The inequality continues to hold if $f^{-\frac{1}{p-1}}$ is not integrable, in which case $c_p(K) = 0$. ∎

Proof (The Case $p = 1$) For a nonnegative $u \in C_o^\infty(\mathbb{R}^N)$

$$\int_{\mathbb{R}^N} |Du|dx = \int_0^\infty \sigma\big([\partial[u > t]\big)dt$$

$$\geq \int_0^1 \sigma\big([\partial[u > t]\big)dt \geq \inf_{E \in \mathcal{E}_K} \sigma(\partial E).$$

Let $E \in \mathcal{E}_K$ and, for $0 < \delta < 1$ let E_δ and u_δ be defined as in (3.2) and (3.3). Since $K \subset E$ is compact, there is δ sufficiently small that $K \subset [u_\delta = 1]$. For such a u_δ,

$$c_1(K) \leq \int_E |\nabla u_\delta|dx$$

up to a mollification process. Letting $\delta \to 0$ and proceeding as in the proof of Proposition 3.1, gives $c_1(K) \leq \sigma(\partial E)$, for all $E \in \mathcal{E}_K$. ∎

6   Lower Estimates of $c_p(K)$ for $1 \leq p < N$

Lemma 6.1 *Let K be a compact subset of $\mathbb{R}^N$. Then for all $1 \leq p < N$,*

$$c_p(K) \geq N\left(\frac{\omega_N}{N}\right)^{\frac{p}{N}}\left(\frac{N-p}{p-1}\right)^{p-1}\mu(K)^{\frac{N-p}{N}} \tag{6.1}$$

where for $p = 1$ the term in round brackets is meant to be 1.

Proof (The Case $1 < p < N$) Let $u \in C_o^\infty(\mathbb{R}^N)$ be nonnegative and let $0 \leq s < t$. By Hölder's inequality

$$\left(\int_{[s<u<t]} u^{p-1}|Du|dx\right)^{\frac{p}{p-1}} = \left(\int_{[u>s]-[u\geq t]} u^{p-1}|Du|dx\right)^{\frac{p}{p-1}}$$

$$\leq \left(\int_{[u>s]-[u\geq t]} u^p dx\right)\left(\int_{[s<u<t]} |Du|^p dx\right)^{\frac{1}{p-1}}.$$

From this, by the co-area formula (2.1), for smooth functions,

$$\left[\int_s^t \tau^{p-1}\left(\int_{\partial[u>\tau]} d\sigma\right)d\tau\right]^{\frac{p}{p-1}} \leq t^p\mu\big([u > s] - [u > t]\big)$$

$$\times \left[\int_s^t \left(\int_{\partial[u>\tau]} |Du|^{p-1}d\sigma\right)d\tau\right]^{\frac{1}{p-1}}.$$

Divide by $(t - s)^{\frac{p}{p-1}}$ and let $s \to t$ to obtain

$$\left[\sigma(\partial[u > t])\right]^{\frac{p}{p-1}} \le -\frac{d}{dt}\mu([u > t])\left(\int_{\partial[u>t]}|Du|^{p-1}d\sigma\right)^{\frac{1}{p-1}}$$

for a.e. $t > 0$. The various limits are justified by the Lebesgue-Besicovitch differentiation theorem, and the monotonicity of the function $t \to \mu([u > t])$ (see Theorems 11.1 and 3.1 of Chap. 5). From this and (5.3) of Theorem 5.1

$$
\begin{aligned}
c_p(K) &= \inf_{u \in \Gamma(K)}\left[\int_0^1\left(\int_{\partial[u>t]}|Du|^{p-1}d\sigma\right)^{\frac{1}{1-p}}dt\right]^{1-p} \\
&\ge \inf_{u \in \Gamma(K)}\left(-\int_0^1\frac{\frac{d}{dt}\mu([u > t])}{\sigma(\partial[u > t])^{\frac{p}{p-1}}}dt\right)^{1-p}.
\end{aligned}
\tag{6.2}
$$

By the isoperimetric inequality (3.1)

$$\sigma(\partial[u > t]) \ge N^{\frac{N-1}{N}}\omega_N^{\frac{1}{N}}\mu([u > t])^{\frac{N-1}{N}}$$

Therefore, from (6.2) for $1 < p < N$,

$$
\begin{aligned}
c_p(K) &\ge N^{\frac{p(N-1)}{N}}\omega_N^{\frac{p}{N}}\inf_{u \in \Gamma(K)}\left(-\int_0^1\frac{\frac{d}{dt}\mu([u > t])}{\mu([u > t])^{\frac{p}{p-1}\frac{N-1}{N}}}dt\right)^{1-p} \\
&\ge N^{\frac{N-p}{N}}\omega_N^{\frac{p}{N}}\left(\frac{N-p}{p-1}\right)^{p-1}\inf_{u \in \Gamma(K)}\left(\int_0^1\frac{d}{dt}\mu([u > t])^{-\frac{N-p}{N(p-1)}}dt\right)^{1-p} \\
&\ge N^{\frac{N-p}{N}}\omega_N^{\frac{p}{N}}\left(\frac{N-p}{p-1}\right)^{p-1}\mu(K)^{\frac{N-p}{N}}.
\end{aligned}
$$
∎

Proof (The Case $p = 1$) The proof for $p = 1$ follows by combining (5.2) of Theorem 5.1 with the isoperimetric inequality (3.1) of Proposition 3.1. ∎

6.1 A Simpler Proof of Lemma 6.1 with a Coarser Constant

Let u be in anyone of the classes $\Gamma(K), \Gamma_o(K), \Gamma_1(K)$ introduced in (4.1). Then by the Gagliardo embedding of $W_o^{1,p}(\mathbb{R}^N)$ into $L^{p^*}(\mathbb{R}^N)$ as stated in (1.6) of Corollary 1.1 of Chap. 10.

$$\mu(K)^{\frac{N-p}{N}} \le \left(\int_{\mathbb{R}^N}u^{\frac{Np}{N-p}}dx\right)^{\frac{N-p}{N}} \le p^p\left(\frac{N-1}{N-p}\right)^p\int_{\mathbb{R}^N}|Du|^p dx$$

for $N \geq 2$ and $1 \leq p < N$. Minimizing the right-hand side over $u \in \Gamma(K)$, yields

$$c_p(K) \geq p^{-p} \left(\frac{N-p}{N-1}\right)^p \mu(K)^{\frac{N-p}{N}}. \tag{6.3}$$

This estimate is analogous to (6.1), except for the different value of the constant. One verifies that

$$N \left(\frac{\omega_N}{N}\right)^{\frac{p}{N}} \left(\frac{N-p}{p-1}\right)^{p-1} > p^{-p} \left(\frac{N-p}{N-1}\right)^p \tag{6.4}$$

for all $1 \leq p < N$. Thus (6.1) is a more precise lower bound of $c_p(\cdot)$ in terms of $\mu(\cdot)$.

6.2   p-Capacity of a Closed Ball $\bar{B}_\rho \subset \mathbb{R}^N$, for $1 \leq p < N$

Corollary 6.1 *Let B_ρ be an open ball of radius ρ in $\mathbb{R}^N$. Then,*

$$c_p(\bar{B}_\rho) = \omega_N \left(\frac{N-p}{p-1}\right)^{p-1} \rho^{N-p} \qquad \text{for } 1 < p < N. \tag{6.5}$$

When $p = 1$ the power in round brackets is meant to be 1.

Proof If $p = 1$ the statement follows from (5.3) of Theorem 5.1. If $1 < p < N$, from the previous lower estimate with $K = \bar{B}_\rho$ one computes

$$c_p(\bar{B}_\rho) \geq \left(\frac{N-p}{p-1}\right)^{p-1} \omega_N \rho^{N-p}.$$

On the other hand, from the definition (4.2) the p-capacity of B_ρ is majorized if the infimum is taken over radially symmetric functions. Then in (4.2) take

$$u(x) = \begin{cases} 1 & \text{for } |x| \leq 1 : \\ \left(\frac{\rho}{|x|}\right)^{\frac{N-p}{p-1}} & \text{for } |x| > \rho \end{cases}$$

up to a density argument.   ∎

Remark 6.1 From the first of (4.9) it follows that $c_p(\bar{B}_\rho) = \rho^{N-p} c_p(\bar{B}_1)$. It is the precise lower estimate (6.1) that permits one to compute $c_p(\bar{B}_1)$.

6.3 $c_p(\bar{B}_\rho) = c_p(\partial B_\rho)$

For $u \in \Gamma_1(\partial B_\rho)$ define

$$v_u = \begin{cases} u & \text{for } |x| \ge \rho; \\ 1 & \text{for } |x| \le \rho. \end{cases}$$

One verifies that $v_u \in \Gamma_*(\bar{B}_\rho)$ and that

$$\int_{\mathbb{R}^N} |Du|^p dx \ge \int_{\mathbb{R}^N} |Dv_u|^p dx \ge \inf_{v \in \Gamma_*(\bar{B}_\rho)} \int_{\mathbb{R}^N} |Dv|^p dx.$$

Therefore,

$$c_p(\partial B_\rho) = \inf_{u \in \Gamma_1(\partial B_\rho)} \int_{\mathbb{R}^N} |Du|^p dx \ge c_p(\bar{B}_\rho).$$

On the other hand $c_p(\partial B_\rho) \le c_p(\bar{B}_\rho)$ and hence $c_p(\partial B_\rho) = c_p(\bar{B}_\rho)$. An almost identical argument establishes the following

Corollary 6.2 *Let E be a bounded open set in $\mathbb{R}^N$ with smooth boundary ∂E and let $1 \le p < N$. Then $c_p(\bar{E}) = c_p(\partial E)$.*

7 The Norm $\|Du\|_p$, for $1 \le p < N$, in Terms of the p-Capacity Distribution Function of $u \in C_o^\infty(\mathbb{R}^N)$

The p-capacity distribution function of a nonnegative function $u \in C_o^\infty(\mathbb{R}^N)$ is defined by

$$\mathbb{R}^+ \ni t \to c_p([u \ge t]).$$

This is a nonincreasing function of t, vanishing for t sufficiently large.

Theorem 7.1 *Let $u \in C_o^\infty(\mathbb{R}^N)$ be nonnegative. Then for all $1 \le p < N$,*

$$\int_0^\infty t^{p-1} c_p([u \ge t]) dt \le \left(\frac{p}{p-1}\right)^{p-1} \int_{\mathbb{R}^N} |Du|^p dx. \tag{7.1}$$

When $p = 1$ the coefficient on the right-hand side of (7.1) is meant to be 1.

7.1 Some Auxiliary Estimates for $1 < p < N$

For a nonnegative $u \in C_o^\infty(\mathbb{R}^N)$ set

$$t_u = \inf\{t > 0 \mid c_p([u \ge t]) = 0\}.$$

Let f be defined as in (5.4) and set

$$(0, t_u) \ni t \to s(t) = \int_0^t f^{-\frac{1}{p-1}}(\tau)d\tau. \tag{7.2}$$

This integral is well defined for $t \in (0, t_u)$. Indeed from (5.2) of Theorem 5.1, for all such t

$$c_p([u \geq t]) \leq \left[\int_0^1 \left(\int_{\partial[\frac{u}{t} > \lambda]} \left|D\frac{u}{t}\right|^{p-1} d\sigma\right)^{\frac{1}{1-p}} d\lambda\right]^{1-p}$$

$$= \left[\int_0^t \left(\int_{\partial[u > \tau]} |Du|^{p-1} d\sigma\right)^{\frac{1}{1-p}} d\tau\right]^{1-p} \tag{7.3}$$

$$= \left(\int_0^t f^{-\frac{1}{p-1}}(\tau)d\tau\right)^{1-p} = s(t)^{1-p}.$$

From this,

$$s(t) \leq \left[c_p([u \geq t])\right]^{-\frac{1}{p-1}} < \infty \quad \text{for a.e. } t \in (0, t_u).$$

This estimate and the definition (5.4) of $f(\cdot)$ imply that

$$0 < f(t) < \infty \qquad \text{for a.e. } t \in (0, t_u).$$

It follows that $s(\cdot)$ is strictly increasing from 0 to $s_u = s(t_u)$. The inverse of $s(\cdot)$, denoted by $h(\cdot)$, is also strictly increasing

$$(0, s_u) \ni s \to h(s) \in (0, t_u), \quad \text{with} \quad h[s(t)] = t \quad \text{for all} \quad t \in (0, t_u).$$

Therefore, $h(\cdot)$ is of bounded variation in $(0, s_u)$ and its derivative

$$h'(s) = f(t)^{\frac{1}{p-1}}$$

is integrable in $(0, s_u)$.

Lemma 7.1 $h' \in L^p(0, s_u)$ and,

$$\int_0^{s_u} h'^p(s)ds \leq \int_{\mathbb{R}^N} |Du|^p dx. \tag{7.4}$$

Proof Let

$$0 = s_o < s_1 < \cdots < s_{n-1} < s_n = s_u \quad \text{and}$$
$$0 = t_o < t_1 < \cdots < t_{n-1} < t_n = t_u$$

be a partition of $(0, s_u)$ and the corresponding partition of $(0, t_u)$ by which $h(s_j) = t_j$. By the reverse Hölder inequality applied with conjugate powers $\frac{1}{1-p}$ and $\frac{1}{p}$,

$$\int_{t_j}^{t_{j+1}} f(\tau)d\tau \ge \left(\int_{t_j}^{t_{j+1}} f^{-\frac{1}{p-1}}(\tau)d\tau \right)^{1-p} \left(\int_{t_j}^{t_{j+1}} 1\, d\tau \right)^p$$

$$= \frac{(t_{j+1} - t_j)^p}{\left(\int_{t_j}^{t_{j+1}} f^{-\frac{1}{p-1}}(\tau)d\tau \right)^{p-1}}.$$

From this

$$\sum_{j=0}^{n-1} \frac{\left(h(s_{j+1}) - h(s_j) \right)^p}{(s_{j+1} - s_j)^{p-1}} = \sum_{j=0}^{n-1} \frac{(t_{j+1} - t_j)^p}{\left(\int_{t_j}^{t_{j+1}} f^{-\frac{1}{p-1}}(\tau)d\tau \right)^{p-1}}$$

$$\le \sum_{j=0}^{n-1} \int_{t_j}^{t_{j+1}} f(\tau)d\tau \le \int_{\mathbb{R}^N} |Du|^p dx. \qquad \blacksquare$$

7.2 Proof of Theorem 7.1

In the integral below we effect the change of variable $t = h(s)$, where $s(\cdot)$ is defined in (7.2) and $h(\cdot)$ is the inverse of $s(\cdot)$. We also use the inequality (7.3) with no further mention.

$$\int_0^\infty t^{p-1} c_p([u \ge t])dt \le \int_0^{s_u} \left(\frac{h(s)}{s} \right)^{p-1} h'(s)ds$$

$$\le \left(\int_0^{s_u} \left(\frac{h(s)}{s} \right)^p ds \right)^{\frac{p-1}{p}} \left(\int_0^{s_u} h'^p(s)ds \right)^{\frac{1}{p}}.$$

By Hardy's inequality (§ 15 of Chap. 9)

$$\int_0^{s_u} \left(\frac{h(s)}{s} \right)^p ds \le \left(\frac{p}{p-1} \right)^p \int_0^{s_u} h'^p(s)ds.$$

Therefore,

$$\int_0^\infty t^{p-1} c_p([u \ge t])dt \le \left(\frac{p}{p-1} \right)^{p-1} \int_0^{s_u} h'^p(s)ds$$

$$\le \left(\frac{p}{p-1} \right)^{p-1} \int_{\mathbb{R}^N} |Du|^p dx,$$

by virtue of (7.4) of Lemma 7.1. $\blacksquare$

8 Relating Gagliardo Embeddings, Capacities, and the Isoperimetric Inequality

Proposition 8.1 *Let $p \geq 1$ and $q > p$. Then the Gagliardo type embedding inequality*

$$\|u\|_q \leq \gamma \|Du\|_p \quad \text{for all } u \in W_o^{1,p}(\mathbb{R}^N) \tag{8.1}$$

for a constant $\gamma > 0$ depending only on p, q, and N, holds if and only if, there exists a positive constant $\overline{\gamma}$, depending only upon N, p, and q, such that for all compact sets $K \subset \mathbb{R}^N$

$$\mu(K)^{\frac{p}{q}} \leq \overline{\gamma}^p c_p(K), \tag{8.2}$$

Proof Let (8.1) hold. Having fixed a compact set $K \subset \mathbb{R}^N$, let u be in the class $\Gamma_o(K)$ introduced in (4.1). Then from (8.1)

$$\mu(K)^{\frac{p}{q}} \leq \gamma^p \int_{\mathbb{R}^N} |Du|^p d\mu.$$

minimizing the right-hand side over $\Gamma_o(K)$ implies (8.2) with $\overline{\gamma} = \gamma$. Conversely, assuming the latter, for a nonnegative $u \in C_o^\infty(\mathbb{R}^N)$, compute and estimate

$$\|u\|_q^p = \left(\int_{\mathbb{R}^N} [u^p]^{\frac{q}{p}} d\mu \right)^{\frac{p}{q}} = \sup_{\|v\|_{\frac{q}{q-p}} = 1} \int_{\mathbb{R}^N} u^p v \, d\mu$$

$$= \sup_{\|v\|_{\frac{q}{q-p}} = 1} \int_0^\infty p t^{p-1} \left(\int_{[u>t]} v \, d\mu \right) dt$$

$$\leq \int_0^\infty p t^{p-1} \left(\sup_{\|v\|_{\frac{q}{q-p}} = 1} \int_{\mathbb{R}^N} v \chi_{[u \geq t]} d\mu \right) dt$$

$$= \int_0^\infty p t^{p-1} \|\chi_{[u \geq t]}\|_{\frac{q}{p}} dt = \int_0^\infty p t^{p-1} \mu([u \geq t])^{\frac{p}{q}} dt$$

$$\leq \overline{\gamma}^p \int_0^\infty p t^{p-1} c_p([u \geq t]) dt \leq p \overline{\gamma}^p \left(\frac{p}{p-1} \right)^{p-1} \|Du\|_p^p,$$

where, in the last inequality we have used (7.1) of Theorem 7.1. The proof is concluded by writing $u = u^+ - u^-$ and by density. ∎

The proposition asserts that Gagliardo's embedding inequalities are equivalent to a lower estimate of the capacity of compact sets $K \subset \mathbb{R}^N$ in terms of their corresponding Lebesgue measure. By Lemma 6.1 this occurs for $1 \leq p < N$, and with parameters

$$\frac{p}{q} = \frac{N-p}{N} \quad \text{and hence} \quad q = p^* = \frac{Np}{N-p}. \tag{8.3}$$

Remark 8.1 Assume $q = p^*$. The constant $\overline{\gamma}^p$ in (8.2) is computed by the lower estimate (6.1) of Lemma 6.1. Hence the constant γ in embedding (8.1) is

$$\gamma = \left(\frac{p}{N}\right)^{\frac{1}{p}} \left(\frac{N}{\omega_N}\right)^{\frac{1}{N}} \left(\frac{p}{N-p}\right)^{\frac{p-1}{p}}. \tag{8.4}$$

This improves the constant of the Gagliardo embedding (1.6) of Corollary 1.1 of Chap. 10.

9 Relating $\mathcal{H}^{N-p}(K)$ to $c_p(K)$ for $1 < p < N$

The "size" of a compact set $K \subset \mathbb{R}^N$ can be "measured" by its Lebesgue measure, or its capacity or its Hausdorff dimension (§ 5.1c of Chap. 3). The lower estimate of Lemma 6.1, is a first coarse relation between $c_p(K)$ and the Lebesgue measure of K. If $c_p(K) = 0$ then also $\mu(K) = 0$. There exist sets of positive p-capacity, whose Lebesgue measure is zero. An example is in § 6.3.

For a Borel set $E \subset \mathbb{R}^N$ and $k > 0$ denote by $\mathcal{H}^k(E)$ the k-dimensional Hausdorff measure of E. The next theorem provides a relation between the p-capacity of a compact set $K \subset \mathbb{R}^N$ and its $(N - p)$-dimensional Hausdorff measure.

Theorem 9.1 *Let K be a compact subset of $\mathbb{R}^N$ and let $1 < p < N$. Then*

$$\mathcal{H}^{N-p}(K) < \infty \quad \Longrightarrow \quad c_p(K) = 0. \tag{9.1}$$

9.1 An Auxiliary Proposition

Proposition 9.1 *Let $K \subset \mathbb{R}^N$ be compact and such that $\mathcal{H}^{N-p}(K) < \infty$. For every open set $\mathcal{O}_o \supset K$, there exists a bounded open set $\mathcal{O}_1$ containing K, and such that $\overline{\mathcal{O}}_1 \subset \mathcal{O}_o$, and a nonnegative function $\zeta \in W_o^{1,p}(\mathcal{O}_o)$, such that*

$$\mathcal{O}_1 \subset [\zeta = 1] \quad and \quad \int_{\mathbb{R}^N} |D\zeta|^p dy \leq 2^{N-p} \frac{\omega_N}{N} \mathcal{H}^{N-p}(K).$$

Proof Let $\mathcal{H}_{N-p,\varepsilon}(K)$ be the Hausdorff outer measure defined in (5.1) of Chap. 3, with

$$0 < \varepsilon < \tfrac{1}{8} \operatorname{dist}\{K; \partial\mathcal{O}_o\},$$

Since

$$\mathcal{H}_{N-p,\varepsilon}(K) \leq \mathcal{H}^{N-p}(K) < \infty,$$

having fixed a positive number ϵ, there exists a countable collection of sets $\{E_n\}$ each of diameter not exceeding ε such that

$$K \subset \bigcup E_n \quad \text{and} \quad \sum \operatorname{diam}(E_n)^{N-p} \leq \mathcal{H}^{N-p}(K) + \epsilon.$$

We may assume that each of the E_n intersects K. Pick $x_n \in K \cap E_n$ and consider the open ball

$$B_{\rho_n}(x_n) \quad \text{with} \quad \rho_n = 2 \operatorname{diam}(E_n).$$

By construction $E_n \subset B_{\rho_n}(x_n)$ and

$$\operatorname{dist}\left\{B_{2\rho_n}(x_n); \partial \mathcal{O}_o\right\} > \tfrac{1}{2} \operatorname{dist}\{K; \partial \mathcal{O}_o\}.$$

The family $\{B_{\rho_n}(x_n)\}$ is an open cover for K, from which we extract a finite one $\{B_{\rho_1}(x_1), \ldots, B_{\rho_k}(x_k)\}$ for some $k \in \mathbb{N}$. As a set $\mathcal{O}_1$ takes

$$\mathcal{O}_1 = \bigcup_{j=1}^{k} B_{\rho_j}(x_j).$$

By construction

$$K \subset \mathcal{O}_1 \subset \overline{\mathcal{O}}_1 \subset \mathcal{O}_o.$$

For each $B_{2\rho_j}(x_j)$ construct a nonnegative, piecewise smooth cutoff function ζ_j, which equals one on $B_{\rho_j}(x_j)$, vanishes for $|x - x_j| \geq 2\rho_j$ and such that $|D\zeta_j| \leq \rho_j^{-1}$. For such functions

$$\int_{\mathbb{R}^N} |D\zeta_j|^p dy \leq \frac{\omega_N}{N} \rho_j^{N-p}$$

The function ζ claimed by the proposition can be taken to be

$$\zeta = \max_{1 \leq j \leq k} \zeta_j.$$

By construction $\mathcal{O}_1 \subset [\zeta = 1]$ and ζ is compactly supported in $\mathcal{O}_o$. Moreover

$$\int_{\mathbb{R}^N} |D\zeta|^p dy \leq \sum_{j=1}^{k} \int_{\mathbb{R}^N} |D\zeta_j|^p dy$$

$$\leq \frac{\omega_N}{N} \sum_{j=1}^{k} \rho_j^{N-p}$$

$$\leq 2^{N-p} \frac{\omega_N}{N} \sum_{j=1}^{k} \operatorname{diam}(E_n)^{N-p}$$

$$\leq 2^{N-p} \frac{\omega_N}{N} \left[\mathcal{H}^{N-p}(K) + \epsilon\right]. \qquad \blacksquare$$

9.2   Proof of Theorem 9.1

Fix an open set $\mathcal{O}_o$ containing K and apply the previous proposition recursively to construct a sequence of nested open sets $\mathcal{O}_j$ and a sequence of functions $\{v_j\} \in W_o^{1,p}(\mathcal{O}_{j-1})$ satisfying

$$K \subset \mathcal{O}_j \subset \overline{\mathcal{O}}_j \subset [v_j = 1] \subset \mathcal{O}_{j-1}$$

and in addition

$$\int_{\mathbb{R}^N} |Dv_j|^p dy \leq 2^{N-p}\frac{\omega_N}{N}\mathcal{H}^{N-p}(K) \qquad \text{for all } j = 1, 2, \dots.$$

Consider now the sequence of weighted sums

$$u_n = \frac{1}{w_n}\sum_{j=1}^{n}\frac{v_j}{j}, \qquad \text{where} \qquad w_n = \sum_{j=1}^{n}\frac{1}{j}.$$

By construction $\mathcal{O}_n \subset [u_n = 1]$. Therefore,

$$c_p(K) \leq \int_{\mathbb{R}^N} |Du_n|^p dy \qquad \text{for all } n = 1, 2, \dots.$$

The last integral is computed and estimated by observing that the construction of the v_j implies that

$$\text{supp}\{|Dv_j|\} \subset \mathcal{O}_{j-1} - \overline{\mathcal{O}}_j \quad \text{for all } j = 1, 2, \dots.$$

From this

$$c_p(K) \leq \int_{\mathbb{R}^N} |Du_n|^p dy$$

$$= \frac{1}{w_n^p}\sum_{j=1}^{n}\frac{1}{j^p}\int_{\mathbb{R}^N} |Dv_j|^p dy$$

$$\leq 2^{N-p}\frac{\omega_N}{N}\mathcal{H}^{N-p}(K)\frac{1}{w_n^p}\sum_{j=1}^{n}\frac{1}{j^p}.$$

Letting $n \to \infty$ the right hand side goes to zero since $p > 1$. ∎

Remark 9.1 As a consequence, segments in $\mathbb{R}^3$ have positive, and finite 1-dimensional Hausdorff measure and zero 2-capacity.

10   Relating $c_p(K)$ to $\mathcal{H}^{N-p+\varepsilon}(K)$ for $1 \leq p < N$

Theorem 10.1 *Let K be a compact subset of $\mathbb{R}^N$ and let $1 \leq p < N$. Then*

$$c_p(K) = 0 \quad \Longrightarrow \quad \mathcal{H}^{N-p+\varepsilon}(K) = 0 \quad \text{for all } \varepsilon > 0. \tag{10.1}$$

Proof For every $j \in \mathbb{N}$ there exists a nonnegative function $u_j \in C_o^\infty(\mathbb{R}^N)$ such that $u_j \geq 1$ in a open neighborhood of K, and

$$\int_{\mathbb{R}^N} |Du_j|^p dy \leq \frac{1}{2^j}.$$

Set $u = \sum u_j$. Then $|Du| \in L^p(\mathbb{R}^N)$ and $u \in L^{p^*}(\mathbb{R}^N)$, where $p^* = \frac{Np}{N-p}$. Indeed

$$\|Du\|_p \leq \sum \|Du_j\|_p \leq \sum \frac{1}{2^{\frac{j}{p}}}.$$

Moreover by the embedding of the Corollary 1.1, of Chap. 10, for all $n \in \mathbb{N}$,

$$\left\| \sum_{j=1}^n u_j \right\|_{p^*} \leq \sum_{j=1}^n \|u_j\|_{p^*} \leq \frac{p(N-1)}{N-p} \sum \|Du_j\|_p < \infty.$$

The set K is contained in a open neighborhood of $[u \geq k]$ for all $k \in \mathbb{N}$. Therefore, having fixed $x \in E$, for all $k \in \mathbb{N}$ there exists $\rho_{x,k} > 0$ small enough that

$$(u)_{x,\rho} \stackrel{\text{def}}{=} \fint_{B_\rho(x)} u \, dy \geq k \quad \text{for all } \rho > \rho_{x,k}.$$

Hence,

$$\lim_{\rho \to 0} \fint_{B_\rho(x)} u \, dy = \infty \quad \text{for all } x \in K. \tag{10.2}$$

We next establish that K is included in the set

$$E_\varepsilon = \left\{ x \in \mathbb{R}^N \ \middle| \ \limsup_{\rho \to 0} \rho^{p-\varepsilon} \fint_{B_\rho(x)} |Du|^p \, dy = \infty \right\} \tag{10.3}$$

for all $\varepsilon > 0$. Let $x \in E_\varepsilon$ be such that

$$\limsup_{\rho \to 0} \rho^{p-\varepsilon} \fint_{B_\rho(x)} |Du|^p \, dy < \infty$$

for some $\varepsilon > 0$. Then there exists a positive constant γ_x such that

$$\rho^{p-\varepsilon} \fint_{B_\rho(x)} |Du|^p \, dy \leq \gamma_x^p \quad \text{for all } \rho \in (0, 1).$$

From this by the Poincarè inequality in the form (10.5) of Chap. 10

$$\fint_{B_\rho(x)} |u - (u)_{x,\rho}| dy \le C'\left(\rho^p \fint_{B_\rho(x)} |Du|^p dy\right)^{\frac{1}{p}} \le C'\gamma_x \rho^{\frac{\varepsilon}{p}}.$$

For $\rho \in (0, 1]$, compute and estimate

$$|(u)_{x,\frac{1}{2}\rho} - (u)_{x,\rho}| = \frac{1}{\mu(B_{\frac{1}{2}\rho})}\left|\fint_{B_\rho(x)}\left[u - (u)_{x,\rho}\right]dy\right|$$

$$\le 2^N \fint_{B_\rho(x)} |u - (u)_{x,\rho}| dy \le C_x \rho^{\frac{\varepsilon}{p}}$$

where $C_x = 2^N C'\gamma_x$. Fix any two positive integers $m < n$. Applying this estimate recursively, gives

$$\left|(u)_{x,\frac{1}{2^n}} - (u)_{x,\frac{1}{2^m}}\right| \le \sum_{j=m+1}^{n} \left|(u)_{x,\frac{1}{2^j}} - (u)_{x,\frac{1}{2^{j-1}}}\right| \le C_x \sum_{j>m} \frac{1}{2^{\frac{\varepsilon}{p}j}}.$$

The right-hand side is the tail of a convergent series and hence it tends to zero as $m \to \infty$. Therefore $\{(u)_{x,\frac{1}{2^n}}\}$ is a Cauchy sequence, and hence convergent. This contradicts (10.2) and establishes the inclusion $K \subset E_\varepsilon$. The proof of Theorem 10.1 is concluded by the following lemma, whose proof is a special case of Proposition 15.1. ∎

Lemma 10.1 $\mathcal{H}^{N-p+\varepsilon}(E) = 0$.

Remark 10.1 As a consequence, for $1 \le p < N$, the Hausdorff dimension of compact sets of zero p-capacity, does not exceed $(N - p)$.

11 The p-Capacity of a Set $E \subset \mathbb{R}^N$ for $1 \le p < N$

Let $\mathcal{O}$ be an open subset of $\mathbb{R}^N$. The inner p-capacity of $\mathcal{O}$ is defined by

$$\underline{c}_p(\mathcal{O}) = \sup_{\substack{K \subset \mathcal{O} \\ K \text{ compact}}} c_p(K). \tag{11.1}$$

For a set $E \subset \mathbb{R}^N$ the inner p-capacity $\underline{c}_p(E)$ and outer p-capacity $\overline{c}_p(E)$ of E are defined as

$$\underline{c}_p(E) = \sup_{\substack{K \subset E \\ K \text{ compact}}} c_p(K); \qquad \overline{c}_p(E) = \inf_{\substack{\mathcal{O} \supset E \\ \mathcal{O} \text{ open}}} \underline{c}_p(\mathcal{O}). \tag{11.2}$$

A set $E \subset \mathbb{R}^N$ is p-capacitable if its inner and outer p-capacities coincide. For a p-capacitable set $E \subset \mathbb{R}^N$, we set

$$c_p(E) = \underline{c}_p(E) = \overline{c}_p(E). \tag{11.3}$$

The definition implies that compact sets and open sets in $\mathbb{R}^N$ are p-capacitable.

Proposition 11.1 *Let E and F be subsets of $\mathbb{R}^N$ and let $p \geq 1$. Then*

$$\overline{c}_p(E \cup F) + \overline{c}_p(E \cap F) \leq \overline{c}_p(E) + \overline{c}_p(F). \tag{11.4}$$

Remark 11.1 This is called *strong sub-additivity* of the set function $\overline{c}_p(\cdot)$.

Proof (of Proposition 11.1) May assume that the right hand side of (11.4) is finite. Assume first that E and F are compact and let $u \in \Gamma_1(E)$ and $v \in \Gamma_1(F)$. Then $(u \vee v)$ and $(u \wedge v)$ are Lipschitz continuous, compactly supported in $\mathbb{R}^N$, and

$$(u \vee v) = 1 \quad \text{in an open neighborhood of } E \cup F;$$
$$(u \wedge v) = 1 \quad \text{in an open neighborhood of } E \cap F.$$

Moreover

$$|D(u \vee v)|^p + |D(u \wedge v)|^p = |Du|^p + |Dv|^p \quad \text{a.e. in } \mathbb{R}^N.$$

Also,

$$(u \vee v) \in \Gamma_*(E \cup F) \quad \text{and} \quad (u \wedge v) \in \Gamma_*(E \cap F)$$

where $\Gamma_*(\cdot)$ is the class introduced in (4.4). From this

$$c_p(E \cup F) + c_p(E \cap F) \leq \int_{\mathbb{R}^N} |D(u \vee v)|^p dx + \int_{\mathbb{R}^N} |D(u \wedge v)|^p dx$$
$$= \int_{\mathbb{R}^N} |Du|^p dx + \int_{\mathbb{R}^N} |Dv|^p dx.$$

This implies (11.4) for E and F compact. Assume next that E and F are open. As such, they are the countable union of compact sets (see for example Proposition 1.2 and Remark 1.1 of Chap. 3). For $\varepsilon > 0$ there exists compact sets $H \subset E$ and $K \subset F$, such that

$$\overline{c}_p(E \cup F) \leq c_p(H \cup K) + \tfrac{1}{2}\varepsilon. \quad \text{and} \quad \overline{c}_p(E \cap F) \leq c_p(H \cap K) + \tfrac{1}{2}\varepsilon.$$

From this and (11.4), valid for compact sets,

$$\overline{c}_p(E \cup F) + \overline{c}_p(E \cap F) \leq c_p(H) + c_p(K) + \varepsilon.$$

This establishes (11.4) for open sets. Next, let E and F be sets in $\mathbb{R}^N$ of finite outer p-capacity and no further topological restriction. Let $\mathcal{O}_E$ and $\mathcal{O}_F$ be open sets containing E and F respectively. Writing down (11.4) for $\mathcal{O}_E$ and $\mathcal{O}_F$ gives

$$\overline{c}_p(E \cup F) + \overline{c}_p(E \cap F) \le c_p(\mathcal{O}_E) + c_p(\mathcal{O}_F).$$

Taking the infimum of the right-hand side for all open sets $\mathcal{O}_E$ containing E and $\mathcal{O}_F$ containing F proves the proposition. ∎

Proposition 11.2 *Let $\{E_n\}$ and $\{F_n\}$ be countable collections of sets in $\mathbb{R}^N$, each of finite outer p-capacity, and such that $E_n \subset F_n$. Then for all $n \in \mathbb{N}$,*

$$\overline{c}_p\left(\bigcup_{j=1}^n F_j \right) - \overline{c}_p\left(\bigcup_{j=1}^n E_j \right) \le \sum_{j=1}^n \left(\overline{c}_p(F_j) - \overline{c}_p(E_j) \right). \tag{11.5}$$

Proof It suffices to prove (11.5) for $n = 2$. Apply (11.4) to the pair of sets

$$E = E_1 \cup F_1 \quad \text{and} \quad F = E_1 \cup F_2$$

to get

$$\overline{c}_p(F_1 \cup F_2) + \overline{c}_p\big(E_1 \cup (F_1 \cap F_2)\big) \le \overline{c}_p(F_1) + \overline{c}_p(E_1 \cup F_2).$$

From this, by the monotonicity of $\overline{c}_p(\cdot)$

$$\overline{c}_p(F_1 \cup F_2) + \overline{c}_p(E_1) \le \overline{c}_p(F_1) + \overline{c}_p(E_1 \cup F_2). \tag{11.6}$$

Next, apply (11.4) to the pair of sets

$$E = E_2 \cup F_2 \quad \text{and} \quad F = E_1 \cup E_2.$$

This gives

$$\overline{c}_p(E_1 \cup F_2) + \overline{c}_p\big(E_2 \cup (E_1 \cap F_2)\big) \le \overline{c}_p(F_2) + \overline{c}_p(E_1 \cup E_2).$$

From this, by the monotonicity of $\overline{c}_p(\cdot)$

$$\overline{c}_p(E_1 \cup F_2) + \overline{c}_p(E_2) \le \overline{c}_p(F_2) + \overline{c}_p(E_1 \cup E_2). \tag{11.7}$$

Combining (11.6) and (11.7) the proposition follows. ∎

12   Limits of Sets and Their Outer p-Capacities

Proposition 12.1 *Let* $\{K_n\}$ *be a countable collection of compact sets such that* $K_{n+1} \subset K_n$. *Then*

$$c_p\left(\bigcap K_n\right) = \lim c_p(K_n). \tag{12.1}$$

Remark 12.1 This is called *right continuity* of the set function $\overline{c}_p(\cdot)$ when acting on compact sets. Compare this with the notion of right continuity given by (4.7).

Proof (of Proposition 12.1) We may assume that $c_p(K_1) < \infty$. The set $K = \cap K_n$ is compact and hence p-capacitable. By the monotonicity of $c_p(\cdot)$

$$c_p(K) \le \lim c_p(K_n).$$

To establish the converse inequality, let $\Gamma_o(K)$ be the class of functions introduced in (4.1). For each $\varepsilon > 0$ there exists $\zeta \in \Gamma_o(K)$ such that

$$\int_{\mathbb{R}^N} |D\zeta|^p dx \le c_p(K) + \varepsilon.$$

Let $\mathcal{O}$ be the open neighborhood of K where $\zeta \ge 1$. Since K is compact and $\{K_n\}$ is nested, there exists $n_\varepsilon \in \mathbb{N}$ such that $K_n \subset \mathcal{O}$ for all $n \ge n_\varepsilon$. From this

$$\lim c_p(K_n) \le \int_{\mathbb{R}^N} |D\zeta|^p dx \le c_p(K) + \varepsilon. \qquad\blacksquare$$

Proposition 12.2 *Let* $\{E_n\}$ *be a countable collection of sets in* $\mathbb{R}^N$ *such that* $E_n \subset E_{n+1}$, *and with* $c_p(\bigcup E_n) < \infty$. *Then*

$$\overline{c}_p\left(\bigcup E_n\right) = \lim \overline{c}_p(E_n). \tag{12.2}$$

Proof Set $E = \bigcup E_n$. By monotonicity of $\overline{c}_p(\cdot)$

$$\overline{c}_p(E) \ge \lim \overline{c}_p(E_n).$$

To establish the converse inequality, assume first that all E_n are open. Having fixed $\varepsilon > 0$, let K be a compact set contained in E such that

$$c_p(E) \le c_p(K) + \varepsilon.$$

Since the sets E_n are open and nested, there exists $n_\varepsilon \in \mathbb{N}$ such that $K \subset E_n$ for all $n \ge n_\varepsilon$. From this

$$c_p(E) \le c_p(K) + \varepsilon \le \lim c_p(E_n) + \varepsilon.$$

Thus (12.2) holds for a countable collection of open sets. Let now $\{E_n\}$ be nested, and have no further topological restriction. Having fixed $\varepsilon > 0$, for all $n \in \mathbb{N}$, there exists an open set $\mathcal{O}_n$ such that

$$E_n \subset \mathcal{O}_n, \quad \text{and} \quad \overline{c}_p(E_n) \geq c_p(\mathcal{O}_n) - \frac{\varepsilon}{2^n}.$$

Applying (11.5) to the pair of collections $\{E_n\}$ and $\mathcal{O}_n$, gives

$$c_p\left(\bigcup_{j=1}^{n} \mathcal{O}_n\right) - \overline{c}_p\left(\bigcup_{j=1}^{n} E_n\right) \leq \varepsilon.$$

From this, since $E_n \subset E_{n+1}$

$$c_p\left(\bigcup_{j=1}^{n} \mathcal{O}_n\right) \leq \lim \overline{c}_p(E_n) + \varepsilon.$$

The sets $\bigcup_{j=1}^{n} \mathcal{O}_n$ are open and nested, and as such, (12.2) holds for them. Then

$$\overline{c}_p(E) \leq c_p(\cup \mathcal{O}_n) = \lim c_p\left(\bigcup_{j=1}^{n} \mathcal{O}_n\right) \leq \lim \overline{c}_p(E_n) + \varepsilon. \qquad \blacksquare$$

Corollary 12.1 *The set function $\overline{c}_p(\cdot)$ is countably sub-additive.*

Proof By Proposition 11.1 the outer p-capacity is finitely sub-additive. Let $\{E_n\}$ be a countable collection of sets in $\mathbb{R}^N$. It suffices to consider the case when each E_n and $\bigcup E_n$ are of finite outer p-capacity. The collection of sets $\bigcup_{j=1}^{n} E_j$ satisfies the assumptions of Proposition 12.2. Therefore,

$$\overline{c}_p\left(\bigcup E_n\right) = \lim \overline{c}_p\left(\bigcup_{j=1}^{n} E_j\right) \leq \sum \overline{c}_p(E_n). \qquad \blacksquare$$

13 Capacitable Sets

Proposition 13.1 *Sets of the type of $\mathcal{F}_{\sigma\delta}$ are p-capacitable*

Proof Any such set is of the form

$$E = \bigcap_i \bigcup_j E_{i,j}$$

where the sets $E_{i,j}$ are closed. We may assume that for each fixed i, the sets $E_{i,j}$ are compact and $E_{i,j} \subset E_{i,j+1}$. This is effected by rewriting E as

$$E = \bigcap_i \bigcup_j E'_{i,j}, \quad \text{where} \quad E'_{i,j} = \bigcup_{\ell=1}^{j} E_{i,\ell} \cap \bar{B}_j$$

where $\bar{B}_j$ is the closed ball in $\mathbb{R}^N$ centered at the origin and radius j. We may also assume that $\bar{c}_p(E) < \infty$. To proof the proposition, for each $\varepsilon > 0$ we will exhibit a compact set $K_\varepsilon \subset E$, such that

$$c_p(K_\varepsilon) \geq \bar{c}_p(E) - \varepsilon.$$

Set

$$H_{1,j} = E \cap E_{1,j}.$$

By construction

$$E = \bigcup H_{1,j}, \quad \text{and} \quad H_{1,j} \subset H_{1,j+1}.$$

From this, by Proposition 12.2,

$$\bar{c}_p(E) = \lim \bar{c}_p(H_{1,j}).$$

Therefore, having fixed $\varepsilon > 0$ there is an index j_1 such that

$$\bar{c}_p(E) - \bar{c}_p(H_{1,j_1}) \leq \tfrac{1}{2}\varepsilon. \tag{13.1}$$

Set

$$H_1 = H_{1,j_1}, \quad \text{and} \quad K_1 = E_{1,j_1}.$$

Consider the sets

$$H_{2,j} = H_1 \cap E_{2,j}.$$

By construction

$$H_1 = \bigcup H_{2,j}, \quad \text{and} \quad H_{2,j} \subset H_{2,j+1}.$$

From this, by Proposition 12.2,

$$\bar{c}_p(H_1) = \lim \bar{c}_p(H_{2,j}).$$

Therefore, for the same fixed $\varepsilon > 0$ there is an index j_2 such that

$$\bar{c}_p(H_1) - \bar{c}_p(H_{2,j_2}) \leq \tfrac{1}{4}\varepsilon.$$

Set

$$H_2 = H_{2,j_2}, \quad \text{and} \quad K_2 = E_{2,j_2}.$$

Construct inductively countable collections of sets $\{H_n\}$ and $\{K_n\}$ as follows. If H_n an K_n have been constructed, set

$$H_{n+1,j} = H_n \cap E_{n+1,j}.$$

By construction

$$H_{n+1} = \bigcup H_{n+1,j}, \quad \text{and} \quad H_{n+1,j} \subset H_{n+1,j+1}.$$

From this, by Proposition 12.2,

$$\overline{c}_p(H_{n+1}) = \lim \overline{c}_p(H_{n+1,j}).$$

Therefore, for the same fixed $\varepsilon > 0$ there is an index j_{n+1} such that

$$\overline{c}_p(H_n) - \overline{c}_p(H_{n+1,j_{n+1}}) \le \frac{1}{2^{n+1}}\varepsilon. \tag{13.2}$$

Set

$$H_{n+1} = H_{n+1,j_{n+1}}, \quad \text{and} \quad K_{n+1} = E_{n+1,j_{n+1}}.$$

Having constructed $\{H_n\}$ and $\{K_n\}$, set

$$H_\varepsilon = \bigcap H_n, \quad \text{and} \quad K_\varepsilon = \bigcap K_n.$$

By construction

$$K_\varepsilon \subset E, \quad \text{and} \quad H_\varepsilon = E \cap \left(\bigcap K_n\right) = E \cap K_\varepsilon.$$

Therefore K_ε is a compact subset of E. Moreover

$$H_\varepsilon = K_\varepsilon \quad \text{and} \quad H_\varepsilon = \lim \bigcap_{j=1}^{n} K_j.$$

Since the sets $\bigcap_{j=1}^{n} K_j$ are compact and nested, by Proposition 12.1

$$\overline{c}_p(H_\varepsilon) = \lim \overline{c}_p\left(\bigcap_{j=1}^{n} K_j\right).$$

Also by construction

$$H_\varepsilon \subset H_n = \bigcap_{j=1}^{n} H_j \subset \bigcap_{j=1}^{n} K_j.$$

Therefore the limit of $\bar{c}_p(H_n)$ exists and

$$c_p(K_\varepsilon) = \bar{c}_p(H_\varepsilon) \leq \lim \bar{c}_p(H_n) \leq \lim \left(\bigcap_{j=1}^{n} K_j \right) = c_p(K_\varepsilon).$$

From this, taking into account (13.1) and (13.2)

$$c_p(K_\varepsilon) = \lim \bar{c}_p(H_n) = \bar{c}_p(H_1) + \lim \sum_{j=1}^{n} \left[\bar{c}_p(H_{j+1}) - \bar{c}_p(H_j) \right]$$

$$= \bar{c}_p(E) - \left[\bar{c}_p(E) - \bar{c}_p(H_1) \right] + \lim \sum_{j=1}^{n} \left[\bar{c}_p(H_{j+1}) - \bar{c}_p(H_j) \right]$$

$$\geq \bar{c}_p(E) - \varepsilon. \qquad\qquad \blacksquare$$

Corollary 13.1 *Sets of the type of $\mathcal{F}_\sigma$ and $\mathcal{G}_\delta$ are p-capacitable.*

14   Capacities Revisited and p-Capacitability of Borel Sets

Let $\varphi(\cdot)$ be a nonnegative set function, defined on compact subsets of $\mathbb{R}^N$, such that $\varphi(\emptyset) = 0$, and satisfying:

(a) $\varphi(\cdot)$ is monotone increasing in the sense of (4.8);
(b) $\varphi(\cdot)$ is strongly sub-additive, in the sense of (11.4);
(c) $\varphi(\cdot)$ is right-continuous, in the sense of Proposition 12.1;

Using such a $\varphi(\cdot)$, for an open set $\mathcal{O}$ in $\mathbb{R}^N$ define

$$\varphi(\mathcal{O}) = \sup_{\substack{K \subset \mathcal{O} \\ K \text{ compact}}} \varphi(K). \qquad\qquad (14.1)$$

For a set $E \subset \mathbb{R}^N$ define the inner and outer φ-capacity of E as

$$\underline{\varphi}(E) = \sup_{\substack{K \subset E \\ K \text{ compact}}} \varphi(K); \qquad \overline{\varphi}(E) = \inf_{\substack{\mathcal{O} \supset E \\ \mathcal{O} \text{ open}}} \varphi(\mathcal{O}). \qquad (14.2)$$

A set $E \subset \mathbb{R}^N$ is φ-capacitable if its inner and outer φ-capacities coincide. For a φ-capacitable set $E \subset \mathbb{R}^N$ we set

$$\varphi(E) = \underline{\varphi}(E) = \overline{\varphi}(E). \qquad\qquad (14.3)$$

The definition implies that compact sets and open sets in $\mathbb{R}^N$ are φ-capacitable. The proofs of Propositions 11.1 and 11.2, only use properties (a) and (b) of the set function $\varphi(\cdot) = c_p(\cdot)$. Likewise the proof of Proposition 12.2 only uses properties (a)–(c) of the set function $\varphi(\cdot) = c_p(\cdot)$ on compact subsets of $\mathbb{R}^N$. Thus these Propositions, continue to hold for any such set function $\varphi(\cdot)$. We summarize the following:

Proposition 14.1 *Let $\varphi(\cdot)$ be a nonnegative set function defined on compact subsets of $\mathbb{R}^N$, such that $\varphi(\emptyset) = 0$, and satisfying* (a)–(c) *above. Then*

(i) *For any two sets $E, F \subset \mathbb{R}^N$*

$$\overline{\varphi}(E \cup F) + \overline{\varphi}(E \cap F) \leq \overline{\varphi}(E) + \overline{\varphi}(F). \tag{14.4}$$

(ii) *Let $\{E_n\}$ and $\{F_n\}$ be countable collections of sets in $\mathbb{R}^N$, each of finite outer φ-capacity, and such that $E_n \subset F_n$. Then for all $n \in \mathbb{N}$,*

$$\overline{\varphi}\left(\bigcup_{j=1}^{n} F_j\right) - \overline{\varphi}\left(\bigcup_{j=1}^{n} E_j\right) \leq \sum_{j=1}^{n} \left(\overline{\varphi}(F_j) - \overline{\varphi}(E_j)\right). \tag{14.5}$$

(iii) *Let $\{E_n\}$ be a countable collection of sets in $\mathbb{R}^N$ such that $E_n \subset E_{n+1}$, and with $\overline{\varphi}\left(\bigcup E_n\right) < \infty$. Then*

$$\overline{\varphi}\left(\bigcup E_n\right) = \lim \overline{\varphi}(E_n). \tag{14.6}$$

(iv) *The set function $\overline{\varphi}(\cdot)$ is countably sub-additive.*

Next the proof of Proposition 13.1 only uses properties (a)–(c) of a φ-capacity and its consequences as stated in Proposition 14.1. Hence we conclude:

Proposition 14.2 *Let $\varphi(\cdot)$ be a nonnegative set function defined on compact subsets of $\mathbb{R}^N$, such that $\varphi(\emptyset) = 0$, and satisfying* (a)–(c) *above. Then sets of the type of $\mathcal{F}_{\sigma\delta}$ in $\mathbb{R}^N$ are φ-capacitable. In particular sets of the type of $\mathcal{F}_\sigma$ and $\mathcal{G}_\delta$ are φ-capacitable.*

14.1 The Borel Sets in $\mathbb{R}^N$ Are p-Capacitable

Continue to assume that $1 \leq p < N$. For a set $E \subset \mathbb{R}^{N+1}$, denote by $P_N(E)$ the projection of E into $\mathbb{R}^N$. If $\mathcal{O}$ is open in $\mathbb{R}^{N+1}$ then $P_N(\mathcal{O})$ is open in $\mathbb{R}^N$. Likewise if K is compact in $\mathbb{R}^{N+1}$, then $P_N(K)$ is compact in $\mathbb{R}^N$. If E is closed in $\mathbb{R}^{N+1}$ then $P_N(E)$ need not be closed in $\mathbb{R}^N$.

A Borel set in $\mathbb{R}^N$ is the projection into $\mathbb{R}^N$ of some set of the type of $\mathcal{G}_\delta$ in $\mathbb{R}^{N+1}$ [71]. For a compact set $K \subset \mathbb{R}^{N+1}$ set

$$\varphi(K) = c_p\left(P_N(K)\right).$$

One verifies that such a $\varphi(\cdot)$ satisfies (a)–(c) above. Thus by Proposition 14.2 sets of the type of $\mathcal{G}_\delta$ in $\mathbb{R}^{N+1}$ are φ-capacitable in the sense of (14.3). Hence all Borel sets in $\mathbb{R}^N$ are p-capacitable.

The same reasoning applies to a larger class of sets in $\mathbb{R}^N$, called *Suslin* sets, or *analytic* sets. This is an algebra of sets closed under continuous transformations of $\mathbb{R}^N$. Every analytic set is the projection into $\mathbb{R}^N$ of a set of the type of a $\mathcal{G}_\delta$ in $\mathbb{R}^{N+1}$ [71]. Thus, by the same reasoning, analytic sets are p-capacitable.

14.2 Generating Measures by p-Capacities

The set function $\bar{c}_p(\cdot)$ satisfies the requirements (i)–(iv) of § 4 of Chap. 3, to be an outer measure in $\mathbb{R}^N$. The countable sub-additivity of $\bar{c}_p(\cdot)$ is in Corollary 12.1. Thus, by the Carathéodory procedure outlined in § 6 of Chap. 3, and leading to Proposition 6.1, of the same chapter, it generates a σ-algebra $\mathcal{A}_p$ and a measure $\mathcal{C}_p$ defined on $\mathcal{A}_p$. A set $E \subset \mathbb{R}^N$ is $\mathcal{A}_p$ if and only if

$$\bar{c}_p(A) \geq \bar{c}_p(A \cap E) + \bar{c}_p(A - E) \tag{14.7}$$

for all sets $A \subset \mathbb{R}^N$. If $\bar{c}_p(E) = 0$ then $E \in \mathcal{A}_p$. Let now E be a bounded set in $\mathbb{R}^N$ such that $\bar{c}_p(E) > 0$. Let $\bar{B}_\rho$ be the closed ball centered at the origin and large enough radius ρ, so that $E \subset B_\rho$. By the arguments of § 6.3 and Corollary 6.2, $c_p(\bar{B}_\rho) = c_p(\partial B_\rho)$. Writing down (14.7) with $A = \bar{B}_\rho$ gives

$$c_p(\partial B_\rho) = c_p(\bar{B}_\rho) \geq \bar{c}_p(E) + c_p(\partial B_\rho).$$

Since $\bar{c}_p(E) > 0$, this is a contradiction and hence $E \notin \mathcal{A}_p$. We conclude that no bounded set of positive p-capacity is $\mathcal{C}_p$-measurable. In particular, no bounded Borel set of positive p-capacity is $\mathcal{C}_p$-measurable. Hence $\mathcal{C}_p$ is not a Borel measure. As a consequence $\bar{c}_p(\cdot)$ is not a metric outer measure (§ 5.1 of Chap. 3). If it were, $\mathcal{A}_p$ would contain the Borel sets (Remark 8.2 of Chap. 3).

15 Precise Representatives of Functions in $L^1_{\text{loc}}(\mathbb{R}^N)$

Let $u \in L^1_{\text{loc}}(\mathbb{R}^N)$ with respect to the Lebesgue measure μ in $\mathbb{R}^N$, and set

$$u_*(x) = \begin{cases} \lim_{\rho \to 0} \fint_{B_\rho(x)} u\, dy & \text{if the limit exists;} \\ 0 & \text{otherwise.} \end{cases} \tag{15.1}$$

This is called the L^1_{loc} precise representative of u. By the Lebesgue-Besicovitch Theorem 11.1 of Chap. 5, the limit exists μ–a.e. in $\mathbb{R}^N$ and u is precisely defined everywhere in $\mathbb{R}^N$, except for a set of measure zero. If u has a continuous representative then u_* it is unambiguously well defined everywhere in $\mathbb{R}^N$. A point where the limit exists is a differentiability point of u, as opposed to a Lebesgue point (§ 11.2 of Chap. 5). For $0 < s < N$ set

$$E_s = \left\{ x \in \mathbb{R}^N \;\middle|\; \limsup_{\rho \to 0} \rho^s \fint_{B_\rho(x)} |u|\,dy > 0 \right\}. \tag{15.2}$$

The set E_s contains points where, roughly speaking, u is unbounded. If x is a Lebesgue point, then the limit in (15.2) is zero and therefore $x \notin E_s$. Hence E_s is a subset of the complement of the Lebesgue points of u. Since the latter has measure zero, and since the Lebesgue σ-algebra is complete, E_s is measurable and $\mu(E_s) = 0$. In this sense the set E_s is "small". The next proposition further quantifies the "smallness" of E_s in terms of its $\mathcal{H}^{N-s}$ Hausdorff measure.

Proposition 15.1 $\mathcal{H}^{N-s}(E_s) = 0$.

Proof For $t > 0$ introduce the set

$$E_{s,t} = \left\{ x \in \mathbb{R}^N \;\middle|\; \limsup_{\rho \to 0} \rho^s \fint_{B_\rho(x)} |u|\,dy > t \right\}.$$

Since $E_{s,t} \subset E_s$, one also has $\mu(E_{s,t}) = 0$. Therefore for all $\eta > 0$ there exists an open set $\mathcal{O} \supset E_{s,t}$ and $\mu(\mathcal{O}) < \eta$.

By the Vitali theorem on the absolute continuity of the integral (Theorem 11.1 of Chap. 4), for all $\varepsilon > 0$ there exists $\eta > 0$ such that

$$\int_{\mathcal{E}} |u|\,d\mu < \varepsilon \qquad \begin{array}{l} \text{for all Lebesgue measurable sets} \\ \mathcal{E} \subset \mathbb{R}^N \text{ such that } \mu(\mathcal{E}) < \eta \end{array} .$$

Next, fix $\delta > 0$ and define an arbitrary function

$$E_{s,t} \ni x \to \rho(x) \in (0, \delta]$$

such that the corresponding balls $B\big(x, \rho(x)\big)$ centered at $x \in E_{s,t}$ and radii $0 < \rho(x) \le \delta$ satisfy

$$B\big(x, \rho(x)\big) \subset \mathcal{O} \qquad \text{and} \qquad \int_{B(x,\rho(x))} |u|\,dy > t\rho^{N-s}.$$

By Proposition 18.2c of the Complements of Chap. 3, there exists a countable collection of disjoint, closed balls $\{B_{\rho_n}(x_n)\}$ with $\rho_n = \rho(x_n)$ such that

$$E_{s,t} \subset \bigcup B_{3\rho_n}(x_n).$$

From this

$$\mathcal{H}_{N-s,6\delta}(E_{s,\varepsilon}) \leq \sum (6\rho_n)^{N-s}$$

$$\leq \frac{6^{N-s}}{t} \sum \int_{B_{\rho_n}(x_n)} u\, dy$$

$$\leq \frac{6^{N-s}}{t} \int_{\mathcal{O}} u\, dy \leq \frac{6^{N-s}}{t} \varepsilon.$$

To prove the proposition let $\delta \to 0$ and then $\varepsilon \to 0$. ∎

Let now $u \in W^{1,p}_{\mathrm{loc}}(\mathbb{R}^N)$ for some $p \in [1, \infty)$. If $p > N$ then by the embedding Theorem 8.1 of Chap. 10, u has a Hölder continuous representative and hence it is unambiguously well defined by (15.1) for all $x \in \mathbb{R}^N$. If $1 \leq p \leq N$ the Poincarè inequality, in the form (10.5) of Chap. 10 implies that u is unambiguously well defined by (15.1) except possibly in the set

$$E_p = \left\{ x \in \mathbb{R}^N \mid \limsup_{\rho \to 0} \rho^p \int_{B_\rho(x)} |Du|^p dy > 0 \right\}. \tag{15.3}$$

For $p = N$, the set E_N is empty and hence u, although not necessarily continuous, is unambiguously well defined by (15.1) for all $x \in \mathbb{R}^N$. If $1 \leq p < N$ the set E_p is a subset of the complement of the Lebesgue points of $|Du|^p \in L^1_{\mathrm{loc}}(\mathbb{R}^N)$. Therefore E_p is measurable and $\mu(E_p) = 0$. Moreover by Proposition 15.1 also $\mathcal{H}^{N-p}(E_p) = 0$. We summarize:

Proposition 15.2 *Let $u \in W^{1,p}_{\mathrm{loc}}(\mathbb{R}^N)$ for some $p \in [1, \infty)$. There exists a Lebesgue measurable set $E_p \subset \mathbb{R}^N$ such that $\mu(E_p) = 0$ and $\mathcal{H}^{N-p}(E_p) = 0$, and u is unambiguously well defined by (15.1) for all $x \in \mathbb{R}^N - E_p$.*

It turns out that for $1 \leq p < N$ a function $u \in W^{1,p}_{\mathrm{loc}}(\mathbb{R}^N)$, can be unambiguously defined except for a Borel set of p-capacity zero. This is the content of the next sections.

16 Estimating the p-Capacity of $[u > t]$ for $t > 0$

Let $u \in L^p(\mathbb{R}^N)$ for some $1 \leq p < \infty$ be non-negative. Then for all $t > 0$,

$$\mu([u_* > t]) \leq \frac{1}{t^p} \int_{\mathbb{R}^N} u^p d\mu$$

where u_* is the L^1 precise representative of u introduced in (15.1). The analog of this estimate, in terms of outer p-capacity for functions $u \in W^{1,p}_{p^*}(\mathbb{R}^N)$, for $1 \leq p < N$, is

$$\overline{c}_p([u_* > t]) \leq \frac{\gamma}{t^p} \int_{\mathbb{R}^N} |Du|^p d\mu$$

for a constant $\gamma = \gamma(N, p)$. While correct, this estimate is coarse. Indeed the set $[u_* > t]$ avoids the complement of the set of differentiability of u. Such a set has measure zero, but it might have positive capacity. A more precise estimate can be given in terms of averages of u

$$(u)_{x,\rho} = \fint_{B_\rho(x)} u \, dy.$$

For $t > 0$ set

$$\begin{aligned}[u > t]_* &= \bigcup_{\rho>0} [(u)_{\cdot,\rho} > t] \\ &= \{x \in \mathbb{R}^N \mid (u)_{x,\rho} > t \text{ for some } \rho > 0\}.\end{aligned} \tag{16.1}$$

If $u_*(x) > t$, then there exists some $\rho > 0$ such that $(u)_{x,\rho} > t$, and hence $[u_* > t] \subset [u > t]_*$ with strict inclusion. Let $W^{1,p}_{p^*}(\mathbb{R}^N)$ be the class of functions introduced in § 24 of Chap. 10, for $E = \mathbb{R}^N$.

Proposition 16.1 *Let $u \in W^{1,p}_{p^*}(\mathbb{R}^N)$ for some $1 \le p < \infty$ be non-negative. There exists a constant γ depending only on N and p, such that, for all $t > 0$,*

$$\bar{c}_p([u > t]_*) \le \frac{\gamma}{t^p} \int_{\mathbb{R}^N} |Du|^p d\mu \tag{16.2}$$

Proof The closed balls $B_\rho(x)$ for $x \in [u > t]_*$ form a Besicovitch covering $\mathcal{F}$, of $[u > t]_*$, and their radius is uniformly bounded. Indeed

$$t \, \omega_N \rho^N \le \int_{B_\rho(x)} u \, d\mu \le (\omega_N \rho^N)^{1-\frac{1}{p^*}} \|u\|_{p^*}.$$

Therefore

$$\{\text{the supremum of the radii of the balls in } \mathcal{F}\} = R \le \left(\frac{1}{t^{p^*}\omega_N} \|u\|_{p^*}^{p^*}\right)^{\frac{1}{N}}.$$

By the Besicovitch covering theorem, in its general form of § 18.1c of Chap. 3, there exist a finite collection $\{\mathcal{B}_1, \ldots, \mathcal{B}_{c_N}\}$ of countable collections of disjoint closed balls $B_{i,j} \in \mathcal{B}_j$, such that

$$[u > t]_* \subset \bigcup_{j=1}^{c_N} \bigcup_{i \in \mathbb{N}} B_{i,j}$$

with $B_{i,j} \cap B_{i',j} = \emptyset$ for $i \ne i'$, for all $j \in \{1, \ldots, c_N\}$. Moreover

$$(u)_{B_{ij}} > t \quad \text{for all } i \in \mathbb{N}, \text{ and all } j \in \{1, \ldots, c_N\}.$$

Set

$$u_{ij} = \max\left\{(u)_{B_{ij}} - u \, ; \, 0\right\} \quad \text{on } B_{ij}.$$

By construction $u_{ij} \in W^{1,p}(B_{ij})$. Extend u_{ij} to functions $v_{ij} \in W_o^{1,p}(\mathbb{R}^N)$, defined in the whole $\mathbb{R}^N$, and such that

$$\int_{\mathbb{R}^N} |Dv_{ij}|^p dy \leq \bar\gamma(N, p) \int_{B_{ij}} |Du_{ij}|^p dy$$

for a constant $\gamma(N, p)$ depending only on N and p and independent of i, j, and u. The extension is effected by means of Corollary 10.2 of Chap. 10. Set

$$v = \sup_{i,j} v_{ij}$$

and estimate

$$\int_{\mathbb{R}^N} |Dv|^p dy \leq \sum_{j=1}^{c_N} \sum_{i \in \mathbb{N}} \int_{\mathbb{R}^N} |Dv_{ij}|^p dy \leq \bar\gamma \sum_{j=1}^{c_N} \sum_{i \in \mathbb{N}} \int_{B_{ij}} |Du_{ij}|^p dy$$

$$\leq \bar\gamma \sum_{j=1}^{c_N} \sum_{i \in \mathbb{N}} \int_{B_{ij}} |Du|^p dy \leq \bar\gamma \sum_{j=1}^{c_N} \int_{\mathbb{R}^N} |Du|^p dy$$

$$= \bar\gamma c_N \int_{\mathbb{R}^N} |Du|^p dy < \infty.$$

Therefore $v \in W_{p^*}^{1,p}(\mathbb{R}^N)$, by Proposition 24.2 of Chap. 10. By construction

$$u_* + v > t \quad \text{in a open neighborhood of } [u > t]_*.$$

From this and the previous estimate,

$$t^p \bar c_p([u > t]_*) \leq \int_{\mathbb{R}^N} |D(u + v)|^p dy \leq \gamma(N, p) \int_{\mathbb{R}^N} |Du|^p dy. \qquad \blacksquare$$

17   Precise Representatives of Functions in $W_{loc}^{1,p}(\mathbb{R}^N)$

Theorem 17.1 *Let $u \in W_{loc}^{1,p}(\mathbb{R}^N)$. There exists a set $\mathcal{E} \subset \mathbb{R}^N$ such that $\bar c_p(\mathcal{E}) = 0$ and u_* is well defined by the limit in (15.1) for all $x \in \mathbb{R}^N - \mathcal{E}$. Moreover, for all $\varepsilon > 0$ there exists an open set $\mathcal{E}_\varepsilon$ such that $c_p(\mathcal{E}_\varepsilon) < \varepsilon$ and u_* restricted to $\mathbb{R}^N - \mathcal{E}_\varepsilon$ is continuous. Finally all points of $\mathbb{R}^N - \mathcal{E}$ are Lebesgue points for u.*

Proof Assume first $u \in W^{1,p}(\mathbb{R}^N)$ and let E_p be the set introduced in (15.3). The limit in (15.1) might not exist if $x \in E_p$. By Proposition 15.2 and Theorem 9.1, $c_p(E_p) = 0$. Construct a sequence $\{u_n\} \subset W^{1,p}(\mathbb{R}^N) \cap C^\infty(\mathbb{R}^N)$ such that

$$\|u - u_n\|_p + \|Du - Du_n\|_p \to 0 \quad \text{as } n \to \infty.$$

Then for $j \in \mathbb{N}$ select a subsequence $\{u_{n(j)}\} \subset \{u_n\}$ such that

$$\int_{\mathbb{R}^N} |Du - Du_{n(j)}|^p dy \le \frac{1}{2^{(p+1)j}} \quad \text{for } j = 1, 2 \ldots .$$

Set

$$\mathcal{E}_j = \bigcup_{\rho > 0} \left[(|u - u_{n(j)}|)_{\cdot,\rho} > \frac{1}{2^j} \right]$$

$$= \left\{ x \in \mathbb{R}^N \mid (|u - u_{n(j)}|)_{x,\rho} > \frac{1}{2^j} \quad \text{for some } \rho > 0 \right\}$$

Set also

$$\mathcal{E}^h = E_p \cup \bigcup_{j \ge h} \mathcal{E}_j$$

The set where the limit in (15.1) fails to exist is contained in $\mathcal{E}^\infty$. By Proposition 16.1 and the construction of $\{u_{n(j)}\}$

$$\overline{c}_p(\mathcal{E}_j) \le \gamma 2^{jp} \int_{\mathbb{R}^N} |Du - Du_{n(j)}|^p dy \le \frac{\gamma}{2^j} \quad \text{for } j = 1, 2 \ldots .$$

Therefore ,

$$\overline{c}_p(\mathcal{E}^h) \le \overline{c}_p(E_p) + \sum_{j \ge h} \overline{c}_p(\mathcal{E}_j) \le \frac{1}{2^{h-1}}.$$

Moreover

$$\limsup_{\rho \to 0} |(u)_{x,\rho} - u_{n(j)}(x)| \le \frac{1}{2^j} \quad \text{for all } x \in \mathbb{R}^N - \mathcal{E}_j.$$

For all $x \in \mathbb{R}^N - \mathcal{E}^h$ and all $i, j > h$

$$|u_{n(i)}(x) - u_{n(j)}(x)| \le \limsup_{\rho \to 0} |u_{n(i)}(x) - (u)_{x,\rho}|$$

$$+ \limsup_{\rho \to 0} |(u)_{x,\rho} - u_{n(j)}(x)| \le \frac{1}{2^i} + \frac{1}{2^j}.$$

Therefore $\{u_{n(j)}\}$ is a Cauchy sequence converging uniformly, as $j \to \infty$, to some continuous function u^* in $\mathbb{R}^N - \mathcal{E}^h$. Next

$$\limsup_{\rho \to 0} |u^*(x) - (u)_{x,\rho}| \leq \limsup_{\rho \to 0} |u^*(x) - u_{n(j)}(x)|$$

$$+ \limsup_{\rho \to 0} |(u)_{x,\rho} - u_{n(j)}(x)|.$$

Therefore $u_* = u^*$ in $\mathbb{R}^N - \mathcal{E}^h$. As a consequence for all $\varepsilon > 0$ there exists a set $\mathcal{E}^h$ of outer p-capacity not exceeding ε such that u_* is well defined and continuous in $\mathbb{R}^N - \mathcal{E}^h$. Since $h \in \mathbb{N}$ is arbitrary, u_* is well defined by the limit in (15.1) in $\mathbb{R}^N - \mathcal{E}^\infty$ with $\overline{c}_p(\mathcal{E}^\infty) = 0$. Thus every point of $\mathbb{R}^N - \mathcal{E}^\infty$ is a differentiability point of u. It is also a Lebesgue point since

$$\lim_{\rho \to 0} \fint_{B_\rho(x)} |u - u_*(x)| dy \leq \lim_{\rho \to 0} |(u)_{x,\rho} - u^*(x)|$$

$$+ \lim_{\rho \to 0} \left(\fint_{B_\rho(x)} |u - (u)_{x,\rho}|^{p^*} dy \right)^{\frac{1}{p^*}}$$

$$\leq \lim_{\rho \to 0} \rho \left(\fint_{B_\rho(x)} |Du|^p dy \right)^{\frac{1}{p}}.$$

The last term is zero since $x \notin E_p$. The proof for $u \in W^{1,p}_{\text{loc}}(\mathbb{R}^N)$ is concluded by standard truncations and approximations. ∎

17.1 Quasi-Continuous Representatives of Functions $u \in W^{1,p}_{loc}(\mathbb{R}^N)$

A function $u : \mathbb{R}^N \to \mathbb{R}$ is quasi-continuous if for every $\varepsilon > 0$ there exists an open set $\mathcal{E}_\varepsilon$ such that $c_p(\mathcal{E}_\varepsilon) < \varepsilon$ and the restriction of u to $\mathbb{R}^N - \mathcal{E}_\varepsilon$ is continuous.

The notion is analogous to that of Lusin's theorem in § 5 of Chap. 4. In that context, the measure was any Borel regular measure μ, and continuity of a μ-measurable function was claimed, except for a "small" set, quantified in terms of its measure. The set E where the function was defined, had to be of finite measure.

Here $\mathbb{R}^N$ is endowed with the Lebesgue measure, and the domain of definition of u need not be of finite measure. Then continuity is sought everywhere in $\mathbb{R}^N$, except possibly for a "small" set quantified in terms of its outer p-capacity.

References

1. R.A. Adams, *Sobolev Spaces* (Academic Press, New York, 1975)
2. L. Alaoglu, Weak topologies of normed linear spaces. Ann. Math. **41**, 252–267 (1940)
3. A.D. Alexandrov, On the extension of a Hausdorff space to an H-closed space. C. R. (Doklady) Acad. Sci. USSR., N.S. **37**, 118–121 (1942)
4. R. Arens, Note on convergence in topology. Math. Mag. **23**, 229–234 (1950)
5. C. Arzelà, Sulle funzioni di linee. Mem. Accad. Sc. Bologna **5**(5), 225–244 (1894/5)
6. G. Ascoli, Le curve limiti di una varietà data di curve. Rend. Accad. Lincei **18**, 521–586 (1884)
7. R. Baire, Sur les fonctions de variables réelles. Annali di Mat. **3**(3), 1–124 (1899)
8. J.M. Ball, F. Murat, Remarks of Chacon's biting lemma. Proc. Am. Math. Soc. **107**(3), 655–663 (1989)
9. S. Banach, Sur les fonctions dérivées des fonctions mesurables. Fundam. Math. **3**, 128–132 (1922)
10. S. Banach, Sur un théorème de Vitali. Fundam. Math. **5**, 130–136 (1924)
11. S. Banach, Sur les lignes rectifiables et les surfaces dont l'aire est finie. Fundam Math. **7**, 225–237 (1925)
12. S. Banach, Über die Baire'sche Kategorie gewisser Funktionenmengen. Studia Math. **3**, 174 (1931)
13. S. Banach, *Théorie des Opérations Linéaires*, 2nd edn. (Monografie Matematyczne, Warsaw, 1932); reprinted by Chelsea Publishing Co., New York, 1963
14. S. Banach, S. Saks, Sur la convergence forte dans les espaces L^p. Studia Math. **2**, 51–57 (1930)
15. S. Banach, H. Steinhaus, Sur le principe de la condensation des singularités. Fundam. Math. **9**, 50–61 (1927)
16. A.S. Besicovitch, A General form of the covering principle and relative differentiation of additive functions. Proc. Cambridge Philos. Soc. **I**(41), 103–110 (1945); ibidem **II**(42), (1946), 1–10
17. H.F. Bohnenblust, A. Sobczyk, Extensions of functionals on complex linear spaces. Bull. Am. Math. Soc. **44**, 91–93 (1938)
18. E. Borel, Théorie des Fonctions. Ann. École Norm., Serie 3 **12**, 9–55 (1895)
19. E. Borel, *Leçons sur la Théorie des Fonctions* (Gauthier-Villars, Paris, 1898)
20. A.P. Calderón, A. Zygmund, On the existence of certain singular integrals. Acta. Math. **88**, 85–139 (1952)
21. G. Cantor, Über unendliche, lineare Punktmannigfaltigkeiten. Math. Ann. **20**, 113–121 (1882)
22. G. Cantor, De la puissance des ensembles parfaits de points. Acta Math. **4**, 381–392 (1884)

© Springer Science+Business Media New York 2016
E. DiBenedetto, *Real Analysis*, Birkhäuser Advanced
Texts Basler Lehrbücher, DOI 10.1007/978-1-4939-4005-9

23. G. Cantor, Über verschiedene Theoreme aus der Theorie der Punktmengen in einem n-fach ausggedehnten stetigen Raum G_n. Acta Math. **7**, 105–124 (1885)
24. H. Cartan, *Fonctions Analytiques d'une Variable Complexe* (Dunod, Paris, 1961)
25. C. Carathéodory, *Vorlesungen über reelle Funktionen* (Chelsea Publishing Co., New York, 1946) (reprint)
26. C. Carathéodory, *Algebraic Theory of Measure and Integration* (Chelsea Publishing Co., New York, 1963) (reprint)
27. G. Chiti, Rearrangement of functions and convergence in Orlicz spaces. Appl. Anal. **9**, 23–27 (1979)
28. J.A. Clarkson, Uniformly convex spaces. Trans. AMS **40**, 396–414 (1936)
29. M.G. Crandall, L. Tartar, Some relations between nonexpansive and order preserving mappings. Proc. Am. Math. Soc. **78**, 385–390 (1980)
30. M.M. Day, *Normed Linear Spaces*, 3rd edn. (Springer, New York, 1973)
31. M.M. Day, The spaces L^p with $0 < p < 1$. Bull. Am. Math. Soc. **46**, 816–823 (1940)
32. E. DeGiorgi, Sulla differenziabilità e l'analiticità delle estremali degli integrali multipli regolari. Mem. Acc. Sc. Torino, Cl. Sc. Mat. Fis. Nat. **3**(3), 25–43 (1957)
33. E. DeGiorgi, Sulla proprietà isoperimetrica dell'ipersfera, nella classe degli insiemi aventi perimetro finito. Rend. Accad. Lincei, Mem. Cl. Sci. Fis. Mat. Natur., Sez.I **5**(8), 33–44 (1958)
34. E. DiBenedetto, *Partial Differential Equations* (Birkhäuser, Boston, 1995)
35. E. DiBenedetto, D.J. Diller, On the rate of drying in a photographic film. Adv. Diff. Equ. **1**(6), 989–1003 (1996)
36. E. DiBenedetto, D.J. Diller, a new form of the sobolev multiplicative inequality and applications to the asymptotic decay of solutions to the neumann problem for quasilinear parabolic equations with measurable coefficients. Proc. Nat. Acad. Sci. Ukraine **8**, 81–88 (1997)
37. U. Dini, Fondamenti per la Teorica delle Funzioni di una Variabile Reale, Pisa. *reprint of Unione Matematica Italiana* (Bologna, Italy, 1878)
38. N. Dunford, J.T. Schwartz, *Linear Operators, Part I* (Wiley-Inter-science, New York, 1958)
39. D.F. Egorov, Sur les suites des fonctions mesurables. C. R. Acad. Sci. Paris **152**, 244–246 (1911)
40. L. Ehrenpreis, Solutions of some problems of division. Am. J. Math. **76**, 883–903 (1954)
41. L.C. Evans, R.F. Gariepy, *Measure Theory and Fine Properties of Functions* (CRC Press, Boca Raton, 1992)
42. P.J. Fatou, Séries trigonométriques et séries de Taylor. Acta Math. **30**, 335–400 (1906)
43. H. Federer, *Geometric Measure Theory* (Springer, New York, 1969)
44. C. Fefferman, E.M. Stein, H^p spaces of several variables. Acta Math. **129**, 137–193 (1972)
45. E. Fischer, Sur la convergence en moyenne. C. R. Acad. Sci. Paris **144**, 1148–1150 (1907)
46. W.H. Fleming, R. Rischel, An integral formula for total gradient variation. Arch. Math. **XI**, 218–222 (1960)
47. M. Fréchet, Des familles et fonctions additives d'ensembles abstraits. Fund. Math. **5**, 206–251 (1924)
48. I. Fredholm, Sur une classe d'équations fonctionnelles. Acta Math. **27**, 365–390 (1903)
49. A. Friedman, *Foundations of Modern Analysis* (Dover, New York, 1982)
50. K.O. Friedrichs, The identity of weak and strong extensions of differential operators. Trans. AMS **55**, 132–151 (1944)
51. G. Fubini, Sugli integrali multipli. Rend. Accad. Lincei, Roma **16**, 608–614 (1907)
52. G. Fubini, Sulla derivazione per serie. Atti Accad. Naz. Lincei Rend. **16**, 608–614 (1907)
53. E. Gagliardo, Proprietà di alcune funzioni in n variabili. Ricerche Mat. **7**, 102–137 (1958)
54. E. Giusti, *Functions of Bounded Variation* (Birkhäuser, Basel, 1983)
55. C. Goffman, G. Pedrick, *First Course in Functional Analysis* (Chelsea Publishing Co., New York, 1983)
56. H.H. Goldstine, Weakly complete banach spaces. Dukei Math J. **4**, 125–131 (1938)
57. H. Hahn, *Theorie der reellen Funktionen* (Berlin 1921)

58. H. Hahn, Über Folgen linearer operationen. Monatshefte für Mathematik un Physik **32**(1), 3–88
59. H. Hahn, Über lineare Gleichungen in linearen Räumen. J. für Math. **157**, 214–229 (1927)
60. H. Hahn, Über die Multiplikation total-additiver Menegnfunktionen. Ann. Sc. Norm. Sup. Pisa **2**, 429–452 (1933)
61. P.R. Halmos, *Measure Theory* (Springer, New York, 1974)
62. P.R. Halmos, *Introduction to Hilbert Spaces* (Chelsea Publishing Co., New York, 1957)
63. O. Hanner, On the uniform convexity of L^p and ℓ^p. Ark. Math. **3**, 239–244 (1956)
64. G.H. Hardy, J.E. Littlewood, G. Pólya, The maximum of a certain bilinear form. Proc. London Math. Soc. **2**(25), 265–282 (1926)
65. G.H. Hardy, J.E. Littlewood, Some properties of fractional integrals I. Math. Zeitschr. **27**, 565–606 (1928)
66. G.H. Hardy, J.E. Littlewood, On certain inequalities connected with the calculus of variations. J. London Math. Soc. **5**, 34–39 (1930)
67. G.H. Hardy, J.E. Littlewood, A maximal theorem with function-theoretical applications. Acta Math. **54**, 81–116 (1930)
68. G.H. Hardy, J.E. Littlewood, G. Pólya, *Inequalities* (Cambridge Univ, Press, 1963)
69. H.H. Hardy, Note on a theorem of Hilbert. Math. Zeitschr. **6**, 314–317 (1920)
70. F. Hausdorff, Dimension und äusseres Mass. Math. Ann. **79**, 157–179 (1919)
71. F. Hausdorff, *Grundzüge der Mengenlehre*, Leipzig 1914 (Chelsea Publishing Co., New York, 1955) (reprint)
72. E. Helly, Über lineare Funktionaloperationen. Österreich. Akad. Wiss. Natur. Kl., S.-B. IIa **121**, 265–297 (1912)
73. E. Hewitt, K.A. Ross, *Abstract Harmonic Analysis* (Springer, Berlin, 1963)
74. O. Hölder, Über einen Mittelwertsatz. Göttinger Nachr. 38–47 (1889)
75. J. Jensen, Sur les fonctions convexes et les inegalités entre les valeurs moyennes. Acta Math. **30**, 175–193 (1906)
76. F. John, L. Nirenberg, On functions of bounded mean oscillation. Comm. Pure Appl. Math. #**14**, 415–426 (1961)
77. C. Jordan, Sur la Série de Fourier. C. R. Acad. Sci. Paris **92**, 228–230 (1881)
78. C. Jordan, *Course d'Analyse* (Gauthier Villars, Paris, 1893)
79. J.L. Kelley, *General Topology* (Van Nostrand, New York, 1961)
80. O.D. Kellogg, *Foundations of Potential Theory* (Dover, New York, 1953)
81. A. Khintchine, A.N. Kolmogorov, Über Konvergenz von Reihen, deren Glieder durch den Zufall bestimmt werden Math. Sbornik **32**, 668–677 (1925)
82. M.D. Kirzbraun, Über die zusammenziehenden und Lipschitzschen Transformationen. Fundam. Math. **22**, 77–108 (1934)
83. A.N. Kolmogorov, Über Kompaktheit der Funktionenmengen bei der Konvergenz im Mittel. Nachrr. Ges. Wiss. Göttingen **9**, 60–63 (1931)
84. V.I. Kondrachov, Sur certaines propriétés des fonctions dans l'espace L^p. C. R. (Doklady), Acad. Sci. USSR (N.S.) **48**, 535–539 (1945)
85. M. Krein, D. Milman, On extreme points of regularly convex sets. Studia Math. **9**, 133–138 (1940)
86. H. Lebesgue, Sur une généralisation de l'intégrale définite. C. R. Acad. Sci. Paris **132**, 1025–1028 (1901)
87. H. Lebesgue, Sur les Fonctions représentables analytiquement. J. de Math. Pures et Appl. **6**(1), 139–216 (1905)
88. H. Lebesgue, Sur l'intégrale de Stieltjes et sur les opérations fonctionnelles linéaires. C. R. Acad. Sci. Paris **150**, 86–88 (1910)
89. H. Lebesgue, Sur l'intégration des fonctions discontinues. Ann. Sci. École Norm. Sup. **27**, 361–450 (1910)
90. H. Lebesgue, *Leçons sur l'Integration et la Recherche des Fonctions Primitives*, 2nd edn. (Gauthier–Villars, Paris, 1928)

91. A. Legendre, Mémoire sur l'integration de quelques équations aux différences partielles. Mémoires de l'Académie des Sciences 309–351 (1787)
92. L. Lichenstein, Eine elementare Bemerkung zur reellen Analyis. Math. Z. **30**, 794–795 (1929)
93. E.H. Lieb, M. Loss, *Analysis*, Graduate Studies in Mathematics, Vol. 14 (AMS, Providence R.I. 1996)
94. J.E. Littlewood, *Lectures on the Theory of Functions* (Oxford, 1944)
95. L.A. Liusternik, V.J. Sobolev, *Elements of Functional Analysis* (Fredrick Ungar Publishing Co., New York, 1967)
96. G. Lorentz, *Approximation of Functions* (Chelsea Publishing Co., New York, 1986)
97. G.G. Lorentz, *Bernstein Polynomials* (University of Toronto Press, Toronto, Canada, 1953)
98. N. Lusin, Sur les propriété des functions measurables. C. R. Acad. Sci. Paris **154**, 1688–1690 (1912)
99. B. Malgrange, Existence at approximation des solutions des équations aux dérivée partielles et des équations de convolution. Ann. Inst. Fourier **6**, 271–355 (1955–56)
100. J. Marcinkiewicz, Sur les series de Fourier. Fundam. Math. **27**, 38–69 (1936)
101. J. Marcinkiewicz, Sur quelques integrales du type de Dini. Ann. Soc. Pol. Math. **17**, 42–50 (1938)
102. J. Marcinkiewicz, Sur l'interpolation d'opérations. C. R. Acad. Sc. Paris. **208**, 1272–1273 (1939)
103. V.G. Mazja, *Sobolev Spaces* (Springer, New York, 1985)
104. S. Mazur, Über konvexe Mengen in linearen normierten Räumen. Studia Math. **4**, 70–84 (1933)
105. E.J. McShane, Extension of range of functions. Bull. AMS **40**, 837–842 (1934)
106. N. Meyers, J. Serrin, $H = W$. Proc. Nat. Acad. Sci. **51**, 1055–1056 (1964)
107. S.G. Mikhlin, *Integral Equations*, Vol. 4 (Pergamon Press, New York, 1957)
108. H. Minkowski, *Geometrie der Zahlen*, Leipzig 1896 and 1910 (reprint Chelsea 1953)
109. C.B. Morrey, On the solutions of quasilinear elliptic partial differential equations. Trans. AMS **43**, 126–166 (1938)
110. C.B. Morrey, *Multiple Integrals in the Calculus of Variations* (Springer, New York, 1966)
111. A.P. Morse, A theory of covering and differentiation. Trans. Am. Math. Soc. **55**, 205–235 (1944)
112. O.M. Nikodým, Sur les fonctions d'ensembles, Comptes Rendus du 1er Congrès de Mathematciens des Pays Slaves, Warsaw, pp. 304–313 (1929)
113. O.M. Nikodým, Sur une généralisation des intégrales de M. J. Radon Fund. Math. **15**, 131–179 (1930)
114. O.M. Nikodým, Sur les suites des fonctions parfaitement additive d'ensembles abstraits. C. R. Acad. Sci. Paris **192**, 727–728 (1931) (Monatshefte für Mathematik un Physik **32**(1), A23)
115. L. Nirenberg, On elliptic partial differential equations. Ann. Sc. Norm. Sup. Pisa **3**(13), 115–162 (1959)
116. W.F. Osgood, Non-uniform convergence and the integration of series term by term. Am. J. Math. **19**, 155–190 (1897)
117. H. Rademacher, Über partielle und totale Differenzierbarkeit. Math. Ann. **79**, 340–359 (1919)
118. J. Radon, Theorie und Anwendungen der absolut additiven Mengenfunktionen. Sitzungsber. Akad. Wiss. Wien **122**, 1295–1438 (1913)
119. A. Rajchman, S. Saks, Sur la dérivabilité des fonctions monotones. Fundam. Math. **4**, 204–213 (1923)
120. R. Rellich, Ein Satz über mittlere Konvergenz. Nachr. Akad. Wiss. Göttingen, Math.–Phys. Kl., 30–35 (1930)
121. F. Riesz, Sur un théoréme de M. Borel. C. R. Acad. Sci. Paris **144**, 224–226 (1905)
122. F. Riesz, Sur les systèmes ortogonaux de fonctions. C. R. Acad. Sci. Paris **144**, 615–619 (1907)
123. F. Riesz, Sur les suites des fonctions mesurables. C. R. Acad. Sci. Paris **148**, 1303–1305 (1909)

124. F. Riesz, Sur les opérations fonctionnelles linéaires. C. R. Acad. Sci. Paris **149**, 974–977 (1909)

125. F. Riesz, Über lineare Funktionalgleichungen. Acta. Math. **41**, 71–98 (1917)

126. F. Riesz, Sur les fonctions subharmoniques et leur rapport à la théorie du potentiel. Acta Math. **48**, 329–343 (1926)

127. F. Riesz, Sur les fonctions subharmoniques et leur rapport à la théorie du potentiel. Acta Math. **54**, 162–168 (1930)

128. F. Riesz, Sur une inégalité intégrale. J. London Math. Soc. **5**, 162–168 (1930)

129. F. Riesz, Sur les ensembles compacts de fonctions sommables. Acta Szeged Sect. Math. **6**, 136–142 (1933)

130. F. Riesz, Zur Theorie des Hilbertschen Raumes. Acta Sci. Math. Szeged. **7**, 34–38 (1934)

131. F. Riesz, B. Nagy, *Leçons d'Analyse Fontionnelle*, 6th edn. (Akadé-miai Kiadó, Budapest, 1972)

132. H.L. Royden, *Real Analysis*, 3rd edn. (Macmillan, New York, 1988)

133. W. Rudin, *Real and Complex Analysis*, 3rd edn. (McGraw–Hill, 1986)

134. S. Saks, *Theory of the Integral*, Vol. 7, 2nd edn. (Monografje Matematyczne, Warsaw, 1933); (Hafner Publishing Co. New York, 1937)

135. S. Saks, On some functionals. Trans. Am. Math. Soc. **35**(4), 549–556 (1933)

136. S. Saks, Addition to the note on some functionals. Trans. Am. Math. Soc. **35**(4), 965–970 (1933)

137. A. Sard, The measure of the critical values of differentiable maps. Bull. AMS **48**(12), 883–890 (1942)

138. L. Schwartz, *Théorie des Distributions* (Hermann & Cie, Paris, 1966)

139. E. Schmidt, Entwicklung willkürlicher Funktionen nach Systemen vorgeschriebener. Math. Ann. **63**, 433–476 (1907)

140. J. Schur, Über lineare Transformationen in der unendlichen Reihen. Crelle J. Reine Ange-wandte Mathematik **151**, 79–111 (1921)

141. C. Severini, Sopra gli sviluppi in serie di funzioni ortogonali. Atti Accad. Gioenia **3**(5), Mem. XI, 1–7 (1910)

142. C. Severini, Sulle successioni di funzioni ortogonali. Atti Accad. Gioenia **3**(5), Mem. XIII, 1–10 (1910)

143. W. Sierpiński, Sur un problème concernant les ensembles measurables superficiellement. Fundam. Math. **1**, 112–115 (1920)

144. W. Sierpiński, Sur les fonctions dérivées des fonctions discontinues. Fundam. Math. **3**, 123–127 (1922)

145. S.L. Sobolev, On a theorem of functional analysis. Mat. Sbornik **46**, 471–496 (1938)

146. S.L. Sobolev, S.M. Nikol'skii, Embedding theorems. Leningrad, Izdat. Akad. Nauk SSSR, 227–242 (1963)

147. R.H. Sorgenfrey, On the topological product of paracompact spaces. Bull. AMS **53**, 631–632 (1947)

148. G. Soukhomlinoff, Über Fortsetzung von linearen Funktionalen in linearen komplexen Räu-men und linearen Quaternionräumen. Recueil (Sbornik) Math. Moscou, N.S. **3**, (1938), 353–358

149. E. Stein, *Singular Integrals and Differentiability Properties of Functions* (Princeton University Press, Princeton, NJ, 1970)

150. J. Steiner, Einfacher Beweis der isoperimetrischen Hauptsätze. Crelle J. Reine Angewandte Mathematik **18** (1838); reprinted in *Gesammelte Werke 2, Berlin 1882, 77-91*

151. T.J. Stieltjes, Note sur l'intégrale $\int_a^b f(x)G(x)dx$. Nouv. Ann. Math. Paris, série **3**(7), 161–171 (1898)

152. M.H. Stone, Generalized Weierstrass approximation theorem. Math. Mag. **21**, 167–184, and 237–254 (1947/1948)

153. G. Talenti, *Calcolo delle Variazioni*, Quaderni dell'Unione Mat. Italiana, Pitagora Ed. (Roma, 1977)

154. G. Talenti, Best constants in sobolev inequalities. Ann. Mat. Pura Appl. **110**, 353–372 (1976)

155. H. Tietze, Über Funktionen, die auf einer abgeschlossenen Menge stetig sind. J. für Math. **145**, 9–14 (1915)

156. E.C. Titchmarsh, *The Theory of Functions*, 2nd edn. (Oxford University Press, London, 1939)

157. L. Tonelli, Sull'integrazione per parti. Atti Accad. Naz. Lincei (5) **18**(2), 246–253 (1909)

158. L. Tonelli, Successioni di curve e derivazione per serie. Atti Accad. Lincei **25**, 85–91 (1916)

159. F. Tricomi, *Integral Equations* (Dover, New York, 1957)

160. N.S. Trudinger, On embeddings into Orlicz spaces and some applications. J. Math. Mech. **17**, 473–483 (1967)

161. J.W. Tukey, Some notes on the separation of convex sets. Portugaliae Math. **3**, 95–102 (1942)

162. A. Tychonov, Über einen Funktionenräum. Math. Ann. **111**, 762–766 (1935)

163. P. Urysohn, Über die Mächtigkeit der zusammenhängenden Mengen. Math. Ann. **94**, 290 (1925)

164. W.T. van Est, H. Freudenthal, Trennung durch stetige Funktionen in topologischen Räumen. Indag. Math. **13**, 305 (1951)

165. B.L. Van der Waerden, Ein einfaches Beispiel einer nichtdifferenzierbaren stetigen Funktion. Math. Zeitschr. **32**, 474–475 (1930)

166. G. Vitali, Sul problema della misura dei gruppi di punti di una retta, *Memorie Accad. delle Scienze di Bologna, 1905.*

167. G. Vitali, Sulle funzioni integrali. Atti R. Accad. Sci. Torino **40**, 1021–1034 (1905)

168. G. Vitali, Sull'Integrazione per Serie. Rend. Circ. Mat. Palermo **23**, 137–155 (1907)

169. G. Vitali, Sui gruppi di punti e sulle funzioni di variabili reali. Atti Accad. Sci. Torino **43**, 75–92 (1908)

170. J. von Neumann, Zur Algebra der Funktionaloperationen und Theorie der Normalen Operatoren. Math. Ann. **102**, 370–427 (1930)

171. J. von Neumann, Zur Operatorenmethode der klassichen Mechanik. Ann. Math. **33**, 587–642 (1932)

172. K. Weierstrass, *Über die analytische Darstellbarkeit sogenannter willkürlicher Funktionen einer reellen Veränderlichen* (Kön. Preussichen Akad, Wissenschaften, 1885)

173. R.L. Wheeden, A. Zygmund, *Measure and Integral* (Marcel Dekker, New York, 1977)

174. H. Whitney, Analytic extensions of functions defined in closed sets. Trans. Am. Math. Soc. **36**, 63–89 (1934)

175. N. Wiener, The ergodic theorem. Duke Math. J. **5**, 1–18 (1939)

176. K. Yoshida, *Functional Analysis* (Springer, New York, 1974)

177. E. Zermelo, Neuer Beweis für die Wohlordnung. Math. Ann. **65**, 107–128 (1908)

178. A. Zygmund, On an integral inequality. J. London Math. Soc. **8**, 175–178 (1933)

Index

© Springer Science+Business Media New York 2016
E. DiBenedetto, *Real Analysis*, Birkhäuser Advanced
Texts Basler Lehrbücher, DOI 10.1007/978-1-4939-4005-9

Printed in the United States
By Bookmasters